Lecture Notes in Computational Science and Engineering

80

Editors:

For further volumes:
http://www.springer.com/series/3527

Peter H. Lauritzen • Christiane Jablonowski
Mark A. Taylor • Ramachandran D. Nair
Editors

Numerical Techniques for Global Atmospheric Models

Springer

Editors

Peter H. Lauritzen
Climate and Global Dynamics Division
National Center for Atmospheric Research
1850 Table Mesa Drive
Boulder, CO 80305
USA
pel@ucar.edu

Mark A. Taylor
Sandia National Laboratories, MS 0370
Albuquerque, NM 87185
USA
mataylo@sandia.gov

Christiane Jablonowski
University of Michigan
Department of Atmospheric, Oceanic
and Space Sciences
2455 Hayward St.
Ann Arbor, MI 48109
USA
cjablono@umich.edu

Ramachandran D. Nair
Institute for Mathematics Applied
to Geosciences
National Center for Atmospheric Research
1850 Table Mesa Drive
Boulder, CO 80305
USA
rnair@ucar.edu

ISBN 978-3-642-26761-1 ISBN 978-3-642-11640-7 (eBook)
DOI 10.1007/978-3-642-11640-7
Springer Heidelberg Dordrecht London New York

Mathematical Subject Classification (2010): 76, 76-06, 76U05, 76R50, 76M10, 76M12, 76M20, 76M22, 76M25, 35, 35L65, 35R05, 35Q30

Cover design: deblik, Berlin. Background visualization courtesy of Jamison Daniel at Oak Ridge National Laboratory.

Printed on acid-free paper

Springer is part of Springer Science+Business Media (www.springer.com)

Foreword

This book appears at a time of rapid change in the field of global atmospheric modeling. The field is being transformed, and the authors of this volume are driving many of those rapid changes.

As always, the models and the computing systems that run them are co-evolving. Processor speeds have (almost) stopped increasing, but parallelism is exploding, and systems with tens of millions of processors are expected in the next few years. These technology trends are driving atmospheric models towards much higher spatial resolution and more local discretization schemes.

The trend to higher model resolution is forcing a healthy re-examination of familiar methods that have been accepted, for decades, as standards of global modeling practice. Perhaps the most obvious point is that the quasi-static approximation is not applicable with high resolution. Depending on the approach, non-hydrostatic models must use short time steps, or else methods that avoid the need for short time steps. This is motivating the design of new time-differencing schemes. With fine horizontal grids, realistically steep topography can influence the choice of vertical coordinate. A variety of new horizontal grids is being very actively explored. At the same time, the horizontal and vertical staggering of the variables is also being revisited. Conservation principles are now widely recognized as key to successful long-term integrations. Vorticity dynamics is gaining a higher profile. Energy and enstrophy spectra present new challenges at high resolution, and this is motivating increased attention to dissipation parameterizations. Parameterized processes, especially those associated with clouds, are highly scale dependent, so that the parameterizations of high-resolution global models must behave very differently from their counterparts in low-resolution models.

In short, the field is in turmoil. This is good. Our rate of progress has accelerated, and new capabilities are being realized at a rapid pace. The book you are holding in your hands is an exciting report on progress from the front lines of research.

Fort Collins, USA — *Prof. David A. Randall*
June 2010

Preface

Approximating the solution to the partial differential equations for atmospheric flows using numerical algorithms implemented on a computer has been intensively researched since the pioneering work of Prof. John von Neuman in the late 1940s and 1950s. Since von Neuman's numerical experimentation on the first general purpose computer, the processing power of computers has increased at a breath-taking pace. While global models used for climate modeling a decade ago used horizontal grid spacings of order hundreds of kilometers, computing power now permits horizontal resolutions near the kilometer scale. Hence, the range of the scales of motion that next-generation global models will resolve spans from thousands of kilometers (planetary and synoptic scale) to the kilometer scale (meso-scale). Hence, the distinction between global climate models and global weather forecast models is starting to disappear due to the closing of the resolution gap that has historically existed between the two. For anyone interested in the dynamics of the weather and climate problem, this is a significant milestone since two branches of modeling, previously considered two separate disciplines, have started to merge.

Making effective use of massively parallel supercomputers, that are necessary for running global models at high resolution, has forced model developers back to the drawing board. Many current numerical methods are not scalable and therefore not amenable for massively parallel processing. This has forced the community to consider novel spherical grids (in the context of atmospheric global climate/weather modeling) where the grid-cell size is globally quasi-uniform in contrast to the highly nonuniform geographical longitude–latitude grid that has been the preferred choice for decades. The higher resolutions also affect which equation set is appropriate as a basis for the numerical discretizations. Model users now also expect the numerical method to preserve key integral invariants in discretized space, demand the accurate maintenance of balances in the flow, and request a truthful representation of waves on many scales as well as realistic scale interactions. Needless to say, the breadth of the choices of the computational grids and numerical schemes that should fulfill all these requirements is daunting, to say the least, and requires insight into the multi-scale nature of the problem and the properties of the chosen numerical methods.

The NCAR[1] ASP Colloquium 2008

To start tackling the significant challenges that lie ahead in global modeling, the Editors organized a colloquium on the latest developments in numerical methods for the dynamical cores of atmospheric General Circulation Models (GCMs). Dynamical cores are the central component of every climate and weather model. Loosely speaking, they solve the equations of motion on the resolved scales and determine not only the choice of the computational grid but also the predicted variables. Research in dynamical cores faces many scientific and computational challenges as was briefly outlined above.

On 1–13 June 2008, the colloquium entitled *Numerical Techniques for Global Atmospheric Models* was held at the National Center for Atmospheric Research (NCAR) in Boulder, Colorado. The colloquium was hosted by NCAR's Advanced Study Program (ASP) that hosts colloquia on an annual basis. The colloquium had two main objectives.

First, it introduced a multidisciplinary group of graduate students to the science of dynamical cores for global weather and climate models through lectures and hands-on tutorials. The chapters of this book are based on the lectures given at the colloquium by leaders in the field of numerical techniques for global atmospheric models. Second, the colloquium brought together the global modeling community by having the GCM modeling groups port their models to NCAR supercomputers, configure the models for idealized test cases defined by the colloquium organizers and to have the students exercise their models on these test cases during the colloquium. Nine international modeling groups accepted our invitation to participate in the colloquium, and each group had at least one modeling mentor present during the entire duration of the colloquium.

The modeling groups were as follows:

- Colorado State University (CSU) with the CSU-GCM
- Max Planck Institute for Meteorology (MPI-M) with the ICON (ICOsahedral Non-hydrostatic) model
- Goddard Institute for Space Studies (GISS) and Goddard Space Flight Center (GSFC) both part of National Aeronautics and Space Administration (NASA) with ModelE
- NCAR with the CAM (Community Atmosphere Model)
- NCAR and Sandia National Laboratories with the HOMME (High-Order Method Modeling Environment) model
- Massachusetts Institute of Technology (MIT) with the MIT-GCM
- Duke University, Earth System Science Interdisciplinary Center (ESSIC, University of Maryland) with the OLAM (Ocean-Land-Atmosphere Model)
- German Weather Service (DWD) with GME[2] (Global Model for Europe)

[1] The National Center for Atmospheric Research is sponsored by the National Science Foundation.

[2] Before the GME became operational, GME was an acronym for Global Model 'Ersatz' (which means 'replacement' in German) as the GME was a replacement for the spectral transform Global Model (GM).

- NASA GSFC joint with Geophysical Fluid Dynamics Laboratory (GFDL) run by National Oceanic and Atmospheric Administration (NOAA) with the GEOS5 (Goddard Earth Observing System model version 5)
- Joint Center for Earth Systems Technology (University of Maryland) with the GEF (Global Eta Framework) model

Some groups participated with several model versions.

A total of six test cases with several variants were used. Two of the test cases are described in Jablonowski and Williamson (2006, Quarterly Journal of the Royal Meteorological Society) and Lauritzen et al. (2010, Journal of Advances in Modeling Earth Systems), and the remaining four in Jablonowski et al. (submitted, Geoscientific Model Development). These papers also show results from the model simulations.

ASP Summer Colloquium June 1-13, 2008
Numerical Techniques for Global Atmospheric Models

Fig. 1 NCAR ASP 2008 summer colloquium group picture behind NCAR's Mesa Laboratory. From left to right (in order of increasing x-coordinate if photo was overlaid by a Cartesian coordinate system): Svetlana Dubinkina, Oksana Guba, Mark A. Taylor, Peter Hjort Lauritzen, Ramachandran D. Nair, Paul Ullrich, Dale Durran, Christiane Jablonowski, Jin-Young Kim, Richard Rood, Jasper Kok, Jung-Eun Kim, Todd Ringler, Lucas Harris, Matthew Long, Detlev Majewski, Hajoon Song, Dustin Williams, Sean Crowell, Junsu Kim, Jairo Gomes, Jochen Förstner, Aneesh Subramanian, Atul Kapur, David Devlin, Willian Sawyer, Verica Savic-Jovcic, Alberto Casado, Angela Marie Zalucha, Robert Walko, Marcia DeLonge, Matthew Norman, Guan Song, Qiang Deng, Colm Clancy, Almut Gassmann, Lin Su, Priscilla Mooney, Lee Murray, Jared Pierce Whitehead, Joakim R. Nielsen, Benjamin Kravitz, Ole-Kristian Kvissel, Lantao Sun, Brian Sørensen, Ayoe Buus Hansen, Cheng Zhou, Prabhakar Shrestha, Allan Christensen. Photo courtesy of Kathleen Barney (ASP)

About This Book

The chapters in this book collectively address almost every step in the development of dynamical cores for global atmospheric models. The 16 chapters have been divided into three parts: (1) equations of motion and basic ideas on discretizations, (2) conservation laws and traditional finite-volume as well as emerging numerical methods, and (3) practical considerations for dynamical cores in weather and climate models.

In the first chapter, Prof. J. Thuburn gives an introduction to the equations of motion for the atmosphere and commonly applied assumptions that are used to render the equations numerically more tractable and/or understand the types of waves supported by the equations of motion. Also the multiscale nature of atmospheric dynamics is introduced. Dr. J. Tribbia continues the theoretical discussion on the three-dimensional equations of motion through a mode decomposition analysis. In Chaps. 3 and 4, we leave the continuous equations behind and start exploring the properties of some basic horizontal and vertical numerical discretizations, and discuss the consequences of colocating and staggering prognostic variables. Thereafter some basic ideas on time-discretizations are introduced in Chap. 5 followed by a discussion on how to control fast waves through appropriate time-differencing (Chap. 6). The latter two chapters were written by Prof. D. R. Durran and conclude part I of this book.

In part II, Dr. T. D. Ringler discusses in detail the finite-volume advection of momentum and its relationship with other kinematic relationships such as conservation of vorticity (Chap. 7). Momentum advection is a key to the overall accuracy of any dynamical core as it determines the transport of mass and tracers. Chapter 8 focuses on transport, in particular finite-volume transport schemes, and reviews them from a semi-Lagrangian perspective. It presents an in-depth discussion on desirable properties for transport operators intended for global atmospheric models (Dr. P. H. Lauritzen). While most global models today use the spectral transform method or the finite-volume method, emerging new algorithms that are local but posses spectral convergence properties are at the time of writing being tested and integrated into atmospheric models. Such methods are being reviewed in Chap. 9 by Dr. R. D. Nair. To conclude part II, Prof. L. Ju gives an introduction to Voronoi diagrams that may be used to construct global spherical meshes with very flexible options for variable resolution.

After the discussion of the continuous equations of motion and basic discretization techniques in part I and the discussion of some classes of numerical schemes and spherical meshes in part II, we turn our attention to the properties of the dynamical core that are considered important in global atmospheric models (part III). Prof. J. Thuburn discusses conservation issues in Chap. 11 followed by a discussion on how to enforce key integral invariants numerically on unstructured grids (Dr. M. A. Taylor's Chap. 12). Almost all models need some level of filtering or damping to render the computed solutions physically realizable and smooth. Although these are rarely documented in the literature, they are paramount in model applications. Prof. C. Jablonowski reviews the pros and cons of these diffusion mechanisms, filters, and fixers in Chap. 13 and provides many illustrating examples from GCM runs. Continuing the filtering discussion, Dr. W. C. Skamarock focuses on the kinetic energy spectra in atmospheric models and how the tail of such spectra is influenced by discretization techniques and filtering. In Chap. 15 Prof. R. B. Rood gives a perspective on the dynamical core and its place in full model systems that include parameterizations of sub-grid-scale processes, data-assimilation, surface models, and others. Finally, Dr. J. M. Dennis discusses the many challenges in designing and implementing models for massively parallel supercomputers with concrete examples from NCAR's Coupled Climate System Model (CCSM).

The complex topic of dynamical cores, which includes choices between hundreds of numerical methods and half a dozen spherical grids as well as variable staggering options, offers an endless set of combinations and choices. Exploring all options is simply not feasible, and it is therefore necessary to make intelligent selections among the many choices. In the research community, there is, however, no consensus regarding a particular numerical method or spherical grid being superior for all applications (or even for a single application). The careful reader will find such differences among some chapters in this book, as different authors advocate particular approaches. It is deliberate that such diversity, which was discussed intensively during the 2008 ASP colloquium, is represented in this book as it depicts state-of-the-art knowledge in the field of dynamical cores. Despite this lack of collective agreement on numerical methods and grids, there seems to be broad consensus regarding dynamical core properties such as conservation, consistency, scalability, accuracy, energy spectra, and capabilities. In other words, the goal seems clear, but the optimal avenue to get there remains an open research question. We hope this book can contribute to this quest and enlighten the interested reader in the many deliberations that are an integral part of dynamical core development.

Acknowledgments

We thank the authors and coauthors of the chapters who generously agreed not only to participate in the colloquium but also to write-up their lectures for this book. All chapters have undergone a peer-review process and the comments by the many anonymous reviewers are gratefully acknowledged. This book would not have been written without the encouragement of Dr. Martin Peters at Springer-Verlag, and the generous funding and support provided by NCAR's Advanced Study Program lead by Dr. Maura Hagan and her team (Ms Paula Fisher, Mr Scott Briggs, Ms Kathleen Barney). Computing time and support was generously provided by NCAR's Computational and Information Systems Laboratory. Partial funding for the colloquium was also provided by NASA, the U.S. Department of Energy and the University of Michigan, Ann Arbor.

Boulder
December, 2010

Peter H. Lauritzen
Christiane Jablonowski
Mark A. Taylor
Ramachandran D. Nair

Contents

Contributors

John Thuburn School of Engineering, Computing and Mathematics, University of Exeter, North Park Road, Exeter, EX4 4QF, UK, j.thuburn@ex.ac.uk

Joseph Tribbia National Center for Atmospheric Research, 1850 Table Mesa Drive, Boulder, CO 80305, USA, tribbia@ucar.edu

Roger Temam Institute for Scientific Computing and Applied Mathematics, Indiana University, Rawles Hall, Bloomington, IN 47405-5701, USA, temam@indiana.edu

Dale R. Durran Department of Atmospheric Sciences, Box 351640, University of Washington, Seattle, WA, 98195, USA, durrand@atmos.washington.edu

Todd Ringler T-3 Fluid Dynamics Group, Theoretical Division, Los Alamos National Laboratory, Los Alamos, NM 87545, USA, ringler@lanl.gov

Peter H. Lauritzen National Center for Atmospheric Research, 1850 Table Mesa Drive, Boulder, CO 80305, USA, pel@ucar.edu

Paul A. Ullrich University of Michigan, 2455 Hayward St., Ann Arbor, MI 48109, USA, paullric@umich.edu

Ramachandran D. Nair National Center for Atmospheric Research, 1850 Table Mesa Drive, Boulder, CO 80305, USA, rnair@ucar.edu

Mike N. Levy Sandia National Laboratory, Albuquerque, NM 87185, USA, mnlevy@sandia.gov

Lili Ju Department of Mathematics, University of South Carolina, Columbia, SC 29208, USA, ju@math.sc.edu

Max Gunzburger Department of Scientic Computing, Florida State University, Tallahassee, FL 32306, USA, gunzburger@fsu.edu

Mark A. Taylor Sandia National Laboratories, Albuquerque, NM 87185, USA, mataylo@sandia.gov

Christiane Jablonowski University of Michigan, 2455 Hayward St., Ann Arbor, MI 48109, USA, cjablono@umich.edu

David L. Williamson National Center for Atmospheric Research, 1850 Table Mesa Drive, Boulder, CO 80305, USA, wmson@ucar.edu

William C. Skamarock National Center for Atmospheric Research, 3450 Mitchell Lane, Boulder, CO 80307, USA, skamaroc@ucar.edu

Richard B. Rood University of Michigan, 2455 Hayward St., Ann Arbor, MI 48109, USA, rbrood@umich.edu

John M. Dennis Computational & Information Systems Laboratory, National Center for Atmospheric Research, 1850 Table Mesa Drive, Boulder, CO 80305, USA, dennis@ucar.edu

Richard D. Loft Computational & Information Systems Laboratory, National Center for Atmospheric Research, 1850 Table Mesa Drive, Boulder, CO 80305, USA, loft@ucar.edu

Part I
Equations of Motion and Basic Ideas on Discretizations

Chapter 1
Some Basic Dynamics Relevant to the Design of Atmospheric Model Dynamical Cores

John Thuburn

Abstract The dynamics of the global atmosphere is highly complex and multiscale. In this chapter a few aspects are discussed that are considered especially important for the design of numerical models of the atmosphere. Commonly used approximations to the governing equations are discussed. The dynamics of fast acoustic and inertio-gravity waves is briefly explained along with their role in maintaining the atmosphere close to hydrostatic and geostrophic balance. The balanced dynamics is exemplified through quasigeostrophic theory, which embodies the key ideas of advection and invertibility of potential vorticity. Finally, some important effects of nonlinearity are discussed, in particular the interaction between different scales and the transfer of energy and potential enstrophy across scales.

1.1 Introduction

Geophysical Fluid Dynamics is a huge and complex subject, and we can barely scratch the surface of it in this pair of introductory lectures. Therefore, I have tried to pick out a set of topics that are most relevant to the design of atmospheric model dynamical cores. There are several excellent introductory and graduate level textbooks that cover these topics and many more in greater depth (e.g., Gill 1982; Pedlosky 1987; Salmon 1998; Holton 2004; Vallis 2006).

On large scales, the dynamics of the atmosphere is approximately *balanced*, and it is important for numerical solutions to be approximately balanced in the same sense. In this lecture we will discuss the nature of this balance, and the linear dynamics of the fast *acoustic* and *inertio-gravity waves* responsible for the *adjustment* towards balance. We will also discuss the *quasigeostrophic equations*, which approximately describe the slow, balanced dynamics, and the *Rossby waves* that

J. Thuburn
School of Engineering, Computing and Mathematics, University of Exeter, North Park Road, Exeter, EX4 4QF, UK
e-mail: j.thuburn@ex.ac.uk

P.H. Lauritzen et al. (eds.), *Numerical Techniques for Global Atmospheric Models*, Lecture Notes in Computational Science and Engineering 80,
DOI 10.1007/978-3-642-11640-7_1,

these equations support. Some aspects of atmospheric dynamics are strongly nonlinear, and numerical models must handle various nonlinear processes in a satisfactory way. In this context we will mention *Eulerian and Lagrangian timescales* for atmospheric dynamics, *conservation properties*, and *turbulent cascades*. Conservation properties and turbulent cascades will be discussed again in Chap. 11. We begin here by emphasizing the complex and multiscale nature of atmospheric dynamics.

1.2 The Multiscale Nature of Atmospheric Dynamics

Figure 1.1 indicates schematically the time scales and horizontal spatial scales of a range of atmospheric phenomena. On the largest spatial scales (comparable to the Earth's radius) and seasonal timescales are large scale circulations such as that associated with the Asian summer monsoon. Undulations in the jet stream and pressure patterns associated with the largest scale Rossby waves (called *planetary waves*) also have length scales of order 10^4 km. Cyclones and anticyclones have length scales of a few thousand kilometers and timescales of order 10 days. The transition zones between relatively warm and cool air masses can collapse in scale to form fronts with widths a few tens of kilometers. Convection can be organized on a huge range of different scales, from the tropical intraseasonal oscillation on scales of thousands of kilometers and a timescale of months, through supercell complexes and squall lines of order 10 km across with lifetimes of several hours, down to individual small cumulus clouds on scales of a few hundred meters and a few minutes. These small cumulus clouds are formed when the turbulent eddies in the boundary layer lift and cool air far enough for condensation to occur. The boundary layer is the lowest few hundred meters of the atmosphere, where the dynamics is dominated by turbulent transports. The turbulent eddies range in scale from a few hundred meters (the boundary layer depth) down to the millimeter scale at which molecular diffusion becomes significant.

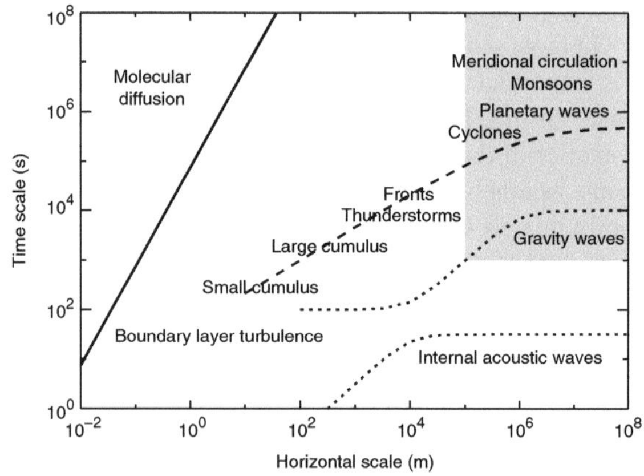

Fig. 1.1 Schematic showing the range of time and horizontal scales of different atmospheric phenomena

The atmospheric spectrum of horizontal kinetic energy is observed to have a slope very close to k^{-3} on large scales and $k^{-5/3}$ on small scales, where k is the horizontal wavenumber, with a gradual transition between the two at scales of a few hundred kilometers (Nastrom and Gage 1985). The dashed line in Fig. 1.1 is consistent with this observed spectrum, re-expressed in terms of length and time scales. The dynamically important phenomena mentioned above are those that dominate the atmospheric energy spectrum, and all lie close to this dashed line. Molecular diffusion, in contrast, is only significant to the left of the continuous line; thus it is completely negligible for atmospheric dynamics until we reach scales of order 1 mm (see Chap. 2).

All of the phenomena along the dashed line in Fig. 1.1 are important for weather and climate, and so need to be represented in numerical models. Important phenomena occur at all scales – there is no significant *spectral gap*. Moreover, there are strong interactions between the phenomena at different scales, and these interactions need to be represented. However, computer resources are finite and so numerical models must have a finite resolution. The shaded region in the figure shows the resolved space and time scales in a typical current day climate model. The important unresolved processes cannot be neglected and so must be represented by *sub-grid models* or *parameterizations*. The lack of any spectral gap makes this task more challenging. The emphasis in this series of lectures is on how we model the resolved dynamics; however, it should be borne in mind that equally important is how we represent the unresolved processes, and how we represent the interactions between resolved and unresolved processes. There are significant research challenges in all three areas.

Also shown in Fig. 1.1 are two dotted curves. These correspond to the dispersion relations for internal inertio-gravity waves and internal acoustic waves (see Sect. 1.4). The fact that the dotted lines lie significantly below the energetically dominant processes on the dashed line indicate that inertio-gravity waves and acoustic waves are relatively fast processes. One consequence of this is that inertio-gravity waves and acoustic waves are energetically weak compared to the dominant processes along the dashed curve. The fact that these waves are fast puts strong constraints on the size of timestep that can be used in numerical models with explicit time schemes. At the same time, the fact that they are energetically weak means that we do relatively little damage if we distort their propagation by using a semi-implicit time scheme in order to avoid the timestep restriction. See Chaps. 5 and 6 for a detailed discussion.

1.3 Governing Equations

The governing equations for a compressible fluid in a frame of reference rotating with angular velocity $\boldsymbol{\Omega}$ may be written in the form

$$\frac{\partial \rho}{\partial t} + \nabla \cdot (\rho \mathbf{u}) = 0, \tag{1.1}$$

$$\frac{D\theta}{Dt} = Q, \tag{1.2}$$

$$\frac{D\mathbf{u}}{Dt} + 2\boldsymbol{\Omega} \times \mathbf{u} = -\frac{1}{\rho}\nabla p - \nabla \Phi + \mathbf{F}. \tag{1.3}$$

Here, ρ is the fluid density, $\mathbf{u}$ is the fluid velocity vector, θ is the potential temperature, p is pressure, and Φ is the geopotential. D/Dt represents the derivative following a fluid parcel. Q is the diabatic source term for potential temperature and $\mathbf{F}$ represents any forces not already accounted for, for example molecular viscosity.

Equation (1.1) describes conservation of mass of the fluid. For simplicity, here we restrict attention to a single phase fluid of fixed composition. The real atmosphere contains varying amounts of water vapor and condensed water, and this complicates the governing equations.

Equation (1.2) is one form of the thermodynamic equation; θ is related to the other thermodynamic variables through

$$\theta = T\left(\frac{p_0}{p}\right)^{\kappa}, \tag{1.4}$$

(T is temperature, p_0 is a constant reference pressure, often taken to be 10^5 Pa, $\kappa = R/C_p$ where R is the gas constant for dry air and C_p is the specify heat capacity at constant pressure), along with the equation of state for an ideal gas

$$p = RT\rho. \tag{1.5}$$

In adiabatic flow the source term Q vanishes, so that the θ of an air parcel is conserved. If an air parcel of potential temperature θ were moved adiabatically from its current pressure p to the reference pressure p_0 its final temperature would be $T = \theta$. The potential temperature is closely related to the specific entropy η:

$$\eta = C_p \ln\theta + \text{const.} \tag{1.6}$$

Equation (1.3) is the momentum equation; it expresses Newton's second law of motion for a fluid. Because we are in a rotating frame, two new terms with the appearance of 'virtual' forces enter the equation of motion. One is the Coriolis term $2\boldsymbol{\Omega} \times \mathbf{u}$. The other is the centrifugal term $\boldsymbol{\Omega} \times (\boldsymbol{\Omega} \times \mathbf{u})$. However, the centrifugal term may be written as the gradient of a certain potential; this potential is then combined with the gravitational potential to obtain the geopotential Φ. The centrifugal term, therefore, does not appear explicitly.

For the flow regime of the Earth's atmosphere, rotation is extremely important. On synoptic scales, the Coriolis term is one of the dominant terms in the horizontal components of the momentum equation. Along with stratification effects, rotation gives atmospheric flow a distinctive character that is qualitatively quite different from other flows.

1.3.1 Approximate Equation Sets

Almost no approximations were made in writing (1.1)–(1.5). However, it is often desirable to work with approximate versions of the governing equations. These may be conceptually simpler, for example by filtering out certain kinds of motion; they may be analytically more tractable; or they may be easier to solve numerically, for example by removing certain terms or types of motion that are difficult to handle numerically.

Some of the most common approximations are the following (e.g., Durran 1999; Gill 1982; White 2002; White et al. 2005, 2008 and references therein).

- *Spherical geoid.* It is common to approximate the geopotential Φ as a function only of r, the distance from the centre of the Earth. As a result the effective gravity $\nabla\Phi$ acts only in the vertical component of the momentum equation in the usual spherical coordinate system. This is a good approximation for the Earth's atmosphere, where the true gravitational acceleration is much stronger than the centrifugal acceleration. But it would not be a good approximation for Jupiter, for example.
- *Quasi-hydrostatic approximation.* This involves neglecting the acceleration term Dw/Dt in the vertical component of the momentum equation. This is a good approximation on horizontal scales greater than about 10 km.
- *Anelastic approximation.* There are several flavours of anelastic or pseudo-incompressible approximation. They involve neglecting the elasticity of the fluid by approximating the mass continuity equation as something like

$$\nabla \cdot (\rho_0 \mathbf{u}) = 0, \tag{1.7}$$

  where ρ_0 is a reference density profile that depends only on height z. The anelastic approximation is a good approximation on horizontal scales smaller than about 10 km.
- *Shallow atmosphere approximation.* This is a collection of several approximations, but they must all be made together so that the resulting approximate equations retain conservation laws for energy and angular momentum. The Coriolis terms involving the horizontal components of $\boldsymbol{\Omega}$ are neglected; factors of $1/r$ in the spherical coordinate component form of the equations are replaced by $1/a$ where a is a constant equal to the Earth's mean radius; and certain other 'metric' terms are neglected.

It is often considered desirable for numerical models to use equation sets that do not support acoustic modes. The high frequency of acoustic modes would make it expensive or complicated to retain them in the numerical solution; on the other hand, because they are energetically very weak we lose little by leaving them out. The anelastic equations do not support acoustic modes. The hydrostatic equations do not support internal acoustic modes, only horizontally propagating external acoustic modes which, because of the anisotropic grids used in global atmospheric models, impose less of a restriction on the time step. Many past and present climate models make the hydrostatic and shallow atmosphere approximations (leading to the so-called hydrostatic primitive equations). Many models of small-scale dynamics use some form of anelastic equations. Unfortunately neither the hydrostatic nor the anelastic approximation is valid on all horizontal scales. Consequently, several recently developed atmospheric models, designed to work from global scales down to kilometer scales, use the fully compressible equations. (Very recently, some progress has been made towards acoustically filtered equation sets valid on all horizontal scales: Durran 2008; Arakawa and Konor 2009).

The different approximate equation sets can be arranged systematically into a hierarchy. Figure 1.2 shows part of that hierarchy. Some of these approximate equation sets have been discussed already above. The quasigeostrophic, planetary geostrophic, and semi-geostrophic equation sets filter inertio-gravity waves as well as acoustic waves. The quasigeostrophic equations will be introduced briefly

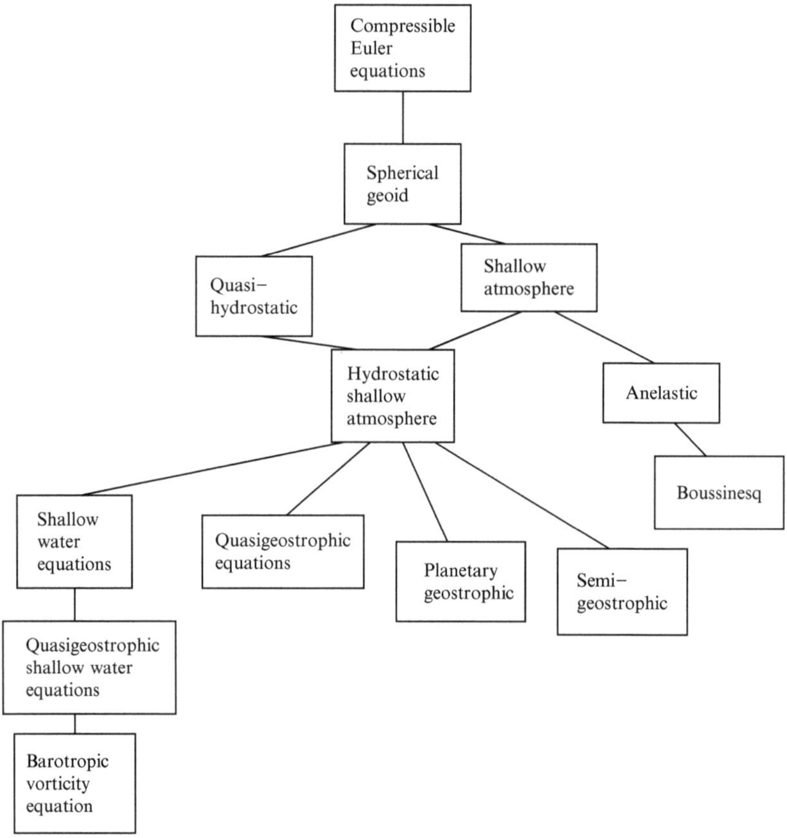

Fig. 1.2 Part of the hierarchy of frequently used approximate equation sets for atmospheric dynamics

in Sect. 1.6. The shallow water equations, their quasigeostropic version, and the barotropic vorticity equation all describe a single-layer two-dimensional fluid. They are too inaccurate for weather forecasting or climate modeling, but they are still used for idealized studies and are useful for testing numerical algorithms before applying them to more complete equation sets (Williamson et al. 1992). White (2002) presents a thorough and readable survey of various approximate equation sets used for atmospheric modeling.

1.4 Fast Waves

We noted in Sect. 1.1 that the fast acoustic and inertio-gravity waves are observed to be energetically weak. It might be tempting to think that it is therefore not necessary to treat these fast waves accurately in atmospheric model dynamical cores. However, the weakness of these fast waves corresponds to certain kinds of approximate balance between other terms in the governing equations, discussed more in Sect. 1.5 below. This balance is a leading order feature of atmospheric dynamics and it is essential to capture it accurately in numerical models. The atmosphere is continually being perturbed away from balance by a variety of mechanisms, including flow over orography, convective instability, and the nonlinear nature of the balanced dynamics. The mechanism by which the atmosphere adjusts back towards balance involves the radiation and ultimate dissipation of the fast acoustic and inertio-gravity waves. Thus, an accurate representation of balance in numerical models requires a causally correct representation of the adjustment mechanism involving the fast waves. In practice this means that some artificial slowing of the fast waves, for example by a semi-implicit time scheme, is usually considered acceptable provided the group velocity – see below – retains the correct sign. With this motivation in mind, we will now look at the dynamics of acoustic waves and inertio-gravity waves. For this purpose we will use the simplest equation sets that contain the essential dynamical ingredients.

1.4.1 Acoustic Waves

Consider a compressible fluid, but neglect rotation effects, gravity, and non-conservative processes. The mass and momentum equations may be written as

$$\frac{D\rho}{Dt} = \left(\frac{\partial\rho}{\partial p}\right)_\theta \frac{Dp}{Dt} + \left(\frac{\partial\rho}{\partial\theta}\right)_\rho \frac{D\theta}{Dt} = -\rho\nabla\cdot\mathbf{u} = 0, \tag{1.8}$$

$$\frac{D\mathbf{u}}{Dt} = -\frac{1}{\rho}\nabla p. \tag{1.9}$$

Now linearize these equations about a reference state at rest with constant density ρ_0 and temperature T_0, noting that $D\theta/Dt = 0$, to obtain

$$\frac{1}{c^2}\frac{\partial p}{\partial t} = -\rho_0 \nabla \cdot \mathbf{u}, \tag{1.10}$$

$$\frac{\partial \mathbf{u}}{\partial t} = -\frac{1}{\rho_0}\nabla p, \tag{1.11}$$

where $c^2 = \partial p/\partial \rho|_\theta = RT_0/(1-\kappa)$ and ρ and p are now perturbations from the reference state. Hence $\mathbf{u}$ may be eliminated to leave a wave equation for p:

$$\frac{\partial^2 p}{\partial t^2} - c^2 \nabla^2 p = 0. \tag{1.12}$$

Equation (1.12) has solutions

$$p \propto \exp\{\mathrm{i}(\mathbf{k}\cdot\mathbf{x} - \omega t)\}, \tag{1.13}$$

where $\mathbf{x}$ is the position vector and where the frequency ω is related to the wavenumber $\mathbf{k}$ by the *dispersion relation*

$$\omega^2 = c^2 |\mathbf{k}|^2. \tag{1.14}$$

Thus acoustic waves all propagate at speed c, independent of the wave vector; they are said to be *non-dispersive*. Typical values of c are around 315–350 ms^{-1}.

Acoustic waves are *longitudinal*, that is, velocity perturbations are parallel to the wave vector $\mathbf{k}$. The physical mechanism for acoustic waves involves the interaction of compressibility and flow divergence: convergence of fluid locally leads to an increase in density and hence pressure; the resulting pressure gradient then drives fluid acceleration leading to new convergence displaced from the original convergence.

1.4.2 Inertio-Gravity Waves

To simplify the governing equations we will make the *Boussinesq approximation*: assume the fluid to be incompressible and neglect variations in density from its reference value ρ_0 except where they appear in a buoyancy term, i.e., multiplied by the gravitational acceleration g. We will also neglect the Coriolis terms involving the horizontal component of $\boldsymbol{\Omega}$ (one element of the shallow atmosphere approximation), and work in Cartesian coordinates. The governing equations become

$$\frac{Du}{Dt} - fv = -\frac{1}{\rho_0}\frac{\partial p}{\partial x}, \tag{1.15}$$

$$\frac{Dv}{Dt} + fu = -\frac{1}{\rho_0}\frac{\partial p}{\partial y}, \tag{1.16}$$

$$\frac{Dw}{Dt} = -\frac{1}{\rho_0}\frac{\partial p}{\partial z} + b, \tag{1.17}$$

$$\frac{\partial u}{\partial x} + \frac{\partial v}{\partial y} + \frac{\partial w}{\partial z} = 0, \tag{1.18}$$

$$\frac{Db}{Dt} + wN^2 = 0, \tag{1.19}$$

where

$$b = -g\frac{\rho - \overline{\rho}}{\rho_0} \tag{1.20}$$

and

$$N^2 = -\frac{g}{\rho_0}\frac{d\overline{\rho}}{dz}. \tag{1.21}$$

Here $f = 2|\boldsymbol{\Omega}|\sin\phi$ at latitude ϕ; f is called the Coriolis parameter. We are interested in motions on scales much smaller than the Earth's radius so we can take f to be a constant. There are two reference densities: ρ_0 is a constant while $\overline{\rho}$ is a function only of z. N^2 is called the buoyancy frequency or Brunt-Väisälä frequency.

Now linearize these equations about a hydrostatically balanced state of rest. (Hydrostatic balance means that the reference buoyancy and vertical pressure gradient terms exactly cancel implying no vertical acceleration; see Sect. 1.5 below.)

$$\frac{\partial u}{\partial t} - fv = -\frac{1}{\rho_0}\frac{\partial p}{\partial x}, \tag{1.22}$$

$$\frac{\partial v}{\partial t} + fu = -\frac{1}{\rho_0}\frac{\partial p}{\partial y}, \tag{1.23}$$

$$\frac{\partial w}{\partial t} = -\frac{1}{\rho_0}\frac{\partial p}{\partial z} + b, \tag{1.24}$$

$$\frac{\partial u}{\partial x} + \frac{\partial v}{\partial y} + \frac{\partial w}{\partial z} = 0, \tag{1.25}$$

$$\frac{\partial b}{\partial t} + wN^2 = 0. \tag{1.26}$$

Because these equations are linear and have constant coefficients they will have solutions in which all variables are proportional to $\exp\{\mathrm{i}(kx + ly + mz - \omega t)\} = \exp\{\mathrm{i}(\mathbf{k}\cdot\mathbf{x} - \omega t)\}$. Substituting a solution of this form allows us to replace all derivatives by algebraic factors. Then, systematically eliminating the velocity and thermodynamic variables leads to the inertio-gravity wave dispersion relation

$$\omega^2 = \frac{(k^2 + l^2)N^2 + m^2 f^2}{k^2 + l^2 + m^2}. \tag{1.27}$$

Inertio-gravity waves are *transverse* waves: the velocity is perpendicular to the wave vector (which can be seen by considering (1.25)). Two interesting limiting cases are that of very deep waves $m^2/(k^2+l^2) \ll 1$, for which $\omega^2 \approx N^2$, and that of very shallow waves $(k^2+l^2)/m^2 \ll 1$, for which $\omega^2 \approx f^2$. More generally, ω^2 lies between f^2 and N^2.

There are two basic physical mechanisms underlying inertio-gravity waves. At the inertial end of the spectrum, i.e., shallow waves, an air parcel displaced from its equilibrium position experiences a restoring force provided by the Coriolis effect. At the gravity wave end of the spectrum, i.e., deep waves, an air parcel displaced from its equilibrium position has a density different from that of the reference profile at that height and so experiences a restoring force due to buoyancy, i.e., the imbalance between the gravitational force on the parcel and the vertical pressure gradient force. In intermediate parts of the spectrum both mechanisms operate to some degree.

1.4.3 Phase Velocity and Group Velocity

Two quantities are often used to describe the propagation of a wave or of a wave packet: the *phase velocity* and the *group velocity*. The phase velocity $\mathbf{c}_p$ is the velocity at which wave crests and troughs propagate. Suppose a wave has a structure proportional to $e^{i\phi(\mathbf{x},t)}$, where

$$\phi = \mathbf{k} \cdot \mathbf{x} - \omega(\mathbf{k})t; \tag{1.28}$$

ϕ is called the phase. The phase velocity is therefore the velocity at which surfaces of constant ϕ move. So let

$$\phi = \mathbf{k} \cdot \mathbf{x} - \omega t = \mathbf{k} \cdot (\mathbf{x} - \mathbf{c}_p t). \tag{1.29}$$

This relation is not enough to uniquely determine $\mathbf{c}_p$, but if we also demand that $\mathbf{c}_p$ be parallel to $\mathbf{k}$ (the most natural choice), then

$$\mathbf{c}_p = \frac{\omega \mathbf{k}}{|\mathbf{k}|^2}. \tag{1.30}$$

(However, the reader should be warned that $\mathbf{c}_p$ does not behave like a standard velocity vector, for example under transformation to a moving frame of reference.)

The group velocity is the velocity at which a packet or group of waves of approximately the same frequency propagates. It is, therefore, the velocity at which waves of that frequency transport energy. One of the simplest derivations of the mathematical expression for group velocity is the following. Consider a one-dimensional wave field Φ that is a superposition of waves at two nearby frequencies and wavenumbers (both of which satisfy the dispersion relation):

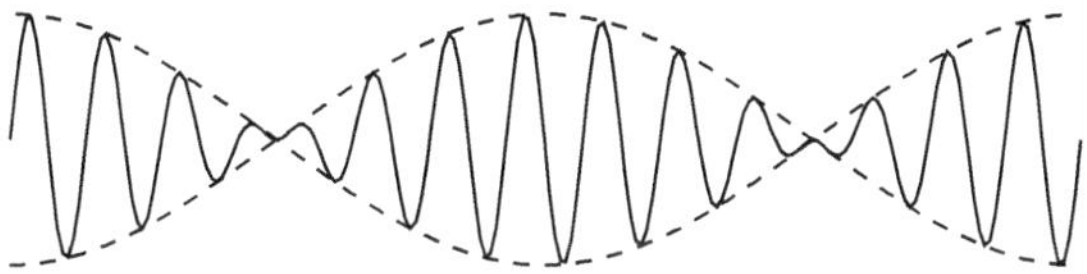

Fig. 1.3 Schematic showing the formation of wave packets from the superposition of two waves of similar wavenumber, given by the real part of (1.31)

$$\begin{aligned}\Phi &= \frac{1}{2}\left(e^{\mathrm{i}[(k+\delta k)x-(\omega+\delta\omega)t]} + e^{\mathrm{i}[(k-\delta k)x-(\omega-\delta\omega)t]}\right)\\ &= \cos(\delta kx - \delta\omega t)e^{\mathrm{i}(kx-\omega t)};\end{aligned} \tag{1.31}$$

see Fig. 1.3. The field Φ consists of a series of wave packets. The individual wave crests and troughs, described by the $e^{\mathrm{i}(kx-\omega t)}$ factor, propagate at the phase speed ω/k. The wave packets, whose envelope is defined by the $\cos(\delta k\, x - \delta\omega\, t)$ factor, propagate at group velocity $c_g = \delta\omega/\delta k$. Taking the limit as δk and $\delta\omega$ tend to zero, we have

$$c_g = \frac{d\omega}{dk}. \tag{1.32}$$

The generalization to three dimensions is

$$\mathbf{c}_g = \nabla_k\omega = \left(\frac{\partial\omega}{\partial k}, \frac{\partial\omega}{\partial l}, \frac{\partial\omega}{\partial m}\right), \tag{1.33}$$

where (k, l, m) are the components of the wave vector **k**.

Both the phase velocity and the group velocity can be computed from the dispersion relation. For acoustic waves, from (1.14) we find

$$\mathbf{c}_p = \mathbf{c}_g = \pm c\frac{\mathbf{k}}{|\mathbf{k}|}. \tag{1.34}$$

Acoustic waves are unusual in that they are non-dispersive (their phase speed is independent of the magnitude of the wave vector) and their phase and group velocities are equal.

For inertio-gravity waves, (1.27) implies

$$\mathbf{c}_p = \frac{\omega}{|\mathbf{k}|^2}(k, l, m) \tag{1.35}$$

and

$$\mathbf{c}_g = \frac{N^2 - f^2}{\omega|\mathbf{k}|^4}\left(km^2, lm^2, -m(k^2 + l^2)\right). \tag{1.36}$$

Among other things, these results show that the vertical components of the phase and group velocities have opposite sign (provided $N^2 > f^2$, which is usually the

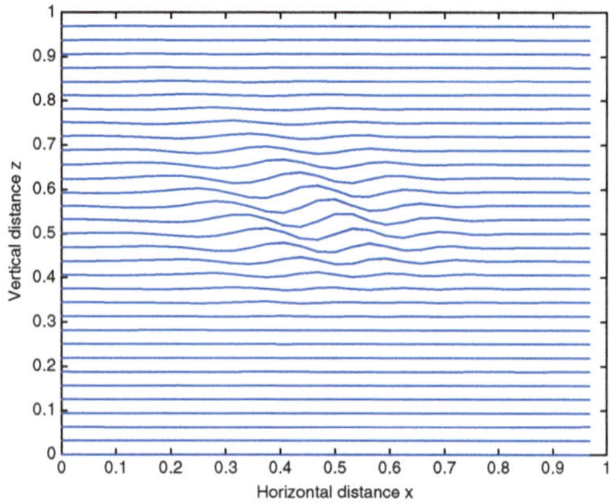

Fig. 1.4 Vertical slice showing the displacement of material lines in the presence of a packet of inertio-gravity waves. The wave crests and troughs are oriented top-left to bottom-right. The individual crests and troughs move towards the bottom left at the phase velocity, while the packet as a whole moves towards the top left at the group velocity

case), and that $\mathbf{c}_p.\mathbf{c}_g = 0$, i.e., group velocity is perpendicular to phase velocity. See Fig. 1.4.

An important measure of the accuracy of any numerical method is how well it captures the phase velocity and group velocity of different kinds of waves. We will look at dispersion relations and phase and group velocity for some example numerical schemes in Chaps. 3 and 4.

1.5 Balance

Atmospheric dynamics is characterized by being close to certain kinds of balance, at least on large enough horizontal scales, namely hydrostatic balance in the vertical, and geostrophic balance in the horizontal.

1.5.1 Hydrostatic Balance

Table 1.1 (closely following Holton 2004) shows typical scalings and typical values for the terms in the vertical momentum equation written in spherical polar coordinates for mid-latitude synoptic scale motions. Here L and H are typical horizontal and vertical length scales, U and W are typical horizontal and vertical velocity scales, a is the Earth's radius, f_0 is a typical value of the Coriolis parameter, and P_0 is a typical pressure value. It is clear that the dominant terms in the vertical momentum equation are

$$g + \frac{1}{\rho}\frac{\partial p}{\partial r} \approx 0, \tag{1.37}$$

i.e., *hydrostatic balance*.

Table 1.1 Typical scales of terms in the vertical momentum equation for synoptic scale midlatitude flow

w-equation	$\frac{Dw}{Dt}$	$-\frac{u^2+v^2}{r}$	$-2\Omega u\cos\phi$	g	$\frac{1}{\rho}\frac{\partial p}{\partial r}$
Scales	UW/L	U^2/a	$f_0 U$	g	$P_0/\rho H$
Values ms^{-2}	10^{-7}	10^{-5}	10^{-3}	10	10

Table 1.2 Typical scales of terms in the eastward and northward component momentum equations for synoptic scale midlatitude flow

u-equation	$\frac{Du}{Dt}$	$-\frac{uv\tan\phi}{r}$	$\frac{uw}{r}$	$-2\Omega v\sin\phi$	$2\Omega w\cos\phi$	$\frac{1}{\rho r\cos\phi}\frac{\partial p}{\partial\lambda}$
v-equation	$\frac{Dv}{Dt}$	$-\frac{u^2\tan\phi}{r}$	$\frac{vw}{r}$	$2\Omega u\sin\phi$		$\frac{1}{\rho r}\frac{\partial p}{\partial\phi}$
Scales	U^2/L	U^2/a	UW/a	$f_0 U$	$f_0 W$	$\delta P/\rho L$
Values ms^{-2}	10^{-4}	10^{-5}	10^{-8}	10^{-3}	10^{-6}	10^{-3}

1.5.2 *Geostrophic Balance*

Table 1.2 (also closely following Holton 2004) shows typical scalings and typical values for the terms in the horizontal momentum equation written in spherical polar coordinates for mid-latitude synoptic scale motions. The same typical scales are used as in Table 1.1, except that δP is a typical horizontal variation in pressure, and $\delta P \ll P_0$. Clearly the dominant terms are

$$u \approx -\frac{1}{f_0 \rho r}\frac{\partial p}{\partial \phi} \equiv u_g, \qquad v \approx \frac{1}{f_0 \rho r\cos\phi}\frac{\partial p}{\partial \lambda} \equiv v_g, \tag{1.38}$$

i.e., *geostrophic balance.*

A useful dimensionless number that measures the relative importance of the inertial term $D\mathbf{u}/Dt$ and the Coriolis term $2\boldsymbol{\Omega}\times\mathbf{u}$ is the Rossby number

$$Ro = U/(f_0 L). \tag{1.39}$$

Geostropic balance will be a good approximation provided $Ro \ll 1$.

1.5.3 *Conditions for Hydrostatic Balance to be a Good Approximation*

Hydrostatic balance is a good approximation on synoptic scales, but not necessarily on smaller horizontal scales. We can employ scale analysis to determine the conditions under which it will be a good approximation, i.e., under which we can neglect Dw/Dt compared to the other terms in the vertical momentum equation.

First note that we can define a global horizontal mean density $\rho_m(r)$ and a pressure field $p_m(r)$ in hydrostatic balance with it; these mean fields are dynamically uninteresting and we can subtract $g\rho_m + dp_m/dr = 0$ from the vertical momentum equation. Thus, the vertical acceleration will be negligible compared with the pressure gradient term provided

$$\frac{UW}{L} \ll \frac{\delta P}{\rho H}. \tag{1.40}$$

From the horizontal momentum equation

$$\frac{\delta P}{\rho} \sim U^2 \quad \text{or} \quad f_0 LU, \tag{1.41}$$

depending on whether the inertial term dominates (large Ro) or the Coriolis term dominates (small Ro). So we require

$$\frac{WH}{UL} \ll 1 \quad \text{or} \quad \frac{WH}{UL} Ro \ll 1. \tag{1.42}$$

From the mass continuity equation we obtain a relationship between the velocity scales and the length scales

$$\frac{W}{U} \sim \frac{H}{L} \quad \text{or} \quad \frac{W}{U} \sim \frac{H}{L} Ro.$$

The first case arises when $\partial w/\partial r$ is comparable to horizontal velocity gradients. The second case arises when there is a strong cancellation between the two horizontal components of the divergence. This happens when the Rossby number is small; the horizontal flow is then approximately non-divergent, and the divergence and hence $\partial w/\partial r$ are smaller by a factor Ro than suggested by the most obvious scaling. See Sect. 1.6.

Hence, hydrostatic balance will be a good approximation when

$$\frac{H^2}{L^2} \ll 1 \quad \text{or} \quad \frac{H^2}{L^2} Ro^2 \ll 1. \tag{1.43}$$

In practice this means L greater than about 10 km (a typical H); for smaller L the Rossby number is typically not small, so the second criterion in (1.43) is no more likely to be satisfied than the first.

1.5.4 Balance and Nonlocality

When the atmosphere is perturbed away from hydrostatic balance, it adjusts back towards hydrostatic balance through the radiation and ultimately dissipation of internal acoustic waves and inertio-gravity waves. Making the quasi-hydrostatic approximation in the governing equations (i.e., crossing out the Dw/Dt term) corresponds to filtering internal acoustic waves from the governing equations (and modifying the dynamics of inertio-gravity waves). More precisely, it corresponds to taking the limit in which the propagation speed of internal acoustic waves becomes infinite, so that the adjustment to hydrostatic balance is instantaneous. In the unapproximated equations all information propagates at finite speed; these are *hyperbolic* equations. The hydrostatic approximation introduces a certain nonlocality. Mathematically, this is reflected in the appearance of a one-dimensional boundary value problem. For example, in height coordinates we must solve a one-dimensional boundary value problem known as Richardson's equation for the vertical velocity (e.g., White 2002). If instead we use pressure as the vertical coordinate we must still solve two vertical integrals in order to compute the time tendencies of the prognostic fields.

Similar ideas apply in the case of geostrophic balance. The atmosphere adjusts towards geostrophic balance (or a nonlinear generalization of geostrophic balance such as *gradient wind balance*, e.g., Holton 2004) through the radiation and dissipation of inertio-gravity waves. The quasi-geostrophic approximation (see the next section) filters inertio-gravity waves from the governing equations, or, rather, corresponds to the limit in which inertio-gravity waves propagate infinitely fast so that the geostrophic adjustment process is instantaneous. This nonlocality is reflected mathematically in the appearance of a three-dimensional elliptic equation that must be solved in order to compute the time tendency of the prognostic field, in this case the potential vorticity.

Hydrostatic and geostrophic balance are physically relevant asymptotic limits of the governing equations. Even if we are solving the unapproximated (i.e., hyperbolic) governing equations, balance and the implied nonlocality are important. However, the solution of elliptic equations requires quite different numerical techniques from the solution of hyperbolic equations, particularly on massively parallel computers. Model developers therefore face an important choice between inherently local explicit time stepping techniques and inherently nonlocal implicit time stepping techniques.

1.6 Sketch of Quasigeostrophic Theory

A very brief sketch of quasigeostrophic theory is given here to lead up to a discussion of the dynamics of Rossby waves. See any of Gill (1982), Pedlosky (1987), Holton (2004), or Vallis (2006) for a fuller and more rigorous discussion.

We will work in Cartesian β-plane geometry, where $f = f_0 + \beta y$, f_0 is a constant mid-latitude value of the Coriolis parameter and $\beta = \partial f/\partial y$, and use a log-pressure vertical coordinate $\tilde{z} = -H_\rho \ln(p/p_{00})$, where $H_\rho = RT_{\text{ref}}/g$ is a constant density scale height related to a constant reference temperature T_{ref}, and p_{00} is a constant reference pressure. Now make four key assumptions:

- The flow is in hydrostatic balance. In terms of the geopotential Φ, $\partial\Phi/\partial\tilde{z} = RT/H_\rho$.

- $Ro \ll 1$ so that the flow is close to geostrophic balance.
- Thermodynamic quantities are close to reference profiles that are functions only of $\tilde{z}$. Reference profiles are indicated by subscript 0 and departures from reference profiles by a$'$.
- $\beta L/f_0 \ll 1$, i.e., fractional changes in the Coriolis parameter are small over the horizontal scales of interest.

With these assumptions, the leading order terms in the horizontal momentum equations simply state that the flow is close to geostrophic balance

$$u \approx u_g \equiv -\frac{1}{f_0}\frac{\partial \Phi}{\partial y}; \qquad v \approx v_g \equiv \frac{1}{f_0}\frac{\partial \Phi}{\partial x}. \tag{1.44}$$

It is convenient to introduce the geostrophic stream function $\psi = \Phi'/f_0$, so that

$$u_g = -\frac{\partial \psi}{\partial y}; \quad v_g = \frac{\partial \psi}{\partial x}; \quad \frac{\theta'}{\theta_{\mathrm{ref}}} = \frac{f_0}{g}\frac{\partial \psi}{\partial \tilde{z}}, \tag{1.45}$$

where $\theta_{\mathrm{ref}} = T_{\mathrm{ref}}(p_{00}/p)^{\kappa}$.

In order to say anything about the time evolution of the flow we need to go to next order. So define the *ageostrophic* velocity u_a, v_a by

$$u = u_g + u_a; \qquad v = v_g + v_a. \tag{1.46}$$

Then the next order terms in the momentum equations bring in the time derivatives of u_g and v_g. The two component equations may be combined to give a vorticity equation

$$\frac{D_g \zeta_g}{Dt} = \frac{f_0}{\rho_0}\frac{\partial}{\partial \tilde{z}}(\rho_0 \tilde{w}). \tag{1.47}$$

Here $\zeta_g = f + \partial v_g/\partial x - \partial u_g/\partial y$ is the geostrophic approximation to the vertical component of absolute vorticity, $\tilde{w} = D\tilde{z}/Dt$ is the vertical velocity in the log-pressure coordinate system, and ρ_0 is a reference density profile. $D_g/Dt \equiv \partial/\partial t + u_g \partial/\partial x + v_g \partial/\partial y$ is the derivative following the geostrophic flow. The thermodynamic equation at the same order becomes

$$\frac{D_g \theta'}{Dt} + \tilde{w}\frac{\partial \theta_0}{\partial \tilde{z}} = 0, \tag{1.48}$$

and this may be combined with the vorticity equation to obtain the potential vorticity equation

$$\frac{D_g q}{Dt} = 0, \tag{1.49}$$

where

$$q = f_0 + \beta y + \nabla_{\tilde{z}}^2 \psi + \frac{1}{\rho_0}\frac{\partial}{\partial \tilde{z}}\left(\rho_0 \frac{f_0^2}{N_{\mathrm{ref}}^2}\frac{\partial \psi}{\partial \tilde{z}}\right), \tag{1.50}$$

with $N_{\mathrm{ref}}^2 = (g/\theta_{\mathrm{ref}})\partial\theta_0/\partial\tilde{z}$.

The two equations (1.49) and (1.50), together with the diagnostic relations (1.45) and suitable boundary conditions, represent a closed set of equations for the evolution of the flow. Equation (1.49) embodies the advection or material conservation of potential vorticity. Equation (1.50) embodies the *invertibility* of potential vorticity, the idea that if we are given the three dimensional distribution of potential vorticity, along with suitable boundary conditions and the condition that the flow be in hydrostatic and geostrophic balance, then we can infer everything else about the wind and thermodynamic fields (e.g. Hoskins et al. 1985). Many phenomena in geophysical fluid dynamics can be understood in terms of the twin properties of advection and invertibility of potential vorticity. The potential benefits of respecting material conservation of potential vorticity in numerical models are discussed further in Chap. 11.

1.6.1 Rossby Waves

We can use quasigeostrophic theory, and the ideas of advection and invertibility of potential vorticity, to understand the dynamics of Rossby waves. Linearize (1.49) and (1.50) about a state of rest:

$$\frac{\partial q}{\partial t} + \beta v_g = 0; \tag{1.51}$$

$$q = \nabla_{\tilde{z}}^2 \psi + \frac{1}{\rho_0} \frac{\partial}{\partial \tilde{z}} \left(\rho_0 \frac{f_0^2}{N_{\text{ref}}^2} \frac{\partial \psi}{\partial \tilde{z}} \right). \tag{1.52}$$

(q is now the potential vorticity perturbation.) Seek solutions

$$\psi = Re \left\{ \hat{\psi}(\tilde{z}) \exp\left[\mathrm{i}(kx + ly + m\tilde{z} - \omega t) \right] \right\} \tag{1.53}$$

that are wavelike in the horizontal and in time but may have a more complicated vertical structure expressed through $\hat{\psi}(\tilde{z})$. By expressing (1.51) in terms of $\hat{\psi}$ and eliminating $\hat{\psi}$ we obtain the dispersion relation

$$\omega = -\frac{\beta k}{k^2 + l^2 + (m^2 + 1/(4H_\rho^2)) f_0^2 / N_{\text{ref}}^2}. \tag{1.54}$$

Figure 1.5 shows schematically the horizontal propagation of a Rossby wave packet. The background potential vorticity increases towards the North. The displacement of the potential vorticity contours (which are material contours) indicates how the potential vorticity has been advected by the wind field. The potential vorticity anomalies in turn determine the wind field through invertibility: positive potential vorticity anomalies have cyclonic circulation while negative potential vorticity anomalies have anticyclonic circulation. The wind field then further

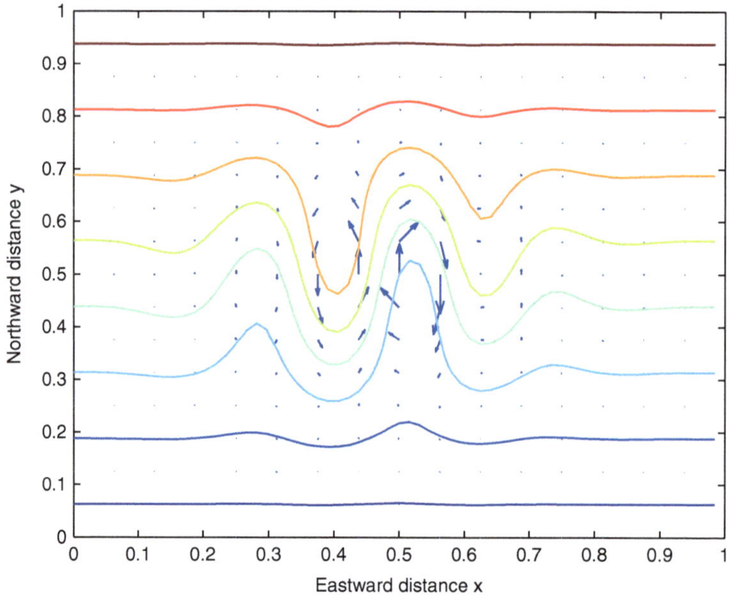

Fig. 1.5 Schematic showing the propagation of a packet of Rossby waves in the longitude-latitude plane. The *contours* indicate potential vorticity values; the *arrows* indicate the wind field

advects the potential vorticity. In this case it is clear that the wind field acts to displace the pattern of potential vorticity crests and troughs towards the west, consistent with the negative values of ω given by (1.54).

1.7 Eulerian and Lagrangian Timescales

The Eulerian view of fluid mechanics looks at the evolution of the fluid fields at fixed locations in space as the fluid moves past. When a feature of length scale L or wavenumber k is advected past at a velocity of scale U, the timescale for its rate of change is

$$\tau_{\mathrm{Eul}} \sim \frac{L}{U} \sim \frac{1}{kU}. \tag{1.55}$$

The Lagrangian view of fluid mechanics looks at the evolution of the fluid fields following fluid parcels. Some quantities (χ, say) are approximately materially conserved ($D\chi/Dt \approx 0$), so they have long Lagrangian timescales. Other quantities, like the pressure or vorticity, evolve on a timescale determined by the velocity gradients or strain field S experienced by the fluid parcel

$$\tau_{\mathrm{Lag}} \sim \frac{1}{S}. \tag{1.56}$$

For the large-scale, balanced, atmospheric flow, the energy spectrum is relatively steep, close to k^{-3}, which implies that the strain field is dominated by the largest scales (or smallest wavenumbers, say k_0) of the flow

$$\tau_{\text{Lag}} \sim \frac{1}{S} \sim \frac{1}{k_0 U}. \tag{1.57}$$

Thus, the Lagrangian timescale for atmospheric flow is typically significantly longer than the Eulerian timescale. This fact may be exploited through the use of semi-Lagrangian time discretizations in atmospheric models; the slow Lagrangian evolution can be captured more accurately (for a given time step) than the faster Eulerian evolution. However, this disparity in timescales is less clear cut for smaller scales of motion or when departures from balance (i.e., fast waves) become important. Another important exception is flow over orography; in this case τ_{Eul} becomes very long, because the flow is quasi-steady from the Eulerian point of view, while

$$\tau_{\text{Lag}} \sim \frac{L}{U} \sim \frac{1}{kU}, \tag{1.58}$$

where L and k are now the length scale and wavenumber of the orography and the flow perturbations it induces. In this case, semi-Lagrangian schemes, using the long time steps permitted by a semi-implicit treatment of the fast waves, can suffer from spurious *orographic resonance* (e.g., Rivest et al. 1994).

1.8 Turbulence and Cascades

The nonlinearity of the governing equations implies that there is an interaction between different scales of motion. A numerical model must be able to handle appropriately these nonlinear scale-interactions. In particular, even for an initially smooth and well resolved initial condition, the dynamics will attempt to generate variability near the grid scale, which may be poorly represented, and below the grid scale, which cannot be resolved at all. In this section we will look at some idealized models of turbulence and the nonlinear scale interactions that they describe. Space permits only the very briefest of introductions here; see, for example, Salmon (1998) for an excellent fuller discussion.

1.8.1 Three Dimensional Turbulence

Consider three-dimensional, statistically steady, homogeneous and isotropic turbulence in an incompressible constant density fluid. Assume that the fluid is stirred, and energy is input, on some large scale, and that energy is dissipated by viscosity

Fig. 1.6 Schematic indicating the downscale energy cascade in three-dimensional turbulence

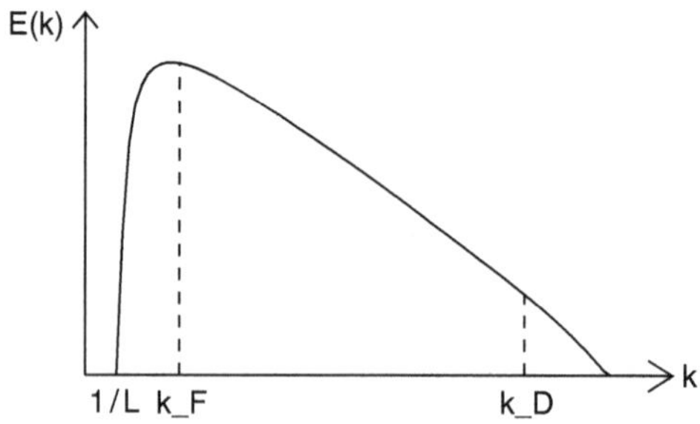

at some small scale; there must therefore be a systematic transfer of energy from the forcing scale to the dissipation scale. When this transfer occurs through a succession of gradually smaller eddies it is referred to as a *cascade*. Assume, also, that there is some range of scales in between the forcing and dissipation scales – the *inertial range* – that is statistically independent of the details of the forcing and dissipation. The rate of energy production ε must equal the rate of energy dissipation. Moreover, the rate of transfer of energy from wavenumbers smaller than k to wavenumbers greater than k, for any k in the inertial range, must also equal ε. See Fig. 1.6.

The following dimensional argument (Kolmogorov 1941) then implies a particular form for the energy spectrum. The dimensions of the spectral energy density $\hat{E}(k)$, i.e., the energy per unit wavenumber of the spectrum, are

$$\left[\hat{E}(k)\right] = \mathrm{L}^3\mathrm{T}^{-2}, \tag{1.59}$$

where L stands for length and T stands for time. In the inertial range at wavenumber k, the only dimensional quantities are k itself and ε.

$$[k] = \mathrm{L}^{-1} \quad \text{and} \quad [\varepsilon] = \mathrm{L}^2\mathrm{T}^{-3}, \tag{1.60}$$

so the only way to construct a quantity with the same dimensions as $\hat{E}(k)$ is

$$\hat{E}(k) = C_1\varepsilon^{2/3}k^{-5/3} \tag{1.61}$$

for some universal C_1 of order 1.

1.8.2 *Two-dimensional Turbulence*

Now consider two-dimensional, statistically steady, homogeneous and isotropic turbulence in an incompressible constant density fluid. In two dimensions we have another conservable quantity, the enstrophy, and therefore a cascade of enstrophy at a rate η.

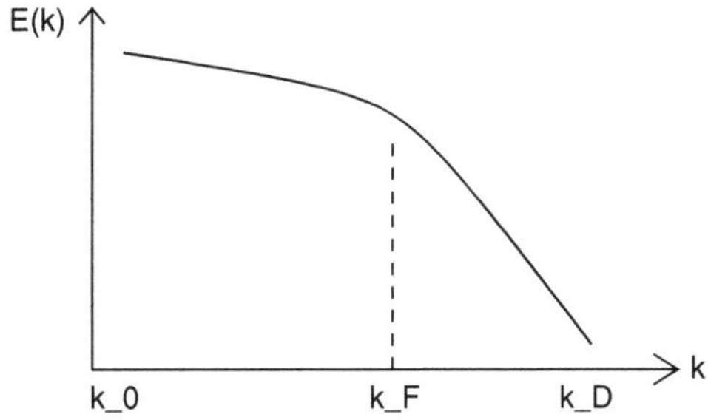

Fig. 1.7 Schematic indicating the upscale energy cascade and downscale enstrophy cascade in two-dimensional turbulence

Typically energy now cascades upscale while enstrophy cascades downscale (Fig. 1.7). The argument for the $k^{-5/3}$ spectrum given above did not depend on the number of space dimensions, nor on the direction of the energy cascade. We therefore expect to see a $k^{-5/3}$ spectrum on scales larger than the forcing scale, provided there is a mechanism to provide a sink of energy at very large scales.

In the inertial range on the small-scale side of the forcing, again the dimensions of $\hat{E}(k)$ are given by (1.59), but now the only dimensional quantities are

$$[k] = \mathrm{L}^{-1} \quad \text{and} \quad [\eta] = \mathrm{T}^{-3}. \tag{1.62}$$

Hence, the only way to construct a quantity with the same dimensions as $\hat{E}(k)$ is

$$\hat{E}(k) = C_2 \eta^{2/3} k^{-3} \tag{1.63}$$

for some universal C_2 of order 1.

1.8.3 *Energy Upscale, Enstrophy Downscale*

The above arguments, based on statistically steady flow, suggest that, in two dimensions, energy will cascade predominantly upscale while enstrophy will cascade predominantly downscale. Another argument, leading to the same conclusion, is given by considering an initial value problem.

Let E and Z be the total energy and enstrophy per unit area:

$$E = \int \hat{E}(k)\, dk \quad \text{and} \quad Z = \int \hat{Z}(k)\, dk. \tag{1.64}$$

The enstrophy spectrum is related to the energy spectrum by $\hat{Z}(k) = k^2 \hat{E}(k)$. Suppose energy is initially concentrated near wavenumber k_1 and subsequently spreads out, so that

$$\frac{d}{dt} \int (k - k_1)^2 \hat{E}(k)\, dk > 0. \tag{1.65}$$

Expanding the integral and substituting from (1.64), and using the fact that E and Z are conserved (neglecting viscosity) leads to

$$\frac{d}{dt}\left(\frac{\int k\hat{E}(k)\,dk}{\int \hat{E}(k)\,dk}\right) < 0. \tag{1.66}$$

However, the quantity under the time derivative here is a representative wavenumber for energy, implying that, in some mean sense, energy moves to larger scales.

Similarly, if we assume that

$$\frac{d}{dt}\int (k^2 - k_1^2)^2 \hat{E}(k)\,dk > 0, \tag{1.67}$$

then, again expanding the integral and substituting from (1.64), conservation of E and Z implies

$$\frac{d}{dt}\left(\frac{\int k^2\hat{Z}(k)\,dk}{\int \hat{Z}(k)\,dk}\right) > 0. \tag{1.68}$$

Thus, a representative wavenumber for the enstrophy increases in time, implying that enstrophy, in some mean sense, moves to small scales.

Figures 1.8 and 1.9 show an example numerical solution of the barotropic vorticity equation

$$\frac{D\zeta}{Dt} = 0, \tag{1.69}$$

where the velocity field used to calculate the material derivative is given by

$$u = -\frac{\partial\psi}{\partial y}; \quad v = \frac{\partial\psi}{\partial x}; \quad \nabla^2\psi = \zeta. \tag{1.70}$$

The barotropic vorticity equation is one form of the equations describing two-dimensional incompressible flow. It bears some resemblance to quasigeostrophic theory as it embodies the advection and invertibility of vorticity. In this example the domain is square and doubly periodic. The initial condition is a not-quite-regular array of vortices of alternating sign. The numerical solution is calculated using a Spectral method (e.g., Williamson and Laprise 2000) based on Fourier transforms. The maximum resolved wavenumber in the x- and y-directions is 85. A suitably tuned $\kappa\nabla^4\zeta$ is added to the right hand side of (1.69) to dissipate enstrophy that cascades towards the resolution limit; here κ is the dissipation coefficient. See Chap. 11 for a discussion of what happens when this term is not included.

The right hand panel of Fig. 1.9 shows the solution after a few vortex turnover times. Several mergers between vortices of the same sign have taken place, and some are clearly in the process of taking place at this instant. This tendency for like-signed vortices to merge is one of the physical space manifestations of the upscale energy cascade discussed above.

At the same time, fluid has been stripped from the edges of most vortices and drawn out into long thin filaments that fill the space between the vortices. This

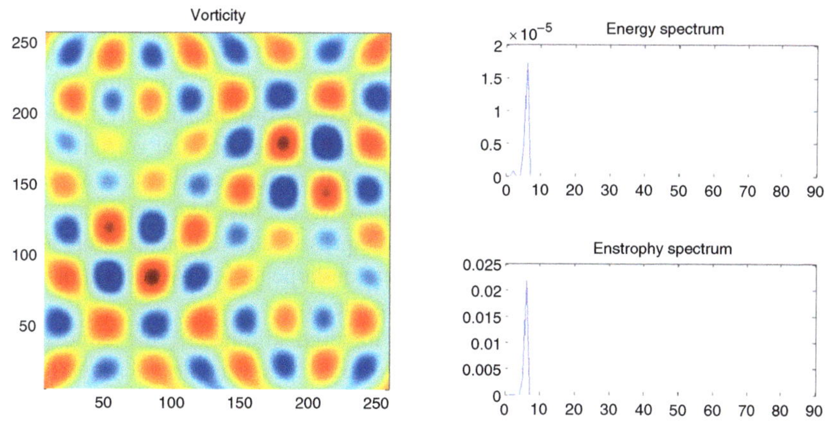

Fig. 1.8 Initial condition for a numerical solution of the barotropic vorticity equation. The *left hand panel* shows the initial vorticity field; *red* is positive vorticity, *blue* is negative vorticity. The *right hand panels* show the initial energy and enstrophy spectra

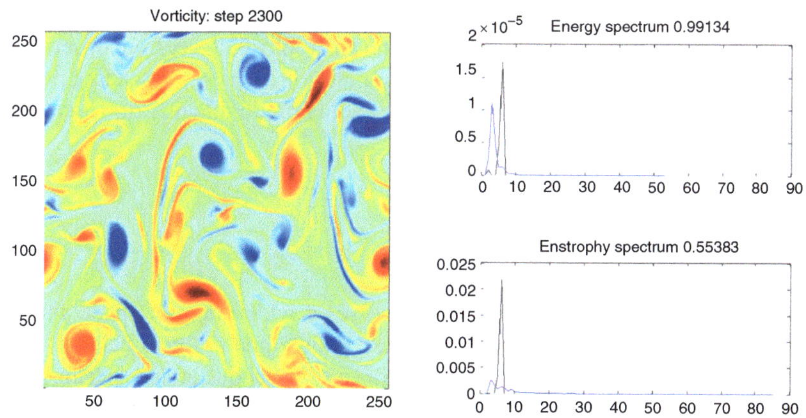

Fig. 1.9 As in Fig. 1.8 but after a few vortex turnover times. The *right hand panels* show both the initial spectra (*black*) and the spectra at the current time (*blue*)

process is the physical space manifestation of the downscale enstrophy cascade discussed above.

1.8.4 Application to the Real Atmosphere

There are a number of caveats associated with these arguments, besides their extreme idealization, including the fact that they neglect intermittency, and the fact that a spectrum as steep as k^{-3} is just barely consistent with the idea of an inertial range because the large scales will begin to dominate the strain rate and interactions will cease to be local in spectral space. Furthermore, the atmosphere is not a two-dimensional incompressible fluid. However, much of the atmosphere is stably stratified and moves approximately layerwise two-dimensionally. Moreover, the atmosphere has an approximate material invariant, the potential vorticity, somewhat analogous to the vorticity in two-dimensional incompressible flow, and hence has a quadratic invariant, the potential enstrophy (see Chap. 11), somewhat analogous to the enstrophy in two-dimensional incompressible flow. It is therefore argued that the turbulent behaviour of the atmosphere on large scales will be qualitatively similar to that of two-dimensional incompressible flow.

On horizontal scales larger than a few hundred kilometers, the atmospheric kinetic energy spectrum is observed to be close to k^{-3}, as in an inertial range (potential) enstrophy cascade. However, analysis of global datasets implies that there are significant sources and sinks of energy across a wide range of scales, which is inconsistent with the idea of an inertial range. Furthermore, the observed kinetic energy spectrum makes a transition to something close to $k^{-5/3}$ on scales smaller than a few hundred kilometers; this transition is quite different from the prediction of two-dimensional turbulence theory and there is currently no widely accepted explanation for it. However, careful analysis of energy and enstrophy budgets from observations and global datasets implies that the general conclusion of energy cascading predominantly upscale and (potential) enstrophy cascading predominantly downscale does indeed hold.

1.9 Conclusion

Atmospheric dynamics is complex and involves a wide range of space and time scales. The energetically dominant dynamics is slow and close to balance, and it may be wavelike, vortical, or strongly nonlinear. Fast acoustic and inertio-gravity waves represent departures from balance, but are also the mechanism by which the atmosphere continuously adjusts towards balance. Nonlinearity implies interactions between the different space and time scales. Particularly important are energy and potential enstrophy transfers across scales; for any practical global atmospheric model there will inevitably be important dynamics occurring near the resolution limit. The need to capture all of these processes with sufficient accuracy make numerical modeling of the atmosphere one of the most challenging branches of computational fluid dynamics.

References

Arakawa A, Konor CS (2009) Unification of the anelastic and quasi-hydrostatic systems of equations. Mon Wea Rev 137:710–726

Durran DR (1999) Numerical Methods for Wave Equations in Geophysical Fluid Dynamics. Springer-Verlag

Durran DR (2008) A physically motivated approach for filtering acoustic waves from the equations governing compressible stratified flow. J Fluid Mech 601:365–379
Gill AE (1982) Atmosphere-Ocean Dynamics. Academic Press, New York
Holton JR (2004) An Introduction to Dynamic Meteorology, fourth edition edn. Elsevier Academic Press, Amsterdam
Hoskins BJ, McIntyre ME, Robinson AW (1985) On the use and significance of isentropic potential vorticity maps. Quart J Roy Meteorol Soc 111:877–946
Kolmogorov AN (1941) Dissipation of energy in locally isotropic turbulence. Dokl Akad Nauk SSSR 32:16–18, (reprinted in Proc. Roy. Soc. Lond. A, **434**, 15–17 (1991))
Nastrom GD, Gage KS (1985) A climatology of atmospheric wavenumber spectra of wind and temperature observed by commercial aircraft. J Atmos Sci 42:950–960
Pedlosky J (1987) Geophysical Fluid Dynamics. Springer, New York
Rivest C, Staniforth A, Robert A (1994) Spurious resonant response of semi-Lagrangian discretizations to orographic forcing: Diagnosis and solution. Mon Wea Rev 122:366–376
Salmon R (1998) Lectures on Geophysical Fluid Dynamics. Oxford University Press
Vallis GK (2006) Atmospheric and Oceanic Fluid Dynamics. Cambridge University Press, Cambridge
White AA (2002) Large-Scale Atmosphere-Ocean Dynamics I: Analytical Methods and Numerical Models, Cambridge University Press, chap A view of the equations of meteorological dynamics and various approximations, pp 1–100
White AA, Hoskins BJ, Roulstone I, Staniforth A (2005) Consistent approximate models of the global atmosphere: Shallow, deep, hydrostatic, quasi-hydrostatic and non-hydrostatic. Quart J Roy Meteorol Soc 131:2081–2107
White AA, Staniforth A, Wood N (2008) Spheroidal coordinate systems for modelling global atmospheres. Quart J Roy Meteorol Soc 134:261–270
Williamson DL, Drake JB, Hack JJ, Jakob R, Swarztrauber PN (1992) A standard test set for numerical approximations to the shallow water equations in spherical geometry. J Comput Phys 102:211–224
Williamson DL, Laprise R (2000) Numerical Modeling of the Global Atmosphere in the Climate System, Kluwer, chap Numerical approximations for global atmospheric GCMs., pp 127–219

Chapter 2
Waves, Hyperbolicity and Characteristics

Joseph Tribbia and Roger Temam

Abstract This lecture describes the basics of hyperbolic systems as needed to solve the initial boundary value problem for hydrostatic atmospheric modeling. We examine the nature of waves in the hydrostatic primitive equations and how the modal decomposition can be used to effect a complete solution in the interior of an open domain. The relevance of the open boundary problem for the numerical problem of static and adaptive mesh refinement is discussed.

2.1 Introduction

The most comprehensive dynamical model of the atmosphere is the Navier Stokes equation for a compressible gas. Because of the viscous stress term this system of equations is parabolic, i.e., formally similar to the diffusion equation. However, on the length scales which we currently numerically model the atmosphere for weather prediction and climate simulation the dissipation time scale is quite large. For example, using the molecular viscosity of dry air, $\nu = 1.5 \times 10^{-6}\,\mathrm{m}^2/\mathrm{s}$, and a length scale, $L = 1$ km, the e-folding time for viscous decay is 2,000 years. Note that if L corresponds to the grid length in a numerical model this scale is currently beyond our computational capability for a global weather or climate model. But even for much smaller length scales, $L = 1$ m, the e-folding time is greater than a half day which, as will be shown later, is still much longer than the relevant propagation time scale of many atmospheric waves which are represented in the compressible Navier–Stokes equations. Thus, for the purpose of understanding the behavior of numerical weather and climate models, the atmosphere can be considered a hyperbolic system

J. Tribbia (✉)
National Center for Atmospheric Research, 1850 Table Mesa Drive, Boulder, CO 80305, USA
e-mail: tribbia@ucar.edu

R. Temam
Institute for Scientific Computing and Applied Mathematics, Indiana University, Rawles Hall, Bloomington, IN 47405-5701, USA
e-mail: temam@indiana.edu

P.H. Lauritzen et al. (eds.), *Numerical Techniques for Global Atmospheric Models*,
Lecture Notes in Computational Science and Engineering 80,
DOI 10.1007/978-3-642-11640-7_2,

of equations and the (molecular) dissipation terms may be neglected to a very good first approximation.

A classical mathematical treatment of hyperbolic systems uses the method of characteristics to simplify and formally solve the governing partial differential equations. In the solution of the PDEs in this manner, the question of boundary conditions for open spatial domains is elucidated. And while the main focus of this volume of lecture notes is global modeling, the challenges associated with static and adaptive grid refinement can be shown to be related to the issues surrounding the open boundary value problem. Thus the remainder of this contribution will be presented as follows; Sect. 2.2 will give a very general introduction to the method of characteristics and give the simplest example of its use; Sect. 2.3 will develop the normal mode structure of the hydrostatic primitive equations in Cartesian geometry discussing the role of the hydrostatic balance and the resultant modified oscillations in the reduced (hydrostatic) system. We will then use the methods of Sect. 2.2 to solve the open boundary problem for the hydrostatic system in this simplest context. Section 2.4 will examine the equivalent problem in spherical geometry and Sect. 2.5 will conclude with a discussion of the utility of these results within the context of global non-hydrostatic weather and climate models.

2.2 The Method of Characteristics

In this section the basics of the method of characteristics is presented in the simplest context for the solution of a first order partial differential equation in two variables (x, t) with the dependent variable to be solved for given as $u(x, t)$. The governing equation is then:

$$a(x,t,u)u_t + b(x,t,u)u_x = c(x,t,u), \tag{2.1}$$

and treating t as the time variable, the initial value problem for (2.1) can be posed by specifying $u(x, 0) = F(x)$. The solution via the characteristic method is then forged by solving the auxiliary set of ordinary differential equations in the variable s taken to be the distance along a characteristic curve in $(x, t) : s = s(x, t)$:

$$\frac{dt}{ds} = a(x,t,u), \qquad \frac{dx}{ds} = b(x,t,u), \qquad \frac{du}{ds} = c(x,t,u). \tag{2.2}$$

That this results in a solution to (2.1) is easily seen by rewriting u as a function of s, i.e., $u(s) = u(x(s), t(s))$ and using the chain rule.

As an example of the method, let $a(x,t,u) = 1$, $b(x,t,u) = U_0$ with U_0 a constant, and $c(x,t,u) = 0$. The solution of (2.2) is then $t = s$, $x(s) = x(0) + b_0 s$ and $u = const$ along each characteristic curve $x = x(0) + U_0 t$. Using the initial value of $u(x, 0)$ gives the result that $u(x,t) = F(x - U_0 t)$. If, rather than specifying $c(x,t,u) = 0$, $c(x,t,u) = -ru$ with $r = const$ is given, then the solution above would be modified to $u(x,t) = \exp(-rt)F(x - U_0 t)$. Note that above the general partial differential equation (2.1) can, in fact, be nonlinear since the coefficients of u_t, u_x and u can depend on u. The method of characteristics is useful for this special type of nonlinearity which restricts the dependence of these coefficients to be only on u with no dependence on partial derivatives of u. Such equations are termed quasi-linear equations due to this restriction.

The solution of the initial value problem above is determined for all $t > 0$ and the entire x axis. If we wish to limit the domain of the solution to the strip in (x, t) such that $t > 0$ and $0 < x < L$, then the solution given informs us as to how this must be done. Assuming $U_0 > 0$, the characteristic curves carry the solution from left to right in x. For the solution to the limited domain problem it is clear that $u(0, t)$ must be given in order to update $u(x, t)$ in the interior. However, $u(L, t)$ should not be specified as the solution is carried by the characteristics from the interior to this boundary. The method of characteristics is then the appropriate analysis technique for determining the boundary conditions that lead to well-posed initial boundary value problems (IBVPs).

The extension of the method to two space and one time variable is straightforward. The general form of the governing equation is then:

$$a(x, y, t, u)u_t + b(x, y, t, u)u_x + c(x, y, t, u)u_y = d(x, y, u, t), \tag{2.3}$$

and the characteristic curves in (x, y, t) are now determined by the solution to:

$$\frac{dt}{ds} = a(x, y, t, u), \quad \frac{dx}{ds} = b(x, y, t, u), \quad \frac{dy}{ds} = c(x, y, t, u),$$

$$\frac{du}{ds} = d(x, y, t, u). \tag{2.4}$$

The spatially two-dimensional generalization of the constant coefficient case, i.e., letting $a(x, y, t, u) = 1$, $b(x, y, t, u) = U_0$, $c(x, y, t, u) = V_0$ and $d(x, y, t, u) = -ru$, with U_0, V_0, and r all constants and initial condition $u(x, y, 0) = F(x, y)$, has as the solution: $u(x, y, t) = \exp(-rt)F(x - U_0t, y - V_0t)$. The analysis of the limited area IBVP proceeds in the same way as in the case of one space dimension, leading to the specification of u on boundaries for which the characteristic curves point inward as time increases and allowing the solution to evolve at boundaries where the characteristic curves are directed outward. This leads to a well-posed IBVP. It should be noted that in any number of space dimensions a singular case exists for the IBVP, where the boundary corresponds precisely to a characteristic curve. In this case the characteristic curves are neither inward nor outward and so no specification leads to a well-posed problem and no solution is possible. In this singular case the IBVP is ill-posed.

The method can be used to solve the IBVP for a system of quasi-linear PDEs of the general form:

$$\mathbf{A}(\mathbf{U}, \mathbf{x}, t)\mathbf{U}_t + \mathbf{B}(\mathbf{U}, \mathbf{x}, t) \cdot \nabla\mathbf{U} + \mathbf{C}(\mathbf{U}, \mathbf{x}, t)\mathbf{U} = \mathbf{D}(\mathbf{U}, \mathbf{x}, t), \tag{2.5}$$

where **U** is an N-dimensional vector dependent variable, **A**, **B**, **C** are N by N matrices and **D** is an N-dimensional vector function of $(\mathbf{U}, \mathbf{x}, t)$. This form is general enough that the Euler equations for a perfect gas can be seen to be one of the system of PDEs for which the method of characteristics can be used. In the vector case, with multiple space dimensions, the needed mathematical trick is to diagonalize the system so that (2.5) is equivalent to multiple scalar equations and reduce the problem to one similar to solving (2.1) or (2.3). For the primitive equations a preliminary step is needed to reach the above form This step and the diagonalization of the atmospheric equations is the topic of the next section.

2.3 The Normal Modes of the Hydrostatic Equations

In this section, the diagonalization of the hydrostatic equations is taken up. As noted in the previous section, the fully compressible Euler equations are of the form of (2.5) above and so are amenable to solution using the method of characteristics. In addition, as noted in the lecture on basic atmospheric dynamics (Chap. 1), global modeling efforts are increasingly giving up the use of model formulations which impose balance conditions within their formulation. Why, one might ask, are the hydrostatic equations the topic of this section? There are two primary reasons for this: First, only one of the global models represented at the colloquium is based on the non-hydrostatic, fully compressible equations. The remaining ten models studied are formulated using the hydrostatic balance assumption. Second, even in the global and regional non-hydrostatic models that currently exist or are planned for the future, the issues that arise in the actual application of the method of characteristics to such models are very much related to the difficulties that exist in the solution of the hydrostatic primitive equations. This latter point will be elaborated upon below.

The diagonalization of the hydrostatic system will be discussed in two parts. First, the simpler problem of diagonalization in Cartesian geometry will be developed because of its easy connection with the presentation given in Chap. 1 and because of its relation to the local problems to be discussed in the next section on well-posedness. In addition to hydrostatic balance the equations used also make the further approximation of incompressibility and are thus more applicable to ocean than the atmosphere. However, a change in vertical coordinate for the atmosphere can bring about a strong similarity to these equations. After the development in Cartesian geometry a brief detour will be made to demonstrate the similarities and differences caused by more realistic spherical geometry and the restoration of compressibility.

In the presence of viscosity, well-posedness of the full primitive equations has been established by Lions et al. (1992a,b), for both the atmosphere and the ocean. Because of the very long time scale associated with viscous dissipation noted above, in this article we are interested in the zero viscosity case. We restrict ourselves in this Cartesian geometry analysis to the primitive equations linearized around a constant flow velocity U_0 in the x direction with no dependence in y. Note that the essence of the of the difficulties discussed are not changed by the restriction to the linear form assumed here since linearization can be used as a guide to the solution of the full nonlinear equations. The equations are then

$$u_t - fv + \phi_x + U_0 u_x = 0, \tag{2.6}$$
$$v_t + fu + U_0 v_x = 0, \tag{2.7}$$
$$\theta_t + U_0\theta_x + N^2\frac{\theta_0}{g}w = 0, \tag{2.8}$$
$$u_x + w_z = 0, \tag{2.9}$$
$$\phi_z = \frac{g\theta}{\theta_0}. \tag{2.10}$$

Where g is the gravity constant, θ_0 a reference potential temperature, f is the (constant) Coriolis parameter and $N = N(z)$ is the Brunt–Väisälä frequency for the unperturbed flow and lower case variables (u, v, ϕ, w, θ) are perturbations from the reference values. Equations (2.8) and (2.10) can be combined to eliminate θ and yield an equation for ϕ:

$$\phi_{zt} + U_0\phi_{zx} + N^2 w = 0. \tag{2.11}$$

Attempting separation of variables, we look for a solution of (2.6)–(2.10) in the form[1]

$$\begin{pmatrix} u \\ v \\ \phi \end{pmatrix} = \mathscr{U}(z)\begin{pmatrix} \hat{u} \\ \hat{v} \\ \hat{\phi} \end{pmatrix}, w = \mathscr{W}(z)\hat{w}, \tag{2.12}$$

where $\hat{u}$, $\hat{v}$, $\hat{w}$, and $\hat{\phi}$ depend only on x and t. By substitution in (2.7) and (2.9) we find

$$\frac{\mathscr{U}'}{N^2\mathscr{W}} = -\frac{\hat{w}}{\hat{\phi}_t + U_0\hat{\phi}_x} \; (= c_1), \frac{\mathscr{U}}{\mathscr{W}'} = -\frac{\hat{w}}{\hat{u}_x} \; (= c_2). \tag{2.13}$$

The quantities above are constant $(= c_1, c_2)$, since the left-hand sides of the equations depend on z alone and the right-hand sides depend only on x and t.

Combining these equations we obtain:

$$\left(\frac{\mathscr{U}_z}{N^2}\right)_z + \lambda^2\mathscr{U} = 0, \quad \text{and} \quad \mathscr{W}_{zz} + \lambda^2 N^2\mathscr{W} = 0, \tag{2.14}$$

[1] Note that if a solution of the form $u = \mathscr{U}\hat{u}, v = \mathscr{V}\hat{v}, \phi = \varphi\hat{\phi}$, is assumed then (2.6) and (2.7) imply that $\mathscr{U}, \mathscr{V}, \varphi$ are proportional to each other, and therefore without loss of generality may be taken to be equal.

with $\lambda^2 = -c_1/c_2$, and we now solve each (2.14) as an eigenvalue problem taking the boundary conditions into consideration. Since $w = 0$ on top and bottom, we have

$$\mathscr{W}(0) = \mathscr{W}(H) = \mathscr{U}'(0) = \mathscr{U}'(H) = 0. \tag{2.15}$$

Equation (2.14) with boundary conditions (2.15) for $\mathscr{U}$ ($N > 0$ bounded from above and from below) is solved and we denote by λ_n^2 the corresponding eigenvalues and write

$$\lambda_n^2 = \frac{1}{gH_n}, \text{ then, as known from Sturm–Liouville theory, } H_1 \geq H_2 \geq \cdots, \text{ and}$$
$$H_n \to 0 \text{ as } n \to \infty.$$

A particular simple example of this is that of N^2 equal to a positive constant. For this case:

$$\lambda_n = \frac{n\pi}{NH} \text{ and the corresponding eigenfunctions are } \sin\left(\frac{n\pi z}{NH}\right) \text{ or } \cos\left(\frac{n\pi z}{NH}\right)$$

as shown by Thuburn et al. (2002).

Denoting by $\mathscr{U}_n, \mathscr{W}_n$ the corresponding vertical normal modes and by $\hat{u}_n, \hat{v}_n, \hat{\phi}_n, \hat{w}_n$ the corresponding (x, t) dependent variables, we eliminate $\hat{w}_n$ and obtain a system identical to the linearized shallow water equations.

$$\begin{aligned} &\hat{u}_t - f\hat{v} + \hat{\phi}_x + U_0\hat{u}_x = 0, \\ &\hat{v}_t + f\hat{u} + U_0\hat{v}_x = 0, \\ &\hat{\phi}_t + U_0\hat{\phi}_x + gH_n\hat{u}_x = 0. \end{aligned} \tag{2.16}$$

(Note that in the system above the subscript n has been dropped on the variables $\hat{u}, \hat{v}, \hat{\phi}$ leaving the dependence on n to be indicated through the coefficient H_n.) The characteristic/eigen values are given as $U_0 \pm \sqrt{gH_n}$ and U_0. Now, if $gH_n < U_0^2$, three characteristics enter the $x-$ domain $(0, L)$ and three boundary conditions are needed at $x = 0$. If $gH_n > U_0^2$, only two characteristics enter the domain $(0, L)$ and only two boundary conditions are needed at $x = 0$ (and one at $x = L$). Analogous comments are valid at $x = L$.

For solving (2.6)–(2.10) in the general case, since from Sturm–Liouville theory the vertical eigenfunctions form a complete set, we can expand all functions in the basis of vertical normal modes that we have just determined:

$$\begin{pmatrix} u \\ v \\ \phi \end{pmatrix}(x, z, t) = \sum_n \mathscr{U}_n(z) \begin{pmatrix} \hat{u}_n \\ \hat{v}_n \\ \hat{\phi}_n \end{pmatrix}(x, t), \tag{2.17}$$

$$w(x, z, t) = \sum_n \mathscr{W}_n(z)\hat{w}_n(x, t), \tag{2.18}$$

and the analysis above holds for each n. Because for $gH_n < U_0^2$ we need three boundary conditions for each mode at $x = 0$ and for $gH_n > U_0^2$, we need two boundary conditions for each mode at $x = 0$, the number of boundary conditions to be applied depends on n, and thus the vertical transform of the prognostic variables. The index n is determined from a vertical integration of the variables and the vertical normal modes and is thus a non-local property of the each dependent variable. Using an argument similar to this, Oliger and Sundström (1978) concluded that there is no set of local (i.e., pointwise) boundary conditions at $x = 0$ which makes the system (2.6)–(2.10) well-posed.

To remedy this problem, as shown in Temam and Tribbia (2003), we can modify the primitive equations by the addition of a Newtonian damping term on the vertical velocity and add this to the hydrostatic balance equation so that (2.10) becomes:

$$\delta\tilde{w} + \phi_z = \frac{g\theta}{\theta_0}. \tag{2.19}$$

With the addition of this term it can then be shown through the conservation of energy constraint that the solutions to this system are unique and have continuous dependence on the data for local boundary conditions. In this way the addition of dissipation (even of a rather mild type) can regularize the ill-posed nature of the hydrostatic primitive equations for the Initial Boundary Value Problem (IVBP).

The efficacy of this dissipation term can be seen in Figs. 2.1–2.5 which depict the numerical solution of the IBVP for both the standard and the δ modified hydrostatic systems above with the initial conditions shown in Fig. 2.1. The lateral boundary conditions correspond to upwinding for both versions of hydrostatic system. One can easily see the effects of ill-posedness in Fig. 2.2 where the wind is subcritical for the first internal mode (i.e., $gH_1 > U_0^2$) and the solution is thus over-specified through the use of upstream boundary conditions at the inflow boundary. On the

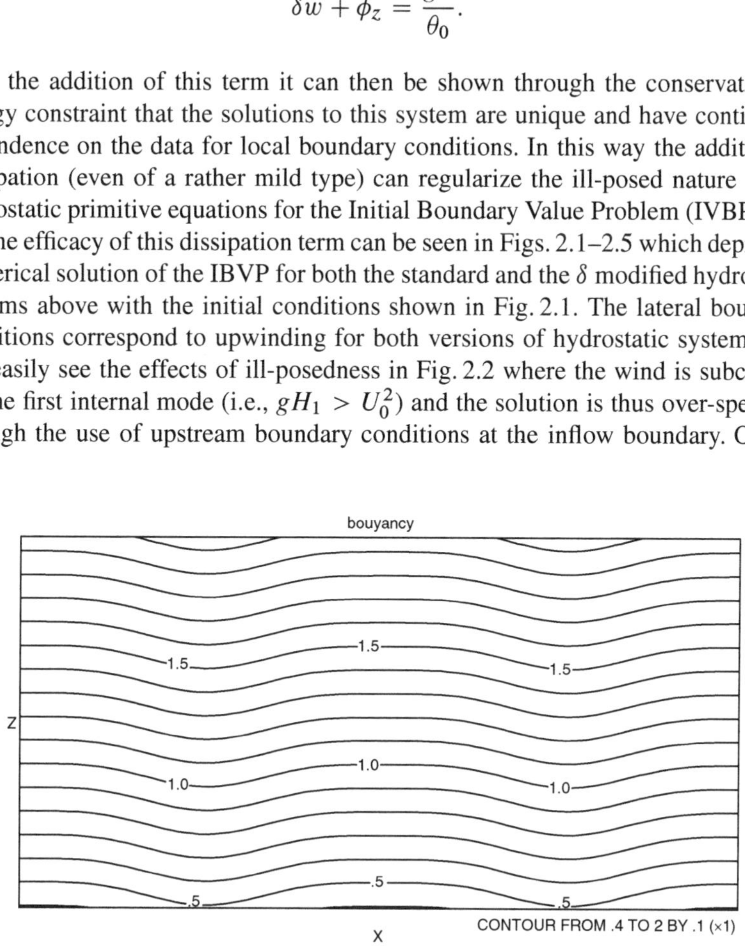

Fig. 2.1 Initial state buoyancy, $b = \phi_z$, as a function of x and z

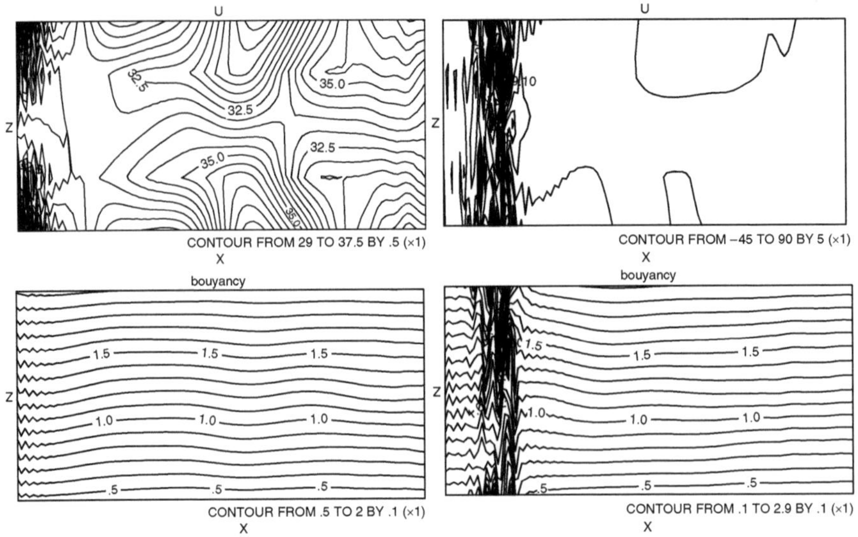

Fig. 2.2 *Top row*: horizontal velocity, U, at $t = 28\,\mathrm{h}$ (*left panel*) and $t = 56\,\mathrm{h}$ (*right panel*) for the traditional hydrostatic limited area model ($\delta = 0$) with subcritical U_0. *Bottom row*: as in the *upper row*, but for the Buoyancy, b

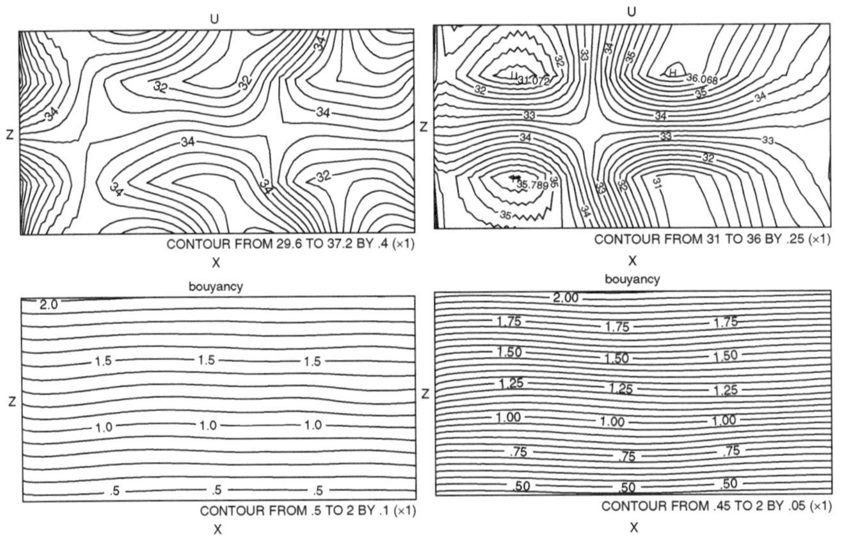

Fig. 2.3 *Top row*: horizontal velocity, U, at $t = 28\,\mathrm{h}$ (*left panel*) and $t = 56\,\mathrm{h}$ (*right panel*) for the modified hydrostatic limited area model ($\delta = 0.3$) with subcritical U_0. *Bottom row*: as in the *top row*, but for the buoyancy, b

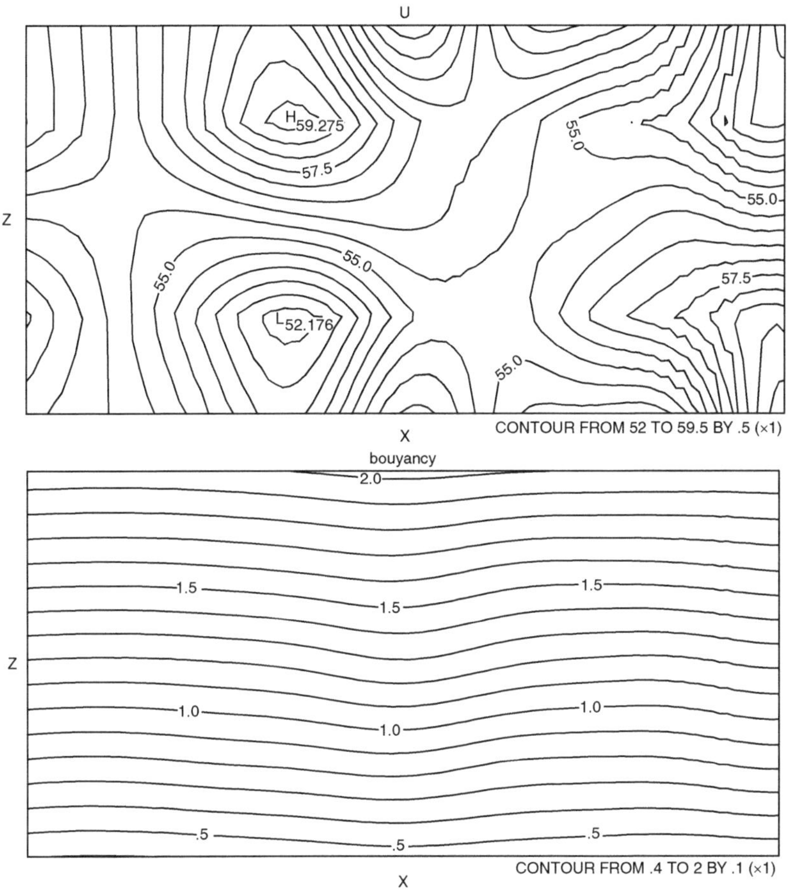

Fig. 2.4 *Top panel:* horizontal velocity, U, at $t = 28$ h for the traditional hydrostatic limited area model ($\delta = 0$) with supercritical U_0. *Bottom panel:* as in the *top panel* but for the buoyancy, b

other hand the solution in Fig. 2.4 is much smoother in space and time despite the identical subcritical zonal flow U_0.

2.4 The Modes of the Primitive Equations on the Sphere

The Cartesian geometry analysis above gives a clear example of the well-posedness issues that exist locally in open boundary models based on the hydrostatic equations. The purpose of this section is to demonstrate that the vertical modal analysis above survives nearly intact for the global, spherical hydrostatic system and thus well-posedness across limited area boundaries is generally feasible for a less simplified

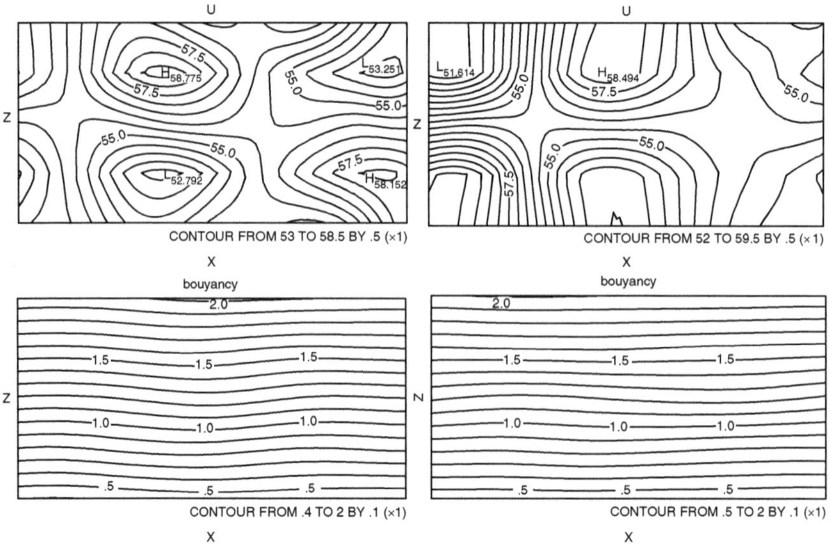

Fig. 2.5 *Top row*: horizontal velocity, U, at $t = 28\,\mathrm{h}$ (*left panel*) and $t = 56\,\mathrm{h}$ (*right panel*) for the modified hydrostatic limited area model ($\delta = 0.3$) with supercritical U_0. *Bottom row*: as in the *top row*, but for the buoyancy, b

set of equations but only if one is willing to deal with non-locality or add artificial dissipation.

For many reasons it is advantageous when using the hydrostatic equation to define a vertical coordinate which differs from the geometric height coordinate z. Some commonly used vertical coordinates in meteorology are pressure, p, and the terrain following counterpart, sigma, where $\sigma \equiv p/p_s$, with p_s being the surface pressure, or a hybrid combination of these two. We use here the equations written with pressure as vertical coordinate because of the relative dynamical simplicity of the equations in this form and since orographic forcing is not the main concern, here. The governing equations become:

$$U_t - fV + \frac{1}{a\cos\varphi}\Phi'_\lambda = NLT_U \tag{2.20}$$

$$V_t + fU + \frac{1}{a}\Phi'_\varphi = NLT_V \tag{2.21}$$

$$\Phi'_p = -\frac{RT'}{p} \tag{2.22}$$

$$\nabla_h \cdot \mathbf{V}_h + \omega_p = 0 \tag{2.23}$$

$$\Phi'_{pt} + S(p)\omega = NLT_\theta \tag{2.24}$$

In the above equations, $\Phi \equiv gz$, where z is the height of the surface of constant pressure, NLT stands for the nonlinear advection terms and curvature terms in spherical geometry, $\omega \equiv \frac{dp}{dt}$, the 'vertical' component of the velocity and U and V are the horizontal components of the velocity (i.e., $\mathbf{V}_h$) in the easterly and northerly directions respectively. Lastly, Φ' and T' are deviations from a resting, stratified basic state so that $\Phi = \overline{\Phi}(p) + \Phi'(\lambda, \varphi, p, t)$ and $T = \overline{T}(p) + T'(\lambda, \varphi, p, t)$. When these are substituted into the first law of thermodynamics $S(p) \equiv \frac{\kappa}{p}\left(\frac{R\overline{T}}{p} - C_p\frac{d\overline{T}}{dp}\right)$, which is positive for a stably stratified fluid in which the entropy increases with height. (A full derivation of the transformation to arbitrary vertical coordinate in a hydrostatic atmosphere may be found in Kasahara (1974) and in Staniforth and Wood (2003) for the deep nonhydrostatic case. Because the effects of a mean zonal velocity lead to complications in the case of spherical geometry, for simplicity we set the mean wind to zero along with NLT. We then (as previously) try a separation of variables in the vertical :

$$\begin{bmatrix} U(\lambda, \varphi, p, t) \\ V(\lambda, \varphi, p, t) \\ \Phi'(\lambda, \varphi, p, t) \end{bmatrix} = G(p) \begin{bmatrix} \widetilde{U}(\lambda, \varphi, t) \\ \widetilde{V}(\lambda, \varphi, t) \\ \widetilde{\Phi}(\lambda, \varphi, t) \end{bmatrix} \tag{2.25}$$

As in the example in Cartesian geometry, the last two equations in (2.20) above are the keys to the separation of variables. Combining them results in:

$$\frac{\partial}{\partial p}\left(\frac{1}{S}\frac{\partial}{\partial p}\Phi'_t\right) - D = 0, \text{ with } D \equiv \nabla_h \cdot \vec{V}_h = G(p)\nabla_h \cdot \widetilde{\vec{V}_h}. \tag{2.26}$$

Straightforward manipulations then show that separability will demand:

$$\frac{d}{dp}\left(\frac{1}{S}\frac{dG}{dp}\right) = -\lambda^2 G, \tag{2.27}$$

where λ^2 is a constant. The fact that our (model) atmosphere has impenetrability as a lower boundary condition and no loss of mass at the top boundary demands that G satisfy the following conditions:

$$w = 0 \text{ at the bottom demands } \quad \frac{dG}{dp} = \Gamma G \text{ at } p = p_s, \tag{2.28}$$

while

$$\omega = 0 \text{ at } p = p_T \text{ requires } \quad \frac{dG}{dp} = 0 \text{ at } p = p_T. \tag{2.29}$$

Note that in the above, $\Gamma \equiv S(p_s)/\left.\frac{d\overline{\Phi}}{dp}\right|_{p=p_s}$. The equation for G above with the homogeneous boundary conditions is a standard Sturm–Liouville eigenfunction equation and λ^2 is the eigenvalue. As in the vertical expansion which arose in the simpler Cartesian case examined in Sect. 2.3, Sturm–Liouville theory for $G(p)$

shows that if $S(p) > 0$ for all p, then, as noted previously, solutions for $G(p)$ exist for an infinite discrete set of λ^2's which are ordered $\lambda_0^2 < \lambda_1^2 < \ldots < \lambda_n^2 < \ldots$ and associated with each λ_n^2 is an H_n, or equivalent depth. These are ordered inversely to the λ^2's, i.e., $H_0 > H_1 \ldots > H_n > \ldots$. The significance of the term 'equivalent depth' becomes obvious when the vertical structure equation is separated from the full equations leaving the following set of horizontal equations:

$$
\begin{aligned}
&\widetilde{U}_t - f\widetilde{V} + \frac{1}{a\cos\varphi}\widetilde{\Phi}_\lambda = 0 \\
&\widetilde{V}_t + f\widetilde{U} + \frac{1}{a}\widetilde{\Phi}_\varphi = 0 \\
&\widetilde{\Phi}_t + gH_n(\widetilde{D}) = \widetilde{\Phi}_t + \frac{gH_n}{a\cos\varphi}(\widetilde{U}_\lambda + (\widetilde{V}\cos\varphi)_\varphi) = 0
\end{aligned}
\tag{2.30}
$$

These are now the (rotating) linear shallow water equations in spherical coordinates for a fluid with mean depth H_n. Each eigenvalue of the vertical structure equation leads to a set of linear shallow water equations with a different mean depth H_n, which is the equivalent shallow water depth for each eigenfunction. Now, gH_n is also the square of the gravity wave speed in a non-rotating fluid and so gH_0 corresponds to the fastest gravity wave speed in the linear stratified system we are considering. For realistic vertical stratification, $S(p)$, the vertical structure equation results in a largest equivalent depth, $H_0 \cong 10\,\text{km}$ and a corresponding gravity wave speed of 300 m/s. Solutions to the vertical structure equation for each equivalent depth are shown in the figure from Kasahara and Puri (1981). The key aspects of the above for our purposes are that (1) the general form of the modal decomposition remains the same and thus an exchange of vertical mode information is necessary to effect lateral boundary conditions and (2) that the shallow water system arises from this decomposition. The second point, in part, explains the utility and widespread use of the shallow water equations in testing numerical methods, since the essence of the horizontal numerical difficulties remain the same when the hydrostatic approximation is used.

The results of this and the previous section have demonstrated that the study of well posedness for the hydrostatic equations commonly used in meteorology and oceanography can be (approximately) reduced to the examination of proper boundary conditions for the shallow water equations in two space dimensions and the analysis of their characteristics. Because the solution to the general, hyperbolic system in two space dimensions is a technically challenging (though straightforward) problem, we only briefly sketch the highlights here. All the gory details of the solution for the linear problem in plane Cartesian geometry are developed in (Weiyan 1992, Chap. 2). The primary complication that arises is that shallow water gravity waves, in the absence of mean advection by the flow, propagate isotropically in the radial direction. Thus the characteristics associated with gravity waves have circular wavefronts and the solution is carried within cones in space-time. Thus, actually forming a solution to the two dimensional IBVP using characteristics is analogous to utilizing Huygen's principle to solve a diffraction problem in optics, straightforward but computationally inefficient. A final note should be made that the method of characteristics differs from the related traditional normal mode approach in its treatment of the Coriolis terms in rotating flow. In the normal mode approach the Coriolis terms are naturally part of the diagonalization and eigenvalue problem associated with the linear operator. In the shallow water case this leads to a mathematical decomposition in terms of inertia-gravity waves and geostrophic potential vorticity modes. In the method of characteristics the Coriolis terms appear as (linear) source terms in the equations for the characteristics much like the damping terms in the single equation examples in Sect. 2.2. The rotational effects of the Coriolis terms are thus integrated along the characteristic directions as opposed to being accounted for in the dispersive nature of the normal modes.

2.5 Discussion and Conclusions

The focus of this contribution has been problems associated with the implementation of lateral boundary conditions in limited domain, open boundary models of the atmosphere which use the hydrostatic primitive equations. As demonstrated above, a fundamental difficulty arises because of the replacement of prognostic equation for the vertical velocity with the diagnostic equation expressing hydrostatic balance. The resulting loss of a wave type (vertically propagating acoustic waves) in the underlying fully hyperbolic system requires that vertical communication be effected through non-locality in the lateral boundary conditions. We have also shown that artificial dissipation can also ameliorate the problems of non-locality at the expense of accuracy of the solution.

The above concerns will clearly arise when the numerical model being integrated is a limited-area hydrostatic model of the atmosphere or ocean. However, the non-locality ill-posedness issue will also affect the quality of solutions in a global model when mesh refinement is used. These problems are similar in a sense, because there are fewer incoming characteristics going from a coarse mesh region to a fine mesh region and more outgoing characteristics leaving a fine mesh region toward a coarse mesh region, due to the change in resolution. Thus a sharp boundary separating a refined mesh region from a coarser mesh region will be susceptible to computational noise similar to that depicted in Fig. 2.2, since the refined region will in essence be a local limited area model.

It would seem to follow that a more consistent resolution to all of these issues requires one abandon the hydrostatic approximation and embrace the fully compressible system. Indeed there are significant advantages in doing so but there is also a steep price to be paid in terms of time step limitations required by the CFL condition. This is particularly true in the vertical dimension where grid spacings are the smallest, $\Delta z < 1$ km, and vertically propagating acoustic waves, which are filtered using the hydrostatic system, must be resolved. Currently, this problem is avoided through the use of an implicit method in the vertical for any time integration method which is split explicit in the horizontal. While the acoustic and gravity wave characteristics can be accurately accounted for in this method, the wave phase velocities are distorted because of implicit component in the vertical which will again raise the possibility of communication being mis-handled leading to enhanced numerical noise at the boundaries between coarse and fine resolution domains. Thus, the price to tackle the problems discussed above in a physically and mathematically sound fashion remains high and awaits computational platforms a decade or so in the future.

References

Kasahara A (1974) Various vertical coordinate systems used for numerical weather prediction. Mon Wea Rev 102(3):504–522

Kasahara A, Puri K (1981) Spectral representation of three-dimensional global data by expansion in normal mode functions. Mon Wea Rev 109(1):37–51

Lions JL, Temam R, Wang S (1992a) New formulations of the primitive equations of the atmosphere and application. Nonlinearity 5(2):237–288

Lions JL, Temam R, Wang S (1992b) On the equations of the large-scale ocean. Nonlinearity 5(4):1007–1053

Oliger J, Sundström A (1978) Theoretical and practical aspects of some initial boundary value problems in fluid mechanics. SIAM J Appl Math 35(3):419–446

Staniforth A, Wood N (2003) The deep-atmosphere equations in a generalized vertical coordinate. Mon Wea Rev 131(8):1931–1938

Temam R, Tribbia J (2003) Open boundary conditions for the primitive and Boussinesq equations. J Atmos Sci 60(8):2647–2660

Thuburn J, Wood N, Staniforth A (2002) Normal modes of deep atmospheres. II: f–F-plane geometry. Q J R Meteorol Soc 128(6):1793–1806

Weiyan T (1992) Shallow water hydrodynamics. Elsevier Oceanography Series; Elsevier, Holland 55:1–434

Chapter 3
Horizontal Discretizations: Some Basic Ideas

John Thuburn

Abstract This chapter will introduce some key ideas in the construction of horizontal discretizations for atmospheric models. One important topic is the ability of different schemes to capture wave propagation accurately. The von Neumann method for analysing numerical wave propagation is presented and applied to some simple schemes to demonstrate the advantages of staggered grids in finite difference models. Another important topic is whether the discretization respects the conservation properties of the differential equations being solved. An introduction to the topic is given, using energy conservation as an illustrative example.

3.1 Introduction

This lecture will introduce some key, basic ideas related to horizontal discretizations in atmospheric model dynamical cores. We will focus on two topics: wave propagation and the effect of using staggered grids (Sect. 3.2), and energy conservation (Sect. 3.3). We will restrict attention to grid point methods (though in many cases finite volume methods can be looked at in the same way). We will not discuss Galerkin methods (although some of the ideas do carry across to Galerkin methods too), nor spectral methods (e.g., Williamson and Laprise 2000). Also, we will not discuss the treatment of advection. Advection is a large and complicated topic; some discussion is given in Chaps. 7, 8, and 9.

3.2 Wave Propagation and Staggered Grids

Chapter 1 in this volume discussed the role of fast waves (acoustic and inertio-gravity waves) in adjustment towards and maintenance of balance. An accurate representation of balance in atmospheric models therefore requires a sufficiently

J. Thuburn
School of Engineering, Computing and Mathematics, University of Exeter, North Park Road, Exeter, EX4 4QF, UK
e-mail: j.thuburn@ex.ac.uk

P.H. Lauritzen et al. (eds.), *Numerical Techniques for Global Atmospheric Models*,
Lecture Notes in Computational Science and Engineering 80,
DOI 10.1007/978-3-642-11640-7_3,

accurate representation of the propagation of the fast waves. In the next two subsections we will look at a technique that can be used to analyse the wave propagation characteristics of numerical schemes. We will see some examples of poor numerical wave propagation that would be damaging to a model's ability to represent near-balanced flow, and show that in some circumstances improved numerical wave propagation can be obtained through the use of a staggered grid.

Slow, balanced motions, Rossby waves and nonlinear vortex dynamics, are energetically dominant on large scales (e.g., Holton 2004). In Sect. 3.2.3 we will point out that Rossby wave propagation can be sensitive to details of the numerical schemes, particularly the treatment of the Coriolis terms.

3.2.1 Gravity Waves in One-Dimension

The simplest relevant model to illustrate our first point is the linearized, one-dimensional, non-rotating shallow water equations:

$$\begin{aligned}\frac{\partial \Phi}{\partial t} + \Phi_0 \frac{\partial u}{\partial x} &= 0\\ \frac{\partial u}{\partial t} + \frac{\partial \Phi}{\partial x} &= 0.\end{aligned} \tag{3.1}$$

Here u is the velocity perturbation and Φ is the geopotential perturbation. The equations have been linearized about a state of rest with geopotential Φ_0.

Assume the domain is either periodic or infinite, and look for wavelike solutions:

$$\begin{aligned}\Phi &= Re\left\{\hat{\Phi}\exp[\mathrm{i}(kx - \omega t)]\right\}\\ u &= Re\left\{\hat{u}\exp[\mathrm{i}(kx - \omega t)]\right\}.\end{aligned} \tag{3.2}$$

Here, k is the wavenumber and ω is the frequency. (The wavelength L is equal to $2\pi/k$.) Substituting the wavelike solutions in (3.1) and eliminating $\hat{u}$ and $\hat{\Phi}$ leads to the dispersion relation

$$\omega^2 = k^2 \Phi_0. \tag{3.3}$$

There are two solutions: a wave propagating to the right with $\omega = k\Phi_0^{1/2}$ and $\Phi = \Phi_0^{1/2}u$, and a wave propagating to the left with $\omega = -k\Phi_0^{1/2}$ and $\Phi = -\Phi_0^{1/2}u$. If we restrict attention to waves propagating in one direction we find that the phase velocity and group velocity (see lecture 1) are both independent of k and equal to $\Phi_0^{1/2}$; these waves are *non-dispersive*. Consider an arbitrary initial condition satisfying $\Phi = \Phi_0^{1/2}u$. This can be Fourier decomposed into waves of different k. Each Fourier component will propagate at the same velocity $\Phi_0^{1/2}$. The solution at some later time t will be a superposition of waves that have all propagated the same distance $\Phi_0^{1/2}t$; it will therefore look identical to the initial state except for a translation.

Now consider a numerical solution of (3.1). We will leave time continuous and concentrate on the spatial discretization. Suppose, first, that Φ and u are stored at the same locations on a uniform grid with spacing Δx (Fig. 3.1), and approximate the x-derivatives by second order centered differences:

$$\frac{\partial u_j}{\partial t} + \frac{\Phi_{j+1} - \Phi_{j-1}}{2\Delta x} = 0;$$
$$\frac{\partial \Phi_j}{\partial t} + \frac{u_{j+1} - u_{j-1}}{2\Delta x} = 0. \tag{3.4}$$

How well do the solutions of the discrete equations replicate the solutions of the continuous equations? We can address this question using a technique known as *von Neumann analysis*. Again, look for wavelike solutions, but now on the grid:

$$\Phi_j = Re\left\{\hat{\Phi}\exp[\mathrm{i}(kx_j - \omega t)]\right\}$$
$$u_j = Re\left\{\hat{u}\exp[\mathrm{i}(kx_j - \omega t)]\right\}. \tag{3.5}$$

The analysis follows exactly the same steps as in the continuous case, except that the x-derivative is approximated by the difference of two exponentials which, using well known identities, can be expressed as a sine. For example,

$$\begin{aligned}\frac{\Phi_{j+1} - \Phi_{j-1}}{2\Delta x} &= \frac{\Phi_i\left(e^{\mathrm{i}k\Delta x} - e^{-\mathrm{i}k\Delta x}\right)}{2\Delta x}\\ &= \Phi_j\frac{2\mathrm{i}\sin(k\Delta x)}{2\Delta x}\\ &= \mathrm{i}\tilde{k}\Phi_j.\end{aligned} \tag{3.6}$$

Thus k is replaced everywhere by

$$\tilde{k} = \sin(k\Delta x)/\Delta x \tag{3.7}$$

and the dispersion relation becomes

$$\omega^2 = \tilde{k}^2\Phi_0. \tag{3.8}$$

The right panel of Fig. 3.1 shows the resulting dispersion relation for the numerical solutions. A large part of the spectrum has significant artificial reduction of its frequency. In particular, the shortest resolvable wave, which has $k\Delta x = \pi$, has zero frequency and does not propagate at all. A disturbance pattern like this that spuriously fails to propagate is sometimes called a *computational mode*. Furthermore, the short wavelength half of the spectrum has $d\omega/dk < 0$, i.e., it has group velocity of the wrong sign. Waves that have spurious propagation characteristics like this are sometimes called *parasitic modes*. Such poor wave propagation characteristics are

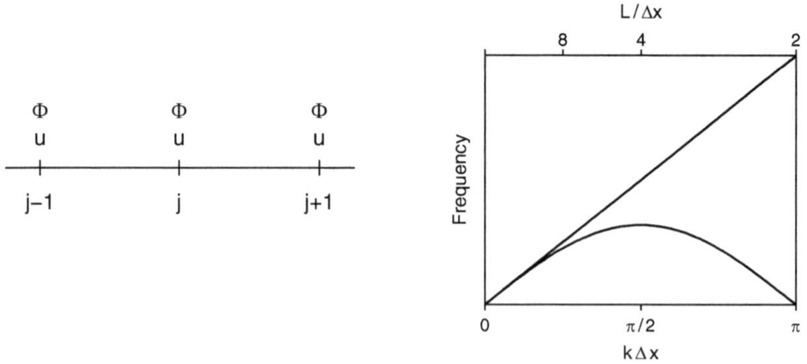

Fig. 3.1 *Left*: Schematic showing the arrangement of variables on a one-dimensional unstaggered grid. *Right*: Dispersion relations for gravity wave solutions of the one-dimensional linearized shallow water equations: the straight line is for the continuous equations (3.3); the curved line is for the discrete equations (3.8)

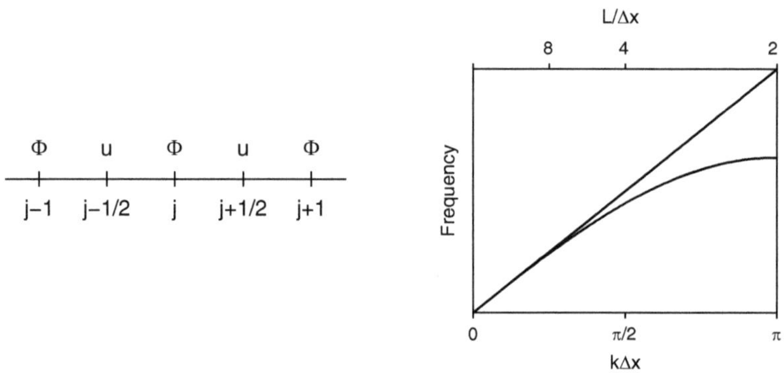

Fig. 3.2 *Left*: Schematic showing the arrangement of variables on a one-dimensional staggered grid. *Right*: Dispersion relations for gravity wave solutions of the one-dimensional linearized shallow water equations: the straight line is for the continuous equations (3.3); the curved line is for the discrete equations (3.11)

likely to lead to a poor representation of geostrophic adjustment and balance in a numerical model.

Now consider an alternative discretization in which Φ and u are stored staggered relative to each other (Fig. 3.2). Again, the x-derivatives are approximated by centred differences, but with a more compact stencil:

$$\frac{\partial u_{j+1/2}}{\partial t} + \frac{\Phi_{j+1} - \Phi_j}{\Delta x} = 0;$$

$$\frac{\partial \Phi_j}{\partial t} + \frac{u_{j+1/2} - u_{j-1/2}}{\Delta x} = 0. \tag{3.9}$$

The analysis follows the same steps as before, except that the x-derivatives are approximated by a more compact difference of exponentials. We find k in the continuous case is replaced everywhere by

$$\tilde{k}' = \sin(k\Delta x/2)/(\Delta x/2). \tag{3.10}$$

and the dispersion relation becomes

$$\omega^2 = \tilde{k}'^2 \Phi_0. \tag{3.11}$$

The right panel of Fig. 3.2 shows the dispersion relation for the staggered grid. There is still significant slowing for large wavenumbers, but much less than for the unstaggered grid. In particular the group velocity always has the correct sign (except for the two-grid length wave $k\Delta x = \pi$ which has zero group velocity).

Figure 3.3 shows an example numerical solution of the linearized one-dimensional shallow water equations using both an unstaggered and a staggered grid. In both cases dispersion errors lead to the main peak lagging behind the true solution, though the lag is worse on the unstaggered grid. And in both cases dispersion errors have led to short wavelength oscillations behind the main peak, though again these are worse on the unstaggered grid.

Another way of viewing the poor behaviour of the unstaggered grid is as follows. A low frequency forcing should lead to the generation of long wavelength waves. However, as can be seen from the right hand panel of Fig. 3.1, on an unstaggered grid

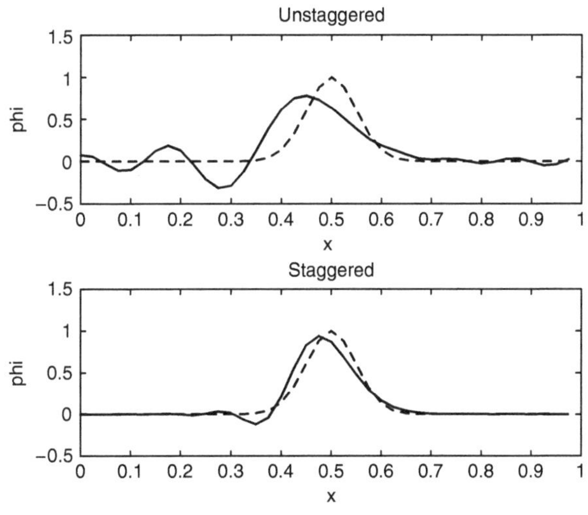

Fig. 3.3 Numerical solution of the linearized shallow water equations on a periodic domain of 40 grid points. *Top*: using an unstaggered grid. *Bottom*: using a staggered grid. At the time shown, the solution should have propagated exactly once around the domain (*left to right*) and returned to its initial position. The initial condition for Φ, shown by the dashed curves, comprises a pulse about 8 grid lengths wide

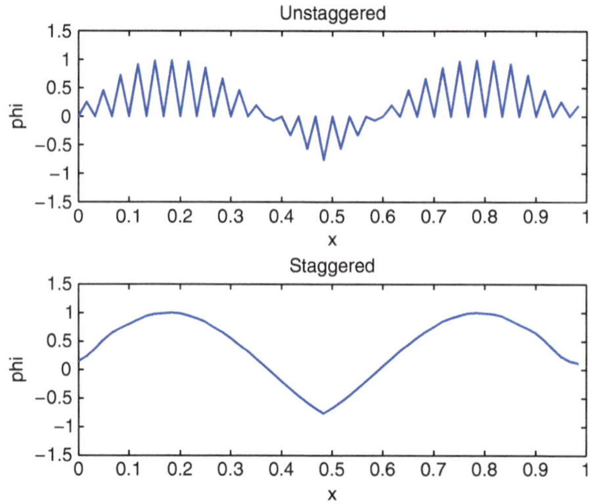

Fig. 3.4 Numerical solution of the linearized shallow water equations where Φ in the center of the domain has been forced to oscillate like $\sin(\omega t)$. The domain shown is 60 grid points across. The initial condition is $u = 0$, $\Phi = 0$ and the solution is shown after less than one forcing period. *Top*: using an unstaggered grid. *Bottom*: using a staggered grid

any resolvable frequency ω corresponds to two different k; there is the possibility that low frequency forcing can generate short wavelength as well as long wavelength waves. Figure 3.4 shows the result of exactly this process. The initial condition was set to $u = 0$, $\Phi = 0$, and the Φ value in the centre of the domain was forced to oscillate like $\sin(\omega t)$. On the staggered grid long wavelength waves have been radiated to the left and to the right, close to the correct solution. However, on the unstaggered grid a superposition of short and long wavelength waves have been radiated, giving a very noisy solution.

3.2.2 Inertio-Gravity Waves in Two-Dimensions

Let us extend the above discussion to the two-dimensional linearized shallow water equations and include the effects of rotation through a constant Coriolis parameter $f = f_0$. The governing equations are

$$\frac{\partial \Phi}{\partial t} + \Phi_0 \left(\frac{\partial u}{\partial x} + \frac{\partial v}{\partial y} \right) = 0,$$

$$\frac{\partial u}{\partial t} - fv + \frac{\partial \Phi}{\partial x} = 0,$$

$$\frac{\partial v}{\partial t} + fu + \frac{\partial \Phi}{\partial y} = 0. \tag{3.12}$$

Seeking wavelike solutions proportional to $\exp\{i(kx + ly - \omega t)\}$ leads to the dispersion relation

$$\omega \left(\omega^2 - f_0^2 - (k^2 + l^2)\Phi_0\right) = 0. \qquad (3.13)$$

The root $\omega = 0$ corresponds to Rossby waves. (In this example Rossby waves do not propagate because we have approximated f as a constant. The effect of spatial variations in f, called the β-effect because f is sometimes approximated as $f = f_0 + \beta y$, causes the Rossby wave frequency to become non-zero; see Chap. 1). The other two roots correspond to left and right propagating inertio-gravity waves.

An important parameter here (and in various other contexts) is the *Rossby radius*

$$\lambda = \Phi_0^{1/2} / f_0 \qquad (3.14)$$

(e.g., Holton 2004). It defines a natural horizontal scale for geostrophically balanced motion, and it can also be interpreted as the distance a gravity wave would propagate (at speed $\Phi_0^{1/2}$) on the inertial timescale $1/f_0$. On length scales significantly shorter than λ the non-zero roots of (3.13) approximately satisfy $\omega^2 - (k^2 + l^2)\Phi_0 = 0$. Pressure gradient forces dominate the dynamics. These are gravity waves. On length scales significantly longer than λ the non-zero roots of (3.13) approximately satisfy $\omega^2 - f_0^2 = 0$. Coriolis terms dominate the dynamics. These are inertial waves. Which regime we are in has implications for the relative accuracy of different numerical methods, as we shall see.

In two dimensions there are more possibilities for staggering than in one-dimension. Arakawa and colleagues (Winninghoff 1968; Arakawa and Lamb 1977; Randall 1994) systematically studied the shallow water wave dispersion properties of a number of staggered quadrilateral grids, and introduced the naming convention that is now universally used (see Fig. 3.5). We will look at three of these grids in some detail.

The A-grid. By analogy with the one-dimensional case, when we look for wave-like solutions of the finite difference equations we find that the k that comes from an x-derivative is replaced by

$$\tilde{k} = \sin(k\Delta x)/\Delta x \qquad (3.15)$$

while the l that comes from a y-derivative is replaced by

$$\tilde{l} = \sin(l\Delta y)/\Delta y. \qquad (3.16)$$

Figure 3.6 shows the ratio of the numerical frequency to the exact frequency as a function of k and l (where $\Delta x = \Delta y$) for two regimes: well-resolved Rossby radius $\Delta x/\lambda = 0.2$ and poorly resolved Rossby radius $\Delta x/\lambda = 5$. The A-grid scheme does well (as do all the others) for well-resolved waves, i.e., when $k\Delta x$ and $l\Delta y$ are small. What distinguishes the various schemes is how they perform for less well resolved waves. The A-grid scheme performs well when the Rossby radius is poorly resolved because near-grid-scale waves are dominated by the Coriolis term,

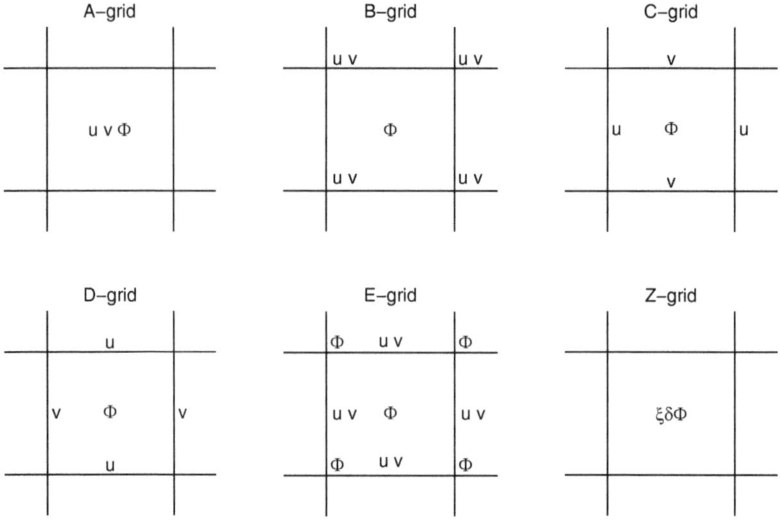

Fig. 3.5 Schematic showing the arrangement of variables on six possible staggered grids for the two-dimensional shallow water equations. On the Z-grid the predicted variables are the (vertical component of) relative vorticity ξ and the horizontal divergence δ

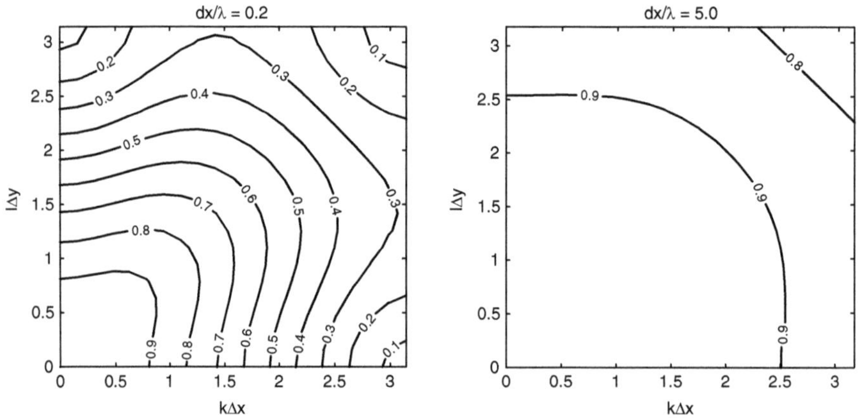

Fig. 3.6 Ratio of numerical frequency to exact frequency versus $k\Delta x$ and $l\Delta y$ for the A-grid. *Left*: $\Delta x/\lambda = 0.2$. *Right*: $\Delta x/\lambda = 5$. The contour interval is 0.1 and the values approach 1 in the bottom left corners

which is accurately represented on the A-grid. However, when the Rossby radius is well-resolved near-grid-scale waves are dominated by the pressure gradient and divergence terms, which are inaccurately represented just as in Sect. 3.2.1.

The B-grid. On the B-grid some of the finite differences are more compact than on the A-grid, but also some averaging is required to obtain values of the variables at the locations where they are needed. For example, u must be averaged in the

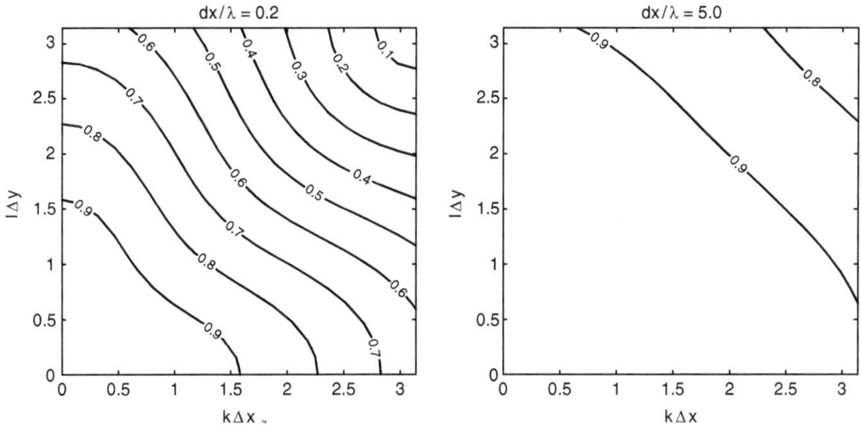

Fig. 3.7 Ratio of numerical frequency to exact frequency versus $k\Delta x$ and $l\Delta y$ for the B-grid. *Left*: $\Delta x/\lambda = 0.2$. *Right*: $\Delta x/\lambda = 5$. The contour interval is 0.1 and the values approach 1 in the bottom left corners

y-direction and differenced in the x-direction in order to approximate $\partial u/\partial x$ in the Φ equation. A similar thing happens for $\partial v/\partial y$ in the Φ equation and for $\partial\Phi/\partial x$ and $\partial\Phi/\partial y$ in the u and v equations. We find that k is replaced by

$$\tilde{k} = \cos(l\Delta y/2)\sin(k\Delta x/2)/(\Delta x/2) \tag{3.17}$$

while l is replaced by

$$\tilde{l} = \cos(k\Delta x/2)\sin(l\Delta y/2)/(\Delta y/2). \tag{3.18}$$

The resulting errors in the dispersion relation are shown in Fig. 3.7. Like the A-grid, it performs well when the Rossby radius is poorly resolved but performs poorly (though slightly better than the A-grid) when the Rossby radius is well resolved.

The C-grid. On the C-grid the variables are ideally placed for calculating the spatial derivatives that arise. However, u and v are no longer located at the same points; u must be averaged in both the x and y-directions to approximate the $f_0 u$ term in the v equation, and similarly for the $f_0 v$ term in the u equation. We find that k is replaced by

$$\tilde{k} = \sin(k\Delta x/2)/(\Delta x/2) \tag{3.19}$$

and l is replaced by

$$\tilde{l} = \sin(l\Delta y/2)/(\Delta y/2) \tag{3.20}$$

while f_0 is replaced by

$$\tilde{f_0} = f_0\cos(k\Delta x/2)\cos(l\Delta y/2). \tag{3.21}$$

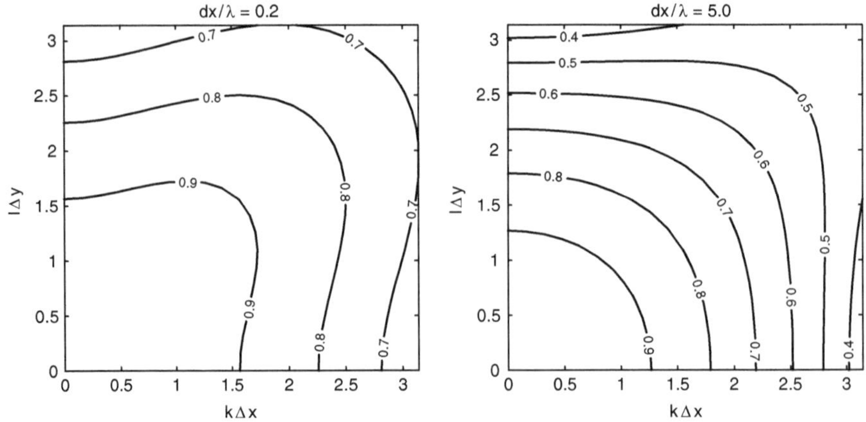

Fig. 3.8 Ratio of numerical frequency to exact frequency versus $k\Delta x$ and $l\Delta y$ for the C-grid. *Left*: $\Delta x/\lambda = 0.2$. *Right*: $\Delta x/\lambda = 5$. The contour interval is 0.1 and the values approach 1 in the bottom left corners

Consequently, the C-grid performs well when the Rossby radius is well-resolved and pressure gradient and divergence terms dominate near-grid-scale waves, but performs poorly when the Rossby radius is poorly resolved and Coriolis terms dominate near-grid-scale waves (Fig. 3.8).

The Rossby radius in the atmosphere (for the deepest modes–in three dimensions the Rossby radius depends on the vertical scale and is greater when the vertical scale is greater) is of the order of 1,000 km, which is well-resolved in any practical atmospheric dynamical core. For this reason, the C-grid has often been the grid of choice for grid point atmospheric models. In the ocean the Rossby radius is typically of order 10 km; historically this has not been well resolved, so the B-grid has often been used for ocean models. As computer power increases and it begins to become practical to resolve the Rossby radius, some ocean modelers are beginning to turn to the C-grid.

The discussion here has concentrated on grid staggering options for quadrilateral grids. However, analogues exist for other grid cell shapes such as triangles and hexagons. See Chap. 10.

3.2.3 Rossby Wave Propagation on the C-grid

The Coriolis terms play a crucial role in the Rossby wave propagation mechanism. Given the need for some averaging in evaluating the Coriolis terms on a C-grid, we might expect the propagation of near-grid-scale Rossby waves to be poorly captured. However, when f is a function of position there are a variety of options for exactly how the averaging is done, e.g., should we multiply by f before averaging or after averaging? The Rossby wave propagation turns out to be sensitive to these details. The following *f-at-Φ-points* scheme turns out to work quite well:

$$\partial_t u - \overline{f\overline{v}^y}^x + \delta_x \Phi = 0, \tag{3.22}$$

$$\partial_t v + \overline{f\overline{u}^x}^y + \delta_y \Phi = 0. \tag{3.23}$$

(Here an overline indicates an average, with the superscript indicating the direction of the average, and δ_x and δ_y indicate centred finite difference approximations to x and y partial derivatives.) This scheme captures the Rossby wave frequency quite accurately even for short north–south wavelengths, though not for short east–west wavelengths. See Thuburn (2007) for details.

In spherical geometry it is important to include appropriate geometrical factors in the averaging of the Coriolis terms, for consistency with the mass continuity equation:

$$\frac{\partial u}{\partial t} - \overline{\frac{f}{\cos\phi}\overline{v\cos\phi}^\phi}^\lambda + \frac{1}{a\cos\phi}\delta_\lambda \Phi = 0 \tag{3.24}$$

$$\frac{\partial v}{\partial t} + \overline{f\overline{u}^\lambda}^\phi + \frac{1}{a}\delta_\phi \Phi = 0. \tag{3.25}$$

Here a is the Earth's radius, λ is longitude and ϕ is latitude. When this is done, normal mode calculations show that the dispersion relations for both Rossby modes and inertio-gravity modes are captured quite accurately (Fig. 3.9). Otherwise, a significant part of the Rossby mode spectrum is lost and replaced by spurious grid-scale modes with positive (eastward) frequency (Thuburn and Staniforth 2004); see Fig. 3.10.

3.3 Conservation Properties

It is often considered desirable for a dynamical core to possess analogues of some of the conservation properties of the continuous adiabatic and frictionless governing equations. Energy is a particularly interesting quantity in this respect, because it is a nonlinear quantity, it can be decomposed into available and unavailable contributions, and it is subject to both upscale and downscale nonlinear transfers (see Chap. 11). Even if we choose to formulate the nonlinear advection terms in a way that does not conserve energy (either to allow for transfers to unresolved scales, or purely for numerical efficiency as in semi-Lagrangian schemes) a strong argument can be made for formulating the linear terms in the equations, the Coriolis and pressure gradient terms, in an energy conserving way. The following two subsections illustrate some of the kinds of techniques that have been used to obtain such conservation properties.

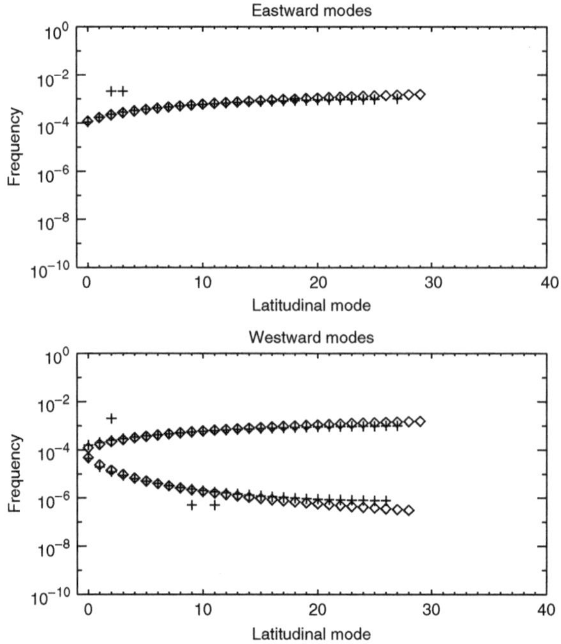

Fig. 3.9 Numerical dispersion relation (*crosses*) for a latitudinal discretization given by (3.24), (3.25) and the corresponding discrete linearized mass equation on the sphere; (the fields are assumed to be proportional to $\exp(im\lambda)$ with zonal wavenumber $m = 2$, and east–west derivatives are handled analytically). Frequency is plotted against increasing latitudinal mode index–smaller index corresponds to greater north–south wavelength. Analytical approximations to the exact frequencies are given by the diamonds. The eastward modes and the higher frequency branch of westward modes are inertio-gravity modes. The lower frequency branch of westward modes are Rossby modes. The Rossby modes are handled quite accurately, despite the averaging of the Coriolis terms. (A small number of modes are handled poorly; this is a result of the polar singularity)

3.3.1 Energy Conservation: Coriolis Terms

The Coriolis terms should cancel when we take u times

$$\frac{Du}{Dt} - fv = \ldots \tag{3.26}$$

plus v times

$$\frac{Dv}{Dt} + fu = \ldots . \tag{3.27}$$

This is achieved very straightforwardly on an A-grid or B-grid for which the u and v points coincide. On a C-grid, however, the cancellation is non-trivial.

Arakawa and Lamb (1981) presented a systematic way of achieving the desired cancellation on a spherical C-grid. They work with mass flux variables

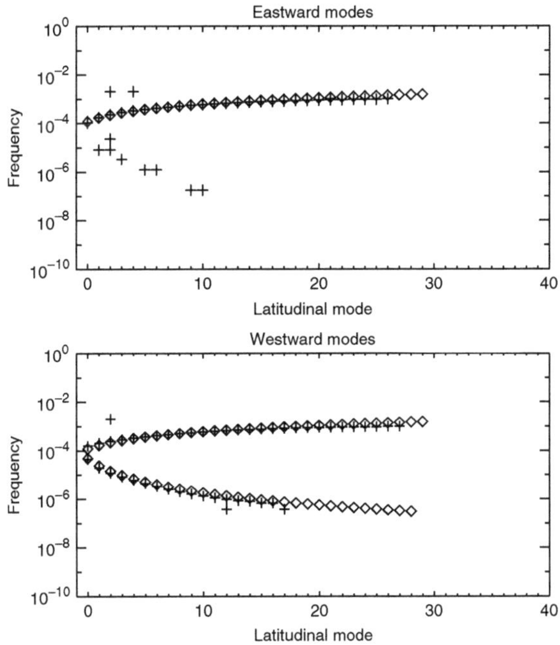

Fig. 3.10 As in Fig. 3.9 except that the $\cos\phi$ factors do not appear in the Coriolis term in (3.24). The Rossby wave spectrum is now badly distorted

$$u^* = u\Phi a\Delta\phi, \tag{3.28}$$

$$v^* = v\Phi a\cos\phi\Delta\lambda, \tag{3.29}$$

($\Delta\lambda$ and $\Delta\phi$ are the longitudinal and latitudinal grid spacing) and express the Coriolis terms using four sets of coefficients α, β, γ and δ:

$$\begin{aligned}&\tfrac{\partial}{\partial t}(ua\cos\phi\Delta\lambda)_{i,j+1/2}\\ &-\alpha_{i,j+1/2}\, v^*_{i+1/2,j+1} - \beta_{i,j+1/2}\, v^*_{i-1/2,j+1}\\ &-\gamma_{i,j+1/2}\, v^*_{i-1/2,j} - \delta_{i,j+1/2}\, v^*_{i+1/2,j} = \dots\end{aligned} \tag{3.30}$$

$$\begin{aligned}&\tfrac{\partial}{\partial t}(va\Delta\phi)_{i+1/2,j}\\ &+\alpha_{i,j-1/2}\, u^*_{i,j-1/2} + \beta_{i+1,j-1/2}\, u^*_{i+1,j-1/2}\\ &+\gamma_{i+1,j+1/2}\, u^*_{i+1,j+1/2} + \delta_{i,j+1/2}\, u^*_{i,j+1/2} = \dots\end{aligned} \tag{3.31}$$

Here the grid indexing convention is that Φ is stored at points labelled with subscripts $i + 1/2,\ j + 1/2$, etc., u points are labelled $i,\ j + 1/2$, etc., and v points are labelled $i + 1/2,\ j$, etc. It may be verified that all terms involving α, β, γ and δ do indeed cancel when we take $u_{i,j+1/2}$ times (3.30) plus $v_{i+1/2,j}$ times (3.31) and sum globally. There is still considerable freedom available in choosing the exact values of α, β, γ and δ. For example, the scheme (3.24), (3.25) is of this form and achieves second order accuracy and good Rossby wave propagation.

3.3.2 Energy Conservation: Pressure Gradient Terms

In order for the pressure gradient terms to conserve energy we require a discrete analogue of

$$\mathbf{v}\Phi.\nabla\Phi + \Phi\nabla.(\mathbf{v}\Phi) = \nabla.(\mathbf{v}\Phi^2) \tag{3.32}$$

or, at least,

$$\int \mathbf{v}\Phi.\nabla\Phi \, dA + \int \Phi\nabla.(\mathbf{v}\Phi) \, dA = 0. \tag{3.33}$$

On an A-grid this is relatively straightforward to achieve. On a C-grid the required cancellation can again be achieved by working with mass flux variables u^* and v^* (Arakawa and Lamb 1981). Then the discrete analogue of (3.33) is

$$\begin{aligned}
&\Sigma_{ij}\, u^*_{i,j+1/2}\left(\Phi_{i+1/2,j+1/2} - \Phi_{i-1/2,j+1/2}\right) + \\
&\Sigma_{ij}\, v^*_{i+1/2,j}\left(\Phi_{i+1/2,j+1/2} - \Phi_{i+1/2,j-1/2}\right) + \\
&\Sigma_{ij}\, \Phi_{i+1/2,j+1/2}\left(u^*_{i+1,j+1/2} - u^*_{i,j+1/2}\right) + \\
&\Sigma_{ij}\, \Phi_{i+1/2,j+1/2}\left(v^*_{i+1/2,j+1} - v^*_{i+1/2,j}\right) = 0,
\end{aligned} \tag{3.34}$$

which does indeed hold.

3.4 Conclusions

A sufficiently accurate representation of the propagation of fast waves is required for a numerical model of the atmosphere to capture the near-balanced large-scale flow. The von Neumann method for analysing the numerical dispersion relation of a discretization has been presented and used to illustrate the behaviour of some simple, well-known schemes. The method shows that staggered grids can be advantageous in some circumstances. Incidentally, the method can be applied to more complex schemes such as higher-order schemes (e.g., Leslie and Purser 1991) or schemes that avoid averaging (e.g., McGregor 2005), and even to more exotic grids such as hexagons (e.g., Ničković et al. 2002; Thuburn 2008), though the analysis becomes more laborious.

The numerical solution could become inaccurate if the discretization introduces spurious sources or sinks of energy. One approach to avoiding this problem is to design the discretization to mimic certain cancellation properties of the continuous equations. This approach has been illustrated for discretizations of the Coriolis and pressure gradient terms.

References

Arakawa A, Lamb VR (1977) Computational design and the basic dynamical processes of the UCLA general circulation model. Methods in Computational Physics 17:172–265

Arakawa A, Lamb VR (1981) A potential enstrophy and energy conserving scheme for the shallow water equations. Mon Wea Rev 109:18–36

Holton JR (2004) An Introduction to Dynamic Meteorology, fourth edn. Elsevier Academic Press, Amsterdam

Leslie LM, Purser RJ (1991) High-order numerics in an unstaggered 3-dimensional time-split semi-Lagrangian forecast model. Mon Wea Rev 119:1612–1623

McGregor JL (2005) Geostrophic adjustment for reversibly staggered grids. Mon Wea Rev 133:1119–1128

Ničković S, Gavrilov MB, Tosić IA (2002) Geostrophic adjustment on hexagonal grids. Mon Wea Rev 130:668–683

Randall DA (1994) Geostrophic adjustment and the finite-difference shallow-water equations. Mon Wea Rev 122:1371–1377

Thuburn J (2007) Rossby wave propagation on the C-grid. Atmos Sci Lett 8:37–42

Thuburn J (2008) Numerical wave propagation on the hexagonal C-grid. J Comput Phys 227:5836–5858

Thuburn J, Staniforth A (2004) Conservation and linear Rossby-mode dispersion on the spherical C grid. Mon Wea Rev 132:641–653

Williamson DL, Laprise R (2000) Numerical Modeling of the Global Atmosphere in the Climate System, Kluwer, chap Numerical approximations for global atmospheric GCMs., pp 127–219

Winninghoff FJ (1968) On the adjustment toward a geostrophic balance in a simple primitive equation model with application to the problems of initialization and objective analysis. PhD thesis, Department of Meteorology, UCLA

Chapter 4
Vertical Discretizations: Some Basic Ideas

John Thuburn

Abstract This chapter introduces some key ideas in the design of vertical discretizations for atmospheric models. Various choices of vertical coordinate are possible, and the most widely used ones are introduced. The requirement to retain certain conservation properties can constrain or determine aspects of the discretization: this is illustrated using the Simmons and Burridge angular momentum and energy conserving scheme for hydrostatic models. Another important set of issues surrounds the ability to capture hydrostatic balance and wave dispersion accurately and to avoid computational modes: some implications for the vertical discretization are discussed.

4.1 Introduction

This lecture will introduce some key, basic ideas related to vertical discretizations in atmospheric model dynamical cores. We will first discuss the choice of vertical coordinate and its relation to the bottom and top boundary conditions. We will then look at how the details of the vertical discretization can influence conservation properties and wave propagation.

4.2 Alternative Vertical Coordinates

Systematic derivation of the governing equations usually involves writing their components in some orthogonal coordinate system, such as spherical polars (λ, ϕ, r). However, for numerical solution of the equations there may be advantages to

J. Thuburn
School of Engineering, Computing and Mathematics, University of Exeter, North Park Road, Exeter, EX4 4QF, UK
e-mail: j.thuburn@ex.ac.uk

P.H. Lauritzen et al. (eds.), *Numerical Techniques for Global Atmospheric Models*, Lecture Notes in Computational Science and Engineering 80,
DOI 10.1007/978-3-642-11640-7_4,

writing the equations in terms of some alternative vertical coordinate. The following transformation rules (e.g., Kasahara 1974; Staniforth and Wood 2003) allow us to re-express the horizontal and vertical derivatives and hence transform the equations to an arbitrary vertical coordinate $\eta(\lambda, \phi, r, t)$:

$$\frac{\partial \psi}{\partial r} = \frac{\partial \eta}{\partial r}\frac{\partial \psi}{\partial \eta}, \tag{4.1}$$

$$\left(\frac{\partial \psi}{\partial s}\right)_r = \left(\frac{\partial \psi}{\partial s}\right)_\eta + \left(\frac{\partial \psi}{\partial \eta}\right)\left(\frac{\partial \eta}{\partial s}\right)_r, \tag{4.2}$$

$$\frac{D\psi}{Dt} = \frac{\partial \psi}{\partial t} + \mathbf{v}.\nabla_{\mathrm{H}}\psi + \dot{\eta}\frac{\partial \psi}{\partial \eta}. \tag{4.3}$$

Here s may be λ, ϕ or t, and ∇_{H} is the horizontal gradient at constant η.

In transforming to a different vertical coordinate it is usual to continue to express vectors in terms of their components in the original orthogonal coordinate system, rather than transform to covariant or contravariant components in the new coordinate system. In particular, it is usual to retain the velocity components $u = \dot{\lambda} r \cos\phi$, $v = \dot{\phi} r$, $w = \dot{r}$ (though $\dot{\eta}$ may be needed too).

4.2.1 Examples

- *Height* $\eta = r$ or $\eta = z$. This is the most obvious choice, requiring no transformation of the governing equations.
- *Pressure* $\eta = p$. A pressure-based coordinate is particularly attractive in hydrostatic models because the mass continuity equation becomes purely diagnostic, and because the pressure difference across a layer is proportional to the mass per unit area in that layer (under the shallow atmosphere approximation), making it easier to formulate schemes with desired conservation properties.
- *Mass* $\eta = \int_z^\infty \rho\, dz'$ for Cartesian geometry with height z or $\eta = \int_r^\infty \rho r'^2\, dr'$ with distance r from Earth's centre. This is the natural generalization of the pressure coordinate to non-hydrostatic models.
- *Terrain-following variants*. It is possible to modify the three coordinate systems mentioned above so that the ground becomes a coordinate surface (e.g., Phillips 1957; Gal-Chen and Somerville 1975, Fig. 4.1); this greatly simplifies the application of the bottom boundary condition (see Sect. 4.3). Some examples are $\eta = z - z_s$ where z_s is the height of the ground, or $\eta = p/p_s$ where p_s is the surface pressure. This latter is sometimes called a σ coordinate.
- *Hybrid terrain-following variants*. To avoid numerical artefacts at high altitudes resulting from a terrain following coordinate, it is possible to use a hybrid coordinate that is terrain-following near the ground but returns to a height or pressure coordinate at high altitude (Fig. 4.1). One well known example is the hybrid σ-p coordinate introduced by Simmons and Burridge (1981). A value of a and a value of b are defined on each model level. Then the pressure on each model level is

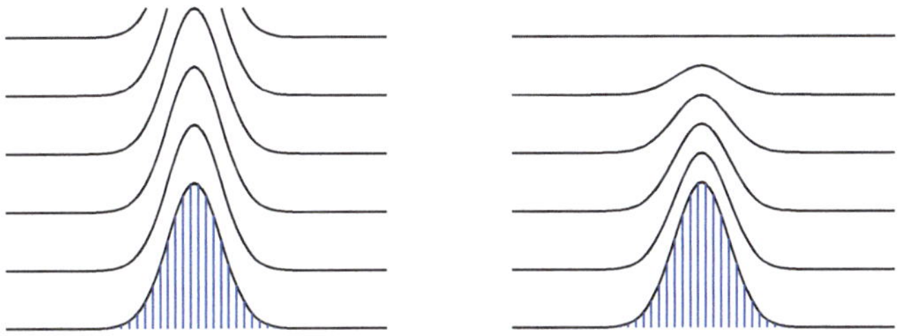

Fig. 4.1 Schematics showing a terrain following coordinate (*left*) and a hybrid terrain following coordinate (*right*)

given by $p = ap_0 + bp_s$ where p_0 is a constant reference pressure and p_s is again the surface pressure. Near the ground a is chosen to be zero or small so that the coordinate looks like a σ coordinate; at high altitude b is chosen to be zero or small so that the coordinate looks like a pressure coordinate. The coefficients are chosen to give a smooth transition in between. (A value of η is given by $\eta = a + b$, though in fact the scheme can be formulated without explicit reference to the value of η.)

- *Isentropic coordinate* $\eta = f(\theta)$. There are a number of potential advantages of using an isentropic vertical coordinate that make it attractive for atmospheric modeling (e.g., Hsu and Arakawa 1990). Diabatic heating is generally weak, so an isentropic coordinate is approximately Lagrangian, leading to improved Lagrangian conservation properties and conservation of entropy-related quantities and perhaps potential vorticity. The handling of moist processes may also be improved. And in some dynamical situations the coordinate automatically adapts to give extra resolution where it is needed. On the other hand, the bottom boundary is difficult to handle, the coordinate cannot handle situations where $N^2 < 0$, and experience suggests it is more difficult to obtain a robust numerical formulation. (A hybrid vertical coordinate can help with all of these issues, e.g., Konor and Arakawa 1997). Also, in regions such as the tropical upper troposphere, where N^2 is close to zero, vertical resolution becomes relatively poor.
- *Lagrangian coordinate*. A Lagrangian vertical coordinate (apparently first suggested by Starr 1945) is defined by $\dot{\eta} = 0$. Like the isentropic coordinate, it is expected to give improved Lagrangian conservation properties. However, over time Lagrangian coordinate surfaces will bend and fold, making them inaccurate or unusable as a vertical coordinate. To circumvent this, the Lagrangian coordinate must be periodically re-initialized and the solution remapped to the re-initialized coordinate system (e.g., Lin 2004).

4.3 Bottom and Top Boundary Conditions

The normal component of velocity at the bottom boundary must vanish. If η is a terrain following coordinate then the boundary condition may be expressed particularly simply as $\dot{\eta} = 0$. In terms of velocity components we must have $w = \mathbf{v}.\nabla_H z_s$, where $\mathbf{v} = (u, v)$ is the horizontal velocity. If the ground is flat then we have $w = 0$, but not in general. Typically w will be stored at the bottom boundary, but u and v will be staggered in the vertical relative to w (see Sect. 4.5). Thus, some means of evaluating $\mathbf{v}$ at the gound will be needed. If the model includes a boundary layer scheme then it is appropriate to apply a no-slip boundary condition $\mathbf{v} = 0$, and it again follows that $w = 0$. However, for a frictionless dynamical core a free-slip boundary condition, which imposes no constraint on $\mathbf{v}$, is more appropriate; then $\mathbf{v}$ must be extrapolated to the ground in a way consistent with the free-slip condition.

A disadvantage of terrain-following coordinates, particularly at high horizontal resolution, is that the coordinate system becomes far from orthogonal near steep orography. Numerical methods can then lose accuracy. To avoid this problem, an alternative is not to use a terrain-following coordinate but to retain a height coordinate and allow the coordinate surfaces to intersect the terrain.

In the simplest version of this idea the orography appears step-like, with the top of each step coincident with a model coordinate surface (Fig. 4.2). This has been found to be too inaccurate. However, the idea can be extended (e.g., Adcroft et al. 1997) by allowing the bottom face of grid cells adjacent to the gound to be at any height, not necessarily coincident with a model coordinate surface (*fractional cells*), or even allowing them to slope (*cut cells* or *shaved cells*). A disadvantage remains that vertical resolution in the boundary layer becomes reduced at mountain tops as model grids are typically vertically stretched at higher altitudes.

The real atmosphere has no top boundary, but in a practical numerical model of the atmosphere we must impose a boundary somewhere. Practical choices include the following (e.g., Staniforth and Wood 2003).

- *Rigid lid*: $w = 0$ is imposed at some constant height $z = z_T$. This is most easily done if the vertical coordinate is height (or a hybrid coordinate reducing to height near the top boundary). Conservation of energy and angular momentum are maintained in the governing equations.
- *Elastic lid*: $Dp/Dt = 0$ is imposed on some surface of constant pressure $p = p_T$. (p_T may equal 0.) This is most easily done if the vertical coordinate is pressure (or a hybrid coordinate reducing to pressure near the top boundary). The governing equations then conserve angular momentum and enthalpy.

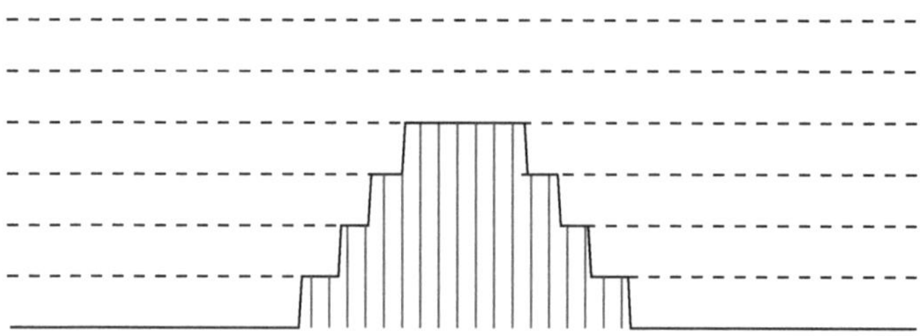

Fig. 4.2 Schematic showing the simplest form of terrain intersecting vertical coordinate

Both a rigid lid and an elastic lid are artificial and may cause spurious reflection of upward propagating waves. To reduce the problem it is common to include a scale-independent damping on model fields near the model top, but note that the strength and depth of the damping layer must be chosen carefully. Moreover, it is recommended that such damping only be applied to departures from the zonal mean to avoid an unrealistic sink of angular momentum and spurious feedbacks (Shaw and Shepherd 2007). An alternative is to apply a wave radiation condition at the model top (e.g., Durran 1999). However, this approach is more complex and some approximation is usually required.

4.4 The Simmons and Burridge Energy and Angular Momentum Conserving Scheme

In this section we use the well-known Simmons and Burridge (1981) vertical discretization to illustrate the kinds of considerations that come into play to obtain properties such as conservation of energy and angular momentum. It is assumed that the hydrostatic primitive equations are being solved, and a hybrid $\sigma - p$ coordinate is used. Figure 4.3 shows the vertical arrangement of variables: the pressure, and the vertical coordinate η if needed, are defined on 'half-levels', while the prognostic variables u, v and T are defined at the 'full-levels'. We suppose there are N full-levels, numbered from the top (where $\eta = 0$) to the bottom (where $\eta = 1$). Surface pressure (or log of surface pressure) is predicted at the ground, which is the half-level with index $N + 1/2$.

4.4.1 *Hydrostatic Equation*

We first look at the discretization of the hydrostatic equation

$$\frac{\partial \Phi}{\partial \eta} = -\frac{RT}{p}\frac{\partial p}{\partial \eta}. \tag{4.4}$$

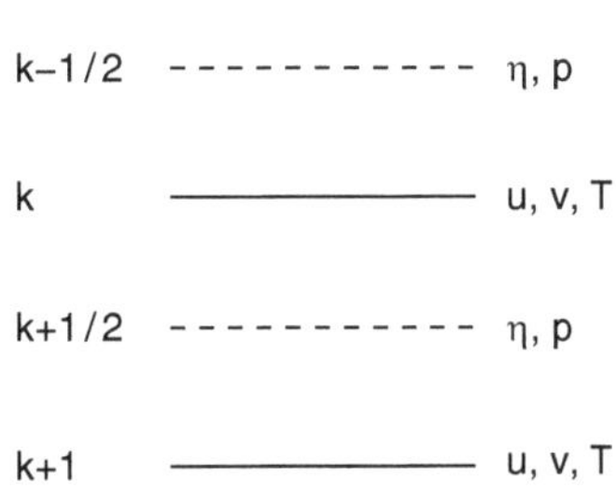

Fig. 4.3 Schematic showing the vertical arrangement of variables for the Simmons and Burridge scheme. k is the level index

Here Φ is the geopotential and R is the gas constant for dry air. This is naturally discretized as

$$\Phi_{k+1/2} - \Phi_{k-1/2} = -RT_k \ln \frac{p_{k+1/2}}{p_{k-1/2}}. \tag{4.5}$$

Since Φ at the ground is given, if we know the half-level pressures and the full-level temperatures then we can easily integrate (4.5) to obtain Φ at any half-level.

However, Φ is required in the momentum equations at full-levels. Therefore a further contribution proportional to T_k is added to obtain full-level values of Φ:

$$\Phi_k = \Phi_{k+1/2} + \alpha_k RT_k. \tag{4.6}$$

We have some freedom in exactly how the α's are specified; see Simmons and Burridge (1981) for a specific example. Here we will not specify α but keep the discussion as general as possible. Note that α_k may depend on $p_{k-1/2}$ and $p_{k+1/2}$ and hence on p_s.

Note that this scheme supports a *computational mode*: for any given hydrostatically balanced profile of T_k, p_s, and Φ_k, we can find a pattern of T and p_s perturbations that, when added to the original profile, has no effect on the Φ_k. Such a perturbation profile is therefore invisible to the dynamics. See Sect. 4.5.2 for further discussion.

4.4.2 Angular Momentum Conservation

The vertical coordinate defines the pressure at the half-levels. However, for the momentum equation we require the horizontal pressure gradient at the full-levels. Demanding angular momentum conservation tells us how we should define the full-level pressure gradient.

In spherical coordinates, the equation for the eastward velocity component is

$$\frac{Du}{Dt} - \frac{uv\tan\phi}{a} - fv + \frac{1}{a\cos\phi}\frac{\partial\Phi}{\partial\lambda} + \frac{RT}{p}\frac{1}{a\cos\phi}\frac{\partial p}{\partial\lambda} = 0, \tag{4.7}$$

where a is the Earth's radius. Multiplying by $a\cos\phi$ and using $a\,D\phi/Dt = v$ gives an equation for the angular momentum density $m = ua\cos\phi + a^2\Omega\cos^2\phi$:

$$\frac{Dm}{Dt} + \frac{\partial\Phi}{\partial\lambda} + \frac{RT}{p}\frac{\partial p}{\partial\lambda} = 0. \tag{4.8}$$

The net source of angular momentum, integrated over a latitudinal slice, is

$$\int_0^{2\pi}\int_0^1 \left(\frac{\partial\Phi}{\partial\lambda} + \frac{RT}{p}\frac{\partial p}{\partial\lambda}\right)\frac{\partial p}{\partial\eta}\,d\eta\,d\lambda$$
$$= \int_0^{2\pi}\int_0^1 \left(\frac{\partial\Phi}{\partial\lambda}\frac{\partial p}{\partial\eta} - \frac{\partial\Phi}{\partial\eta}\frac{\partial p}{\partial\lambda}\right)d\eta\,d\lambda$$

$$= \int_0^{2\pi} \int_0^1 \frac{\partial}{\partial \eta}\left(-\Phi \frac{\partial p}{\partial \lambda}\right) + \frac{\partial}{\partial \lambda}\left(\Phi \frac{\partial p}{\partial \eta}\right) d\eta\, d\lambda$$
$$= -\int_0^{2\pi} \Phi_s \frac{\partial p_s}{\partial \lambda} d\lambda, \tag{4.9}$$

where Φ_s is the surface geopotential, and we have used the hydrostatic relation and the boundary conditions to simplify the integral. Repeating this derivation for the finite difference scheme, we find that a finite difference analogue of this formula will hold provided

$$\sum_{k=1}^{N} \Phi_k \frac{\partial}{\partial \lambda} \Delta p_k = \Phi_s \frac{\partial p_s}{\partial \lambda} + \sum_{k=1}^{N} R \left(\frac{T}{p}\frac{\partial p}{\partial \lambda}\right)_k \Delta p_k, \tag{4.10}$$

which, in turn, will hold provided we define the full-level pressure gradient via

$$\left(\frac{RT}{p}\nabla p\right)_k = \frac{RT_k}{\Delta p_k}\left[\left(\ln \frac{p_{k+1/2}}{p_{k-1/2}}\right)\nabla p_{k-1/2} + \alpha_k \nabla(\Delta p_k)\right], \tag{4.11}$$

where α_k is the same as used in (4.6) to define the full-level Φ.

4.4.3 Energy Conservation

Taking $\mathbf{v}$ times the momentum equation gives

$$\frac{D}{Dt}\left(\frac{\mathbf{v}^2}{2}\right) = -\mathbf{v}.\nabla\Phi - \frac{RT}{p}\mathbf{v}.\nabla p, \tag{4.12}$$

while the thermodynamic equation may be written

$$\frac{D}{Dt}c_p T = \frac{RT\omega}{p}, \tag{4.13}$$

where

$$\omega \equiv \frac{Dp}{Dt} = -\int_0^{\eta} \nabla.\left(\mathbf{v}\frac{\partial p}{\partial \eta}\right) d\eta + \mathbf{v}.\nabla p. \tag{4.14}$$

The terms on the right hand sides of (4.12) and (4.13) represent conversions between kinetic and potential or internal energy. The global integral of the sum of the conversion terms vanishes, implying energy conservation.

For the discretization we need to define ω at full-levels using a finite-difference analogue of (4.14). It may be verified that if we evaluate RT/p times the vertical integral term as

$$\frac{RT_k}{\Delta p_k}\left[\left(\ln \frac{p_{k+1/2}}{p_{k-1/2}}\right)\sum_{r=1}^{k-1} \nabla.(\mathbf{v}_r \Delta p_r) + \alpha_k \nabla.(\mathbf{v}_k \Delta p_k)\right] \quad (4.15)$$

and evaluate $(RT/p)\nabla p$ using (4.11) as in the momentum equation then all contributions to the global integral of the conversion terms do indeed cancel and so the scheme preserves energy conservation.

The expressions (4.11) and (4.15) are not the most obvious finite difference discretizations of the corresponding continuous expressions; it is typically non-trivial to obtain such conservation properties.

4.5 Wave Dispersion and Balance

In Chap. 3 we saw how different choices of horizontal grid staggering (and prognostic variables) can affect the accuracy with which we capture the propagation of different classes of waves, particularly short wavelength waves that are marginally resolved. Accurate representation of the propagation of fast waves is important for capturing adjustment towards balance, and hence for capturing balanced motions themselves. Similar issues arise when considering vertical discretizations.

4.5.1 The Lorenz and Charney–Phillips Grids

For models solving the hydrostatic equations we have three-dimensional fields of three prognostic variables: usually the two horizontal wind components (or some equivalent information in terms of vorticy and divergence) and a thermodynamic variable. (In some formulations we also have a surface pressure field.) Two classes of vertical grid are widely used for hydrostatic models: those in which the thermodynamic variable is stored at the same levels as the wind variables, and those in which the thermodynamic variable is staggered in the vertical relative to the wind variables. These are commonly referred to as the Lorenz grids and Charney–Phillips grids, respectively, (Fig. 4.4) after Lorenz (1960) and Charney and Phillips (1953).

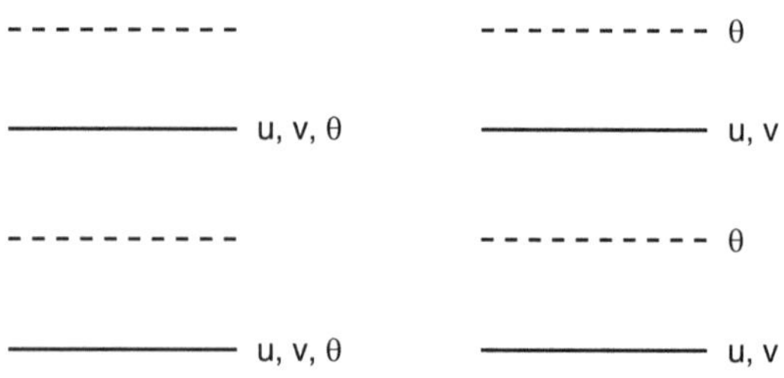

Fig. 4.4 Schematic showing the vertical arrangement of variables for the Lorenz (*left*) and Charney–Phillips (*right*) grids

4.5.2 Lorenz Grid Computational Mode

One well-known drawback of the Lorenz grids is that they support a *computational mode*. Consider, for example, the Simmons and Burridge scheme discussed above. Suppose we have vertical profiles of T, p_s and Φ satisfying hydrostatic balance (4.5). Now consider making some perturbations T'_k to the temperature values and p'_s to the surface pressure; through the linearized version of (4.5), these will imply corresponding perturbations Φ'_k in the geopotential. The geopotential perturbation at the lowest full-level is

$$\Phi'_N = \alpha_N R T'_N + R T_N \frac{d\alpha_N}{dp_s} p'_s, \tag{4.16}$$

while the difference between successive full-level geopotential perturbations is

$$\begin{aligned} \Phi'_k - \Phi'_{k-1} &= R\left\{\alpha_k + \ln\left(\frac{p_{k+1/2}}{p_{k-1/2}}\right)\right\} T'_k \\ &\quad - R\alpha_{k-1} T'_{k-1} \\ &\quad + \left\{R T_k \left(\frac{d\alpha_k}{dp_s} + \frac{b_{k+1/2}}{p_{k+1/2}} - \frac{b_{k-1/2}}{p_{k-1/2}}\right) - R T_{k-1} \frac{d\alpha_{k-1}}{dp_s}\right\} p'_s, \end{aligned} \tag{4.17}$$

where $b_{k+1/2} = dp_{k+1/2}/dp_s$.

For an arbitrary p'_s, we can ensure that Φ'_N vanishes by a suitable choice of T'_N. But then we can ensure that Φ'_{N-1} vanishes by a suitable choice of T'_{N-1}, and so on. In this way we can find a profile of T'_k such that all Φ'_k vanish. Such a p'_s and T'_k profile will therefore have no effect on the momentum equation; it will be invisible to the dynamics and will not propagate (Tokioka 1978; Arakawa and Moorthi 1988). The key point here is that there is one more degree of freedom in $\{T_k, k = 1, ..., N; p_s\}$ than there is in $\{\Phi_k, k = 1, ..., N\}$; basic linear algebra then implies that there exists a family of nonzero solutions for $\{T'_k, k = 1, ..., N; p'_s\}$ that make $\{\Phi'_k, k = 1, ..., N\}$ vanish.

Related to the existence of the computational mode is the possibility of a spurious resonant response to a steady thermal forcing. If the forcing projects onto the computational mode then the response will grow linearly with time (until nonlinear effects come into play) rather than reaching a steady response (Schneider 1987).

4.5.3 Compressible Euler Equations

The compressible Euler equations do not make the hydrostatic approximation or any kind of incompressibility approximation; they therefore support acoustic waves as well as inertio-gravity and Rossby waves. For the compressible Euler equations we have five prognostic variables, usually three velocity variables and two thermodynamic variables. There are therefore many different possibilities for choosing staggered grids. There are also different possible choices for which thermodynamic variables are predicted, e.g., any two from ρ, p, T, θ, etc. In this subsection we will restrict attention to the use of height z as the vertical coordinate with a uniform grid spacing Δz, but similar reasoning applies to other vertical coordinates (Thuburn and Woollings 2005; Thuburn 2006).

Numerical exploration of a large number of possible configurations (Thuburn and Woollings 2005) shows that:

- Accurate representation of acoustic waves is necessary (but not sufficient) for an accurate representation of inertio-gravity waves
- Which in turn is necessary (but not sufficient) for an accurate representation of Rossby waves

Here, by considering the dispersion relations for different kinds of waves, we attempt to give heuristic explanations for the kinds of configuration that give the best representation of wave propagation. Just as we found when considering horizontal discretizations, we want to avoid or minimize taking averages, and we want to avoid or minimize taking differences over $2\Delta z$, since these approximations introduce large errors in the propagation of short waves.

In what follows we will consider wavelike solutions of the governing equations with wavevector (k, l, m) and frequency ω. Also define the total horizontal wavenumber squared $K^2 = k^2 + l^2$, the gravitational acceleration g, the Coriolis parameter f, and the buoyancy frequency N.

4.5.3.1 Acoustic Waves

The acoustic wave dispersion relation is

$$\omega^2 \approx (m^2 + K^2)c^2, \tag{4.18}$$

where c is the speed of sound. Here, one factor of m comes from the vertical derivative of p appearing in the w equation, and the other comes from the vertical derivative of w appearing in the p equation. Thus, we will capture the dispersion relation as accurately as possible if we capture these two vertical derivatives as accurately as possible in the limit of short vertical wavelength. We therefore require:

- $\delta_z p$ at the same level as w
- $\delta_z w$ at the same level as p

where δ_z represents a finite difference approximation to $\partial/\partial z$. This implies that p should be staggered with respect to w to obtain the most compact finite difference approximations.

If p is not predicted but ρ is, then, for vertical-grid-scale waves, p perturbations (expressed in terms of the two predicted thermodynamic variables) will be dominated by ρ perturbations provided $\Delta z \ll g/N^2$, which will always hold in practice; this then implies that ρ should be staggered with respect to w.

4.5.3.2 Inertio-Gravity Waves

The inertio-gravity wave dispersion relation is

$$\omega^2 \approx \frac{m^2 f^2 + K^2 N^2}{m^2 + K^2}. \tag{4.19}$$

The denominator arises in the same way as the $m^2 + K^2$ factor in the acoustic wave dispersion relation and so again will be captured as accurately as possible provided p (or ρ) is staggered with respect to w. The m^2 term in the numerator also arises in the same way, yet again requiring p (or ρ) staggered with respect to w. Depending on the horizontal wavelength and on the relative sizes of f and N, it is possible that the K^2N^2 term in the numerator could dominate even for the shortest resolved vertical wavelengths. To capture the K^2N^2 term accurately requires that u and v be stored at the same levels as p in order to capture the pressure gradient term in the horizontal momentum equations without averaging, and also requires that the buoyancy variable (e.g., the potential temperature θ) be stored at the same levels as w in order to capture the buoyancy source due to vertical advection and the effect of buoyancy in the w equation without averaging.

If we do not predict θ but predict, say, T and p or T and ρ then there are comparable contributions to the θ perturbation from the two predicted thermodynamic variables. Optimal wave propagation would then require both p or ρ staggered with respect to w (to capture the m^2f^2 term) *and* p or ρ collocated with w (to capture the K^2N^2 term), which is obviously not possible. In other words, optimal wave propagation requires that we predict θ or some function of θ.

4.5.3.3 Rossby Waves

The Rossby wave dispersion relation is

$$\omega \approx -\frac{k\beta N^2}{m^2 f^2 + K^2 N^2}. \tag{4.20}$$

The denominator arises in the same way as the $m^2f^2 + K^2N^2$ factor in the inertio-gravity wave dispersion relation. It will be captured as accurately as possible provided p is staggered with respect to w and, if K^2N^2 can be large, provided u and v are stored at the same levels as p and θ is stored at the same levels as w. The numerator will be captured accurately provided θ is stored at the same levels as w.

4.5.3.4 Numerical Dispersion Relations for Some Example Configurations

Figure 4.5 shows two plausible grid configurations that are natural extensions to the compressible Euler equations of the Lorenz and Charney–Phillips grids. According to our heuristic reasoning above, the Charney–Phillips grid should be as accurate as possible for all types of waves. This does indeed turn out to be the case; Fig. 4.6

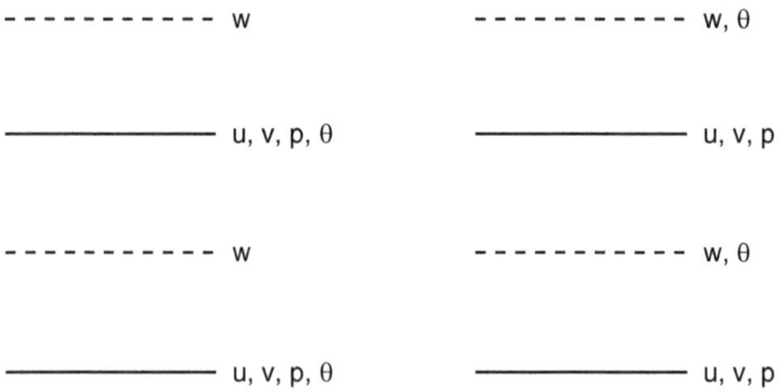

Fig. 4.5 Schematic showing the vertical arrangement of variables for compressible Euler versions of the Lorenz (*left*) and Charney–Phillips (*right*) grids

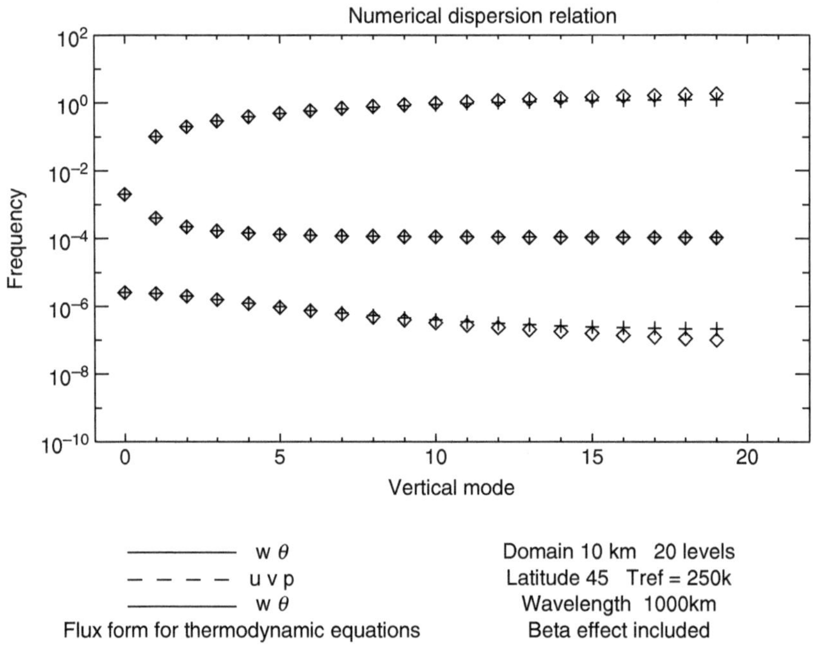

Fig. 4.6 Numerical dispersion relation for the optimal vertical configuration shown in the *right panel* of Fig. 4.5 (*crosses*) and exact dispersion relation (*diamonds*). The domain depth is 10^4 m with a rigid lid, horizontal wavelength is 1,000 km, and the geometry is that for a β-plane at 45°N. The reference state is resting and in hydrostatic balance with a uniform temperature of 250 K. The numerical dispersion relation was calculated for a uniform grid with 20 full-levels. The *upper curve* corresponds to internal acoustic modes, the *middle curve* corresponds to the external acoustic mode (mode number zero) and internal inertio-gravity modes, and the *lower curve* corresponds to Rossby modes. Only westward propagating modes are shown. There are also eastward propagating acoustic and inertio-gravity mode branches almost identical to the westward branches shown

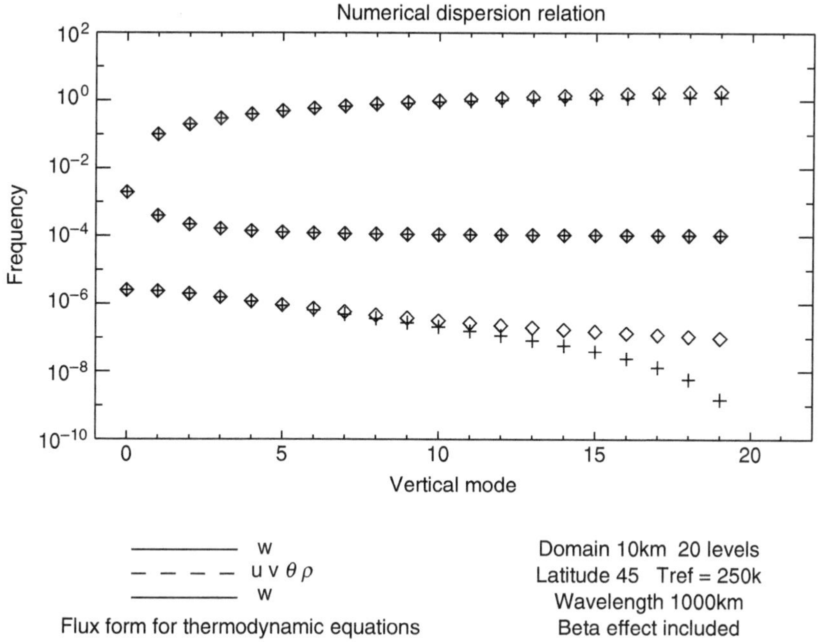

Fig. 4.7 As in Fig. 4.6 but for the vertical configuration shown in the *left panel* of Fig. 4.5

shows an example numerical dispersion relation, computed numerically, for this configuration.

The Lorenz grid should be accurate for acoustic and inertio-gravity waves provided K^2N^2 does not dominate m^2f^2. Figure 4.7 shows the numerical dispersion relation for this configuration when this condition holds. The acoustic and inertio-gravity wave dispersion relations are indeed captured accurately, but the Rossby modes are retarded (compare Fig. 4.6). Also, as in the hydrostatic case, the Lorenz grid supports a zero-frequency computational mode, which is not visible in Fig. 4.7.

There are some subtleties in exactly how the pressure gradient term should be evaluated, particularly if we wish to predict ρ rather than p to facilitate mass conservation (Thuburn 2006). Figure 4.8 shows the numerical dispersion relation when we predict ρ instead of p on the Charney–Phillips grid, assuming that the pressure gradient term is written as $(1/\rho)\nabla p$, discretized in the obvious way, with p diagnosed from ρ and a vertically averaged θ. In this calculation the buoyancy effect of θ is, in effect, vertically averaged, with the result that short-vertical-wavelength Rossby waves are retarded. Figure 4.9 shows the numerical dispersion relation for the same configuration if we use the alternative form $c_p\theta\nabla\Pi$ for the pressure gradient term, where c_p is the specific heat capacity at constant pressure and $\Pi = (p/p_0)^\kappa$. This calculation feels the full buoyancy effect of θ, and all waves are handled as accurately as possible.

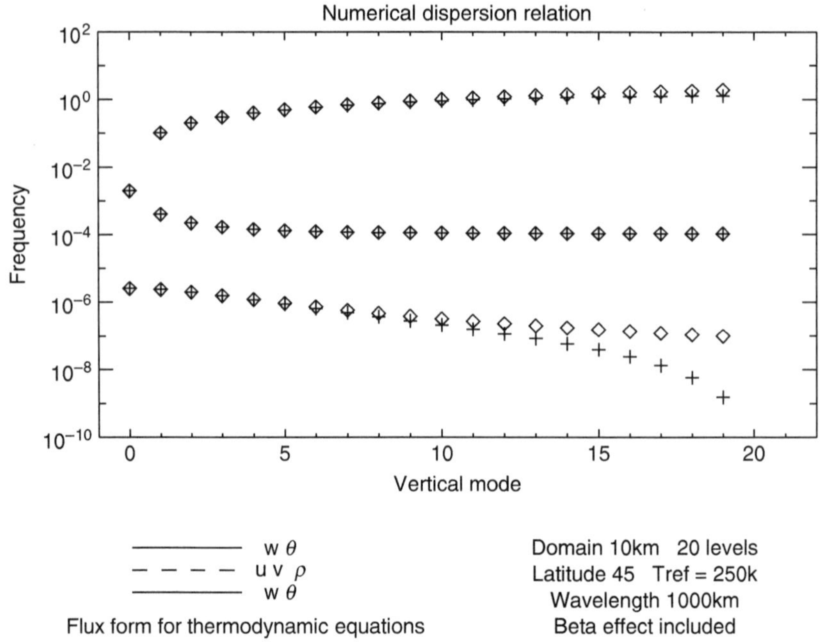

Fig. 4.8 As in Fig. 4.6 but for the vertical configuration predicting ρ instead of p and using the $(1/\rho)\nabla p$ form of the pressure gradient

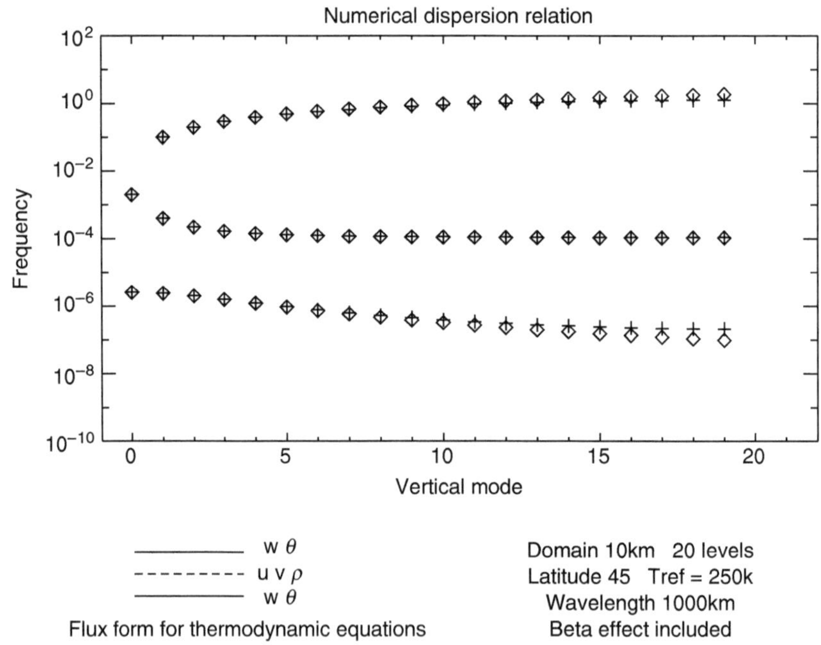

Fig. 4.9 As in Fig. 4.6 but for the vertical configuration predicting ρ instead of p and using the $c_p\theta\nabla\Pi$ form of the pressure gradient

4.6 Conclusion

The main choices of vertical coordinate have been introduced. They each have advantages and disadvantages, and indeed there is ongoing research and development pursuing several of the options.

Two of the main issues in the design of vertical discretizations are conservation properties and wave dispersion properties, and we have touched on both topics. Better wave dispersion properties can be obtained with the Charney–Phillips family of grids, particularly if careful attention is paid to the formulation of the pressure gradient term. On the other hand, conservation properties are more easily obtained using the Lorenz family of grids. There is ongoing debate over the relative importance of these two factors, and new models are being developed with both Charney–Phillips and Lorenz grids.

Incidentally, further issues and complexity arise when considering the coupling of the dynamical core to the physical parameterizations. For example, with a Charney–Phillips grid, should one store moisture at density levels, facilitating conservation of moisture, or at θ-levels, facilitating the calculation of important thermodynamic quantities like relative humidity? There is clearly great scope for further research.

References

Adcroft A, Hill C, Marshall J (1997) Representation of topography by shaved cells in a height coordinate ocean model. Mon Wea Rev 125:2293–2315

Arakawa A, Moorthi S (1988) Baroclinic instability in vertically discrete systems. J Atmos Sci 45:1688–1707

Charney JG, Phillips NA (1953) Numerical integration of the quasi-geostrophic equations for barotropic and simple baroclinic flow. J Meteorol 10:71–99

Durran DR (1999) Numerical Methods for Wave Equations in Geophysical Fluid Dynamics. Springer-Verlag

Gal-Chen T, Somerville RC (1975) On the use of a coordinate transformation for the solution of Navier-Stokes equations. J Comput Phys 17:209–228

Hsu YJG, Arakawa A (1990) Numerical modeling of the atmosphere with an isentropic vertical coordinate. Mon Wea Rev 118:1933–1959

Kasahara A (1974) Various vertical coordinate systems used for numerical weather prediction. Mon Wea Rev 102(7):509–522

Konor CS, Arakawa A (1997) Design of an atmospheric model based on a generalized vertical coordinate. Mon Wea Rev 125(7):1649–1673

Lin SJ (2004) A 'vertically Lagrangian' finite-volume dynamical core for global models. Mon Wea Rev 132:2293–2307

Lorenz EN (1960) Energy and numerical weather prediction. Tellus 12:364–373

Phillips NA (1957) A coordinate system having some special advantage for numerical forecasting. J Meteorol 14:184–185

Schneider EK (1987) An inconsistency in vertical discretization in some atmospheric models. Mon Wea Rev 115:2166–2169

Shaw TA, Shepherd TG (2007) Angular momentum conservation and gravity wave drag parametrization: Implications for climate models. J Atmos Sci 64:190–203

Simmons AJ, Burridge DM (1981) An energy and angular-momentum conserving vertical finite-difference scheme and hybrid vertical coordinates. Mon Wea Rev 109(4):758–766

Staniforth A, Wood N (2003) The deep-atmosphere Euler equations in a generalized vertical coordinate. Mon Wea Rev 131:1931–1938

Starr VP (1945) A quasi-Lagrangian system of hydrodynamical equations. J Atmos Sci 2:227–237

Thuburn J (2006) Vertical discretizations giving optimal representation of normal modes: Sensitivity to the form of the pressure gradient term. Quart J Roy Meteorol Soc 132:2809–2825

Thuburn J, Woollings TJ (2005) Vertical discretizations for compressible Euler equation atmospheric models giving optimal representation of normal modes. J Comput Phys 203:386–404

Tokioka T (1978) Some considerations on vertical differencing. J Meteorol Soc Japan 56:98–111

Chapter 5
Time Discretization: Some Basic Approaches

Dale R. Durran

Abstract The basic concepts of stability, consistency and convergence are introduced. Additional measures of stability, such as A- and L-stability are discussed, along with desirable stability properties for the time integration of partial differential equations. The family of Runge–Kutta methods is reviewed, including both classical schemes and more recently developed strong-stability preserving and diagonally implicit methods. The chapter concludes with a brief discussion of linear multistep methods.

5.1 Introduction

Although the fundamental equations governing the evolution of geophysical fluids are partial differential equations, ordinary differential equations arise in several contexts. The trajectories of individual fluid parcels in an inviscid flow are governed by simple ordinary differential equations, and systems of ordinary differential equations may describe chemical reactions or highly idealized dynamical systems. Since basic methods for the numerical integration of ordinary differential equations are simpler than those for partial differential equations, and since the time-differencing formulae used in the numerical solution of partial differential equations are closely related to those used for ordinary differential equations, this chapter is devoted to the analysis of methods for the approximate solution of ordinary differential equations. Nevertheless some approaches to the solution of partial differential equations, such finite-volume methods, arise from fully discretized approximations in both space and time that cannot be correctly analyzed by considering the spatial and temporal differencing in isolation (Chap. 8).

D.R. Durran
Department of Atmospheric Sciences, Box 351640, University of Washington, Seattle, WA, 98195, USA
e-mail: durrand@atmos.washington.edu

P.H. Lauritzen et al. (eds.), *Numerical Techniques for Global Atmospheric Models*, Lecture Notes in Computational Science and Engineering 80,
DOI 10.1007/978-3-642-11640-7_5,

Most of this chapter will focus on methods potentially suitable for use in the numerical integration of time dependent partial differential equations. In comparison with typical ordinary differential equation solvers, the methods used to integrate partial differential equations are relatively low order. Low-order schemes are used for two basic reasons. First, the approximation of the time derivative is not the only source of error in the solution of partial differential equations; other errors arise through the approximation of the spatial derivatives. In many circumstances the largest errors in the solution are introduced through the numerical evaluation of the spatial derivatives, so it is pointless to devote additional computational resources to higher-order time differencing. The second reason for using low-order methods is that practical limitations on computational resources often leave no other choice.

Consider the initial value problem

$$\frac{d\psi}{dt} = F(\psi, t), \quad \psi(0) = \psi_0; \tag{5.1}$$

which will have a unique solution provided the function F is sufficiently smooth (in particular, F must satisfy a Lipschitz condition.)[1] Numerical approximations ϕ_n to the true solution at some set of discrete time levels $t_n = n\Delta t$, $n = 0, 1, 2, \ldots$ may be obtained by setting $\phi_0 = \psi_0$ and repeatedly stepping the solution forward by solving algebraic equations in which ϕ_n depends only on the approximate solution at previous time levels.

It is often helpful to consider the algebraic equations used to create these approximate solutions as arising from one of two approaches. In the first approach, the time derivative in (5.1) is replaced with a finite difference. In the second approach (5.1) is integrated over a time interval Δt

$$\psi(t_{n+1}) = \psi(t_n) + \int_{t_n}^{t_{n+1}} F\big(\psi(t), t\big)\, dt, \tag{5.2}$$

and the algebraic equations that constitute the numerical method provide an approximate formula for evaluating the integral of F.

In the following, we will focus on the how the various stability conditions satisfied by simple ordinary differential equation solvers influence their suitability for the solution of partial differential equations. We then take a close look at Runge–Kutta methods, which include a wide selection of schemes with various desirable properties, many of which are not familiar to the atmospheric-science community. The chapter concludes with a brief discussion of another popular family of schemes, the linear multistep methods.

[1] The *Lipschitz condition* is that $|F(x,t) - F(y,t)| \leq L|x - y|$ for all x and y, and all $t \geq 0$; where $L > 0$ is a real constant. One way to satisfy this condition is if $|\partial F/\partial x|$ is bounded.

5.2 Stability, Consistency and Convergence

The basic goal when computing a numerical approximation to the solution of a differential equation is to obtain a result that indeed approximates the true solution. In this section we will examine the relationship between three fundamental concepts characterizing the quality of the numerical solution in the limit where the separation between adjacent nodes on a numerical mesh approaches zero: consistency, stability and convergence.

5.2.1 *Truncation Error*

The derivative of a function $\psi(t)$ at time t_n could be defined either as

$$\frac{d\psi}{dt}(t_n) = \lim_{\Delta t \to 0} \frac{\psi(t_n + \Delta t) - \psi(t_n)}{\Delta t}, \tag{5.3}$$

or as

$$\frac{d\psi}{dt}(t_n) = \lim_{\Delta t \to 0} \frac{\psi(t_n + \Delta t) - \psi(t_n - \Delta t)}{2\Delta t}. \tag{5.4}$$

If the derivative of $\psi(t)$ is continuous at t_n, both expressions produce the same unique answer. In practical applications, however, it is impossible to evaluate these expressions with infinitesimally small Δt. The approximations to the true derivative obtained by evaluating the algebraic expressions on the right side of (5.3) and (5.4) using finite Δt are known as *finite differences*. When Δt is finite, the preceding finite-difference approximations are not equivalent; they differ in their accuracy, and when they are substituted for derivatives in differential equations they generate different algebraic equations. The differences in the structure of these algebraic equations can have a great influence on the stability of the numerical solution.

Which of the preceding finite-difference formula is likely to be more accurate when Δt is small but finite? If $\psi(t)$ is a sufficiently smooth function of t, this question can be answered by expanding the terms $\psi(t_n \pm \Delta t)$ in Taylor series about t_n and substituting these expansions into the finite-difference formula. For example, substituting

$$\psi(t_n + \Delta t) = \psi(t_n) + \Delta t \frac{d\psi}{dt}(t_n) + \frac{(\Delta t)^2}{2}\frac{d^2\psi}{dt^2}(t_n) + \frac{(\Delta t)^3}{6}\frac{d^3\psi}{dt^3}(t_n) + \ldots$$

into (5.3), one finds that

$$\frac{\psi(t_n + \Delta t) - \psi(t_n)}{\Delta t} - \frac{d\psi}{dt}(t_n) = \frac{\Delta t}{2}\frac{d^2\psi}{dt^2}(t_n) + \frac{(\Delta t)^2}{6}\frac{d^3\psi}{dt^3}(t_n) + \ldots . \tag{5.5}$$

The right side of the preceding is the *truncation error* of the finite difference. The lowest power of Δt in the truncation error determines the *order of accuracy* of the finite difference. Inspection of (5.5) shows that the one-sided difference is first-order accurate. In contrast, the truncation error associated with the centered difference (5.4) is

$$\frac{(\Delta t)^2}{6}\frac{d^3\psi}{dt^3}(t_n) + \frac{(\Delta t)^4}{120}\frac{d^5\psi}{dt^5}(t_n) + \dots ,$$

and the centered difference is therefore second-order accurate. If the higher-order derivatives of ψ are bounded in some interval about t_n, (i.e., if ψ is "smooth") and Δt is repeatedly reduced, the error in the second-order difference (5.4) will approach zero more rapidly than the error in the first-order difference (5.3). The fact that the truncation error of the centered difference is higher order does not, however, guarantee that it will always generate a more accurate estimate of the derivative. If the function is sufficiently rough and Δt sufficiently coarse, neither formula is likely to produce a good approximation, and the superiority of one over the other will be largely a matter of chance.

Euler's method (sometimes called forward-Euler) approximates the derivative in (5.1) with the forward difference (5.3) to give the formula

$$\frac{\phi_{n+1} - \phi_n}{\Delta t} = F(\phi_n, t_n). \tag{5.6}$$

Clearly this formula can be used to obtain ϕ_1 from the initial condition $\phi_0 = \psi_0$, and then be applied recursively to obtain ϕ_{n+1} from ϕ_n. How well does this simple method perform?

One way to characterize the accuracy of this method is through its *truncation error, defined as the residual by which smooth solutions to the continuous problem fail to satisfy the discrete approximation* (5.6). Let τ_n denote the truncation error at time t_n, then from (5.5),

$$\frac{\psi(t_{n+1}) - \psi(t_n)}{\Delta t} - F\big(\psi(t_n), t_n\big) = \frac{d\psi}{dt}(t_n) + \tau_n - F\big(\psi(t_n), t_n\big) = \tau_n, \tag{5.7}$$

where the second equality holds because ψ is a solution to the continuous problem and

$$\tau_n = \frac{\Delta t}{2}\frac{d^2\psi}{dt^2}(t_n) + O\big[(\Delta t)^2\big].$$

It is not necessary to explicitly consider the higher-order terms in the truncation error to bound $|\tau_n|$; if ψ has continuous second derivatives, the mean-value theorem may be used to show

$$|\tau_n| \le \frac{\Delta t}{2} \max_{t_n \le s \le t_{n+1}} \left| \frac{d^2\psi}{dt^2}(s) \right|. \tag{5.8}$$

Euler's method is *consistent of order one* because the lowest power of Δt appearing in τ_n is unity. If the centered difference (5.4) were used to approximate the time derivative in (5.6), the resulting method would be consistent of order two.

5.2.2 Convergence

A *consistent* method is one for which the truncation error approaches zero as $\Delta t \to 0$. The order of the consistency determines the rate at which the solution of a *stable* finite-difference method *converges* to the true solution as $\Delta t \to 0$. To examine the relation between consistency and convergence, define the *global error* at time t_n as $E_n = \phi_n - \psi(t_n)$. From (5.7),

$$\psi(t_{n+1}) = \psi(t_n) + \Delta t F\big(\psi(t_n), t_n\big) + \Delta t\, \tau_n, \tag{5.9}$$

which implies that if we start with the true solution at t_n, the *local* or *one-step* error generated by Euler's method in approximating the solution at t_{n+1} is $\Delta t\, \tau_n$, which is one power of Δt higher that the truncation error itself. One might suppose that the global error in the solution at time T is bounded by the maximum local error times the number of times steps $(\max_n |\Delta t\, \tau_n|)(T/\Delta t)$ which, like τ_n itself, is $O(\Delta t)$. This would be a welcome result because it would imply the error becomes arbitrarily small as the time step approaches zero, but such reasoning is incorrect because it does not account for the difference between ϕ_n and $\psi(t_n)$ arising from the accumulation of local errors over the preceding time steps. The increase in the global error generated over a single step satisfies

$$E_{n+1} = E_n + \Delta t \left[F(\phi_n, t_n) - F\big(\psi(t_n), t_n\big)\right] - \Delta t\, \tau_n, \tag{5.10}$$

which may be obtained by solving (5.6) for ϕ_{n+1} and subtracting (5.9). As apparent from (5.10), the numerical solution will converge to the true solution provided $F(\phi_n, t_n) - F\big(\psi(t_n), t_n\big)$ remains finite in the limit $\Delta t \to 0$.

It is easy to show that Euler's method converges for the special case where

$$F(\psi, t) = \lambda\psi + g(t), \tag{5.11}$$

where λ is a constant.[2] We will examine this special case because it reveals the relatively weak stability condition required to assure convergence to the true solution in the limit $\Delta t \to 0$. Substituting (5.11) into (5.10) gives

$$E_{n+1} = (1 + \lambda\Delta t)E_n - \Delta t\, \tau_n. \tag{5.12}$$

[2] More general conditions sufficient to guarantee the convergence of Euler's are that F is an analytic function (Iserles 1996, p. 7) or that the first two derivatives of ψ are continuous (Hundsdorfer and Verwer 2003).

Note that $g(t)$, the part of $F(\psi, t)$ that is independent ψ, drops out and has no impact on the growth of the global error. Suppose that $N = T/\Delta t$ is the number of time steps required to integrate from the initial condition at $t = 0$ to some fixed time T.

From (5.12)

$$\begin{aligned} E_N &= (1 + \lambda \Delta t) E_{N-1} - \Delta t\, \tau_{N-1} \\ &= (1 + \lambda \Delta t)\,[1 + \lambda \Delta t) E_{N-2} - \Delta t\, \tau_{N-2}] - \Delta t\, \tau_{N-1} \end{aligned}$$

and by induction,

$$E_N = (1 + \lambda \Delta t)^N E_0 - \Delta t \sum_{m=1}^{N} (1 + \lambda \Delta t)^{N-m} \tau_{m-1}.$$

Let

$$\tau_{\max} = \frac{\Delta t}{2} \max_{0 \le s \le t_N} \left| \frac{d^2 \psi}{dt^2}(s) \right|,$$

which from (5.8) is an upper bound on $|\tau_n|$ for all n independent of the choice of time step used to divide up the interval $[0, T]$. Assuming the initial error E_0 is zero (although an $O(\Delta t)$ error would not prevent convergence), and noting that for $\Delta t > 0$, $1 + |\lambda| \Delta t \le e^{|\lambda| \Delta t}$, one obtains

$$|E_N| \le N \Delta t\, (1 + |\lambda| \Delta t)^N \tau_{\max} = T e^{|\lambda| T} \tau_{\max}. \tag{5.13}$$

Since $T e^{|\lambda| T}$ has some finite value independent of the numerical discretization and $\tau_{\max}$ is $O(\Delta t)$, the global error at time T must approach zero in proportion to the first power of Δt.

5.2.3 Stability

The foundation for the theory of numerical methods for differential equations is built on the theorem that *consistency of order p and stability imply convergence of order p* (Dalhquist 1956; Lax and Richtmyer 1956) . Evidently Euler's method satisfies some type of stability condition since it is consistent and is convergent of order unity. The relation (5.12), which states that previous global errors amplify by a factor of $(1 + \lambda \Delta t)$ over each individual time step, provides the key for bounding the growth of the global error over a finite time interval. Define the *amplification factor A* as the ratio of the approximate solution at two adjacent time steps,

$$A = \phi_{n+1}/\phi_n. \tag{5.14}$$

A two-time-level method is stable in the sense that, if it is also consistent, it will converge in the limit $\Delta t \rightarrow 0$ provided that for some constant η (independent of the properties of the numerical discretization)

$$|A| \leq 1 + \eta \Delta t. \tag{5.15}$$

In the previous simple example, $\eta = \lambda$ is just the coefficient of ψ in the forcing $F(\psi, t)$. When Euler's method is applied to more general problems, η is a constant associated with the Lipschitz condition on $F(\psi, t)$. Essentially all consistent *two-time-level* ordinary differential equation solvers satisfy this stability condition, but as discussed in Sect. 5.3.4, bounds similar to (5.15) are not satisfied by many potentially reasonable approximations to time-dependent partial differential equations.

5.3 Additional Measures of Stability and Accuracy

Although Euler's method is sufficiently stable to converge in the limit $\Delta t \rightarrow 0$, it may nevertheless generate a sequence $\phi_0, \phi_1, \ldots$ that blows up in a completely nonphysical manner when the computations are performed with finite values of Δt. Again suppose $F(\psi, t) = \lambda\psi$, if $\lambda < 0$, the true solution $\psi_0 e^{\lambda t}$ is bounded by ψ_0 for all time and approaches zero as $t \rightarrow \infty$. Yet if $\lambda \Delta t < -2$, then $A = 1 + \lambda \Delta t < -1$, and the numerical solution changes sign and amplifies geometrically every time step, diverging wildly from the true solution.

5.3.1 A-Stability

How can we characterize the stability of a consistent numerical method to give an indication whether a solution computed using finite Δt is likely to blow up in such an "unstable" manner? Clearly there are many physical problems where the true solution does amplify rapidly with time, and of course any convergent numerical method must be able to capture such amplification. On the other hand, there are also many problems in which the norm of the solution is bounded or decays with time. It is not practical to consider every possible case individually, but it is very useful to evaluate the behavior of schemes on the simple test problem

$$\frac{d\psi}{dt} = \gamma\psi, \qquad \psi(0) = \psi_0, \tag{5.16}$$

where in contrast to our previous examples, ψ and γ are complex-valued. Breaking γ into its real and imaginary parts, such that $\gamma = \lambda + i\omega$ with λ and ω real, the solution to (5.16) is

$$\psi(t) = \psi_0 e^{\lambda t} e^{i\omega t},$$

showing that $\Re\{\gamma\}$ determines rate of change of the magnitude (or modulus) of ψ, while $\Im\{\gamma\}$ governs the rate of change of its phase (or argument).

Despite its simplicity, (5.16) is prototypical of the time variations found in many important fluid-dynamical problems. For example the concentration χ of a passive tracer in a flow moving at speed c and diffusing with a diffusivity M along one spatial dimension is given by the partial differential equation

$$\frac{\partial \chi}{\partial t} + c\frac{\partial \chi}{\partial x} = M\frac{\partial^2 \chi}{\partial x^2}. \tag{5.17}$$

Suppose the spatial domain is $|x| \leq 1$ and periodic, then the solution may be determined as the superposition of Fourier modes, each of which may be expressed in the form $b_k(t)e^{ikx}$, where b_k is a complex number determining the amplitude and phase of each mode and $k = n\pi,\ n = 0, \pm 1, \pm 2, \ldots$ is the *wavenumber*. The wavenumber is inversely proportional to the *wavelength*, $L = 2\pi/k$, which is the distance over which a wave's shape repeats. Substituting an arbitrary Fourier mode into (5.17) yields the following ordinary differential equation of the form (5.16):

$$\frac{db_k}{dt} = -\left(Mk^2 + ick\right) b_k. \tag{5.18}$$

Note that in the context of the advection-diffusion problem, $\Re\{\gamma\}$ determines the changes in amplitude produced by diffusion and $\Im\{\gamma\}$ governs changes in phase produced by advection.

Numerical solutions to (5.16) computed with some specific value of Δt are *absolutely stable* if $|\phi_n| \leq |\phi_0|$ for all n, or equivalently, if $|A| \leq 1$. The amplification factor for Euler's-method solutions to (5.16) is $1+\gamma\Delta t$, so the values of $(\lambda\Delta t, \omega\Delta t)$ for which $|A| \leq 1$ satisfy the inequality

$$(1 + \lambda\Delta t)^2 + (\omega\Delta t)^2 \leq 1.$$

This region of absolute stability, which is the interior of a unit circle centered at (-1,0) in the $\lambda\Delta t$-$\omega\Delta t$ plane, is plotted in Fig. 5.1a.

The true solution to (5.16) is non-amplifying for all $\lambda \leq 0$. This behavior is captured by numerical methods that are A-stable. *A numerical method that is absolutely stable for all $\lambda\Delta t \leq 0$ is A-stable*. Forward differencing is not A-stable, but as will be discussed in Sect. 5.3.3, the other methods whose absolute stability regions are shown in Fig. 5.1 are A-stable. A less restrictive variant of A-stability, known as A(α) stability, is discussed in Iserles (1996), Hundsdorfer and Verwer (2003) and LeVeque (2007).

5.3.2 Phase-Speed Errors

When $M = 0$, the prototype (5.18) reduces to the *oscillation equation*

$$\frac{d\psi}{dt} = i\omega\psi, \tag{5.19}$$

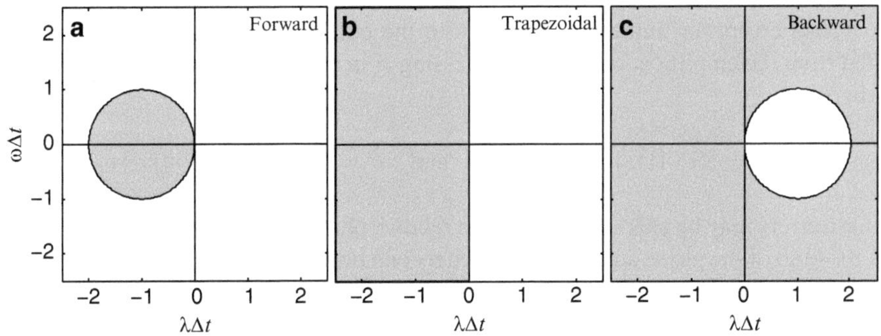

Fig. 5.1 Absolute stability regions (*shaded*) for (**a**) forward Euler, (**b**) trapezoidal differencing, and (**c**) backward Euler

which serves as important model for many non-dissipative dynamical systems. The oscillation equation may also be derived from a two-component real-valued system of ordinary differential equations such as those representing Coriolis accelerations,

$$\frac{du}{dt} - fv = 0$$

$$\frac{dv}{dt} + fu = 0,$$

by setting $\psi = u + iv$ and $\omega = -f$.

Integrating the oscillation equation over a time Δt yields,

$$\psi(t_0 + \Delta t) = e^{i\omega\Delta t}\psi(t_0) \equiv A_e\psi(t_0). \tag{5.20}$$

Here the last relation defines an "exact amplification factor" A_e, which in the case of the oscillation equation, is a complex number of modulus one. According to (5.20), over the time interval Δt, ψ moves $\omega\Delta t$ radians around a circle in the complex plane of radius $|\psi(t_0)|$ centered at the origin.

Hundreds of papers have been written investigating techniques for solving (5.17) with $M = 0$ (see for example the extensive review in Rood (1987)). The vastness of this body of literature is a testament to the subtle tradeoffs involved in the selection of the "best" numerical method for even very simple equations. It might be supposed that the relative accuracy of different methods for the $M = 0$ problem could be easily determined by comparing their respective truncation error. The analysis of truncation error is, however, most effective at predicting the behavior of well resolved features which oscillate over periods at least an order of magnitude larger than a single time step. The most serious errors are, however, often found in the poorly resolved features oscillating over periods between $2\Delta t$ and $4\Delta t$. These errors typically appear in both the phase and amplitude of the solution.

Phase errors for numerical solutions to the oscillation equation can be evaluated from the amplification factor. Expressing A in modulus-argument form $|A|e^{i\theta}$, where

$$|A| = (\Re\{A\}^2 + \Im\{A\}^2)^{1/2}, \quad \text{and} \quad \theta = \arctan\left(\Im\{A\}/\Re\{A\}\right).$$

phase errors may be characterized by the relative phase change, $R = \theta/\omega\Delta t$, which is the ratio of the phase advance produced by one time step of the numerical scheme, divided by the change in phase experienced by the true solution over the same time interval. If $R > 1$, the method is *accelerating*; if $R < 1$, the scheme is *decelerating*. Phase errors accumulate over the period of integration and can become quite large over long time periods.

In a non-dissipative system, amplitude errors represent spurious sinks or sources of energy. Amplitude errors arise from the difference between the magnitude of the approximate amplification factor $|A|$ and the correct value of unity. When $|A| = 1$, the scheme is *neutral*. If $|A| < 1$, the scheme is *damping*; and if $|A| > 1$, it is *amplifying*. The range of values of Δt for which a given approximation to the oscillation equation is not amplifying are given by the intersection of the absolute stability region for the scheme and the imaginary $(\omega\Delta t)$ axis, which in the case of Euler's method (Fig. 5.1a) is just the origin.

5.3.3 Single-Stage, Single-Step Schemes

The simplest techniques for the solution of the ordinary differential equation (5.1) are members of the general family of single-stage single-step schemes, which may be written in the form

$$\frac{\phi_{n+1} - \phi_n}{\Delta t} = (1-\alpha)F(\phi_n, t_n) + \alpha F(\phi_{n+1}, t_{n+1}). \tag{5.21}$$

Euler's method is obtained by setting $\alpha = 0$; the backward-Euler method corresponds to the case $\alpha = 1$, and the trapezoidal method is obtained when $\alpha = 1/2$. Substituting the true solution ψ into (5.21), expanding all terms in Taylor series about t_n, and using

$$F(\psi(t_{n+1}), t_{n+1}) = \frac{d\psi}{dt}(t_{n+1}) = \frac{d\psi}{dt}(t_n) + \Delta t \frac{d^2\psi}{dt^2}(t_n) + \frac{(\Delta t)^2}{2}\frac{d^3\psi}{dt^3}(t_n) + \ldots,$$

one may show the truncation error for all members of this family of schemes is $O(\Delta t)$, except for the trapezoidal method which is $O\left[(\Delta t)^2\right]$.

Application of (5.21) to the test equation for absolute stability (5.16) yields

$$A = \frac{\phi_{n+1}}{\phi_n} = \frac{1 + (1-\alpha)\gamma\Delta t}{1 - \alpha\gamma\Delta t}. \tag{5.22}$$

For backward Euler, $A = (1 - \gamma \Delta t)^{-1} = (1 - \lambda \Delta t - i\omega \Delta t)^{-1}$. Multiplying A by its complex conjugate gives

$$|A|^2 = \frac{1}{(1 - \lambda \Delta t)^2 + (\omega \Delta t)^2},$$

implying that backward-Euler differencing will produce absolutely stable solutions for all $(\lambda \Delta t, \omega \Delta t)$ *outside* the circle

$$(1 - \lambda \Delta t)^2 + (\omega \Delta t)^2 \leq 1. \tag{5.23}$$

This region is shown in Fig. 5.1c, and since it includes the region $\lambda \Delta t \leq 0$, backward-Euler differencing is A-stable. Although it generates physically appropriate solutions for $\lambda < 0$, the backward-Euler method can produce large errors if $\lambda > 0$. If $\lambda > 0$ and $(\lambda \Delta t, \omega \Delta t)$ is not inside the circle (5.23), the numerical solution will decay but the true solution should grow exponentially with time.

The amplification factor for the trapezoidal method is

$$A = \frac{1 + \gamma \Delta t/2}{1 - \gamma \Delta t/2}, \tag{5.24}$$

from which

$$|A|^2 = \frac{(1 + \lambda \Delta t/2)^2 + (\omega \Delta t)^2}{(1 - \lambda \Delta t/2)^2 + (\omega \Delta t)^2}.$$

Thus, the absolute stability region for the trapezoidal method (shown in Fig. 5.1b) is the half-plane $\lambda \Delta t \leq 0$ and it is A-stable.

Now consider the behavior of these schemes in the purely oscillatory case; then $\gamma = i\omega$, and from (5.22)

$$|A|^2 = \frac{1 + (1 - \alpha)^2 (\omega \Delta t)^2}{1 + \alpha^2 (\omega \Delta t)^2} = 1 + (1 - 2\alpha) \frac{(\omega \Delta t)^2}{1 + \alpha^2 (\omega \Delta t)^2}. \tag{5.25}$$

Inspection of the preceding shows that the scheme is amplifying when $\alpha < 1/2$, neutral when $\alpha = 1/2$; damping when $\alpha > 1/2$. These results are consistent with the locations of the boundaries of the absolute stability regions in Fig. 5.1.

The amplitude and phase errors in the approximate solution are functions of the *numerical resolution*. The solution to the governing differential equation (5.19) oscillates with a period $T = 2\pi/\omega$. An appropriate measure of numerical resolution is the number of time steps per oscillation period, $T/\Delta t$. The numerical resolution is improved by decreasing the step size. In the limit of very good numerical resolution, $T/\Delta t \to \infty$ and $\omega \Delta t \to 0$. Assuming good numerical resolution, the Taylor series expansion

$$(1 + x)^{1/2} = 1 + \frac{x}{2} - \frac{x^2}{8} + \ldots, \quad \text{for} \quad |x| < 1,$$

may be used to reduce (5.25) to

$$|A| \approx 1 + \tfrac{1}{2}(1-2\alpha)(\omega\Delta t)^2.$$

It follows that

$$|A|_{\text{forward}} \approx 1 + \tfrac{1}{2}(\omega\Delta t)^2 \qquad \text{and} \qquad |A|_{\text{backward}} \approx 1 - \tfrac{1}{2}(\omega\Delta t)^2, \tag{5.26}$$

indicating that the spurious amplitude changes introduced by both the forward and backward-Euler methods are $O[(\omega\Delta t)^2]$.

The relative phase change in the family of single-stage two-level schemes is

$$R = \frac{1}{\omega\Delta t}\arctan\left(\frac{\omega\Delta t}{1-\alpha(1-\alpha)(\omega\Delta t)^2}\right).$$

Thus,

$$R_{\text{forward}} = R_{\text{backward}} = \frac{\arctan\omega\Delta t}{\omega\Delta t}, \tag{5.27}$$

which ranges between 0 and 1, implying that both the forward and backward-Euler schemes are decelerating. Assuming, once again, that the numerical solution is well resolved, the preceding expression for the phase-speed error may be approximated using the Taylor series expansion

$$\arctan x = x - \frac{x^3}{3} + \frac{x^5}{5} - \ldots \quad \text{for} \quad |x| < 1,$$

to obtain

$$R_{\text{forward}} = R_{\text{backward}} \approx 1 - \frac{(\omega\Delta t)^2}{3}.$$

The phase-speed error, like the amplitude error, is $O[(\Delta t)^2]$. The relative phase change for the trapezoidal scheme is

$$R_{\text{trapezoidal}} = \frac{1}{\omega\Delta t}\arctan\left(\frac{\omega\Delta t}{1-\omega^2\Delta t^2/4}\right).,$$

which for small values of $\omega\Delta t$ is approximately,

$$R_{\text{trapezoidal}} \approx \frac{1}{\omega\Delta t}\arctan\left(\omega\Delta t\left(1+\frac{\omega^2\Delta t^2}{4}\right)\right) \approx 1 - \frac{\omega^2\Delta t^2}{12}.$$

As with the forward and backward Euler methods, the trapezoidal scheme retards the phase change of well resolved oscillations. However, the deceleration is only $\frac{1}{4}$ as great as that produced by the other schemes.

Although the trapezoidal scheme is second-order accurate and A-stable, it has the disadvantage in that it requires the evaluation of $F(\phi_{n+1})$ during the computation of ϕ_{n+1}. A scheme such as the trapezoidal method, in which the calculation of ϕ_{n+1} depends on $F(\phi_{n+1})$, is an *implicit* method. If the calculation of ϕ_{n+1} does not depend on $F(\phi_{n+1})$, the scheme is *explicit*. In the case of the test problem (5.16), implicitness is a trivial complication. However, if F is a nonlinear function, any implicit finite-difference scheme will convert the differential equation into a nonlinear algebraic equation for ϕ_{n+1}. In the general case, the solution to this nonlinear equation must be obtained by some iterative technique. Thus, implicit finite-difference schemes generally require much more computation per individual time step than do similar explicit methods. Nevertheless, in problems where accuracy considerations do not demand a short time step, the extra computation per implicit time step can be more than compensated by using a much larger time step than that required to maintain the stability of comparable explicit schemes.

5.3.4 Application to PDEs

Consider once again the advection-diffusion equation (5.17) that motivated the selection of (5.16) as a prototype ODE. According to (5.18), each individual Fourier coefficient b_k oscillates at the frequency ck. The highest frequency resolved by any completely discrete approximation to (5.17) will be that of the highest-wavenumber, or equivalently, the shortest-horizontal-wavelength disturbance captured by the discretization. As a concrete example, suppose the spatial derivatives are replaced by finite differences, then the maximum resolved k scales like $(\Delta x)^{-1}$. Let us temporarily suppose that the physical viscosity M is zero, and that the finite difference approximation to $\partial\psi/\partial x$ does not introduce "numerical diffusion."[3] Then if Euler's method is used to approximate the time derivative, the frequency of the most rapidly varying Fourier component $\omega_{\max}$ will be $O(c/\Delta x)$, and over each time step its Fourier coefficient $b_{k_{\max}}$ will change by a factor $A_{k_{\max}} = 1 + iO(c\Delta t/\Delta x)$.

When attempting to obtain converged solutions to partial differential equations, the spatial and temporal resolution are typically reduced simultaneously, keeping $\Delta t/\Delta x$ constant. But if Δt and Δx are both repeatedly halved and Euler's method is used to integrate the numerical solution over a fixed physical time $T = N\Delta t$, the inequality,

$$|A_{k_{\max}}| \leq 1 + |\omega_{\max}\Delta t| = 1 + O(|c\Delta t/\Delta x|),$$

cannot be used to bound $|A_{k_{\max}}|^N$. Thus, the approach used to prove the convergence of Euler's method for ODEs in Sect. 5.2.2 fails, and as may be shown rigorously (Durran 1999, p. 90), forward-in-time, centered-in-space approximations to the pure advection problem are unstable. Those time stepping schemes suitable

[3] Such diffusion can be avoided by using centered spatial differences (Durran 1999, p. 80).

for use with centered-in-space approximations to the advection equation are ones for which the point $(0, \omega_{\max}\Delta t)$ lies in the scheme's region of absolute stability whenever $|c\Delta t/\Delta x|$ is less than some constant.[4]

Now consider the case of pure diffusion, for which (5.18) reduces to

$$\frac{db_k}{dt} = -Mk^2 b_k. \tag{5.28}$$

If the time-derivative is approximated by Euler's method, the amplification factor for the Fourier coefficient of the shortest wavelength, most rapidly decaying component of the solution becomes $1 - O[M\Delta t/(\Delta x)^2]$, which approaches negative infinity if Δt and Δx are both repeatedly halved in an effort to obtain a convergent approximation. In most practical applications involving diffusion dominated problems, $\Delta t/(\Delta x)^2$ becomes unbounded as the numerical resolution is refined, and it is therefore advantageous to approximate their temporal evolution using A-stable schemes, all of which are implicit.

Explicit time differences may, nevertheless, yield good results in the special case where M represents an "eddy diffusivity" M_e rather than a true physical diffusivity. Eddy diffusivities are designed to parameterize the effects of mixing by fluid motions whose scale is too small to be captured on the numerical mesh, and M_e is typically proportional to the spatial grid interval. Thus $M_e\Delta t/(\Delta x)^2$ remains constant as Δt, $\Delta x \to 0$ with $\Delta t/\Delta x$ fixed, and it becomes practical to satisfy conditions such as $0 \leq M_e\Delta t/(\Delta x)^2 \leq 1$, which would allow Euler's to be used to stably integrate those terms representing parameterized diffusion.

5.3.5 L-Stability

A-stability is not always sufficient to guarantee good behavior in practical applications involving systems of equations in which the individual components decay at very different rates. When A-stable trapezoidal time differencing is used in conjunction with finite-difference approximations to the spatial derivative in the diffusion equation, the amplification factor for the Fourier coefficient of the shortest wavelength mode may be obtained by replacing $\lambda/2$ in (5.24) with $-\sigma M/(\Delta x)^2$, to give

$$A_{k_{\max}} = \frac{1 - \sigma M\Delta t/(\Delta x)^2}{1 + \sigma M\Delta t/(\Delta x)^2},$$

where σ is a positive constant determined by the exact finite difference formulation.

[4] When choosing a time step for the numerical solution of time-dependent PDEs, one must also satisfy the Courant–Friedrichs–Levy condition that the numerical domain of dependence include the domain of dependence of the true solution (see, e.g., Durran 1999).

In some applications it is not necessary to follow the precise behavior of the most rapidly decaying, shortest wavelength modes, and a time step appropriate for the accurate and efficient simulation other aspects of the problem (for example the slower diminution of the longer wavelength components) can make $M\Delta t/(\Delta x)^2 \gg 1$. Yet in the limit $M\Delta t/(\Delta x)^2 \to \infty$, $A_{k_{\max}} \to -1$, in which case the short-wavelength components of the trapezoidal integration will flip sign every time step without significant loss of amplitude. Although large time steps will not produce an unstable amplification of the shortest wavelength modes, sufficiently large steps do prevent those modes from properly decaying.

The correct behavior in the limit $M\Delta t/(\Delta x)^2 \to \infty$ is recovered if backward-Euler differencing is used to approximate the time derivative. Then the amplification factor for the Fourier coefficient of the shortest-wavelength mode becomes

$$A_{k_{\max}} = \frac{1}{1 + 2\sigma M\Delta t/(\Delta x)^2},$$

and the amplification factor approaches zero as $\Delta t/(\Delta x)^2$ becomes arbitrarily large. Backward-Euler differencing is an example of an *L-stable* method. L-stable methods are defined in the context of the prototype ODE (5.16) as those schemes that are A-stable and satisfy the additional property that $A \to 0$ as $\Re\{\gamma\}\Delta t \to -\infty$. L-stable methods are of great use in simulation of systems in which chemical reactions occur over a broad range of time scales (Hundsdorfer and Verwer 2003; LeVeque 2007).

5.4 Runge–Kutta (Multi-Stage) Methods

Definite integrals are often evaluated numerically through quadrature formulae

$$\int_a^b f(t)\,dt \approx \sum_{j=1}^{s} b_j f(c_j), \tag{5.29}$$

where the *weights* b_j and the *nodes* c_j are independent of the function f (Iserles 1996, p. 33). A similar strategy may be used to step the solution of an ordinary differential equation forward over a time interval Δt by approximating the integral in (5.2) such that

$$\psi(t_{n+1}) \approx \psi(t_n) + \Delta t \sum_{j=1}^{s} b_j F(\psi(t_n + c_j\Delta t), t_n + c_j\Delta t). \tag{5.30}$$

In contrast to the situation with the simple quadrature formula (5.29), however, the values of $\psi(t_n + c_j\Delta t)$ required for the evaluation of (5.30) are not known at time t_n, and must therefore be estimated numerically through a series of preliminary calculations, or *stages*. An *explicit* s-stage Runge–Kutta scheme iteratively builds an approximation to (5.30) as follows

$$\xi_1 = \phi_n \tag{5.31}$$

$$\xi_2 = \phi_n + \Delta t\, a_{2,1} F(\xi_1, t_n) \tag{5.32}$$

$$\xi_3 = \phi_n + \Delta t\, [a_{3,1} F(\xi_1, t_n) + a_{3,2} F(\xi_2, t_n + c_2 \Delta t)] \tag{5.33}$$

$$\vdots$$

$$\xi_s = \phi_n + \Delta t \sum_{j=1}^{s-1} a_{s,j} F(\xi_j, t_n + c_j \Delta t) \tag{5.34}$$

$$\phi_{n+1} = \phi_n + \Delta t \sum_{j=1}^{s} b_j F(\xi_j, t_n + c_j \Delta t) \tag{5.35}$$

By convention, we ensure that ξ_j is at least a first order approximation to $\psi(t_n + c_j \Delta t)$ by setting $c_1 = 0$ and

$$c_j = \sum_{k=1}^{j-1} a_{j,k} \qquad j = 2, 3, \ldots, s. \tag{5.36}$$

In explicit Runge–Kutta schemes $a_{j,k} = 0$ for $k \geq j$. Implicit s-stage Runge–Kutta schemes are obtained by replacing (5.31)–(5.34) with

$$\xi_j = \phi_n + \Delta t \sum_{k=1}^{s} a_{j,k} F(\xi_k, t_n + c_k \Delta t), \tag{5.37}$$

where in general all the $a_{j,k}$ may be non-zero. The order conditions given above (and in the next two sections) apply both to implicit and explicit Runge–Kutta methods.

5.4.1 Explicit Two-Stage Schemes

Taylor series expansions may be used to arrive at the additional conditions Runge–Kutta methods must satisfy to achieve a given level of accuracy. First-order accuracy requires

$$\sum_{j=1}^{s} b_j = 1. \tag{5.38}$$

For a single-stage method, the unique solution to (5.38) is $b_1 = 1$ and (5.31)–(5.35) reduce to the forward Euler method. Second order accuracy requires (5.36), (5.38) and

$$\sum_{j=1}^{s} b_j c_j = \frac{1}{2}. \tag{5.39}$$

For an explicit two-stage scheme, these accuracy requirements reduce to

$$c_2 = a_{2,1}, \quad b_1 + b_2 = 1, \quad b_2 c_2 = 1/2,$$

which is a system of three equations in four unknowns whose solution is not unique, but may be expressed in terms of the free parameter $a_{2,1}$. One well-known second-order two-stage scheme is the *Heun* method, for which $a_{2,1} = 1$ (and therefore $b_1 = b_2 = 1/2$, $c_2 = 1$). The Heun method creates a trapezoidal-like approximation to the integral of F, but differs from the true trapezoidal method because $F(\phi_{n+1}, t_{n+1})$ is replaced by the estimate $F(\xi_2, t_{n+1})$. Another second-order two-stage scheme is the *midpoint* method, in which $a_{2,1} = 1/2$. Also of note is the *first-order* two-stage forward-backward scheme (Matsuno 1966) in which $a_{2,1} = b_2 = c_2 = 1$ and $b_1 = 0$.

One important difference among the basic explicit Runge–Kutta schemes is whether they generate non-amplifying solutions in purely oscillatory problems. If the oscillation equation, is approximated using a two-stage scheme of at least first order, the result may be written as

$$\phi_{n+1} = \phi_n + b_2 i\omega\Delta t\,(\phi_n + a_{2,1} i\omega\Delta t\,\phi_n) + (1 - b_2) i\omega\Delta t\,\phi_n. \tag{5.40}$$

The amplification factor is

$$A = 1 + i\omega\Delta t - a_{2,1} b_2 (\omega\Delta t)^2,$$

and

$$|A|^2 = 1 + (1 - 2a_{2,1} b_2)(\omega\Delta t)^2 + (a_{2,1} b_2)^2 (\omega\Delta t)^4, \tag{5.41}$$

which shows that the set of second-order schemes, (*i.e.,* those schemes for which $a_{2,1} b_2 = 1/2$) have $O[(\Delta t)^4]$ amplitude error, whereas the amplitude error in first-order two-stage schemes is $O[(\Delta t)^2]$. Unfortunately, all the second-order two-stage explicit Runge–Kutta schemes are amplifying, since in the limit of good numerical resolution,

$$|A|_{\text{RKe2}} \approx 1 + \tfrac{1}{8}(\omega\Delta t)^4.$$

Although these schemes are amplifying, the growth is $O[(\Delta t)^4]$. At a given step size, the erroneous amplification produced by a second-order two-stage scheme will be much weaker than the $O[(\Delta t)^2]$ growth produced by forward time-differencing (or equivalently the first-order one-stage Runge–Kutta method, see (5.26)).

Many physical systems contain several different modes, each oscillating at a different frequency. When simulating these systems, the highest frequency components of the numerical solution are likely to be most seriously in error because of their poor numerical resolution. It is precisely these poorly resolved features that are amplified most rapidly by the second-order two-stage methods. In contrast, non-amplifying solutions in which the high frequency components are strongly damped can be obtained using Matsuno's forward-backward scheme, for which

$$|A|^2_{\text{Matsuno}} = 1 - (\omega \Delta t)^2 + (\omega \Delta t)^4. \tag{5.42}$$

The Matsuno scheme damps the solution whenever $0 < \omega \Delta t < 1$. Differentiation of (5.42), with respect to $\omega \Delta t$, shows that the maximum damping occurs when $\omega \Delta t = 1/\sqrt{2}$. Thus, if the time step is chosen such that $0 \le \omega \Delta t \le 1/\sqrt{2}$ for all frequencies ω in the physical system, Matsuno time differencing will preferentially damp the highest frequency waves. The damping properties of the Matsuno scheme have been exploited to eliminate high-frequency gravity waves generated during the initialization of weather prediction models. The standard Matsuno scheme produces too much damping, however, for most nonspecialized applications. The fourth-order Runge–Kutta scheme (see Sect. 5.4.2) may also be used to preferentially damp high frequency modes, and in most instances it would be a better choice than the Matsuno scheme because it is more efficient and far more accurate.

The amplitude errors generated by the preceding Runge–Kutta schemes are compared those produced by backward Euler and trapezoidal differencing in Fig. 5.2. The strong damping associated with the backward Euler and Matsuno schemes is evident, along with the rapid amplification produced by forward Euler differencing. These relatively large errors may be contrasted with the significantly weaker amplification produced by the second-order Runge–Kutta methods, and the neutral amplification of the trapezoidal method.

The relative phase change associated with the general two-stage explicit Runge–Kutta method (5.40) is

$$R = \frac{1}{\omega \Delta t} \arctan\left(\frac{\omega \Delta t}{1 - a_{2,1} b_2 (\omega \Delta t)^2}\right).$$

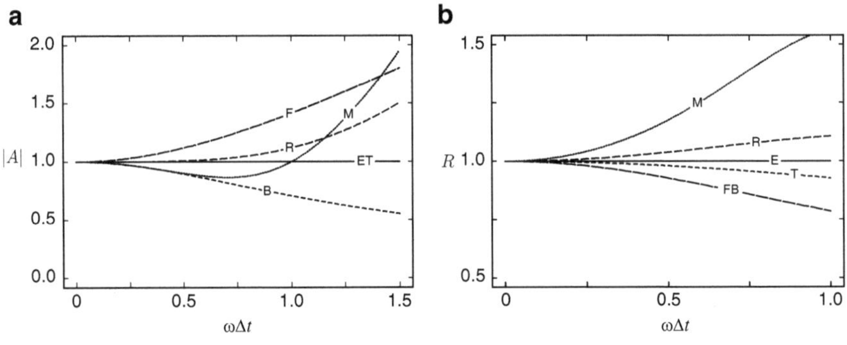

Fig. 5.2 The modulus of the amplification factor (**a**) and the relative phase change (**b**) as a function of temporal resolution $\omega \Delta t$ for the true solution and five two-level schemes: exact solution (E) and trapezoidal method (T), forward Euler (F), backward Euler (B), second-order Runge–Kutta (R), and Matsuno (M)

In the limit of good numerical resolution, the relative phase changes produced by second-order schemes and the Matsuno method scheme are

$$R_{\mathrm{RKe2}} \approx 1 + \tfrac{1}{6}(\omega \Delta t)^2. \qquad R_{\mathrm{Matsuno}} \approx 1 + \tfrac{2}{3}(\omega \Delta t)^2.$$

The relative phase change for these schemes is plotted as a function of temporal resolution in Fig. 5.2, along with that for forward Euler, backward Euler, and trapezoidal differencing. The Matsuno and second-order Runge–Kutta schemes are accelerating, whereas the forward Euler, backward Euler, and trapezoidal schemes are decelerating.

5.4.2 Explicit Three- and Four-Stage Schemes

Runge–Kutta schemes satisfying

$$\sum_{j=1}^{s} b_j c_j^2 = \frac{1}{3} \quad \text{and} \quad \sum_{j=1}^{s}\sum_{k=1}^{s} b_j a_{j,k} c_k = \frac{1}{6}, \tag{5.43}$$

as well as (5.36), (5.38) and (5.39) are third-order accurate. For explicit three-stage Runge–Kutta schemes, (5.43) reduces to

$$b_2 c_2 + b_3 c_3 = \frac{1}{3} \quad \text{and} \quad b_3 a_{3,2} c_2 = \frac{1}{6}.$$

As with the second-order methods there is no unique choice for the coefficients of a three-stage third-order scheme. On example is Heun's third-order method,

$$\xi_1 = \phi_n, \quad \xi_2 = \phi_n + \frac{\Delta t}{3} F(\xi_1, t_n), \quad \xi_3 = \phi_n + \frac{2\Delta t}{3} F(\xi_2, t_n + \tfrac{\Delta t}{3}),$$

$$\phi_{n+1} = \phi_n + \frac{\Delta t}{4}\Big[F(\xi_1, t_n) + 3F(\xi_3, t_n + \tfrac{2\Delta t}{3})\Big]$$

Another possibility is the low storage variant recommended by Williamson (1980) which may be written as

$$\begin{aligned}
q_1 &= \Delta t F(\phi_n, t_n) & \phi_{(1)} &= \phi_n + q_1/3 \\
q_2 &= \Delta t F(\phi_{(1)}, t_n + \tfrac{\Delta t}{3}) - 5q_1/9 & \phi_{(2)} &= \phi_{(1)} + 15q_2/16 \\
q_3 &= \Delta t F(\phi_{(2)}, t_n + \tfrac{5\Delta t}{12}) - 153q_2/128 & \phi_{n+1} &= \phi_{(2)} + 8q_3/15.
\end{aligned}$$

In practical applications involving time-dependent partial differential equations, ϕ_n may be an extremely long vector of unknown variables (e.g., the velocity, temperature, pressure and mixing ratio of chemical species at every node on a large three-dimensional mesh). It may, therefore, be difficult to store several copies of ϕ and $F(\phi)$ in the in-core memory of a digital computer. If m is the number of unknowns in ϕ, the Williamson–Runge–Kutta scheme economizes on storage by allowing the integration to proceed using only 2 m storage locations, divided between the arrays q and ϕ, which are overwritten three times during each integration step.

In addition to (5.36)–(5.39) and (5.43), fourth-order Runge–Kutta methods must satisfy four additional equations (Hundsdorfer and Verwer 2003, p. 141). Once again, the solutions for the coefficients of a four-stage explicit method are not unique. The most well-known four-stage fourth-order method is the classical Runge–Kutta formulation,

$$\xi_1 = \phi_n, \qquad \xi_2 = \phi_n + \frac{\Delta t}{2} F(\xi_1, t_n),$$
$$\xi_3 = \phi_n + \frac{\Delta t}{2} F(\xi_2, t_n + \tfrac{\Delta t}{2}), \quad \xi_4 = \phi_n + \Delta t\, F(\xi_3, t_n + \tfrac{\Delta t}{2}), \tag{5.44}$$

$$\phi_{n+1} = \phi_n + \frac{\Delta t}{6}\Big[F(\xi_1, t_n) + 2F(\xi_2, t_n + \tfrac{\Delta t}{2}) + 2F(\xi_3, t_n + \tfrac{\Delta t}{2}) + F(\xi_4, t_{n+1})\Big].$$

Low-storage variants also exist for fourth-order schemes (Blum 1962), but in contrast to the third-order methods, they require 3 m storage locations to advance an m-dimensional vector of unknowns forward in time.

Fifth- or higher-order explicit Runge–Kutta schemes are relatively unattractive because the number of stages required to achieve order s exceeds s for all $s > 4$. Nevertheless, the simple s-stage scheme

$$\xi_0 = \phi_n; \qquad \xi_j = \phi_n + \frac{\Delta t}{s - j + 1} F(\xi_{j-1}), \quad 1 \le j \le s; \qquad \phi_{n+1} = \xi_s, \tag{5.45}$$

is accurate to order s when $F(\psi)$ is linear in ψ (as would be the case in many applications involving time-dependent partial differential equations). When F is nonlinear, (5.45) is only second-order accurate.

Figure 5.3 shows the amplification factor for third- and fourth-order Runge–Kutta solutions to the oscillation equation (5.19) plotted as a function of temporal resolution. As shown in Fig. 5.3, once the time step exceeds the maximum stable time step for the third-order scheme, the fourth-order method becomes highly damping. In some circumstances it may be desirable to selectively damp the highest frequency modes, and in such cases the fourth-order Runge–Kutta method would be clearly preferable to the first-order Matsuno method. On the other hand, if one wishes to avoid excessive damping of the high-frequency components, it will not be possible to use the full stable time step of the fourth-order Runge–Kutta scheme.

As was the case for the two-stage first-order Matsuno method, the stability of explicit Runge–Kutta solutions to the oscillation equation may be enhanced by adding extra stages if one is willing to settle for first- or second-order accuracy. In particular, the stability condition $\max|\omega \Delta t| = s - 1$ may be obtained for an

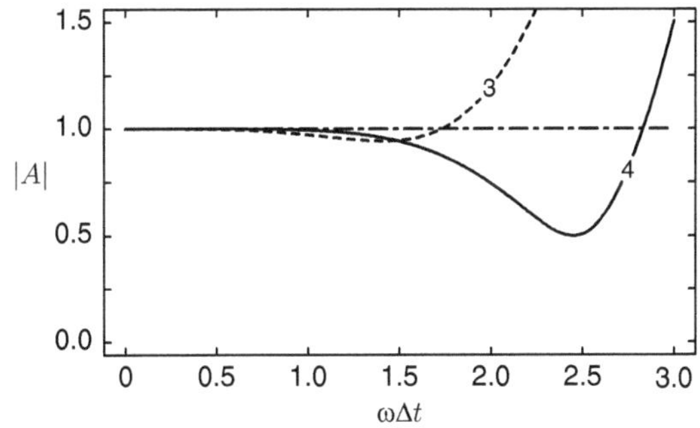

Fig. 5.3 Modulus of the amplification factor plotted as a function of temporal resolution $\omega \Delta t$ for third-order three-stage (*dashed*) and fourth-order four-stage (*solid*) explicit Runge–Kutta solutions to the oscillation equation

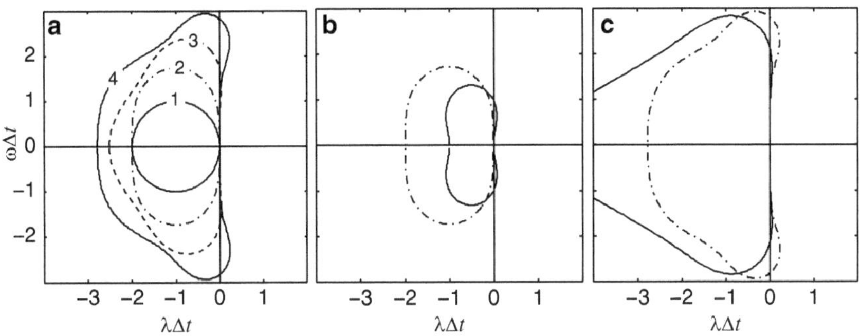

Fig. 5.4 Absolute stability regions for explicit Runge–Kutta schemes: (**a**) of equal orders and stages, 1 through 4; (**b**) two-stage methods: Matsunno (*solid*) and second-order (*dashed*); (**c**) four-stage methods: Spiteri and Ruuth's third-order SSPRK scheme (*solid*) and fourth-order (*dashed*). In each case, the region of absolute stability lies inside the curve. When $\omega = 0$, the absolute stability region for the Spiteri–Ruuth scheme extends to roughly $\lambda \Delta t = -5.15$

s-stage scheme that will be second-order accurate if s is odd, and first-order accurate when s is even (Hundsdorfer and Verwer 2003, p. 150). Note that despite their high-order accuracy, explicit fourth-order four-stage Runge–Kutta methods are stable for max $|\omega \Delta t| < 2.82$ which is very close to the optimal limit of max $|\omega \Delta t| = 3$ obtainable using a *first-order* four-stage explicit method.

Absolute stability regions for explicit Runge–Kutta schemes of orders one through four are plotted in Fig. 5.4a. Consistent with the behaviors of the amplification factors for the oscillation equation shown in Figs. 5.2 and 5.3, the third- and fourth-order methods are the only ones for which the absolute stability regions includes a finite segment of the imaginary axis. None of these methods, and indeed no explicit Runge–Kutta scheme is A-stable.

Figure 5.4b compares the absolute stability region for a pair of explicit two-stage methods, the first-order Matsuno method and any second-order scheme. The increase in absolute stability along the real axis in the Matsuno scheme is achieved not only by sacrificing accuracy, but also by considerably reducing the overall region of absolute stability relative to the second-order schemes.

5.4.3 Strong-Stability Preserving Methods

Many methods for the numerical integration of conservation laws avoid the generation of spurious maxima and minima through the use of some type of flux limiter. The time differencing associated with such methods is often forward Euler. Strong-stability perserving Runge–Kutta (SSPRK) schemes can be used to obtain higher-order accuracy in time while preserving the beneficial results of the flux limiter. To be more precise, suppose that U is a vector of unknowns at every point on the spatial mesh, and that $\|U\|$ represents a measure such as the maximum of $|U|$ or the total variation of U over all spatial grid points. Let $B(\phi)$ be an approximation to the flux divergences in a conservation law such that

$$U_{n+1} = (I + \Delta t B)\, U_n, \tag{5.46}$$

and suppose that the fluxes are limited so that $\|U_{n+1}\| \leq \|U_n\|$ provided $|c\Delta t/\Delta x| \leq 1$, where c is the phase speed at which signals are propagated by the conservation law. SSPRK methods allow the forward-in-time approximation in (5.46) to be replaced by a higher-order scheme while preserving the *strong-stability condition* that $\|U_{n+1}\| \leq \|U_n\|$.

SSPRK schemes are constructed by forming linear combinations of forward-Euler operators in which the coefficient multiplying each operator is positive. The positivity of the coefficients ensures that a conservation law integrated with the new scheme retains the strong-stability properties of the original forward-Euler approximation (5.46). The precise value of the positive coefficients is chosen to obtain some combination of high-order accuracy and a favorable maximum stable time step. A two-stage second order SSPRK method is

$$\begin{aligned}
\phi_{(1)} &= \phi_n + \Delta t B(\phi_n),\\
\phi_{(2)} &= \phi_{(1)} + \Delta t B(\phi_{(1)}),\\
\phi_{n+1} &= \tfrac{1}{2}\left(\phi_n + \phi_{(2)}\right),
\end{aligned} \tag{5.47}$$

and a three-stage third-order scheme is

$$\begin{aligned}
\phi_{(1)} &= \phi_n + \Delta t B(\phi_n),\\
\phi_{(2)} &= \tfrac{3}{4}\phi_n + \tfrac{1}{4}\left[\phi_{(1)} + \Delta t B(\phi_{(1)})\right],\\
\phi_{n+1} &= \tfrac{1}{3}\phi_n + \tfrac{2}{3}\left[\phi_{(2)} + \Delta t B(\phi_{(2)})\right].
\end{aligned} \tag{5.48}$$

Both of these schemes, which were proposed by Shu and Osher (1988), are strong-stability preserving for $|c\Delta t/\Delta x| \leq 1$.

The schemes (5.47) and (5.48) are optimal in the sense that no second-order two-stage or third-order three-stage SSPRK scheme exists that allows a larger maximum time step. (Gottlieb and Shu 1998). Nevertheless, in some applications the four-stage, third-order SSPRK scheme proposed by Spiteri and Ruuth (2002)

$$\begin{aligned}
\phi_{(1)} &= \phi_n + \tfrac{1}{2}\Delta t B(\phi_n), \\
\phi_{(2)} &= \phi_{(1)} + \tfrac{1}{2}\Delta t B(\phi_{(1)}), \\
\phi_{(3)} &= \tfrac{2}{3}\phi_n + \tfrac{1}{3}\left[\phi_{(2)} + \tfrac{1}{2}\Delta t B(\phi_{(2)})\right], \\
\phi_{n+1} &= \phi_{(3)} + \tfrac{1}{2}\Delta t B(\phi_{(3)}),
\end{aligned} \tag{5.49}$$

may be more efficient because it is strong-stability preserving for $|c\Delta t/\Delta x| \leq 2$, allowing one to double the time step while increasing the computational burden associated with the evaluation of B by only 33% relative to that required by (5.48).

It should be emphasized that these methods are only strong-stability preserving when flux-limiting ensures that the forward step (5.46) yields a strongly-stable result. Amplifying solutions are produced if (5.47) is applied directly to the oscillation equation (5.19). Since (5.47) is an explicit two-stage second-order method and since (5.48) is an explicit three-stage third-order scheme, their absolute stability regions are exactly those shown for the second- and third-order methods in Fig. 5.4a. On the other hand, as shown in Fig. 5.4c, the four-stage third-order scheme (5.49) has a different, and generally larger, region of absolute stability than the family of four-stage, fourth-order Runge–Kutta methods. More information about SSP time-differencing schemes may be found in the reviews by Gottlieb et al. (2001) and Gottlieb (2005).

5.4.4 Diagonally Implicit Runge–Kutta Methods

Diagonally implicit Runge Kutta schemes are obtained when the implicit coupling in (5.37) is limited by requiring $a_{j,k} = 0$ whenever $k > j$. In comparison to methods with more extensive implicit coupling, the relative efficiency of diagonally implicit schemes make them more attractive for applications involving PDEs or large systems of ODEs. Backward Euler differencing is a first-order accurate single-stage diagonally implicit Runge–Kutta scheme. The implicit midpoint method,

$$\begin{aligned}
\xi_1 &= \phi_n + \tfrac{1}{2}\Delta t \; F(\xi_1, t + \tfrac{1}{2}\Delta t) \\
\phi_{n+1} &= \phi_n + \Delta t \; F(\xi_1, \tfrac{1}{2}\Delta t),
\end{aligned} \tag{5.50}$$

is a second-order accurate single-stage scheme. The implicit midpoint method is A-stable; its amplification factor is identical to that for the trapezoidal method (5.24).

A family of two-stage diagonally implicit Runge–Kutta schemes of at least second-order accuracy may be written in terms of a single free parameter α as

$$\begin{aligned}\xi_1 &= \phi_n + \alpha \Delta t F(\xi_1, t_n + \alpha \Delta t),\\ \xi_2 &= \phi_n + (1-2\alpha)\Delta t F(\xi_1, t_n + \alpha\Delta t) + \alpha \Delta t F(\xi_2, t_n + (1-\alpha)\Delta t)\\ \phi_{n+1} &= \phi_n + \tfrac{1}{2}\Delta t\left[F(\xi_1, t_n + \alpha\Delta t) + F(\xi_2, t_n + (1-\alpha)\Delta t)\right]. \end{aligned} \tag{5.51}$$

Third order accuracy is obtained if $\alpha = 1/2 \pm \sqrt{3}/6$.

If one of the schemes defined by (5.51) is applied to the test problem (5.16), the resulting amplification factor is

$$A = \frac{1 + (1-2\alpha)\gamma\Delta t + (\frac{1}{2} - 2\alpha + \alpha^2)(\gamma\Delta t)^2}{(1-\alpha\gamma\Delta t)^2}. \tag{5.52}$$

These schemes are A-stable if and only if $\alpha \geq 1/4$, as may be easily appreciated in the particular case for which $\gamma\Delta t \to (-\infty, 0)$; then the leading order behavior of $|A|$ is $(\frac{1}{2} - 2\alpha + \alpha^2)/\alpha^2$ which is bounded by unity for $\alpha \geq 1/4$. The $(\gamma\Delta t)^2$ term in the numerator of (5.51) is zero, and the scheme is L-stable if $\alpha = 1 \pm \frac{1}{2}\sqrt{2}$. One attractive way for handling the implicitness in (5.51) is through Runge–Kutta Rosenbrock methods. These are discussed in the context of photochemical air pollution models in (Verwer et al. 1999).

5.5 Multistep Methods

Multistep methods are an alternative to multi-stage methods in which information from several earlier time levels is incorporated into the integration formula. For example, the general form for an explicit two-step method is

$$\phi_{n+1} = \alpha_1\phi_n + \alpha_2\phi_{n-1} + \beta_1 \Delta t F(\phi_n, t_n) + \beta_2 \Delta t F(\phi_{n-1}, t_{n-1}). \tag{5.53}$$

In contrast to multistage methods, the forcing $F(\psi, t)$ is only evaluated at integer time steps and all the required values except $F(\phi_n, t_n)$ have been already calculated at previous time steps. Since the evaluation of $F(\psi, t)$ is often computationally intensive, storing and reusing these values has the potential to increase efficiency, although obviously it may also require more storage. Multistep methods also require special start-up procedures, because an n-step method requires data from the previous n time levels, but initial conditions for well posed physical problems give information about the solution at only one time. Multistage or lower-order multistep methods must therefore be used for the first $n-1$ steps of the integration.

5.5.1 Explicit Two-Step Schemes

A complete discussion of linear multi-step methods is beyond the scope of this chapter. In many geophysical applications, the memory required to store data from each time level is enormous, so we will focus on the family of two-step schemes (5.53). When formulating a two-step scheme, one seeks to improve upon the single-step methods, so it is reasonable to require that the global truncation error be at least second order. The scheme (5.53), will be at least second order if

$$\alpha_1 = 1 - \alpha_2, \qquad \beta_1 = \tfrac{1}{2}(\alpha_2 + 3), \qquad \beta_2 = \tfrac{1}{2}(\alpha_2 - 1), \tag{5.54}$$

where the coefficient α_2 remains a free parameter. Choosing $\alpha_2 = 5$ gives a third-order scheme, but this method is useless because it is highly unstable (Durran 1999). Since it is not practical to choose the coefficients in (5.54) to minimize the truncation error, the most important explicit two-step schemes are obtained by choosing α_2 to minimize the amount of data that must be stored and carried over from the $n-1$ time level, i.e., by setting $\alpha_2 = 1$, in which case $\beta_2 = 0$, or by setting $\alpha_2 = 0$. If α_2 is set to one, (5.53) becomes the *leapfrog* scheme. The choice $\alpha_2 = 0$ gives the two-step *Adams–Bashforth* method. The remainder of this section will be devoted to an examination of the performance of these two schemes in problems with purely oscillatory solutions.

The leapfrog and two-step Adams–Bashforth methods must be initialized using a single-step scheme to compute ϕ_1 from ϕ_0. In most instances, a simple forward step is adequate. Although forward differencing is amplifying, the amplification produced by a single step will generally not be large. Moreover, even though the truncation error of a forward-difference is $O(\Delta t)$, the execution of a single forward time step does not reduce the $O\left[(\Delta t)^2\right]$ global accuracy of leapfrog and Adams–Bashforth integrations. The basic reason that $O\left[(\Delta t)^2\right]$ accuracy is preserved is that forward differencing is only used over a Δt-long portion of the total integration. The contribution to the total error produced by the accumulation of $O\left[(\Delta t)^2\right]$ errors in a stable scheme over a finite time interval is the same order as the error arising from the accumulation of $O(\Delta t)$ errors over a time Δt.

5.5.2 *The Leapfrog Scheme*

If the leapfrog scheme,

$$\phi_{n+1} = \phi_{n-1} + 2\Delta t F(\phi_n, t_n), \tag{5.55}$$

is used to integrate the oscillation equation (5.19), its amplification factor satisfies

$$A^2 - 2i\omega\Delta t A - 1 = 0,$$

whose two roots are

$$A_\pm = i\omega\Delta t \pm \left(1 - \omega^2\Delta t^2\right)^{1/2}. \tag{5.56}$$

Evidently, the numerical solution is capable of behaving in two very different ways or *modes*. The mode associated with A_+ is known as the *physical mode* because it approximates the solution to the original differential equation. The mode associated with A_- is referred to as the *computational mode* since it arises solely as an artifact of the numerical computation. If $|\omega \Delta t| \leq 1$, the second term in (5.56) is real and $|A_+| = |A_-| = 1$, i.e., both the physical and the computational modes are stable and neutral. In the case $\omega \Delta t > 1$,

$$|A_+| = \left| i\omega \Delta t + i\left(\omega^2 \Delta t^2 - 1\right)^{1/2} \right| > |i\omega \Delta t| > 1,$$

and the scheme is unstable. When $\omega \Delta t < -1$, a similar argument shows that $|A_-| > 1$.

The complete leapfrog solution can typically be written as a linear combination of the physical and computational modes. An exception occurs if $\omega \Delta t = \pm 1$, in which case $A_+ = A_- = i\omega \Delta t$, and the physical and computational modes are not linearly independent. In such circumstances, the general solution to the leapfrog approximation to the oscillation equation has the form

$$\phi_n = C_1 (i\omega \Delta t)^n + C_2 n (i\omega \Delta t)^n.$$

Since the magnitude of the preceding solution grows as a function of time step, the leapfrog scheme is *not* stable when $|\omega \Delta t| = 1$. Nevertheless, the $O(n)$ growth of the solution that occurs when $\omega \Delta t = \pm 1$ is far slower than the $O(A^n)$ amplification that is produced when $|\omega \Delta t| > 1$.

The source of the computational mode is particularly easy to analyze in the trivial case of $\omega = 0$; then the analytic solution to the oscillation equation (5.19) is $\psi(t) = C$, where C is a constant determined by the initial condition at $t = t_0$. Under these circumstances, the leapfrog scheme reduces to

$$\phi_{n+1} = \phi_{n-1}, \tag{5.57}$$

and the amplification factor has the roots $A_+ = 1$; $A_- = -1$. The initial condition requires $\phi_0 = C$, which, according to the difference scheme (5.57), also guarantees that $\phi_2 = \phi_4 = \phi_6 = \ldots = C$. The odd time levels are determined by a second, computational initial condition imposed on ϕ_1. In practice ϕ_1 is often obtained from ϕ_0 by taking a single time step with a single-step method, and the resulting approximation to $\psi(t_0 + \Delta t)$ will contain some error E. It is obvious that in our present example, the correct choice for ϕ_1 is C, but in order to mimic the situation in a more general problem, suppose that $\phi_1 = C + E$. Then the numerical solution at any subsequent time will be the sum of two modes

$$\phi_n = (A_+)^n \phi_+ + (A_-)^n \phi_- = (C + E/2) - (-1)^n (E/2).$$

Here, the first term represents the physical mode and the second term represents the computational mode. The computational mode oscillates with a period of $2\Delta t$, and does not decay with time.

In the previous example, the amplitude of the computational mode is completely determined by the error in the specification of the computational initial condition ϕ_1. Since there is no coupling between the physical and computational modes in solutions to linear problems, the errors in the initial conditions also govern the amplitude of the computational mode in leapfrog solutions to most linear equations. If the governing equations are nonlinear, however, the nonlinear terms introduce a coupling between ϕ_+ and ϕ_- that often amplifies the computational mode until it eventually dominates the solution. This spurious growth of the computational mode can be avoided by periodically discarding the solution at ϕ_{n-1} and taking a single time step with a two-level scheme, or by filtering the high-frequency components of the numerical solution (Asselin 1972; Williams 2009). Various techniques for controlling the leapfrog scheme's computational mode are discussed in Durran (1999).

5.5.3 The Two-Step Adams–Bashforth Scheme

Another limitation of the leapfrog scheme is that is region of absolute stability is just a line segment on the imaginary axis. That is, solutions to (5.16) will undergo spurious amplification unless $\Re\{\gamma\} = 0$. A larger region of absolute stability can be obtained using the two-step Adams–Bashforth method,

$$\phi_{n+1} = \phi_n + \Delta t \left(\tfrac{3}{2} F(\phi_n, t_n) - \tfrac{1}{2} F(\phi_{n-1}, t_{n-1})\right). \tag{5.58}$$

although as will become apparent shortly, that method is not suitable for the simulation of purely oscillatory systems.

Applying (5.58) to the oscillation equation, one obtains

$$\phi_{n+1} = \phi_n + i\omega\Delta t \left(\tfrac{3}{2}\phi_n - \tfrac{1}{2}\phi_{n-1}\right).$$

The amplification factor associated with this scheme is given by the quadratic

$$A^2 - \left(1 + \frac{3i\omega\Delta t}{2}\right) A + \frac{i\omega\Delta t}{2} = 0,$$

in which case

$$A_\pm = \frac{1}{2}\left(1 + \frac{3i\omega\Delta t}{2} \pm \left(1 - \frac{9(\omega\Delta t)^2}{4} + i\omega\Delta t\right)^{1/2}\right). \tag{5.59}$$

As the numerical resolution increases, $A_+ \to 1$ and $A_- \to 0$. Thus, the Adams–Bashforth method damps the computational mode., which of course is highly desirable. Unfortunately the physical mode is weakly amplifying, as revealed if (5.59) is approximated under the assumption that $\omega\Delta t$ is small; then

$$A_+ = \left(1 - \frac{(\omega\Delta t)^2}{2} - \frac{(\omega\Delta t)^4}{8} - \ldots\right) + i\left(\omega\Delta t + \frac{(\omega\Delta t)^3}{4} + \ldots\right),$$

$$A_- = \left(\frac{(\omega\Delta t)^2}{2} + \frac{(\omega\Delta t)^4}{8} + \ldots\right) + i\left(\frac{\omega\Delta t}{2} - \frac{(\omega\Delta t)^3}{4} - \ldots\right),$$

and

$$|A_+|_{\mathrm{A-B2}} \approx 1 + \tfrac{1}{4}(\omega\Delta t)^4, \qquad |A_-|_{\mathrm{A-B2}} \approx \tfrac{1}{2}\omega\Delta t.$$

The modulus of the amplification factor of the physical mode exceeds unity by an $O[(\omega\Delta t)^4]$ term, as was the case for the two-stage second-order Runge–Kutta methods. The dependence of $|A_+|$ and $|A_-|$ upon temporal resolution is plotted in Fig. 5.5.

Although the two-step Adams–Bashforth method generates unstable amplification, the three-step variant,

$$\phi_{n+1} = \phi_n + \frac{\Delta t}{12}\left[23F(\phi_n) - 16F(\phi_{n-1}) + 5F(\phi_{n-2})\right], \tag{5.60}$$

gives non-amplifying solutions to the oscillation equation for $\omega\Delta t < 0.724$, and this third-order method is a better choice for the time integration of problems with oscillatory solutions (Durran 1999).

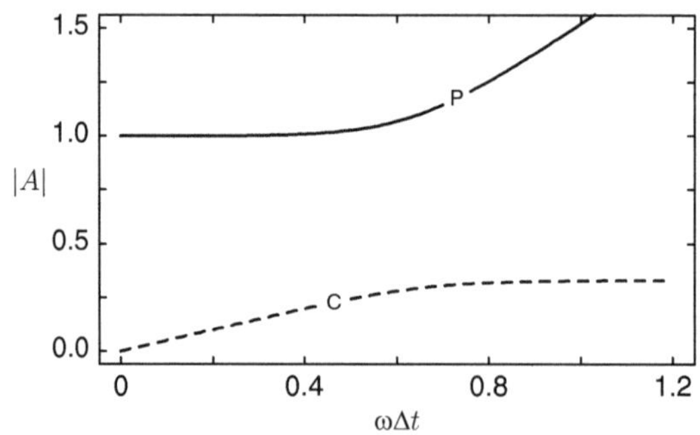

Fig. 5.5 Modulus of the amplification factors for the second order Adams–Bashforth scheme as a function of temporal resolution $\omega\Delta t$. The *solid* and *dashed lines* represent the physical and the computational modes, respectively

5.6 Summary Discussion

In this chapter we have investigated the performance of basic two-time-level, single-step schemes to illustrate the various stability properties that might be satisfied by numerical approximations to ordinary differential equations. A scheme that is only stable enough to ensure convergence in the limit $\Delta t \rightarrow 0$, such as the forward-Euler method, will prove unsatisfactory when used with non-dissipative approximations to spatial derivatives in problems like advective scalar transport. Stable results for non-dissipative approximations to the advection problem may be obtained using ODE solvers whose region of absolute stability includes a finite segment of the imaginary axis, but all two-time-level, single-step schemes that meet this criterion are implicit.

The leapfrog scheme is an explicit three-time-level scheme that can be used to obtain stable, efficient second-order integrations to linear systems with oscillatory solutions. The attractiveness of the leapfrog scheme is reduced by the behavior of its undamped computational mode, which can become unstable through interactions with the physical mode in nonlinear problems. As a consequence, in most practical applications the leapfrog algorithm must be modified in some manner that reduces it to first-order accuracy.

One possible alternative to the leapfrog scheme is the three-step (four-time-level) Adams–Bashforth method. Other possibilities may be found among the family of Runge–Kutta methods, which provide a large and flexible framework for creating suitable solvers for many atmospheric applications. Classical three-step, third-order and four-step, fourth-order Runge–Kutta schemes provide accurate and efficient methods that are absolutely stable over a significant segment of the imaginary axis and therefore suitable for use with non-dissipative approximations to spatial derivatives in transport problems. Strong stability preserving Runge–Kutta schemes offer a way to increase the accuracy of the time integration of flux-limited approximations to conservation laws. Unlike classical linear multistep methods, diagonally implicit Runge–Kutta methods can be A-stable and higher than second-order accurate.

Acknowledgments Thanks to the editors and to an anonymous reviewer for comments that helped improve this chapter. Support for this research was provided by NSF grant ATM-0836316.

References

Asselin R (1972) Frequency filter for time integrations. Mon Wea Rev 100:487–490
Blum E (1962) A modification of the Runge-Kutta fourth-order method. Math Comp 16:176–187
Dalhquist G (1956) Numerical integration of ordinary differential equations. Math Scandinavica 4:33–53
Durran DR (1999) Numerical Methods for Wave Equations in Geophysical Fluid Dynamics. Springer-Verlag, New York
Gottleib S (2005) On high order strong stability preserving Runge-Kutta and multi step time discretizations. J Sci Comput 25:105–128

Gottleib S, Shu CW, Tadmor E (2001) Strong stability-preserving high-order time discretization. SIAM Review 43:89–112

Gottlieb S, Shu CW (1998) Total variation diminishing Runge-Kutta schemes. Math Comp 67: 73–85

Hundsdorfer W, Verwer J (2003) Numerical Solution of Time-Dependent Advection-Diffusion-Reaction Equations. Springer, Berlin, 471 p.

Iserles A (1996) A First Course in the Numerical Analysis of Differential Equations. Cambridge University Press, Cambridge, 378 p.

Lax P, Richtmyer RD (1956) Survey of the stability of linear finite difference equations. Comm Pure Appl Math 9:267–293

LeVeque RJ (2007) Finite Difference Methods for Ordinary and Partial Differential Equations. SIAM, Philadelphia, 341 p.

Matsuno T (1966) Numerical integrations of the primative equations by a simulated backward difference method. J Meteor Soc Japan, Ser 2 44:76–84

Rood RB (1987) Numerical advection algorithms and their role in atmospheric transport and chemistry models. Reviews of Geophysics 25:71–100

Shu CW, Osher S (1988) Efficient implementation of essentially nonoscillatory shock-capturing schemes. J Comp Phys 77:439=471

Spiteri RJ, Ruuth SJ (2002) A new class of optimal high-order strong-stability-preserving time discretization methods. SIAM J Numer Anal 40:469–491

Verwer JG, Spee EJ, Blom JG, Hundsdorfer W (1999) A second-order Rosenbrock method applied to photochemical dispersion problems. SIAM J Sci Comput 20:1456–1480

Williams PD (2009) A proposed modification to the Robert–Asselin time filter. Mon Wea Rev 137:2538–2546

Williamson J (1980) Low-storage Runge-Kutta schemes. J Comp Phys 35:48–56

Chapter 6
Stabilizing Fast Waves

Dale R. Durran

Abstract The atmosphere transmits wavelike signals at a wide range of speeds. Rapidly moving, physically insignificant waves can impose very strict time-step limitations on numerical methods in order to ensure the integrations remain stable. Sound waves, for example, travel very rapidly but are of essentially no meteorological significance, and it is not practical to simulate most atmospheric circulations using the very short time steps required for the accurate and stable integration of the sound waves. This chapter reviews techniques for circumventing such time step restrictions, thereby allowing the step size to be chosen to ensure the accuracy and stability of the physically significant components of the solution.

6.1 Introduction

One reason that explicit time differencing is widely used in the simulation of wave-like flows is that accuracy considerations and stability constraints often yield similar criteria for the maximum time step in numerical integrations of systems that support a single type of wave motion. Many fluid systems, however, support more than one type of wave motion, and in such circumstances accuracy considerations and stability constraints can yield very different criteria for the maximum time step. If explicit time differencing is used to construct a straightforward numerical approximation to the equations governing a system that supports several types of waves, the maximum stable time step will be limited by the Courant number associated with the most rapidly propagating wave, yet in some cases that rapidly propagating wave may be of little physical significance.

As an example, consider the earth's atmosphere which supports sound waves, gravity waves and Rossby waves. Rossby waves propagate more slowly than gravity

D.R. Durran
Department of Atmospheric Sciences, Box 351640, University of Washington, Seattle, WA, 98195, USA
e-mail: durrand@atmos.washington.edu

P.H. Lauritzen et al. (eds.), *Numerical Techniques for Global Atmospheric Models*, Lecture Notes in Computational Science and Engineering 80,
DOI 10.1007/978-3-642-11640-7_6,

waves which in turn move more slowly than sound waves. The maximum stable time step with which an explicit numerical method can integrate the full equations governing atmospheric motions will be limited by the Courant number associated with sound wave propagation. If finite differences are used in the vertical and the vertical grid spacing is 300 m, the maximum stable time step will be on the order of one second. Since sound waves have no direct meteorological significance, they need not be accurately simulated in order to obtain a good weather forecast. The quality of the weather forecast depends solely on the ability of the model to accurately simulate atmospheric disturbances that evolve on much slower time scales. Gravity waves can be accurately simulated with time steps on the order of 10–100 s; Rossby waves require a time step on the order of 500–5,000 s. To obtain a reasonably efficient numerical model for the simulation of atmospheric circulations, it is necessary to circumvent the stability constraint associated with sound wave propagation and bring the maximum stable time step into closer agreement with the time step limitations arising from accuracy considerations.

There are two basic approaches for circumventing the time step constraint imposed by a rapidly moving, physically insignificant wave. The first approach is to approximate the full governing equations with set of "filtered" equations that do not support the rapidly moving wave. As an example, the full equations for stratified compressible flow might be approximated by the Boussinesq equations for incompressible flow. In this approach fundamental approximations are introduced to the original continuous equations prior to any numerical approximations that may be subsequently employed to generate finite-difference or spectral solutions to the filtered governing equations. The use of the filtered equation set may be motivated entirely by numerical considerations, or it may arise naturally from the standard approximations used in the study of a given physical phenomena. Gravity waves, for example, are often studied in the context of Boussinesq incompressible flow to simplify and streamline the mathematical description of the problem.

The second approach for circumventing the time step constraint imposed by a rapidly moving, physically insignificant wave leaves the continuous governing equations unmodified and relies on numerical techniques to stabilize the fast moving wave. These numerical techniques achieve efficiency by sacrificing the accuracy with which the fast moving wave is represented. Note that although the fast waves are retained, this approach is not appropriate in applications where the fast moving wave needs to be accurately simulated.

Approximate equation sets that filter sound waves include the Boussinesq, anelastic (Ogura and Phillips 1962; Lipps and Hemler 1982), and pseudo-incompressible (Durran 2008) systems. Of these three approaches, the Boussinesq equations are the most concise mathematically, but the least accurate quantitatively because they do not account for the decrease in atmospheric density with height. To reveal the essential properties of the filtered nonhydrostatic equations with a minimum of mathematically complexity, we will focus on the Boussinesq equations.

This chapter begins by examining techniques for the numerical solution of the Boussinesq equations via the projection method. Numerical methods for stabilizing the solution to problems that simultaneously support fast- and slow-moving waves are then considered including the semi-implicit method in Sect. 6.3 and fractional step methods in Sect. 6.4. Section 6.5 contains a summary discussion of these methods.

6.2 The Projection Method

The Boussinesq system for adiabatic inviscid flow can be expressed in a compact form involving the pressure potential P, the buoyancy b and the Brunt–Väisälä frequency N such that

$$\frac{\partial \mathbf{v}}{\partial t} + \nabla P = \mathbf{F}(\mathbf{v}, b) \equiv -\mathbf{v} \cdot \nabla \mathbf{v} + b\mathbf{k}, \tag{6.1}$$

$$\frac{Db}{1} Dt + N^2 w = 0, \tag{6.2}$$

$$\nabla \cdot \mathbf{v} = 0, \tag{6.3}$$

where

$$P = \frac{p'}{\rho_0}, \qquad b = -g\frac{\rho'}{\rho_0}, \qquad \text{and} \qquad N^2 = -\frac{g}{\rho_0}\frac{d\overline{\rho}}{dz}. \tag{6.4}$$

Here ρ_0 is a constant reference density, p' and ρ' are the deviations of the pressure and density from their values in a hydrostatically balanced reference state,[1] $\overline{p}(z)$ and $\overline{\rho}(z)$,

$$\frac{D}{Dt} = \frac{\partial}{\partial t} + \mathbf{v} \cdot \nabla,$$

and $\mathbf{v}$, w and $\mathbf{k}$ are the full velocity vector, the vertical velocity component and the unit vector directed opposite to the gravitational acceleration.

Alternatively, if the fluid in question is an ideal gas, a similar set of approximations can be invoked in which P, b and N are expressed in terms of the Exner function pressure $\pi = (p/p_0)^{R/c_p}$ and the potential temperature $\theta = T\pi^{-1}$, where T is the temperature, c_p is the specific heat at constant pressure, R is the gas constant and p_0 a constant reference pressure. As noted by Durran and Arakawa (2007), (6.1)–(6.3) is then recovered with

$$P = c_p\theta_0\pi', \qquad b = g\frac{\theta'}{\theta_0}, \qquad \text{and} \qquad N^2 = \frac{g}{\theta_0}\frac{d\overline{\theta}}{dz}. \tag{6.5}$$

As before the over-bars denote horizontally uniform reference-state fields in hydrostatic balance[2] and primes are the deviations from those reference values; θ_0 is a constant reference potential temperature.

---

[1] To satisfy hydrostatic balance $d\overline{p}/dz = -\overline{\rho}g$.

[2] In the π–θ formulation, hydrostatic balance requires $c_p\overline{\theta}d\overline{\pi}/dz = -g$.

The unknown variables are the three velocity components, the perturbation density and the perturbation pressure. In contrast to the full compressible system, there is no prognostic equation available to determine the time tendency of P. The perturbation pressure field at a given instant can, however, be diagnosed from the instantaneous velocity and perturbation density fields by solving the Poisson equation

$$\nabla^2 P = \nabla \cdot \mathbf{F}, \tag{6.6}$$

which can be derived by taking the divergence of (6.1) and then using (6.3). The perturbation pressure satisfying (6.6) is the instantaneous pressure distribution that will keep the evolving velocity field non-divergent.

6.2.1 Forward-in-Time Implementation

The *projection method* (Chorin 1968; Témam 1969) is a classical technique that may be used to obtain numerical solutions to the Boussinesq system. Suppose the momentum equation is integrated over a time interval Δt to yield

$$\int_{t^n}^{t^{n+1}} \frac{\partial \mathbf{v}}{\partial t}\, dt = -\int_{t^n}^{t^{n+1}} \nabla P\, dt + \int_{t^n}^{t^{n+1}} \mathbf{F}(\mathbf{v}, b)\, dt, \tag{6.7}$$

where $t^n = n\Delta t$. Define the quantity $\tilde{P}^{n+1}$ such that

$$\Delta t\, \nabla \tilde{P}^{n+1} = \int_{t^n}^{t^{n+1}} \nabla P\, dt.$$

Note that $\tilde{P}^{n+1}$ is not necessarily equal to the actual perturbation pressure at any particular time. Using the definition of $\tilde{P}^{n+1}$, (6.7) may be written as

$$\mathbf{v}^{n+1} - \mathbf{v}^n = -\Delta t\, \nabla \tilde{P}^{n+1} + \int_{t^n}^{t^{n+1}} \mathbf{F}(\mathbf{v}, b)\, dt. \tag{6.8}$$

Define $\tilde{\mathbf{v}}$ such that

$$\tilde{\mathbf{v}} = \mathbf{v}^n + \int_{t^n}^{t^{n+1}} \mathbf{F}(\mathbf{v}, b)\, dt. \tag{6.9}$$

As noted by Orszag et al. (1986), the preceding integral can be conveniently evaluated using an explicit finite-difference scheme such as the third-order Adams-Bashforth method ((5.60) in Chap. 5). Equations (6.8) and (6.9) imply that

$$\mathbf{v}^{n+1} = \tilde{\mathbf{v}} - \Delta t\, \nabla \tilde{P}^{n+1}, \tag{6.10}$$

which provides a formula for updating $\tilde{\mathbf{v}}$ to obtain the new velocity field $\mathbf{v}^{n+1}$ once $\tilde{P}^{n+1}$ has been determined.

A Poisson equation for $\tilde{P}^{n+1}$ that is analogous to (6.6) is obtained by taking the divergence of (6.10) and noting that $\nabla \cdot \mathbf{v}^{n+1} = 0$, in which case

$$\nabla^2 \tilde{P}^{n+1} = \frac{\nabla \cdot \tilde{\mathbf{v}}}{\Delta t}. \tag{6.11}$$

Boundary conditions for this equation are obtained by computing the dot product of the unit vector normal to the boundary ($\mathbf{n}$) with each term of (6.10) to yield

$$\frac{\partial \tilde{P}^{n+1}}{\partial n} = -\frac{1}{\Delta t}\, \mathbf{n} \cdot (\mathbf{v}^{n+1} - \tilde{\mathbf{v}}). \tag{6.12}$$

If there is no flow normal to the boundary, the preceding reduces to

$$\frac{\partial \tilde{P}^{n+1}}{\partial n} = \frac{\mathbf{n} \cdot \tilde{\mathbf{v}}}{\Delta t}, \tag{6.13}$$

which eliminates the implicit coupling between $\tilde{P}^{n+1}$ and $\mathbf{v}^{n+1}$ that is present in the general boundary condition (6.12). In this particularly simple case, in which an inviscid fluid is bounded by rigid walls, the projection method is implemented by first updating (6.9), which accounts for the time-tendencies produced by advection and buoyancy forces, and then solving (6.11) subject to the boundary conditions (6.13). As the final step of the algorithm, $\mathbf{v}^{n+1}$ is obtained by projecting $\tilde{\mathbf{v}}$ onto the subspace of non-divergent vectors using (6.10).

The preceding algorithm loses some of its simplicity when the computation of $\mathbf{v}^{n+1}$ is coupled with that of $\tilde{P}^{n+1}$, as would be the case if a wave-permeable boundary condition replaced the rigid wall condition that $\mathbf{n} \cdot \mathbf{v}^{n+1} = 0$. In practice, the coupling between $\mathbf{v}^{n+1}$ and $\tilde{P}^{n+1}$ is eliminated by imposing some approximation to the full, implicitly coupled boundary condition. Coupling between $\mathbf{v}^{n+1}$ and $\tilde{P}^{n+1}$ may also occur when the projection method is applied to viscous flows with a no-slip condition at the boundary. The no-slip condition that $\mathbf{v} = 0$ at the boundary reduces (6.12) to

$$\frac{\partial \tilde{P}^{n+1}}{\partial n} = \frac{1}{\Delta t}\, \mathbf{n} \cdot \int_{t^n}^{t^{n+1}} b\mathbf{k} + \nu \nabla^2 \mathbf{v}\, dt, \tag{6.14}$$

where viscous forcing is now included in the momentum equations and ν is the kinematic viscosity. High spatial resolution is often required to resolve the boundary layer in no-slip viscous flow. In order to maintain numerical stability in the high-resolution boundary layer without imposing an excessively strict limitation on the time step, the viscous terms are often integrated using implicit differencing[3] (Karniadakis et al. 1991). When the time integral of $\mathbf{F}(\mathbf{v}, b)$ includes viscous terms

[3] Explicit time differencing can still be used for the advection terms because the wind speed normal to the boundary decreases as the fluid approaches the boundary.

that are approximated using implicit finite differences, (6.14) is an implicit relation between $\tilde{P}^{n+1}$ and $\mathbf{v}^{n+1}$ whose solution is often computed via a fractional step method (see discussion of (6.58) and (6.59)). As noted by Orszag et al. (1986), the accuracy with which this boundary condition is approximated can significantly influence the accuracy of the overall solution. The design of optimal approximations to (6.14) has been the subject of considerable research, however, the emphasis in this chapter is not on viscous flow, and especially not on highly viscous flow in which the diffusion terms need to be treated implicitly for computational efficiency. The reader is referred to Boyd (1989) for further discussion of the use of the projection method in viscous no-slip flow.

6.2.2 Leapfrog Implementation

In atmospheric science the projection method is often implemented using leapfrog time differences, in which case (6.8) becomes

$$\mathbf{v}^{n+1} = \mathbf{v}^{n-1} - 2\Delta t\, \nabla P^n + 2\Delta t\, \mathbf{F}\left(\mathbf{v}^n, b^n\right).$$

The solution procedure is very similar to the algorithm described in the preceding section. The velocity field generated by advection and buoyancy forces acting over the time period $2\Delta t$ is defined as

$$\tilde{\mathbf{v}} = \mathbf{v}^{n-1} + 2\Delta t\, \mathbf{F}\left(\mathbf{v}^n, b^n\right);$$

then the Poisson equation for P^n is

$$\nabla^2 P^n = \frac{\nabla \cdot \tilde{\mathbf{v}}}{2\Delta t},$$

and the velocity field is updated using the relation

$$\mathbf{v}^{n+1} = \tilde{\mathbf{v}} - 2\Delta t\, \nabla P^n.$$

Some technique, such as time filtering (Asselin 1972; Williams 2009), must also be used to prevent time splitting instability in the leapfrog solution to nonlinear problems.

One difference between this approach and the standard projection method is that by virtue of the leapfrog time difference, the pressure field that insures the non-divergence of $\mathbf{v}^{n+1}$ is the actual pressure at time $n\Delta t$. The pressure must, nevertheless, be updated at the same point in the integration cycle at which $\tilde{P}^{n+1}$ is obtained in the standard projection method, i.e., part way through the calculation of $\mathbf{v}^{n+1}$. Thus, the same problems with implicit coupling between the pressure and $\mathbf{v}^{n+1}$ arise in both the standard and the leapfrog projection methods. If viscosity is included in the momentum equations, stability considerations require that the contribution of viscosity to $\mathbf{F}(\mathbf{v}, b)$ be evaluated at time level $n-1$ so that the viscous terms are treated using forward differencing over a time interval of $2\Delta t$. This is not a particularly accurate way to represent the viscous terms and is not suitable for highly viscous flow in which the viscous terms are more efficiently integrated using implicit time differences.

6.2.3 Solving the Poisson Equation for Pressure

Suppose that the Boussinesq equations are to be solved in a two-dimensional x-z domain and that the velocity and pressure variables are staggered as indicated in Fig. 6.1. Approximating the diagnostic pressure equation (6.11) using the standard five-point finite-difference stencil for the two-dimensional Laplacian, one obtains

$$\delta_x^2 \tilde{P}^{n+1} + \delta_z^2 \tilde{P}^{n+1} = \frac{1}{\Delta t}\left(\delta_x \tilde{u} + \delta_z \tilde{w}\right), \tag{6.15}$$

where the finite-difference operator δ_t is defined such that

$$\delta_{nx} f(x) = \frac{f(x + n\Delta x/2) - f(x - n\Delta x/2)}{n\Delta x}. \tag{6.16}$$

This is an implicit algebraic relation for the $\tilde{P}_{i,j}^{n+1}$. If pressure is defined at M grid points in x and N points in z, an $M \times N$ system of linear algebraic equations must be solved in order to determine the pressure. Let the unknown grid-point values of the pressure be ordered such that

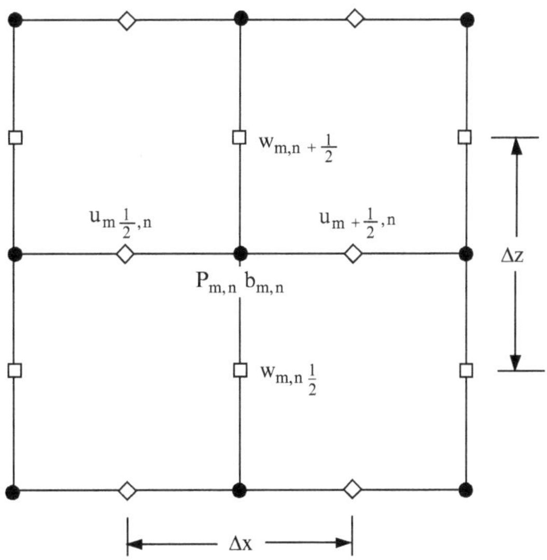

Fig. 6.1 Distribution of the dependent variables on a staggered mesh (the Arakawa C-grid) for the finite-difference approximation of the two-dimensional Boussinesq system

$$\mathbf{p} = (\tilde{P}^{n+1}_{1,1}, \tilde{P}^{n+1}_{1,2}, \ldots, \tilde{P}^{n+1}_{1,N}, \tilde{P}^{n+1}_{2,1}, \ldots, \tilde{P}^{n+1}_{M,N}),$$

then the system may be written as the matrix equation

$$\mathbf{A}\mathbf{p} = \mathbf{f}, \tag{6.17}$$

in which $\mathbf{f}$ is an identically ordered vector containing the numerically evaluated divergence of $\tilde{\mathbf{v}}$. The matrix $\mathbf{A}$ is very sparse with only five non-zero diagonals. In practical applications the number of unknown pressures can easily exceed one million, and to solve (6.17) efficiently it is important to take advantage of the sparseness of $\mathbf{A}$. Direct methods based on some variant of Gaussian elimination are, therefore, not appropriate. Direct methods for band matrices are also not suitable because the bandwidth of $\mathbf{A}$ is not 5, but $2N + 1$ and direct methods for band matrices do not preserve sparseness within the band.

Direct solutions to (6.17) can, nevertheless, be efficiently obtained by exploiting the block structure of $\mathbf{A}$. For simplicity, suppose that (6.15) is to be solved subject to Dirichlet boundary conditions, then the diagonal of $\mathbf{A}$ contains M copies of the $N \times N$ tridiagonal sub-matrix

$$\begin{pmatrix} -4 & 1 & & & & \\ 1 & -4 & 1 & & & \\ & \ddots & \ddots & \ddots & & \\ & & \ddots & \ddots & \ddots & \\ & & & 1 & -4 & 1 \\ & & & & 1 & -4 \end{pmatrix},$$

and the super- and sub-diagonals are made up of $M - 1$ copies of the $N \times N$ identity matrix. This system can be efficiently solved using block cyclic reduction (Golub and van Loan 1996, p. 177). Numerical codes for the solution of two- and three-dimensional Poisson equations subject to the most common types of boundary conditions may be accessed through the Internet at several cites including the National Institute of Standards and Technology's Guide to Available Mathematical Software (NIST/GAMS, http://gams.nist.gov), the National Center for Atmospheric Research's Mathematical and Statistical Libraries (NCAR, http://www.cisl.ucar.edu/softlib/mathlib.html) and the Netlib Repository at the Oak Ridge National Laboratory (ORNL, http://www.netlib.org).

6.3 The Semi-Implicit Method

As an alternative to filtering the governing equations to eliminate insignificant fast waves, one can retain the unapproximated governing equations and use numerical techniques to stabilize the simulation of the fast moving waves. One common way to improve numerical stability is through the use of implicit time differences such as the backward or the trapezoidal methods. Implicit methods can, however, produce rather inaccurate solutions when the time step is too large. It is therefore useful to analyze the effect of the time step on the accuracy of fully implicit solutions to wave propagation problems before discussing the true semi-implicit method.

6.3.1 Large Time Steps and Poor Accuracy

Suppose that a differential-difference approximation to the one-dimensional advection equation

$$\frac{\partial \psi}{\partial t} + c\frac{\partial \psi}{\partial x} = 0 \tag{6.18}$$

is constructed in which finite differences are used to represent the time derivative and the spatial derivative is not discretized. If the time derivative is approximated using leapfrog differencing such that

$$\frac{\phi^{n+1} - \phi^{n-1}}{2\Delta t} + c\left(\frac{\partial \phi}{\partial x}\right)^n = 0.$$

wave solutions of the form

$$\phi^n(x) = e^{i(kx - \omega n t)} \tag{6.19}$$

must satisfy the semi-discrete dispersion relation

$$\omega = \frac{1}{\Delta t}\arcsin(ck\Delta t). \tag{6.20}$$

The phase speed of the leapfrog-differenced solution is

$$c_l = \frac{\omega}{k} = \frac{\arcsin(ck\Delta t)}{k\Delta t}. \tag{6.21}$$

The stability constraint, $|ck\Delta t| < 1$, associated with the preceding leapfrog scheme can be avoided by switching to trapezoidal differencing. Many semi-implicit formulations use a combination of leapfrog and trapezoidal differencing, and in those formulations the trapezoidal time difference is computed over an interval of $2\Delta t$. To facilitate the application of this analysis to these semi-implicit formulations, and to more directly compare the trapezoidal and leapfrog schemes, (6.18) will be approximated using trapezoidal differencing over a $2\Delta t$-wide stencil such that

$$\frac{\phi^{n+1} - \phi^{n-1}}{2\Delta t} + \frac{c}{2}\left[\left(\frac{\partial \phi}{\partial x}\right)^{n+1} + \left(\frac{\partial \phi}{\partial x}\right)^{n-1}\right] = 0.$$

Wave solutions to this scheme must satisfy the dispersion relation

$$\omega = \frac{1}{\Delta t} \arctan(ck\Delta t). \tag{6.22}$$

The phase speed of the trapezoidally differenced solution is

$$c_t = \frac{\arctan(ck\Delta t)}{k\Delta t}.$$

The phase-speed errors generated by the leapfrog and $2\Delta t$ trapezoidal methods are compared in Fig. 6.2. The phase speed at a fixed Courant number is plotted as a function of both spatial wavenumber (bottom axis) and wavelength (top axis) in units of Δx. These curves give the phase speed that would be obtained if the spatial dependence of the numerical solution was represented by a Fourier spectral method with a cutoff wavelength of $2\Delta x$. When $c\Delta t/\Delta x = 1/\pi$ the errors in wavelengths greater than $2.5\Delta x$ generated by the leapfrog and the $2\Delta t$-trapezoidal methods are similar in magnitude and opposite in sign. The leapfrog scheme is unstable for Courant numbers greater than $1/\pi$, but solutions can still be obtained using the trapezoidal scheme. The phase-speed errors in the $2\Delta t$-trapezoidal solution computed with $c\Delta t/\Delta x = 5/\pi$ are, however, rather large. Even modes with relatively good spatial resolution, such as a $10\Delta x$ wave, are in significant error.

The deceleration generated by $2\Delta t$-trapezoidal differencing may be alternatively expressed in terms of the reduced phase speed

$$\hat{c} = c\cos(\omega\Delta t).$$

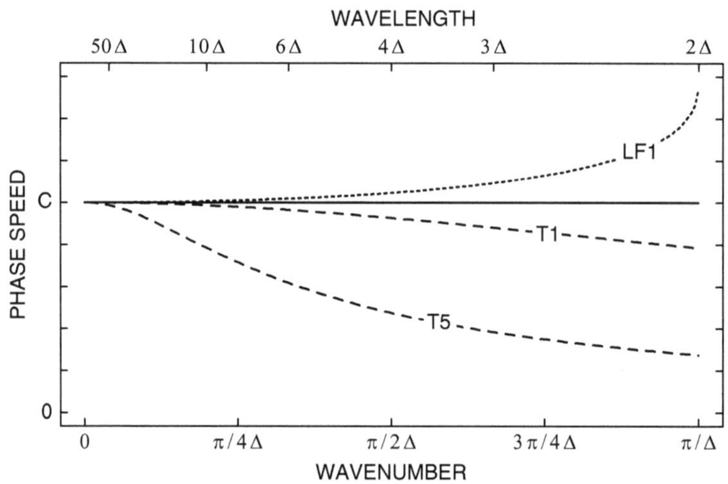

Fig. 6.2 Phase speed of leapfrog (*dotted*) and $2\Delta t$-trapezoidal (*dashed*) approximations to the advection equation when $c\Delta t/\Delta x = 1/\pi$ (LF1 and T1), and for the trapezoidal solution when $c\Delta t/\Delta x = 5/\pi$ (T5)

Then the $2\Delta t$-trapezoidal dispersion relation (6.22) assumes the form

$$\omega = \frac{1}{\Delta t} \arcsin(\hat{c}k\Delta t).$$

and the phase speed of the $2\Delta t$-trapezoidal solution becomes

$$c_t = \frac{\omega}{k} = \frac{\arcsin(\hat{c}k\Delta t)}{k\Delta t}.$$

The preceding differ from the corresponding expressions for the leapfrog scheme (6.20) and (6.21) in that the true propagation speed, c, has been replaced by the reduced speed $\hat{c}$. As the time step increases, $\hat{c}$ decreases so that $|\hat{c}k\Delta t|$ remains less than one and the numerical solution remains stable, but the relative error in $\hat{c}$ can become arbitrarily large. As a consequence, it is not possible to take advantage of the unconditional stability of the trapezoidal method by using very large time steps to solve wave-propagation problems unless one is willing to tolerate a considerable decrease in the accuracy of the solution.

6.3.2 A Prototype Problem

The loss of accuracy associated with poor temporal resolution that can occur using implicit numerical methods is not a problem if the poorly resolved waves are not physically significant. If the fastest moving waves are insignificant, the accuracy constraints imposed on the time step by these waves can be ignored and, provided the scheme is unconditionally stable, a good solution can be obtained using any time step that adequately resolves the slower moving features of primary physical interest. A simple but computationally inefficient way to insure the unconditional stability of a numerical scheme is to use trapezoidal time differencing throughout the approximate equations. It is, however, more efficient to implicitly evaluate only those terms in the governing equations that are crucial to the propagation of the fast wave and to approximate the remaining terms with some explicit time-integration scheme. This is the fundamental strategy in the "semi-implicit" approach which gains efficiency relative to a "fully implicit" method by reducing the complexity of the implicit algebraic equations that must be solved during each integration step. Semi-implicit differencing is particularly attractive when all the terms that are evaluated implicitly are linear functions of the unknown variables.

In order to investigate the stability of semi-implicit time-differencing schemes consider a prototype ordinary differential equation of the form

$$\frac{d\psi}{dt} + i\omega_H\psi + i\omega_L\psi = 0. \tag{6.23}$$

This is simply a version of the oscillation equation (see (5.19) in Chap. 5) in which the oscillatory forcing is divided into high-frequency (ω_H) and low-frequency

(ω_L) components. The division of the forcing into two terms may appear to be rather artificial, but the dispersion relation associated with wave-like solutions to more complex systems of governing equations (such as the shallow-water system discussed in the next section) often has individual roots of the form

$$\omega = \omega_H + \omega_L,$$

and (6.23) serves as the simplest differential equation describing the time-dependence of such waves.

The simplest semi-implicit approximation to (6.23) is

$$\frac{\phi^{n+1} - \phi^n}{\Delta t} + i\omega_H \phi^{n+1} + i\omega_L \phi^n = 0.$$

The stability and the accuracy of this scheme have already been analyzed in connection with ((5.21) in Chap. 5); it is first order accurate and is stable whenever $|\omega_L| < |\omega_H|$. Since $|\omega_L| < |\omega_H|$ by assumption, the method is stable for all Δt. The weakness of this scheme is its low accuracy. A more accurate second-order method can be obtained using the centered-in-time formula

$$\frac{\phi^{n+1} - \phi^{n-1}}{2\Delta t} + i\omega_H \left(\frac{\phi^{n+1} + \phi^{n-1}}{2} \right) + i\omega_L \phi^n = 0. \tag{6.24}$$

The stability of this method may be investigated by considering the behavior of oscillatory solutions of the form $\exp(-i\omega n \Delta t)$, which satisfy (6.24) when

$$\sin \tilde{\omega} = \tilde{\omega}_H \cos \tilde{\omega} + \tilde{\omega}_L, \tag{6.25}$$

where

$$\tilde{\omega} = \omega \Delta t, \quad \tilde{\omega}_H = \omega_H \Delta t, \quad \text{and} \quad \tilde{\omega}_L = \omega_L \Delta t.$$

To solve for $\tilde{\omega}$, let $\tan \beta = \tilde{\omega}_H$, then (6.25) becomes

$$\sin \tilde{\omega} = \tan \beta \cos \tilde{\omega} + \tilde{\omega}_L,$$

or equivalently

$$\sin \tilde{\omega} \cos \beta - \sin \beta \cos \tilde{\omega} = \tilde{\omega}_L \cos \beta.$$

By the Pythagorean theorem, $\cos \beta = (1 + \tilde{\omega}_H^2)^{-1/2}$, and the preceding reduces to

$$\sin(\tilde{\omega} - \beta) = \tilde{\omega}_L (1 + \tilde{\omega}_H^2)^{-1/2},$$

or equivalently,

$$\tilde{\omega} = \arctan(\tilde{\omega}_H) + \arcsin\left(\tilde{\omega}_L (1 + \tilde{\omega}_H^2)^{-1/2} \right).$$

The semi-implicit scheme (6.24) will be stable when the $\tilde{\omega}$ satisfying this equation are real and distinct, which is guaranteed when

$$\tilde{\omega}_L^2 \leq 1 + \tilde{\omega}_H^2. \tag{6.26}$$

Since, by assumption $|\omega_L| \leq |\omega_H|$, (6.24) is stable for all Δt. Note that (6.26) will also be satisfied whenever $|\omega_L \Delta t| \leq 1$, implying that semi-implicit differencing permits an increase in the maximum stable time step relative to that for a fully explicit approximation even in those cases where $|\omega_L| > |\omega_H|$ because the terms approximated with the trapezoidal difference do not restrict the maximum stable time step.

6.3.3 Semi-Implicit Solution of the Shallow-Water Equations

The shallow-water equations for motion in a rotating reference frame with Coriolis parameter f may be expressed

$$\frac{Du}{Dt} - fv + g\frac{\partial h}{\partial x} = 0, \qquad \frac{Dv}{Dt} + fu + g\frac{\partial h}{\partial y} = 0, \tag{6.27}$$

$$\frac{Dh}{Dt} + h\left(\frac{\partial u}{\partial x} + \frac{\partial v}{\partial y}\right) = 0, \tag{6.28}$$

where u and v are the eastward and northward components of the velocity and h is the fluid depth. This system supports rapidly moving gravity waves. If there are spatial variations in the potential vorticity of the undisturbed system, f/h, the shallow-water equations can also support slowly propagating potential-vorticity (or Rossby) waves. In many large-scale atmospheric and oceanic models the Rossby waves are of greater physical significance than the faster moving gravity waves and the Rossby-waves can be efficiently simulated using semi-implicit time-differencing to accommodate the CFL stability condition associated with gravity-wave propagation.

The simplest example in which to illustrate the influence of semi-implicit differencing on the CFL condition can be obtained by examining a one-dimensional system without the Coriolis force that is linearized about a reference state with a constant fluid velocity U and fluid depth H,

$$\frac{\partial u}{\partial t} + U\frac{\partial u}{\partial x} + g\frac{\partial h}{\partial x} = 0, \tag{6.29}$$

$$\frac{\partial h}{\partial t} + U\frac{\partial h}{\partial x} + H\frac{\partial u}{\partial x} = 0. \tag{6.30}$$

If the mean-flow velocity is less than the phase speed of a shallow-water gravity wave $c = \sqrt{gH}$, the numerical integration can be stabilized by evaluating those terms responsible for gravity-wave propagation with trapezoidal differencing;

leapfrog differencing can be used for the remaining terms (Kwizak and Robert 1971). The terms essential to gravity-wave propagation are the hydrostatic pressure gradient ($g\partial h/\partial x$) in (6.29) and the velocity divergence in (6.30), so the semi-implicit approximation to the linearized shallow-water system is

$$\delta_{2t}u^n + U\frac{\partial u^n}{\partial x} + g\left\langle\frac{\partial h^n}{\partial x}\right\rangle^{2t} = 0, \tag{6.31}$$

$$\delta_{2t}h^n + U\frac{\partial h^n}{\partial x} + H\left\langle\frac{\partial u^n}{\partial x}\right\rangle^{2t} = 0, \tag{6.32}$$

where the finite-difference operator δ_t is defined by (6.16) and the averaging operator $\langle\ \rangle^t$ is given by

$$\langle f(x)\rangle^{nx} = \frac{f(x+n\Delta x/2)+f(x-n\Delta x/2)}{2}. \tag{6.33}$$

Solutions to (6.31) and (6.32) exist of the form $e^{i(kx-\omega j\Delta t)}$ provided k and ω satisfy the semi-discrete dispersion relation

$$\sin\omega\Delta t = Uk\Delta t \pm ck\Delta t\cos\omega\Delta t.$$

This dispersion relation has the same form as (6.25), so as demonstrated in the preceding section, the method will be stable provided that $|U| \leq c$, or equivalently, whenever the phase speed of shallow-water gravity waves exceeds the speed of the mean flow. The Coriolis force has been neglected in the preceding shallow-water system, and as a consequence, there are no Rossby wave solutions to (6.31) and (6.32). In a more general two-dimensional system that includes the Coriolis force semi-implicit time differencing leads to a system that is stable whenever the CFL condition for the Rossby waves is satisfied.

6.3.4 Semi-implicit Solution of the Compressible Governing Equations

Now consider how semi-implicit differencing can be used to eliminate the stability constraint imposed by sound waves in the numerical solution of the Euler equations for stratified flow. To streamline the discussion we will focus on the compressible Boussinesq system, which supports both sound and gravity wave propagation while eliminating small terms reflecting the decrease in mean density produced by the decrease in pressure with height.[4] The compressible Boussinesq system consists of

[4] See Sect. 7.2.4 of (Durran 1999) for details about the difference between the Euler equations and the compressible Boussinesq system.

the relations

$$\frac{d\mathbf{v}}{dt} + \nabla P = b\mathbf{k}, \tag{6.34}$$

$$\frac{db}{dt} + N^2 w = 0, \tag{6.35}$$

$$\frac{dP}{dt} + c_s^2 \nabla \cdot \mathbf{v} = 0. \tag{6.36}$$

Note that (6.34) and (6.35) are identical to the buoyancy and momentum equations in the standard Boussinesq system (6.1) and (6.2), while the incompressible continuity equation (6.3) has been replaced by (6.36) and is recovered in the limit $c_s \to \infty$.

Suppose the flow is confined to the x–z plane and linearize (6.34)–(6.36) about a basic state with uniform horizontal velocity U and zero means for the other fields. Letting (u, w, b, P) now denote the perturbations, the linear system becomes

$$\left(\frac{\partial}{\partial t} + U\frac{\partial}{\partial x}\right) u + \frac{\partial P}{\partial x} = 0, \tag{6.37}$$

$$\left(\frac{\partial}{\partial t} + U\frac{\partial}{\partial x}\right) w + \frac{\partial P}{\partial z} = b, \tag{6.38}$$

$$\left(\frac{\partial}{\partial t} + U\frac{\partial}{\partial x}\right) b + N^2 w = 0, \tag{6.39}$$

$$\left(\frac{\partial}{\partial t} + U\frac{\partial}{\partial x}\right) P + c_s^2 \left(\frac{\partial u}{\partial x} + \frac{\partial w}{\partial z}\right) = 0. \tag{6.40}$$

As in the standard Boussinesq approximation, the compressible Boussinesq system neglects the influence of density variations on inertia while retaining the influence of density variations on buoyancy and assumes that buoyancy is conserved following a fluid parcel. In contrast to the standard Boussinesq system, the compressible Boussinesq system does retain the influence of density fluctuations on pressure and thereby allows the formation of the prognostic pressure equation (6.40).

Suppose that the simplified compressible system (6.37)–(6.40) is approximated using leapfrog time differencing and that the spatial derivatives are computed using a Fourier pseudo-spectral method. Waves of the form

$$(u, w, b, P) = (u_0, w_0, b_0, P_0)e^{i(kx + \ell z - \omega n \Delta t)}$$

are solutions to this system provided that ω, k and ℓ satisfy the dispersion relation

$$\hat{\omega}^4 - c_s^2 \left(k^2 + \ell^2 + N^2/c_s^2\right) \hat{\omega}^2 + N^2 k^2 c_s^2 = 0,$$

where

$$\hat{\omega} = \frac{\sin \omega \Delta t}{\Delta t} - Uk.$$

This dispersion relation is quadratic in $\hat{\omega}^2$ and has solutions

$$\hat{\omega}^2 = \frac{c_s^2}{2}\left(k^2 + \ell^2 + \frac{N^2}{c_s^2} \pm \left[\left(k^2 + \ell^2 + \frac{N^2}{c_s^2}\right)^2 - \frac{4N^2k^2}{c_s^2}\right]^{1/2}\right). \tag{6.41}$$

The positive root yields the dispersion relation for sound waves; the negative root yields the dispersion relation for gravity waves.[5] The individual dispersion relations for sound and gravity waves may be greatly simplified whenever the last term inside the square root in (6.41) is much smaller than the first term. One condition sufficient condition for this simplification, which is easily satisfied in most atmospheric applications, is that $N^2/c_s^2 \ll \ell^2$. If $N^2/c_s^2 \ll \ell^2$, then

$$\frac{4N^2k^2}{c_s^2} \ll \frac{2N^2k^2}{c_s^2} + 2k^2\ell^2 \leq \left(k^2 + \ell^2 + \frac{N^2}{c_s^2}\right)^2, \tag{6.42}$$

and therefore the sound-wave dispersion relation is well approximated by

$$\hat{\omega}^2 = c_s^2\left(k^2 + \ell^2 + N^2/c_s^2\right). \tag{6.43}$$

Dividing the terms inside the square root in (6.41) by $\left(k^2 + \ell^2 + N^2/c_s^2\right)^2$ and again using (6.42), the gravity wave-dispersion relation may be well approximated as

$$\hat{\omega}^2 = \frac{N^2k^2}{k^2 + \ell^2 + N^2/c_s^2}. \tag{6.44}$$

Consider the time-step limitation imposed by sound wave propagation. Using the definition of $\hat{\omega}$, (6.43) may be expressed as

$$\sin \omega \Delta t = \Delta t\left(Uk \pm c_s\left(k^2 + \ell^2 + N^2/c_s^2\right)^{1/2}\right).$$

Stable leapfrog solutions are obtained when the right side of this expression is a real number whose absolute value is less than unity. A necessary condition for stability is that

$$\left(|U|k_{\max} + c_s(k_{\max}^2 + \ell_{\max}^2)^{1/2}\right)\Delta t < 1, \tag{6.45}$$

where $k_{\max}$ and $\ell_{\max}$ are the largest horizontal and vertical wavenumbers retained in the truncation. In many applications the vertical resolution is much higher than the horizontal resolution and the most severe restriction on the time step is associated with vertically propagating sound waves; (6.45) is also a good approximation to the sufficient condition for stability since the term involving N^2/c_s^2 is typically insignificant for the highest frequency waves.

[5] In the limit $N \to 0$, the positive root gives $\hat{\omega}^2 = c_s^2\left(k^2 + \ell^2\right)$; the negative root gives $\hat{\omega}^2 = 0$.

The dispersion relation for gravity waves (6.44) may be written as

$$\sin \omega \Delta t = \Delta t \left(Uk \pm \frac{Nk}{\left(k^2 + \ell^2 + N^2/c_s^2\right)^{1/2}} \right). \tag{6.46}$$

Since

$$\frac{N|k|}{\left(k^2 + \ell^2 + N^2/c_s^2\right)^{1/2}} \leq c_s|k|,$$

the necessary condition for sound-wave stability (6.45) is sufficient to insure the stability of the gravity waves. Although (6.45) guarantees the stability of the gravity wave modes, it is far too restrictive. Since

$$\frac{N|k|}{\left(k^2 + \ell^2 + N^2/c_s^2\right)^{1/2}} \leq N,$$

(6.46) implies the gravity waves will be stable provided that

$$(|U|k_{\max} + N)\Delta t < 1.$$

This is also a good approximation to the necessary condition for stability because the term involving N^2/c_s^2 is usually dominated by $k_{\max}^2$.

In most geophysical applications

$$c_s(k_{\max}^2 + \ell_{\max}^2)^{1/2} >> |U|k_{\max} + N$$

and the maximum stable time step with which the gravity waves can be integrated is, therefore, far larger than the time step required to maintain stability in the sound wave modes. In such circumstances, the sound waves can be stabilized using a semi-implicit approximation in which the pressure gradient and velocity divergence terms are evaluated using trapezoidal differencing (Tapp and White 1976). The resulting semi-implicit system is

$$\delta_{2t} u^n + U \frac{\partial u^n}{\partial x} + \left\langle \frac{\partial P^n}{\partial x} \right\rangle^{2t} = 0, \tag{6.47}$$

$$\delta_{2t} w^n + U \frac{\partial w^n}{\partial x} + \left\langle \frac{\partial P^n}{\partial z} \right\rangle^{2t} = b^n, \tag{6.48}$$

$$\delta_{2t} b^n + U \frac{\partial b^n}{\partial x} + N^2 w^n = 0, \tag{6.49}$$

$$\delta_{2t} P^n + U \frac{\partial P^n}{\partial x} + c_s^2 \left(\left\langle \frac{\partial u^n}{\partial x} \right\rangle^{2t} + \left\langle \frac{\partial w^n}{\partial z} \right\rangle^{2t} \right) = 0. \tag{6.50}$$

Let $\hat{c}_s = c_s \cos(\omega \Delta t)$, then the dispersion relation for the semi-implicit system is identical to that obtained for leapfrog differencing except that c_s is replaced by $\hat{c}_s$ throughout (6.41). The dispersion relation for the sound-wave modes is

$$\hat{\omega}^2 = \hat{c}_s^2 \left(k^2 + \ell^2 + N^2/\hat{c}_s^2\right),$$

or

$$\sin \omega \Delta t = \Delta t \left(Uk \pm \hat{c}_s \left(k^2 + \ell^2 + N^2/\hat{c}_s^2\right)^{1/2} \right). \tag{6.51}$$

The most severe stability constraints are imposed by the shortest waves for which the term $N^2/\hat{c}_s^2$ can be neglected in comparison with $k^2 + \ell^2$. Neglecting $N^2/\hat{c}_s^2$, (6.51) becomes

$$\sin \omega \Delta t = Uk\Delta t \pm c_s \Delta t \left(k^2 + \ell^2\right)^{1/2} \cos \omega \Delta t,$$

which has the same form as (6.25) implying that the sound wave modes are stable whenever

$$|Uk| \leq c_s \left(k^2 + \ell^2\right)^{1/2}.$$

A sufficient condition for the stability of the sound waves is simply that the flow be sub-sonic ($|U| \leq c_s$), or equivalently, that the Mach number be less than unity.

Provided that the flow is sub-sonic, the only constraint on the time step required to keep the semi-implicit scheme stable is that associated with gravity wave propagation. The dispersion relation for the gravity waves in the semi-implicit system is

$$\hat{\omega}^2 = \frac{N^2 k^2}{k^2 + \ell^2 + N^2/\hat{c}_s^2} \tag{6.52}$$

which differs from the result for leapfrog differencing only in the small term $N^2/\hat{c}_s^2$. Stable gravity wave solutions to the semi-implicit system are obtained whenever

$$(|U| k_{\max} + N)\Delta t < 1,$$

which is the same condition obtained for the stability of the gravity waves using leapfrog differencing. Thus, as suggested previously, the semi-implicit scheme allows the compressible equations governing low Mach-number flow to be integrated with a much larger time step than that allowed by fully explicit schemes. This increase in efficiency comes at a price; whenever the time step is much larger than that allowed by the CFL condition for sound waves, the sound waves are artificially decelerated by a factor of $\cos(\omega \Delta t)$. This error is directly analogous to that considered in Sect. 6.3.1 in which spurious decelerations were produced by fully implicit schemes using very large time steps. Nevertheless, in many practical applications the errors in the sound waves are of no consequence and the quality of the solution is entirely determined by the accuracy with which the slower moving waves are approximated.

6.3.5 Numerical Implementation

The semi-implicit approximation to the compressible Boussinesq system discussed in the preceding section generates a system of implicit algebraic equations that must be solved every time step. First consider the situation where only the sound waves are stabilized by semi-implicit differencing and suppose that the spatial derivatives are not discretized. Then (6.34)–(6.36) take the form

$$\mathbf{v}^{n+1} + \Delta t \nabla P^{n+1} = \mathbf{G}, \tag{6.53}$$

$$b^{n+1} = b^{n-1} - 2\Delta t \left(\mathbf{v}^n \cdot \nabla b^n + N^2 w^n\right), \tag{6.54}$$

$$P^{n+1} + c_s^2 \Delta t \nabla \cdot \mathbf{v}^{n+1} = h. \tag{6.55}$$

Here

$$\mathbf{G} = \mathbf{v}^{n-1} - \Delta t \left[\nabla P^{n-1} - 2b^n \mathbf{k} + 2\mathbf{v}^n \cdot \nabla \mathbf{v}^n\right],$$

and

$$h = P^{n-1} - \Delta t \left[c_s^2 \nabla \cdot \mathbf{v}^{n-1} + 2\mathbf{v}^n \cdot \nabla P^n\right].$$

A single Helmholtz equation for P^{n+1} can be obtained by substituting the divergence of (6.53) into (6.55) to yield

$$\nabla^2 P^{n+1} - \frac{P^{n+1}}{(c_s \Delta t)^2} = \frac{\nabla \cdot \mathbf{G}}{\Delta t} - \frac{h}{(c_s \Delta t)^2}. \tag{6.56}$$

The numerical solution of this Helmholtz equation is trivial if the Fourier spectral method is employed in a rectangular domain or if spherical harmonic expansion functions are used in a global spectral model. If the spatial derivatives are approximated by finite differences, (6.56) yields a sparse linear algebraic system that can be solved using the techniques described in Sect. 6.2.3. After solving (6.56) for P^{n+1}, the momentum equations can be stepped forward and the buoyancy equation (6.54), which is completely explicit, can be updated to complete the integration cycle.

This implementation of the semi-implicit method is closely related to the projection method for incompressible Boussinesq flow. Indeed in the limit $c_s \to \infty$ the preceding approach will be identical to the leapfrog projection method (described in Sect. 6.2.2) if $(P^{n+1} + P^{n-1})/2$ is replaced by P^n in (6.56). Although the leapfrog projection method and the semi-implicit method yield algorithms involving very similar algebraic equations, these methods are derived via very different approximation strategies. The projection method is an efficient way to solve a set of continuous equations that is obtained by filtering the exact Euler equations to eliminate sound waves. In contrast, the semi-implicit scheme is obtained by directly approximating the full compressible equations and using implicit time differencing to stabilize the sound waves. Neither approach allows one to correctly simulate sound waves, but both approaches allow the accurate and efficient simulation of the slower moving gravity waves.

6.4 Fractional-Step Methods

The semi-implicit method requires the solution of an elliptic equation for the pressure during each step of the integration. This can be avoided by splitting the complete problem into fractional steps and using a smaller time step to integrate the subproblem containing the terms responsible for the propagation of the fast-moving wave. Consider a general partial differential equation of the form

$$\frac{\partial \psi}{\partial t} + \mathscr{L}(\psi) = 0, \tag{6.57}$$

where $\mathscr{L}(\psi)$ contains the spatial derivatives and other forcing terms. Assuming for simplicity in the following analysis that $\mathscr{L}$ is time-independent, the exact solution to (6.57) may be written in the form $\psi(t) = \exp(t\mathscr{L})\psi(0)$, where the exponential of the operator $\mathscr{L}$ is defined by the infinite series

$$\exp(t\mathscr{L}) = I + t\mathscr{L} + \frac{t^2}{2}\mathscr{L}^2 + \frac{t^3}{6}\mathscr{L}^3 + \cdots,$$

and I is the identity operator. The change in ψ over one time step is therefore

$$\psi(t+\Delta t) = \exp[(\Delta t+t)\mathscr{L}]\psi(0) = \exp(\Delta t\mathscr{L})\exp(t\mathscr{L})\psi(0) = \exp(\Delta t\mathscr{L})\psi(t).$$

Suppose that $\mathscr{L}(\psi)$ can be split into two parts

$$\mathscr{L}(\psi) = \mathscr{L}_1(\psi) + \mathscr{L}_2(\psi),$$

such that $\mathscr{L}_1$ and $\mathscr{L}_2$ contain those terms responsible for the propagation of slow- and fast-moving waves, respectively. Each of these individual operators can also be formally integrated over an interval Δt to obtain

$$\psi(t+\Delta t) = \exp(\Delta t\mathscr{L}_1)\psi(t), \quad \psi(t+\Delta t) = \exp(\Delta t\mathscr{L}_2)\psi(t).$$

Let $\mathscr{F}_1(\Delta t)$ and $\mathscr{F}_2(\Delta t)$ be numerical approximations to the exact operators $\exp(\Delta t\mathscr{L}_1)$ and $\exp(\Delta t\mathscr{L}_2)$.

6.4.1 Complete Operator Splitting

In the standard fractional-step approach, the approximate solution is stepped forward over a time interval Δt using

$$\phi^s = \mathscr{F}_1(\Delta t)\phi^n, \tag{6.58}$$

$$\phi^{n+1} = \mathscr{F}_2(\Delta t)\phi^s, \tag{6.59}$$

but it is not necessary to use the same time step in each subproblem. If the maximum stable time step with which the approximate slow-wave operator (6.58) can be integrated is M times that with which the fast-wave operator (6.59) can be integrated, the numerical solution could be evaluated using the formula

$$\phi^{n+1} = [\mathscr{F}_2(\Delta t/M)]^M \, \mathscr{F}_1(\Delta t)\phi^n. \tag{6.60}$$

This approach can be applied to the linearized one-dimensional shallow water system by writing (6.29) and (6.30) in the form

$$\frac{\partial \mathbf{r}}{\partial t} + \mathscr{L}_1(\mathbf{r}) + \mathscr{L}_2(\mathbf{r}) = 0, \tag{6.61}$$

where

$$\mathbf{r} = \begin{pmatrix} u \\ h \end{pmatrix}, \quad \mathscr{L}_1 = \begin{pmatrix} U\partial_x & 0 \\ 0 & U\partial_x \end{pmatrix}, \quad \mathscr{L}_2 = \begin{pmatrix} 0 & g\partial_x \\ H\partial_x & 0 \end{pmatrix},$$

and ∂_x denotes the partial derivative with respect to x. The first fractional step, which is an approximation to

$$\frac{\partial \mathbf{r}}{\partial t} + \mathscr{L}_1(\mathbf{r}) = 0,$$

involves the solution of two decoupled advection equations. Since this is a fractional step method, it is generally preferable to approximate the preceding with a two-time level method. In order to avoid using implicit, unstable or Lax–Wendroff methods the first step can be integrated using the Runge–Kutta scheme

$$\mathbf{r}^* = \mathbf{r}^n + \Delta t/3\, \mathscr{L}_1(\mathbf{r}^n), \tag{6.62}$$

$$\mathbf{r}^{**} = \mathbf{r}^n + \Delta t/2\, \mathscr{L}_1(\mathbf{r}^*), \tag{6.63}$$

$$\mathbf{r}^{n+1} = \mathbf{r}^n + \Delta t\, \mathscr{L}_1(\mathbf{r}^{**}). \tag{6.64}$$

This Runge–Kutta method is third-order accurate for linear problems and is stable and damping for $|U|k_{\max}\Delta t < 1.73$, where $k_{\max}$ is the maximum retained wavenumber.

The second fractional step, which approximates

$$\frac{\partial \mathbf{r}}{\partial t} + \mathscr{L}_2(\mathbf{r}) = 0,$$

can be efficiently integrated using forward-backward differencing. Defining $\Delta\tau = \Delta t/M$ as the length of a small time step, the forward-backward scheme is

$$\frac{u^{m+1} - u^m}{\Delta\tau} + g\frac{\partial h^m}{\partial x} = 0, \tag{6.65}$$

$$\frac{h^{m+1} - h^m}{\Delta\tau} + H\frac{\partial u^{m+1}}{\partial x} = 0. \tag{6.66}$$

This scheme is stable for $ck_{\max}\Delta\tau < 2$ and is second order accurate in time. Since the operators used in each fractional step commute,[6] the complete method will be $O\left[(\Delta t)^2\right]$ accurate and stable whenever each of the individual steps are stable.

Although the preceding fractional step scheme works fine for the linearized one-dimensional shallow water system, it does not generalize as nicely to problems in which the operators do not commute. As an example, consider the compressible two-dimensional Boussinesq equations, which could be split into the form (6.61) by defining

$$\mathbf{r} = \left(u\ w\ b\ P\right)^T,$$

$$\mathscr{L}_1 = \begin{pmatrix} \mathbf{v}\cdot\nabla & 0 & 0 & 0 \\ 0 & \mathbf{v}\cdot\nabla & 0 & 0 \\ 0 & 0 & \mathbf{v}\cdot\nabla & 0 \\ 0 & 0 & 0 & \mathbf{v}\cdot\nabla \end{pmatrix}, \quad \mathscr{L}_2 = \begin{pmatrix} 0 & 0 & 0 & \partial_x \\ 0 & 0 & -1 & \partial_z \\ 0 & N^2 & 0 & 0 \\ c_s^2\partial_x & c_s^2\partial_z & 0 & 0 \end{pmatrix},$$

where $\mathbf{v}$ is the two-dimensional velocity vector and $\nabla = (\partial/\partial x, \partial/\partial z)$. Suppose that N and c_s are constant and that the full nonlinear system is linearized about a reference state with a mean horizontal wind $U(z)$. The operators associated with this linearized system will not commute unless dU/dz is zero.

As in the one-dimensional shallow-water system, the advection operator $\mathscr{L}_1$ can be approximated using the third-order Runge–Kutta method (6.62)–(6.64). The second fractional step may be integrated using trapezoidal differencing for the terms governing the vertical propagation of sound waves and forward-backward differencing for the terms governing horizontal sound-wave propagation and buoyancy oscillations. The resulting scheme is

$$\frac{u^{m+1} - u^m}{\Delta\tau} + \frac{\partial P^m}{\partial x} = 0, \tag{6.67}$$

$$\frac{w^{m+1} - w^m}{\Delta\tau} + \frac{\partial}{\partial z}\left(\frac{P^{m+1} + P^m}{2}\right) - b^m = 0, \tag{6.68}$$

$$\frac{b^{m+1} - b^m}{\Delta\tau} + N^2 w^{m+1} = 0, \tag{6.69}$$

$$\frac{P^{m+1} - P^m}{\Delta\tau} + c_s^2\frac{\partial u^{m+1}}{\partial x} + c_s^2\frac{\partial}{\partial z}\left(\frac{w^{m+1} + w^m}{2}\right) = 0, \tag{6.70}$$

[6] The operators $\mathscr{L}_1$ and $\mathscr{L}_2$ commute if $\mathscr{L}_1(\mathscr{L}_2(\mathbf{r})) = \mathscr{L}_2(\mathscr{L}_1(\mathbf{r}))$. See Durran (1999, Sect. 3.3) for a discussion of the impact of operator commutativity on the performance of fractional-step schemes.

This approximation to $\exp(\Delta\tau\mathscr{L}_2)$ is stable and non-damping if $\max(c_s k_{\max}, N)\,\Delta\tau < 2$. Note that if the spatial derivatives are replaced by finite differences, the trapezoidal approximation of the terms involving vertical derivatives will not significantly increase the computations required on each small time step because it leads to a tridiagonal system of algebraic equations for the w^{m+1} throughout each vertical column within the domain. If the horizontal resolution is very coarse, so that $k_{\max} << N/c_s$ further efficiency can be also obtained by treating the terms involving buoyancy oscillations with trapezoidal differencing. Since these terms do not involve derivatives, the resulting implicit algebraic system remains tridiagonal.

As an alternative to the trapezoidal method, the terms involving the vertical pressure gradient and the divergence of the vertical velocity could be integrated using forward-backward differencing, in which case the stability criteria for the small time step would include an additional term proportional to $c_s \ell_{\max}\Delta\tau$ where $\ell_{\max}$ is the maximum resolvable vertical wavenumber. It may be appropriate to use forward-backward differencing instead of the trapezoidal scheme in applications with identical vertical and horizontal grid spacing, but if the vertical resolution is much finer than the horizontal resolution the additional stability constraint imposed by vertical sound-wave propagation will reduce efficiency by requiring an excessive number of small time steps.

The performance of the preceding scheme is evaluated in simulation of two dimensional compressible Boussinesq flow past a compact gravity-wave generator. The wave generator is modeled by including forcing terms in the momentum equations such that the non-discretized versions of (6.67) and (6.68) take the form

$$\frac{du}{dt} + \frac{\partial P}{\partial x} = -\frac{\partial \Psi}{\partial z}, \tag{6.71}$$

$$\frac{dw}{dt} + \frac{\partial P}{\partial z} - b = \frac{\partial \Psi}{\partial x}, \tag{6.72}$$

where

$$\Psi(x, z, t) = E(x, z)\sin\omega t \sin k_1 x \cos \ell_1 z,$$

and

$$E(x, z) = \begin{cases} \alpha\,(1 + \cos k_2 x)\,(1 + \cos \ell_2 z) & \text{if } |x| \le \pi/k_2 \text{ and } |z| \le \pi/\ell_2, \\ 0 & \text{otherwise}. \end{cases}$$

This forcing has no influence on the time tendency of the divergence, and as a consequence it does not excite sound waves. The spatial domain is periodic at $x = \pm 50$ km and bounded by rigid horizontal walls at $z = \pm 5$ km. In the following tests $\Delta x = 250$ m, $\Delta z = 50$ m, $N = 0.01\ \mathrm{s}^{-1}$, $c_s = 350\ \mathrm{ms}^{-1}$, and the parameters defining the wave generator are $\alpha = 0.2$, $2\pi/k_1 = 10$ km, $2\pi/\ell_1 = 2.5$ km, $2\pi/k_2 = 11$ km, $2\pi/\ell_2 = 1.5$ km, and $\omega = 0.002\ \mathrm{s}^{-1}$. The forcing is evaluated every $\Delta\tau$ and applied to the solution on the small time step, $\Delta x = 250$ m, $\Delta z = 50$ m, $N = 0.01\ \mathrm{s}^{-1}$, and $c_s = 350\ \mathrm{ms}^{-1}$.

The spatial derivatives are approximated using centered differencing on a staggered grid identical to that shown in Fig. 6.1 except that b is co-located with the w points rather than the P points. As a consequence of the mesh staggering, the horizontal wavenumber obtained from the finite-difference approximations to the pressure gradient and velocity divergence is $(2/\Delta x)(\sin k\Delta x/2)$, and the small-step stability criteria is max $(2c_s/\Delta x+N)\Delta\tau < 2$. The horizontal wavenumber generated by the finite-difference approximation to the advection operator is $(\sin k\Delta x)/\Delta x$, so the large time step is stable when $|U|\Delta t/\Delta x < 1.73$. Strang splitting,

$$\phi^{n+1} = [\mathscr{F}_2(2\Delta t/M)]^{(M/2)}\, \mathscr{F}_1(\Delta t)\, [\mathscr{F}_2(2\Delta t/M)]^{(M/2)}\, \phi^n,$$

is used in preference to (6.60) to preserve $O\left[(\Delta t)^2\right]$ accuracy in those cases where $\mathscr{F}_1$ and $\mathscr{F}_2$ don't commute.

In the first simulation $\Delta t = 12.5$ s, there are twenty small time steps per large time step, and $U = 10\,\mathrm{ms}^{-1}$ throughout the domain. In this case $(2c_s\Delta x+N)\Delta\tau = 1.76$ so the small time step is being integrated using time steps near the stability limit. The horizontal velocity field and the pressure field obtained from this simulation are plotted in Fig. 6.3. The velocity field is essentially identical to that obtained using the full compressible equations. Very small errors are detectable in the pressure field, but the overall accuracy of the solution is excellent.

Now consider a second simulation that is identical to the first in every respect except that the mean wind U increases linearly from 5 to $15\,\mathrm{ms}^{-1}$ between the bottom and the top of the domain. The pressure perturbations that develop in this simulation are shown in Fig. 6.4a, along with streamlines for the forcing function Ψ.

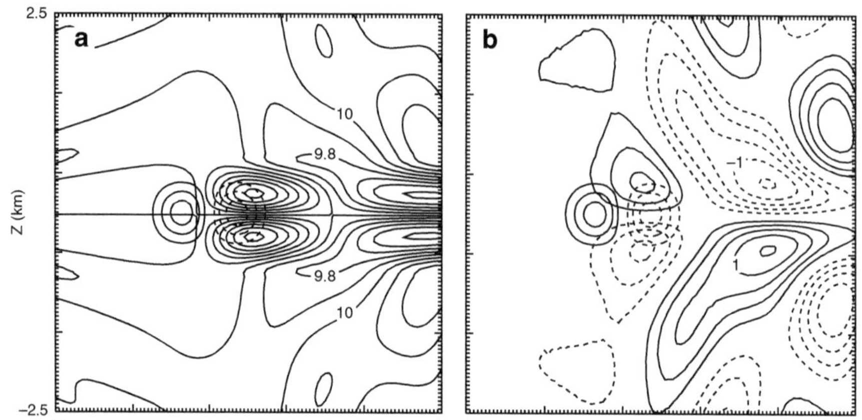

Fig. 6.3 (**a**) contours of $U+u$ at intervals of $0.1\,\mathrm{ms}^{-1}$ and Ψ at intervals of $0.1\,\mathrm{s}^{-1}$ at $t = 8000$ s. (**b**) as in (**a**) except that P is contoured at intervals of $0.25\,\mathrm{m}^2\mathrm{s}^{-2}$. No zero contour is shown for the P and Ψ fields. *Minor tick marks* indicate the location of the P points on the numerical grid. Only the central portion of the total computational domain is shown

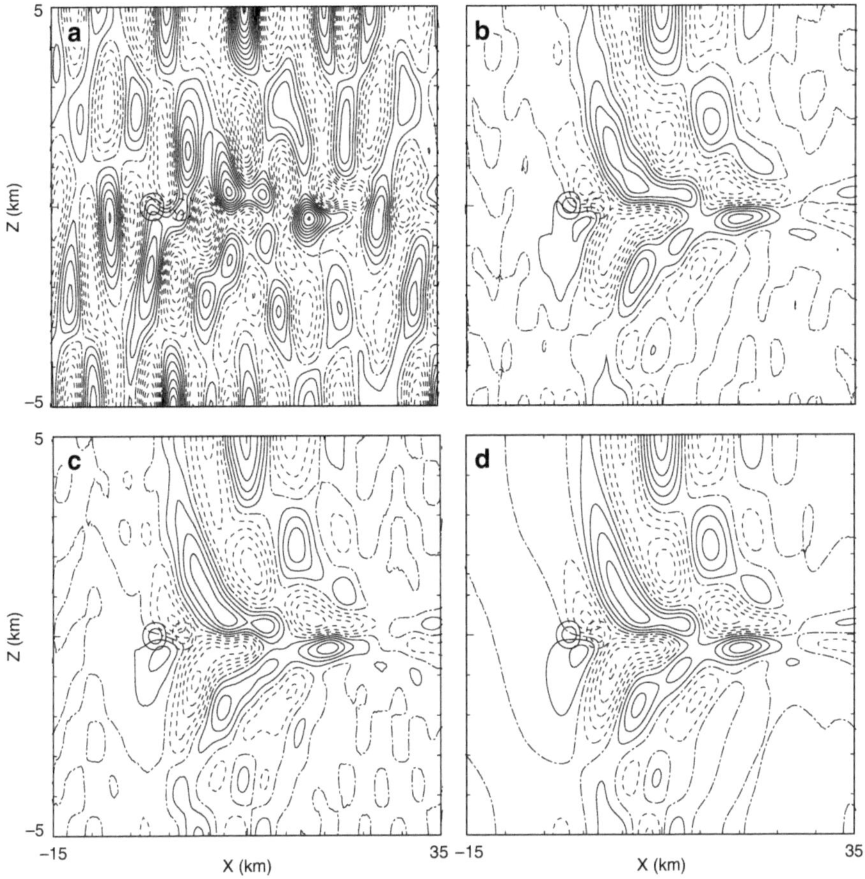

Fig. 6.4 Contours of P at intervals of $0.25\,\mathrm{m^2s^{-2}}$ (the zero contour is *dot-dashed*) and Ψ at intervals of $0.15\,\mathrm{s^{-1}}$ at $t = 3000$ s for the case with vertical shear in the mean wind and (**a**) $\Delta t = 12.5$ s, $M = 20$, (**b**) $\Delta t = 6.25$ s, $M = 20$, (**c**) $\Delta t = 6.25$ s, $M = 10$, (**d**) the solution is computed using the partial splitting method described in the next section with $\Delta t = 12.5$ s, $M = 20$. *Tick marks* appear every 20 grid intervals

Spurious pressure perturbations appear throughout the domain. The correct pressure field is shown in Fig. 6.4d, which was computed using a scheme that will be described in the next subsection. Although the pressure field in Fig. 6.4a is clearly in error, most of the spurious signal in the pressure field relates to sound waves whose velocity perturbations are very weak. The velocity fields associated with all the solutions shown in Fig. 6.4 are essentially identical. The extrema in the pressure perturbations shown in Fig. 6.4a are approximately twice those in the other panels and are growing very slowly suggesting that the solution is subject to a weak instability. Since the operators for each fractional step do not commute, the stability of each individual operator no longer guarantees the stability of the overall scheme.

Nevertheless, the fundamental problem with the completely split method seems to be one of inaccuracy arising from inadequate temporal resolution. Cutting Δt by a factor of 2, while leaving $M = 20$ so that $\Delta\tau$ is also reduced by a factor of 2, gives the pressure distribution shown in Fig. 6.4b, which is clearly a significant improvement over that obtained using the original time step, but still contains spurious perturbations of the same spatial scale shown in Fig. 6.4a. Similar results are obtained if both Δt and M are cut in half, as shown in Fig. 6.4c, which demonstrates that it is the decrease in Δt, rather than $\Delta\tau$, that is responsible for the improvement. Further discussion of the source of the error in the completely split method is provided in Sect. 6.5.

6.4.2 Partially-Split Operators

The first task involved in implementing the fractional-step methods discussed in the previous section is to identify those terms in the governing equations that need to be updated on a shorter time step. Having made this identification, it is possible to leave all the terms in the governing equations coupled together and to update those terms governing the slowly evolving processes less frequently than those terms responsible for the propagation of high frequency physically insignificant waves. This technique will be referred to as a partial splitting, since the individual fractional steps are never completely decoupled in the conventional manner given by (6.58) and (6.59).

Once again the linearized one-dimensional shallow-water system provides a simple context in which to illustrate partial splitting. As before, it is assumed that the gravity-wave phase speed is much larger than the velocity of the mean flow U. Klemp and Wilhelmson (1978) and (Tatsumi 1983) suggested a partial splitting in which the terms on the right side of the following

$$\frac{\partial u}{\partial t} + g\frac{\partial h}{\partial x} = -U\frac{\partial u}{\partial x}, \tag{6.73}$$

$$\frac{\partial h}{\partial t} + H\frac{\partial u}{\partial x} = -U\frac{\partial h}{\partial x}, \tag{6.74}$$

are updated as if the time derivative were being approximated using a leapfrog difference, but rather than advancing the solution from time level $t - \Delta t$ to $t + \Delta t$ in a single step of length $2\Delta t$, the solution is advanced through a series of $2M$ "small time steps." During each small time step the terms on the right side of (6.73) and (6.74) are held constant at their value at time level t and the remaining terms are updated using forward-backward differencing. Let m and n be time indices for the small and large time steps respectively and define $\Delta\tau = \Delta t/M$ as the length of a small time step. The solution is advanced from time level $n-1$ to $n+1$ in $2M$ small time steps of the form

$$\frac{u^{m+1} - u^m}{\Delta\tau} + g\frac{\partial h^m}{\partial x} = -U\frac{\partial u^n}{\partial x},$$

$$\frac{h^{m+1} - h^m}{\Delta\tau} + H\frac{\partial u^{m+1}}{\partial x} = -U\frac{\partial h^n}{\partial x}.$$

Note that the left sides of the preceding equations are identical to those appearing in the completely split scheme (6.65) and (6.66).

The complete small-step large-step integration cycle for this problem can be written as a four-dimensional linear system as follows. Define $\hat{u}^m = u^n$, $\hat{h}^m = h^n$, and let

$$\mathbf{r} = (u, h, \hat{u}, \hat{h})^T.$$

Then an individual small time step can be expressed in the form

$$\mathbf{r}^{m+1} = \mathbf{A}\mathbf{r}^m,$$

where

$$\mathbf{A} = \begin{pmatrix} 1 & -\tilde{g}\partial_x & -\tilde{U}\partial_x & 0 \\ -\tilde{H}\partial_x & 1 + \tilde{c}^2\partial_{xx}^2 & \tilde{U}\tilde{H}\partial_{xx}^2 & -\tilde{U}\partial_x \\ 0 & 0 & 1 & 0 \\ 0 & 0 & 0 & 1 \end{pmatrix},$$

and the tilde denotes multiplication of the parameter by $\Delta\tau$ (e.g., $\tilde{c} = c\Delta\tau$). At the beginning of the first small time step in an complete big-step, small-step integration cycle

$$\mathbf{r}^{m=1} = (u^{n-1}, h^{n-1}, u^n, h^n)^T.$$

At the end of the $2M$-th small step

$$\mathbf{r}^{m=2M} = (u^{n+1}, h^{n+1}, u^n, h^n)^T.$$

Thus, if $\mathbf{S}$ is a matrix interchanging the first pair and second pair of elements in $\mathbf{r}$,

$$\mathbf{S} = \begin{pmatrix} 0\,0\,1\,0 \\ 0\,0\,0\,1 \\ 1\,0\,0\,0 \\ 0\,1\,0\,0 \end{pmatrix},$$

the complete big-step, small-step integration cycle is given by

$$\mathbf{r}^{n+1} = \mathbf{S}\mathbf{A}^{2M}\mathbf{r}^n.$$

Since the individual operators commute, the completely split approximation to this problem is stable whenever both of the individual fractional steps are stable. One might hope that the stability of the partially-spilt method could also be guaranteed whenever the large- and small-step sub-problems are stable. Unfortunately, there are

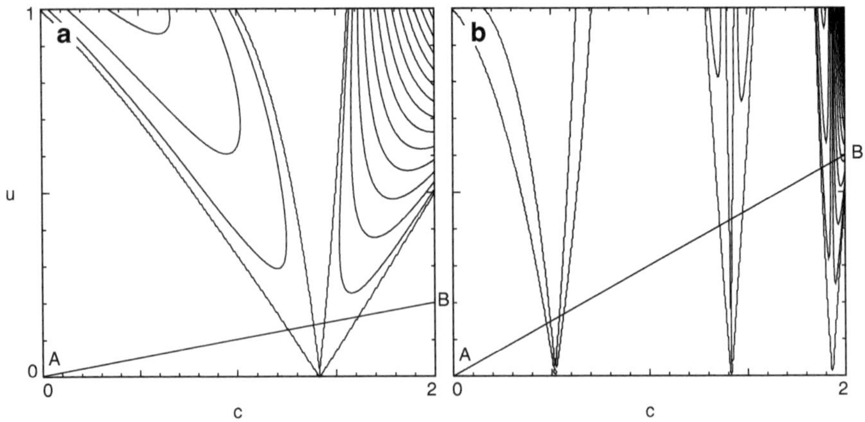

Fig. 6.5 Spectral radius of the amplification matrix for the partially-split method contoured as a function of $\hat{c}$ and $\hat{u}$ for (**a**) $M = 1$ (**b**) $M = 3$. Unstable regions are enclosed in the *wedged-shaped* areas. Contour intervals are 1.0 (*heavy line*), 1.2, 1.4, ... *Line* AB indicates the possible combinations of $\hat{c}$ and $\hat{u}$ that can be realized when $U/c = 1/10$ and M is specified as 1 or 3

many combinations of Δt and $\Delta\tau$ for which the partially-split method is unstable even though the sub-problems obtained by setting either U or c to zero are both stable (Tatsumi 1983; Skamarock and Klemp 1992). Suppose that the partially-split scheme is applied to an individual Fourier mode with horizontal wavenumber k, then the amplification matrix for an individual small time step is given by a matrix in which the partial derivative operators in $\mathbf{A}$ are replaced by ik; let this matrix be denoted $\hat{\mathbf{A}}$.

Consider the case $M = 1$ for which the amplification matrix is $\mathbf{S}\hat{\mathbf{A}}^2$. The magnitude of the maximum eigenvalue, or spectral radius ρ_{m}, of $\mathbf{S}\hat{\mathbf{A}}^2$ is plotted in Fig. 6.5a as a function of $\hat{c} = ck\Delta\tau$ and $\hat{u} = Uk\Delta t$. The domain over which ρ_{m} is contoured, $0 \leq \hat{c} \leq 2$ and $0 \leq \hat{u} \leq 1$, is that for which the individual small- and large-step problems are stable. When $M = 1$, ρ_{m} exceeds unity and the partially-split scheme is unstable throughout two regions of the $\hat{c}$-$\hat{u}$ plane whose boundaries intersect at $(\hat{c}, \hat{u}) = (\sqrt{2}, 0)$. If $U << c$, only a limited subset the $\hat{c}$-$\hat{u}$ plane shown in Fig. 6.5a is actually relevant to the solution of the shallow-water problem. Once the number of small time steps per large time step is fixed, the possible combinations of $\hat{u}$ and $\hat{c}$ will lie along a straight line of slope

$$\frac{\hat{u}}{\hat{c}} = \frac{U\Delta t}{c\Delta\tau} = M\frac{U}{c}.$$

Suppose that $U/c = 1/10$, then if the partial splitting method is used with $M = 1$, the only possible combinations of $\hat{u}$ and $\hat{c}$ are those lying along line AB in Fig. 6.5a. The maximum stable value of $\Delta\tau$ is determined by the intersection of the line AB and the left boundary of the leftmost region of instability. Thus, for $U/c = 1/10$ and $M = 1$, the stability requirement is that $\hat{c}$ be less than approximately 1.25.

As demonstrated in Fig. 6.5b, which shows contours of the spectral radius of $\mathbf{S}\tilde{\mathbf{A}}^6$, the restriction on the maximum stable time step becomes more severe as M increases to 3. The regions of instability are narrower and the strength of the instability in each unstable region is reduced, but additional regions of instability appear and the distance from the origin to the nearest region of instability decreases. When $M = 3$ and $U/c = 1/10$ the maximum stable value of $\hat{c}$ is roughly 0.48. Further reductions in the maximum stable value for $\hat{c}$ occur as M is increased, and as a consequence, the gain in computational efficiency that one might expect to achieve by increasing the number of small time steps per large time step is eliminated by a compensating decrease in the maximum stable value for $\Delta\tau$.

The partial splitting method has, nevertheless, been used extensively in many practical applications. The method has proved useful because in most applications it is very easy to remove these instabilities by using a filter. As noted by Tatsumi (1983) and Skamarock and Klemp (1992), the instability is efficiently removed by time filtering (Asselin 1972), which is often used in conjunction with leapfrog time differencing to prevent the divergence of the solution on the odd and even time steps. Other filtering techniques have also been suggested and will be discussed after considering a partial splitting approximation to the compressible Boussinesq system.

The equations evaluated each small time step in a partial splitting approximation to the two-dimensional compressible Boussinesq equations linearized about a basic-state flow with Brunt-Väisälä frequency N and horizontal velocity U are

$$\frac{u^{m+1} - u^m}{\Delta\tau} + \frac{\partial P^m}{\partial x} = -U\frac{\partial u^n}{\partial x} - w^n\frac{\partial U}{\partial z}, \tag{6.75}$$

$$\frac{w^{m+1} - w^m}{\Delta\tau} + \frac{\partial}{\partial z}\left(\frac{P^{m+1} + P^m}{2}\right) - b^m = -U\frac{\partial w^n}{\partial x}, \tag{6.76}$$

$$\frac{b^{m+1} - b^m}{\Delta\tau} + N^2 w^{m+1} = -U\frac{\partial b^n}{\partial x}, \tag{6.77}$$

$$\frac{P^{m+1} - P^m}{\Delta\tau} + c_s^2\frac{\partial u^{m+1}}{\partial x} + c_s^2\frac{\partial}{\partial z}\left(\frac{w^{m+1} + w^m}{2}\right) = -U\frac{\partial P^n}{\partial x}, \tag{6.78}$$

where as before m and n are the time indices associated with the small and large time steps. The left sides of these equations are identical to the small-time step equations in the completely split method (6.67)–(6.70). The right sides are updated every large time step.

This method is applied to the problem previously considered in connection with Fig. 6.3, in which fluid flows past a compact gravity-wave generator. The forcing from the wave generator appears in the horizontal and vertical momentum equations as per (6.71) and (6.72) and is updated on the small time step. In this test U is a constant $10\,\mathrm{ms}^{-1}$, $\Delta t = 12.5$ s and $\Delta\tau = 0.625$. The horizontal velocity field and the pressure field from this simulation are plotted in Fig. 6.6. The horizontal velocity field is very similar, though slightly noisier than that shown in Fig. 6.3a.

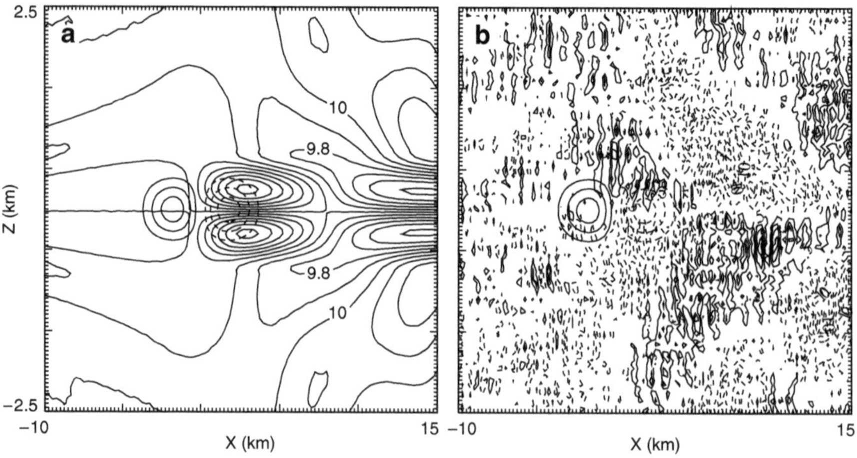

Fig. 6.6 (**a**) contours of $U + u$ at intervals of $0.1\ \mathrm{ms}^{-1}$ and Ψ at intervals of $0.1\ \mathrm{s}^{-1}$ at $t = 8000$ s. (**b**) as in (**a**) except that P is contoured at intervals of $0.5\ \mathrm{m}^2\mathrm{s}^{-2}$

The pressure field is, however, complete garbage. Indeed, it is surprising that errors of the magnitude shown in Fig. 6.6b can exist in the pressure field without seriously degrading the velocity field. These pressure perturbations are growing with time (the contour interval in Fig. 6.6b is twice that in Fig. 6.3b)); the velocity field eventually becomes very noisy, and the solution eventually blows up.

This instability can be prevented by applying an Asselin time filter (Asselin 1972) at the end of each big-step small-step integration cycle. Skamarock and Klemp (1992) have shown that filtering coefficients on the order of $\gamma = 0.1$ may be required to stabilize the partially-split solution to the one-dimensional shallow water system. A value of $\gamma = 0.1$ is sufficient to completely remove the noise in the pressure field and to eliminate the instability in the preceding test. Nevertheless, Asselin-filtering reduces the accuracy of the leapfrog scheme to $O(\Delta t)$ so it is best not to rely exclusively on the Asselin filter to stabilize the partially-split approximation. Other techniques for stabilizing the preceding partially-split approximation include divergence damping and forward biasing the trapezoidal integral of the vertical derivative terms (6.76) and (6.78). Forward biasing the trapezoidal integration is accomplished without additional computational effort by replacing those terms of the form $(\phi^{m+1} + \phi^m)/2$ with

$$\left(\frac{1+\epsilon}{2}\right)\phi^{m+1} + \left(\frac{1-\epsilon}{2}\right)\phi^m,$$

where $0 \le \epsilon \le 1$. A value of $\epsilon = 0.2$ provides an effective filter that does not noticeably modify the gravity waves (Durran and Klemp 1983).

Since trapezoidal time differencing is only used to approximate the vertical derivatives, forward-biasing those derivatives will not damp horizontally propagating sound waves. Skamarock and Klemp (1992) recommended including a "divergence damper" in the momentum equations such that the system of equations that is integrated on the small time step becomes

$$\frac{\partial u}{\partial t} + \frac{\partial P}{\partial x} - \alpha_x \frac{\partial \delta}{\partial x} = F_u,$$

$$\frac{\partial w}{\partial t} + \frac{\partial P}{\partial z} - b - \alpha_z \frac{\partial \delta}{\partial z} = F_w,$$

$$\frac{\partial b}{\partial t} + N^2 w = F_b,$$

$$\frac{\partial P}{\partial t} + c_2^2 \delta = F_p, \tag{6.79}$$

where

$$\delta = \frac{\partial u}{\partial x} + \frac{\partial w}{\partial z},$$

and F_u, F_w, F_b, and F_p represent the forcing terms that are updated every Δt. Damping coefficients of $\alpha_x = 0.001(\Delta x)^2/\Delta\tau$ and $\alpha_z = 0.001(\Delta z)^2/\Delta\tau$ removed all trace of noise and instability in the test problem shown in Fig. 6.6 without a supplemental Asselin-filter.

The role played by divergence damping in stabilizing the small-time-step integration in the partial splitting method can be appreciated by noting that if a single damping coefficient α is used in all components of the momentum equation, the divergence satisfies

$$\frac{\partial \delta}{\partial t} + \nabla^2 P - \alpha \nabla^2 \delta = G. \tag{6.80}$$

where $G = -\nabla \cdot (\mathbf{v} \cdot \nabla \mathbf{v}) + \partial b/\partial z$. Eliminating the pressure between (6.79) and (6.80), one obtains

$$\frac{\partial^2 \delta}{\partial t^2} - \alpha \nabla^2 \frac{\partial \delta}{\partial t} - c_s^2 \nabla^2 \delta = \frac{\partial G}{\partial t} - \nabla^2 F_p.$$

The forcing on the right side of this equation will tend to produce divergence in an initially non-divergent flow. Substituting a single Fourier mode into the homogeneous part of this equation, one obtains the classic equation for a damped harmonic oscillator

$$\frac{d^2 \tilde{\delta}}{dt^2} + \alpha \kappa^2 \frac{d\tilde{\delta}}{dt} + c_s^2 \kappa^2 \tilde{\delta} = 0, \tag{6.81}$$

where $\tilde{\delta}(t)$ is the amplitude and $\kappa = \sqrt{k^2 + \ell^2}$. The damping increases with wavenumber and is particularly effective in eliminating the high wavenumber modes at which the instability in the partial splitting method occurs. Gravity waves, on the other hand, are not significantly impacted by the divergence damper because the velocity field in internal gravity waves is almost non-divergent. Skamarock and Klemp (1992) have shown that divergence damping slightly reduces the amplitude of the gravity waves.

At this point it might appear that the partial splitting approach is inferior to the complete splitting method considered previously since filters are required to stabilize the partially-split approximation in situations where the completely split scheme performs quite nicely. Recall, however, that the completely split method does not generate usable solutions to the compressible Boussinesq equations when there is a vertical shear in the basic-state horizontal velocity impinging on the gravity-wave generator. The same filtering strategies that stabilize the partially-split method in the no-shear problem remain effective in the presence of vertical wind shear. This is demonstrated in Fig. 6.4d which shows the pressure perturbations in the test case with vertical shear as computed by the partially-split method using a divergence damper with the values of α_x and α_z given previously. Results similar to those in Fig. 6.4d may also be obtained using Asselin time filtering with $\alpha = 0.1$ in lieu of the divergence damper. The advantages of the partial splitting method are not connected with its performance in the simplest test cases, for which it can indeed be inferior to a completely split approximation, but in its adaptability to more complex problems.

One might inquire whether divergence damping can also be used to stabilize the completely-split approximation to the test case with vertical shear in the horizontal wind. The norm of the amplification matrix for the large-time-step third-order Runge-Kutta integration (6.62)–(6.64) is strictly less than unity for all sufficiently small Δt. Divergence damping makes the norm of the amplification matrix for the small time step strictly less than unity for all sufficiently small $\Delta\tau$ and thereby stabilizes the completely split scheme by guaranteeing that the norm of the amplification matrix for the overall scheme will be less than unity. Nevertheless, divergence damping only modestly improves the solution obtained with the completely split scheme; the pressure field remains very noisy and completely unacceptable.[7] The fundamental problem with the completely split method appears to be one of inaccuracy, not instability. This will be discussed further in the next section.

The linearly third-order Runge–Kutta scheme (6.62)–(6.64) can provide a simple accurate alternative to leapfrog time differencing for use on the large time step in partially split integrations (Wicker and Skamarock 2002), and it has replaced the leapfrog scheme in several operational codes. To clarify how (6.62)–(6.64) are modified for use as the large-time-step integrator in a partially split problem, let the small time step again be defined such that $\Delta\tau = \Delta t/M$, where M must now be a multiple of 6. Let ${}_1\mathbf{r}^m$ be the vector of unknowns at the start of the mth small time step during the first Runge–Kutta iteration, which is initialized by setting ${}_1\mathbf{r}^1 = \mathbf{r}^n$. The mth small time step of the this iteration has the form

$$ {}_1\mathbf{r}^{m+1} = {}_1\mathbf{r}^m + \Delta\tau\left(\mathscr{L}_1(\mathbf{r}^n) + \mathscr{L}_2({}_1\mathbf{r}^m, {}_1\mathbf{r}^{m+1})\right). \tag{6.82} $$

[7] One way to appreciate the difference in the effectiveness of divergence damping in the completely- and partially-split schemes is to note the difference in wavelength at which spurious pressure perturbations appear in each solution. The partially split scheme generates errors at much shorter wavelengths than those produced by the completely split method (compare Figs. 6.4a and 6.6b), and the short-wavelength features are removed more rapidly by the divergence damper.

As before, $\mathscr{L}_1$ and $\mathscr{L}_2$ contain the terms responsible for the low- and high-frequency forcing, respectively. After $M/3$ small time steps, the solution to (6.82) is projected forward to time $t^n + \Delta t/3$. The low-frequency forcing is then evaluated using this new estimated solution, and the second Runge–Kutta iteration is stepped forward from time t^n to $t^n + \Delta t/2$ in $M/2$ steps, beginning with ${}_2\mathbf{r}^1 = \mathbf{r}^n$. The mth small time step of this iteration is

$$ {}_2\mathbf{r}^{m+1} = {}_2\mathbf{r}^m + \Delta\tau \left(\mathscr{L}_1({}_1\mathbf{r}^{M/3+1}) + \mathscr{L}_2({}_2\mathbf{r}^m, {}_2\mathbf{r}^{m+1}) \right). $$

Following a similar update of the large-time-step forcing with the estimated solution at $t^n + \Delta t/2$, the mth small time step of the final Runge–Kutta iteration, which integrates from t^n to t^{n+1} in M steps, becomes

$$ {}_3\mathbf{r}^{m+1} = {}_3\mathbf{r}^m + \Delta\tau \left(\mathscr{L}_1({}_2\mathbf{r}^{M/2+1}) + \mathscr{L}_2({}_3\mathbf{r}^m, {}_3\mathbf{r}^{m+1}) \right), $$

where ${}_3\mathbf{r}^1 = \mathbf{r}^n$, and $\mathbf{r}^{n+1} = {}_3\mathbf{r}^{M+1}$. Several other alternatives to leapfrog-based partial splitting have also been recently been proposed (Gassmann 2005; Park and Lee 2009; Wicker 2009).

6.5 Summary Discussion

One way to compare the preceding methods for increasing efficiency when modeling fluids that support physically insignificant sound waves is to compare the way each approximation treats the velocity divergence. The pressure and the divergence in the compressible Boussinesq system satisfy

$$ \frac{\partial P}{\partial t} + c_s^2 \delta = F_p, \tag{6.83} $$

$$ \frac{\partial \delta}{\partial t} + \nabla^2 P = G, \tag{6.84} $$

where $\delta = \nabla \cdot \mathbf{v}$, $F_p = -\mathbf{v} \cdot \nabla P$ and $G = -\nabla \cdot (\mathbf{v} \cdot \nabla \mathbf{v}) + \partial b/\partial z$. The semi-implicit method approximates the left sides of the preceding equations with a stable trapezoidal time difference. Sound waves are artificially slowed when large time steps are used in this trapezoidal difference, but the gravity wave modes are accurately approximated. The implicit coupling in the trapezoidal difference leads to a Helmholtz equation for the pressure that must be solved at every time step.

The prognostic pressure equation (6.83) is discarded in the incompressible Boussinesq approximation and the local time derivative of the divergence is set to zero in (6.84). This leads to a Poisson equation for pressure that must be solved at every time step. The computational effort required to evaluate the pressure is similar to that required by the semi-implicit method. The Boussinesq system does, however,

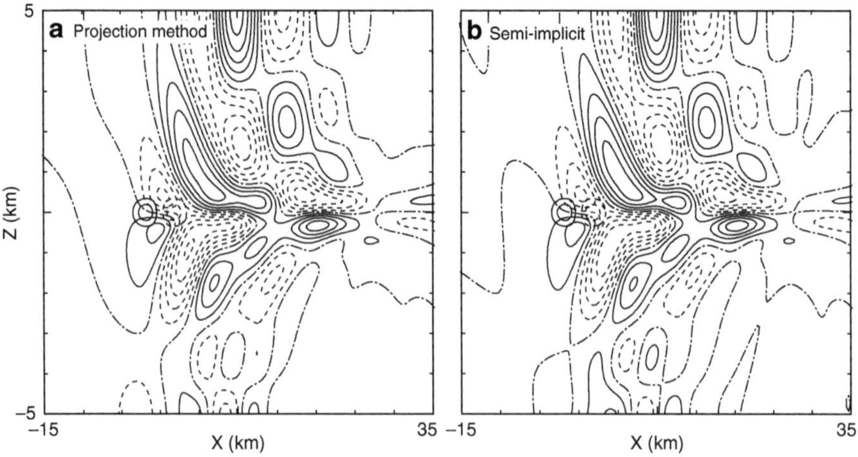

Fig. 6.7 As in Fig. 6.4: contours of P at intervals of $0.25\,\mathrm{m^2 s^{-2}}$ and Ψ at intervals of $0.15\ \mathrm{s^{-1}}$ at $t = 3000$ s. Solutions are obtained using (**a**) the Boussinesq projection method, (**b**) the semi-implicit method

have the advantage of allowing a wider choice of methods for the integration of the remaining oscillatory forcing terms, which are approximated using leapfrog differencing in the conventional semi-implicit method.

The pressure fields generated by the Boussinesq projection method and the semi-implicit method for the test problem (6.71)–(6.72) are compared in Fig. 6.7. As in Fig. 6.4 the basic-state horizontal flow is vertically sheared from $U = 5\,\mathrm{m\ s^{-1}}$ at the bottom to $U = 15\,\mathrm{m\ s^{-1}}$ at the top of the domain. In the projection method, the integral (6.9) is evaluated using the third-order Adams-Bashforth method with a time step of 10 s. The semi-implicit method is integrated using a 12.5 s time step. The pressure fields generated by both of these methods look very similar to that produced by the partially split method (Fig. 6.4d) and show no evidence of the noise produced using the completely split method (Figs. 6.4a–c).

The elliptic pressure equations that appear when using the semi-implicit or projection methods are most efficiently solved by sophisticated algorithms such as block-cyclic reduction, conjugate gradient, or multi-grid methods. One may think of the small-time-step procedure used in the fractional step methods as a sort of specialized iterative solver for the Helmholtz equation obtained using the conventional the semi-implicit method. The difference in the character of the solution obtained by the complete and the partial splitting methods can be appreciated by considering the behavior of the divergence during the small-time step integration.

During the small-time-step portion of the completely-split method the divergence satisfies

$$\frac{\partial^2 \delta}{\partial t^2} - c_s^2 \nabla^2 \delta = \frac{\partial^2 b}{\partial t \partial z}.$$

The initial conditions for δ are those at the beginning of each small-time-step cycle, and since divergence is typically generated by the operators evaluated on the large time step, the initial δ is non-zero. This divergence is propagated without loss during the small-time-step integration (except for minor modification by the buoyancy forcing) and tends to accumulate over a series of large-step, small-step cycles. The test in which the completely split scheme performs well is the case where the basic-state horizontal velocity is uniform throughout the fluid. When U is constant, the linearized advection operator merely produces a Galilean translation of the fluid that does not generate any divergence. (Recall that the forcing from the wave generator was computed on the small time step.) Nonlinear advection can, of course, generate divergence as can the linearized advection operator when there is vertical shear in the basic-state wind, and these are the circumstances in which the complete splitting method produces spurious sound waves.

In contrast, the divergence is almost zero at the start of the first small time step of the partially split method and only small changes in the divergence are forced during each individual small step. Moreover, the divergence forcing on each small time step closely approximates that which would appear in an explicit small-time-step integration of the full compressible equations *provided* that the amplitude of all the sound waves is negligible in comparison to slower modes. The divergence damper insures that the amplitude of the sound waves remains small and thereby preserves the stability and accuracy of the solution.

In summary, the projection, semi-implicit and partially split fractional step methods all provide viable ways to model atmospheric circulations in which sound waves are of no significance. Assuming that one wishes to capture nonhydrostatic motions, there does not appear to be a clear-cut best approach and the choice of method may be dictated by a number of additional considerations such as compatibility with larger-scale models, the complexity introduced by any proposed coordinate transformations, or the ease with which the method can be adapted to particular computer architectures. If the focus is on larger scales, in which the all circulations of interest are approximately hydrostatic, the semi-implicit method has generally been the method of choice. For example, the semi-implicit method, is frequently used to integrate the primitive equations in applications where the phenomena of primary interest are slow-moving Rossby waves. In such applications the numerical integration is stabilized with respect to two different types of physically insignificant, rapidly moving waves. Sound waves are filtered by the hydrostatic approximation, and the most rapidly moving gravity waves (and the horizontally propagating Lamb wave) are stabilized by the semi-implicit time integration. (cross reference primitive equations and Lamb wave)

References

Asselin, R. (1972). Frequency filter for time integrations. *Mon. Wea. Rev. 100*, 487–490.

Boyd, J. P. (1989). *Chebyshev and Fourier Spectral Methods*. Berlin: Springer-Verlag. 798 p.

Chorin, A. J. (1968). Numerical solution of the Navier-Stokes equations. *Math. Comp. 22*, 745–762.

Durran, D. R. (1999). *Numerical Methods for Wave Equations in Geophysical Fluid Dynamics.* New York: Springer-Verlag. 465 p.
Durran, D. R. (2008). A physically motivated approach for filtering acoustic waves from the equations governing compressible stratified flow. *J. Fluid Mech. 601*, 365–379.
Durran, D. R. and A. Arakawa (2007). Generalizing the boussinesq approximation to stratified compressible flow. *Comptes Rendus Mécanique 335*, 655–664.
Durran, D. R. and J. B. Klemp (1983). A compressible model for the simulation of moist mountain waves. *Mon. Wea. Rev. 111*, 2341–2361.
Gassmann, A. (2005). An improved two-time-level split-explicit integration scheme for non-hydrostatic compressible models. *Meteorol. Atmos. Phys. 88*, 23–38.
Golub, G. H. and C. F. van Loan (1996). *Matrix Computations* (Third ed.). Baltimore: John-Hopkins University Press. 694 p.
Karniadakis, G. E., M. Israeli, and S. Orszag (1991). High-order splitting methods for the incompressible Navier-Stokes equations. *J. Comp. Phys. 97*, 414–443.
Klemp, J. B. and R. Wilhelmson (1978). The simulation of three-dimensional convective storm dynamics. *J. Atmos. Sci. 35*, 1070–1096.
Kwizak, M. and A. J. Robert (1971). A semi-implicit scheme for grid point atmospheric models of the primitive equations. *Mon. Wea. Rev. 99*, 32–36.
Lipps, F. and R. Hemler (1982). A scale analysis of deep moist convection and some related numerical calculations. *J. Atmos. Sci. 29*, 2192–2210.
Ogura, Y. and N. Phillips (1962). Scale analysis for deep and shallow convection in the atmosphere. *J. Atmos. Sci. 19*, 173–179.
Orszag, S. A., M. Israeli, and M. O. Deville (1986). Boundary conditions for incompressible flows. *J. Scientific Comp. 1*, 75–111.
Park, S.-H. and T.-Y. Lee (2009). High-order time-integration schemes with explicit time- splitting methods. *Mon. Wea. Rev. 137*, 4047–4060.
Skamarock, W. C. and J. B. Klemp (1992). The stability of time-split numerical methods for the hydrostatic and nonhydrostatic elastic equations. *Mon. Wea. Rev. 120*, 2109–2127.
Tapp, M. C. and P. W. White (1976). A non-hydrostatic mesoscale model. *Quart. J. Roy. Meteor. Soc. 102*, 277–296.
Tatsumi, Y. (1983). An economical explicit time integration scheme for a primitive model. *J. Meteor. Soc. Japan 61*, 269–287.
Témam, R. (1969). Sur l'approximation de la solution des équations Navier-Stokes par la méthode des pas fractionnaires. *Archiv. Ration. Mech. Anal. 33*, 377–385.
Wicker, L. J. (2009). A two-step Adams–Bashforth–Moulton split-explicit integrator for compressible atmospheric models. *Mon. Wea. Rev. 137*, 3588–3595.
Wicker, L. J. and W. C. Skamarock (2002). Time-splitting methods for elastic models using forward time schemes. *Mon. Wea. Rev. 130*, 2088–2097.
Williams, P. D. (2009). A proposed modification to the Robert–Asselin time filter. *Mon. Wea. Rev. 137*, 2538–2546.

Part II
Conservation Laws, Finite-Volume Methods, Remapping Techniques and Spherical Grids

Chapter 7
Momentum, Vorticity and Transport: Considerations in the Design of a Finite-Volume Dynamical Core

Todd D. Ringler

Abstract This chapter provides an end-to-end discussion of issues related to the design and construction of dynamical cores. The governing equations of motion are derived from basic principles cast in the Lagrangian frame of motion. The Reynolds Transport Theorem is derived so that these conservation statements can be recast in their weak, integral form in the Eulerian reference frame. Special attention is given to the relationship between the momentum equation and vorticity dynamics. The principles of conservation of circulation and vorticity are derived in the continuous system. It is demonstrated that the kinematic principles related to circulation and vorticity can be carried over exactly into the discrete system. The analysis is conducted in an idealized, two-dimensional setting that is meant to serve as a prototype system for the consideration of the full three-dimensional general circulation of the atmosphere and ocean.

7.1 Introduction

More than 40 years after the first global models for the simulation of the fluid motion in the atmosphere and ocean appeared, research into the construction of atmosphere and ocean "dynamical cores" has never been more vibrant. The dynamical core refers to the fluid-dynamic core of an atmosphere or ocean general circulation model; the part of the model that evolves the distribution of mass, momentum and tracer constituents forward in time. The diversity of approaches that are being explored to simulate the evolution of mass, momentum and tracers in the atmosphere and ocean systems points to both the richness and complexity of the problem.

The motivation for this chapter is to present an"end-to-end" view in the design of numerical models used for the simulation of fluid motion in the atmosphere and ocean. The process starts with a rigorous construction and description of the

T.D. Ringler
Theoretical Division, Los Alamos National Laboratory, Los Alamos, NM 87545, USA
e-mail: ringler@lanl.gov

P.H. Lauritzen et al. (eds.), *Numerical Techniques for Global Atmospheric Models*, Lecture Notes in Computational Science and Engineering 80,
DOI 10.1007/978-3-642-11640-7_7,

underlying continuous system. The process ends with the specification of a numerical model that is suitable for its target application. Both the beginning and end are essentially applied math activities, with the former manipulating continuous equations and the latter manipulating discrete equations. In between these ends is the "art" of constructing dynamical cores. If the process were as simple as discretizing a set of continuous equations, we would not see the vibrancy in dynamical core development that we see today. A host of subtle, yet profound, questions such as "which form of a continuous equations should be the starting point for the discrete model?" fall squarely in the middle of the end-to-end design process. This chapter explores some of those questions in order to illuminate the intricacies of the decisions that have to be made in the design process.

The price-to-be-paid for this end-to-end view is scope. Many relevant aspects of the design process have been omitted in order to contain the discussion to an appropriate length. The discussion is focused primarily on one important component of a dynamical core: the prediction of momentum. This proves to be an important and rich topic for several reasons. First, since the velocity that is derived from momentum acts as the transport velocity for the mass and tracers fields, a robust simulation of velocity is a prerequisite for any viable dynamical core. Furthermore, as the velocity field responds to changes in the applied forces it must also satisfy certain kinematic conditions, such as conservation of circulation and absolute vorticity. Satisfying the desire to accurately model $\mathbf{F} = m\,\mathbf{a}$, where $\mathbf{F}$ is the vector force, m symbolizes the mass and $\mathbf{a}$ stands for the vector acceleration, while also accommodating important kinematic constraints is a challenge for any numerical model. And finally, the majority of the nonlinearity in dynamical core simulations arises from the simulation of the evolving velocity field. In many ways, getting the evolution of momentum "right" is the hardest part in the design and construction of a dynamical core.

The analysis presented below is conducted in a very simple, two-dimensional framework and is, in some ways, quite removed from the global three-dimensional motions that compose the atmosphere and ocean general circulations. As such, it is important to address the relevance of this chapter to the modeling of the more complicated three-dimensional systems. First and foremost, the analysis conducted here is a prerequisite for the construction of a robust three-dimensional model. In that, what follows below could be considered a set of necessary, but not sufficient, properties of robust three-dimensional models of atmosphere and ocean circulations. Since the general circulation of the atmosphere and ocean occurs primarily along a vertical stack of two-dimensional sheets, it is folly to suppose that a numerical method that performs poorly in the solution of the two-dimensional system will perform acceptably in the solution of the three-dimensional system. Second, while the two-dimensional system might seem trivial in some respects, many numerical methods used in the modeling of geophysical fluid dynamics fall short when viewed from the perspective of vorticity dynamics. Vorticity dynamics largely represent the "slow modes" of these system where relatively small truncation errors can accumulate and, eventually, completely corrupt the simulation. The struggle to control the form of truncation error with respect to vorticity dynamics is as important today as it was when Arakawa (1966) wrote the seminal paper on the topic (see also the reprint Arakawa 1997). And finally, this chapter is meant as an introduction to the concept of designing numerical methods that respect the continuous system in some relevant aspects. For this goal, the very simple, two-dimensional framework is perfectly appropriate.

The omissions in the discussion are sometimes glaring. For example, the importance of accurately simulating transport phenomena in dynamical cores is largely omitted (see, e.g. the discussion in Chap. 8). The notable exception is the detailed discussion on the relationship between fluid acceleration and absolute vorticity transport. The next glaring omission is the lack of discussion of potential vorticity and its relationship to the velocity field; the discussion below is limited to an analysis of the absolute vorticity field. While absolute vorticity is connected only to the velocity field, potential vorticity is connected to both the velocity field and to the mass field. The analysis below can (and has) been extended from absolute vorticity to potential vorticity (Ringler et al. 2010). The choice was made based on the belief that a firm grasp of the absolute vorticity dynamics is a prerequisite to understanding the potential vorticity dynamics. And finally, while the primary target geometry of a dynamical core is the surface of the sphere, the *f-plane* approximation is made throughout. All of the analysis carries over to the sphere, the simplification to the *f-plane* is for the sake of conciseness in presentation. And finally, while the focus is on the relationship between the evolution of velocity and its relationship to vorticity dynamics, we need to be sure to understand that the velocity equation is derived from $\mathbf{F} = m\,\mathbf{a}$ and that the system cannot be closed without the knowledge of the density field and an equation, such as the ideal gas law, that relates density to pressure.

The discussion unfolds in the following manner. First, the relevant evolution equations are constructed from the Lagrangian perspective. These conservation statements are then transferred to an Eulerian reference frame through the use of the Reynolds Transport Theorem (RTT). Since a full discussion of RTT is rarely found in texts related to geophysical fluid dynamics, RTT is derived from first principles for completeness. Following the development of the evolution equations appropriate for an Eulerian reference frame, a qualitative analysis is conducted of the various "flavors" of the momentum equation that can be used as the basis for a numerical solution. The discussion then moves into the setting of discrete numerics by asking the most basic question of "How do we begin the process of discretization?" And finally, a numerical model is developed that meets the criteria developed throughout the entire discussion. The numerical model is constructed in such a way that it can easily be implemented in development environments such as MATLAB.

7.2 Reference Frames and Conceptual Constructs

When we consider the motions of the atmosphere or ocean, we expend considerable effort on the phenomena of transport, such as the transport of fluid density from one region to another, or the transport of tracer substance from a source region to a sink region, or the transport of momentum from one area to another. In almost all cases, the most natural setting to consider transport is the *Lagrangian* reference frame where we, as the observer, move with the fluid.

To start, let us define a volume of fluid, V, composed of a set of particles, $\mathbf{R}$, enclosed at all times by a surface, S. Each particle in the set of $\mathbf{R}$ is denoted by its vector position $\mathbf{X}(t) = X_1\mathbf{e}_1 + X_2\mathbf{e}_2 + X_3\mathbf{e}_3$. As indicated, $\mathbf{X}$ is only a function of time. Also, $\mathbf{e}_{1,2,3}$ is the set of orthogonal unit vectors spanning the $\mathbb{R}^3$ space (see Fig. 7.1). The idea of constructing the volume as a set of particles is entirely a conceptual construct; the particles are simply the most basic "element" that is used to define all other features; lines, surfaces and volumes can be "built" from sets of particles. Each particle is accompanied by an arbitrarily long list of labels representing such things as the particle position ($\mathbf{X}$), density (ρ) and the vector velocity ($\mathbf{u}$). The validity of such an approach is that the particles can be made arbitrarily small and, thus, approach the continuum limit.

The amount of mass, M, or tracer substance, Q, within the boundary surface can be expressed as

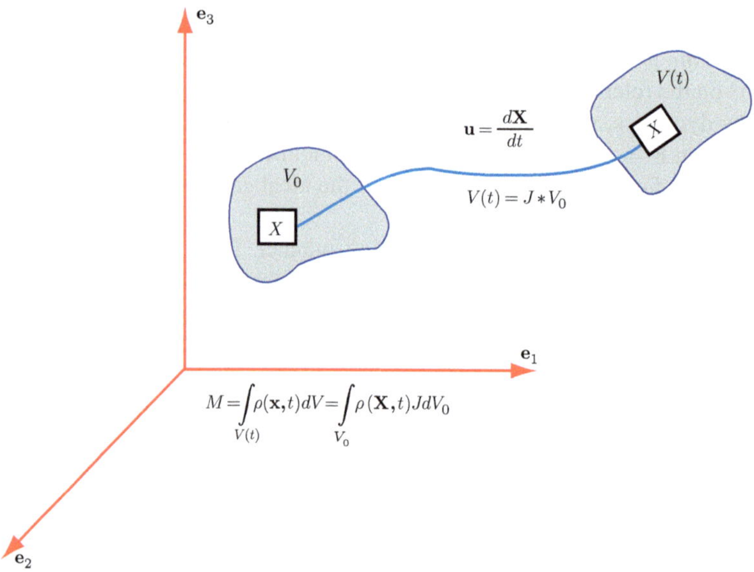

Fig. 7.1 The Lagrangian perspective. At *time* = 0 a volume of fluid, V_0, is identified. The volume is composed of a set of particles, $\mathbf{R}$, with each particle identified by its vector position $\mathbf{X}$. Even though the volume is sheared, rotated and dilated as it moves through space, it is always composed of the same set of particles $\mathbf{R}$. Thus, the boundary surrounding V is impermeable. The Jacobian, J, integrates the time-rate-of-change of V and represents the fractional change in the volume between *time* = 0 and *time* = t. The volume of fluid at any time t is equal to its volume at some initial time, V_0, times the fractional change in volume, J. Since the boundary of V is impermeable, the mass, M, within V is a constant in time

$$M = \int_{V(t)} \rho(\mathbf{x}, t)\, dV \tag{7.1}$$

$$Q = \int_{V(t)} \rho(\mathbf{x}, t)\, q(\mathbf{x}, t)\, dV \tag{7.2}$$

where the limits of integration span the positions $\mathbf{x}$ inside the volume $V(t)$. The dependence of V on time is retained to make clear that the limits of integration, in general, change in time. ρ is the fluid density with units of $mass$ per $volume$ and q has units of concentration, such as kg of Q per kg of fluid.

Assume that no mass or tracer substance is exchanged across the boundary S such that

$$\frac{dM}{dt} = 0 \tag{7.3}$$

and

$$\frac{dQ}{dt} = 0. \tag{7.4}$$

Equations (7.3) and (7.4) define the material derivative as measured in the Lagrangian reference frame of motion by stating that the amount of M and Q is invariant in time when following a volume $V(t)$ that is always composed of the same set of particles included in $\mathbf{R}$.

Another reference frame of great utility is the *Eulerian* reference frame where the observer remains at a fixed position in space, as opposed to moving in space along particle trajectories. The material derivative (of, say, Q) is expressed in the Eulerian reference frame as

$$\left.\frac{dQ}{dt}\right|_{fluid\ particle} \equiv \frac{DQ}{Dt} = \frac{\partial Q}{\partial t} + \mathbf{u} \cdot \nabla Q \tag{7.5}$$

where, as shown in Fig. 7.1, $\mathbf{u}$ is the particle velocity vector defined as

$$\mathbf{u} = \frac{d\mathbf{X}}{dt}. \tag{7.6}$$

The gradient in (7.5) is defined as

$$\nabla Q = \frac{\partial Q}{\partial x_1}\mathbf{e}_1 + \frac{\partial Q}{\partial x_2}\mathbf{e}_2 + \frac{\partial Q}{\partial x_3}\mathbf{e}_3. \tag{7.7}$$

The terms on the right-hand side (RHS) of (7.5) are evaluated at a fixed point and at a fixed time, respectively. Even when the material derivative is identically zero, a non-zero time-rate of change, $\frac{\partial\{\}}{\partial t}$, can be observed at a fixed location due to the differential transport, $\mathbf{u} \cdot \nabla\{\}$, into and out-of a specific region. An Eulerian observer essentially balances $\frac{dM}{dt} = 0$ by measuring the differential transport at one location, then setting the local time tendency to the value required to make the material derivative sum to zero.

The blending of the Lagrangian and Eulerian reference frames through the use of Arbitrary Lagrangian Eulerian (ALE) (Hirt et al. 1997) methods is increasingly popular in climate system modeling. While the full discussion of ALE methods is beyond the scope of this chapter, the analysis of the continuous system given in the following section can be extended to ALE frameworks.

7.3 Evolution Equations from a Lagrangian Perspective

The elegance and simplicity of the Lagragrian reference frame is clearly apparent in (7.1)–(7.4). In a model of the global atmosphere or ocean we could envision decomposing the domain into a set of Lagrangian volumes where each volume is separated by an invisible, yet impermeable, barrier. The numerical algorithms would then track the "blobs" as they move through space being pushed, squeezed and rotated due to their contact with neighboring blobs. In such a model the phenomena of transport would be remarkably well simulated; no mass or tracer substance would be erroneously exchanged between the Lagrangian volumes. In fact, ideas along these lines are under development by Haertel et al. (2009).

The primary reason that no robust climate model is constructed entirely in a Lagrangian reference frame is due to the rapid deformation of the Lagrangian control volumes. As seen in Fig. 7.1, while the mass within the volume V is constant in time, the volume itself can evolve in time through rotation, dilation and shearing. Figure 7.2 demonstrates what happens to control volumes in typical geophysical flows. Initially compact control volumes are stretched due to shearing. The stretching creates long filaments that are folded. Tracking these rapidly distorting control volumes poses a tremendous challenge for numerical models.

So while the Lagrangrian reference frame proves exceptionally useful for the construction of the evolution equations, numerical models are currently restricted to reference frames that are essentially Eulerian. As a result, we require a robust means of transforming conservation laws and evolution equations between the Lagrangian and Eulerian frames of motion. While several methodologies are available for

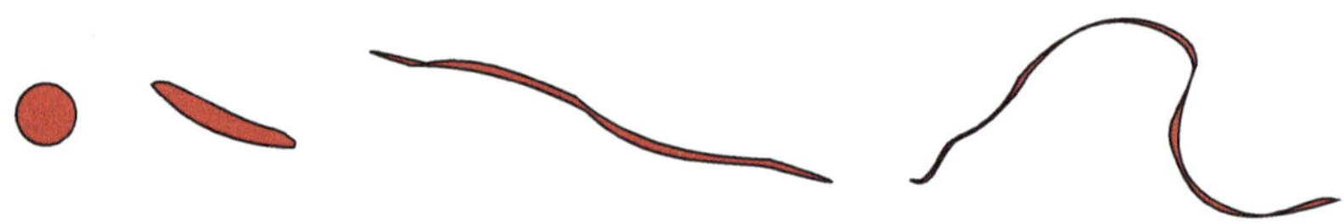

Fig. 7.2 In the highly nonlinear flows that characterize fluid motion in the atmosphere and ocean, Lagrangian control volumes are rapidly distorted due the presence of strong shear, rotation and dilation. The rapid distortion of Lagrangian control volumes makes the formulation of numerical models within the Lagrangian reference frame an extremely difficult challenge

transforming between these reference frames, an approach based on the Reynold's Transport Theorem (RTT) is particularly appealing for two reasons. First, the RTT is formulated in an integral form that leads naturally to equations suitable to finite-volume models that will be discussed in Sects. 7.5 and 7.6. Second, a generalization of the RTT allows for the seamless transformation between the Lagrangian reference frame and any other reference frame that falls between the Lagragian (moving) and Eulerian (fixed) reference frame. Thus, the emerging type of models based on ALE methods are fully accommodated in approaches based on the RTT; this chapter serves as a useful waypoint on the path to developing numerical models in the ALE reference frame. A full analysis of RTT and its generalizations can be found in F. White's Fluid Mechanics textbook (White 2008).

7.3.1 The Reynolds Transport Theorem

Let F be any intensive property of the fluid. Examples of F include ρ with units of mass per unit volume, ρq with units of tracer mass per unit volume or $\rho\,\mathbf{u}$ with units of momentum per unit volume. The conservation statement for F in the Lagrangian reference frame in the absence of sources and sinks is expressed as

$$\frac{d}{dt}\left[\int\limits_{V_L} F(\mathbf{x},t)\,dV\right] = 0. \tag{7.8}$$

Note that (7.3) is included as a specific example of (7.8). In general the RHS of (7.8) need not be zero. A source term for F can be placed on the RHS of (7.8). The proper evaluation of this source term is along the volume trajectory.

The subscript L on the volume V in (7.8) has been added to denote that the volume is being viewed by an observer moving in the Lagrangian reference frame. The goal is to move the time derivative inside the volume integral and, thereby, allow for the integration to occur over the same volume V but with respect to an observer in a different reference frame. This is somewhat problematic since the limits of integration, V_L, are a function of time.

The way around this difficulty is to make use of the fact that the volume V_L is composed of the same set of particles $\mathbf{R}$ at every instant in time. Thus, as shown in Fig. 7.1, the differential volume element dV at some time t is related to its value at time $t = 0$ as

$$dV = J\,dV_0 \tag{7.9}$$

where J accounts for the fractional change in the volume element between time 0 and time t. Conceptually we can consider each of these differential fluid elements dV_0 as being associated with a single particle. Thus, (7.8) can be transformed to

$$\frac{d}{dt}\left[\int_{V_L} F(\mathbf{x},t)\,dV\right] = \frac{d}{dt}\left[\int_{V_0} F(\mathbf{X},t)\,J\,dV_0\right] = 0. \tag{7.10}$$

Note that both sides of (7.10) integrate over the same group of particles **R**, but do so in different ways. The LHS indirectly sums over the particles by integrating over V_L, which is identical to the spatial extent spanned by **R** at time t. The RHS explicitly sums over the particle positions **X** at time t included in V_L and weights each particle by its initial volume, V_0, times the fraction change in V_0 between $time = 0$ and $time = t$. Now that the limits of integration on the RHS are not a function of time, the order of integration and differentiation can be exchanged. In particular, we can write

$$\frac{d}{dt}\int_{V_0} F(\mathbf{X},t)\,J\,dV_0 = \int_{V_0}\left[J\frac{D}{Dt}F(\mathbf{X},t) + F(\mathbf{X},t)\frac{D}{Dt}J\right]dV_0 = 0. \tag{7.11}$$

Just as J accounts for the time-integrated fractional change in the size of the volume elements, $\frac{DJ}{Dt}$ accounts for the instantaneous rate-of-change in the size of the volume elements, namely

$$\frac{DJ}{Dt} = J\ \nabla\cdot\mathbf{u}. \tag{7.12}$$

Equation (7.12) states that the rate-of-change of a Lagrangian volume (JV_0) is equal to its present volume ($J\ V_0$) times the divergence of the fluid; since V_0 is not a function of time it cancels in (7.11). Using (7.12) we can simplify (7.11) to

$$\int_{V_0}\left[\frac{D}{Dt}F(\mathbf{X},t) + F(\mathbf{X},t)\nabla\cdot\mathbf{u}\right]dV_0 = 0. \tag{7.13}$$

We can expand the first term in (7.13) using the definition of the material derivative (7.5) and combine terms to obtain

$$\int_{V_0}\left[\frac{DF}{Dt} + F\nabla\cdot\mathbf{u}\right]dV_0 = \int_{V_0}\left[\frac{\partial F}{\partial t} + \nabla\cdot(F\,\mathbf{u})\right]dV_0 = 0. \tag{7.14}$$

The broad utility and analytic power of (7.14) is in the choice of V_0. Note that the only requirements on V_0 are the following: V_0 is coincident with V_L at some instant in time and V_0 is fixed in space. Of particular interest is when V_L and V_0 *span the same volume of space* at the instant $time = 0$. At this instant in time, we can see that V_0 is the Eulerian representation of V_L, in that it spans the same volume but is not moving with the fluid. The volumes V_0 and V_L only differ in the reference frame of the observer, with the former in the Eulerian reference frame and the latter in the Lagrangian reference frame. Relabeling V_0 as V_E to emphasis this point we can now write

$$\frac{d}{dt}\left[\int_{V_L} F(x,t)\,dV\right] = \int_{V_E}\left[\frac{\partial F}{\partial t} + \nabla\cdot(F\,\mathbf{u})\right]dV$$

$$= \int_{V_E}\left[\frac{DF}{Dt} + F\nabla\cdot\mathbf{u}\right]dV = 0. \tag{7.15}$$

Note that the Eulerian volume, V_E, is often referred to as a "control volume" when discussed in the context of finite-volume methods

Equation (7.15) is the Reynolds Transport Theorem (RTT). The term "Reynolds Transport Theorem" is most commonly used when the volume V_L is transported with the fluid, as is the case for the first term in (7.15). When the volume is not being observed in the Lagrangian reference frame, a generalization of RTT still holds and that theorem is commonly referred to as the "Generalized Transport Theorem". The only way to satisfy (7.15) for any V_E is to guarantee that

$$\frac{\partial F}{\partial t} + \nabla\cdot(F\,\mathbf{u}) = 0. \tag{7.16}$$

A more useful form of (7.15) is obtained by applying the divergence theorem to the $\nabla\cdot(F\,\mathbf{u})$ term to yield

$$\frac{d}{dt}\left[\int_{V_L} F(\mathbf{x},t)\,dV\right] = \int_{V_E}\frac{\partial F}{\partial t}dV + \int_{S_E} F\;\mathbf{u}\cdot\mathbf{n}\,dS = 0 \tag{7.17}$$

where S_E is the surface bounding V_E and $\mathbf{n}$ is the unit vector normal to S_E directed outward. The RTT states that the time-rate-of-change of any intensive quantity F inside a volume V_L following the fluid motion can be computed at any instant in time as the sum of the time-rate-of-change of F inside V_E and the net flux of F across the surface bounding V_E (see Fig. 7.3). The RTT allows for conservation statements to be naturally cast in an integral form as shown in (7.17). The *integral form* is also referred to as the *weak form* since, in general, the statements hold only for a compact region of integration. With the machinery of the RTT in place, we can easily apply it to any conservation statement to obtain an analytic expression of the dynamical core expressed in integral form.

7.3.2 Conservation of Mass and Tracer Substance

Applying (7.17) to the conservation of mass and tracer expressions in (7.3) and (7.4), we obtain

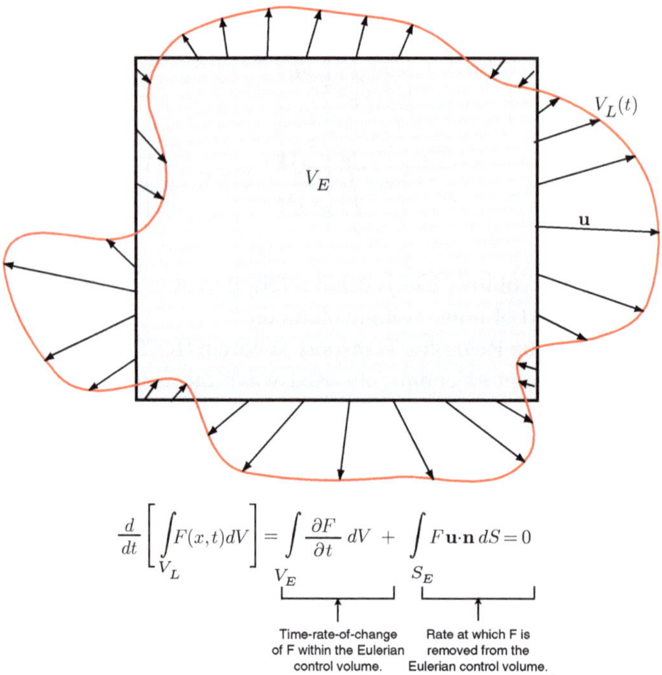

Fig. 7.3 An illustration of the Reynolds Transport Theorem. At some time $t = 0$, the volume V_L is coincident with the volume V_E. The Eulerian volume V_E remains fixed in place while the Lagrangian volume V_L deforms to $V_L(t)$ at time t. The conservation statement for F is that the integral of FdV over V_L is constant for all time. The Reynolds Transport Theorem allows for the computation of the time-rate-of-change for F within V_E by computing the transport of F across the surface of V_E over time t

$$\frac{d}{dt}\left[\int_{V_L} \rho dV\right] = \int_{V_E} \frac{\partial\rho}{\partial t} dV + \int_{S_E} \rho\,\mathbf{u}\cdot\mathbf{n}\, dS = 0 \tag{7.18}$$

$$\frac{d}{dt}\left[\int_{V_L} \rho q dV\right] = \int_{V_E} \frac{\partial(\rho q)}{\partial t} dV + \int_{S_E} \rho q\,\mathbf{u}\cdot\mathbf{n}\, dS = 0. \tag{7.19}$$

Equations (7.18) and (7.19) are inextricably coupled and a discussion of the coupling is worthy of its own chapter. A glimpse at this entanglement can be seen by simply defining $G_m = \rho\,\mathbf{u}\cdot\mathbf{n}$ and rewriting (7.18) and (7.19) as

$$\frac{d}{dt}\left[\int_{V_L} \rho dV\right] = \int_{V_E} \frac{\partial\rho}{\partial t} dV + \int_{S_E} G_m\, dS = 0 \tag{7.20}$$

$$\frac{d}{dt}\left[\int_{V_L} \rho q dV\right] = \int_{V_E} \frac{\partial(\rho q)}{\partial t} dV + \int_{S_E} q\, G_m\, dS = 0. \tag{7.21}$$

G_m is the mass flux per unit area across S_E. Equation (7.21) shows that a prerequisite to computing the tracer flux across S_E is the knowledge of the mass flux G_m. In fact, when written in this manner it is clear that tracer transport is meaningless without the underlying mass transport field G_m. Those transport algorithms that fully recognize the relationship between mass and tracer transport are most appropriate for use in climate simulations.

Differential forms of mass and tracer transport can be obtained directly from (7.16) or by letting $V_E \to 0$ in (7.18) and (7.19) to obtain

$$\frac{\partial \rho}{\partial t} + \nabla \cdot (\rho\, \mathbf{u}) = 0 \tag{7.22}$$

and

$$\frac{\partial(\rho q)}{\partial t} + \nabla \cdot (\rho q\, \mathbf{u}) = 0. \tag{7.23}$$

Equations (7.22) and (7.23) can be written in material derivative form as

$$\frac{D\rho}{Dt} + \rho \nabla \cdot \mathbf{u} = 0 \tag{7.24}$$

and

$$\frac{Dq}{Dt} = 0 \tag{7.25}$$

The last two forms will be used in the discussion below.

7.3.3 A Statement of Newton's Second Law

In order to complete the Lagrangian perspective illustrated in Fig. 7.1, we need to describe how the volume evolves in time, i.e. what determines the set of particle velocities $\mathbf{u}$ that will dilate, rotate and shear the volume V_L shown in Fig. 7.1? In this case the intensive quantity is momentum per unit volume

$$\mathbf{P} = \rho\, \mathbf{u}\,. \tag{7.26}$$

In its most basic form, the statement of Newton's Second Law is

$$\frac{d\mathbf{P}}{dt} = \frac{d}{dt}\left[\int_{V_L} \mathbf{P}(\mathbf{x}, t)\, dV\right] = \int_{V_L} \mathbf{F}_b\, dV + \int_{S_L} \mathbf{F}_s\, dS \tag{7.27}$$

where $\mathbf{F}_b$ is a *body force* acting throughout the volume V_L and $\mathbf{F}_s$ is a *surface force* acting on the surface S_L. $\mathbf{F}_b$ has units of force per unit volume and $\mathbf{F}_s$ has units of force per unit area. Applying RTT as expressed in (7.13)–(7.27) yields

$$\int\limits_{V_E} \left[\frac{D}{Dt}(\rho\,\mathbf{u}) + (\rho\,\mathbf{u})\nabla\cdot\mathbf{u} \right] dV = \int\limits_{V_E} \mathbf{F}_b\, dV + \int\limits_{S_E} \mathbf{F}_s\, dS. \tag{7.28}$$

Expanding the material derivative and combining terms results in

$$\int\limits_{V_E} \left[\rho\frac{D\mathbf{u}}{Dt} + \mathbf{u}\left(\frac{D\rho}{Dt} + \rho\nabla\cdot\mathbf{u} \right) \right] dV = \int\limits_{V_E} \mathbf{F}_b\, dV + \int\limits_{S_E} \mathbf{F}_s\, dS. \tag{7.29}$$

The term $\left(\frac{D\rho}{Dt} + \rho\nabla\cdot\mathbf{u}\right)$ is a statement of conservation shown in (7.24) and is identically zero. The momentum equation now has a form that is analogous to $m\mathbf{a} = \mathbf{F}$ with

$$\int\limits_{V_E} \rho\frac{D\mathbf{u}}{Dt}\, dV = \int\limits_{V_E} \mathbf{F}_b\, dV + \int\limits_{S_E} \mathbf{F}_s\, dS \tag{7.30}$$

where $\frac{D\mathbf{u}}{Dt}$ is exactly equal to the particle acceleration. The specific forces that are applied to the RHS can range from the Coriolis force[1] to the pressure gradient force to surface drag to shear stress, just to name a few. The focus here will be on the forces responsible for geostrophic balance: Coriolis and pressure. In addition, the Coriolis force is representative of a body force with the integration over V_E, and the pressure force is representative of a surface force with the integration over S_E. The Coriolis force can be expressed as

$$\int\limits_{V_E} \mathbf{F}_b\, dV = -\int\limits_{V_E} f_o\, \mathbf{k}\times(\rho\,\mathbf{u})\, dV \tag{7.31}$$

where f_o is the Coriolis parameter that is assumed to be a constant (i.e. an *f-plane* approximation has been assumed) and $\mathbf{k}$ is the unit vector pointing in the local vertical direction. The pressure force can be expressed as

$$\int\limits_{S_E} \mathbf{F}_s\, dS = -\int\limits_{S_E} p\,\mathbf{n}\, dS = -\int\limits_{V_E} \nabla p\, dV \tag{7.32}$$

[1] The Coriolis force is an *apparent* force that arises due to casting the equations of motion in a non-inertial, rotating reference frame. Both the Lagrangian and Eulerian reference frames are measured relative to the underlying rotating reference frame. If the system of equations were cast in an inertial reference frame, then the Coriolis force would not be present.

where $\mathbf{n}$ is the outward directed normal vector to S_E. The negative sign on the $p\,\mathbf{n}$ term in (7.32) is because, by definition, pressure p pushes inward on S_E resulting in a force directed in the $-\mathbf{n}$ direction. Equation (7.32) also uses the divergence theorem to transform the pressure force from an integral over S_E to an integral over V_E. Letting $V_E \to 0$ allows (7.30) to be expressed in its differential form as

$$\frac{D\mathbf{u}}{Dt} = -f_o\,\mathbf{k}\times\mathbf{u} - \frac{1}{\rho}\nabla p. \tag{7.33}$$

One numerical method that will be of particular interest below is the "finite-volume approach." In this approach, we retain prognostic equations for *mean values* over discrete regions. As a result, the weak or integral form of (7.33) is more amenable to a finite-volume approach. In order to convert the momentum equation shown in (7.33) into its weak form, we can apply (7.17) to the intensive quantity $P = \rho\,\mathbf{u}$ and obtain

$$\int_{V_E} \frac{\partial(\rho\,\mathbf{u})}{\partial t} dV + \int_{S_E} (\rho\,\mathbf{u})\;\mathbf{u}\cdot\mathbf{n}\,dS = \int_{V_E} \mathbf{F}_b\,dV + \int_{S_E} \mathbf{F}_s\,dS. \tag{7.34}$$

With examples of $\mathbf{F}_b$ and $\mathbf{F}_s$ in place, the integral form of the momentum equation becomes

$$\int_{V_E} \frac{\partial(\rho\,\mathbf{u})}{\partial t} dV + \int_{S_E} (\rho\,\mathbf{u})\;\mathbf{u}\cdot\mathbf{n}\,dS = -\int_{V_E} f_o\,\mathbf{k}\times(\rho\,\mathbf{u})\;dV - \int_{S_E} p\,\mathbf{n}\;dS. \tag{7.35}$$

Figure 7.4 illustrates the various terms involved in (7.35). Allowing $V_E \to 0$ in (7.35) and transforming the second and fourth term using the divergence theorem gives

$$\frac{\partial(\rho\,\mathbf{u})}{\partial t} + \nabla\cdot(\rho\,\mathbf{u}\;\mathbf{u}) = -f_o\,\mathbf{k}\times(\rho\,\mathbf{u}) - \nabla p \tag{7.36}$$

where the notation $(\rho\,\mathbf{u}\;\mathbf{u})$ symbolizes a tensor.

We have developed several different analytic forms of $\mathbf{F} = m\,\mathbf{a}$ in this section. In particular, a particle-based formulation of momentum is shown in (7.33) and a control-volume formulation is shown in (7.35). When constructing a numerical model, each form will have its own advantages and disadvantages. We will return to this discussion in Sect. 7.4.

7.3.4 Dynamics of Vorticity

By using $\mathbf{F} = m\,\mathbf{a}$ to construct the evolution equation for velocity or momentum, we describe how a particle (7.33) or a region of fluid (7.35) responds to applied forces. In addition to the balance-of-forces in the momentum equation, there are

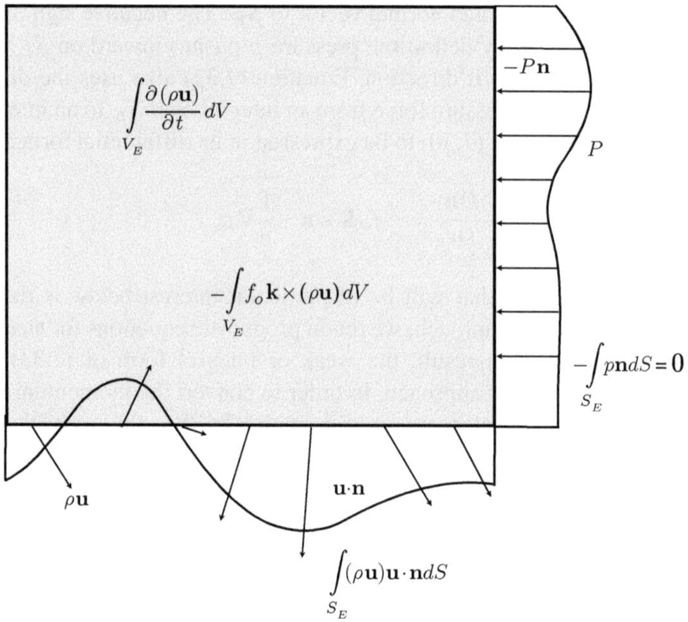

Fig. 7.4 A control volume perspective of conservation of momentum: The time-rate-of-change of momentum, $\rho\mathbf{u}$, within V_E is due to three mechanisms. The first is the apparent body force, $-f_o\,\mathbf{k} \times \rho\mathbf{u}$, acting over the entire control volume V_E. The second is due to the pressure force acting along the surface of V_E. And the last mechanism is the transport of momentum, $\rho\mathbf{u}$, across the surface of V_E. Other mechanisms such as dissipation and external sources can also be included

kinematic constraints on the structure of the velocity field. A vector velocity field can always be described as a sum of two vector velocity fields where one vector field is purely rotational and the other vector field is purely divergent. This is known as the Helmoltz Decomposition.[2] The Helmoltz Decomposition states that we can always decompose a vector field as

$$\mathbf{u} = \mathbf{u}_\delta + \mathbf{u}_\zeta \tag{7.37}$$

with

$$\nabla \cdot \mathbf{u} = \nabla \cdot \mathbf{u}_\delta = \delta, \tag{7.38}$$

and

$$\nabla \times \mathbf{u} = \nabla \times \mathbf{u}_\zeta = \boldsymbol{\zeta}, \tag{7.39}$$

where δ is the scalar divergence field associated with $\mathbf{u}$ and $\boldsymbol{\zeta}$ is the vector vorticity field associated with $\mathbf{u}$. Equations (7.38) and (7.39) show that the divergent

[2] The simplification to singly-connected domains extending to infinity is made here for clarity in presentation, see (Batchelor 1967) page 85 for a full discussion.

component of $\mathbf{u}$ is contained entirely in $\mathbf{u}_\delta$ and the rotational component of $\mathbf{u}$ is contained entirely in $\mathbf{u}_\zeta$. Given a divergence and vorticity field, the velocity field can be determined by first finding the potential fields consistent with δ and ζ as

$$\nabla^2 \phi = \delta, \tag{7.40}$$

and

$$\nabla^2 \boldsymbol{\beta} = \zeta \tag{7.41}$$

and then differentiating the scalar potential field ϕ and vector potential field $\boldsymbol{\beta}$ to obtain the velocities as

$$\nabla \phi = \mathbf{u}_\delta, \tag{7.42}$$

and

$$\nabla \times \boldsymbol{\beta} = \mathbf{u}_\zeta \,. \tag{7.43}$$

Solving (7.40) and (7.41) for the potential fields requires the inversion of the ∇^2 operator.[3] While the Helmoltz Decomposition holds for three-dimensional flows, we will limit the velocity to 2-D planar flows in the following section.

Broadly speaking, the rotational component of the velocity field, $\mathbf{u}_\zeta$, is associated with slow modes, such as Rossby waves, and the divergent component of the velocity field, $\mathbf{u}_\delta$, is associated with fast modes, such as gravity waves. An adequate representation of both the rotational and divergent components of motion is a prerequisite to robust simulations of geophysical fluid dynamics.

From a climate modeling perspective, avoiding the spurious forcing of the rotational component of the velocity field is of great concern. Since the vorticity field tends to evolve slowly in time via transport (i.e. it is a slow mode), errors in the evolution of the rotational component of velocity tend to be advected along with the fluid flow and, thus, accumulate in time. Discrete numerical models with spurious forcing of the vorticity field resort, inevitably, to inappropriately large levels of dissipation in order to control the spurious accumulation of vorticity variance at the model grid-scale.

Throughout the remaining sections of this chapter a tremendous amount of discussion will focus how to design numerical methods that appropriately solve $\mathbf{F} = m\,\mathbf{a}$ while avoiding any spurious forcing of the vorticity field. We will begin this discussion by developing conservation statements in the continuous system regarding how the rotational component of the velocity field *should* evolve in time. Later sections will focus on how to build these conservation statements into the discrete system.

[3] In singly-connected domains, like the entire surface of the sphere, no additional boundary conditions are required to solve (7.40) and (7.41). In multi-connected domains, additional boundary conditions are required to close the system.

7.3.4.1 Conservation of Circulation

Circulation measures the mean rotation around a material contour (see Fig. 7.5). Circulation is essentially the area-weighted representation of vorticity. In the discussion of circulation and vorticity, we will limit the velocity field to two spatial directions, such as the surface of a plane. The reduction in the space spanned by the velocity field means that volume integrals in RTT reduce to surface integrals and surface integrals in RTT reduce to contour integrals. The *relative* circulation is defined as

$$\Gamma^r_{c(t)} = \oint_{c(t)} \mathbf{u}\cdot\mathbf{dr} = \int_{S(t)} \mathbf{k}\cdot(\nabla\times\mathbf{u})\,dS = \int_{S(t)} \zeta\,dS \tag{7.44}$$

where $\Gamma^r_{c(t)}$ measures the mean rotation produced by the velocity field $\mathbf{u}$ around a contour $c(t)$ that moves with the material particles. For the 2D system considered here, $\mathbf{k}$ is the local vertical with ζ measuring the component of vorticity in the vertical direction. The limits of integration are around the contour $c(t)$, or over the area $S(t)$ associated with the contour. The explicit dependence on time has been retained in $c(t)$ and $S(t)$ to emphasize that the limits of integration are a function of time. All analysis in this section will take place in the Lagrangian reference frame; the use of RTT to transform the conservation statements to the more practical Eulerian reference frame will be done in the following section.

The first task is to determine the appropriate conservation statement for circulation within a Lagrangian control area. Note that since the contour of integration in (7.44) moves with the fluid, the contour is composed of the same set of particles for all time. Applying the time derivative to (7.44) yields

$$\frac{d}{dt}\Gamma^r_{c(t)} = \frac{d}{dt}\oint_{c(t)} \mathbf{u}\cdot\mathbf{dr} = \oint_{c(t)} \left[\mathbf{dr}\cdot\left.\frac{d\mathbf{u}}{dt}\right|_{particle} + \mathbf{u}\cdot\frac{d\,(\mathbf{dr})}{dt}\right]. \tag{7.45}$$

Since the element $\mathbf{dr}$ is transported with velocity $\mathbf{u}$, its time-rate-of-change can be expressed as

$$\frac{d\,(\mathbf{dr})}{dt} = \mathbf{dr}\cdot\nabla\,\mathbf{u}\,. \tag{7.46}$$

The RHS of (7.46) measures the deformation and rotation of $\mathbf{dr}$ due to spatial variations in the $\mathbf{u}$ field.[4] Using (7.46) in (7.45) yields

$$\frac{d}{dt}\Gamma^r_{c(t)} = \oint_{c(t)} \left[\frac{D\mathbf{u}}{Dt} + \nabla\left(\frac{|\mathbf{u}|^2}{2}\right)\right]\cdot\mathbf{dr} = \oint_{c(t)} \frac{D\mathbf{u}}{Dt}\cdot\mathbf{dr} \tag{7.47}$$

[4] Equation (7.46) is obtained by noting that $\frac{d(\mathbf{dr})}{dt} = \mathbf{u}(\mathbf{x}+\mathbf{dr}) - \mathbf{u}(\mathbf{x})$, expanding $\mathbf{u}(\mathbf{x}+\mathbf{dr})$ in a Taylor series and retaining the first two terms. The $\nabla\,\mathbf{u}$ term is the gradient of the vector velocity field and is a rank-2 tensor. A detailed explanation of $\nabla\,\mathbf{u}$ is given in DeCaria and Sikora (2010).

where (7.5) is used to recast the time derivative of $\mathbf{u}$ as a material derivative. The relationship between the evolution of circulation and $\mathbf{F} = m\,\mathbf{a}$ is becoming apparent with the appearance of the $\frac{D\mathbf{u}}{Dt}$ in (7.47). If we substitute in the form of the momentum equation defined in (7.33) we obtain

$$\frac{d}{dt}\Gamma^r_{c(t)} = \oint_{c(t)} \left[-f_o\,\mathbf{k}\times\mathbf{u} - \frac{\nabla p}{\rho} \right] \cdot \mathbf{dr}. \tag{7.48}$$

The first source of relative circulation on the RHS of (7.48) is related to the amount of planetary vorticity captured in $c(t)$ due to expansion or contraction of the area associated with $c(t)$. Referring to Fig. 7.5 and under the condition that the Coriolis parameter is constant, we can manipulate this source term as

$$-\oint_{c(t)} [f_o\,\mathbf{k}\times\mathbf{u}]\cdot\mathbf{dr} = -\oint_{c(t)} [\mathbf{u}\times\mathbf{dr}]\cdot f_o\,\mathbf{k} = -f_o\frac{D}{Dt}S(t) = -\frac{D}{Dt}[f_o\,S(t)]\,. \tag{7.49}$$

The term $\mathbf{u}\times\mathbf{dr}$ represents the rate at which area is swept by the transport of element $\mathbf{dr}$ by velocity $\mathbf{u}$. When integrated around the entire contour and multiplied by the planetary vorticity, the result measures the time-rate-of-change in the amount of

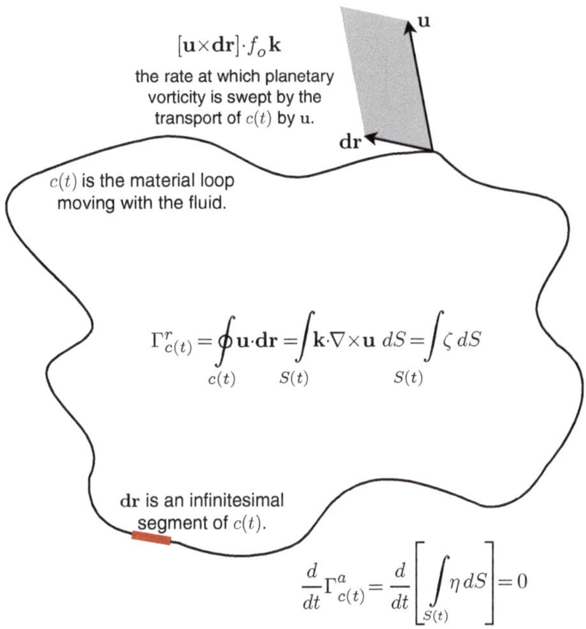

Fig. 7.5 A graphical representation of circulation. The symbol $\eta = \zeta + f_o$ denotes the absolute vorticity

planetary vorticity contained within $c(t)$. If we define the *planetary* circulation as

$$\Gamma^{p}_{c(t)} = f_o\, S(t) \tag{7.50}$$

then we can express the *absolute* circulation as

$$\Gamma^{a}_{c(t)} = \Gamma^{r}_{c(t)} + \Gamma^{p}_{c(t)} = \int\limits_{S(t)} (\zeta + f_o)\, dS = \int\limits_{S(t)} \eta\, dS \tag{7.51}$$

where η is the absolute vorticity defined as the sum of the relative vorticity and the planetary vorticity. We can now rewrite (7.48) as

$$\frac{d}{dt}\Gamma^{a}_{c(t)} = \oint\limits_{c(t)} \left[-\frac{\nabla p}{\rho}\right] \cdot \mathbf{dr} \tag{7.52}$$

where (7.52) is an expression for the rate-of-change of absolute circulation associated with a contour $c(t)$ that is observed in the Lagrangian reference frame. The remaining source term on the RHS of (7.52) is the due to the differential acceleration of particles along $c(t)$ produced by the pressure gradient force when variations in the density field are present. The primary interest here is on the situation when the density field is constant,[5] i.e. $\rho = \rho_o$. In this situation we find

$$\oint\limits_{c(t)} \left[-\frac{\nabla p}{\rho_o}\right] \cdot \mathbf{dr} = \frac{-1}{\rho_o} \oint\limits_{c(t)} \nabla p \cdot \mathbf{dr} = 0. \tag{7.53}$$

The term $\nabla p \cdot \mathbf{dr}$ measures the gradient of the pressure field in the direction of **dr**. So long as the $c(t)$ loop traced out by the differential **dr** elements is closed, the integration of $\nabla p \cdot \mathbf{dr}$ around $c(t)$ is guaranteed to be identically zero. This results holds for any loop and for any scalar field. With the result provided in (7.53), we can end the analysis with

$$\frac{d}{dt}\Gamma^{a}_{c(t)} = \frac{d}{dt}\left[\int\limits_{S(t)} \eta\, dS\right] = 0 \tag{7.54}$$

that states that the absolute circulation contained within contour $c(t)$ as it moves with the fluid will be a constant in time; absolute circulation within $c(t)$ is conserved

[5] When variations in density are present, as in the real atmosphere and ocean, then the RHS of (7.52) serves as a source of circulation and vorticity. When considering the numerical simulation of this process, a critical prerequisite is the guarantee that vorticity is *not* created when these variations are *not* present.

in time. The relationship also makes it clear that, in general, the absolute vorticity *is not* constant within the contour $c(t)$. Only in the special case of non-divergent flow resulting in $\frac{D}{Dt}[S(t)] = 0$ will the mean value of η be a constant within contour $c(t)$.

7.3.4.2 Conservation of Absolute Vorticity

The entire analysis in the section above is conducted in the Lagrangian reference frame. The purpose of this section is to use RTT to transfer the conservation statements into an Eulerian reference frame. Comparing (7.54) to (7.8) shows that the form of conservation of absolute circulation shown in (7.54) is suitable for the application of RTT. Applying RTT as stated (7.15) to (7.54), we find

$$\frac{d}{dt}\Gamma^a_{c(t)} = \frac{d}{dt}\left[\int_{S(t)} \eta\, dS\right] = \int_S \left[\frac{\partial \eta}{\partial t} + \nabla \cdot (\eta\, \mathbf{u})\right] dS = 0. \tag{7.55}$$

The form of (7.55) that is most suitable for finite-volume applications discussed below is

$$\int_S \frac{\partial \eta}{\partial t} dS + \oint_c \eta\, \mathbf{u} \cdot \mathbf{n}\, dr = 0 \tag{7.56}$$

that states that the time-tendency of absolute vorticity in region S is equal and opposite to the rate at which absolute vorticity is being transported into or out of region S. A primary goal in the construction of the numerical model developed below is to guarantee that the velocity field evolves in such a way as to mimic (7.56) exactly.

For the sake of completeness we note that in the limit of $dS \to 0$ and allowing ρ to be nonuniform, (7.55) becomes

$$\frac{\partial \eta}{\partial t} + \nabla \cdot (\eta\, \mathbf{u}) = -\mathbf{k} \cdot \left(\nabla \times \left[\frac{\nabla p}{\rho}\right]\right) \tag{7.57}$$

where the RHS source term shown in (7.52) has been retained. And finally, introducing the material derivative into (7.57) yields

$$\frac{D\eta}{Dt} + \eta \nabla \cdot \mathbf{u} = -\mathbf{k} \cdot \left(\nabla \times \left[\frac{\nabla p}{\rho}\right]\right). \tag{7.58}$$

7.3.5 Summary of Evolution Equations

The analytic analysis of the continuous system is now complete. The approach has been to identify conservation statements in the Lagrangian reference frame and to use the Reynolds Transport Theorem to transfer these conservation statements into an Eulerian reference frame. The value of the Reynolds Transport Theorem is that it provides a machine-like approach to the derivation of evolution equations specified naturally in the integral form conducive to the development of finite-volume methods.

Before turning to the process of discretization, a survey is conducted of the various flavors of $\mathbf{F} = m\,\mathbf{a}$ that can be used as the basis, or starting point, for the discretization process. The specific form of $\mathbf{F} = m\,\mathbf{a}$ that is chosen as the starting point for the numerical model has a tremendous impact on the attributes of that numerical model. Particular attention is paid to the ability of each form to satisfy both $\mathbf{F} = m\,\mathbf{a}$ and conservation of absolute vorticity (7.56).

7.4 The Various Flavors of $\mathbf{F} = m\,\mathbf{a}$

In the continuous system, all forms of the momentum equation are equivalent. The equivalence holds for smooth flows. If singularities develop in the solution, the equivalence between the various forms is more tenuous. Since each form can be manipulated into any other form, there is no difference between the various expressions of $\mathbf{F} = m\,\mathbf{a}$. This is not true in the setting of discrete numerics. Discretizing the continuous system implies an approximation of the continuous fields as a finite set of values that typically exist on a mesh that spans the spatial extent of the system. In addition, the continuous operators such as ∇, $\nabla\cdot$ and $\nabla\times$ are replaced with discrete approximations. One result of discretizing the momentum equation is that the various forms are no longer equivalent; we cannot, in general, manipulate one discrete form of the momentum equation into another discrete form using the discrete operators. As a result, when we choose the form of the momentum equation used in a numerical model, we are saying a great deal about what aspects of $\mathbf{F} = m\,\mathbf{a}$ are most important in the target application. Each form has its own advantages and disadvantages and, thus, each form has its own niche to fill in the modeling of the global atmosphere and ocean systems. This section provides a brief review of the commonly used flavors of $\mathbf{F} = m\,\mathbf{a}$ with a discussion of their respective advantages and disadvantages.

7.4.1 The Advective Form

The advective form of the momentum equation (7.33) is restated here for convenience:

$$\frac{D\mathbf{u}}{Dt} = -f_o\,\mathbf{k}\times\mathbf{u} - \frac{1}{\rho}\nabla p. \tag{7.59}$$

This is essentially an evolution equation for one of the particles in the Lagrangian system, such as a particle $\mathbf{X}$ shown in Fig. 7.1. Assume that the system is discretized on a regular mesh composed of squares, such as the one shown in Fig. 7.6. If at some

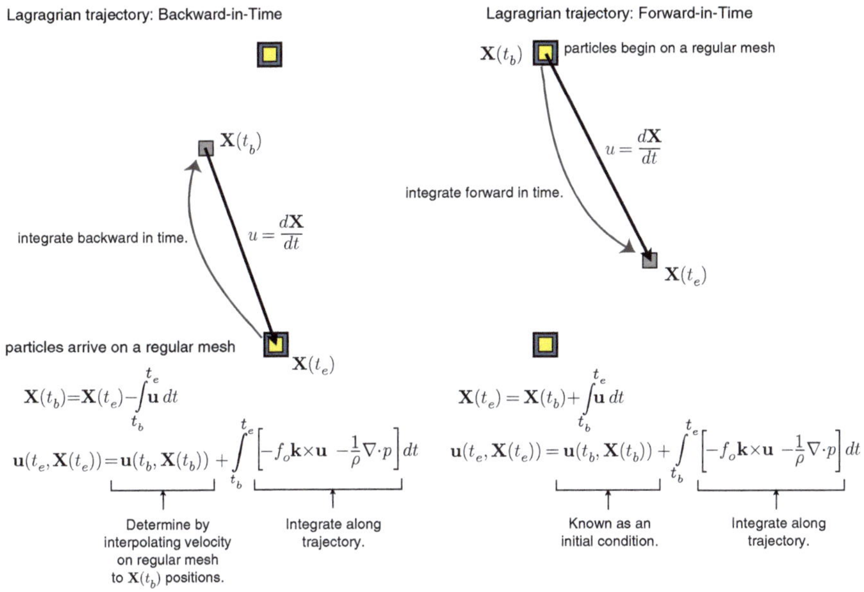

Fig. 7.6 A graphical representation of forward Lagrangian and backward Lagrangian (i.e. the semi-Lagrangian) method

time, say $t = t_b$ one particle is placed at the center of each square shown in Fig. 7.6, then the particle position and velocity at some later time, say $t = t_e$, are determined by integrating (7.59) *along the particle trajectory* as

$$\int_{t_b}^{t_e} \frac{D\mathbf{u}}{Dt} dt = \mathbf{u}(t_e, \mathbf{X}(t_e)) - \mathbf{u}(t_b, \mathbf{X}(t_b)) = \int_{t_b}^{t_e} \left[-f_o\, \mathbf{k} \times \mathbf{u} - \frac{1}{\rho} \nabla p \right] dt. \quad (7.60)$$

Assuming that the particle positions and velocities are known at t_b, the system is solved for $\mathbf{X}(t_e)$ and $\mathbf{u}(\mathbf{X}(t_e), t_e)$ as

$$\mathbf{X}(t_e) = \mathbf{X}(t_b) + \int_{t_b}^{t_e} \mathbf{u}\, dt, \quad (7.61)$$

$$\mathbf{u}(t_e, \mathbf{X}(t_e)) = \mathbf{u}(t_b, \mathbf{X}(t_b)) + \int_{t_b}^{t_e} \left[-f_o\, \mathbf{k} \times \mathbf{u} - \frac{1}{\rho} \nabla p \right] dt. \quad (7.62)$$

It needs to be emphasized that all of the source-term integrals on the RHS of (7.62) are along the particle path starting at time t_b at position $\mathbf{X}(t_b)$ and ending at time t_e at position $\mathbf{X}(t_e)$. While there are certainly challenges with the discrete evaluation of the RHS of (7.62), a more basic problem with the approach is that the particle positions at the end of the time step are, in general, no longer on a regular mesh (see forward-in-time diagram in Fig. 7.6). More forward-in-time steps will lead to a continuous distortion of particle positions due to the same shearing, stretching and deformation mechanisms illustrated in Fig. 7.2. In order to prevent this continuous distortion, (7.60) is generally evaluated *backward in time* in what is commonly known as the *semi-Lagrangian approach* (see Staniforth and Côté 1991 for a complete review).

Instead of assuming that the particles exist on a regular mesh at the beginning of the time step, the particles are assumed to reside on the regular mesh at the end of the time step. In this situation, the particle positions $\mathbf{X}(t_e)$ are required to form the regular mesh shown in Fig. 7.6. The challenge is then to determine $\mathbf{X}(t_b)$ by integrating particle trajectories backward in time, i.e. to determine the starting point of the particles such that the particles arrive on a regular mesh at t_e. In this approach the system is solved for $\mathbf{X}(t_b)$ and $\mathbf{u}(\mathbf{X}(t_e), t_e)$ as

$$\mathbf{X}(t_b) = \mathbf{X}(t_e) - \int_{t_b}^{t_e} \mathbf{u}\, dt, \tag{7.63}$$

$$\mathbf{u}(t_e, \mathbf{X}(t_e)) = \mathbf{u}(t_b, \mathbf{X}(t_b)) + \int_{t_b}^{t_e} \left[-f_o\, \mathbf{k} \times \mathbf{u} - \frac{1}{\rho} \nabla p \right] dt. \tag{7.64}$$

In general, $\mathbf{u}(t_b, \mathbf{X}(t_b))$ is determined by interpolating the velocity values known on the fixed mesh at time t_b to $\mathbf{X}(t_b)$ locations. Equations (7.63) and (7.64) are coupled and need to be solved jointly or iteratively. The challenges of evaluating the RHS along the particle trajectory still remain.

The advantage of this approach is that exceptionally long time steps are possible.[6] Since the integration is occurring along the particle characteristic, traditional advective Courant–Friedrichs–Lewy (CFL) time step constraints do not apply. An additional advantage is the ease with which tracer constituents can be updated. Using (7.25) and integrating $\frac{Dq}{Dt}$ from t_b to t_e, we have

$$q(t_e, \mathbf{X}(t_e)) = q(t_b, \mathbf{X}(t_b)). \tag{7.65}$$

$q(t_b, \mathbf{X}(t_b))$ is determined by interpolating the tracer values from the regular mesh to the departure points $\mathbf{X}(t_b)$. Once this interpolation is complete, the updated tracer values are known immediately since q is conserved along particle trajectories.

The disadvantages in this approach to solving the momentum equation are related to the lack of conservation of mass and tracer substance and the spuri-

[6] While longer time steps reduce the computational expense of a given simulation, longer time steps also often lead to less accurate results. Weighing the relative value of "fast" versus "correct" is important in choosing the time step for a simulation.

ous generation of vorticity. While these disadvantages pose severe problems in the context of long-time simulations typical in climate applications, these disadvantages have been successfully mitigated and/or circumvented for numerical weather prediction applications where the integration time scales are on the order of days to a week or two. Another alternative is to abandon the particle-centric approach of pure semi-Lagrange schemes and move to a cell-based approach (see Chap. 8).

The issues regarding conservation can be readily identified by comparing (7.65)–(7.4). The conservation statement is that the mass-weighted integral of q (i.e. Q) is conserved in time when no sources or sinks are present. Yet (7.65) only sees the tracer concentration q for an isolated number of particles and, furthermore, that concentration is computed at locations $\mathbf{X}(t_b)$ via an interpolation procedure where accuracy is generally much more important than conservation.

The issues regarding spurious vorticity generation are equally problematic in the context of climate system modeling. In general, getting a handle on the evolution of vorticity in a particle-based formulation is extremely difficult. Using (7.58) we could certainly tag each particle with an associated vorticity, but the evolution of absolute vorticity during the time step involves spatial gradients that are difficult to compute. In addition, the same issue regarding lack of conservation occurs in the context of vorticity as occurs in the context of tracer transport. And finally, even if one could manage to evolve vorticity with the particles in a realistic manner, it is not clear how that information could be used to control the evolution of the prognostic velocity field shown in (7.64).

7.4.2 *The Flux Form*

The flux form of the momentum equation is shown in (7.34), illustrated in Fig. 7.4 and rewritten here for 2D planar flow as

$$\int_{S_E} \frac{\partial(\rho\,\mathbf{u})}{\partial t} dS + \int_{c_E} (\rho\mathbf{u})\;\mathbf{u}\cdot\mathbf{n}\,dc = -\int_{S_E} f_o\,\mathbf{k}\times(\rho\,\mathbf{u})\;dS - \int_{c_E} p\,\mathbf{n}\;dc. \tag{7.66}$$

where c stands for a line segment along the contour c_E. The main advantage of the flux-form momentum equation is that it is relatively easy to insure that the transport of momentum (the second term in (7.66)) is conservative, i.e. momentum that exits one cells across c_E enters a neighbor cell. This same conservation property occurs in the evaluation of the pressure force; along a contour c_E the pressure force results in an equal and opposite source of momentum for the surfaces that share c_E. An additional advantage of the flux-form is that density is incorporated into the prognostic variable. When using the flux-form of momentum, the prognostic variable is $\rho\,\mathbf{u}$, whereas all the other forms have $\mathbf{u}$ as the prognostic variable. The merit in retaining $\rho\,\mathbf{u}$ as the prognostic variable is that as $\rho \rightarrow 0$ the prognostic variable goes to zero so long as $\mathbf{u}$ remains bounded. In the emerging class of atmosphere and ocean models, ρ is often related to the vertical layer thickness, so $\rho \rightarrow 0$ is equivalent to a layer collapsing to zero thickness when all of the mass in a given layer at a given position is evacuated (e.g. Konor and Arakawa 1997; Bleck and Smith 1990). This is a common occurrence in numerical models and the flux-form momentum equation provides ample opportunities to insure that the discrete system remains well-behaved even in the presence of massless layers.

The primary disadvantage in the use of the flux-form momentum equation is that the curl of (7.66) does not lead directly to a vorticity equation; vorticity and circulation are purely kinematic quantities that are related to the $\nabla \times \mathbf{u}$ not $\nabla \times (\rho\, \mathbf{u})$. As a result, discrete models based on the flux-form of the momentum equation do not conserve circulation or absolute vorticity. In a discrete formulation of (7.66) every term has the potential to generate spurious vorticity. If no guarantees can be provided in regards to the conservation of circulation or vorticity, in general the only recourse is to increase the level of dissipation to maintain a regular, well-behaved solution. If the level of dissipation required to suppress the spurious generation of vorticity is significantly higher than is physically warranted, one should expect the numerical simulation to be degraded due to the physically-excessive dissipation.

The spurious generation of vorticity is due to errors in the discretization of the system. Assuming smooth flows, these errors approach zero as the order-of-accuracy of the discrete operators is increased and/or as the grid resolution is increased. The possibility certainly exists that these spurious errors are acceptably small, even for climate simulations, when employing high-order numerical methods and/or high-resolution meshes.

7.4.3 *The Vector-Invariant Form*

The vector-invariant form is derived from the advective form (7.59) where the material derivative is expanded into time tendency and transport terms using (7.5) to obtain

$$\frac{\partial \mathbf{u}}{\partial t} + (\mathbf{u} \cdot \nabla)\, \mathbf{u} = -f_o\, \mathbf{k} \times \mathbf{u} - \frac{1}{\rho} \nabla p. \tag{7.67}$$

If the $(\mathbf{u} \cdot \nabla)\, \mathbf{u}$ term is replaced based on the following vector identity

$$(\mathbf{u} \cdot \nabla)\, \mathbf{u} = (\nabla \times \mathbf{u}) \times \mathbf{u} + \nabla \left[\frac{1}{2} |\mathbf{u}|^2 \right], \tag{7.68}$$

we obtain

$$\frac{\partial \mathbf{u}}{\partial t} = -\eta\, \mathbf{k} \times \mathbf{u} - \nabla K - \frac{1}{\rho} \nabla p \tag{7.69}$$

where $\zeta = \mathbf{k} \cdot (\nabla \times \mathbf{u})$, $\eta = \zeta + f_o$ and the kinetic energy is defined as $K = \frac{1}{2} |\mathbf{u}|^2$.

Since the vector-invariant form of the evolution of momentum has no notion of a material derivative, it is a natural expression of the velocity tendency at a *fixed point in space*. The interesting and powerful aspect of (7.69) is that while **u** is defined at a point, the integral of **u** around a closed contour defines an area, a circulation and the area-mean vorticity. This relationship will be fully developed in Sect. 7.5.

The $\eta\,\mathbf{k}\times\mathbf{u}$ term will be referred to as the *nonlinear Coriolis force* because it contains both the linear tendency term $f_o\,\mathbf{k}\times\mathbf{u}$ and a portion of the nonlinear transport term in the form of $\zeta\,\mathbf{k}\times\mathbf{u}$.

When considering the momentum equation we are primarily interested in the velocity field that is needed for the evolution of the mass and tracer fields. Beyond the velocity itself, we are interested in three *derived* quantities: divergence, vorticity and kinetic energy. Two of these three derived quantities appear explicitly in (7.69). The appearance of vorticity and kinetic energy does not necessarily imply that the necessary controls are available to insure that these quantities remain well-behaved and bounded, but it is a step in the right direction.

In the context of climate modeling, it is difficult to find shortcomings in choosing the vector-invariant form of the momentum equation as the basis for a discrete model. This approach was successfully employed on hexagonal grids (Sadourny and Morel 1969) and on latitude-longitude grids (Arakawa and Lamb 1981) decades ago. The primary reason to not choose this form of the momentum equation is that another form of the momentum equation, such as the advective form or flux form, is a more natural choice for the application of interest.

7.4.4 The Vorticity-Divergence Form

Since a great deal of emphasis has been placed on the importance of vorticity in the above discussion, it is reasonable to consider *exchanging* the prediction of the vector velocity for the prediction of the vorticity and divergence. As discussed above, the Helmoltz Decomposition guarantees that vorticity and divergence form a complete description of the vector velocity field, so prognosing ζ and δ is a theoretically-sound approach (e.g. Heikes and Randall 1995; Ringler et al. 2000; Thuburn 1997). In addition, retaining ζ as a prognostic variable leads to a strong control over its evolution.

For 2D planar flow, we generate the evolution equations for ζ and δ by taking $\mathbf{k}\cdot\nabla\times$ and $\nabla\cdot$ of the momentum equation, respectively. As long as we are working with the continuous equations, we can start with any form of the momentum equation and obtain the same resulting vorticity and divergence equation. Starting with the vector-invariant form of the momentum equation expressed in (7.69) and applying the $\mathbf{k}\cdot\nabla\times$ and $\nabla\cdot$ operators yields

$$\mathbf{k}\cdot\left(\nabla\times\frac{\partial\mathbf{u}}{\partial t}\right)=\frac{\partial\zeta}{\partial t}=\mathbf{k}\cdot\left(\nabla\times\left[-\eta\mathbf{k}\times\mathbf{u}-\nabla K-\frac{1}{\rho}\nabla p\right]\right), \tag{7.70}$$

and

$$\nabla \cdot \frac{\partial \mathbf{u}}{\partial t} = \frac{\partial \delta}{\partial t} = \nabla \cdot \left[-\eta \mathbf{k} \times \mathbf{u} - \nabla K - \frac{1}{\rho} \nabla p \right]. \tag{7.71}$$

Focusing on the vorticity equation, we can recover the Eulerian expression derived in (7.57) written as

$$\frac{\partial \eta}{\partial t} + \nabla \cdot (\eta \, \mathbf{u}) = -\mathbf{k} \cdot \left(\nabla \times \left[\frac{\nabla p}{\rho} \right] \right). \tag{7.72}$$

The first important aspect to note in (7.72) is that $\mathbf{k} \cdot (\nabla \times [-\eta \mathbf{k} \times \mathbf{u}]) = -\nabla \cdot (\eta \, \mathbf{u})$. The application of the curl operator to the nonlinear Coriolis force results in the divergence of the absolute vorticity flux. The second important aspect to note in (7.72) is that $\nabla \times \nabla K = 0$; the curl of the gradient is identically zero.

The divergence equation can be expressed as

$$\frac{\partial \delta}{\partial t} + \nabla \cdot \left(\eta \, \mathbf{u}^{\perp} \right) = -\nabla^2 K - \nabla \cdot \left[\frac{1}{\rho} \nabla p \right] \tag{7.73}$$

where $\mathbf{u}^{\perp} = \mathbf{k} \times \mathbf{u}$.

The primary advantage of using the vorticity-divergence form of the velocity evolution equation is the ability to retain (7.72) as a prognostic equation. In the presence of uniform density, the time-rate-of-chance of absolute vorticity is the divergence of the absolute vorticity flux. The absolute vorticity flux can be computed numerically using advanced transport algorithms that can guarantee that η will remain smooth at the grid-scale without the introduction of excessive dissipation.

The primary disadvantage of this formulation can be seen in (7.40) and (7.41). After each time step, two elliptic equations must be inverted in order to compute the velocity field that will be required to compute the tendency terms in (7.72) and (7.73) on the next time step. For simple domains, such as the global atmosphere, inverting (7.40) and (7.41) is straightforward but relatively expensive in regards to the computational effort. In more complicated domains, inverting (7.40) and (7.41) is analytically challenging and, at least to date, computationally prohibitive.

7.5 The Process of Discretization

In this section the continuous equations developed above will be *discretized* in order to obtain a numerical model for the evolution of momentum. The process of discretization truncates the infinite degrees of freedom that are present in the continuous system to a finite number of degrees of freedom in order to produce a computationally-tractable algebraic problem suitable for existing computer architectures. When the numerical methods are based on traditional finite-volume techniques, such as those to be developed below, the spatial extent of the continuous system is decomposed into *cells* and the temporal extent of the continuous system is decomposed into *time steps*. The discussion here will be limited to the spatial discretization of the continuous system.

The possibilities for the specific form the discrete momentum equation can quickly become unwieldy. For example, the optimal way to decompose the sphere into cells is still very much a research topic. Even if we limit the scope to decompositions that attempt to produce quasi-uniform meshes the choices include, at a minimum, the cubed-sphere (Chap. 9), Voronoi tessellations (Chap. 10) and Delaunay triangulations (Chap. 10). Furthermore, once a mesh is chosen there are at least five different staggering arrangements of the prognostic variables: A-grid, B-grid, C-grid, D-grid, and E-grid (Chap. 3). In addition, we can choose one of the four viable flavors of $\mathbf{F} = m\,\mathbf{a}$ to discretize. So three meshes times five grid-staggerings times four momentum forms leads to 60 permutations. And this is before we even consider the specification of the numerical operators.

A "down-select" of the 60 permutations is required. Some of this down-select can be made based on the target application. Some of this down-select can be based on the wealth of experience that has been gained over the last 40 years. And finally, some of this down-select can be made based on an intuition of what method(s) are likely to emerge as the preferred-alternative over the next decade. Furthermore, the selection method should not be made as an *a la carte* process; some choices of grid staggering are clearly inappropriate for certain choices in the form of the momentum equation. Rather, the process is similar to a *table d'hote* where choices are made with the prior knowledge of the other choices and the intention to produce the *best overall product* as opposed to the best single course. The courses in this chapter's *table d'hote* are discussed directly below.

7.5.1 Target Application: Joint Climate-Weather Prediction

The traditional gap between the atmospheric component of climate models and weather prediction models is disappearing. Atmosphere climate models have been used to conduct global cloud resolving simulations (Tomita et al. 2005). Weather prediction models have been used to study regional climate change (Leung et al. 2004). While each model is finding application outside what has been its core mission, these uses are clearly "off-label applications" where, as expected, the quality of the results vary. The criteria driving the choices in model specification (i.e. the choice of mesh, grid staggering and form of momentum) have traditionally been very different in the climate and weather modeling communities. Climate applications have emphasized concepts related to mass, tracer and vorticity conservation, as well as long-time stability of numerical simulations. Weather applications have emphasized concepts related to local accuracy and simulation throughput. The driving need is for a *single* atmosphere model to excel at both climate applications and weather applications. So the target application for this discussion is a joint climate-weather simulation. As a result, the choices made below may differ from the choices made if the target application was solely climate simulation or solely weather prediction. And finally, these same choices will be applicable to a unified ocean model that is appropriate for both global ocean simulations and regional eddy-resolving simulations.

7.5.2 Grid Staggering: C-grid Staggering

The choice of the grid staggering is very much constrained by the target application. Weather prediction models have often used a collocated staggering of variables in order to apply semi-Lagrangian methods to the advective form of the momentum equation (Ritchie et al. 1995). This is a computationally efficient method that is greatly appreciated in operational settings where simulation throughput is often a driving factor in model specification. Other grid staggerings, such as the B-grid (Zhang and Rančić 2007) and C-grid staggering (Skamarock et al. 2008), have been used with success in both weather and climate models. The choice of the C-grid staggering, when paired with the other choices, will also allow for exact conservation of absolute vorticity.[7] And more importantly, the C-grid staggering will allow for the precise control of the evolution of vorticity in time through the use of advanced flux-limiting transport algorithms. In addition, the C-grid staggering excels in the simulation of divergent modes that dominate the cloud-resolving scales of motion (Randall 1994). The principle difficulty with the C-grid staggering is that while the normal component of velocity is retained as a prognostic variable, the tangential component of velocity is needed to compute the nonlinear Coriolis force (Chap. 3). The robustness of numerical schemes built with a C-grid staggering is very much dependent on the method used for the reconstruction of the tangential velocity component.

7.5.3 Mesh: Locally-Orthogonal Meshes

One of the residual benefits of using the C-grid staggering is that it accommodates a wide class of meshes. The critical aspect of the C-grid staggering is that the edge that separates two cells is orthogonal to the line segment connecting the centers of the two associated cells (see the discussion in Chap. 10). The local orthogonality leads to compact numerical operators that are approximately second-order accurate in space (Ringler et al. 2010). The local orthogonality, C-grid staggering

[7] While the target applications involve full 3D simulations of the atmosphere and ocean, the process of discretization is best elucidated in 2D. The 3D system is clearly more complicated and is not a simple extension of the 2D system. Still, the concept of vorticity dynamics and conservation of (potential) vorticity are equally important in the full 3D system.

and vector-invariant form of momentum will lead to a strong connection between acceleration and vorticity transport.

7.5.4 Form of Momentum Equation: The Vector-Invariant Form

The use of the vector-invariant form of the momentum equation has a long and successful track record in climate modeling dating back to at least Arakawa and Lamb (1981). Weather applications have tended to use other forms, such as the flux form in order to conserve momentum and to obtain higher formal accuracy (e.g. the Weather and Research Forecast (WRF) model described in Skamarock et al. 2008) or the advective form in order to employ semi-Lagrangian methods (e.g. the European Center for Medium-Range Weather Forecasts (ECMWF) model documented in Ritchie et al. 1995).

The comparison of the vector-invariant form to the flux form offers an important insight into conservation. Given all of the choices made above (i.e. climate-weather applications, C-grid staggering, and locally-orthogonal meshes), either the vector-invariant form or the flux form is a viable choice. If one chooses the flux form of the momentum equation, then the prognostic variable, $\rho\,\mathbf{u}$, will be conserved in the numerical model. As derived below, if one chooses the vector-invariant form of the momentum equation, then absolute vorticity will be conserved in the numerical model. The choice between the vector-invariant form or the flux form of momentum comes down to the relative importance of conserving absolute vorticity or conserving momentum in the target application. The choice here is to value the former more than the latter.

7.6 Building a Discrete Model

This section will develop the numerical model that uses a C-grid staggering of the vector-invariant form of the momentum equation discretized on a locally-orthogonal mesh. The analysis will focus on the relationship between the time-tendency of the velocity field and the absolute vorticity flux.

7.6.1 Defining the Mesh and Location of Variables

For this discussion we will assume that the domain is decomposed into a set of squares as shown in Fig. 7.7. The scalar function Φ is defined at the center of each cell that are denoted as mass points in Fig. 7.7. The component of velocity in the direction normal to each edge will be integrated in time with a prognostic equation. Vorticity points are defined at the corners of the scalar function cells and will be

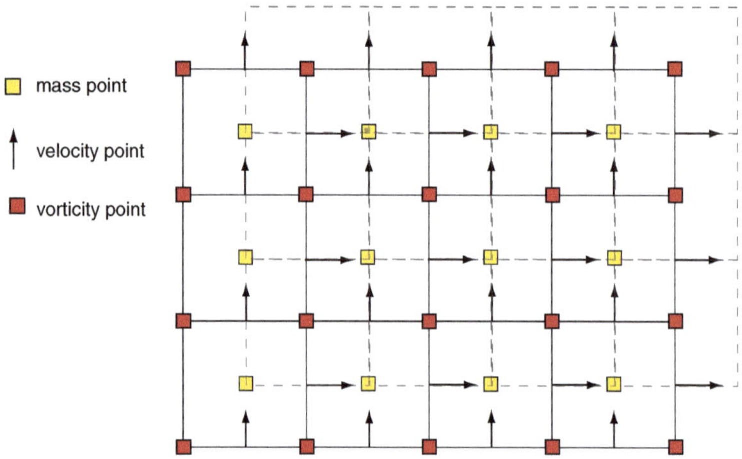

Fig. 7.7 The mesh used in the construction of the discrete system

associated with the mesh denoted by the dashed lines. The assumption is that the mesh continues indefinitely in the horizontal directions.

The choice of squares as the cell shape is based on several reasons. A mesh composed of squares is clearly locally-orthogonal, so it meets the requirement listed in Sect. 7.5. A mesh composed of squares is also the most accessible mesh; the analysis presented here can be easily replicated in development environments such as MATLAB.

While the derivation will be completed for a mesh composed of squares, conformally-mapped cubed-sphere meshes, Voronoi tessellations and Delaunay triangulations (Chap. 10) are all accommodated in the analysis,[8] i.e. the results found for the mesh composed of squares will be applicable to these more practical meshes. In an effort to point the way toward extensions to meshes that are used to discretize the surface of the sphere, an indexing nomenclature will be chosen that is appropriate for any unstructured mesh.

7.6.2 Continuous Prognostic Equation

We discretize the vector-invariant form of the momentum equation as

$$\frac{\partial \mathbf{u}}{\partial t} + \eta \mathbf{k} \times \mathbf{u} = -\nabla \Phi \qquad (7.74)$$

[8] Cubed-sphere grids produced by projections that result in a more uniform distribution of nodes at the expense of orthogonality (e.g. gnomonic-projected cubed-sphere meshes) are not accommodated in this analysis.

where $-\nabla\Phi = -\nabla\left(\frac{p}{\rho_o} + K\right)$ represents the gradient terms on the RHS of (7.69). In the full 3D system, ρ will vary in space and, as a result, the RHS can not be written as the gradient of a potential. Here, the analysis assumes that the density is a constant ρ_o in order to demonstrate that the largest contribution to the RHS of (7.69) (i.e. the ρ_o contribution) does not project onto the vorticity dynamics of the system. The system can be closed by the addition of an equation describing the evolution of fluid pressure, p. For reasons discussed in Sect. 7.1, we will limit the analysis to the evolution of velocity. In addition, special care is required to determine the appropriate discrete form of K in order to avoid numerical instabilities associated with the divergent part of the velocity field (see, for example, Eq. (3.41) of Arakawa and Lamb 1981 or Eq. (63) Ringler et al. 2010). The analysis below focuses on the evolution of the rotational part of the flow and is valid for any definition of K.

The $\mathbf{k}\times\mathbf{u}$ operation acts to rotate the vector velocity by 90° in the counter clockwise (CCW) direction. If we define $\mathbf{u}^{\perp} = \mathbf{k}\times\mathbf{u}$ as in (7.73) then (7.74) is expressed as

$$\frac{\partial \mathbf{u}}{\partial t} + \eta\,\mathbf{u}^{\perp} = -\nabla\Phi. \tag{7.75}$$

7.6.3 Discrete Prognostic Equation

At each cell edge the unit normal vector $\mathbf{e}_N$ is defined to point toward the right or toward the top as appropriate. The choice of the direction of the local normal vector is entirely arbitrary. The choice made here is for the convenience of presentation. In addition, the tangential unit vector is defined as $\mathbf{e}_T = \mathbf{k}\times\mathbf{e}_N$. The discrete version of (7.75) is generated by taking $\mathbf{e}_N\cdot$ (7.75) at each edge to yield

$$\frac{\partial N_k}{\partial t} - \hat{\eta}_k \hat{T}_k = -\left(\mathbf{e}_N\cdot\nabla\Phi\right)_k \tag{7.76}$$

where, as shown in Fig. 7.8, $N_k = \mathbf{e}_N\cdot\mathbf{u}$ represents the component of $\mathbf{u}$ in the normal direction and $\hat{T}_k = -\mathbf{e}_N\cdot\mathbf{u}^{\perp}$ represents the component of $\mathbf{u}$ in the tangential direction. All variables with hats, $(\hat{\cdot})$, require further specification.

The first example of the simplicity afforded by the assumption of a locally-orthogonal mesh is found on the RHS of (7.76). The RHS of (7.76) requires the determination of the component of $\nabla\Phi$ in the $\mathbf{e}_N$ direction. Since $\mathbf{e}_N$ is parallel to the vector connecting the Φ points on either side of the edge, the specification of the $(\mathbf{e}_N\cdot\nabla\Phi)_k$ can be approximated (with second-order accuracy) at velocity point k_1 as simply $\left[\Phi_{i_4} - \Phi_{i_1}\right]/dc_{k_1}$ (see Fig. 7.8). Using this representation of the gradient forcing, (7.76) at velocity point k_1 is rewritten as

$$\frac{\partial N_{k_1}}{\partial t} = \hat{\eta}_{k_1}\hat{T}_{k_1} - \left[\Phi_{i_4} - \Phi_{i_1}\right]/dc_{k_1} \tag{7.77}$$

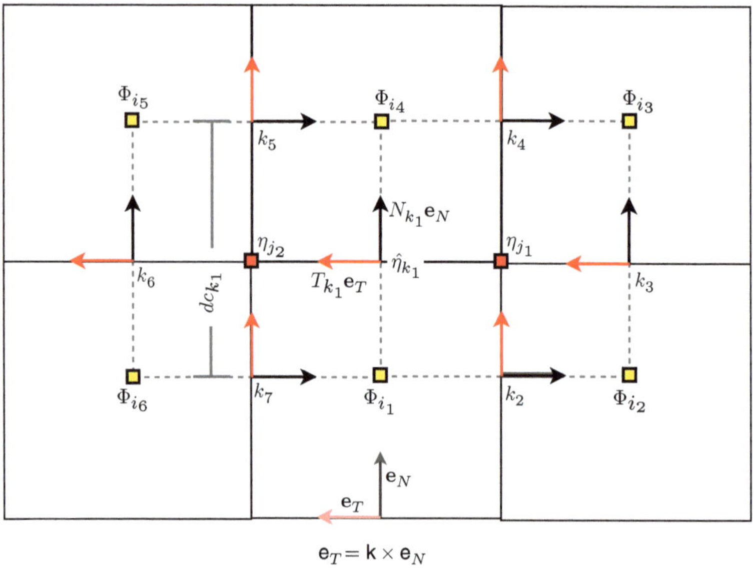

Fig. 7.8 The detailed description of the velocity and vorticity mesh

where dc_{k_1} is the distance between Φ_{i_4} and Φ_{i_1}. While the various ways to specify $\hat{\eta}_{k_1}$ are given in Sect. 7.7, at this point $\hat{\eta}_{k_1}$ can be constrained as

$$\hat{\eta}_{k_1} = f(\eta_{j_1}, \eta_{j_2}). \tag{7.78}$$

The absolute vorticity used to compute the nonlinear Coriolis force, $\hat{\eta}\hat{T}$, at velocity points is only a function of the vorticities defined at the end of the edge. Other approaches to specifying $\hat{\eta}$ are possible and often preferable, e.g. see Sadourny (1975) and Ringler et al. (2010) for a more in-depth discussion of the possible alternatives. In order to complete the specification of (7.77) a definition for $\hat{T}_{k_1}$ is required. The algorithm for computing $\hat{T}_{k_1}$ is also given in Sect. 7.7.

7.6.4 Derived Equation

The importance of derived equations in a discrete representation is frequently overlooked. Attention is more often focused on the analysis of the discrete prognostic equations since these are the variables that are explicitly tracked in the numerical model in time. In practice, an analysis of the derived equations generally provides important insights into the chosen numerical method. The purpose of this section is to demonstrate that the discrete system can mimic the continuous system in terms of the vorticity dynamics. The analysis carried out in Sects. 7.3.4.1 and 7.3.4.2 is repeated here, but in the setting of a discrete system. The primary property of the continuous system that the discrete system needs to mimic is

$$\frac{d}{dt}\Gamma^a_{c(t)} = \frac{d}{dt}\left[\int\limits_{S(t)} \eta\, dS\right] = \int\limits_{S_E} \left[\frac{\partial \eta}{\partial t} + \nabla \cdot (\eta\, \mathbf{u})\right] dS = 0, \tag{7.79}$$

where S_E will span one or more vorticity cells shown in Fig. 7.8. The absolute circulation following a contour $c(t)$ is conserved when the fluid density is constant (as is assumed here) and when no frictional forces are present. The challenge is to demonstrate that absolute circulation is conserved following a contour $c(t)$ *even when the discrete system does not directly prognose circulation or vorticity.* Stated another way, the goal is to demonstrate that the evolution of the discrete velocity field, N_k, is consistent with the kinematic constraints imposed by (7.79). Since the velocity evolution equation is written in an Eulerian reference frame, the analysis is most direct when the focus is on the third part of (7.79). The integration of dS can span a single cell or a collection of cells that are contained in a single loop.

The analysis begins by taking the discrete curl of the velocity tendency equation around the j_1 vorticity cell shown in Fig. 7.8. The discrete circulation operator is shown in Fig. 7.9. As seen in Fig. 7.9 the discrete curl has four terms, one for each edge of a vorticity cell. Using the labels shown in Fig. 7.9, the curl operator at vorticity point j_1 can be expressed as

$$\frac{1}{A}\oint\limits_{c} \mathbf{u} \cdot \mathbf{dr} \approx \frac{1}{A_{j_1}} \sum_{m=1}^{4} N_{k_m}\, \mathbf{e}_N \cdot \mathbf{dr}_{k_m} \tag{7.80}$$

where A_{j_1} is the area of the vorticity cell j_1. The dot product $\mathbf{e}_N \cdot \mathbf{dr}_{k_m}$ accounts for whether or not $N_{k_m}\,\mathbf{e}_N$ points in the same or the opposite direction as $\mathbf{dr}_{k_m}$. In addition, $\left|\mathbf{dr}_{k_m}\right| = dc_{k_m}$ to account for the distance of each segment of the loop around vorticity cell j_1.

A discrete equation for the evolution of absolute vorticity is constructed by applying the curl operator to each term in (7.77). In order to provide a clear representation of the curl operations, we will focus on vorticity point j_1. Beginning with the discrete curl of the time tendency of N_k, we find

$$\frac{1}{A}\oint\limits_{c} \frac{\partial \mathbf{u}}{\partial t} \cdot \mathbf{dr} \approx \frac{1}{A_{j_1}} \sum_{m=1}^{4} \frac{\partial N_{k_m}}{\partial t}\, \mathbf{e}_N \cdot \mathbf{dr}_{k_m} = \frac{\partial \zeta_{j_1}}{\partial t} = \frac{\partial \eta_{j_1}}{\partial t} \tag{7.81}$$

where the curl operator has been moved inside the time derivative and we have used the fact that $\frac{\partial f_o}{\partial t} = 0$. Now moving to the gradient term on the RHS of (7.77) we find

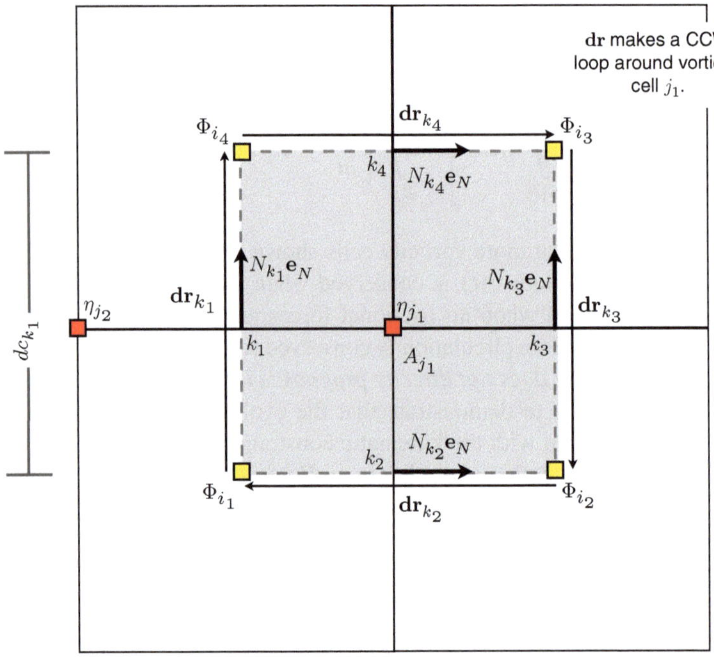

$$|\mathbf{dr}_{k_m}| = dc_{k_m}$$

$$\left[\frac{1}{A}\int_c \mathbf{u}\cdot\mathbf{dr}\right]_{j_1} \approx \frac{1}{A_{j_1}}\sum_{m=1}^{4} N_{k_m}\mathbf{e}_N\cdot\mathbf{dr}_{k_m}$$

Fig. 7.9 A graphical description of the discrete curl operator

$$-\frac{1}{A}\oint_c \nabla\Phi\cdot\mathbf{dr} \approx \frac{1}{A_{j_1}}\left[\frac{(\Phi_2-\Phi_1)}{dc_{k_2}}dc_{k_2} + \frac{(\Phi_3-\Phi_2)}{dc_{k_3}}dc_{k_3} - \frac{(\Phi_3-\Phi_4)}{dc_{k_4}}dc_{k_4} - \frac{(\Phi_4-\Phi_1)}{dc_{k_1}}dc_{k_1}\right] \tag{7.82}$$

where the distance used in the gradient calculation and the distance used in the curl operator cancel on each term. After removing these offsetting terms we find

$$\frac{1}{A}\oint_c \nabla\Phi\cdot\mathbf{dr} \approx \frac{1}{A_{j_1}}\left[(\Phi_2-\Phi_1)+(\Phi_3-\Phi_2)+(\Phi_4-\Phi_3)+(\Phi_1-\Phi_4)\right] = 0. \tag{7.83}$$

Just as in the continuous system, the curl of the gradient is identically zero. This property in the discrete system insures that forces in the velocity tendency equation of the form $\nabla\Phi$, where Φ is any scalar field defined at mass points, do not generate spurious vorticity.

Moving to the final term, the nonlinear Coriolis force, we find

$$\frac{1}{A}\oint_{c} \eta\,\mathbf{u}^{\perp}\cdot\mathbf{dr} \approx -\frac{1}{A_{j_1}}\sum_{m=1}^{4}\hat{\eta}_{k_m}\hat{T}_{k_m}\,\mathbf{e}_{N_{k_m}}\cdot\mathbf{dr}_{k_m}. \tag{7.84}$$

Expanding the summation yields

$$-\frac{1}{A_{j_1}}\sum_{m=1}^{4}\left(\hat{\eta}\hat{T}\,\mathbf{e}_N\right)_{k_m}\cdot\mathbf{dr}_{k_m} = -\frac{1}{A_{j_1}}\left[+\left(\hat{\eta}\hat{T}dc\right)_{k_1} - \left(\hat{\eta}\hat{T}dc\right)_{k_2} - \left(\hat{\eta}\hat{T}dc\right)_{k_3} + \left(\hat{\eta}\hat{T}dc\right)_{k_4}\right]. \tag{7.85}$$

Combining all of the curl operators to produce a discrete equation for the evolution of absolute vorticity yields

$$\frac{\partial\eta_{j_1}}{\partial t} + \frac{1}{A_{j_1}}\left[+\left(\hat{\eta}\hat{T}dc\right)_{k_1} - \left(\hat{\eta}\hat{T}dc\right)_{k_2} - \left(\hat{\eta}\hat{T}dc\right)_{k_3} + \left(\hat{\eta}\hat{T}dc\right)_{k_4}\right] = 0. \tag{7.86}$$

Comparing (7.86) to its continuous counterpart in (7.57), we see that the discrete vorticity evolution equation is an analog to the continuous system when

$$\nabla\cdot(\eta\,\mathbf{u}) \approx \frac{1}{A_{j_1}}\left[+\left(\hat{\eta}\hat{T}dc\right)_{k_1} - \left(\hat{\eta}\hat{T}dc\right)_{k_2} - \left(\hat{\eta}\hat{T}dc\right)_{k_3} + \left(\hat{\eta}\hat{T}dc\right)_{k_4}\right]. \tag{7.87}$$

The RHS of (7.87) is an approximation to the weak form of the divergence operator. The approximation is second-order accurate assuming suitable choices for $\hat{\eta}$ and $\hat{T}$. It is critical to note that in this discrete system *vorticity is transported by the reconstructed, tangential velocity field*. It is useful to recast (7.86) as an expression for the circulation within cell j_1 by moving the area into the time derivative as

$$A_{j_1}\frac{\partial\eta_{j_1}}{\partial t} = \frac{\partial\Gamma^a_{j_1}}{\partial t} = -\left[+\left(\hat{\eta}\hat{T}dc\right)_{k_1} - \left(\hat{\eta}\hat{T}dc\right)_{k_2} - \left(\hat{\eta}\hat{T}dc\right)_{k_3} + \left(\hat{\eta}\hat{T}dc\right)_{k_4}\right]. \tag{7.88}$$

$\Gamma^a_{j_1}$ represents the absolute circulation around the dual cell j_1. This result can be generalized to an arbitrary contour by progressively adding cells. Equation (7.88) represents a contour containing the j_1 vorticity cell. The discrete equation governing the evolution of circulation for the j_2 vorticity cell can be expressed as

$$\frac{\partial\Gamma_{j_2}}{\partial t} = -\left[-\left(\hat{\eta}\hat{T}dc\right)_{k_1} + \left(\hat{\eta}\hat{T}dc\right)_{k_5} + \left(\hat{\eta}\hat{T}dc\right)_{k_6} - \left(\hat{\eta}\hat{T}dc\right)_{k_7}\right]. \tag{7.89}$$

The edge shared by vorticity cells j_1 and j_2 is edge k_1. The term $\left(\hat{\eta}\hat{T}dc\right)_{k_1}$ appears in both (7.88) and (7.89), but with opposite signs. The evolution of absolute circulation formed by the contour containing vorticity cell j_1 and j_2 is thus

$$\frac{\partial(\Gamma_{j_1}+\Gamma_{j_2})}{\partial t} = -\Big[-\left(\hat{\eta}\hat{T}dc\right)_{k_2} - \left(\hat{\eta}\hat{T}dc\right)_{k_3} + \left(\hat{\eta}\hat{T}dc\right)_{k_4} + \left(\hat{\eta}\hat{T}dc\right)_{k_5} + \left(\hat{\eta}\hat{T}dc\right)_{k_6} - \left(\hat{\eta}\hat{T}dc\right)_{k_7}\Big] \tag{7.90}$$

where the shared edge between vorticity cells j_1 and j_2 cancels. The edges that remain all lie on the boundary of the contour and account for the transport of circulation across the boundary of the region. The mean absolute vorticity within the contour can always be determined by dividing the absolute circulation by the area enclosed in the contour. This analysis is sufficient to conclude that the discrete system conserves absolute circulation exactly. By extension, the discrete system conserves the area-mean absolute vorticity exactly. Both of these conservation statements mimic the findings in the continuous system. What is somewhat surprising is that these conservation statements have been proven *without even having to specify* $\hat{\eta}$ *or* $\hat{T}$. In that, the conservation statements hold for any $\hat{\eta}$ and any $\hat{T}$. The two essential ingredients required for these conservation statements to hold in the discrete system are the use of the vector-invariant form of the momentum equation and the discrete analog of the $\nabla \times \nabla\Phi \equiv 0$ identity.

The final and most important conclusion of this section is the following: The time tendency of velocity due to the nonlinear Coriolis force $\left(\hat{\eta}\hat{T}\right)$ **is** the per-unit-length absolute vorticity transport in the direction normal to $\mathbf{e}_N$. This is key to providing a direct handle on the vorticity dynamics of the discrete system via the discrete momentum equation.

7.7 Constraining the Evolution of Velocity Through the Transport of Absolute Vorticity

In the preceding section we were able to accomplish three goals. First, we were able to exhibit that absolute circulation is conserved for any closed loop in the discrete system. Second, the conservation statements related to circulation and vorticity hold exactly in the discrete system, even though neither are retained as prognostic variables. And finally, these conservation statements hold without having to specify the form of the reconstructed tangential velocity or the value of absolute vorticity used to compute the velocity tendency due to the nonlinear Coriolis force. Given this last statement, it should be clear that conservation alone is insufficient in specifying an adequate numerical model. The general framework allows us to specify $\hat{\eta}$ and $\hat{T}$ to meet other constraints that we deem important. The following discussion is meant to demonstrate the flexibility, or lack thereof, in the choice of $\hat{\eta}$ and $\hat{T}$. It turns out that there is some flexibility in the choice of the former and essentially no flexibility in the choice of the latter. As above, constant density is assumed. In addition, the analysis below assumes non-divergent flow in order to illustrate the relationship between vorticity transport and acceleration.

7.7.1 Considerations when Specifying $\hat{\eta}$

The specification of $\hat{\eta}$ should be made with two concerns in mind. The first is that since the nonlinear Coriolis force $\eta\,\mathbf{k}\times\mathbf{u}$ is always orthogonal to $\mathbf{u}$, the nonlinear Coriolis force neither produces nor destroys kinetic energy, i.e. $\mathbf{u}\cdot(\eta\,\mathbf{k}\times\mathbf{u}) = 0$. This is essentially a concern related to the energetics of the discrete system. The second concern is how the specification of $\hat{\eta}$ will influence the structure of the evolving vorticity field. For example, we would like to make some guarantees on the long-time smoothness of the discrete vorticity field. This is essentially a concern related to the vorticity dynamics of the discrete system. The goal, in my view, should be the rigorous guarantee of both of these concerns. In that, the guarantee that the choice of $\hat{\eta}$ neither produces or destroys kinetic energy *and* that this same choice in $\hat{\eta}$ promotes long-term smoothness in the vorticity field. Given the analysis and the anecdotal evidence presented in Ringler et al. (2010), this goal might be possible.

For the discussion presented here, the focus will be on choosing $\hat{\eta}$ such that the evolution of absolute vorticity is monotone in time.[9] In the context of transport, monotonicity implies that the vorticity field at some time t can be determined as a *convex interpolation* of the vorticity field at some previous time (Godunov 1959). Since the interpolation process is *convex*, vorticity values at some previous time are given weights between zero and one. Thus monotonicity implies that the solution of vorticity at any time t is bounded from above and below by the vorticity at any previous time. While it is true that only in the special case of non-divergent flow should we expect absolute vorticity to evolve monotonically in time, extensions of this idea to potential vorticity holds for general 3D flows. If we assume an arbitrary velocity field that is non-divergent, then the continuous vorticity equation (7.58) reduces to

$$\frac{\partial \eta}{\partial t} + \nabla\cdot(\eta\,\mathbf{u}) = \frac{\partial \eta}{\partial t} + \mathbf{u}\cdot\nabla\eta = \frac{D\eta}{Dt} = 0, \tag{7.91}$$

which states that the absolute vorticity attributed to a particle (e.g. Fig. 7.1) is invariant in time. Since we are not in a Lagrangian reference frame where tracking particles is an option, the discrete model will have to attempt to mimic (7.91) in an Eulerian setting. When a property is conserved along particle trajectories it

---

9 Discussing the evolution of potential vorticity, as opposed to absolute vorticity, would be more relevant here. But for the reasons discussed in the Introduction, we will limit the scope to the evolution of absolute vorticity. Only in the special case of non-divergent flow is the evolution of absolute vorticity monotone. In addition, the topic of transport (monotone or otherwise) warrants an entire chapter to itself.

means that the quantity itself (e.g. η) and all moments of that quantity (e.g. η^n where n is any integer) are also conserved along particle trajectories. With only 1° of freedom in the discrete system (i.e. $\hat{\eta}$), we are woefully ill-equipped to mimic the richness contained in the continuous system and, therefore, must make some tough choices regarding how to specify $\hat{\eta}$. The goal here is not to determine an optimal specification of $\hat{\eta}$ but rather to demonstrate that we can *guarantee* a monotone evolution of vorticity even when the only prognostic variable is the normal component of velocity at cell edges.

Assuming that the discrete velocity field is non-divergent, guaranteeing a monotone evolution of the discrete absolute vorticity field is straightforward. Focusing on edge (k_1), we specify $\hat{\eta}_{k_1}$ as

$$\textit{if}\ \hat{T}_{k_1} \geq 0,\ \hat{\eta}_{k_1} = \eta_{j_1} \tag{7.92}$$

$$\textit{if}\ \hat{T}_{k_1} < 0,\ \hat{\eta}_{k_1} = \eta_{j_2} \tag{7.93}$$

in that we always choose the value of $\hat{\eta}$ by picking the vorticity value *upstream of* $\hat{T}$. While this is essential the low-order, monotone solution used in Zalesak (1979), it immediately generalizes to higher-order. Without loss of generality, assume that $\hat{T}_{k_1} \geq 0$ at some instant in time, then the evolution equation of N_{k_1} is written as

$$\frac{\partial N_{k_1}}{\partial t} = \eta_{j_1} \hat{T}_{k_1} - \left[\Phi_{i_4} - \Phi_{i_1}\right] / dc_{k_1}. \tag{7.94}$$

If $\hat{\eta}$ is chosen based on the approach in (7.92), then the absolute vorticity associated with the evolving N_k velocity field will be monotone. To be clear, the donor cell approach results in excessive diffusion and this discussion is in no way meant to advocate for the use of (7.92); it is employed here for demonstration purposes only. In practice, we can apply state-of-the-art transport algorithms for the computation of the absolute vorticity flux, $\hat{\eta}\hat{T}$, and use that flux as the nonlinear Coriolis force in the velocity tendency equation.

7.7.2 Considerations when Specifying $\hat{T}$

It turns out that there is essentially no flexibility in the choice of $\hat{T}$. The mesh used here is essentially identical to that used in Arakawa and Lamb (1981). In that work, the reconstructed velocity is specified as

$$\hat{T}_{k_1} = -\frac{1}{4}\left(N_{k_7} + N_{k_2} + N_{k_4} + N_{k_5}\right). \tag{7.95}$$

(see Fig. 7.8). The reasoning behind this choice is not particularly clear in Arakawa and Lamb (1981). Based on the more recent analysis conducted on general unstructured meshes with C-grid staggerings in Thuburn et al. (2009) and Ringler et al.

(2010), it is clear that the critically important aspect of the reconstructed $\hat{T}$ field is that the $\left[\nabla \cdot \left(\hat{T}\, \mathbf{e}_T\right)\right]_j$ be an interpolation of the neighboring $[\nabla \cdot (N\, \mathbf{e}_N)]_i$ values; the divergence computed at vorticity points based on $\hat{T}_k$ *must* be an interpolation of the divergence computed at mass points based on N_k.

The importance and significance of this requirement can be clearly seen in the following example. Suppose the continuous system is characterized with an initial condition of uniform absolute vorticity field being transported by a non-divergent flow. From (7.58) we see that the solution for all time is simply $\frac{\partial \eta}{\partial t} = 0$. Also suppose that the discrete velocity field N_k is chosen such that it produces a uniform absolute vorticity field and is also non-divergent. The discrete system from (7.86) can be expressed as

$$\frac{\partial \eta_{j_1}}{\partial t} + \frac{\eta_o}{A_{j_1}} \left[\left(\hat{T} dc\right)_{k_1} + \left(\hat{T} dc\right)_{k_2} - \left(\hat{T} dc\right)_{k_3} - \left(\hat{T} dc\right)_{k_4} \right] = 0 \tag{7.96}$$

The only way to reproduce the solution of $\frac{\partial \eta}{\partial t} = 0$ for all time is to require that

$$\left[\left(\hat{T} dc\right)_{k_1} + \left(\hat{T} dc\right)_{k_2} - \left(\hat{T} dc\right)_{k_3} - \left(\hat{T} dc\right)_{k_4} \right] = 0. \tag{7.97}$$

Equation (7.97) requires that the divergence of the reconstructed, tangential velocity at vorticity points is also zero. If one can build a general algorithm for the reconstruction of $\hat{T}$ that produces $\left[\nabla \cdot \left(\hat{T}\, \mathbf{e}_T\right)\right]_j = 0$ when $[\nabla \cdot (N\, \mathbf{e}_N)]_i = 0$, then we have sufficient proof that the divergence computed at vorticity points will be a convex interpolation of the divergence computed at mass points. Unfortunately, the failure of some C-grid staggered model to enforce this essential feature in the reconstruction of the tangential velocity has lead to (sometime severe) limitations in the robustness of the numerical model and the quality of the numerical solutions.

7.8 Final Thoughts

This analysis provided an end-to-end discussion of one aspect in the construction of a dynamical core, namely the derivation and approximation of the equations related to the evolution of momentum. As much as possible, the analysis is developed with the aid of the Reynolds Transport Theorem. In addition to providing a rigorous means to recasting conservation statements made in the Lagrangian reference frame to statements applicable to the Eulerian reference frame, the Reynolds Transport Theorem produces evolution equations cast in a weak, integral form that fit naturally into traditional finite-volume approaches.

The analysis lingered and continually revisited the relationship between the evolution of velocity and vorticity dynamics. The reason for such a strong emphasis on this relationship is that while the evolution of momentum has to be faithful to $\mathbf{F} = m\,\mathbf{a}$, it also has to respect the kinematic constraints implied by conservation statements related to vorticity and circulation. First, the relationship has to be understood in the continuous setting, then the relationship has to be accommodated in the development of the discrete system of equations.

The system of equations that one chooses as the starting point for constructing a discrete model is a critical moment in the construction of a dynamical core. This choice will have a profound impact on the quality of the simulations. Understanding the anticipated use of the numerical model is a prerequisite to making sound, defensible choices for the components of a dynamical core. For this reason, an entire section related to the "process of discretization" is included. While the actual choices made in that section are highly biased, the purpose of the section is to hopefully motivate the extreme importance of choosing numerical methods based on a target application.

Acknowledgments This work was supported by the DOE Office of Science's Climate Change Prediction Program DOE 07SCPF152. I would like to thank Professor Sidney Leibovich for providing a thorough training in the foundation of fluid dynamics. In addition, I would like to thank Professor David Randall for his mentoring and invaluable insights into the construction of robust numerical algorithms. The chapter benefited greatly from a critical reviews provided by John Dukowicz, an anonymous reviewer, and the editors.

References

Arakawa A (1966) Computational design for long-term numerical integration of the equations of fluid motion: Two-dimensional incompressible flow. Part I. J Comput Phys 1:119–143

Arakawa A (1997) Computational design for long-term numerical integration of the equations of fluid motion: Two-dimensional incompressible flow. Part I. J Comput Phys 135:103–114

Arakawa A, Lamb VR (1981) A potential enstrophy and energy conserving scheme for the shallow water equations. Mon Wea Rev 109:18–36

Batchelor GK (1967) Introduction to fluid dynamics. Cambridge Press 636 pp.

Bleck R, Smith LT (1990) A wind-driven isopycnic coordinate model of the North and equatorial Atlantic ocean 1. Model development and supporting experiments. J Geophys Res 95(C3):3273–3285

DeCaria AJ, Sikora TD (2010) Momentum advection and the gradient of a vector field: A discussion of standard notation. J Atmos Sci 67:1287–1291

Godunov SK (1959) A difference method for numerical calculation of discontinuous solutions of the equations of hydrodynamics. Mat Sb 47(89)(3):271–306

Haertel PT, Van Roekel L, Jensen TG (2009) Constructing an idealized model of the North Atlantic Ocean using slippery sacks. Ocean Modelling 27:143–159

Heikes R, Randall DA (1995) Numerical integration of the shallow-water equations on a twisted icosahedral grid. Part I: Basic design and results of tests. Mon Wea Rev 123:1862–1880

Hirt CW, Amsden AA, Cook JL (1997) An arbitrary Lagrangian-Eulerian computing method for all flow speeds. J Comput Phys 135(2):203–216

Konor C, Arakawa A (1997) Design of an atmospheric model based on a generalized vertical coordinate. Mon Wea Rev 125:1649–1673

Leung LR, Qian Y, Bian X, Washington WM, Han J, Roads JO (2004) Mid-century ensemble regional climate change scenarios for the western United States. Climatic Change 62(1–3):75–113

Randall DA (1994) Geostrophic adjustment and the finite-difference shallow-water equations. Mon Wea Rev 122(6):1371–1377
Ringler TD, Heikes R, Randall D (2000) Modeling the atmospheric general circulation using a spherical geodesic grid: A new class of dynamical cores. Mon Wea Rev 128(7):2471–2490
Ringler TD, Thuburn J, Klemp JB, Skamarock WC (2010) Numerical treatment of energy and potential vorticity on arbitrarily structured C-grids. J Comput Phys 229:3065–3090
Ritchie H, Temperton C, Simmons A, Hortal M, Davies T, Dent D, Hamrud M (1995) Implementation of the semi-Lagrangian method in a high-resolution version of the ECMWF forecast model. Mon Wea Rev 123(2):489–514
Sadourny R (1975) The dynamics of finite-difference models of the shallow-water equations. J Atmos Sci 32(4):680–689
Sadourny R, Morel P (1969) A finite-difference approximation of the primitive equations for a hexagonal grid on a plane. Mon Wea Rev 97(6):439–445
Skamarock WC, Klemp JB, Dudhia J, Gill DO, Barker DM, Duda MG, Huang XY, Wang W, Powers JG (2008) A description of the Advanced Research WRF Version 3. NCAR Tech. Note NCAR/TN-475+STR, National Center for Atmospheric Research, Boulder, Colorado, 113 pp., available from http://www.ucar.edu/library/collections/technotes/technotes.jsp
Staniforth A, Côté J (1991) Semi-Lagrangian integration schemes for atmospheric modeling - A review. Mon Wea Rev 119:2206–2223
Thuburn J (1997) A PV-based shallow-water model on a hexagonal–icosahedral grid. Mon Wea Rev 125(9):2328–2347
Thuburn J, Ringler TD, Skamarock WC, Klemp JB (2009) Numerical representation of geostrophic modes on arbitrarily structured C-grids. J Comput Phys 228(22):8321–8335
Tomita H, Miura H, Iga S, Nasuno T, Satoh M (2005) A global cloud-resolving simulation: Preliminary results from an aqua planet experiment. Geophys Res Lett 32(8):L08,805
White F (2008) Fluid mechanics. McGraw-Hill Higher Education 896 pp.
Zalesak ST (1979) Fully multidimensional flux-corrected transport algorithms for fluids. J Comput Phys 31(3):335–362
Zhang H, Rančić M (2007) A global Eta model on quasi-uniform grids. Quart J Roy Meteor Soc 133:517–528

Chapter 8
Atmospheric Transport Schemes: Desirable Properties and a Semi-Lagrangian View on Finite-Volume Discretizations

Peter H. Lauritzen, Paul A. Ullrich, and Ramachandran D. Nair

Abstract This chapter has twofold purpose. After a short introduction to the mass continuity equations in atmospheric models, desirable properties for mass transport schemes intended for meteorological applications are discussed in some detail. This includes a discussion on the complications caused by the non-linearity of most problems of interest that makes it hard to define accuracy and convergence as the 'truth' is not known. Thereafter, some finite-volume schemes from the atmospheric literature are reviewed and discussed. To complement the large existing literature on finite-volume schemes, a less frequently discussed semi-Lagrangian derivation of the finite-volume method is given that focuses on 'remap-type' schemes where the space and time discretizations are combined rather than separated. A discussion on the challenges in deriving accurate schemes intended for global models and non-traditional spherical grids is given as well.

8.1 Introduction

To predict the evolution of air and tracers[1] we solve one of the fundamental laws of physics: namely the equation of mass continuity. This equation is intuitively very simple to understand; perhaps the simplest statement of the equation is that mass of air and tracers is conserved without the presence of sources or sinks. Hence mass in a

[1]A tracer in this context is any quantity that follows the flow of air such as chemical species and water in the atmosphere.

P.H. Lauritzen (✉)
National Center for Atmospheric Research, 1850 Table Mesa Drive, Boulder, CO 80305, USA
e-mail: pel@ucar.edu

P.A. Ullrich
University of Michigan, 2455 Hayward St., Ann Arbor, MI 48109, USA
e-mail: paullric@umich.edu

R.D. Nair
National Center for Atmospheric Research, 1850 Table Mesa Drive, Boulder, CO 80305, USA
e-mail: rnair@ucar.edu

P.H. Lauritzen et al. (eds.), *Numerical Techniques for Global Atmospheric Models*,
Lecture Notes in Computational Science and Engineering 80,
DOI 10.1007/978-3-642-11640-7_8,

volume can only change if there is inflow and/or outflow through surfaces bounding the closed volume or if there are parameterizised sources and/or sinks (e.g., for water vapor, sources and sinks can be evaporation and condensation, respectively)

$$\frac{d}{dt}(mass) = inflow - outflow + sources - sinks. \tag{8.1}$$

The continuity equation is simple, in a strict mathematical sense, and it may appear as a surprise that we use an entire chapter discussing it. However, despite its simplicity finding an accurate and efficient numerical approximation to its solution remains an active research subject and no scheme to date is ideal (and perhaps never will be as long as computing power remains finite). Also, the continuity equation is coupled to the other equations of motion, so a complete discussion of the challenges in air and tracer transport must also consider this coupling. The purpose if this chapter is to convey some of the many deliberations in transport scheme development and to discuss some examples of transport schemes on the sphere.

In the literature there are numerous review articles and books on transport methods in general and specifically on the finite-volume method (e.g., Rood 1987; LeVeque 1996) and this chapter is not an attempt to supersede or replace these reviews. Instead we shall limit the review to space-time (or remap) finite-volume transport schemes used in meteorology. By space-time schemes we refer to schemes where the temporal and spatial discretizations are combined rather than separated. As will become clear one may also refer to space-time (or remap) schemes as *cell-integrated* (or *finite-volume*) *semi-Lagrangian* schemes. Conservative grid-to-grid interpolation (also referred to as remapping), which is usually an integral part of finite-volume schemes, will also be discussed in some detail. Obviously this chapter will only scratch the surface of the enormous literature on transport schemes and we will emphasize the intuitive (and perhaps more physical) derivation of schemes rather than mathematical rigor.

The chapter is organized as follows. Before diving into the nuts and bolts of finite-volume schemes we begin by formulating the transport problem relevant to atmospheric models (Sect. 8.2) and discuss some desirable properties that transport schemes intended for atmospheric applications ideally should possess (Sect. 8.3). In Sect. 8.4 the mathematical foundation for space-time finite-volume schemes is given in Eulerian and Lagrangian forms. The equivalence between the two forms is rarely discussed but useful in gaining more understanding of Eulerian schemes. In Sect. 8.5 the spatial and temporal approximations needed for practical schemes are presented step by step. This includes upstream cell approximation, sub-grid-cell reconstruction and practical integration over cells in space. Section 8.5 is mostly limited to two-dimensional schemes on the Cartesian plane, however, a brief discussion on the extension to spherical geometry is given. In Sect. 8.6 we discuss the extension to three dimensions. Before the final remarks in Sect. 8.8 some practical considerations for the coupling of transport schemes to the continuity equation for air are discussed in Sect. 8.7. This includes the inconsistencies that may arise in the air mass and tracer mass coupling, techniques for sub-cycling the air mass equation with respect to tracers, and coupling a semi-implicit air mass scheme with an explicit tracer mass scheme (Sect. 8.7). For brevity, Sects. 8.6 and 8.7 are cursory while more attention will be given to desirable properties and the space-time scheme derivations on the plane.

8.2 The Continuous Equation

8.2.1 Representation of Mass in Atmospheric Models

Most atmospheric models have at least a handful of continuity equations and, in most cases, many more. From a dynamics point of view the continuity equation for air is the most fundamental and important continuity equation since it is strongly coupled to the momentum equations and the thermodynamic equation. For the representation of moist processes most models have prognostic continuity equations for three water species: Water vapor, cloud liquid water and cloud ice water. Some high resolution models also have resolved-scale continuity equations for rain and snow (if there is no resolved-scale continuity equation for rain and snow the assumption is usually that rain and snow falls to the ground in one time-step). Modern micro-physical parameterizations include prognostic continuity equations for four to eight condensed species. For example, the Morrison and Gettelman (2008) micro-physics package used in NCAR's Community Atmosphere Model (CAM) version five has continuity equations for mass and number concentrations for ice and liquid water. Some microphysics parameterizations also have prognostic continuity equations for mass and number concentrations for ice and liquid precipitation. Modal (and even more for bin) aerosol schemes may have 20 or more prognostic continuity equations for mass and number concentrations of aerosols such as particulate organic matter, dust, sea salt, secondary organic aerosols, number concentrations for different sizes of aerosols, etc. In addition, any prognostic representation of chemical species requires the solution to one continuity equation per species e.g., MOZART (Model of Ozone And Related Tracers, Brasseur et al. 1998). So needless to say, the continuity equations make up a dominant part in atmospheric models at least in terms of the total computational cost of the dynamical core.[2]

First, let us discuss the representation of air mass in atmospheric models as this has fundamental influence on how all other species are treated. The density of well-mixed moist air ρ_m can be separated into a dry and wet part

$$\rho_m = \frac{m_d + m_v}{V} = \rho_d + \rho_v = \rho_d + q_v \rho_m, \tag{8.2}$$

[2] Roughly speaking the *dynamical core* is the part of the model that solves the governing fluid and thermodynamic equations on resolved scales (Thuburn 2008b).

where m_d and m_v are the masses of the dry air and water vapor, respectively, and V is a small volume. The density of dry air and water vapor are denoted ρ_d and ρ_v, respectively, and q_v is the specific humidity,

$$q_v = \frac{m_v}{m_d + m_v}. \tag{8.3}$$

To a very good approximation the mass of dry air is the mass of the dominant well-mixed gases: Nitrogen N_2 (ca. 78.08%), Oxygen O_2 (ca. 20.95%), Argon Ar (ca. 0.93%) and Carbon dioxide CO_2 (at present ca. 0.038%). These gases make up over 99.998% of the volume of dry air and may therefore be considered permanent (although argon and carbon-dioxide are slowly increasing). In addition, small amounts of trace gases are mixed into the air (with sources and sinks varying in space and time), however, the variation in these 'non-permanent' gases is very small compared to the total mass of all the trace gases. Trenberth and Smith (2005) estimated that the dry air mass of the atmosphere corresponds to a surface pressure of approximately 983.05 hPa and it varies less than 0.01 hPa based on changes in atmospheric composition. So the variation in the dry air mass budget is on the order of 0.001%. So to a very good approximation the continuity equation for dry air does not have any source or sink terms, and thus reads

$$\frac{\partial \rho_d}{\partial t} + \nabla \cdot (\rho_d \, \mathbf{v}) = 0, \tag{8.4}$$

where $\mathbf{v}$ is the velocity field and '$\nabla \cdot$' is the divergence operator. The mass of dry air accounts for approximately 99% of the total mass of the atmosphere and the remaining 1% is approximately the mass of water vapor. The continuity equation for humidity (water vapor) is given by

$$\frac{\partial}{\partial t} (\rho_m q_v) + \nabla \cdot (\rho_m q_v \, \mathbf{v}) = P_{q_v \rho_m}, \tag{8.5}$$

where $P_{q_v \rho_m}$ represents sources and sinks (in this case condensation and evaporation processes). Moisture q_v varies significantly (relatively speaking) with values near zero for cold dry air and a few percent in warm moist air. The continuity equation for moist air can be obtained by adding (8.4) and (8.5), and using (8.2) to simplify. The result is

$$\frac{\partial \rho_m}{\partial t} + \nabla \cdot (\rho_m \, \mathbf{v}) = P_{\rho_m}. \tag{8.6}$$

This equation is similar to the equation for dry air (8.4) except for the humidity forcing terms.

The prognostic variables used for tracers are usually defined in terms of mixing ratios. If moist density is prognosed, the mixing ratios for tracers are most conveniently defined in terms of the specific concentration

$$q_m^{(l)} = \frac{m^{(l)}}{m_d + m_v}, \tag{8.7}$$

where $m^{(l)}$ is the mass of constituent (l). So the density of the constituent is $\rho^{(l)} = q_m^{(l)}\rho_m$, where $q_m^{(l)}$ is the 'moist' mixing ratio. However, one may also solve the continuity equation for tracers in terms of the 'dry' mixing ratio $q_d^{(l)}$, defined by

$$q_d^{(l)} = \frac{m_d^{(l)}}{m_d}. \tag{8.8}$$

As discussed in Collins et al. (2004) the advantage of using (8.7) is that the mass of species (l) is obtained by simply multiplying the moist mixing ratio with the moist air density $q_m^{(l)}\rho_m$. However, this approach has the disadvantage of implicitly requiring a change in $q_m^{(l)}$ whenever the water vapor q_v changes. This disadvantage does not exist if (8.8) is used.

8.2.2 Consistency in the Mass Equations

Herein we will respectively use ρ and q to denote air density and mixing ratio (which can be either moist or dry) and we assume no sources or sinks (no forcing terms). Then the continuity equation for air density ρ can be written as

$$\frac{\partial\rho}{\partial t} + \nabla\cdot(\rho\,\mathbf{v}) = 0, \tag{8.9}$$

and similarly for a tracer density $\rho\,q$

$$\frac{\partial(\rho\,q)}{\partial t} + \nabla\cdot(\rho\,q\,\mathbf{v}) = 0, \tag{8.10}$$

where $\mathbf{v}$ is the velocity vector. Note that (8.9) and (8.10) imply

$$\frac{dq}{dt} = 0, \quad \frac{d}{dt} \equiv \frac{\partial}{\partial t} + \mathbf{v}\cdot\nabla, \tag{8.11}$$

which states that q is conserved along trajectories/characteristics of the flow. Note that the continuity equations (8.9) and (8.10) are linked in the sense that ρ appears in both equations. Hence, numerical error introduced in simulating the evolution of air mass ρ may be reflected in the prognosed trace gas mixing ratios when converting from tracer mass $\rho\,q$ to mixing ratio q.

To solve any of the continuity equations given above the flow field $\mathbf{v}$ must be given. The continuity equation for air (8.9) is coupled with the momentum equations and thermodynamic equations. Hence the thermodynamic variables and other prognostic variables feed back on the velocity field which, in turn, feeds back on the solution to the continuity equation. It follows that the continuity equation for air cannot be solved in isolation and one must obey the maximum allowable time-step restrictions imposed by the fastest waves in the system[3] (see Chaps. 1 and 6). The passive tracer transport equation (8.10) or (8.11) can be solved in isolation given prescribed winds and air densities, and is therefore not susceptible to the stricter time-step restrictions imposed by the fastest waves in the system but 'only' to the less restrictive advective velocities.[4] Hence, if for stability relatively short time-steps must be used for the continuity equation for air, one does not necessarily need to use short time-steps for the tracers (at least not for stability reasons). That is, one can solve the tracer transport equations with time-steps longer than what are allowed for stability in (8.9). This technique is referred to as *sub-cycling*, that is, multiple cycles of dynamics (air continuity equation) are performed within one time-step of the tracers. In doing so care must be taken to retain the consistency between tracers and air. For example, if $q = 1$ then (8.10) reduces to (8.9) and additional care must be taken to ensure consistency between these equations in the discretization. Specific examples and details on sub-cycling are given later (Sect. 8.7.2). First, let us consider important design objectives for tracer transport schemes intended for atmospheric applications.

8.3 Desirable Properties

When developing a new transport (or any other) algorithm one is usually striving for a scheme that ensures simulation *veracity*. In other words, a numerical method should be designed so that simulations using it are as truthful as possible. In mathematical literature simulation *veracity* is often synonymous with accuracy which is associated with the absolute truth. Convergence, truncation error and error norms are all associated with quantitative measures of conformity to the truth. In most realistic atmospheric model settings, however, the truth is unknown in an absolute sense (the exact solution is not known). For instance, in most atmospheric applications an increase in resolution will often resolve finer scales and new phenomena appear making it problematic to define convergence in a strict mathematical sense. Adding to the complexity is the fact that the system is chaotic and therefore not deterministic beyond 10 days or so (Lorenz 1982), so any attempt to assess absolute accuracy in simulations beyond the predictability limit must be based on statistical approaches.

In all, simulation veracity in an atmospheric modeling context is more than accuracy in a strict mathematical sense. Perhaps because there is little quantitative knowledge of the true solution a lot of emphasis is placed on physical properties of the solution method. For example, we do know that the numerical solution should ideally obey discretized equivalents of properties we can derive from the

---

[3] Assuming that explicit time-stepping is used.

[4] Although there is a weak coupling between humidity and the thermodynamic/momentum equation.

continuous set of equations such as conservation[5] of mass and higher moments, shape-preservation (including monotonicity, positivity and non-oscillatory property), correlation preservation, and so on. Also, sub-grid-scale parameterizations usually require physical realizable atmospheric states from the resolved scale dynamics. From a computational point of view properties such as parallel efficiency, geometric flexibility, etc. are also very important properties of the final numerical algorithm.

What follows is a list of desirable properties for tracer transport schemes that are all (apart from the properties related to efficiency) essential ingredients of simulation veracity.

8.3.1 Accuracy (Error Norms)

Accuracy describes the degree of closeness of the simulated (numerically computed) solution to its true (exact) state specified in terms of error norms (numeric values). The error measures can either be assessed at a fixed resolution (absolute error) or as a function of resolution (convergence). For linearized equations and approximations a proxy for convergence can be sought by computing the formal order of accuracy of the numerical method through Taylor series expansions. Note that formal order of accuracy does not necessarily guarantee accurate solutions for distributions/flows with near discontinuities (shocks and fronts) nor does it guarantee accuracy at a particular resolution. For many global weather and climate applications absolute accuracy at a particular range of resolutions is perhaps more important than high-order convergence rates. Below is a list of some idealized test cases used to quantitatively assess simulation veracity:

8.3.1.1 Linear Test Cases

Error norms are well defined when the exact solution is known which is usually only the case for linear problems. Commonly used linear test cases, where the analytical solution is known at all times t, can be divided into two categories: Translational and deformational. Here we focus only on global test cases in spherical geometry.

Most test cases are formulated with non-divergent flow fields for which the advective form of the continuity equation for a tracer (8.11), that uses mixing ratio q as the prognostic equation, is equivalent to the flux-form version (8.10) based on tracer mass ρq. That is, q or ρq is set equal to the same spatial distribution and the modeler is implicitly assuming that ρ is one everywhere and since the flow is non-divergent ρ will remain one through-out the simulation at least in the

[5] For a discussion on conservation in the context of the full equation set for the atmosphere see Chap. 11.

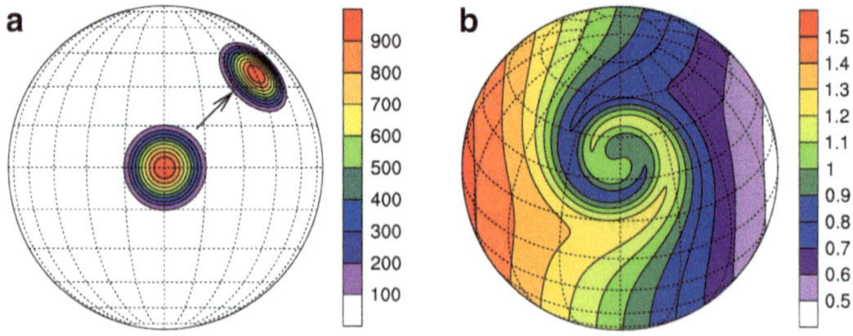

Fig. 8.1 Exact solutions for the (**a**) solid body advection of a cosine bell test case at $t = 0$ (center of plot) and $t = 44$ h (for a 'flow rotation angle' of 45°), and (**b**) the static vortex test case at day 6

analytical case. Hence the modeler is not forced to distinguish between tracer mass ρq and mixing ratio q. However, for a divergent/convergent flow only q is constant along parcel trajectories whereas tracer mass ρq will increase/decrease in areas of convergence/divergence. For a fuller discussion see Nair and Lauritzen (2010).

Translational. Probably the most commonly used idealized test case in the meteorological literature is the solid body rotation of a cosine bell (Fig. 8.1a) (e.g., test case one of the widely used two-dimensional test suite of Williamson et al. 1992 for the shallow-water equations). The exact solution is simply the translation of the initial condition and standard error norms can be computed at every time-step. This two-dimensional test case has been extended to three dimensions in Jablonowski et al. (2011). Another three-dimensional test case on the sphere where the analytic solution is known was proposed by Zubov et al. (1999).

For convergence studies used to assess the formal order of accuracy of a scheme the translated distribution should be sufficiently smooth. For example, the cosine bell distribution may appear smooth but it is only C^1 at the base of the bell. Consequently, schemes that are high-order accurate in terms of a Taylor Series analysis may not show this high-order formal convergence rate when using the cosine bell initial condition. To assess 'ideal' convergence rates it is advised to use C^∞ functions such as Gaussian surfaces (Levy et al. 2007).

Deformational. The translational test case described above has a large degree of symmetry and perhaps is not challenging enough to thoroughly test a numerical algorithm. Real world flows also have deformational, convergent/divergent and rotational components that deform, expand and rotate the initial distribution. A popular purely deformational test case (non-divergent) is the cyclogenesis test case introduced in meteorology by Doswell (1984) and used as a test case for transport schemes by numerious authors (e.g., Rančić 1992; Nair and Machenhauer 2002). The exact solution at day 6 is shown on Fig. 8.1b.[6] As can been seen in the figure the vortex 'curls up' and generates long thin filaments in the process. These, in general, are quite challenging to represent for any numerical scheme.

[6] The dimensionalization of the vortex problem used here follows Nair and Jablonowski (2008).

Fig. 8.2 Exact solutions for the moving vortex test case with a flow orientation of 45° at zero (*left*; day 0), half (*middle*; day 6) and one (*right*; day 12) revolution. The continents are shown for reference purposes only

Another useful application of this test case is to use its velocity field but instead of transporting an initial condition such as shown on Fig. 8.2 in the cyclogenesis test case, instead transport a constant mass field $\rho = \rho_0$. Since the flow is non-divergent any numerical scheme should ideally preserve a constant mass field. Also the solid-body rotation flow field is non-divergent and so for this wind field a constant mass-field should remain constant throughout the simulation. However, the cyclogenesis wind field is much more challenging as a preservation of constancy test since it is deformational (unless the stream function for the velocity field is used to makes sure that the divergence that the scheme 'sees' is zero). Some schemes might preserve a constant mass field for solid body advection but fail to preserve a constant mass field for the deformational wind field. Unfortunately results from such tests are rarely presented in the literature.

Translational and deformational. Although the idealized cyclogenesis test case described above is challenging it lacks a translational component. Nair and Jablonowski (2008) combined the cyclogenesis wind field with the solid body advection wind field on the sphere which makes up the 'moving vortices' test case. Instead of a stationary 'curl up' of the vortex, it is transported as a solid body as it deforms (Fig. 8.2). Obviously such a test case is more challenging and might therefore be more useful to discriminate between schemes than simpler test cases. For example, in the idealized tests of the finite-volume transport scheme in Lauritzen et al. (2010) it was found that the moving vortices test case was more discriminating than the pure translational and stationary cyclogenesis test cases (at least when applied and compared to the Putman and Lin (2007) scheme).

Recently, Nair and Lauritzen (2010) extended LeVeque's test case (LeVeque 1996) to a class of test cases on the sphere. Unlike all the test cases considered so far the wind fields in this test case are time varying. In these cases the wind fields are periodic and reverse so that after one period the initial distribution has returned to its initial position and shape. Hence the analytic solution is known after one period but not throughout the simulation. The flow is swirling (deformational) so the initial condition is highly deformed half way through the simulation. This challenges the numerical scheme since grid-scale features develop from well-resolved initial

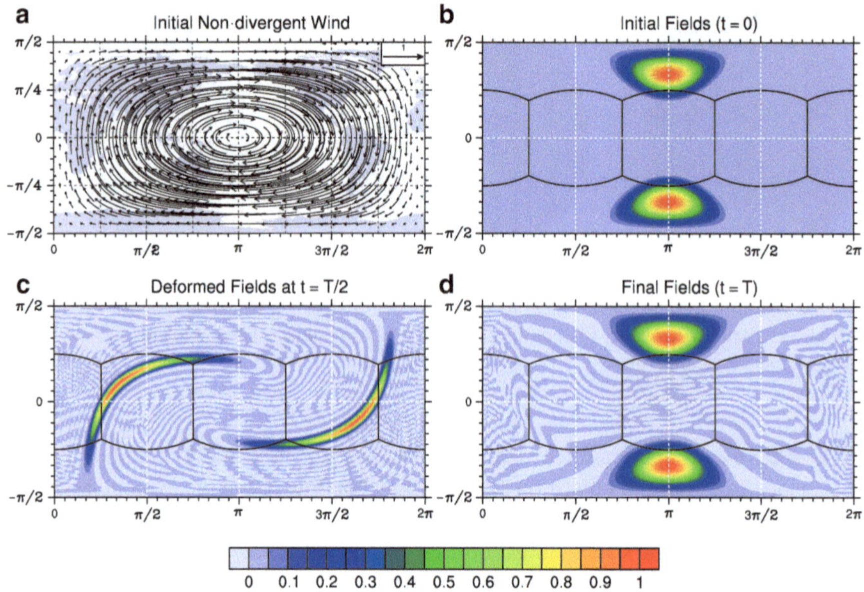

Fig. 8.3 The recently proposed test case by Nair and Lauritzen (2010). (**a**) The initial wind field and (**b**) initial condition and analytical solution after one period. (**c**) and (**d**) show a numerically computed solution after half and full period, respectively. The grid-scale noise in (**c**) and (**d**) are due to the numerical scheme not being monotone/shape-preserving

conditions (Fig. 8.3). And perhaps more importantly, one of the test cases in Nair and Lauritzen (2010) is divergent contrary to most idealized test cases for transport on the sphere which are non-divergent. By introducing divergence the modeler is forced to distinguish between mixing ratio and air mass which is not strictly necessary for non-divergent test cases.

8.3.1.2 Non-Linear Test Cases

In linear test cases for smooth flows the accuracy in terms of error norms is usually improved when the resolution is increased and when the formal order of the numerical method is increased. However, such idealized experiments do not truly quantify the error in realistic atmospheric applications that are far from linear. In general, for non-linear problems the quantification of error is problematic except in very simple cases[7] and, as discussed in Prather et al. (2008), we usually design models with the expectation that a correct solution (truth) exists and that with adequate physical approximations and numerical methods our solutions will converge to a 'true' solution as the resolution is increased.

[7] e.g., the one-dimensional Burgers' equation that has an exact solution although it is non-linear.

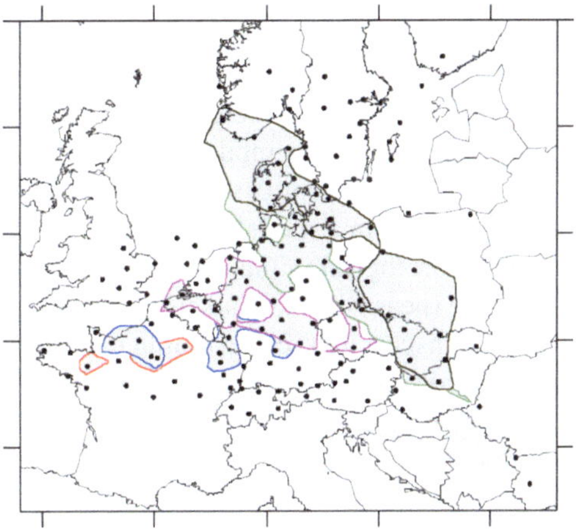

Fig. 8.4 The ETEX sampling stations distribution (*filled black circles*) and 0.1 ngm^{-3} contour of measured cloud at T_0+12 h (*red*), +24 h (*blue*), +36 h (*purple*), +48 h (*green*), +60 h (*black*). Figure courtesy of Stefano Galmarini

In the context of passive tracer transport a non-exhaustive list of non-linear test cases is given below. The examples are meant to give the reader an idea of the 'world' beyond idealized linear test cases that are usually reported on in transport scheme development. All test cases do not have analytical solutions and involve the solution to the entire system of dynamical equations (not just prescribed winds and mass-fields) as well as parameterizations of sub-grid-scale processes making it harder to distinguish numerical errors of the transport scheme from other sources of error.

The ETEX forecast experiment. The worst nuclear power plant disaster in history (the Chernobyl power plant explosion in 1986) generated a radioactive plume that drifted over extensive parts of western Russia and Europe. This is a rude reminder of the importance of having models capable of forecasting long-range transport accurately; at least for emergency management. As a consequence the European Tracer Experiment (ETEX, see, e.g., Girardi et al. 1998; van Dop et al. 1998 and the more recent study of Galmarini et al. 2004) was established in 1994 to evaluate the validity of long-range transport models and to assemble a database which would allow the evaluation of long-range atmospheric dispersion models in general.

ETEX was a controlled experiment where two releases (under different weather conditions) of perfluorocarbon tracers from Western France were tracked across Europe. Perfluorocarbon tracers are non-depositing, non-water-soluble and inert, and therefore a passive tracer for all practical purposes. A large network of samplers deployed eastward on the territory of Central and Eastern Europe collected tracer samples that were later analyzed to determine the concentration levels. That set of measurements was then used to quantitatively evaluate the predictions of the models (Fig. 8.4).

The ETEX experiment and data can be used to evaluate new transport schemes. Obviously this test case indirectly tests more than the transport scheme itself but also parameterizations (such as boundary layer parameterizations, parameterized vertical diffusion etc.) and, in general, the models ability to produce accurate winds and air densities for the tracer transport scheme.[8]

Mixing experiments. There are several experiments in the literature targeting the mixing properties of the model. Probably the simplest was proposed by Rasch et al. (2006). The experiment is setup as follows. The mixing ratio for a tracer is set to one everywhere in a model layer and zero elsewhere. Then the model is run from some meteorological initial conditions for 30 days. The tracer is placed either near the surface (near 800 hPa) and around 200 hPa. The low tracer test serves as an indicator of transport in a region dominated by sub-grid scale transport processes such as convection and turbulence. The high tracer is much more dominated by resolved-scale dynamics at least at middle and high latitudes. The test case also indicates the tropospheric-stratospheric mixing in the model (generally, in the polar and mid-latitude regions stratospheric air is mixed into tropospheric air and in the Equatorial regions deep convection results in a large scale ascent of tropospheric air). For models based on an isentropic vertical coordinate, this test when run adiabatically and with non-zero tracer values in an isentropic layer instead of pressure levels, can be used to indicate the amount of spurious vertical diffusion in the transport scheme since ideally the mixing ratio should remain one in the isentrophic layer for all time (and zero elsewhere).

Another experiment that is probably more widely used is the age-of-air experiment (see, e.g., Waugh and Hall 2002 and references herein). The age of air is the mean transport time from some reference location. For example, stratospheric age of air is the mean transport time from the tropical tropopause to a location in the stratosphere. Monitoring the age of air for species with long lifetimes provides a proxy for the diffusivity (often spurious) of the tracer transport in a particular model. In general, schemes that are too diffusive tend to produce too 'young' air while less diffusive schemes simulate 'older' age of air. Eluszkiewicz et al. (2000) found a large dependency on the choice of advection scheme in age-of-air experiments in addition to the simulated large scale circulation. Even for short-lived tracers with sources and sinks Rasch et al. (2006) found a large dependency on the numerical solution technique. These studies demonstrate that the choice of transport scheme (and driving model) can easily influence the simulation at a level that can strongly modulate the physical signal of interest.

Dynamics/tracer consistency. This test was proposed by D. Johnson (University of Wisconsin) and published in Rasch et al. (2006). It targets the model's ability to simulate transport of conserved tracers consistently and the model's ability to maintain non-linear relationships between six different conserved and non-conserved tracers. It can be shown from the second law of thermodynamics that two points separated in space and time connected by a trajectory should satisfy a non-linear

[8] There have been other controlled tracer transport experiments before ETEX, e.g., ANATEX (The Across north America Tracer Experiment) and CAPTEX (Cross-Appalachian Tracer Experiment) and also more recent experiments such as MEGAPOLI (Emissions, urban, regional and Global Atmospheric POLlution and climate effects, and Integrated tools for assessment and mitigation; http://megapoli.info).

relationship in terms of temperature T, potential temperature θ and pressure p (see Appendix of Rasch et al. 2006):

$$\theta_1 = \left(\frac{\theta_0}{T_0}\right) T_1 \left(\frac{p_0}{p_1}\right)^{R/C_p}, \tag{8.12}$$

where the subscript 0 and 1 refer to the two points, and R and C_p are the gas constant and specific heat constant at constant pressure, respectively. The test case consists of predicting θ, T and p^{R/C_p} separately and then check how well they obey (8.12). The level of agreement between these two ways of computing potential temperature yields a measure of the degree of consistency in the model. See Rasch et al. (2006) for details. It is probably impossible to construct an Eulerian scheme that will exactly fulfill this consistency test, however, it is desirable that schemes strive to be as consistent as possible.

8.3.2 Conservation of Mass

As discussed in Sect. 8.2.1 one of the most fundamental budgets of the global atmosphere is that for the mass of dry air. Since the physical variation in the dry air mass budget is on the order of 0.001% (and usually not modeled) even minor drifts in the dry air mass budget due do numerical errors would be larger than the physical variation in the dry air mass budget (Moorthi et al. 1995).

For the trace gases any spurious non-conservation of mass will effectively correspond to a spurious source or sink for the gas in question. In particular for long-lived trace species such as stratospheric ozone it is paramount that their mass-budgets are well maintained in the models. Even for highly reactive tracers such as reactive chlorine compounds, mass-conservation is important since the sum of all the compounds should be conserved although individual compounds have large sources and sinks (one compound is converted into another).

There are two ways of obtaining mass-conservation in numerical schemes. Either an inherently conservative numerical method is used or mass-fixers (see Chap. 13) can be employed. For the mass of dry air mass-fixers usually operate by increasing or decreasing the mean of the pressure field (mass) by an amount corresponding to the spuriously lost or gained mass caused by the lack of conservation of the numerical method. Note that such a procedure can be done so that it does not alter gradients in the pressure field and was shown by Williamson and Olson (1994) to have minimal effect on the simulation. Mass-fixers are applied in numerious non-conservative models, e.g., the spectral transform versions of NCAR's Community Atmosphere Model (CAM, Collins et al. 2004). Although mass-fixers for the pressure field seem to not adversely affect simulations it is far more problematic to apply mass-fixers for tracers. For example, altering mixing ratios to obtain tracer mass-conservation can lead to unphysical large or small mixing ratios. If that is the case the mass-fixer must do a local adjustment and thereby it might introduce new extrema in the tracer mass fields and gradients are no longer preserved. This may also disrupt tracer correlations (tracer correlations are discussed in Sect. 8.3.7 below) and consistency between tracer and air mass (see Sect. 8.3.4 below). Therefore finite-volume methods that are inherently conservative, have become a popular numerical method in climate and chemistry modeling since ad-hoc adjustments are, in theory, not necessary.[9]

The continuous equations of motion conserve all moments not just mass. However, Thuburn (2008b) argued (see also Chap. 11) it might not be desirable that the advection scheme also preserves higher-order moments.

8.3.3 *Optimal Diffusion and Dispersion Properties*

The linear diffusion and dispersion properties of a linearized scheme can be assessed by performing a von Neumann stability analysis (also known as a Fourier stability analysis). It is a standard analytic analysis technique and is described in many textbooks in the context of grid-point methods (see, e.g., Durran 1999; Haltiner and Williams 1980) and in the context of finite-volume methods in Lauritzen (2007). The analysis consists of assessing analytically how a single Fourier mode is damped and accelerated/decelerated by the numerical scheme during one time-step assuming a constant wind field.

In one dimension the von Neumann analysis is performed by assuming a solution in the form

$$\psi^n(x) = \psi^0\, \Gamma^n \, \exp\left(\hat{\imath}\, \kappa\, x\right), \tag{8.13}$$

where $\hat{\imath}$ is the imaginary unit, ψ^0 the initial amplitude, and $\kappa = 2\pi/L$ is the wavenumber (L is the wavelength), and n is the time-level index. The damping and phase properties of a scheme are assessed by substituting the solution (8.13) into the forecast formula for the finite-volume scheme in question, and subsequently analyzing the complex amplification factor Γ. The stability of a numerical method is governed by the modulus of the complex amplification factor, that is, a particular wave with wavenumber κ is stable if $|\Gamma| \leq 1$. Following Bates and McDonald (1982) the dispersion properties of a scheme is assessed by writing the complex amplification factor as

$$\Gamma = |\Gamma| \, \exp\left(-\hat{\imath}\omega^* \, \Delta t\right), \tag{8.14}$$

where ω^* is the numerical frequency. Define the relative frequency as $R = \omega^*/\omega$ where ω is the exact frequency given by $\kappa\, u_0$ and u_0 is the constant wind. If $R > 1$ the numerical scheme is accelerating and if $R < 1$ the scheme is decelerating compared to the exact solution.

The von Neumann analysis provides useful information about the stability properties of a scheme and may provide new insight into schemes. The limitation of the

[9] We write 'in theory' since if a transport scheme is not strictly monotone local 'ad hoc' adjustments might be necessary even for finite-volume methods.

von Neumann stability analysis is that it is linear. Hence any non-linear operators such as limiters and filters cannot be included in the basic analysis as well as non-linear flows. Usually the spurious numerical diffusion and dispersion decrease rapidly with the formal order of the scheme. So each scheme probably has an optimal order for which the extra computational cost associated with increasing the order of the scheme simply does not pay off in terms of linear diffusion and dispersion properties. For example, Leonard (1991) argued that the reduction in diffusion becomes trivial soon after the order is larger than third.

8.3.4 Tracer and Air Mass Consistency

Tracer and air mass consistency is a stricter concept than simple mass conservation of the individual quantities. It basically states that the discretized tracer transport scheme should reduce to the discretized continuity equation for air when $q = 1$ as is the case for the continuous equations: (8.10) reduces to (8.9) when setting $q = 1$. Tracer-air mass consistency can, for example, be violated if using a numerical method for tracer transport that is different from the scheme used for predicting the evolution of the air density.[10] To achieve a high level of consistency it is usually necessary that the same numerical algorithm is used for the dynamics as well as for tracer transport. For more discussion see Machenhauer et al. (2009), Lee et al. (2004), Jöckel et al. (2001), Zhang et al. (2008).

8.3.5 Divergence Preservation

The transport operator should not be a spurious source of divergence. Usually this property is discussed within the context of non-divergent flow fields. For example, a constant initial mass distribution should remain constant at all time in a non-divergent flow (*preservation of mass-constancy*). This subject has received considerable attention in the magnetohydrodynamics literature since the magnetic flux density is non-divergent and the numerical scheme should ideally retain that property (e.g., Artebrant and Torrilhon (2008) and references therein). A prerequisite for controlling spurious generation of divergence is preservation of mass-constancy as formulated above (see test case suggestion in Sect. 8.3.1.1) for non-divergent flows.

Note that the preservation of constant mixing ratio (and not constant tracer mass field) is trivial in most cases. If the advective form of the transport equation (8.11) is used it is trivial to maintain a constant mixing ratio since q is the prognostic

[10] This discussion applies to online applications where tracer transport is performed in conjunction with the governing fluid and thermodynamic equations. A similar inconsistency appears when driving the tracer transport equation in an offline mode (prescribed winds and mass fields from reanalysis, observations or a different model) in which case the tracer transport scheme with $q = 1$ will not equal the prescribed mass-field unless ad-hoc fixers are applied.

variable and the divergence does not appear explicitly. If the air and tracer equations are solved in flux-form ((8.9) and (8.10), respectively) using the same numerical method, it is usually trivial to preserve a constant mixing ratio field since the mixing ratio q is recovered from (8.10) by dividing the prognosed tracer mass field ρq by ρ from (8.9). So even if the numerical scheme is unable to preserve a constant mass field ρ, it is usually possible to design schemes so that a constant q field is recovered when dividing ρq by the (potentially) non-divergence preserving forecast of ρ.

8.3.6 Physical Realizability (Monotone, Positive-Definite, Non-Oscillatory, Shape-Preserving)

In the absence of sources and sinks the mixing ratio of a Lagrangian parcel being transported by the flow is invariant (8.11). If the numerical solution fulfills this property it is *monotonicity*[11] *preserving*; no new local extrema are generated and the absolute values of pre-existing local extrema is non-increasing. Strict monotonicity preservation can be hard to achieve and enforcing it in numerical schemes is often found to be at the cost of overall accuracy wherefore it is often relaxed somewhat.

The zero-th order shape-preservation property is that the numerical scheme generates physically realizable solutions. Since mixing ratios cannot physically take negative values they should remain non-negative. Schemes that cannot generate negative values are termed *positive definite* and schemes that do not generate wiggles (spurious grid-scale waves as the ones on Fig. 8.3c and d) typically associated with large gradients are referred to as *non-oscillatory*. Obviously a scheme that is monotone is automatically positive-definite and non-oscillatory but not necessarily vice versa. It should be stressed that it is q that should remain monotone and not ρq. For convergent flows ρq can physically take values outside the range of the initial condition whereas q should not. See Nair and Lauritzen (2010) for a discussion and simple illustration of the latter for an idealized flow field.

Note that shape-preservation can be enforced in finite-volume schemes based on (8.9) and (8.10) if these schemes imply some discretized version of (8.11). Schemes that retain such a property are termed *compatible* (Schär and Smolarkiewicz 1996).

8.3.7 Preservation of Pre-Existing Functional Relations Between Species (Correlations)

As described in Plumb (2007): "Relationships between long-lived stratospheric tracers, manifested in similar spatial structures on scales ranging from a few to several thousand kilometers, are displayed most strikingly if the mixing ratio of one is

[11] Atmospheric modelers tend to be a bit loose with the term 'monotone' and normally they do not refer to the careful definition given by Harten (1983).

plotted against another, when the data collapse onto remarkably compact curves." In other words, different longlived trace constituents (such as nitrous oxide N_2O and 'total odd nitrogen' NO_y) seem to be related through rather simple functional relationships in, for example, the polar stratospheric vortex. Such relationships can arise for different reasons (Plumb and Ko 1992), however, it is well-known that transport can establish such relations (e.g., Thuburn and Mclntyre 1997).

In order to accurately simulate such relationships in numerical models, the transport operator should, at least, be able to preserve linear correlations (Lin and Rood 1996; Thuburn and Mclntyre 1997). That is, the transport operator should maintain the relationship in (8.15) throughout the simulation

$$q_1 = \gamma^{(0)} + \gamma^{(1)} q_2, \tag{8.15}$$

where $\gamma^{(i)}$, $i = 0, 1$, are constants, and q_i, $i = 1, 2$, are mixing ratios of two linearly interrelated species. A transport scheme will preserve linear pre-existing functional relations if the transport operator $\mathscr{T}$, that updates q_i, $i = 1, 2$, in time, is 'semi-linear'

$$\mathscr{T}(q_1) = \mathscr{T}\left(\gamma^{(0)} + \gamma^{(1)} q_2\right) = \gamma^{(0)} \mathscr{T}(1) + \gamma^{(1)} \mathscr{T}(q_2) = \gamma^{(0)} + \gamma^{(1)} \mathscr{T}(q_2), \tag{8.16}$$

(Lin and Rood 1996; Thuburn and Mclntyre 1997). As noted by Thuburn and Mclntyre (1997) the successful preservation of linear correlations by a transport operator does not necessarily gurantee an accurate solution since shaping two tracer fields the same way does not necessarily imply shaping them the right way. On the other hand, if a model significantly violates the preservation of linear correlations between chemical constituents, the model is most likely not going to provide truthful simulations of the relation between those constituents.

Since interrelated tracers can also be related non-linearly, it is also of interest to investigate how a transport operator distorts such non-linear relation. For example, consider two tracers that are initially correlated by a fourth-order polynomial

$$q_1 = \gamma^{(0)} + \gamma^{(1)} (q_2)^4, \tag{8.17}$$

(Thuburn and Mclntyre 1997) where the constants $\gamma^{(0)}$ and $\gamma^{(1)}$ should be chosen so that the functional relation is either convex or concave in the range of the initial condition values of q_1 and q_2. Except for fully Lagrangian transport operators, schemes are usually unable to maintain non-linear functional relationships and their degree of non-preservation of correlations effectively translates into numerical mixing of the constituents. Initializing two tracers that are, for example, related through (8.17) and letting the tracers be transported by a challenging flow that develops features that collapse to the near grid-scale, provides physical insight into the numerical mixing that the transport operator introduces (Thuburn and Mclntyre 1997). No practical Eulerian and semi-Lagrangian scheme can preserve (8.17) and will therefore produce scatter points that deviate from the pre-existing functional relationship (8.17). When scatter points deviate from the pre-existing functional relation curve

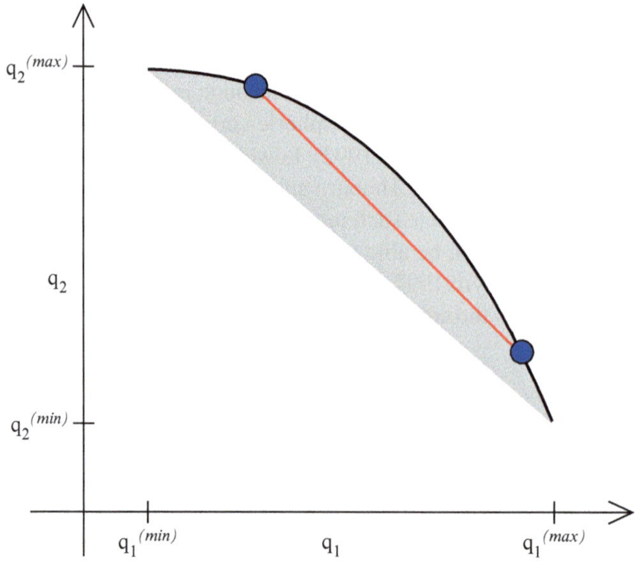

Fig. 8.5 Schematic view of the effect of mixing on a scatter plot where mixing ratio for tracer 1, q_1, is plotted against tracer 2, q_2. The two tracers are initially non-linearly correlated, that is, the scatter points (q_1, q_2) are on the pre-existing functional relation curve (*solid thick line curve*). The initial ranges for the two tracers are $[q_1^{(min)}, q_1^{(max)}]$ and $[q_2^{(min)}, q_2^{(max)}]$, respectively. Partial mixing of two air masses (*two filled circles – scatter points*) will tend to move the two scatter points towards each other along the straight (*red*) line (also referred to as a '*mixing line*'). Hence 'real mixing' occurring in the atmosphere will tend to move points on the scatter plot to the concave side of the pre-existing functional relation curve (also referred to as the '*convex hull*' – *shaded area*)

the transport operator is introducing numerical mixing. The numerical mixing can either be spurious or resemble 'real' mixing. If the scatter values are on the concave side of the pre-existing functional relation, the numerical mixing is similar to 'real mixing' that is observed in the atmosphere (see Fig. 8.5). Mixing in the atmosphere occurs, for example, when the polar stratospheric vortex breaks up (e.g., Waugh et al. 1997). If scatter values appear outside the 'convex hull' (either by producing scatter points on the convex side of the pre-existing functional relationship curve and/or outside the range of the initial condition for q_i, $i = 1, 2$), the model produces numerical unmixing which is unlike 'real mixing'. Thuburn and McIntyre (1997) proved that in order to guarantee only 'real' numerical mixing, the transport operator should be 'semi-linear' and monotone according to Harten (1983). Unfortunately only first-order schemes will meet these requirements. Since first-order schemes are too diffusive for most atmospheric applications, one must accept some level of unmixing. For a more complete discussion of this topic the reader is referred to Thuburn and McIntyre (1997). Recently, Lauritzen and Thuburn (2011) proposed mixing diagnostics that quantifies the amount of numerical mixing that the transport operator introduces for interrelated species.

Another situation relevant to the transport of chemical species is the situation in which more than two species are related through some complicated relation but they add up to a constant (or a smooth spatial field.[12]) With just two species this reduces to preserving a linear correlation but with more than two species it is very challenging to guarantee that the total mixing ratio remains constant, except by transporting the total or using a fully Lagrangian scheme. The transport operators ability to maintain the constant sum is another measure for numerical mixing and has been explored in one dimension by Ovtchinnikov and Easter (2009). Note that maintaining or only perturbing pre-existing functional relations in a 'physical way' is not only important for long-lived stratospheric tracers but also for other parts and processes in the atmosphere such as cloud-aerosol interactions (Ovtchinnikov and Easter 2009). In all, single-tracer testing that has traditionally been used to evaluate transport operators in idealized settings does not provide insight into how well tracer interrelations are maintained although it is important for many atmospheric applications.

8.3.8 Robustness

The numerical method should remain stable and retain simulation veracity throughout the integration. Robustness can be assessed by testing the algorithm for many different flow fields, temporal and spatial resolutions.

8.3.9 Parallel Computational Efficiency

Performance improvements are largely due to increased parallelism rather than improved microprocessor clock frequency. Hence the numerical algorithm should be amenable for execution on massively parallel computing platforms. A way to achieve this is to use local methods with minimal global dependence (for more discussion see Chap. 16).

It is worth noting that although computing power has increased dramatically in the last 20 years or so, these extra computational resources have largely been used to satisfy demands for higher resolution, more advanced physical parameterizations and coupling the atmospheric component to ocean, land, and ice components (i.e., coupled models). Hence it is still desirable to develop efficient dynamical core algorithms, in particular, schemes for efficient tracer transport (see paragraph on *Multi-tracer efficiency* below) even though computing power is increasing.

8.3.10 Multi-Tracer Efficiency

In modern atmospheric models the number of tracers required to be advected continue to increase. For example, the chemistry version of NCAR's CAM model

[12] For example, total reactive chlorine in the stratosphere.

transports over 100 tracers (Lamarque et al. 2008). Given that the dynamical core typically has less than ten prognostic variables defining the state of the fluid flow and thermodynamics, the computational cost of running the dynamical core can primarily be attributed to the transport of tracers. Needless to say, it is highly desirable that the numerical algorithm used for tracer transport be efficient and adaptable for a large number of tracers. A way to achieve multi-tracer efficiency is to design schemes that can reuse information for each additional tracer (Barth and Frederickson 1990; Dukowicz and Baumgardner 2000) and/or transport tracers with longer time-steps than used for the continuity equation for air in the dynamical core (also referred to as 'super-cycling' of tracers with respect to air or, more commonly, 'sub-cycling' of air with respect to tracers; see Sect. 8.7.2).

8.3.11 Geometric Flexibility

It is generally useful to develop numerical methods that can be used on a wide range of spherical grids. Next generation dynamical cores are being developed on spherical grids based on triangles, quadrilateral, pentagonal and/or hexagonal control volumes. It is therefore desirable that a scheme can handle any spherical polygon-based grid. Also models using static or adaptive mesh-refinement benefit from geometrically flexible methods. An example of a geometrically flexible advection scheme is MPDATA (Multidimensional Positive Definite Advection Transport Algorithm); for an overview see Smolarkiewicz (2006).

8.4 Problem Formulation: Discrete Schemes

Finite-volume methods are numerical methods where each prognostic variable is stored as an average quantity over a certain finitely large control volume (also referred to as cell-integrated methods). This choice differs from methods that are based on grid-point values (used in, e.g., finite-difference methods) or weights for expansion functions (e.g., finite-element or spectral method). In order to derive finite-volume discretization schemes the equations of motion, in this case the continuity equation, are integrated over a control volume. This allows for discretizations that keep track exactly of the local mass-budgets and thus provides mass-conservation to machine precision. Note that although finite-volume schemes are designed to conserve mass locally through explicitly tracking mass, conservation of mass can also be achieved in non finite-volume methods (e.g., compatible methods, see Chap. 12). Conservative methods that are not finite-volume methods usually do not conserve mass locally.

Typically finite-volume schemes come in two flavors corresponding to two forms of deriving the equations of motion from first principles: Eulerian and Lagrangian.[13]

[13] The Eulerian and Lagrangian forms are limits of the more general arbitrary Lagrangian–Eulerian (ALE) form (Hirt et al. 1974).

In most textbooks the equations of motion are derived in Eulerian form, that is, as observed from a fixed volume in the atmosphere (stationary to the Earth's surface). Hence there is a flux of mass through the volume boundaries unless the local wind is zero. One may also derive the equations of motion as viewed by a volume not just rotating with the Earth's rotation axis but also moving with the local flow; a.k.a. Lagrangian form. In Lagrangian form there is no flux of mass through the 'walls' of the volume. Both of these forms of the finite-volume discretization of the continuity equation are presented next after the introduction of some notation. For simplicity we consider the two-dimensional problem in Cartesian geometry and defer the discussion of the extension to spherical geometry and three-dimensions to Sects. 8.5.3.3 and 8.6, respectively.

Let the domain of integration be denoted Ω (a Cartesian plane with periodic boundary conditions or no flux through the domain boundaries). The domain Ω is partitioned into N non-overlapping grid cells, A_k, $k = 1, .., N$, so that $\bigcup_{k=1}^{N} A_k$ span Ω. The area of cell A_k is denoted ΔA_k. For now we shall assume a quadrilateral mesh in Cartesian geometry, however, the discussion can trivially be extended to other meshes such as triangular or hexagonal meshes in Cartesian geometry.

As mentioned above the prognostic variable considered is the cell averaged value

$$\overline{\psi}_k = \frac{1}{\Delta A_k} \int_{A_k} \psi(x, y)\, dA, \quad \psi = \rho \text{ or } \rho q, \tag{8.18}$$

where $\psi(x, y)$ is the exact solution. In time we discretize in terms at equidistant time-levels, i.e., superscript n refers to the quantity at time $t = n\,\Delta t$ where Δt is the time-step. So the state of a tracer in cell A_k at time-level n is denoted $\overline{\psi}_k^n$.

8.4.1 (Semi-)Lagrangian Schemes

Consider an arbitrary Lagrangian area $A(t)$. By definition the area $A(t)$ moves with the flow without any flux of mass through its sides and hence it always contains the same material particles. Since there is no flux of mass through the boundaries of $A(t)$, the mass in the area is conserved. In mathematical terms this can be written as

$$\frac{d}{dt} \int_{A(t)} \psi\, dA = 0, \quad \psi = \rho \text{ or } \rho\, q. \tag{8.19}$$

Equation (8.19) is referred to as the Lagrangian finite-volume form of the continuity equation. A temporal discretization of (8.19) reads

$$\int_{A(t+\Delta t)} \psi\, dA = \int_{A(t)} \psi\, dA. \tag{8.20}$$

If the same Lagrangian cell $A(t)$ is tracked throughout the simulation the resulting scheme is referred to as fully Lagrangian. The challenge in such schemes is that for non-trivial flows the areas quickly deform into thin filaments so that the resolution is no longer uniform (see Fig. 7.2 in Chap. 7).

Instead one may consider a different set of areas/parcels at every time-step, for example, enforcing that either $A(t+\Delta t)$ or $A(t)$ is a regular static grid cell. Such an approach is referred to as semi-Lagrangian since it only tracks the same Lagrangian parcels/area for one time-step. The advantage of the semi-Lagrangian approach, as compared to a fully Lagrangian method, is that it retains a quasi uniform resolution as the mesh only deforms for one time-step. However, the grid uniformity is introduced at the expense of having to interpolate variables from a regular static grid to a deformed Lagrangian grid (or vice versa) at every time-step. How this interpolation can be done is discussed in great detail below but for now more mathematical notation is needed.

Assume that $A(t + \Delta t)$ is a regular grid cell resulting in a method referred as upstream semi-Lagrangian.[14] If we consider cell k in the discretized domain then the regular grid cell $(A(t + \Delta t))$ is exactly A_k with area ΔA_k. The corresponding upstream Lagrangian area $(A(t))$ is referred to as a_k with area Δa_k (see Fig. 8.6a). We assume that Δt is chosen such that all the deformed areas a_k are simply connected.

Note that there exists a one-to-one correspondence between A_k and a_k such that the a_k's span Ω without gaps or overlaps between them

$$\bigcup_{k=1}^{N} a_k = \Omega, \text{ and } a_k \cap a_\ell = \emptyset \, \forall \, k \neq \ell. \tag{8.21}$$

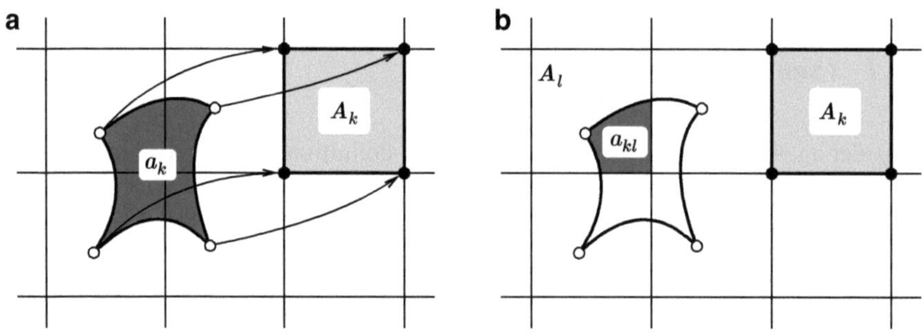

Fig. 8.6 A graphical illustration of the upstream semi-Lagrangian nomenclature. (**a**) The static Eulerian cell A_k (*light shading*) and the corresponding upstream Lagrangian area a_k (*dark shading*) that ends up at A_k after one time-step. For illustration the trajectories of the vertices (*filled circles*) of A_k are depicted with *arrows*. The corresponding upstream vertices (*departure points*) are shown with *open circles*. (**b**) The notation used to define overlap areas between Eulerian cell A_ℓ and upstream Lagrangian area a_k is $a_{k\ell} = A_\ell \cap a_k$ (*dark shaded area*)

[14] Note that one might equally well consider downstream schemes where one considers Eulerian (regular) grid cells at time-level n and let them be transported with the flow for one time-step.

Assume that the evolution of the Lagrangian grid is known analytically so we know the characteristics or trajectories for each fluid parcel at all times. The computation of fluid parcel trajectories is well developed in the semi-Lagrangian literature (e.g., Staniforth and Côté 1991; Staniforth et al. 2003; Hortal 2002) and in the interest of brevity it is not discussed further in this chapter, although accurate trajectories are vital for the accuracy of any Lagrangian method.

With the notation introduced above the forecast (8.20) can be written as

$$\overline{\psi}_k^{n+1}\,\Delta A_k = \overline{\psi}_k^{n}\,\Delta a_k. \tag{8.22}$$

where $\overline{\psi}_k^{n}$ is the average tracer density over the upstream area a_k

$$\overline{\psi}_k^{n} = \frac{1}{\Delta a_k}\int_{a_k} \psi^n(x,y)\,dA, \tag{8.23}$$

and $\overline{\psi}_k^{n+1}$ is the cell averaged value of ψ over the regular area A_k at time-level $n+1$. The function $\psi^n(x,y)$ is the continuous distribution of ψ at time-level n. Obviously, since the prognostic variables are cell averages $\overline{\psi}$ we do not know the variation of ψ at the sub-grid scale and $\psi^n(x,y)$ must be reconstructed from the prognostic cell averages.[15] This procedure is referred to as sub-grid-scale reconstruction. In finite-volume schemes the reconstruction is usually local rather than global. So each cell k will have an associated sub-grid-scale reconstruction function $\psi_k(x,y)$ rather than one global reconstruction function over all cells such as the spherical harmonic functions used in spectral transform models.

Hence the global reconstruction function is a collection of local reconstruction functions

$$\psi(x,y) = \sum_{k=1}^{N} I_{A_k}\psi_k(x,y), \tag{8.24}$$

where I_{A_k} is the indicator function

$$I_{A_k} = \begin{cases} 1, & (x,y) \in A_k, \\ 0, & (x,y) \notin A_k. \end{cases} \tag{8.25}$$

Commonly used methods for computing $\psi_k(x,y)$ from $\overline{\psi}_k$ are discussed in Sect. 8.5.2.

First, we note that $\psi(x,y)$ is not necessary continuous or differentiable across cell boundaries. So if the upstream area a_k covers several Eulerian cells (e.g., Fig. 8.6), the integral on the right-hand side of (8.23) must be broken up into overlap areas between Eulerian cells and a_k. The discretized semi-Lagrangian finite-volume

[15] Unless variables such as gradients are also carried as prognostic variables (e.g., Yabe et al. 2001).

continuity equation (8.22) then reads

$$\overline{\psi}_k^{n+1} \Delta A_k = \sum_{\ell=1}^{L_k} \int_{a_{k\ell}} \psi_\ell^n(x, y)\, dA. \tag{8.26}$$

The number of non-empty overlap areas between the upstream cell (departure cell) a_k and the Eulerian grid cells is denoted L_k. Note that L_k depends on the flow and time-step size, and for time-varying flows it is not necessarily constant. The area $a_{k\ell}$ is the non-empty overlap area between the upstream cell a_k and the Eulerian grid cell A_ℓ (see Fig. 8.6b)

$$a_{k\ell} = a_k \cap A_\ell, \quad a_{k\ell} \neq \emptyset; \quad \ell = 1, \ldots, L_k, \text{ and } 1 \leq L_k \leq N, \tag{8.27}$$

where N is the number of cells in the domain.

Two conditions must be fulfilled to get conservation of mass in Lagrangian finite-volume schemes: Firstly, the upstream cells a_k must be simply connected domains and they must span Ω without gaps or overlaps (8.21). Secondly, the reconstruction function in cell k, $\psi_k(x, y)$, must be conservative in the sense that the integral of $\psi_k(x, y)$ over A_k must yield the cell-average value that is used as prognostic variable,

$$\frac{1}{\Delta A_k} \int_{A_k} \psi_k(x, y)\, dA = \overline{\psi}_k. \tag{8.28}$$

Equation (8.26) is the basic finite-volume form of the continuity equation when using an upstream finite-volume semi-Lagrangian approach. Obviously we do not know the exact Lagrangian trajectory of every parcel in the domain so some approximation to a_k is necessary for the derivation of any practical scheme. This is discussed in Sect. 8.5.1.

In the discussion above ψ generically refers to both ρ and ρq. In the reconstruction of ρq one may chose to reconstruct ρ and q separately and combine them to provide a reconstruction for the product ρq. There are several reasons for choosing this approach. First, it is q and not ρq that is conserved along parcel trajectories (see 8.11) and q should therefore obey monotonicity requirements. Hence one can argue that monotone reconstruction function filters (discussed in Sect. 8.5.2) should be applied to q and not ρq. Second, the consistent coupling of tracers and air density equations in cell-integrated semi-Lagrangian schemes as well as ensuring monotone forecasts of q, is perhaps easier when choosing this approach (Nair and Lauritzen 2010).

The reconstructions for ρ and q can be combined to provide a reconstruction for ρq by simply multiplying the reconstruction functions for ρ and q as done in Dukowicz and Baumgardner (2000). However, in doing so mass-correction terms may be needed to satisfy (8.28) for higher-order reconstructions. The downside of this approach is that if, for example, the reconstruction function for ρ and q are polynomials of ith and jth order the product will be polynomials of $(i + j)$th order which may be computationally intensive to integrate. One may simplify by removing some terms from the product as done in Nair and Lauritzen (2010). The latter also facilitates rendering schemes monotone in q. In Eulerian schemes, discussed next, tracer mixing ratio and air density are usually reconstructed separately for sub-cycling (see Sect. 8.7.2)

8.4.2 Eulerian Scheme

Contrary to the Lagrangian derivations in the previous section, the equations of motion are typically derived in Eulerian form. In the context of the finite-volume form of the continuity equation the Eulerian approach keeps track of the flux of mass through the Eulerian cell walls rather than tracking the mass in a cell moving with the flow. A more formal derivation is given below.

First, integrate (8.9) or (8.10) in space over a grid cell A_k

$$\int_{A_k} \frac{\partial \psi}{\partial t} dA + \int_{A_k} \nabla \cdot (\psi\, \mathbf{v})\, dA = 0, \text{ where } \psi = \rho, \rho q. \tag{8.29}$$

On integrating the first term on the left-hand side of (8.29) to get the area average and applying the divergence theorem to the second term we get

$$\frac{d}{dt}\left(\overline{\psi}_k \Delta A_k\right) + \oint_{\partial A_k} (\psi\, \mathbf{v}) \cdot \mathbf{n}\, dS = 0, \tag{8.30}$$

where ∂A_k is the boundary of A_k and $\mathbf{n}$ the outward normal vector to ∂A_k. The second-term on the left-hand side of (8.29) represents the instantaneous flux of mass through the boundaries of A_k. Temporal integration of (8.30) over one time-step yields

$$\overline{\psi}_k^{n+1} \Delta A_k = \overline{\psi}_k^{n} \Delta A_k - \int_{n\Delta t}^{(n+1)\,\Delta t} \left[\oint_{\partial A_k} (\psi\, \mathbf{v}) \cdot \mathbf{n}\, dS \right] dt, \tag{8.31}$$

after re-arranging terms. The second term on the right-hand side of (8.31) is the flux of mass through the walls of A_k during one time-step. A graphical illustration of the fluxes is given in Fig. 8.7 and discussed in the next paragraph.

Let τ denote the face number and N^f the number of faces of the cells. For simplicity we assume a quadrilateral mesh $N^f = 4$, however, the method can accommodate any kind of mesh (for example, for a triangular and hexagonal mesh N^f would be 3 and 6, respectively). A graphical illustration of the fluxes through the cell walls for Eulerian cell k are shown on Fig. 8.7. As will become clear, the figure also shows the upstream Lagrangian cell although it is not explicitly needed for flux computations. The sides of the Eulerian control volume are numbered counter-clockwise so that sides $\tau = 1, 2, 3, 4$ correspond to the east, north, west and south walls, respectively (using standard compass notation). The flux of mass through the side $\tau = 1$ corresponds to the mass over the shaded area on Fig. 8.7a that is 'swept' through the wall during one time step. The shaded area, referred to as $a_k^{\tau=1}$,

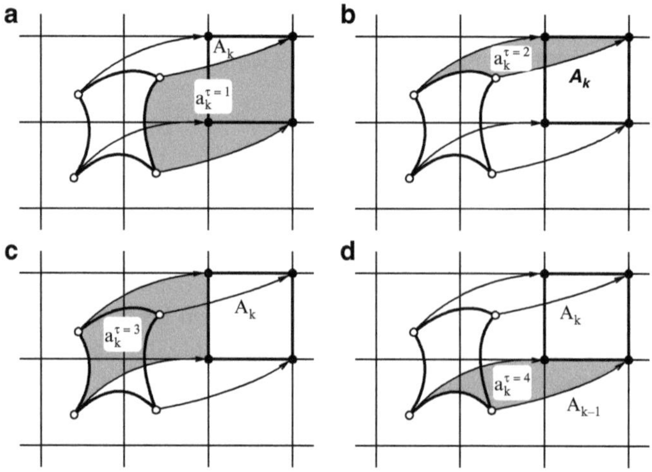

Fig. 8.7 A graphical illustration of the 'flux-areas' associated with Eulerian cell A_k (area in the upper right corner of each plot bounded by *thick lines*). For each vertex of cell A_k (*filled circles*) the upstream trajectories are shown (*curved arrows* departing from *open circles*). The *shaded areas* show the flux-areas for the (**a**) east $a_k^{\tau=1}$, (**b**) north $a_k^{\tau=2}$, (**c**) west $a_k^{\tau=3}$ and (**d**) south $a_k^{\tau=4}$ face, respectively, using standard compass orientation. These areas are swept through each face during one time-step. See text for details

is bounded by the face $\tau = 1$, the two upstream trajectories for the end points of face $\tau = 1$, and the upstream translation of the side $\tau = 1$. We will refer to $a_k^{\tau=1}$ as the 'flux-area' for face $\tau = 1$. Similarly, the fluxes through the remaining cell sides are illustrated in Fig. 8.7b–d.

Using the notation introduced above (8.31) can be written as

$$\overline{\psi}_k^{n+1} \Delta A_k = \overline{\psi}_k^{n} \Delta A_k - \sum_{\tau=1}^{N^f} F_k^\tau, \tag{8.32}$$

where F_k^τ is the flux of mass through face τ during one time-step

$$F_k^\tau = s_k^\tau \int_{a_k^\tau} \psi^n(x, y)\, dA. \tag{8.33}$$

The 'flow-direction' function s_k^τ is used to indicate inflow and outflow

$$s_k^\tau = \operatorname{sgn}(\mathbf{v} \cdot \mathbf{n}), \tag{8.34}$$

where sgn(·) is the sign-function. Hence s_k^τ is 1 for outflow and -1 for inflow.[16] In Fig. 8.7 the flow-direction function s_k^τ is 1 for $\tau = 1, 2$ and -1 for $\tau = 3, 4$.

[16] For simplicity we do not consider the situation where s_k^τ is multi-valued along a particular face. For more details on such a situation see Harris et al. (2011).

Note that the flux of mass through one cell wall is identical, but with opposite sign, to the flux of mass through the neighboring cell that it shares a face with. For the example on Fig. 8.7d

$$a_k^{\tau=4} = a_{k-1}^{\tau=2}, \tag{8.35}$$

where the cell located immediately to the south of the Eulerian cell A_k is A_{k-1}. So in a practical implementation of a scheme based on (8.32) only two fluxes per cell are computed if $N^f = 4$.

Although the scheme outlined above is termed 'Eulerian' it is not Eulerian in the classical sense where the space and time dimensions are separated. In other words, the scheme outlined above could also be termed flux-form semi-Lagrangian since flux-areas that move with the flow are tracked ('remap-type' scheme). It is Eulerian in the sense that we consider the flux of mass through the (stationary or Eulerian) cell walls. When separating the temporal and spatial dimensions, as done in classical Eulerian schemes, there are no trajectory calculations and fluxes are computed using local information and partial derivatives along the coordinate directions at specific times. The temporal discretization is usually based on Runga–Kutta methods (see Chap. 6). One may argue that the classical Eulerian schemes are an approximation of the general Eulerian–Lagrangian concept presented in this chapter where true (along the trajectories) fluxes are approximated with partial fluxes (i.e., the particle path vector can be decomposed into vector components along the coordinate axes).

8.4.3 Equivalence Between the Lagrangian and Eulerian Discretizations

It is interesting to note the equivalence between the Lagrangian finite-volume continuity equation (8.26) and the Eulerian version (8.32): If taking the sum of the flux-areas a_k^τ with weight 1 for outflow and weight -1 for inflow as well as A_k with weight 1 (all areas involved on the right-hand side of (8.32)), the upstream Lagrangian area a_k results (see example on Fig. 8.7). That is, the right-hand side of (8.32) written in terms of areas is

$$\Delta A_k - \sum_{\tau=1}^{N^f} \left(s_k^\tau \, \Delta a_k^\tau \right) = \Delta a_k. \tag{8.36}$$

So the Lagrangian and Eulerian schemes are identical, as expected, since no approximations have been made so far (even if approximations are made and the resulting schemes are applied to the Euler equations, Eulerian and semi-Lagrangian schemes may produce very similar results as shown in Leslie and Dietachmayer 1997). Insights into schemes can be obtained in the light of the equivalence described above. Any Eulerian flux-form scheme should ideally and effectively have an associated upstream cell from which information is fetched (a.k.a. domain of dependence) to produce the forecast. A more detailed discussion is given in Sect. 8.5.1.2.

A significant difference between the Lagrangian and Eulerian formulation is the necessary conditions for mass conservation. Given a mass-conservative reconstruction function (8.28), a necessary condition for the Lagrangian scheme to be conservative is that the upstream areas a_k span the domain Ω without overlap and gaps between them (8.21) and that the reconstruction function is mass-conservative (8.23). For the Eulerian scheme, however, the flux-areas a_k^{τ} need not necessarily to span the domain Ω and the reconstruction function does not need to satisfy (8.23) to produce a mass-conservative scheme. In fact any estimation of the flux will provide an inherently mass-conservative scheme since the flux computed for a particular cell wall is subtracted in the neighboring cell with which it shares that particular face. So the Lagrangian scheme has, in the sense described above, a stricter requirement for mass-conservation than the Eulerian flux-form formulation.

Another significant difference between the Eulerian and Lagrangian formulations is that the Lagrangian formulation requires the upstream areas to be simply-connected domains. The Eulerian formulation does not require that, in fact, even for relatively simple flows the flux-areas can be non-simply connected (see, e.g., Fig. 8.2 in Harris et al. 2011). The Eulerian formulation is therefore more robust in the sense that it can handle non-simply connected flux-areas (and conserve mass simultaneously) whereas the Lagrangian scheme will break down if an upstream area is not simply connected. This difference could be important for an operational application of the scheme.

8.5 Discrete Schemes: Approximations

The Lagrangian and Eulerian finite-volume schemes, given in (8.26) and (8.32) respectively, are exact. Hence we assume the trajectory of every parcel is known exactly (the exact upstream area and flux-areas are known), the sub-grid-cell reconstruction is exact and the integration of the sub-grid-cell reconstruction function over the upstream areas and flux-areas can be done analytically. Now we start to discuss some of the approximations that can be made in order to derive practical numerical schemes that only have a finite number of degrees of freedom. The approximations can be divided into four steps: Computation of parcel trajectories, area approximation (either upstream Lagrangian areas or Eulerian flux-areas), sub-grid-cell reconstruction and integration of $\psi^n(x, y)$ over deformed areas. As already mentioned we will not discuss the computation of trajectories here and therefore simply assume that they are given.

Firstly, the approximation to areas are discussed. Once the areas have been defined, the transport problem has been reduced to a remapping problem, that is, a conservative grid-to-grid interpolation problem. This requires a reconstruction of the sub-grid-cell distribution and an integration over overlap areas. These three steps (area approximation, reconstruction, integration over overlap areas) are discussed separately below.

8.5.1 Approximation to Areas

With only a finite-number of degrees of freedom and therefore only having the capability of tracking a finite number of parcels (typically the same number as cells N) some approximation must be made to the exact upstream Lagrangian area or Eulerian flux-area. The inability of the scheme to approximate the exact areas is referred to as the *geometric error* (Lauritzen and Nair 2008) and is illustrated graphically on Fig. 8.8a. Obviously the geometric error may lead to local mass errors. Another error is due to inexact sub-grid-cell reconstruction. This error, referred to as the *reconstruction error*, is illustrated on Fig. 8.8b and discussed further in Sect. 8.5.2. Strategies for area approximations are the subject of this section.

8.5.1.1 Lagrangian Area Approximations

Fully Two-dimensional Lagrangian Area Approximations

Probably the most rigorous approximation to the exact upstream cell a_k, Fig. 8.9a, is to follow the trajectories of the vertices of A_k upstream and then connect the upstream vertices with straight lines (Fig. 8.9b); (Rančić 1992; Lauritzen et al. 2010). All other approximations involve approximating a_k with line segments parallel to the coordinates axis which, in general, simplifies the overlap-area integration algorithm. Some examples are given on Fig. 8.9. For more details on Lagrangian cell approximations for orthogonal meshes see the comprehensive review by Lauritzen et al. (2006) and Machenhauer et al. (2009).

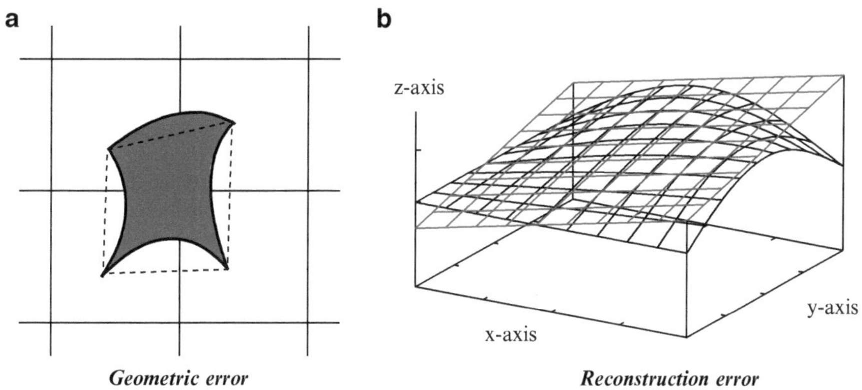

Fig. 8.8 A schematic illustration of the (**a**) geometric error and (**b**) reconstruction (*gradient*) error, respectively, for a cell in two dimensions. (**a**) The geometric error occurs due to the exact region of integration (*shaded area*) being approximated by, for example, *straight line* segments (*dashed lines*). (**b**) The reconstruction error refers to the numerical methods inability to reconstruct the exact sub-grid-scale variation (*black line surface*). The *grey lines* contour the reconstructed sub-grid-scale distribution (in this case a linear approximation)

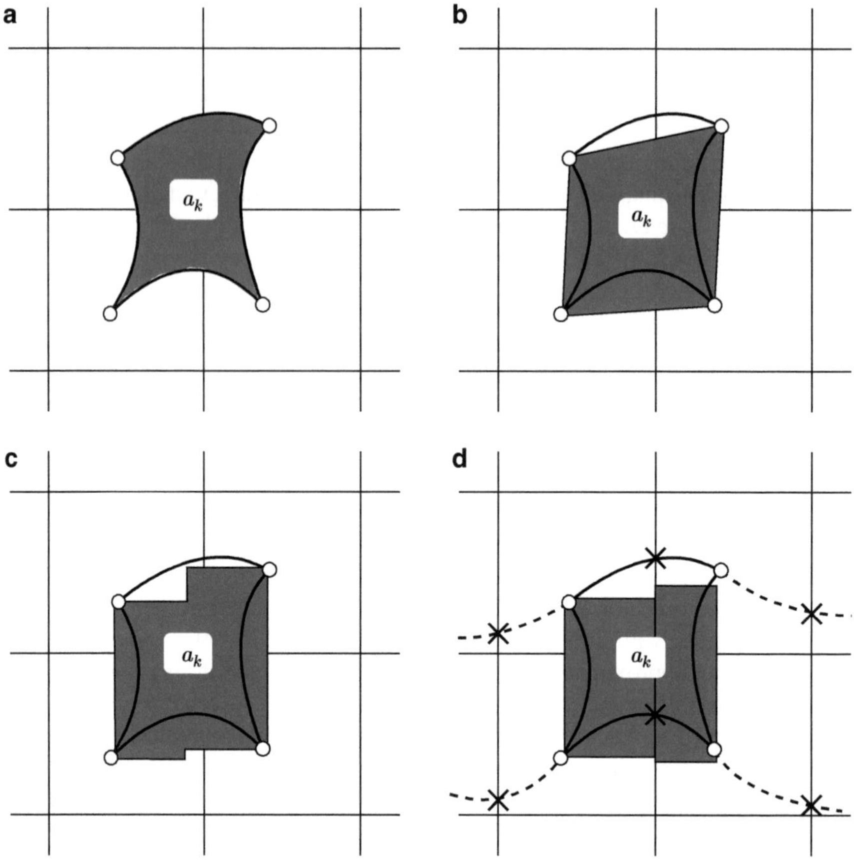

Fig. 8.9 Graphical illustration of approximations to the upstream Lagrangian cell a_k a.k.a. the departure cell. Assume the departure points corresponding to the vertices of the Eulerian grid cell are known (*open circles*). (**a**) Exact departure cell (*shaded area*) with sides depicted with thick lines. (**b**) Sides of the departure cell approximated with *straight lines* by connecting the departure points. (**c**) Departure cell approximation used in Nair and Machenhauer (2002) where the east and west sides are *straight lines* parallel to the Eulerian longitudes (y-axis on the plot) and the north and south sides are approximated with 'step functions'. (**d**) The Lagrangian cell used in the cascade schemes that are based on intersections (*crosses*) between the Lagrangian latitudes (*dashed/solid curved lines*) and the Eulerian longitudes. The 'step' in the step functions used in the cascade schemes always coincides with the Eulerian longitudes (x-isolines on the figure)

Flow-split Lagrangian Area Approximations

More recently the finite-volume cascade[17] approach was suggested by Nair et al. (2002) and Zerroukat et al. (2002) which uses a combination of Eulerian and

[17] The non-conservative cascade interpolation method in Cartesian geometry was introduced by Purser and Leslie (1991).

Lagrangian operators, that is, the one-dimensional operators are successively applied along a coordinate line and a Lagrangian line, respectively. An example is given in Fig. 8.9d where the first one-dimensional operator is applied along the Eulerian longitudinal direction and the second is applied along the deformed Lagrangian latitude (curved solid/dash lines on Fig. 8.9d). So rather than being a fixed direction based splitting method it is flow-based (for a review see Machenhauer et al. 2009). The upstream Lagrangian cell for the cascade scheme is illustrated on Fig. 8.9d. The main difference between the fully two-dimensional area approximation used in Nair and Machenhauer (2002), shown on Fig. 8.9c, and the cascade scheme area approximation, is the location of the 'jump' in the north and south sides of the departure cell. Since the first cascade 'sweep' is along Eulerian longitudes the jump in the north and south sides coincide with an Eulerian longitude. In the Nair and Machenhauer (2002) the jump is located midway between the east and west cell sides.

Approximating the Lagrangian cell with line-segments parallel to the coordinate axis, either with fully two-dimensional or cascade methods, is attractive for orthogonal grids such as a Cartesian rectangular mesh (e.g., Zerroukat et al. 2002) and a regular latitude-longitude grid on the sphere (e.g., Nair and Machenhauer 2002; Nair et al. 2002; Zerroukat et al. 2004). It is less obvious how to extend such approaches to non-orthogonal grids such as triangular or hexagonal grids since the cell sides are no longer orthogonal.

8.5.1.2 Eulerian Flux Area Approximations

The approximation to flux-areas in Eulerian schemes can be divided into two categories: Fully two-dimensional approximations to the flux-areas and dimensionally split area approximations. We remind the reader that only methods that have been extended to global spherical domains are discussed here. We are thereby excluding many transport schemes published in the meteorological literature.

Fully Two-dimensional Flux-area Approximations

The fully two-dimensional flux-area approximations can be divided into two categories. Firstly, one in which one face-centered velocity vector per face is used to trace back the flux-area and, secondly, the approach in which the vertices of the face are traced upstream to compute the flux-area. The first approach only has one degree of freedom for the flux-areas whereas the latter approach has two. Consequently the resulting flux-areas are parallelograms and arbitrary quadrilaterals, respectively, for the two approaches. An elaboration is given below.

Recently, Miura (2007) suggested to approximate the flux-areas from a face-centered wind velocity. So the two vertices of the face would have identical upstream displacements based on the same face-centered velocity vector. The trajectories are therefore parallel and have the exact same length. Hence the flux-areas a_k^{τ},

$\tau = 1, 2, 3, 4$, are parallelograms (Fig. 8.10a,b,d,e, respectively). The fact that the upstream area is a parallelogram may simplify the practical integration of overlap areas at the expense of some potential loss of accuracy if the flow is highly deformational. This is illustrated by computing the effective upstream Lagrangian area for the Miura (2007) scheme using the method outlined in Sect. 8.4.3. That is, by taking the sum of the flux-areas (with signs) shown on Fig. 8.10a,b,d,e and the Eulerian area (Fig. 8.10c), the effective upstream area a_k results (Fig. 8.10f). The upstream area mostly coincides with the exact departure cell, however, there are minor contributions tracing the Eulerian cell vertices that are non-local (not overlapping with the true departure cell). Also, the flux-areas for all cells do not span the domain Ω. If the flow is constant (no deformation) the non-local part of the flux-areas disappear as all the face-centered velocity vectors would be aligned.

If this inability of representing the local flux-areas (geometric error) is a significant source of error has not been investigated (as far as the authors are aware) and the error would only show for challenging test cases with strong deformation. For example, the widely used solid body advection test on the sphere would most likely not expose this potential deficiency. An illustrative example of a highly deformational flow is given on Fig. 8.11 that shows the Lagrangian (upstream) grid for each cubed-sphere panel for one of the test cases in Nair and Lauritzen (2010). Even for a relative short time-step (resulting in a maximum CFL number in each coordinate direction of approximately 0.8) the upstream cells are highly deformed and they might be challenging to approximate accurately using simplified fluxes unless very short time-steps are used. It should, however, be noted that the geometric discussed above will only show if it is larger than the reconstruction error. Consequently, the geometric error is most likely not significant when using low-order reconstruction functions (constant or linear reconstructions).

The potential non-locality problem described above can be resolved by instead of using one face-centered vector (for the trajectories) per face, to use trajectories for the vertices of the cell A_k (Rančić 1992; Lipscomb and Ringler 2005; Yeh 2007). This extra degree of freedom allows the flux-areas to deform into arbitrary quadrilaterals. The equivalent upstream area now equals the Lagrangian area resulting from connecting the upstream points with straight lines. This can be shown as above by taking the sum of the areas involved in the forecast (8.32), Fig. 8.10g,h,i,j,k, with appropriate weights (signs). As for the Eulerian–Lagrangian equivalence in the continuous case, discussed in Sect. 8.4.3, this approximate flux-form scheme is exactly equivalent to the approximate Lagrangian scheme discussed above where the departure points are connected with straight lines (Figs. 8.9b and 8.10, respectively).

Improving the effective approximation to the upstream area further would involve the introduction of more parcels that are tracked (as suggested by Lauritzen et al. 2010) or some approximation to the sides with curved lines. A cursory study addressing the potential benefits of approximating the upstream areas with higher-order polygons was performed in Harris et al. (2011) within the context of a flux-form semi-Lagrangian scheme.

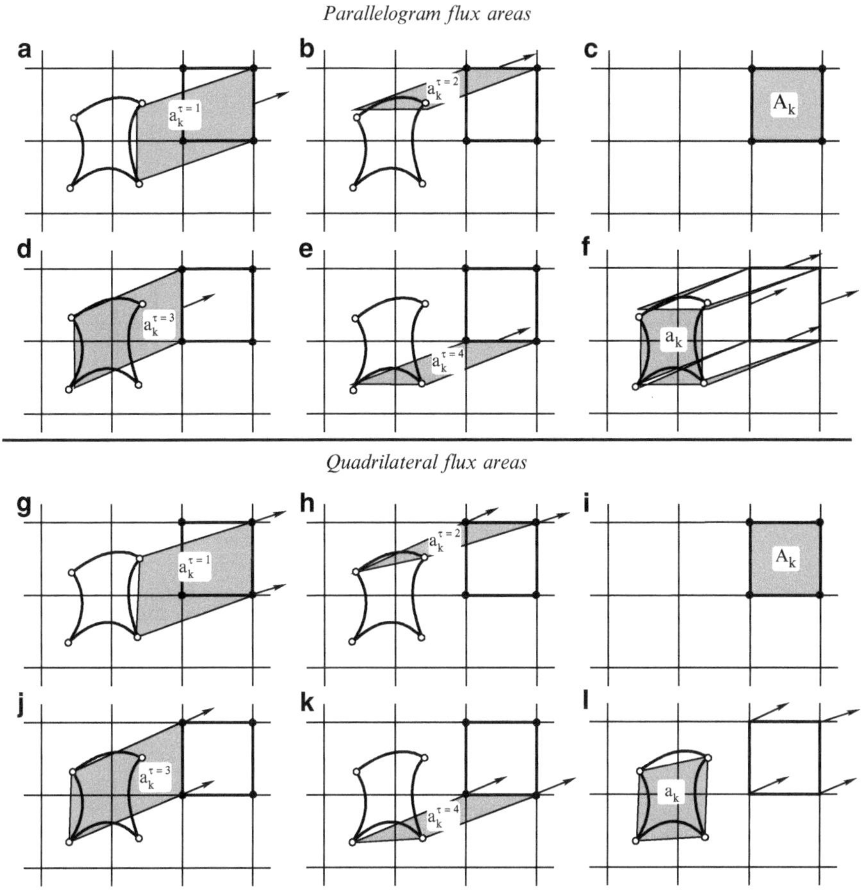

Fig. 8.10 A schematic illustration of different flux approximations, parallelogram (**a,b,d,e**) and quadrilateral (**g,h,j,k**) flux-areas, and the equivalent upstream Lagrangian areas (**f,l**). The equivalent upstream areas are computed by taking the sum of all areas involved in the forecast (**a,b,c,d,e**) or (**g,h,i,j,k**) with appropriate signs (see (8.36)). The velocity vectors used for the flux computations are also shown. The exact upstream Lagrangian cell (*open circles* connected with *curved lines*) is also shown although it is not explicitly used in the flux-form schemes

Dimensionally Split Flux-areas

A popular approach not discussed so far is to use a sequence of one-dimensional operators to approximate the two-dimensional fluxes thereby eliminating the need for solving a fully two-dimensional remapping problem. These methods are also referred to as dimensionally split approaches. A popular scheme based on this strategy is presented in Lin and Rood (1996) and Leonard et al. (1996).

In the present discussion on effective upstream areas, this operator splitting approach was analyzed by Lauritzen (2007) and Machenhauer et al. (2009). When

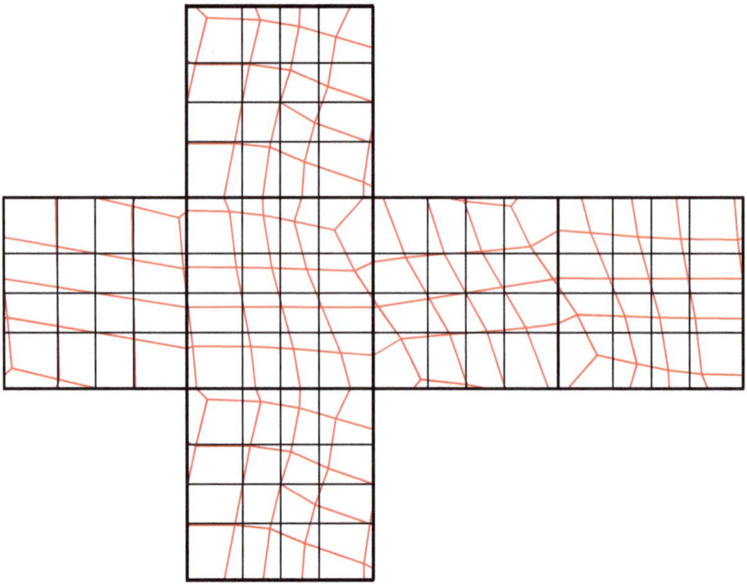

Fig. 8.11 The static Eulerian grid (*thin lines* aligned with coordinate *lines*) and departure grid (deformed *thin lines*) at the first time-step shown on the gnomonic projection on each cubed-sphere panel for test case one of Nair and Lauritzen (2010) illustrated on Fig. 8.3 (time-step was chosen such that the maximum CFL number is approximately 0.8). The departure grid has been constructed by computing trajectories for the cell vertices and then the vertices are connected with *straight lines* (great-circle arcs on the *sphere*)

using dimensionally split approaches the effective upstream area is approximated with a combination of rectangles aligned with the grid lines and with different weights (see Machenhauer et al. 2009). One-dimensional operators cannot represent areas skew to the face in question. As an example of an operator splitting approach the effective departure area for the Lin and Rood (1996) scheme is given on Fig. 8.12 for a flow that has a translational, deformation and rotational component (see Machenhauer et al. 2009 for details).

In dimensional split schemes one can obtain preservation of a constant density field in a non-divergent flow field. This property is harder to obtain with fully two-dimensional semi-Lagrangian schemes but it is possible with cascade semi-Lagrangian schemes (Thuburn et al. 2010).

8.5.1.3 Comment on Area Approximations

One might argue that the errors associated with some of the simplified flux- and upstream- area approximations are not significant at least for orthogonal meshes. For example, for semi-Lagrangian finite-volume schemes Lauritzen et al. (2010) found

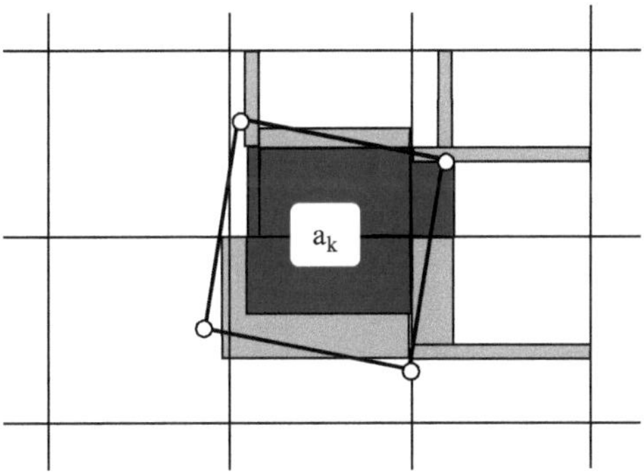

Fig. 8.12 A graphical illustration of the effective departure area a_k for the Lin and Rood (1996) scheme using an analytic wind field which is deformational, rotational and divergent. The exact departure cell is shown with *thick black lines* (and *open circles* as vertices). *Light shading* shows the parts of the departure area where mass is weighted with 1/2 and *dark shaded* areas are weighted with one. See Machenhauer et al. (2009) for details

little difference between the rigorous upstream area approximation and simpler area approximations using line-segments parallel to the coordinate axis.

On non-traditional meshes simplified fluxes might introduce significant inaccuracies. For example, considering a solid-body rotation flow field on the sphere on a non-traditional grid such as the cubed-sphere grid, some of the Lagrangian areas are highly deformed even though the flow field is non-divergent, non-deformational and non-rotational. This is illustrated on Fig. 8.13. The Lagrangian cells entering a cubed-sphere panel from neighboring panels are highly skewed compared to the Lagrangian areas staying within the panel in question. Therefore the need for fully two-dimensional area approximations for non-traditional grid applications seems more evident than for orthogonal quadrilateral grids such as the regular latitude-longitude grid. All of the above is, of course, assuming that the reconstruction error is smaller than the geometric error which will most likely not be the case for first- and second-order methods.

Velocity Staggering and Flux-areas

For the different flux-area approximations described above the velocity components are needed at the center of cell faces (for parallelogram flux areas), cell vertices (for the quadrilateral flux areas) or at multiple locations along the cell sides (for higher-order polygon fluxes). To avoid any interpolation of velocity components Arakawa B and E grid staggering (see Chap. 3) should be used for quadrilateral and parallelogram flux areas, respectively, whereas the higher-order polygon flux inevitably will require interpolation of the velocity components (at least at a subset of the points along the cell sides). The interpolation of velocity components can potentially degrade the overall accuracy of the scheme (McGregor 2005) and the

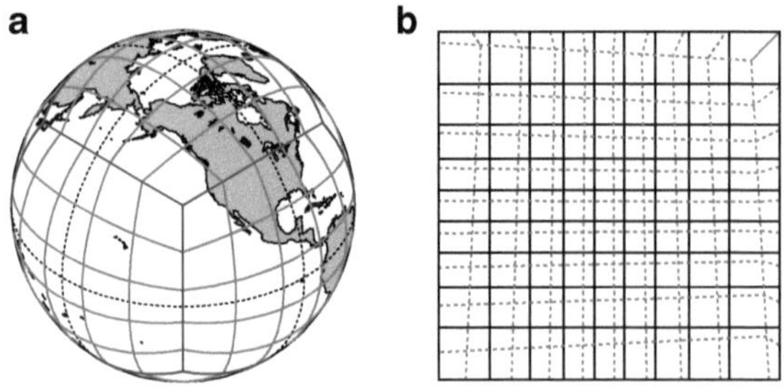

Fig. 8.13 (**a**) Cubed-sphere grid shown with *light shaded lines* and panel edges with *black lines*. (**b**) The upstream/departure grid (*dashed lines*) shown on a local (gnomonic) projection for one of the cubed-sphere panels using the solid-body advection flow field (time-step is so that one revolution is completed in 72 time-steps). The *solid lines* show the Eulerian static grid. The skewed departure cells are cells entering from neighboring panels during the time-step. The parts of the departure cells outside the panel have been 'chopped off'. For an introduction to the cubed-sphere grid see, e.g., Chap. 9

choice of variable staggering impacts wave propagation (when solving the air mass continuity equation with the momentum equations) as discussed in Chap. 3. Hence the choice of flux-area approximation and variable staggering are 'intertwined' and the choices impact not only the accuracy of the transport operator but also wave propagation properties in full models as well as other properties such as the need for filtering etc. (see, e.g., Chaps. 13 and 14). A exhaustive discussion of optimal variable staggering and flux-area approximation is beyond the scope of this chapter.

8.5.2 Sub-Grid-Scale Reconstruction

In the previous sections the geometrical approximation to the upstream areas and flux-areas have been discussed. Next comes the actual integration of $\psi(x, y)$ over these areas, for which a sub-grid-scale reconstruction of the tracer field is needed. We start by discussing reconstruction methods in one spatial dimension and then briefly discuss two dimensional extensions before covering the integration of $\psi(x, y)$ over overlap areas.

8.5.2.1 One-Dimensional Reconstruction Functions

The sub-grid-scale reconstruction is vital for the overall accuracy and efficiency of a scheme, and a thorough discussion is beyond the scope of this chapter. We will, however, discuss some of the most widely used methods. In principle any function could be used for reconstructions, however, the choice of reconstruction function has consequences for any finite-volume scheme. Here are some desirable properties for reconstruction functions that should be considered:

- *Locality.* Locality is generally desirable to maximize parallel efficiency; that is, the stencil (or halo) used for the reconstruction in any cell should use only a limited number of neighboring grid cells. The cells used in the reconstruction of a given cell are referred to as the stencil of that cell.
- *Integrability.* The reconstruction function must later be integrated over overlap areas and it is convenient to use functions that can be integrated exactly. If polynomials are used, polynomials of successively higher degree will lead to more computationally expensive schemes.
- *Conservation.* For Lagrangian finite-volume schemes mass-conservation of the final algorithm requires the reconstruction function to satisfy the so-called cell-averaged property; namely, integration of the reconstruction over the cell (8.28) yields the known cell-average (for each prognostic variable). This requirement is not strictly necessary for Eulerian flux-form schemes but, in general, leads to more accurate reconstructions (Skamarock 2009; personal communication).
- *Filterable.* A scheme can be rendered monotone in the reconstruction step by filtering the reconstruction function so that it is monotone. It may therefore be desirable to use reconstruction functions that are amiable for such filtering. One thing to consider, for example, is that higher-degree polynomial reconstructions are more difficult to filter, since the number of possible extrema increases with the degree of the polynomial. For flux-form Eulerian schemes one may also render the solution monotone *a posteriori* by adequately 'mixing' the (usually low-order) monotone flux with the (usually higher-order) non-monotone flux (Zalesak 1979). In the literature the *a posteriori* filtering is often referred to as limiting. An excellent review on limiting is given in Durran (1999), and we make no effort to try and reproduce it here. Certain reconstructions can also be used that are inherently non-oscillatory by design, such as the class of (W)ENO schemes ((Weighted) Essentially Non-Oscillatory schemes), which generally do not require filtering or limiting.
- *Exactness.* A reconstruction algorithm is referred to as p-exact if it exactly reproduces a global polynomial of degree p (Barth and Frederickson 1990). Generally speaking, strict exactness constraints will lead to an increase in accuracy of the reconstruction function.

Polynomial reconstruction functions, mentioned a couple of times above, are a popular choice in the literature and all properties discussed above can be conveniently dealt with using such a basis. Some work has been done on nonpolynomial-based reconstruction functions (e.g., Norman and Nair 2008; Xiao et al. 2002), however, we will focus on the former here. A comparison of various reconstruction functions in the context of conservative cascade interpolation was tackled by Norman et al. (2009).

Reconstruction Problem Formulation (one Dimension)

The one-dimensional reconstruction problem for a finite-volume scheme utilizing a polynomial basis can be stated as follows: Given discrete cell-averaged values $\overline{\psi}_k$

over cells A_k (here A_k refers to a 1D cell), determine coefficients $c_k^{(i)}, i = 1, \ldots, p$, so that

$$\psi_k(x) = c_k^{(0)} + c_k^{(1)}\, x + c_k^{(2)}\, x^2 + \ldots + c_k^{(p)}\, x^p, \tag{8.37}$$

is an approximation to the underlying field ψ in cell A_k. As mentioned previously, it is desirable that the reconstruction satisfies the cell-averaged property,

$$\int_{A_k} \psi_k(x)\, dx = \overline{\psi}_k\, \Delta x_k, \tag{8.38}$$

where Δx_k is the width of cell A_k. In the context of semi-Lagrangian advection schemes, this property is also referred to as the mass-conservation property.

In the cell-integrated continuity equation (8.18) $\overline{\psi}$ refers to either cell-averaged air density ρ or tracer density ρq, however, in the context of reconstructions it can be desirable to reconstruction ρ and q separately (as mentioned in Sect. 8.4.1). In particular when enforcing shape-preservation it may be convenient to apply the filters/limiters to q and not ρq (e.g., Nair and Lauritzen 2010). Hence, for the discussion on reconstructions ψ can either refer to ρ, ρq or q.

The Piecewise Constant Method (PCoM)

Perhaps the simplest sub-grid-scale representation is the so-called piecewise constant method (PCoM), which simply uses

$$\psi_k(x) = \overline{\psi}_k. \tag{8.39}$$

This approach is attributed to Godunov (1959) and trivially satisfies (8.38), does not need a halo, and is also inherently monotone since it cannot lead to new extrema. This approach is also formally first-order accurate and highly diffusive when used with any scheme over smooth flows and distributions. As a consequence, this choice of reconstruction is considered too diffusive for atmospheric transport problems (unless the flow is 'rough'), and so we must turn our attention to higher-order reconstructions.

Higher-order Reconstructions

Note that by appropriately shifting the polynomial (8.37), we can always map A_k onto the normalized interval $x \in [-\Delta x_k/2, \Delta x_k/2]$ with centerpoint $x = 0$. By doing so, the math behind the reconstruction is dramatically simplified, and so we will hereafter assume that we are working over this domain. Further, we will assume that the grid is *uniform* so that $\Delta x_j = \Delta x$ for all j. Reconstructions based on non-uniform grids are generally a straightforward extension of the uniform case.

Perhaps the most intuitive method for determining the coefficients of (8.37) is to use a Taylor series expansion about the center of the cell ($x = 0$),

$$\psi_k(x)=\psi_k\big|_{x=0}+\left(\frac{\partial\psi_k}{\partial x}\right)\bigg|_{x=0}x+\frac{1}{2}\left(\frac{\partial^2\psi_k}{\partial x^2}\right)\bigg|_{x=0}x^2+\ldots+\frac{1}{p!}\left(\frac{\partial^p\psi_k}{\partial x^p}\right)\bigg|_{x=0}x^p + \mathcal{O}\left[(\Delta x)^{p+1}\right]. \quad (8.40)$$

By pairing terms of equal order, we obtain the association

$$c_k^{(0)} = \psi_k(0), \qquad c_k^{(i)} = \frac{1}{i!}\left(\frac{\partial^i\psi_k}{\partial x^i}\right)\bigg|_{x=0}. \quad (8.41)$$

Since we do not know the exact value of ψ_k or its derivatives, we must approximate these values using, for example, interpolated polynomials through known cell-averaged values.

Note that one must be careful in choosing the correct approximations to these derivatives to preserve high-order accuracy. Specifically, for (8.40) to be formally $\mathcal{O}[(\Delta x)^p]$ accurate, each of the derivatives $\partial^n\psi_k/\partial x^n$ must be approximated to order $\mathcal{O}[(\Delta x)^{p-n}]$, and $\psi_k(0)$ must be approximated to order $\mathcal{O}[(\Delta x)^p]$. The rationale behind this claim is as follows: When evaluating the reconstruction (8.40), each of the derivatives $\partial^n\psi_k/\partial x^n$ is multiplied by x^p, which must satisfy $|x|^p \leq (\Delta x)^p$. Hence, if $\partial^n\psi_k/\partial x^n$ is approximated to $\mathcal{O}[(\Delta x)^{n-p}]$ then each term in the series (8.40) is approximated to $\mathcal{O}[(\Delta x)^p]$. However, since $\psi_k(0)$ is not multiplied by any power of x, it must be approximated to full order-of-accuracy.

Finite-difference Approximations

On averaging the Taylor series (8.40) over a cell A_k, we obtain

$$\frac{1}{\Delta x}\int_{-\Delta x/2}^{\Delta x/2}\psi_k(x)dx = \psi_k(0) + \frac{1}{24}\left(\frac{\partial^2\psi_k}{\partial x^2}\right)\bigg|_{x=0}(\Delta x)^2 + \mathcal{O}\left[(\Delta x)^4\right]. \quad (8.42)$$

The left-hand-side of this expression is simply the cell average $\overline{\psi}_k$, which is known in a finite-volume context. The first term on the right-hand-side is the value of $\psi_k(x)$ evaluated at the cell-centerpoint and it is followed by higher-order terms. Hence, we can conclude that $\overline{\psi}_k$ is a $\mathcal{O}\left[(\Delta x)^2\right]$ approximation to the value of $\psi_k(x)$ evaluated at the centerpoint. This result implies that if we utilize finite-difference approximations to approximate derivatives of any order at $x = 0$, such approximations will only be valid up to $\mathcal{O}\left[(\Delta x)^2\right]$ in a finite-volume context.

The simplest finite-difference approximation is the piecewise-linear method (PLM), given by

$$\psi_k(x) = \overline{\psi}_k + \left(\frac{\partial\psi_k}{\partial x}\right)\bigg|_{x=0}x, \quad (8.43)$$

(van Leer 1977) where $\partial\psi_k/\partial x$ is at least a first-order-accurate approximation to the derivative at $x = 0$. Some choices include an upwind discretization,

$$\left(\frac{\partial\psi_k}{\partial x}\right)\bigg|_{x=0} = \frac{\overline{\psi}_k - \overline{\psi}_{k-1}}{\Delta x} + \mathcal{O}(\Delta x), \tag{8.44}$$

or a centered discretization

$$\left(\frac{\partial\psi_k}{\partial x}\right)\bigg|_{x=0} = \frac{\overline{\psi}_{k+1} - \overline{\psi}_{k-1}}{2\Delta x} + \mathcal{O}\left[(\Delta x)^2\right]. \tag{8.45}$$

Either choice will lead to a scheme which is formally second-order accurate. Larger stencils can be chosen for the approximations to these derivatives, but they can only lead to reconstructions that are at most second-order-accurate. Nonetheless, with larger stencils total accuracy may improve significantly even though the formal order-of-accuracy will not.

The linear reconstruction drastically improves the error measures of finite-volume schemes, when compared to PCoM. This result is illustrated in Fig. 8.14 in terms of a von Neumann stability analysis of a finite-volume scheme based on PCoM and PLM (using the centered approximation (8.45)). In many large-scale atmospheric models PLM is still considered too diffusive and therefore even higher-order reconstructions are often considered.

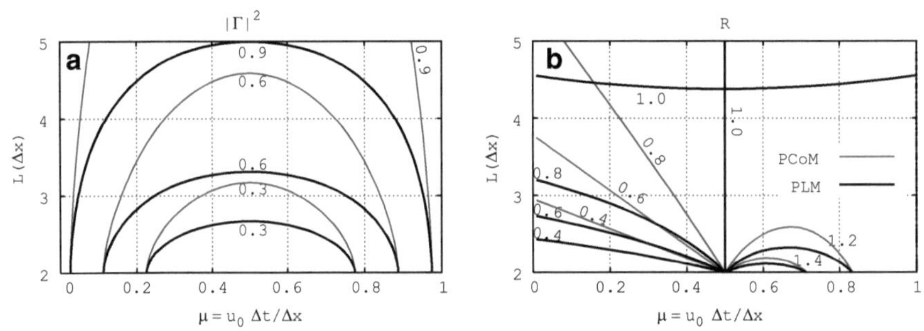

Fig. 8.14 The stability properties (see Sect. 8.3.3 and/or Lauritzen 2007) of a one-dimensional finite-volume scheme based on PCoM (*grey line*) and PLM (*black line*), respectively. Note that in one dimension all finite-volume schemes discussed in this chapter are identical when using the same reconstruction method. (**a**) Squared modulus of the amplification factor ($|\Gamma|^2$) as a function of Courant number (x-axis) and wavelength L (y-axis). Hence (**a**) shows how much each wavelength is damped in one time-step as a function of Courant number. For a fixed Courant number μ the damping decreases monotonically as a function of wavelength L and $\lim_{L\to\infty}|\Gamma|^2 = 1$. For Courant number 0 or 1 the scheme is exact and hence $|\Gamma|^2 = 1$. (**b**) Same as (**a**) but for the relative phase speed (R), that is, how much each Fourier mode is accelerated or decelerated as a function of Courant number

Finite-difference schemes can also be utilized to obtain a third-order reconstruction, even in a finite-volume context. Rearranging (8.42), we can obtain an expression for the centerpoint value $\psi_k(0)$,

$$\psi_k(0) = \overline{\psi}_k - \frac{(\Delta x)^2}{24}\left(\frac{\partial^2 \psi_k}{\partial x^2}\right)\bigg|_{x=0} + \mathscr{O}\left[(\Delta x)^4\right], \tag{8.46}$$

which is a fourth-order-accurate approximation to the pointwise value of $\psi_k(0)$, as long as $\partial^2\psi_k/\partial x^2$ is approximated to at least $\mathscr{O}\left[(\Delta x)^2\right]$. Combining this approximation with (8.40), we obtain a third-order (parabolic) reconstruction

$$\psi_k(x) = \overline{\psi}_k + \left(\frac{\partial \psi_k}{\partial x}\right)\bigg|_{x=0} x + \frac{1}{2}\left(\frac{\partial^2 \psi_k}{\partial x^2}\right)\bigg|_{x=0}\left(x^2 - \frac{(\Delta x)^2}{12}\right) + \mathscr{O}\left[(\Delta x)^3\right], \tag{8.47}$$

when combined with simple finite-difference approximations of the form (8.45) and

$$\left(\frac{\partial^2 \psi_k}{\partial x^2}\right)\bigg|_{x=0} = \frac{\overline{\psi}_{k-1} - 2\overline{\psi}_k + \overline{\psi}_{k+1}}{(\Delta x)^2} + \mathscr{O}\left[(\Delta x)^2\right]. \tag{8.48}$$

In fact, it can be quickly verified that (8.47) also satisfies the cell-averaged property (8.38). This method has the highest formal order of accuracy that can be obtained by treating finite-volume methods in a finite-difference context. This choice of reconstruction was used by Laprise and Plante (1995).

Finite-volume Approximations

To obtain approximations higher than third-order in accuracy, we must first take a step back and understand how finite-volume methods are formulated. First, recall that finite-volume methods use cell-averaged values, which implies that the underlying scalar field is not known point-by-point. Instead, it is cell-averaged values that are known exactly

$$\overline{\psi}_k = \frac{1}{\Delta x}\int_{-\Delta x/2}^{\Delta x/2} \psi_k(x)dx. \tag{8.49}$$

Hence, in the context of finite-volume methods, high-order sub-grid-scale reconstructions cannot be interpolated through specific points (as with finite-difference methods), but must instead satisfy certain cell-averaged properties.

To build a reconstruction that utilizes cell-averages, one generally defines a *cumulative mass function* $W(x)$ via

$$W(x) = \int_{x_{k-j-1/2}}^{x} \psi_k(\tilde{x})d\tilde{x}, \tag{8.50}$$

where $x_{k-j-1/2}$ denotes the left-side edge of cell A_{k-j}. Now, observe

$$\begin{aligned} W(x_{k-j-1/2}) &= 0, \\ W(x_{k-j+1/2}) &= \Delta x \left(\overline{\psi}_{k-j}\right), \\ W(x_{k-j+3/2}) &= \Delta x \left(\overline{\psi}_{k-j} + \overline{\psi}_{k-j+1}\right), \\ &\vdots \end{aligned}$$

Over such a set of consecutive cells one can then define an interpolating polynomial of degree m that approximates the exact cumulative mass function $W(x)$. We denote this approximation by $\tilde{W}(x)$. Finally, we observe that in accordance with the fundamental theorem of Calculus, differentiating (8.50) gives

$$\frac{dW}{dx}(x) = \psi_k(x). \tag{8.51}$$

By evaluating the first derivative of $\tilde{W}(x)$ at a given point, we actually obtain a $\mathcal{O}\left[(\Delta x)^{m-1}\right]$ approximation to the underlying field $\psi_k(x)$ from its cell-averages. This method can then be used to reconstruct $\psi_k(x)$ at any point and, by taking additional derivatives of $\tilde{W}(x)$, its corresponding derivatives.

Alternatively, one can obtain an identical reconstruction by enforcing the cell-averaged constraint on an interpolating polynomial in neighboring cells (Zerroukat et al. 2002). That is, a polynomial $\widehat{\psi}_k(x)$ of degree p that exactly satisfies the mass-conservation constraint not only in cell k but also in p adjacent cells:

$$\int_{x_{j-1/2}}^{x_{j+1/2}} \widehat{\psi}_k(x)\, dx = \overline{\psi}_j \Delta x, \quad j = \left(k - \frac{p}{2}\right) .. \left(k + \frac{p}{2}\right), \tag{8.52}$$

for p even and

$$\int_{x_{j-1/2}}^{x_{j+1/2}} \widehat{\psi}_k(x)\, dx = \overline{\psi}_j \Delta x, \quad j = \left(k - \frac{p+1}{2}\right) .. \left(k + \frac{p-1}{2}\right), \tag{8.53}$$

for p odd.

Either method will yield an identical reconstruction ($\tilde{W}(x) = \widehat{\psi}_k(x)$), although the latter is more adaptable to two dimensions and beyond.

If we utilize the aforementioned procedure over a 3-cell stencil (consisting of cells $k-1, k$ and $k+1$), we will exactly obtain (8.45), (8.46) and (8.48). However, beyond three cells, the finite-difference and finite-volume reconstructions will differ substantially. For instance, over a centered 5-cell stencil, we obtain approximations

$$\psi_k(0) = \frac{9\overline{\psi}_{k-2} - 116\overline{\psi}_{k-1} + 2134\overline{\psi}_k - 116\overline{\psi}_{k+1} + 9\overline{\psi}_{k+2}}{1920} + \mathcal{O}\left[(\Delta x)^6\right],$$

$$\left(\frac{\partial \psi_k}{\partial x}\right)\Bigg|_{x=0} = \frac{5\overline{\psi}_{k-2} - 34\overline{\psi}_{k-1} + 34\overline{\psi}_{k+1} - 5\overline{\psi}_{k+2}}{48\,\Delta x} + \mathscr{O}\left[(\Delta x)^5\right],$$

etc.

High-order reconstructions of this type were adopted for a shallow-water model by Ullrich et al. (2010).

Symmetric Finite-Volume Schemes

In all the methods so far discussed, we have not touched on the issue of continuity between cells. In fact, all of the methods we have described so far do not enforce any sort of continuity between reconstructions in neighboring cells.

As we have seen so far, as the order of the reconstruction polynomial is increased, more options for how to approximate the coefficients $c_k^{(i)}$ are available. Continuity at edges can be enforced (over an arbitrary scalar field) if we adopt a reconstruction that is at least parabolic, i.e.,

$$\psi_k(x) = c_k^{(0)} + c_k^{(1)} x + c_k^{(2)} x^2. \tag{8.54}$$

Since we have three degrees of freedom in this polynomial, we can choose to enforce $\psi_k(-\Delta x/2) = \psi_k^L$ and $\psi_k(\Delta x/2) = \psi_k^R$, where ψ_k^L and ψ_k^R are reconstructed values at the left- and right- edges, respectively. These are purposely chosen to be consistent between neighboring cells, which gives us the desired continuity restriction. With our remaining degree of freedom we enforce the cell-averaged condition (8.38).

This scheme is the well-known piecewise-parabolic method (PPM) of Colella and Woodward (1984). To obtain edgepoint values ψ_k^L and ψ_k^R, PPM makes use of the finite-volume formulation discussed earlier, taken over four cells and evaluated at the cell edgepoint, which gives

$$\psi_k^R = \frac{7}{12}(\overline{\psi}_k + \overline{\psi}_{k+1}) - \frac{1}{12}(\overline{\psi}_{k+2} + \overline{\psi}_{k-1}) + \mathscr{O}\left[(\Delta x)^4\right], \tag{8.55}$$

(also see Zerroukat et al. (2002)) and $\psi_{k+1}^L = \psi_k^R$. In terms of the coefficients $c_k^{(i)}$, this reconstruction can be written as

$$\begin{aligned}
c_k^{(0)} &= \overline{\psi}_k - (\Delta x)^2 \frac{c_k^{(2)}}{12}, \\
c_k^{(1)} &= \frac{1}{\Delta x}\left[\frac{2}{3}\left(\overline{\psi}_{k+1} - \overline{\psi}_{k-1}\right) - \frac{1}{12}\left(\overline{\psi}_{k+2} - \overline{\psi}_{k-2}\right)\right], \\
c_k^{(2)} &= \frac{1}{2\,(\Delta x)^2}\left[-5\overline{\psi}_k + 3\left(\overline{\psi}_{k+1} + \overline{\psi}_{k-1}\right) - \frac{1}{2}\left(\overline{\psi}_{k+2} + \overline{\psi}_{k-2}\right)\right].
\end{aligned}$$

All of the coefficients $c_k^{(i)}$ in this case approximate $\psi_k(x)$ and its derivatives to $\mathcal{O}\left[(\Delta x)^4\right]$.

The approximation to ψ_k^L in (8.55) and ψ_k^R are fourth-order accurate for uniform grids. Obviously, one could also derive second, third, fourth, fifth or sixth-order accurate estimations by fitting a linear, parabolic, cubic, quartic and quintic polynomial so that (8.50) or (8.52–8.53) is satisfied in 2, 3, 4, 5 and 6 adjacent cells, respectively. We will refer to PPM based on second, third, fourth, fifth-order edge value estimates as PPM2, PPM3, PPM4, PPM5, respectively. In this context PPM and PPM4 refer to the same reconstruction.

In terms of p-exactness the edge estimates must be at least third-order for the PPM to exactly reconstruct a global parabola. Hence PPM2 is not p-exact (p = 2) whereas PPM3, PPM4, and PPM5 are. It is noted that PPM4 is significantly more accurate than PPM3 in terms of a Von Neuman stability analysis (Fig. 8.15) whereas PPM5 only gives modest increases in accuracy. Obviously PPM5 needs a larger halo than PPM4. As a consequence, the potential increase in cost associated with the use of larger stencils has likely been a significant factor in determining the widespread adoption of PPM4 over these other schemes. More discussion on edge-value estimates is given in White and Adcroft (2008).

One could naturally ask the question why should one not use the highest-order polynomial that can be approximated with a given halo (stensil)? For example, the cubic polynomial used to estimate the edge value in PPM4 could be used as the reconstruction function, $\psi_k(x) = \widehat{\psi}_k(x)$. While this might improve the accuracy of the scheme, it will make filtering and integration over overlap-areas more cumbersome and computationally expensive, as compared to sticking to a parabola with high-order edge-value estimates (PPM4).

Reconstructions based on polynomials of degree higher than two have been proposed in the literature but have not been widely adopted in transport schemes as of the time of writing. Zerroukat et al. (2002) introduced a symmetric piecewise-cubic method (PCM), along with advanced filtering techniques (Zerroukat et al. 2005),

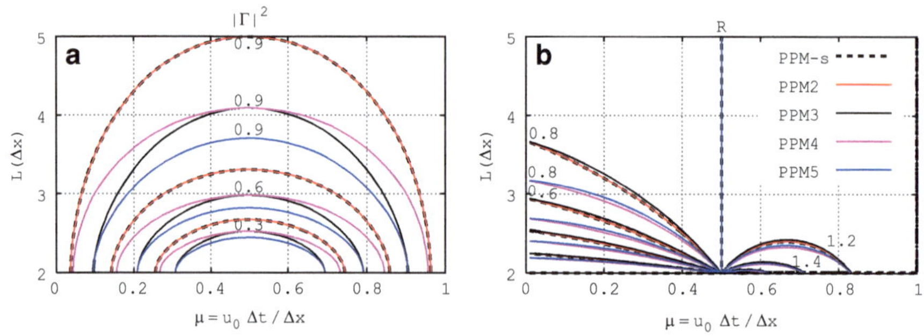

Fig. 8.15 Same as Fig. 8.14 but for PPM using different estimates for the edge values (*solid lines*) as well as PPM-s (*dashed line*). PPM-s is the sub-grid-cell reconstruction method based on the method of Laprise and Plante (1995), that is, using (8.52) with $p = 2$ to determine the sub-grid-cell reconstruction function

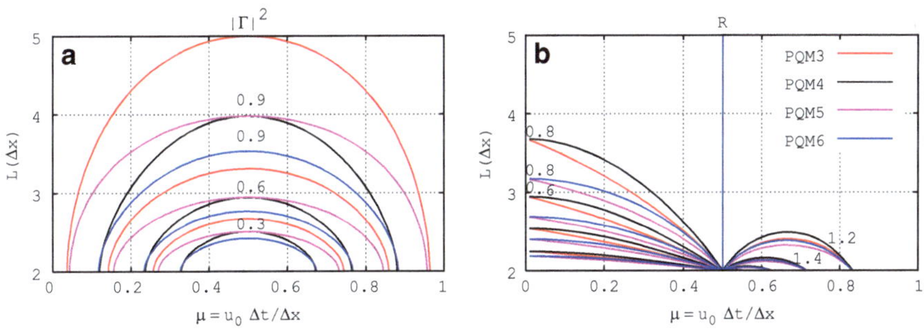

Fig. 8.16 Same as Fig. 8.15 but for PQM

and White and Adcroft (2008) proposed a symmetric piecewise-quartic method (PQM) based on polynomials of degree four. As for the PPM the edge-value estimate is paramount for the accuracy of the scheme (see, e.g., Fig. 8.16). However, even for the most accurate PQM6 the increase in accuracy (in terms of a Von Neumann analysis) is modest compared to PPM4 given the increase in the halo size. Also to consider is that polynomials of degree three (PCM) and four (PQM) can have two and three extrema within a grid cell making it harder to filter such polynomials compared to a (relatively low-order) parabola.

Piecewise Quadratic Splines

An interesting variant on the reconstruction methods discussed so far, and also based on parabolas, is the piecewise quadratic spline method (Zerroukat et al. 2006) and higher-order extensions such as those presented in Zerroukat et al. (2010). Instead of only enforcing C^0 continuity across cell edges also the first derivative of the reconstruction is constrained to be continuous, i.e., the reconstruction is C^1 across cell edges. Enforcing continuity in the derivatives of the reconstruction functions at cell boundaries results in an implicit system of equations for the polynomial coefficients. When written in matrix form, however, the matrix that needs to be inverted has a tri-diagonal form.

In idealized test cases using the scheme of Zerroukat et al. (2002) the piecewise spline reconstruction method is superior to PPM while being 40% more efficient in terms of number of operations (Zerroukat et al. 2007). The price to pay, in terms of a parallel computational environment, is that splines are inherently global since the inversion of a global tri-diagonal matrix is necessary.

Essentially Non-oscillatory (ENO) Reconstructions

Essentially non-oscillatory reconstructions were originally developed by Harten et al. (1987) for shock hydrodynamics problems. This approach is particularly interesting since it leads to a reconstruction that is (under most circumstances) monotonic and positive. The ENO scheme works by applying either a finite-difference or finite-volume approach (as discussed earlier) on a variety of stencils. The reconstruction that satisfies some least-oscillatory property, among all possible stencils, is then chosen to give the 'true' reconstruction. The main drawback of this approach is that it requires a large stencil in order to obtain the same order of accuracy as 'vanilla' finite-difference or finite-volume methods.

Least-squares

The least-squares technique is one of the few approaches available for obtaining approximate reconstructions on unstructured grids. Under this method, we introduce some quantification of the misfit between the reconstruction (8.37) and the known cell-averages, usually given by the square of the difference between the cell-averages of the reconstruction and the known cell-averages. The misfit is then minimized over all possible reconstructions in order to give the "best possible" reconstruction. An example of a Cartesian finite-volume scheme on an unstructured grid based on the least-squares technique can be found in Barth and Frederickson (1990).

One-dimensional Reconstruction Limiters/Filters

As discussed in Sect. 8.3.6, it is desirable that a transport scheme utilizes physically realizable reconstructions. There are two ways to achieve this goal, either a priori by filtering the sub-grid-cell reconstruction function so that it only takes physically realizable values, or a posteriori by limiting prognosed cell averages or by altering the fluxes individually. In the context of an upstream semi-Lagrangian scheme flux-limiting is obviously not an option. For Eulerian schemes one may apply a priori filters or flux-limiters to provide physically realizable solutions. A priori filters are also referred to as slope-limiters as they act directly on the sub-grid-scale reconstruction function.

The PLM, usually based on (8.44) or (8.45), may violate monotonicity as illustrated in Fig. 8.17a. Monotonicity can be enforced by replacing the reconstructed derivative with some weighted average of the upwind and downwind approximations. Many such combinations exist, including MINMOD, Superbee (Roe 1985), and monotone central (van Leer 1977), to name a few (see, for example, Toro 1999). It is beyond the scope of this chapter to provide a comprehensive review of these filters but they all seek to blend the derivative estimates, as hinted above, to obtain the least diffusive monotone solution.

As illustrated on Fig. 8.17b PPM4 is also non-monotone without the application of filters. The seminal paper of Colella and Woodward (1984) constrains the reconstruction so that the entire sub-grid-scale reconstruction is bounded by the cell-averages of the neighboring cells (or is reduced to a constant when the reconstruction is a local extrema). See Fig. 8.17b for an example. This technique for filtering the reconstruction has the tendency to "cut off" or flatten smooth, physical maxima and minima. Several approaches have been proposed to retain physical

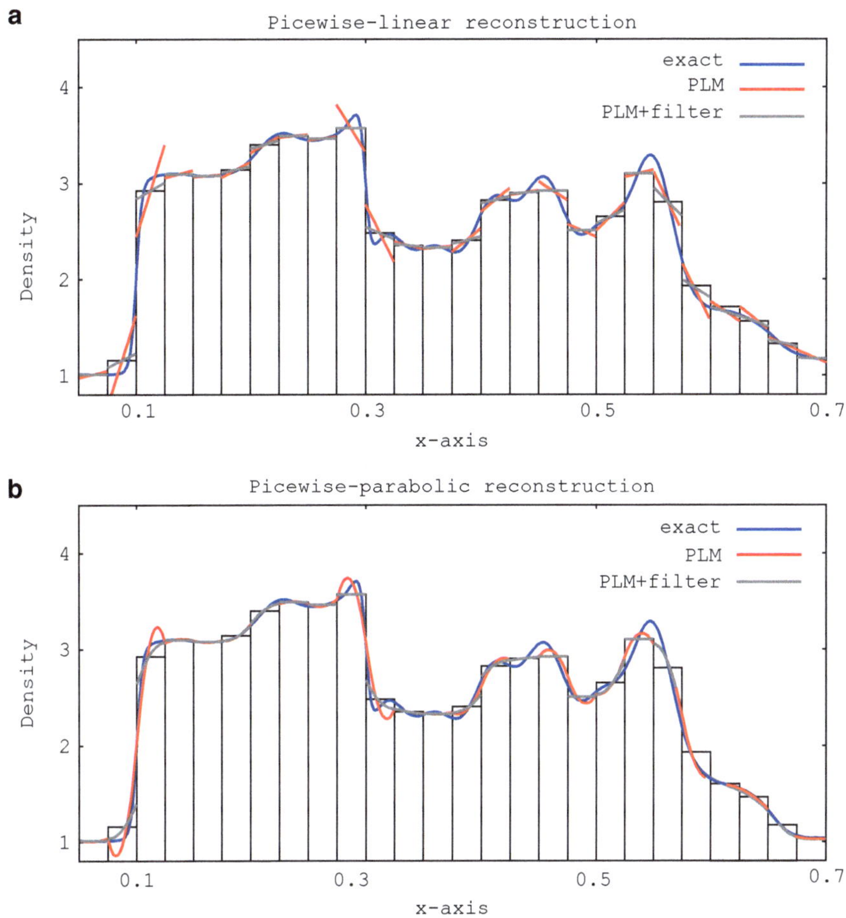

Fig. 8.17 Reconstructions for the irregular signal of Smolarkiewicz and Grabowski (1990, *blue line*) using the (**a**) piecewise-linear method (PLM) and (**b**) the piecewise parabolic method (PPM) with reconstruction function filter (*grey*) and without (*red*). The filter for PLM is the MINMOD limiter (see text) with theta = 1 and the PPM limiter is the original filter presented in Colella and Woodward (1984)

extrema while filtering out spurious grid-scale noise (see, for example, van Albada et al. (1982); Zerroukat et al. (2005); Liu et al. (2007) and Colella and Sekora (2008)). If miniscule over- and undershoots can be tolerated, less invasive filters can be designed using (W)ENO-type methods where the user-specified filter is applied only when a smoothness metric exceeds a certain threshold (Blossey and Durran 2008). Achieving high-order accuracy and physical realizable prognosed values (monotonicity) is very challenging and deserves a chapter on its own for a comprehensive discussion. We will not discuss reconstruction filtering further, although it has profound impact on the diffusion and dispersion properties of a scheme at small scales.

8.5.2.2 Two-Dimensional Reconstruction Functions

Two-dimensional reconstructions can be obtained using nearly direct generalizations of the methods presented in Sect. 8.5.2.1. In fact, for second-order accurate schemes that use a linear reconstruction the linear derivatives can be calculated in each direction independently (dimension-splitting).

Reconstruction Problem Formulation

The two-dimensional reconstruction problem for a finite-volume scheme utilizing a polynomial basis is analogous to the one-dimensional case: Given discrete cell-averaged values $\overline{\psi}_k$ over cells A_k, determine coefficients $c^{(i,j)}$, $i + j \leq p$ (i and j are 0 or positive integers), so that

$$\psi_k(x, y) = \sum_{i+j \leq p} c_k^{(i,j)} x^i y^j, \tag{8.56}$$

is an approximation to the underlying field ψ in cell A_k. The cell-averaged property in two dimensions then reads,

$$\int_{A_k} \psi_k(x, y)\, dA = \overline{\psi}_k\, \Delta A_k. \tag{8.57}$$

Again we can choose to shift the reconstruction so that, for simplicity, A_k has a centroid located at $(x, y) = (0, 0)$.

Piecewise Constant Method (two dimensions)

The extension of the PCoM to two dimensions is trivial, given by

$$\psi_k(x, y) = \overline{\psi}_k. \tag{8.58}$$

This scheme suffers from the same deficiencies as discussed in the one-dimensional case, and so is not discussed further here.

Piecewise Linear Method (two dimensions)

The two-dimensional piecewise linear reconstruction can be written as

$$\psi_k(x, y) = \overline{\psi}_k + c_k^{(1,0)} x + c_k^{(0,1)} y. \tag{8.59}$$

Any choice of $c_k^{(1,0)}$ and $c_k^{(0,1)}$ will yield a mass-conservative reconstruction and, as with the one-dimensional PLM, $c_k^{(1,0)}$ and $c_k^{(0,1)}$ correspond to the components of the gradient along each coordinate direction at the cell centroid,

$$c_k^{(1,0)} = \left(\frac{\partial \psi_k}{\partial x}\right)\bigg|_{(x,y)=(0,0)},$$

$$c_k^{(0,1)} = \left(\frac{\partial \psi_k}{\partial y}\right)\bigg|_{(x,y)=(0,0)}.$$

For this method, the gradient can be limited as in the one-dimensional case (see, e.g., Dukowicz and Baumgardner 2000).

High-order Reconstructions (two dimensions)

True third-order and higher schemes require some method of incorporating cross-derivatives in order to be formally third-order accurate. For example, a true third-order parabolic reconstruction could make use of a reconstruction of the form

$$\psi_k(x,y) = \overline{\psi}_k + c_k^{(1,0)} x + c_k^{(0,1)} y \tag{8.60}$$

$$+c_k^{(2,0)}\left(x^2 - \frac{(\Delta x)^2}{12}\right) + c_k^{(1,1)} xy + c_k^{(0,2)}\left(y^2 - \frac{(\Delta y)^2}{12}\right),$$

where $c_k^{(1,0)}$, $c_k^{(0,1)}$, $c_k^{(2,0)}$, $c_k^{(0,2)}$, and $c_k^{(1,1)}$ are obtained by again approximating the derivatives of ψ. Note that $c^{(0.0)}$ does not equal the cell average $\overline{\psi}$ but includes more terms to ensure the mass-conservation property. Extensions of this form are described in Nair and Machenhauer (2002) and Ullrich et al. (2010). It has been shown that the loss of accuracy attributed to neglecting the cross-derivative term $c_k^{(1,1)}$ can be large, but is less significant on grids of low resolution (Lauritzen et al. 2010).

Piecewise Parabolic Method (Two Dimensions)

A rigorous extension of the PPM method introduced by Colella and Woodward (1984) would require the fully two dimensional biparabolic polynomials to be continuous across cell-borders at selected points and/or in some average sense.

One such extension was developed by Rančić (1992), who chose

$$\psi(x,y) = \phi_2(y)\,x^2 + \phi_1(y)\,x + \phi_0(y), \tag{8.61}$$

where

$$\phi_0(y) = \phi_{02}y^2 + \phi_{01}y + \phi_{00}, \tag{8.62}$$
$$\phi_1(y) = \phi_{12}y^2 + \phi_{11}y + \phi_{10}, \tag{8.63}$$
$$\phi_2(y) = \phi_{22}y^2 + \phi_{21}y + \phi_{20}. \tag{8.64}$$

This reconstruction has nine degrees of freedom, which are restricted by satisfying (a) the cell-averaged constraint (8.57) in cell k, (b) equal average values along each of the four edges of the quadrilateral cells, and (c) continuity at the corner points of each cell. These restrictions lead to 9 constraints, and hence define a unique reconstruction. Note that the reconstruction is not globally continuous. We refer to Rančić (1992) for further details on this algorithm.

Extensions to Irregular Grids

All of the methods described above are tied to quadrilateral (orthogonal) meshes and the extension to triangular, hexagonal and other grids where the cells do not have exactly four vertices, is not obvious. The authors are not aware of any rigorous extensions of PPM to such grids where continuity across cell borders is enforced. In this case enforcement of the cell-average property is more difficult, and requires special treatment of the parabolic terms. For instance, we must have

$$c_k^{(0,0)} = \overline{\psi}_k + c_k^{(2,0)}\left[x^2 - m_k^{(2,0)}\right] + c_k^{(0,2)}\left[y^2 - m_k^{(0,2)}\right] + c^{(1,1)}\left[x\,y - m_k^{(1,1)}\right], \tag{8.65}$$

(Ullrich et al. 2009) where $m_k^{(i,j)}$ are the area-averaged higher-order moments

$$m_k^{(i,j)} = \frac{1}{\Delta A_k}\int_{A_k} x^i\, y^j\, dA.$$

Approximation of the derivative terms may be difficult on irregular grids. For grids where finite-difference approximations to the derivatives are not obvious to compute, as is the case for completely unstructured grids, one might use a two-dimensional extension of the Laprise and Plante (1995) method. That is, enforce the mass-conservation constraint not only in cell k but in a set of adjacent cells. For grids in which cell k has a variable number of adjacent neighbors this approach may not be optimal. In such cases a least-squares approach might be a more natural choice to avoid biases introduced by excluding some adjacent cells and not others. When using a least squares method one may chose to optimize the approximation to the coefficients not just to mass-conservation in adjacent cells but also to p-exactness for example (see, e.g., Barth and Frederickson 1990).

Two-dimensional Limiters / Filters

Reconstruction function filtering in two-dimensions is significantly more complicated than in one dimension, simply because a two-dimensional polynomial of degree two (a parabolic reconstruction) can possess an extrema within a cell, along a cell boundary, or at a corner-point (all of which must be checked). Hence, filtering comes in two flavors: Dimensionally split filtering and fully two-dimensional approaches.

The dimensional split approach simply applies the one-dimensional filters presented in Sect. 8.5.2.1 in each coordinate direction. However, by doing so the entire reconstruction ψ_k is not guaranteed to be monotonic within A_k. Specifically, there are no guarantees of monotonicity at cell corner-points where the reconstruction in each coordinate direction is additive (Lauritzen et al. 2006).

Strict monotonicity at all points within a cell can be guaranteed using the fully two-dimensional approach of Barth and Jespersen (1989), which can also be applied to unstructured grids. This filter guarantees strict monotonicity of linear reconstructions by first determining where a given linear reconstruction has extrema (this is typically the cell corner-points), and then rescaling the linear derivatives so that the linear reconstruction is monotonic with respect to its neighbors. This approach was also extended to parabolic (third-order) reconstructions by Ullrich et al. (2009), which applies rescaling to both linear and high-order derivatives. If strict monotonicity is not necessary, a WENO-type criterion can be used to identify places in which a filter should be applied. An extension of the one-dimensional WENO-filtering in Blossey and Durran (2008) can be found in Harris et al. (2011).

For flux-limiting the most widely used method is flux-corrected transport (FCT) introduced by Zalesak (1979). As in one dimension it seeks to find the optimal "blending" of a monotone flux and a high-order non-monotone flux. FCT is described in detail in Durran (1999) and hence not repeated here.

8.5.3 *Practical Integration Over Areas*

For the approximation of the overlap integrals in (8.26) and (8.33) we have only discussed how to approximate the overlap areas and how to do reconstructions so far. It remains to be shown how to go about integrating the sub-grid-scale reconstruction function over that area. If the sides of the overlap areas are aligned with the coordinate lines, direct integration is usually straightforward since the integrals effectively reduce to one dimension (see, for example, Nair and Machenhauer (2002)). However, if the overlap area is allowed to be an arbitrary polygon the integration is more involved. There are basically two approaches that exactly integrate polynomial functions over polygons. Firstly, direct integration over overlap areas using Gaussian quadrature. Secondly, the area-integrals can be converted into line-integrals via Gauss-Green's theorem.

Both of these approaches are discussed below. We assume that the overlap cell sides are straight lines with arbitrary orientation and that the overlap area $a_{k\ell}$ is simply connected. This is the most general case. Mathematically the problem is stated as follows: Given a reconstruction function in cell A_ℓ which is a polynomial, say of order 3,

$$\psi_\ell(x, y) = \sum_{i+j\leq 2} c_\ell^{(i,j)} x^i y^j, \tag{8.66}$$

where $c_\ell^{(i,j)}$ are reconstruction coefficients, compute the integral

$$\int_{a_{k\ell}} \psi_\ell(x, y) dA. \tag{8.67}$$

8.5.3.1 Direct Integration Using Gaussian Quadrature

For the direct integration using Gaussian quadrature it is often convenient to break up $a_{k\ell}$ into triangles which is the case we will discuss here. So, for simplicity, suppose the overlap area is already an arbitrary triangle[18] with vertices located at $x_{k\ell,h}$, $y_{k\ell,h}$, $h = 1, 2, 3$, and numbered counter-clockwise. Exact integration of the polynomial (8.66) can be achieved using Gaussian quadrature which approximates the integral in terms of a weighted sum of functional evaluations at quadrature points. The quadrature points are

$$(x_{k\ell}^{(a)}, y_{k\ell}^{(a)}) = \tfrac{1}{6}\left(4x_{k\ell,1} + x_{k\ell,2} + x_{k\ell,3}, 4y_{k\ell,1} + y_{k\ell,2} + y_{k\ell,3}\right), \tag{8.68}$$

$$(x_{k\ell}^{(b)}, y_{k\ell}^{(b)}) = \tfrac{1}{6}\left(x_{k\ell,1} + 4x_{k\ell,2} + x_{k\ell,3}, y_{k\ell,1} + 4y_{k\ell,2} + y_{k\ell,3}\right), \tag{8.69}$$

$$(x_{k\ell}^{(c)}, y_{k\ell}^{(c)}) = \tfrac{1}{6}\left(x_{k\ell,1} + x_{k\ell,2} + 4x_{k\ell,3}, y_{k\ell,1} + y_{k\ell,2} + 4y_{k\ell,3}\right). \tag{8.70}$$

(Dukowicz and Baumgardner 2000) and the integral of $\psi_\ell(x, y)$ over the overlap triangle $a_{k\ell}$ is given by

$$\int_{a_{k\ell}} \psi_\ell(x, y)\, dA = \frac{\Delta a_{k\ell}}{3}\left[\psi_\ell(x_{k\ell}^{(a)}, y_{k\ell}^{(a)}) + \psi_\ell(x_{k\ell}^{(b)}, y_{k\ell}^{(b)}) + \psi_\ell(x_{k\ell}^{(c)}, y_{k\ell}^{(c)})\right], \tag{8.71}$$

where $\Delta a_{k\ell}$ is the area of $a_{k\ell}$

$$\Delta a_{k\ell} = \frac{1}{2}\left[(x_{k\ell,2} - x_{k\ell,1})(y_{k\ell,3} - y_{k\ell,1}) - (y_{k\ell,2} - y_{k\ell,1})(x_{k\ell,3} - x_{k\ell,1})\right]. \tag{8.72}$$

Note that the quadrature points only have to be computed once for each overlap area and can then be re-used for each additional tracer (since all tracers follow the

[18] Note that any area with straight line sides can be broken up into triangles.

same trajectories/areas). For flux-form transport schemes efficient algorithms can be designed that decompose the overlap areas into triangles if the flux-areas are confined to nearest neighbors (Dukowicz and Baumgardner 2000). For longer time-steps where the flux-areas may span several Eulerian cells not sharing a face with the flux-face, the decomposition into triangles is more complicated. In such situations it may be more convenient to use the line-integral approach described next.

8.5.3.2 Converting Area-integrals into Line-integrals

This approach was originally introduced by Dukowicz (1984) and Dukowicz and Kodis (1987) in numerical schemes: For the simply connected overlap area $a_{k\ell}$ (not necessarily a triangle but any polygon) the following integral equation holds,

$$\iint_{a_{k\ell}} \psi_\ell(x,y)\, dA = \oint_{\partial a_{k\ell}} \left[P\, dx + Q\, dy\right], \tag{8.73}$$

where $\partial a_{k\ell}$ is the boundary of $a_{k\ell}$. The potentials $P = P(x,y)$ and $Q = Q(x,y)$ are chosen such that they satisfy

$$-\frac{\partial P}{\partial y} + \frac{\partial Q}{\partial x} = \psi_\ell(x,y).$$

The integral of the polynomial reconstruction function $\psi_\ell(x,y)$ in (8.66) can be written as

$$\int_{a_{k\ell}} \psi_\ell(x,y)\, dA = \sum_{i+j\le 2} c_\ell^{(i,j)} w_{k\ell}^{(i,j)}, \tag{8.74}$$

where $c_\ell^{(i,j)}$ are the reconstruction function coefficients and $w_{k\ell}^{(i,j)}$ are weights given by

$$w_{k\ell}^{(0,0)} = \frac{1}{2}\sum_{h=1}^{N_h} \left(x_{k\ell,h} + x_{k\ell,h-1}\right)\left(y_{k\ell,h} - y_{k\ell,h-1}\right), \tag{8.75}$$

$$w_{k\ell}^{(1,0)} = \frac{1}{6}\sum_{h=1}^{N_h} \left(x_{k\ell,h}^2 + x_{k\ell,h}\, x_{k\ell,h-1} + x_{k\ell,h-1}^2\right)\left(y_{k\ell,h} - y_{k\ell,h-1}\right), \tag{8.76}$$

$$w_{k\ell}^{(0,1)} = -\frac{1}{6}\sum_{h=1}^{N_h} \left(y_{k\ell,h}^2 + y_{k\ell,h} y_{k\ell,h-1} + y_{k\ell,h-1}^2\right)\left(x_{k\ell,h} - x_{k\ell,h-1}\right), \tag{8.77}$$

$$w_{k\ell}^{(2,0)} = \frac{1}{12}\sum_{h=1}^{N_h} \left(x_{k\ell,h} + x_{k\ell,h-1}\right)\left(x_{k\ell,h}^2 + x_{k\ell,h-1}^2\right)\left(y_{k\ell,h} - y_{k\ell,h-1}\right), \tag{8.78}$$

$$w_{k\ell}^{(0,2)} = -\frac{1}{12}\sum_{h=1}^{N_h}\left(y_{k\ell,h}+y_{k\ell,h-1}\right)\left(y_{k\ell,h}^2+y_{k\ell,h-1}^2\right)\left(x_{k\ell,h}-x_{k\ell,h-1}\right), \quad (8.79)$$

$$w_{k\ell}^{(1,1)} = \frac{1}{24}\sum_{h=1}^{N_h}\left\{\left[y_{k\ell,h}\left(3\,x_{k\ell,h}^2 + 2\,x_{k\ell,h}\,x_{k\ell,h-1} + x_{k\ell,h-1}^2\right) + y_{k\ell,h-1}\left(x_{k\ell,h}^2 + 2\,x_{k\ell,h}\,x_{k\ell,h-1} + 3\,x_{k\ell,h-1}^2\right)\right]\left(y_{k\ell,h} - y_{k\ell,h-1}\right)\right\}, \quad (8.80)$$

and

$$(x_{k\ell,h}, y_{k\ell,h}), \quad h = 1, \ldots, N_h \quad (8.81)$$

are the coordinates for the sides of the overlap area $a_{k\ell}$ numbered counter-clockwise. So $N_h = 3$ for triangular overlap areas, $N_h = 4$ for quadrilateral $a_{k\ell}$'s etc. Note that $(x_{k\ell,h-1}, y_{k\ell,h-1})$ and $(x_{k\ell,h}, y_{k\ell,h})$ are contiguous points (defining a line segment) and the index h is cyclic so that $h = 0$ equals $h = N_h$.

The weights $w_{k\ell}^{(i,j)}$ given in (8.75–8.80) have been derived by using (8.73) with the following pairs $(P^{(i,j)}, Q^{(i,j)})$

$$\left(P^{(0,0)} = 0, \;\; Q^{(0,0)} = x\right),$$

$$\left(P^{(1,0)} = 0, \;\; Q^{(1,0)} = \tfrac{x^2}{2}\right),$$

$$\left(P^{(0,1)} = -\frac{y^2}{2}, \;\; Q^{(0,1)} = 0\right),$$

$$\left(P^{(2,0)} = 0, \;\; Q^{(2,0)} = \tfrac{x^3}{3}\right),$$

$$\left(P^{(0,2)} = -\frac{y^3}{3}, \;\; Q^{(0,2)} = 0\right),$$

$$\left(P^{(1,1)} = 0, \;\; Q^{(1,1)} = \tfrac{x^2 y}{2}\right).$$

The choice of P and Q is not unique and can often be chosen for convenience. Here we haven chosen P and Q as in Bockman (1989). Note that the integration of the polynomials is exact.

Using the line-integral approach the final discretized transport scheme in Lagrangian and Eulerian form can be written as

$$\overline{\psi}_k^{n+1}\,\Delta A_k = \sum_{\ell=1}^{L_k}\int_{a_{k\ell}}\psi_\ell(x,y)\,dA = \sum_{\ell=1}^{L_k}\left[\sum_{i+j\leq 2} c_\ell^{(i,j)}\,w_{k\ell}^{(i,j)}\right], \quad (8.82)$$

and

$$\overline{\psi}_k^{n+1}\,\Delta A_k = \overline{\psi}_k^{n}\,\Delta A_k + \sum_{\tau=1}^{4}\left[\sum_{\ell=1}^{L_k^\tau} F_{k\ell}^\tau\right]$$
$$= \overline{\psi}_k^{n}\,\Delta A_k + \sum_{\tau=1}^{4}\left\{\sum_{\ell=1}^{L_k^\tau}\left[\sum_{i+j\le 2} c_\ell^{(i,j)} w_{k\ell}^{(i,j)}(\tau)\right]\right\}, \qquad (8.83)$$

respectively, where the individual overlap fluxes are written as

$$F_{k\ell}^\tau = s_{k\ell}^\tau \int_{a_{k\ell}^\tau} \psi_\ell(x,y)\,dA. \qquad (8.84)$$

For each overlap area the sign-function $s_{k\ell}^\tau$ is $+1$ for inflow and -1 for outflow. The subscript ℓ in s_ℓ^τ is added to handle situations where there is both inflow and outflow for a face (see Harris et al. 2011, for details).

It is worth noting the separation of the weights $w_{k\ell}^{(i,j)}$ from the reconstruction coefficients $c_\ell^{(i,j)}$ in (8.82) and (8.83). In practice this separation implies that once the weights have been computed they can be reused for the integral of each additional tracer distribution. Hence the transport of additional tracers reduces to the multiplication of precomputed weights and reconstruction coefficients.

8.5.3.3 Extension to Spherical Geometry

Extending the aforementioned approaches to spherical geometry generally compounds the complexity of the problem, since extra care must be taken when metric terms are present. So instead of having interpolate a polynomial a more complicated function must be integrated

$$\int\int_{a_{k\ell}} g(\alpha,\beta)\,\psi_\ell(\alpha,\beta)d\alpha\,d\beta, \qquad (8.85)$$

where (α,β) is the coordinate for the computational space chosen for the integration[19] and $g(\alpha,\beta)$ is the metric term. For example, if one chooses geographic coordinates $(\alpha,\beta) = (\lambda,\theta)$, where λ is the longitude and θ is latitude, and then the metric term is $g = R^2\cos(\theta)$ where R is the radius of the Earth. So instead of having to integrate a polynomial a much more complicated function must integrated. In general, exact integration is no longer possible as was the case in Cartesian geometry. There are, however, some special cases where direct integration is possible (discussed below).

[19] For simplicity we only consider two-dimensional computational spaces although one may also use three-dimensional Cartesian coordinates for horizontal problems on the sphere.

The choice of coordinate system in which the integration is performed has implications on how the sides of $a_{k\ell}$ are approximated on the sphere and how accurate the reconstruction is. Here we will focus on the former. In Cartesian geometry the most general approximation to cell sides seems to be straight lines. The spherical extension of that is to approximate cell sides with great-circle arcs which seems the most general and accurate approach (at least in the case where the Eulerian cells are constructed from great-circle arcs). Hence, in the following we assume that great-circle arcs are the most accurate approximations to $a_{k\ell}$.

In the widely used Spherical Coordinate Remapping and Interpolation Package (SCRIP) proposed by Jones (1999) the sides of $a_{k\ell}$ are approximated with straight line segments in latitude-longitude coordinates (i.e., line segments of the form $\theta = a\lambda + b$). So for sides that are parallel to longitudes (which are great-circle arcs) and latitudes (which are small circle arcs) the representation of the cell sides is exact. However, for any other orientation it is not. While the error in cell side approximation is small near the Equator the errors may become significant in the polar regions (see Fig. 8.9 in Lauritzen and Nair (2008)). A way to alleviate this problem is to rotate the overlap area to the Equator. Using Gauss–Green's theorem the integration here can be performed exactly whereas direct integration using Gaussian quadrature will not be exact due to the metric term.

An alternative approach is to use the gnomonic coordinate as the computational space. The gnomonic projection was designed so that connecting any two points with a straight line in that computational space will mirror a great-circle arc on the sphere. Another beneficial property of this computational space is that exact integration of (8.85) is possible along coordinate lines in the gnomonic coordinate system when applying the Gauss–Green's theorem (Ullrich et al. 2009). For lines not parallel to the coordinate lines the potentials that need to be integrated in the line-integrals can be evaluated/approximated using one-dimensional Gaussian quadrature (Lauritzen et al. 2010). Again, direct integration will always be inexact due to the gnomonic metric terms.

8.6 Extension to Three Dimensions

The discussion so far has been limited to two spatial dimensions and we will only briefly discuss three-dimensional schemes, as a more thorough discussion would need at least a chapter on its own. There are basically three ways of extending schemes to three dimensions which we will discuss separately below.

8.6.1 Floating Lagrangian Vertical Coordinate

The floating Lagrangian coordinate was introduced in a theoretical context by Starr (1945) and first applied in discretized models over half a century later (e.g., Lin 2004; Lauritzen et al. 2008; Nair et al. 2009). Instead of using vertical coordinates based on height or pressure, a vertical coordinate ζ that is constant along

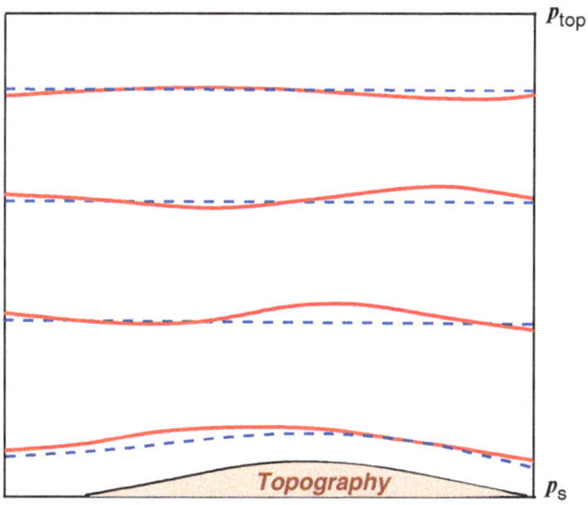

Fig. 8.18 A graphical illustration the floating Lagrangian coordinate. The vertical coordinate is pressure. p_s and p_{top} is the pressure at the surface and model top, respectively. The *dashed lines* is the reference Eulerian grid and *solid lines* are Lagrangian surfaces resulting from letting the Eulerian levels evolve in time, and require a periodic remapping

three-dimensional parcel trajectories is used

$$\frac{d\zeta}{dt} = 0, \tag{8.86}$$

(see Fig. 8.18). The benefit of using such a vertical coordinate is that the vertical advection terms in the equations of motion are eliminated and only two-dimensional transport/advection operators are necessary. The downside, as with any other Lagrangian approach, is that the vertical coordinate deform as the flow evolves. In order to avoid overly deformed vertical coordinates a remapping of the prognostic variables in the vertical to some reference vertical coordinate is necessary. This may be a source of vertical diffusion in the model. Note that isentropic vertical coordinates are a subset of floating Lagrangian vertical coordinates as they are also material surfaces for adiabatic flow.

8.6.2 Operator Splitting

Using a cascade finite-volume scheme (flow based splitting) or Eulerian operator splitting the extension to three dimensions can be made less costly than when using fully three dimensional approaches simply because they require only one-dimensional operators. Eulerian type operator splitting use a combination of operators applied along coordinate lines (see, e.g., Pietrzak 1998). In such approaches errors due to the coordinate splitting (also referred to as splitting error) will appear if care is not taken to alleviate them. Various methods for reducing the splitting error have been proposed (e.g., Strang 1968; Lin and Rood 1996). The traditional Eulerian type operator splitting approach may be referred to as a fixed direction based splitting method as opposed to the flow-based splitting approach discussed below.

More recently the finite-volume cascade approach was suggested by Nair et al. (2002) and Zerroukat et al. (2002) which uses a combination of Eulerian and Lagrangian operators, that is, the operators are successively applied along coordinate lines and Lagrangian lines, respectively. So rather than being a fixed direction based splitting method it is flow-based (for a review see Machenhauer et al. 2009). Since the splitting is flow-based the splitting error is reduced. Note that one may use the cascade approach to extend fully two-dimensional methods to three dimensions by applying a cascade sweep in the vertical based on the horizontally transported values.

8.6.3 Rigorous Three-dimensional Approach

Fully three-dimensional schemes based on the space-time finite-volume approach discussed in this chapter are rather complex. Instead of having to deal with overlap areas (as discussed in this chapter) one has to compute overlap volumes which is significantly complicating the problem. Examples of fully three-dimensional remapping algorithms are given in, e.g., Garimella et al. (2007) for Cartesian geometry. The authors are not aware of any fully three-dimensional finite-volume remapping schemes on the sphere.

8.7 Time-integration and Tracer Transport

If all models would use the same numerical method for tracer transport as used for the continuity equation for air, and if those would always be solved by using the same time-step, then this section would be irrelevant. Most models, however, use one of the following three approaches: Either they use different schemes for air and tracers, use different time-step size for air and tracers (but explicit time-stepping for both) or semi-implicit time-stepping is used for air and explicit time-stepping for tracers (and both use the same time-step). All of these approaches potentially have consistency problems as discussed separately for each approach below.

8.7.1 Different Schemes Air and Tracers

If different schemes are used for air and tracers consistency cannot be achieved other than with fixers that enforce consistency in a 'ad hoc' and somewhat arbitrary manner. See Sect. 8.3.4 and references therein.

8.7.2 Different Time-steps for Air and Tracers (Sub-Cycling, Super-Cycling)

Given the increase in the number of prognostic tracers in atmospheric models, significant computational cost savings can be obtained by using a longer time-step for tracers than for the solution of the air continuity equation. As discussed in Sect. 8.2.2 the maximum allowable time-step that can be used for the solution of the equation for air density (when using explicit time-stepping) is determined by the fastest wave in the system, since the continuity equation for air is directly coupled to the other equations of motion. The continuity equations for tracers, however, are not directly coupled (at least in terms of stability) to the momentum and thermodynamic equations and therefore have less restrictive time-step limitations. So a stable and more efficient integration scheme can be designed by sub-cycling the solution of the air density equation with respect to the tracer equations. In doing so it is important to retain the consistency discussed in Sect. 8.2.2, that is, for a constant mixing ratio ($q = 1$) the tracer transport equation should yield the same solution as the continuity equation for air (a.k.a. 'free-stream preserving'). A scheme possessing the 'free-stream preserving' property can be designed as described below.

A conceptual explanation of sub-cycling is given with the aid of Fig. 8.19. For simplicity assume one spatial dimension, flow from left to right and that the wind at the right cell wall is zero (no mass flux through that boundary). The number of times the integration of the air density equation is sub-cycled with respect to the tracer equations is referred to as *ksplit*. In Fig. 8.19 *ksplit* is 4. At time $t = n\Delta t$ the mass in the cell is ρ^n, where we have assumed that the cell width is one (grey area on Fig. 8.19a). We then integrate the full dynamical system of equations (continuity equation for air, momentum equations and thermodynamic equation) forward in time to $t = (n + 1/ksplit)\Delta t$. The flux of mass into the cell during this forward integration corresponds to the red area 'swept' through the left cell wall, on Fig. 8.19a (left column) and hence the air mass in the cell increases by the red area in cell k (Fig. 8.19a right column). This procedure is repeated three, or ($ksplit - 1$), more time-steps during which the blue, yellow and green areas are 'swept' through the left cell wall and adding to the total air mass in the cell (Fig. 8.19b,c,d respectively). The total flow of mass into the cell is the sum of all the areas on Fig. 8.19a,b,c,d corresponding to an average flux into the cell given by the brown area on Fig. 8.19e.

Since we are updating tracers on the long time-step we use the transport scheme to estimate the average mixing ratio over the full time-step Δt, that is, the average of q^n over the brown area in Fig. 8.19e denoted $\langle q^n \rangle$. Then the final forecast for the tracer is given by the product between the background flow of mass and an estimate of the mixing ratio over the long time-step

$$(\rho q)^{n+1} = (\rho q)^n + \langle q^n \rangle \left[\sum_{i=1}^{ksplit} \Delta\rho^{n+i/ksplit} \right], \tag{8.87}$$

where $\delta\rho^{n+i/ksplit}$ is the flux of air mass into the cell during one sub-cycled time-step $\Delta t/ksplit$. If $q = 1$ then (8.87) reduces to the equation for air mass and consequently the scheme is free-stream preserving. Note that updating the tracers on the short time-step will not yield the same result.

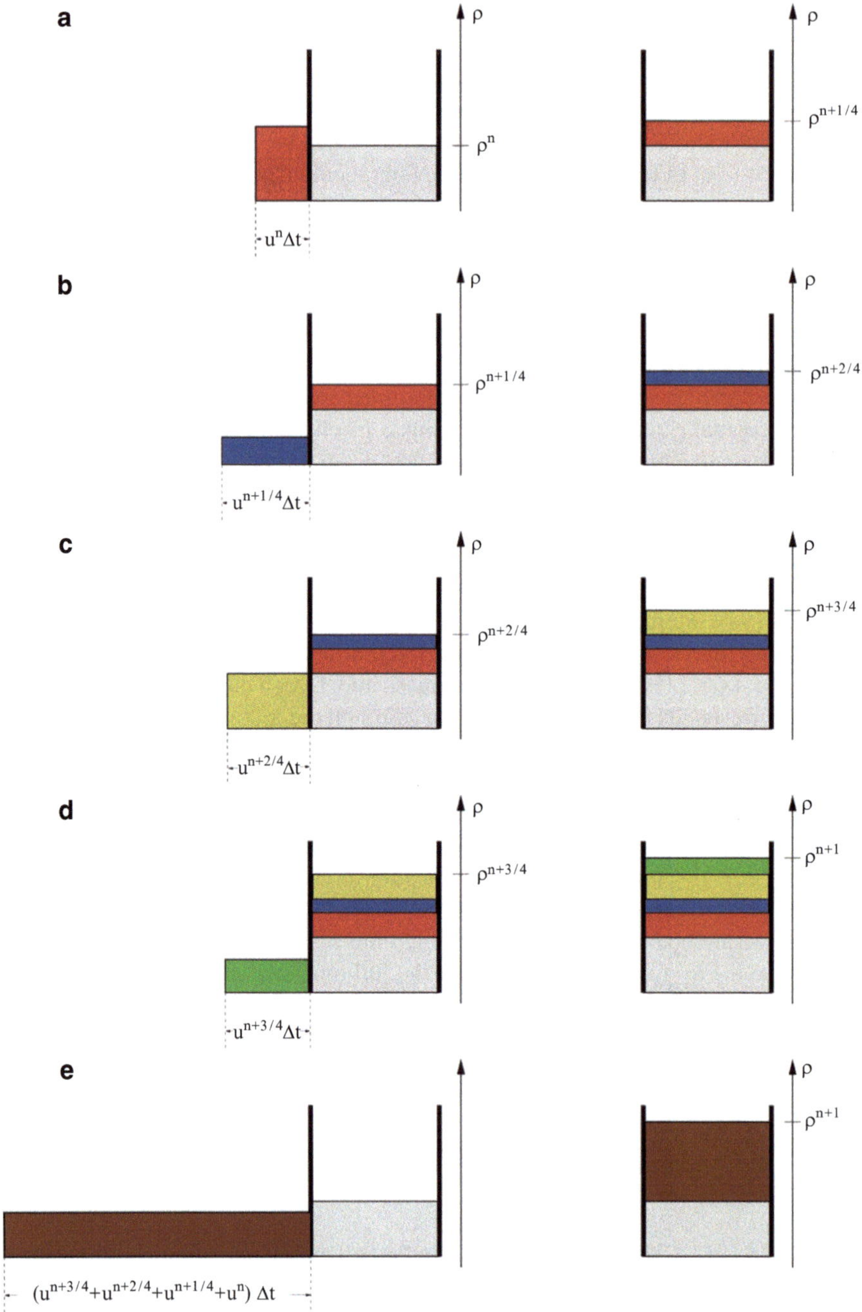

Fig. 8.19 A graphical illustration of sub-cycling the continuity for air mass with respect to tracers. Details and explanations are given in the text

8.7.3 *Semi-Implicit Time-Stepping for Air and Explicit for Tracers*

If semi-implicit time-stepping is used (see Chap. 6) then the prognostic equation for air density can be written as

$$\rho^{n+1} = \rho_{exp}^{n+1} + \frac{\Delta t}{2}\rho^{ref}\left(D^{n+1} - \tilde{D}^{n+1}\right), \tag{8.88}$$

(e.g., Lauritzen et al. 2006) where ρ_{exp}^{n+1} is the explicit prediction, ρ^{ref} is a constant reference density, D is the divergence and $\tilde{D}$ is the divergence extrapolated to time-level $n+1$. The terms on the right-hand side of (8.88) involving D are referred to as the semi-implicit correction terms and represent the implicit coupling to the momentum equations. If the tracer transport equation is solved explicitly, as is usually done, then the scheme is not 'free-stream preserving' because of the semi-implicit correction terms (although they are usually small).

So for consistency, one should also solve the tracer transport equation semi-implicitly

$$(\rho q)^{n+1} = (\rho q)_{exp}^{n+1} + \frac{\Delta t}{2}(q\rho)^{ref}\left(D^{n+1} - \tilde{D}^{n+1}\right), \tag{8.89}$$

(e.g., Lauritzen et al. 2008), however, that seems problematic. For example, if q is zero in some area and the semi-implicit correction terms are non-zero in that area, then tracer mass will be produced in an area where q should be zero.

Thuburn et al. (2010) present a method where they discretize an alternative form of the semi-implicit continuity equation. Through a series of iterations the semi-implicit correction terms cancel and consistency between air mass and tracer transport is obtained. For more details see Thuburn et al. (2010).

8.8 Final Remarks

In this chapter a detailed discussion of desirable properties for transport schemes intended for meteorological applications has been presented. The finite-volume method for tracer transport (in two-dimensional Cartesian geometry) has been introduced and discussed using a remap approach which conceptually introduces the finite-volume method through following characteristics of the flow. This conceptual framework has been used to explain and analyze several schemes from the literature. Practical considerations related to the coupling of air mass equations and tracer mass equations has been discussed in some detail as well as brief introductions to extensions to spherical geometry and three dimensions. The authors hope to have communicated some of the aspects that go into modeling transport accurately in large modeling systems. Although physical parameterizations that represent sub-grid-scale processes are probably among the largest sources of uncertainty in weather and climate models, the accurate representation of transport is very important. Errors in resolved-scale transport can change scientific results (e.g., Rasch et al. 2006; Wild and Prather 2006).

Acknowledgments Thanks to Dr. S. Galmarini (Institute for Environment and Sustainability, European Commission, Joint Research Center) and Dr. A. Baklanov (Danish Meteorological Institute) for details on the ETEX experiment. Many fruitful discussions with Dr. D.L. Williamson (NCAR), Dr. P. Rasch (PNNL), Dr. A.Gettelman (NCAR), Dr. W. Skamarock (NCAR), Dr. M. Taylor (Sandia National Laboratories) and Dr. R. Mittal (NCAR) are acknowledged as well as the internal review performed by Dr. D.L. Williamson and Dr. C. Erath, and the anonymous and Editor (Dr. C. Jablonowski and Dr. M.A. Taylor) reviews. The authors gratefully acknowledge Prof. J.Thuburn's suggestions on numerical mixing tests. The first and third authors were partially supported by the DOE BER Program under award DE-SC0001658. The National Center for Atmospheric Research is sponsored by the National Science Foundation.

References

van Albada GD, van Leer B, Roberts WW (1982) A comparative study of computational methods in cosmic gas dynamics. Astronomy and Astrophysics 108:76–84

Artebrant R, Torrilhon M (2008) Increasing the accuracy in locally divergence-preserving finite volume schemes for MHD. J Comput Phys 227(6):3405–3427

Barth T, Frederickson P (1990) Higher-order solution of the Euler equations on unstructured grids using quadratic reconstruction. In: AIAA Paper 90-0013

Barth T, Jespersen D (1989) The design and application of upwind schemes on unstructured meshes. Proc AIAA 27th Aerospace Sciences Meeting, Reno

Bates JR, McDonald A (1982) Multiply-upstream, semi-Lagrangian advective schemes: Analysis and application to a multi-level primitive equation model. Mon Wea Rev 110(12):1831–1842

Blossey PN, Durran DR (2008) Selective monotonicity preservation in scalar advection. J Comput Phys 227(10):5160–5183

Bockman SF (1989) Generalizing the formula for areas of polygons to moments. The American Mathematical Monthly 96:131–132

Brasseur GP, Hauglustaine DA, Walters S, Rasch PJ, Muller JF, Granier C, Tie XX (1998) MOZART, a global chemical transport model for ozone and related chemical tracers: 1. model description. J Geophys Res 103:28,265–28,289

Colella P, Sekora MD (2008) A limiter for ppm that preserves accuracy at smooth extrema. J Comput Phys 227:7069–7076

Colella P, Woodward PR (1984) The piecewise parabolic method (PPM) for gas-dynamical simulations. J Comput Phys 54:174–201

Collins WD, Rasch PJ, Boville BA, Hack JJ, McCaa JR, Williamson DL, Kiehl JT, Briegleb B (2004) Description of the NCAR Community Atmosphere Model (CAM 3.0). NCAR Tech. Note, NCAR/TN-464+STR

van Dop H, Addis R, Fraser G, Girardi F, Graziani G, Inoue Y, Kelly N, Klug W, Kulmala A, Nodop K, Pretel J (1998) Etex: A European tracer experiment; observations, dispersion modelling and emergency response. Atmospheric Environment 32:4089–4094

Doswell CAI (1984) A kinematic analysis of frontogenesis associated with a nondivergent vortex. J Atmos Sci pp 1242–1248

Dukowicz JK (1984) Conservative rezoning (remapping) for general quadrilateral meshes. J Comput Phys 54:411–424

Dukowicz JK, Baumgardner JR (2000) Incremental remapping as a transport/advection algorithm. J Comput Phys 160:318–335

Dukowicz JK, Kodis JW (1987) Accurate conservative remapping (rezoning) for arbitrary Lagrangian-Eulerian computations. SIAM Journal on Scientific and Statistical Computing 8(3):305–321

Durran DR (1999) Numerical Methods for Wave Equations in Geophysical Fluid Dynamics. Springer-Verlag

Eluszkiewicz J, Hemler RS, Mahlman JD, Bruhwiler L, Takacs LL (2000) Sensitivity of age-of-air calculations to the choice of advection scheme. J Atmos Sci 57:3185–3201

Galmarini S, Bianconi R, Addis R, Andronopoulos S, Astrup P, Bartzis JC, Bellasio R, Buckley R, Champion H, Chino M, D'Amours R, Davakis E, Eleveld H, Glaab H, Manning A, Mikkelsen T, Pechinger U, Polreich E, Prodanova M, Slaper H, Syrakov D, Terada H, der Auwera LV (2004) Ensemble dispersion forecasting–part II: Application and evaluation. Atmospheric Environment 38:4619–4632

Garimella R, Kucharik M, Shashkov M (2007) An efficient linearity and bound preserving conservative interpolation (remapping) on polyhedral meshes. Computers & Fluids 36:224–237

Girardi F, Graziani G, van Veltzen D, Galmarini S, Mosca S, Bianconi R, Bellasio R, Klug W (1998) The ETEX project. EUR report 181-43 en., Office for official publication of the European Communities, Luxembourg, 108 pp.

Godunov SK (1959) A difference scheme for numerical computation of discontinuous solutions of equations in fluid dynamics. Math Sb 47:271, also: Cornell Aero. Lab. translation

Haltiner GJ, Williams RT (1980) Numerical Prediction and Dynamic Meteorology. John Wiley & Sons, 477 pp.

Harris LM, Lauritzen PH, Mittal R (2011) A flux-form version of the conservative semi-Lagrangian multi-tracer transport scheme (CSLAM) on the cubed sphere grid. J Comput Phys 230(4):1215–1237

Harten A (1983) On the symmetric form of systems of conservation laws with entropy. J Comput Phys 49:151–164

Harten A, Engquist B, Osher S, Chakravarthy SR (1987) Uniformly high order accurate essentially non-oscillatory schemes iii. J Comput Phys 71:231–303

Hirt CW, Amsden AA, Cook JL (1974) An arbitrary Lagrangian-Eulerian computing method for all flow speeds. J Comput Phys 14(3):227–253

Hortal M (2002) The development and testing of a new two-time-level semi-Lagrangian scheme (SETTLS) in the ecmwf forecast model. Q J R Meteorol Soc 128(583):1671–1687

Jablonowski C, Lauritzen PH, Taylor MA, Nair RD (2011) Idealized test cases for the dynamical cores of atmospheric general circulation models. Geoscientific Model Development Submitted. http://esse.engin.umich.edu/admg/publications.php

Jöckel P, von Kuhlmann R, Lawrence MG, Steil B, Brenninkmeijer C, Crutzen PJ, Rasch PJ, Eaton B (2001) On a fundamental problem in implementing flux-form advection schemes for tracer transport in 3-dimensional general circulation and chemistry transport models. QJR Meteorol Soc 127(573):1035–1052

Jones PW (1999) First- and second-order conservative remapping schemes for grids in spherical coordinates. Mon Wea Rev 127:2204–2210

Lamarque JF, Kinnison DE, Hess PG, Vitt F (2008) Simulated lower stratospheric trends between 1970 and 2005: Identifying the role of climate and composition changes. J Geophys Res 113(D12301)

Laprise JP, Plante A (1995) A class of semi-Lagrangian integrated-mass (SLIM) numerical transport algorithms. Mon Wea Rev 123:553–565

Lauritzen PH (2007) A stability analysis of finite-volume advection schemes permitting long time steps. Mon Wea Rev 135:2658–2673

Lauritzen PH, Nair RD (2008) Monotone and conservative cascade remapping between spherical grids (CaRS): Regular latitude-longitude and cubed-sphere grids. Mon Wea Rev 136: 1416–1432

Lauritzen PH, Thuburn J (2011) Evaluating advection/transport schemes using scatter plots and numerical mixing diagnostics. Quart J Roy Met Soc Submitted

Lauritzen PH, Kaas E, Machenhauer B (2006) A mass-conservative semi-implicit semi-Lagrangian limited area shallow water model on the sphere. Mon Wea Rev 134:1205–1221

Lauritzen PH, Kaas E, Machenhauer B, Lindberg K (2008) A mass-conservative version of the semi-implicit semi-Lagrangian HIRLAM. Q J R Meteorol Soc 134

Lauritzen PH, Nair RD, Ullrich PA (2010) A conservative semi-Lagrangian multi-tracer transport scheme (CSLAM) on the cubed-sphere grid. J Comput Phys 229:1401–1424

Lee SM, Yoon SC, Byun DW (2004) The effect of mass inconsistency of the meteorological field generated by a common meteorological model on air quality modeling. Atmospheric Environment 38(18):2917–2926

van Leer B (1977) Towards the ultimate conservative difference scheme. IV: A new approach to numerical convection. J Comput Phys 23:276–299

Leonard BP (1991) The ULTIMATE conservative difference scheme applied to unsteady one-dimensional advection. Comput Methods Appl Mech Eng 88:17–74

Leonard BP, Lock A, MacVean M (1996) Conservative explicit unrestricted-time-step multidimensional constancy-preserving advection schemes. Mon Wea Rev 124:2588–2606

Leslie LM, Dietachmayer GS (1997) Comparing schemes for integrating the Euler equations. Mon Wea Rev 125(7):1687–1691

LeVeque RJ (1996) High-resolution conservative algorithms for advection in incompressible flow. SIAM Journal on Numerical Analysis 33:627–665

Levy MN, Nair RD, Tufo HM (2007) High-order Galerkin method for scalable global atmospheric models. Comput Geosci 33:1022–1035

Lin SJ (2004) A 'vertically Lagrangian' finite-volume dynamical core for global models. Mon Wea Rev 132:2293–2307

Lin SJ, Rood RB (1996) Multidimensional flux-form semi-Lagrangian transport schemes. Mon Wea Rev 124:2046–2070

Lipscomb WH, Ringler TD (2005) An incremental remapping transport scheme on a spherical geodesic grid. Mon Wea Rev 133:2335–2350

Liu Y, wang Shu C, Tadmor E, Zhang M (2007) Central discontinuous Galerkin methods on overlapping cells with a non-oscillatory hierarchical reconstruction. SIAM J Numer Anal pp 45–2442

Lorenz EN (1982) Atmospheric predictability experiments with a large numerical model. Tellus pp 505–513

Machenhauer B, Kaas E, Lauritzen PH (2009) Finite volume methods in meteorology, in: R. Temam, J. Tribbia, P. Ciarlet (Eds.), Computational methods for the atmosphere and the oceans. Handbook of Numerical Analysis 14, Elsevier, 2009, pp.3-120

McGregor JL (2005) Geostrophic adjustment for reversibly staggered grids. Mon Wea Rev 133:1119–1128

Miura H (2007) An upwind-biased conservative advection scheme for spherical hexagonal-pentagonal grids. Mon Wea Rev 135:4038–4044

Moorthi S, Higgins RW, Bates JR (1995) A global multilevel atmospheric model using a vector semi-Lagrangian finite-difference scheme. Part II: Version with physics. Mon Wea Rev 123(5):1523–1541

Morrison H, Gettelman A (2008) A new two-moment bulk stratiform cloud microphysics scheme in the community atmosphere model, version 3 (CAM3). Part I: Description and numerical tests. J Climate 21:3642–3659

Nair RD, Jablonowski C (2008) Moving vortices on the sphere: A test case for horizontal advection problems. Mon Wea Rev 136:699–711

Nair RD, Lauritzen PH (2010) A class of deformational flow test cases for linear transport problems on the sphere. J Comput Phys 229:8868–8887

Nair RD, Machenhauer B (2002) The mass-conservative cell-integrated semi-Lagrangian advection scheme on the sphere. Mon Wea Rev 130(3):649–667

Nair RD, Scroggs JS, Semazzi FHM (2002) Efficient conservative global transport schemes for climate and atmospheric chemistry models. Mon Wea Rev 130(8):2059–2073

Nair RD, Choi HW, Tufo HM (2009) Computational aspects of a scalable high-order discontinuous Galerkin atmospheric dynamical core. Computers & Fluids 38:309–319

Norman MR, Nair RD (2008) Inherently conservative nonpolynomial-based remapping schemes: Application to semi-Lagrangian transport. Mon Wea Rev 126:5044–5061

Norman MR, Semazzi FHM, Nair RD (2009) Conservative cascade interpolation on the sphere: An intercomparison of various non-oscillatory reconstructions. Quart J Roy Met Soc 135:795–805

Ovtchinnikov M, Easter RC (2009) Nonlinear advection algorithms applied to interrelated tracers: Errors and implications for modeling aerosol-cloud interactions. Mon Wea Rev 137:632–644

Pietrzak J (1998) The use of TVD limiters for forward-in-time upstream-biased advection schemes in ocean modeling. Mon Wea Rev 126:812–830

Plumb RA (2007) Tracer interrelationships in the stratosphere. Rev Geophys 45(RG4005)

Plumb RA, Ko M (1992) Interrelationships between mixing ratios of long-lived stratospheric constituents. J Geophys Res 97:10,145–10,156

Prather MJ, Zhu X, Strahan SE, Steenrod SD, Rodriguez JM (2008) Quantifying errors in trace species transport modeling. Proceedings of the National Academy of Science pp 19,617–19,621

Purser RJ, Leslie LM (1991) An efficient interpolation procedure for high-order three-dimensional semi-Lagrangian models. Mon Wea Rev 119:2492–2498

Putman WM, Lin SJ (2007) Finite-volume transport on various cubed-sphere grids. J Comput Phys 227(1):55–78

Rančić M (1992) Semi-Lagrangian piecewise biparabolic scheme for two-dimensional horizontal advection of a passive scalar. Mon Wea Rev 120:1394–1405

Rasch PJ, Coleman DB, Mahowald N, Williamson DL, Lin SJ, Boville BA, Hess P (2006) Characteristics of atmospheric transport using three numerical formulations for atmospheric dynamics in a single GCM framework. J Climate 19:2243–2266

Roe PL (1985) Lecture Notes in Applied Mathematics, vol 22, New York: Springer-Verlag, chap Some contributions to modeling of discontinuous flows, pp 163–193

Rood RB (1987) Numerical advection algorithms and their role in atmospheric transport and chemistry models. Rev Geophys 25:71–100

Schär C, Smolarkiewicz PK (1996) A synchronous and iterative flux-correction formalism for coupled transport equations. J Comput Phys 128:101–120

Smolarkiewicz PK (2006) Multidimensional positive definite advection transport algorithm: An overview. Int J Numer Methods Fluids 50:1123–1144

Smolarkiewicz PK, Grabowski WW (1990) The multidimensional positive definite advection transport algorithm: Nonoscillatory option. J Comput Phys 86:355–375

Staniforth A, Côté J (1991) Semi-Lagrangian integration schemes for atmospheric models-a review. Mon Wea Rev 119:2206–2223

Staniforth A, White A, Wood N (2003) Analysis of semi-Lagrangian trajectory computations. Q J R Meteorol Soc 129(591):2065–2085

Starr VP (1945) A quasi-Lagrangian system of hydrodynamical equations. J Atmos Sci 2:227–237

Strang G (1968) On the construction and comparison of difference schemes. SIAM J Numer Anal 5:506–517

Thuburn J (2008) Some conservation issues for the dynamical cores of NWP and climate models. J Comput Phys 227:3715–3730

Thuburn J, McIntyre M (1997) Numerical advection schemes, cross-isentropic random walks, and correlations between chemical species. J Geophys Res 102(D6):6775–6797

Thuburn J, Zerroukat M, Wood N, Staniforth A (2010) Coupling a mass conserving semi-Lagrangian scheme (SLICE) to a semi-implicit discretization of the shallow-water equations: Minimizing the dependence on a reference atmosphere. Q J R Meteorol Soc 136:146–154

Toro EF (1999) Riemann Solvers and Numerical Methods for Fluid Dynamics, Second edn. Springer, ISBN-10: 3540659668, 624 pp.

Trenberth KE, Smith L (2005) The mass of the atmosphere: A constraint on global analyses. J Climate 18:864–875

Ullrich PA, Lauritzen PH, Jablonowski C (2009) Geometrically exact conservative remapping (GECoRe): Regular latitude-longitude and cubed-sphere grids. Mon Wea Rev 137(6): 1721–1741

Ullrich PA, Jablonowski C, van Leer BL (2010) Riemann-solver-based high-order finite-volume models for the shallow-water equations on the sphere. J Comput Phys 229:6104–6134

Waugh DW, Hall TM (2002) Age of stratospheric air: Theory, observations, and models. Rev Geophys 40

Waugh DW, Plumb RA, Elkins JW, Fahey DW, Boering KA, Dutton GS, Volk CM, Keim E, Gao RS, Daube BC, Wofsy SC, Loewenstein M, Podolske JR, Chan KR, Proffitt MH, Kelly KK, Newman PA, Lait LR (1997) Mixing of polar vortex air into middle latitudes as revealed by tracer-tracer scatterplots. J Geophys Res 120(D11):119–134

White L, Adcroft A (2008) A high-order finite volume remapping scheme for nonuniform grids: The piecewise quartic method (PQM). J Comput Phys 227:7394–7422

Wild O, Prather MJ (2006) Global tropospheric ozone modeling: Quantifying errors due to grid resolution. J Geophys Res 111(D11305)

Williamson DL, Olson J (1994) Climate simulations with a semi-Lagrangian version of the NCAR Community Climate Model. Mon Wea Rev 122(7):1594–1610

Williamson DL, Drake JB, Hack JJ, Jakob R, Swarztrauber PN (1992) A standard test set for numerical approximations to the shallow water equations in spherical geometry. J Comput Phys 102:211–224

Xiao F, Yabe T, Peng X, Kobayashi H (2002) Conservative and oscillation-less atmospheric transport schemes based on rational functions. J Geophys Res 107(D22):4609

Yabe T, Tanaka R, Nakamura T, Xiao F (2001) An exactly conservative semi-Lagrangian scheme (cip csl) in one dimension. Mon Wea Rev 129(2):332–334

Yeh KS (2007) The streamline subgrid integration method: I. quasi-monotonic second-order transport schemes. J Comput Phys 225:1632–1652

Zalesak ST (1979) Fully multidimensional flux-corrected transport algorithms for fluids. J Comput Phys 31:335–362

Zerroukat M, Wood N, Staniforth A (2002) SLICE: A semi-Lagrangian inherently conserving and efficient scheme for transport problems. Q J R Meteorol Soc 128:2801–2820

Zerroukat M, Wood N, Staniforth A (2004) SLICE-S: A semi-Lagrangian inherently conserving and efficient scheme for transport problems on the sphere. Q J R Meteorol Soc 130:2649–2664

Zerroukat M, Wood N, Staniforth A (2005) A monotonic and positive-definite filter for a semi-Lagrangian inherently conserving and efficient (SLICE) scheme. Q J R Meteorol Soc 131(611):2923–2936

Zerroukat M, Wood N, Staniforth A (2006) The parabolic spline method (PSM) for conservative transport problems. Int J Numer Meth Fluids 51:1297–1318

Zerroukat M, Wood N, Staniforth A (2007) Application of the parabolic spline method (PSM) to a multi-dimensional conservative semi-Lagrangian transport scheme (SLICE). J Comput Phys 225:935–948

Zerroukat M, Staniforth A, Wood N (2010) The monotonic quartic spline method (QSM) for conservative transport problems. J Comput Phys 229:1150–1166

Zhang K, Wan H, Wang B, Zhang M (2008) Consistency problem with tracer advection in the atmospheric model GAMIL. Adv Atmos Sci 25(2)

Zubov VA, Rozanov EV, Schlesinger ME (1999) Hybrid scheme for three-dimensional advective transport. Mon Wea Rev 127(6):1335–1346

Chapter 9
Emerging Numerical Methods for Atmospheric Modeling

Ramachandran D. Nair, Michael N. Levy, and Peter H. Lauritzen

Abstract This chapter discusses the development of discontinuous Galerkin (DG) schemes for the hyperbolic conservation laws relevant to atmospheric modeling. Two variants of the DG spatial discretization, the modal and nodal form, are considered for the one- and two-dimensional cases. The time integration relies on a second- or third-order explicit strong stability-preserving Runge–Kutta method. Several computational examples are provided, including a solid-body rotation test, a deformational flow problem and solving the barotropic vorticity equation for an idealized cyclone. A detailed description of various limiters available for the DG method is given, and a new limiter with positivity-preservation as a constraint is proposed for two-dimensional transport. The DG method is extended to the cubed-sphere geometry and the transport and shallow water models are discussed.

9.1 Introduction

Atmospheric numerical modeling has undergone radical changes over the past decade. One major reason for this trend is the recent paradigm change in scientific computing, triggered by the arrival of petascale computing resources with core counts in the range of tens to hundreds of thousands. Due to these changes, modelers must develop or adapt grid systems and numerical algorithms which facilitate an unprecedented level of scalability on these modern highly parallel computer

R.D. Nair (✉)
National Center for Atmospheric Research, 1850 Table Mesa Drive, Boulder, CO 80305, USA
e-mail: rnair@ucar.edu

M.N. Levy
Sandia National Laboratories, Albuquerque, NM 87185, USA
e-mail: mnlevy@sandia.gov

P.H. Lauritzen
National Center for Atmospheric Research, 1850 Table Mesa Drive, Boulder, CO 80305, USA
e-mail: pel@ucar.edu

P.H. Lauritzen et al. (eds.), *Numerical Techniques for Global Atmospheric Models*, Lecture Notes in Computational Science and Engineering 80,
DOI 10.1007/978-3-642-11640-7_9,

architectures. The numerical algorithms which can address these challenges should have *local* properties such as a high on-processor operation (floating-point operation or flop) count and a minimal parallel communication footprint.

With the increased amount of computing resources available to modelers, it is now possible to develop global models with resolution of the order of a few kilometers. This capability bridges the gap between traditional weather and climate modeling efforts, which operate on different spatial and temporal scales, and sets the stage for the development of a unified weather and climate model. However, this opens up another challenge – switching the governing equations from the hydrostatic to non-hydrostatic regime. The equation set generally used in traditional global climate models (hydrostatic equations of motion) is not adequate at the non-hydrostatic scale. In the very high-resolution regime viable options for the governing equations of motion are the compressible (or quasi-compressible) Euler equations or Navier–Stokes equations. Also, it is highly desirable that the underlying model equations follow the physical laws of conservation for integral invariants such as mass, energy, enstrophy, etc. In order to comply with these constraints and address new computational challenges, the next generation of atmospheric models should be based on robust numerical methods which satisfy the following set of criteria:

- Inherent local and global conservation
- High-order accuracy
- Computational efficiency
- Geometric flexibility (any type of grid system, suitable for adaptive mesh refinement)
- Non-oscillatory advection (monotonic, positivity preservation)
- High parallel efficiency (local method, petascale capability)

There are several successful numerical methods, particularly in the finite-volume (FV) literature, which satisfy most of the above-mentioned properties. The FV schemes are inherently conservative but mostly low-order accurate (third-order or less). High-order extensions of the FV method are possible at the cost of wider halo regions. For example, the weighed essentially non-oscillatory (WENO) method (Shu 1997) is a powerful approach; however, a $(k + 1)$th-order accurate WENO scheme in 1D requires $2k + 1$ cells (control volumes). Thus, as the order of accuracy grows the WENO scheme requires a wider computational stencil (halo region) which can seriously impede the parallel efficiency. A local method like the spectral element (SE) method has the local domain decomposition property of the finite-element (FE) method combined with high-order accuracy and the weak numerical dispersion and low numerical dissipation of spectral methods. The SE method offers excellent parallel efficiency and has become the method of choice for many practical applications. The classical SE method is not necessarily based on hyperbolic conservation laws and is not inherently conservative. Nevertheless, the conservation properties can be engineered in the SE discretization (Chap. 12) much as they were in the conservative finite-difference discretization developed by Arakawa and Lamb (1977) and Simmons and Burridge (1981).

The discontinuous Galerkin (DG) method retains all the nice properties of the SE method, plus it is inherently conservative. The DG method has the potential to address all of the above-listed properties. DG algorithms for solving partial differential equations are becoming very popular in a wide range of applications in computational science and engineering. The primary focus of this chapter is on the development of the DG method for atmospheric modeling applications.

The DG method may be viewed as a hybrid approach, combining the ideas of classical FV and FE methods into a unified framework to exploit the merits of both. As a FV method, DG discretizations employ discontinuous elements (local control volumes) and flux integrals along its boundaries, guaranteeing local conservation. Similar to the FV method, DG schemes can incorporate slope limiters for controlling spurious oscillations in the solution. However, in contrast to FV methods, the DG method avoids the reconstruction process (often requiring wider stencil). The FE or SE structure (element-wise Galerkin approach) makes the DG method high-order accurate and provides the ability to handle complex geometries such as the Earth's surface or boundary conditions. However, as opposed to the FE/SE methods, the elements used for the DG methods are discontinuous, which leads to a localized discretization. This feature offers excellent parallel efficiency as well as efficient adaptive mesh refinement (AMR) capability, even with non-conforming elements.

The DG method was first introduced by Reed and Hill (1973) and later analyzed by Lesaint and Raviart (1974) for linear advection equation. A rigorous mathematical foundation for the DG method was laid by Cockburn and Shu (1989) and Cockburn et al. (1990), where high-order accurate explicit Runge–Kutta (RK) time integration schemes combined with DG spatial discretizations for nonlinear systems of conservation laws were developed. The resulting RKDG method has become widely popular in different computational science and engineering disciplines (Cockburn et al. 2000; Remacle et al. 2003).

The remainder of the chapter is organized as follows: in Sect. 9.2 we describe the basic DG discretization in 1D, and the extension to 2D is given in Sects. 9.3 and 9.4 describes various limiters for the DG method with examples. An extension of the DG method onto the sphere is given in Sect. 9.5, where the shallow water model for the cubed-sphere is described. Section 9.6 offers some concluding remarks.

9.2 The DG Method

Although the DG method is applicable to a variety of parabolic and elliptic problems (Rivière 2008), our primary focus is on the DG method applied to hyperbolic conservation laws which are relevant to atmospheric numerical modeling. Before detailing the DG discretization procedure we briefly review conservation laws.

9.2.1 Conservation Laws

Systems of conservation laws are very important mathematical models for a variety of physical phenomena that appear in fluid mechanics and several other areas including atmospheric sciences. A large class of atmospheric equations of motion for compressible and incompressible flows can be written in conservation form. Conservation laws are systems of nonlinear partial differential equations (PDEs) most readily expressed in flux form and can be written:

$$\frac{\partial}{\partial t}U(\mathbf{x},t) + \sum_{j=1}^{3} \frac{\partial}{\partial x_j} F_j(U,\mathbf{x},t) = S(U), \tag{9.1}$$

where $\mathbf{x}$ is the 3D space coordinate and time $t > 0$. $U(\mathbf{x},t)$ is the state vector representing conserved quantities (e.g. mass, momentum or energy). $F_j(U)$ are components of $\mathbf{F}$, a prescribed flux vector which accounts for diffusive and convective effects, and $S(U)$ is the source term representing exterior forces. The system of Euler and Navier–Stokes equations, widely used for modeling fluid motion, can be cast in this form. The mass continuity equation is an example of scalar conservation law and is a special case of (9.1), which is obtained by applying the physical principle of conservation of mass in a fluid flow:

$$\frac{\partial \rho}{\partial t} + \nabla \cdot (\rho \mathbf{V}) = 0,$$

where ρ is the fluid density, $\mathbf{V}$ is the velocity of the fluid, and '$\nabla\cdot$' denotes the divergence operator. Note that discretizing the equations in flux-form is important because application of the divergence theorem is straightforward and the conservation can be maintained numerically. We consider several hyperbolic conservation laws based on (9.1) in this chapter and numerically solve them by using the DG method.

9.2.2 The DG Method for 1D Problems

The basic ideas of the DG discretization may be understood in a simple 1D framework. In order to introduce the DG discretization and notations, we first consider the one-dimensional scalar conservation law:

$$\frac{\partial U}{\partial t} + \frac{\partial F(U)}{\partial x} = 0 \quad \text{in} \quad \Omega \times (0,T], \tag{9.2}$$

where $U = U(x,t)$ is the conservative variable evolving in time with a known initial condition $U(x,t=0) = U_0(x)$, $\forall x \in \Omega$, and $F(U)$ is the flux function. For a linear advection problem the flux function is $F(U) = cU$, where c is the velocity;

for the inviscid Burgers' equation, a simple non-linear problem, the flux function is $F(U) = U^2/2$.

9.2.3 Galerkin Formulation

The DG discretization consists of partitioning the global domain Ω into N_{elm} non-overlapping elements such that $\Omega = \cup_{j=1}^{N_{elm}} \Omega_j$ with $\Omega_j \equiv [x_{j-1/2}, x_{j+1/2}]$, $j = 1, \ldots, N_{elm}$. With this setup the width of the jth element is $\Delta x_j = x_{j+1/2} - x_{j-1/2}$ and the midpoint is defined by $x_j = (x_{j+1/2} + x_{j-1/2})/2$. Note that the edges (interface) $x_{j\pm1/2}$ of the element Ω_j are shared by the adjacent elements in this partition, as shown schematically in Fig. 9.1.

The next step is to cast the problem (9.2) into the *weak* Galerkin formulation. This is done by multiplying (9.2) by a test (weight) function $\varphi(x)$ and integrating over the element Ω_j:

$$\int_{\Omega_j} \left[\frac{\partial U}{\partial t} + \frac{\partial F(U)}{\partial x} \right] \varphi(x) dx = 0. \tag{9.3}$$

The term *weak* refers to the fact that the formulation (9.3) admits a larger class of solutions as opposed to the *strong* or classical form (9.2). Integrating the second term of (9.3) by parts (Green's method) yields

$$\begin{aligned} &\int_{\Omega_j} \frac{\partial U(x,t)}{\partial t} \varphi(x) dx - \int_{\Omega_j} F(U(x,t)) \frac{\partial \varphi(x)}{\partial x} dx \\ &\quad + [F(U(x,t))\, \varphi(x)]_{x_{j-1/2}^{+}}^{x_{j+1/2}^{-}} = 0, \end{aligned} \tag{9.4}$$

where $x_{j+1/2}^{-}$ is the left limit at the edge $x_{j+1/2}$, and $x_{j-1/2}^{+}$ is the right limit at the edge $x_{j-1/2}$ of the element Ω_j, as indicated in Fig. 9.1. While the Galerkin formulation procedure (9.4) is the same for each element Ω_j, special attention must be paid to the evaluation of fluxes at the edges because this flux is the only *connection* between the elements.

Each element Ω_j has its own approximate local solution, allowing the global solution on Ω to be discontinuous at the element interfaces $x_{j\pm1/2}$. This leads to

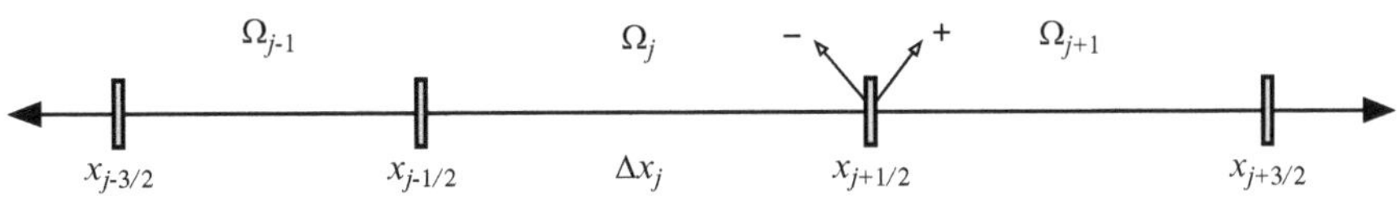

Fig. 9.1 Partition of the 1D domain Ω into non-overlapping elements $\Omega_j = [x_{j+1/2}, x_{j-1/2}]$, with element width Δx_j and edges $x_{j\pm1/2}$. The signs $(-)$ and $(+)$ indicate the *left* and *right* limits of the edge point (interface) $x_{j+1/2}$, respectively. The global solution is discontinuous at these points

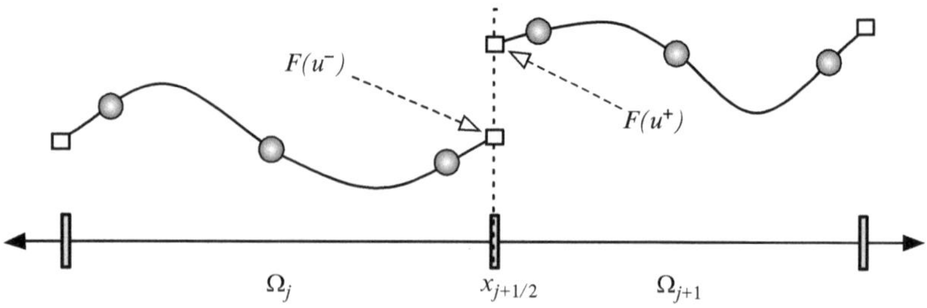

Fig. 9.2 Schematic diagram illustrating the discontinuity of the solution $U(x,t)$ and the flux function $F(U)$ at the element interface (*edge*) $x_{j+1/2}$. *Filled circles* on the smooth curves are the element-wise solution points and the open squares at the edges are the flux points. At the interface the flux function has two contributions, one from the left $F(U^-)$, and one from the right $F(U^+)$. The discontinuity of $F(U)$ at the interfaces is resolved by employing a numerical flux formula

two different values for the flux functions at each interface $x_{j+1/2}$: $F(U(x^-_{j+1/2},t))$ on the left and $F(U(x^+_{j+1/2},t))$ on the right. This discontinuity at the element edges must be addressed by employing a numerical flux (or approximate Riemann solver) $\hat{F}(U^-,U^+) = \hat{F}[U(x^-_{j+1/2},t), U(x^+_{j+1/2},t)]$, which provides the crucial coupling between the elements. Figure 9.2 describes schematically the discontinuity of the flux function at the element interface $x_{j+1/2}$.

The upwind based numerical fluxes used for DG applications are in fact identical to those developed for the finite-volume methods. A variety of numerical flux formulae are available with varying complexity, however, the Lax–Friedrichs (LF) numerical flux is cost-effective and widely used for many applications (Qiu et al. 2006). The LF flux formula is defined as follows:

$$\hat{F}(U^-,U^+) = \frac{1}{2}\left[F(U^-) + F(U^+) - \alpha_{\max}(U^+ - U^-)\right] \tag{9.5}$$

where $\alpha_{\max}$ is the upper bound of $|F'(U)|$, the flux Jacobian, over the entire domain Ω (for scalar problems). If $\alpha_{\max}$ is evaluated only at the local element edges then (9.5) is known as the local Lax–Friedrichs or Russanov flux. For a linear advection problem $\alpha_{\max} = |c|$ and for the inviscid Burgers' equation $\alpha_{\max} = \max(|U^-|,|U^+|)$.

9.2.4 Space Discretization

In order to solve the weak Galerkin formulation (9.4), we assume that the approximate (numerical) solution $U_h \approx U(x,t)$ and the corresponding test function φ_h are polynomial functions belonging to a finite-dimensional space V_h. This space may be formally defined as $V_h = \{p : p\,|_{\Omega_j} \in \mathbb{P}_N(\Omega_j)\}$ where $\mathbb{P}_N$ is the space of polynomials in Ω_j with degree $\leq N$.

For the approximate solution $U_h(x,t)$, the DG spatial discretization based on the weak formulation (9.4) combined with (9.5) can now be written as follows:

$$\int_{\Omega_j} \frac{\partial U_h(x,t)}{\partial t}\varphi_h(x)dx = \int_{\Omega_j} F(U_h(x,t))\frac{\partial \varphi_h(x)}{\partial x}dx - \left[\hat{F}(U_h^-,U_h^+)_{j+1/2}(t)\,\varphi_h(x_{j+1/2}^-) - \hat{F}(U_h^-,U_h^+)_{j-1/2}(t)\,\varphi_h(x_{j-1/2}^+)\right], \tag{9.6}$$

where $U_h, \varphi_h \in V_h$ for all Ω_j; $j = 1,\ldots,N_{elm}$. This completes the DG formulation of problem (9.2).

In order to solve (9.6) accurately and efficiently, we need to make some judicious choices for the integrals and polynomial functions employed in (9.6). The integrals can be accurately computed using the high-order Gaussian quadrature rules. Moreover, choosing orthogonal polynomials as a basis for U_h and φ_h in (9.6) significantly enhances computational efficiency. This is because the coefficients of the time derivative in (9.6) reduce to a diagonal matrix when U_h and φ_h are orthogonal polynomials. The orthogonal basis set which spans V_h may be based on either *modal* or *nodal* expansions. We consider these two cases separately in the following sections.

9.2.4.1 Modal Formulation

The modal basis set consists of orthogonal polynomials of degree k monotonically increasing from 0 to N, and each basis function represents the moment of order k (or, equivalently, each order contributes an extra moment in the expansion (Karniadakis and Sherwin 2005)). The Legendre polynomials $P_k(\xi), k = 0,1,\ldots,N$, $\xi \in [-1,1]$ provide an excellent choice for the orthogonal basis function in V_h. A major advantage of this choice is that the computations in (9.6) can be significantly simplified by exploiting the properties of Legendre polynomials. The first few Legendre polynomials are tabulated in Table 9.1.

Higher degree $P_k(\xi)$ can be generated by the following recurrence relation:

$$P_k(\xi) = \left[\frac{2k-1}{k}\right]\xi\, P_{k-1}(\xi) - \left[\frac{k-1}{k}\right] P_{k-2}(\xi), \quad k = 2,3,4,\cdots. \tag{9.7}$$

At the edges of the interval $[-1,1]$, $P_k(-1) = (-1)^k$ and $P_k(1) = 1$, for any $k \geq 0$. In Fig. 9.3 the left panel shows the Legendre polynomials of degree up to $k = 4$.

Table 9.1 Legendre polynomials $P_k(\xi)$ of degree up to $k = 4$

Degree (k)	0	1	2	3	4
$P_k(\xi)$	1	ξ	$(3\xi^2-1)/2$	$\xi(5\xi^2-3)/2$	$(35\xi^4-30\xi^2+3)/8$

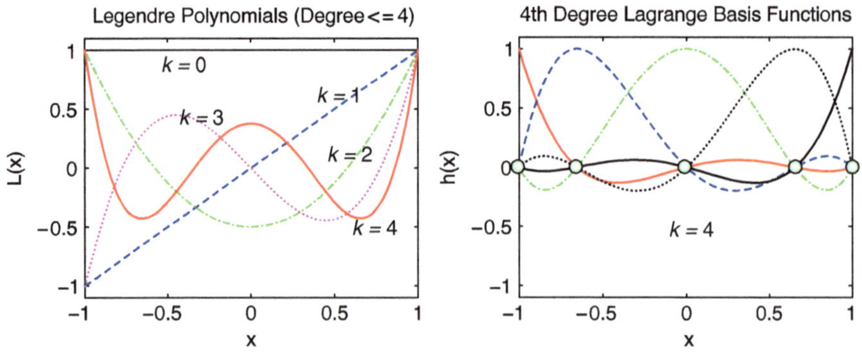

Fig. 9.3 The *left panel* shows Legendre polynomials of degree from $k = 0$ to 4, which can be used as basis functions for the modal DG method. The *right panel* shows Lagrange–Legendre polynomials of fixed degree $k = N = 4$, whose zeros are at the Gauss–Lobatto–Legendre (GLL) quadrature points. The nodal version of DG employs GLL quadrature points, which are in the interval $[-1, 1]$ and marked as *filled circles*

The orthogonality of $P_k(\xi)$ implies that

$$\int_{-1}^{1} P_k(\xi)\, P_\ell(\xi)\, d\xi = \frac{2}{2k+1}\, \delta_{k\ell}, \quad \xi \in [-1, 1], \tag{9.8}$$

where $\delta_{k\ell}$ is the Kronecker delta function ($\delta_{k\ell} = 1$ if $k = \ell$, and $\delta_{k\ell} = 0$ if $k \neq \ell$).

To adopt an orthogonal basis set $\{P_k(\xi)\}_{k=0}^{N}$ for the DG discretization (9.6), we first need to introduce a mapping between x on each element Ω_j and the local variable $\xi \in [-1, 1]$. Irrespective of the physical length Δx_j, each element Ω_j can be mapped onto a unique reference (or standard) element $Q \equiv [-1, 1]$ such that

$$\xi = \frac{2(x - x_j)}{\Delta x_j}. \tag{9.9}$$

Figure 9.4 illustrates schematically the mapping between each Ω_j and the reference element Q. In terms of the new local variable $\xi = \xi(x)$, we denote the approximate solution in any element Ω_j by $U_j = U_j(\xi, t)$ and it can be expressed as

$$U_j(\xi, t) = \sum_{k=0}^{N} U_j^k(t)\, P_k(\xi) \quad \text{for} \quad \xi \in [-1, 1], \tag{9.10}$$

where the expansion coefficients, $U_j^k(t)$, are the *moments* or degrees of freedom (*dof*) evolving in time. The explicit form of $U_j^k(t)$ is derived using (9.8) and given by

$$U_j^k(t) = \frac{2k+1}{2} \int_{-1}^{1} U_j(\xi, t)\, P_k(\xi)\, d\xi, \quad \text{where} \quad k = 0, 1, \ldots, N. \tag{9.11}$$

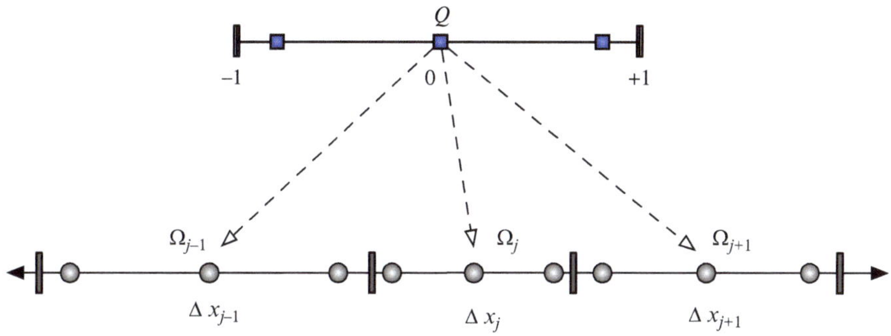

Fig. 9.4 A schematic diagram of the mapping between the unique reference element $Q = [-1, 1]$ and each element Ω_j in the physical domain Ω. The *filled squares* on Q indicate the Gaussian quadrature points in the interval $[-1, 1]$ and the *filled circles* are the corresponding quadrature points on the elements. All the integral and differential operations required for DG discretization are computed on Q

Note that (9.11) may be interpreted as a transformation (or a projection operation) from the physical space to the spectral (Legendre) space with inverse transformation (9.10). It is clear from (9.11) that the zeroth moment,

$$U_j^0(t) = \overline{U}_j = \frac{1}{2}\int_{-1}^{1} U_j(\xi, t)\, d\xi, \tag{9.12}$$

is the average value $\overline{U}_j$. Similarly the first, second, and higher moments are responsible for the linear, quadratic, and higher-order variations of $U(\xi)$ in the element. The left panel in Fig. 9.3 shows the Legendre polynomials of degree up to $N = 4$; each polynomial corresponds to the kth moment in the modal formulation.

We can simplify (9.6) by substituting $U_j(\xi, t)$ for $U_h(x, t)$ and $P_k(\xi)$ for $\varphi_h(x)$, however, this requires a change of variable from x to ξ in (9.6) with the new domain of integration $[-1, 1]$. By using the summation (9.10) and the following relations from (9.9)

$$dx = \frac{\Delta x_j}{2} d\xi, \quad \frac{\partial}{\partial x} = \frac{2}{\Delta x_j}\frac{\partial}{\partial \xi},$$

the weak Galerkin form (9.6) can be written in the semi-discrete form as given below:

$$\frac{\Delta x_j}{2}\sum_{\ell=0}^{N} \frac{d}{dt} U_j^{\ell}(t) \int_{-1}^{1} P_k(\xi) P_\ell(\xi) d\xi = \int_{-1}^{1} F(U_j(\xi, t))\, P_k'(\xi)\, d\xi - \left[\hat{F}_{j+1/2}(t)\, P_k(1) - \hat{F}_{j-1/2}(t)\, P_k(-1)\right], \tag{9.13}$$

where $P_k'(\xi)$ is the derivative of the Legendre polynomials (9.7). The above equation can be further simplified by employing the orthogonality relation (9.8) and the property $P_k(\pm 1) = (\pm 1)^k$ as follows:

$$\frac{1}{2k+1}\frac{d}{dt}U_j^k(t) = \frac{1}{\Delta x_j}\int_{-1}^{1} F(U_j(\xi,t))\, P_k'(\xi)\, d\xi - \frac{1}{\Delta x_j}\left[\hat{F}_{j+1/2}(t) - \hat{F}_{j-1/2}(t)(-1)^k\right], \tag{9.14}$$

where $k = 0, 1, \ldots, N$.

The integral appearing in (9.14) is evaluated using a high-order ($(N+1)$-node) Gaussian quadrature rule. Usually a Gauss–Legendre (GL) quadrature, which is exact for polynomials of degree $2N+1$, or a Gauss–Lobatto–Legendre (GLL) quadrature, which is exact for polynomials of degree $2N-1$, is employed; the choice of a specific quadrature is somewhat application dependent. For a given number of quadrature points, the GL quadrature is more accurate than the GLL quadrature but the former does not place nodes at the end points of the interval $[-1, 1]$ (see the marked points on the reference element Q in Fig. 9.4). We further discuss the relative merits of GL and GLL quadrature rules in a 2D context in Sect. 9.3.1.3.

In order to compute $\hat{F}_{j\pm 1/2}$ in (9.14), the flux $F(U(\xi))$ at the element edges $\xi = \pm 1$ must be known. In the GL case, this means that one must interpolate the solution $U(\xi)$ using (9.10). However, the GLL quadrature includes the edges where values of $U(\xi)$ are readily available, and makes the edge flux computation easy.

Regardless of the choice of quadrature, the DG solution procedure for the conservation law (9.2) on an element Ω_j reduces to solving a system of decoupled ordinary differential equations (ODEs) (9.14), which may be written in the following form.

$$\mathbf{M}_j \frac{d}{dt}\mathbf{U}_j = \mathbf{R}(\mathbf{U}_j), \tag{9.15}$$

where $\mathbf{M}_j$ is the coefficient matrix associated with the time derivative in (9.14) and formally referred to as the *mass matrix*, $\mathbf{U}_j$ is a column vector containing the moments $U^k(t), k = 0, 1, \ldots, N$, and $\mathbf{R}$ is the residual vector corresponding to the right-hand side of (9.14). By virtue of the orthogonality of the Legendre polynomials, the mass matrix $\mathbf{M}_j$ is strictly a diagonal matrix with non-zero entries $\{1/(2k+1)\}_{k=0}^{N}$.

This diagonal structure has great computational advantage because $\mathbf{M}_j$ can be inverted trivially and simplifies the solution process in (9.15). For the DG discretization considered here each element Ω_j relies on the same polynomial bases, therefore the mass matrix $\mathbf{M}_j = \mathbf{M}$ is identical for each element in the domain Ω. Pre-multiplying (9.15) by $\mathbf{M}^{-1}$ for each element results in the following system of ODEs corresponding to the problem (9.2),

$$\frac{d}{dt}\mathbf{U} = \mathbf{L}(\mathbf{U}) \quad \text{in} \quad (0, T], \tag{9.16}$$

where **U** is the global vector of degrees of freedom which evolves in time, **L** is a generic operator combining all the spatial discretizations. A DG method employing $N + 1$ moments (or with polynomial bases up to degree N) is often referred to as a P^N method (Cockburn and Shu 2001). We will consider the time discretization procedure for (9.16) in the following section.

In order to see the close link between DG and FV approaches, we consider the first few moments $k = 0, 1, 2$ in (9.14) as follows:

$$\frac{d}{dt}U_j^0(t) = \frac{1}{\Delta x_j}\left[\hat{F}_{j+1/2}(t) - \hat{F}_{j-1/2}(t)\right], \tag{9.17}$$

$$\frac{1}{3}\frac{d}{dt}U_j^1(t) = \frac{1}{\Delta x_j}\int_{-1}^{1} F(U_j(\xi,t))d\xi - \frac{1}{\Delta x_j}\left[\hat{F}_{j+1/2}(t) + \hat{F}_{j-1/2}(t)\right], \tag{9.18}$$

$$\frac{1}{5}\frac{d}{dt}U_j^2(t) = \frac{6}{\Delta x_j}\int_{-1}^{1} F(U_j(\xi,t))\xi d\xi - \frac{1}{\Delta x_j}\left[\hat{F}_{j+1/2}(t) - \hat{F}_{j-1/2}(t)\right]. \tag{9.19}$$

The mass matrix associated with the above system is $\mathbf{M} = \mathrm{diag}\,[1, 1/2, 1/5]$, and the moments $U_j^k(t)$ can be used for constructing the solution at a known time $t = t_n$ via (9.10) such that

$$U_j(\xi, t_n) = U_j^0(t_n)P_0(\xi) + U_j^1(t_n)P_1(\xi) + U_j^2(t_n)P_2(\xi). \tag{9.20}$$

For the simplest DG formulation, the P^0 case, (9.17) is the only equation to solve in time. In this case $U_j^0(t)$, the moment (*dof*) evolving in time, is nothing more than the cell-average $\overline{U}_j$ given in (9.12), which is an element-wise (or piecewise) constant. Thus the DG P^0 case reduces to the classical piecewise constant Godunov FV method (Toro 1999, Chap. 8). In a similar manner one can show the DG P^1 and P^2 methods are related to the piecewise linear method (PLM, van Leer 1974) and the piecewise parabolic method (PPM, Colella and Woodward 1984), respectively.

Nevertheless, there are subtle differences between regular FV and DG methods. In FV methods such as PLM or PPM there is only one *dof* per control volume evolving in time, namely $\overline{U}_j$, irrespective of the spatial order of accuracy of the method or the dimension of the problem. On the other hand, the DG method carries more *dofs* per element (the cell or the control volume in an FV sense) and the number of *dofs* grows with both the order of accuracy and the dimension (see Fig. 9.5). In other words, a DG method packs more information into each cell than the FV method. For example, in (9.20) three moments are required to construct the solution $U_j(\xi)$ with a P^2 method, and the moments depend only on the element Ω_j resulting in a compact computational stencil. The PPM method requires the reconstruction of parabolas of the form (9.20) by utilizing the averages $\overline{U}_j$ from the neighboring cells, resulting in a wider stencil. In both methods $U_j(\xi)$ essentially represents the sub-grid scale distribution of the solution – even though the underlying discretizations are different. However, as compared to PPM, the high accuracy and compactness of the DG P^2

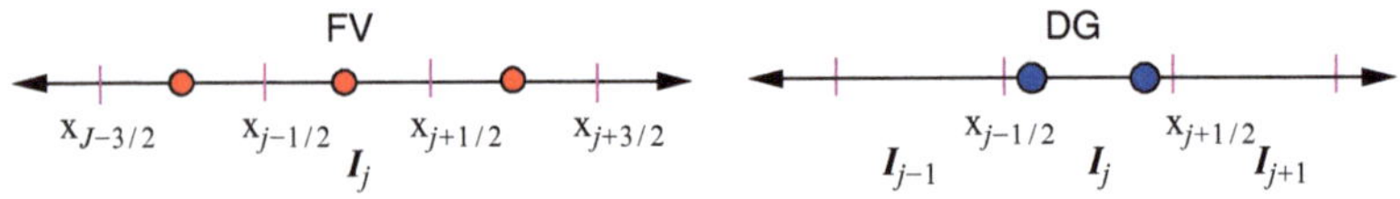

Fig. 9.5 A schematic showing a comparison between the classical 1D finite-volume (FV) and DG methods. $I_j = [x_{j-1/2}, x_{j+1/2}]$ may be interpreted as a cell in the FV method or an element in the DG (P^1) method. For the FV method, the cell-average (shown as *filled circles* in the *left panel*) is the only degree of freedom per cell evolving in time. The DG method has more degrees of freedom (marked as *filled circles* in the *right panel*) per element evolving in time, however both methods employ the same procedure to address the discontinuities at the cell boundaries $x_{j\pm1/2}$

method comes with additional computational cost. The DG method presented here may be viewed as a high-order compact FV method. A P^2 transport scheme is also similar to the multi-moment transport schemes developed by Prather (1986).

9.2.4.2 Nodal Formulation

The nodal expansion is based on Lagrange polynomials with roots at a set of *nodal* points, which may include the edge points. The nodal bases are widely popular in high-order spectral element methods (Karniadakis and Sherwin 2005). An important aspect of the DG discretization is the choice of an efficient basis set (polynomials) that span V_h. Because of the inherent computational advantages associated with nodal bases, they are adopted in DG discretization for many applications (Hesthaven and Warburton 2008). The nodal DG scheme is potentially more computationally efficient because it relies on solutions in physical (grid point) space, obviating the need to transform between spectral and physical space, which is required for the modal DG scheme (9.14).

The nodal basis set is constructed using the Lagrange polynomials $h_k(\xi)$, $\xi \in [-1, 1]$, with roots at the Gauss quadrature points. The nodal points may be based on the Gauss–Legendre (GL) or the Gauss–Lobatto–Legendre (GLL) quadrature rule. However, we consistently employ the GLL quadrature for the nodal formulation considered herein. The $N + 1$ GLL points $\{\xi_l\}_{l=0}^{N}$ (i.e., the nodal points including the edge point ±1), can be generated from the relation $(1 - \xi^2)P'_N(\xi) = 0$, where $P_N(\xi)$ is the Legendre polynomial of degree N. The basis functions are defined by

$$h_k(\xi) = \frac{(\xi - 1)(\xi + 1)\, P'_N(\xi)}{N(N+1)\, P_N(\xi_k)\, (\xi - \xi_k)}, \tag{9.21}$$

where $P_N(\xi)$ is the Legendre polynomial of degree N. In Fig. 9.3 the right panel shows the fourth degree nodal bases $h_k(\xi)$, and $N + 1 = 5$ GLL points are marked as filled circles. Since $h_k(\xi)$ is a Lagrange polynomial, the following property holds at the nodes ξ_l:

$$h_k(\xi_l) = \delta_{kl} = \begin{cases} 1 \text{ if } k = l, \\ 0 \text{ if } k \neq l. \end{cases} \tag{9.22}$$

The discrete orthogonality of $h_\ell(\xi)$ can be established through the GLL quadrature rule, given by

$$\int_{-1}^{1} f(\xi)d\xi \approx \sum_{k=0}^{N} f(\xi_k)w_k, \tag{9.23}$$

where $f(\xi)$ is an arbitrary function with known values at the nodes (quadrature points) and w_k are the weights associated with the GLL quadrature rule, defined to be

$$w_k = \frac{2}{N(N+1)\,[P_N(\xi_k)]^2}.$$

As mentioned earlier, the GLL quadrature rule (9.23) is *exact* for polynomials of degree up to $2N-1$. The discrete orthogonality of the basis function $h_\ell(\xi)$ can be derived using (9.22) and (9.23) as follows:

$$\int_{-1}^{1} h_k(\xi)\,h_l(\xi)d\xi \approx \sum_{\ell=0}^{N} h_k(\xi_\ell)\,h_l(\xi_\ell)\,w_\ell = w_k\,\delta_{kl}. \tag{9.24}$$

Note that the integrand $h_k(\xi)\,h_l(\xi)$ is a polynomial of degree $2N$, so the orthogonality does not strictly hold under exact integration. In other words, the orthogonality of the nodal expansion given in (9.24) is not as rigorous as the continuous orthogonality employed in the modal case (9.8). Fortunately, the error incurred in discrete orthogonality is of the same order as the nodal expansion so the discretization is consistent. Moreover, it is shown in Canuto et al. (2007) that the discrete norm is uniformly equivalent to the continuous norm.

In the nodal expansion, the approximate solution $U_j(\xi,t)$ for an element Ω_j can be written in terms of $h_k(\xi)$ as given below:

$$U_j(\xi,t) = \sum_{k=0}^{N} U_{j,k}(t)\,h_k(\xi), \quad \xi \in [-1,1], \tag{9.25}$$

where $U_{j,k}(t) = U_j(\xi_k,t)$ are the known values of $U_j(\xi,t)$ at the GLL grid points. Also, from (9.25) it is evident that the approximate solution is expressed as a Lagrange interpolation polynomial. Analogous to the modal case, the weak Galerkin formulation (9.6) can be simplified as follows: substitute (9.25) for the approximate solution and $h_k(\xi)$ for the test function, employing the properties (9.24) and $h_k(\pm 1) = 1$. This yields the equation

$$\frac{w_k}{2}\frac{d}{dt}U_{j,k}(t) = \frac{1}{\Delta x_j}\int_{-1}^{1} F(U_j(\xi,t))\,h'_k(\xi)\,d\xi - \frac{1}{\Delta x_j}\left[\hat{F}_{j+1/2}(t) - \hat{F}_{j-1/2}(t)\right], \tag{9.26}$$

where $k = 0,1,\ldots,N$. The right-hand side involves the derivative of the Lagrange polynomial $h'_k(\xi)$, which needs to be calculated and stored at each of the quadrature points in order to evaluate the integral in (9.26). The resulting matrix, known as the differentiation matrix, has the following explicit form (Karniadakis and Sherwin 2005; Canuto et al. 2007):

$$h_k'(\xi_l) = \begin{cases} \frac{L_N(\xi_k)}{L_N(\xi_l)} \frac{1}{(\xi_k - \xi_l)} & \text{if } k \neq l, \\ -\frac{(N+1)N}{4} & \text{if } k = l = 0, \\ \frac{(N+1)N}{4} & \text{if } k = l = N, \\ 0 & \text{otherwise.} \end{cases} \tag{9.27}$$

The mass matrix associated with (9.26) is a diagonal matrix $\mathbf{M}$ with non-zero entries $\{w_k/2\}_{k=0}^{N}$ and, by virtue of the GLL grids, the numerical fluxes $\hat{F}_{j\pm1/2}$ are readily available at the edges $\xi = \pm 1$. The system of ODEs (9.26) can be generalized for the whole domain Ω exactly as in (9.16),

$$\frac{d}{dt}\mathbf{U} = \mathbf{L}(\mathbf{U}) \quad \text{in} \quad (0, T],$$

where $\mathbf{U}$ is the global vector of grid point values $U_{j,k}$, $j = 1, 2, \ldots, N_{elm}$ and $k = 0, 1, \ldots, N$.

A remarkable difference between the nodal version (9.26) and the corresponding modal version (9.14) of the DG discretization is the absence of the spectral coefficients. In other words, the *dofs* to evolve in time in (9.26) are just the grid point values of the approximate solution $U_{j,k}(t)$, not the spectral coefficients as in the modal case. Hence, there is no need to transform between spectral and physical spaces at every time step, and this feature makes the nodal discretization computationally more efficient (Levy et al. 2007).

9.2.5 *Time Integration*

The modal and nodal DG discretization both reduce the one-dimensional scalar conservation law to a system of ODEs (9.16) which can be solved using a variety of time integration techniques (Chap. 5). In fact, the DG discretization reduces conservation law PDEs to a system of ODEs irrespective of the spatial dimension. Therefore we consider the following general form of the ODE system:

$$\frac{d}{dt}\mathbf{U} = \mathbf{L}(\mathbf{U}) \quad \text{in} \quad (0, T].$$

The most widely used explicit time integration technique for the DG method is based on the Runge–Kutta (RK) scheme; a combination of these space and time discretization approaches is often referred to as the RKDG method (Cockburn and Shu

2001). For the DG discretization considered in this Chapter we employ the strong stability-preserving (SSP) RK scheme, also known as the total variation diminishing RK scheme (Cockburn et al. 1997); a detailed account of SSP-RK methods is given in Gottlieb et al. (2001).

The second-order (two-stage) SSP-RK is

$$\begin{aligned} \mathbf{U}^{(1)} &= \mathbf{U}^{n} + \Delta t\, \mathbf{L}(\mathbf{U}^{n}) \\ \mathbf{U}^{n+1} &= \frac{1}{2}\mathbf{U}^{n} + \frac{1}{2}[\mathbf{U}^{(1)} + \Delta t\, \mathbf{L}(\mathbf{U}^{(1)})], \end{aligned} \tag{9.28}$$

and the third-order (three-stage) SSP-RK is

$$\begin{aligned} \mathbf{U}^{(1)} &= \mathbf{U}^{n} + \Delta t\, \mathbf{L}(\mathbf{U}^{n}) \\ \mathbf{U}^{(2)} &= \frac{3}{4}\mathbf{U}^{n} + \frac{1}{4}[\mathbf{U}^{(1)} + \Delta t\, \mathbf{L}(\mathbf{U}^{(1)})] \\ \mathbf{U}^{n+1} &= \frac{1}{3}\mathbf{U}^{n} + \frac{2}{3}[\mathbf{U}^{(2)} + \Delta t\, \mathbf{L}(\mathbf{U}^{(2)})]. \end{aligned} \tag{9.29}$$

In both (9.28) and (9.29), $\mathbf{U}^{(1)}$ and $\mathbf{U}^{(2)}$ are intermediate stages of the RK method while the superscripts n and $n+1$ denote time levels t and $t+\Delta t$, respectively. The overall accuracy of the numerical scheme is dictated by the order of accuracy of both the spatial and temporal discretizations. For example, the DG method using polynomials of degree N along with an $N+1$ stage RK method results in an $(N+1)$th-order accurate method (Cockburn and Shu 2001).

Higher-order RK schemes provide a wider stability region (Butcher 2008), so a longer time step may be used in the numerical integration. Unfortunately, a high-order RK time discretization has multiple stages of function (right-hand side) evaluations and flux communications, resulting in a computationally expensive scheme (especially in a parallel computing environment). Therefore, many practical applications use a fourth- or lower-order RK scheme (Nair et al. 2005a, 2009).

The linear stability analysis for the modal DG method discussed in Cockburn and Shu (2001) may be used as a guideline for choosing the time steps. For an $(N+1)$-th -order accurate RKDG method, the CFL (Courant-Friedrichs-Lewy) stability limit is given by $c\Delta t/\Delta x \leq 1/(2N+1)$, where Δx is the element width and c is the velocity. This has been proven to be true when $N=1$, however, no theoretical proof exists when $N>1$. For $N \gg 1$ the explicit DG method is very time step restrictive, in such cases a semi-implicit or implicit time integration strategy may be desirable (Chap. 5). We also note that when $N>1$, the grid spacing Δx used in calculating the CFL limit should be the *minimum* distance between the non-uniformly distributed quadrature points (see the *right panel* of Fig. 9.3). A detailed discussion of the CFL stability limit for advection problems for high-order Galerkin methods can be found in Chap. 6 of Karniadakis and Sherwin (2005).

9.2.6 DG 1D Computational Examples

Here we illustrate the DG method by solving two examples of the 1D conservation law (9.2). The first one is a simple linear problem involving the advection of both a Gaussian profile (smooth case) and a rectangular wave (non-smooth case). The second example is the solution of the inviscid Burgers equation, a nonlinear problem. Numerical solutions are computed using the 1D DG schemes discussed earlier. Both the modal and nodal versions of the scheme are used for the simulations. However, the results produced by these schemes are almost identical, and we show only modal or nodal solution for each test.

For the linear advection problem, the domain is $\Omega = [-1, 1]$ with periodic boundary conditions. The initial condition for the smooth problem is $U_0(x) = \exp(-8\,x^2)$, a Gaussian hill with unit height, and the wind velocity is $c = 1$. In this case the flux function in (9.2) is simply $F(U) = U$. The domain is partitioned into $N_{elm} = 40$ elements, each with $N_v = 5$ GLL quadrature points, and the nodal DG formulation (9.26) is used for the discretization. The resulting time-dependent ODE is solved with the third-order SSP-RK (9.29). 400 time steps are required for a complete revolution along the domain. Figure 9.6 shows the Gaussian hill (*left panel, dashed line*) after ten revolutions; the reference solution is also plotted with a solid line but it is visually indistinguishable from the numerical solution.

For the non-smooth advection case the initial condition is a rectangular wave pattern located at the center of the domain with unit height and width of 0.5 units; other than this the boundary conditions and discretization are exactly the same as in the smooth case. The right panel in Fig. 9.6 shows the numerical solution after ten revolutions, and the reference solution (initial condition) is also displayed (solid

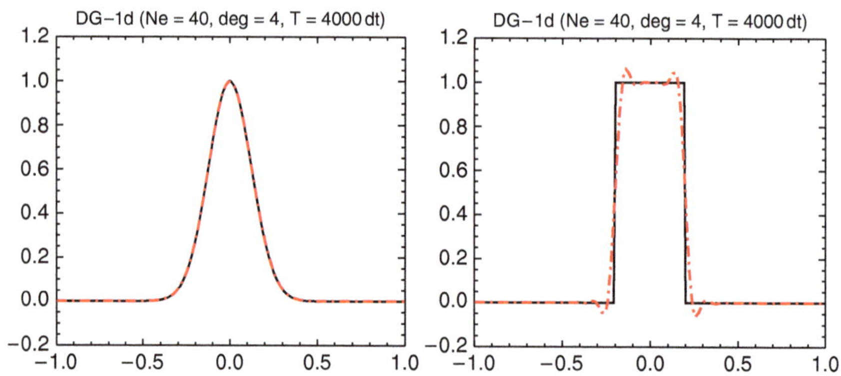

Fig. 9.6 Numerical solution (after ten revolutions) of the 1D advection problem (9.2) with the high-order nodal DG scheme. The *left panel* shows the solution for the smooth case, where a Gaussian hill is used as the initial condition. The *right panel* shows the solution for the non-smooth case, for which a rectangular wave is used as the initial condition. The computational domain $[-1, 1]$ consists of 40 elements, each with 5 GLL quadrature points

line). The DG solution suffers from oscillations at the non-smooth edges. The steep gradients at these point produces the Gibbs phenomena, however, the oscillations are confined (or local) to a narrow region even after ten revolutions. This is a remarkable property of the DG method; other high-order approaches, such as the spectral element method, propagate the noise along the entire domain.

The inviscid Burgers equation, $U_t + (U^2/2)_x = 0$, is a special case of (9.2) with $F(U) = U^2/2$. The initial condition for this problem is $U_0(x) = 1/2 + \sin(\pi x)$ over a periodic domain $\Omega = [0, 2]$. The domain is partitioned into 80 elements, and a modal version DG scheme employing 4 GLL quadrature points is used for the simulations. Time integration is performed with the third-order SSP-RK (9.29), for which a small time step of $\Delta t = 0.0015/\pi$ is used. The exact solution is known for this problem and is shown as solid narrow lines in Fig. 9.7, and the DG solution is marked as diamond points (one value for each element). The left panel in Fig. 9.7 shows the smooth solution time $t = 3/(4\pi)$ (500 time steps). Clearly, the DG solution is in good agreement with the analytic solution. However, at time $t = 9/(8\pi)$ (750 time steps) the numerical solution develops a *shock* at the steep gradient, leading to oscillations, as seen in the right panel of Fig. 9.7. As time evolves the oscillations become severe and they can pollute the numerical solution. As in the non-smooth advection case, the generation of unphysical oscillations in the numerical solution at contact discontinuities or shocks are due to the Gibbs phenomenon. Any linear numerical method higher than first-order is subject to this problem (Godunov 1959), unless there is some measure to control or eliminate the spurious oscillations by *limiting* or *filtering* the numerical solution. We discuss the limiting procedure for DG methods in the following Section.

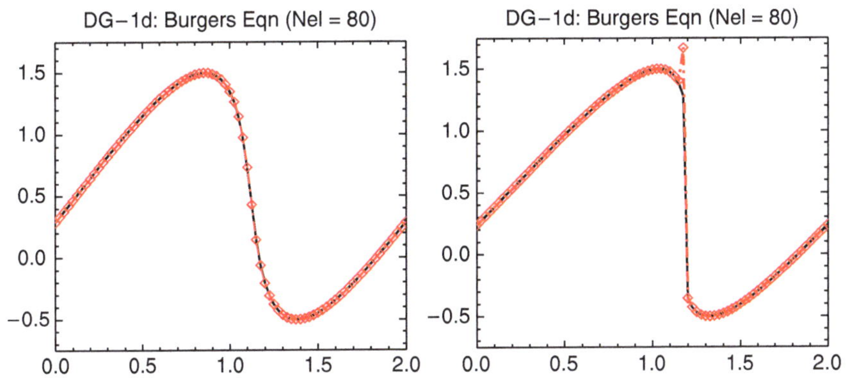

Fig. 9.7 Numerical solution for the inviscid Burgers equation with the modal DG scheme. The *solid line* indicates the exact solution and diamond points show the DG solution. The domain consists of 80 elements; only one value per element is plotted for clarity. The *left panel* shows the solution at time $t = 3/(4\pi)$; at this time the solution is still smooth and free from shocks. The *right panel* shows the solution at time $t = 9/(8\pi)$, at which point shocks have developed

9.3 DG for 2D Cartesian Problems

Although the DG method can be adapted to any type of domain or mesh, we choose a rectangular domain D with quadrilateral elements for simplicity. Consider the two-dimensional (2D) scalar conservation law,

$$\frac{\partial U}{\partial t} + \nabla \cdot \mathbf{F}(U) = S(U), \quad \text{in} \quad D \times (0, T), \tag{9.30}$$

where $U = U(x, y, t)$ is the conservative variable such that $(x, y) \in D$, the 2D gradient operator ∇ on D is defined as $\nabla = (\partial/\partial x, \partial/\partial y)$, $\mathbf{F} = (F_1, F_2)$ is the flux function, and $S(U)$ is the source term (if any). The initial condition for the problem is specified as $U(x, y, t = 0) = U_0(x, y)$ and we assume that the rectangular domain D is periodic in both the x- and y-directions.

Following the steps used in the previous section the 2D extension of the DG discretization is straightforward. The domain D is partitioned into $N_{elm} = N_x \times N_y$ rectangular non-overlapping elements Ω_e such that

$$\Omega_e = \{(x, y) \,|\, x \in [x_{i-1/2}, x_{i+1/2}],\ y \in [y_{j-1/2}, y_{j+1/2}]\}, \quad D = \cup_{e=1}^{N_{elm}} \Omega_e, \tag{9.31}$$

where $e = e(i, j)$ is the element index and $i = 1, 2, \ldots, N_x$, $j = 1, 2, \ldots, N_y$. Figure 9.8 shows a simple partition of D and a general element Ω_e.

We first introduce some basic formal notations required for the discretization. Let $\mathcal{V}_h$ be a finite-dimensional space of polynomials of degree up to $k = N$ such that

$$\mathcal{V}_h = \{\varphi \in L^2(D) : \varphi|_{\Omega_e} \in \mathbb{P}_N(\Omega_e), \forall\, \Omega_e \in D\}, \tag{9.32}$$

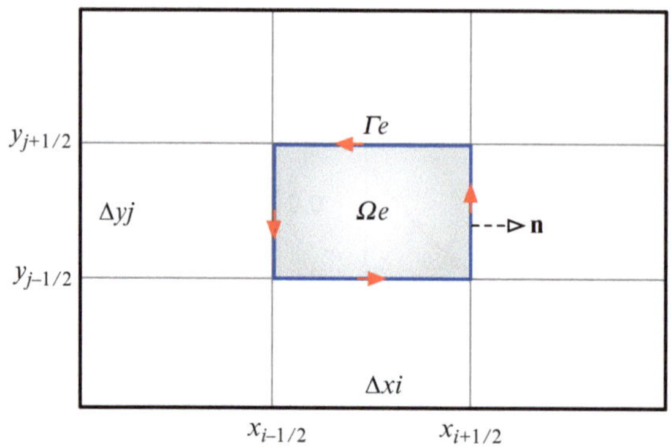

Fig. 9.8 A schematic of a 2D domain with rectangular elements. Ω_e is a generic element with boundary Γ_e and its width in the x- and y-directions are $\Delta x_i = (x_{i+1/2} - x_{i-1/2})$ and $\Delta y_j = (y_{j+1/2} - y_{j-1/2})$, respectively. The outward-facing unit normal vector is denoted by $\mathbf{n}$ and the flux integrals (*line integrals*) are performed along the boundary Γ_e as indicated by the *arrows*

where

$$\mathbb{P}_N = \text{span}\{x^m\, y^n \ : 0 \le m, n \le N\}.$$

The first step for the DG discretization is the weak Galerkin formulation of the problem (9.30). In general, this is achieved by multiplying (9.30) with a test function and integrating by parts (Green's method) over the domain, where both the approximate solution and test function belong to $\mathcal{V}_h$. Since the discretization procedure is the same for each element, it is only necessary to consider a generic element Ω_e with boundary Γ_e in D (as in Fig. 9.8). Thus, to find the approximate solution $U_h \in \mathcal{V}_h$, (9.30) is multiplied by a test function $\varphi_h(x, y) \in \mathcal{V}_h$ and then integrated over the element Ω_e. This results in the following integral equation (i.e., the weak formulation), analogous to (9.4):

$$\int_{\Omega_e} \frac{\partial U_h(x, y, t)}{\partial t} \varphi_h(x, y)\, d\Omega - \int_{\Omega_e} \mathbf{F}[U_h(x, y, t)] \cdot \nabla\varphi(x, y)\, d\Omega$$
$$+ \int_{\Gamma_e} \mathbf{F}[U_h(x, y, t)] \cdot \mathbf{n}\,\varphi_h(x, y)\, d\Gamma = \int_{\Omega_e} S[U_h(x, y, t)]\, \varphi(x, y)\, d\Omega, \tag{9.33}$$

where $\mathbf{n}$ is the outward-normal unit vector on the element boundary Γ_e as shown in Fig. 9.8. A major difference between the weak formulations (9.6) of 1D and (9.33) of 2D cases is the appearance of the flux integral in the 2D case (the last term on the left-hand side of (9.33)). The flux integration should be performed along the element boundary Γ_e. The analytic flux $\mathbf{F}(U_h) \cdot \mathbf{n}$ in (9.33) is discontinuous because the solution itself is discontinuous at the element edges. Therefore, $\mathbf{F}(U_h)\cdot\mathbf{n}$ should be replaced by a numerical flux $\hat{\mathbf{F}}(U_h^-, U_h^+)$. This is addressed by employing a suitable flux formula (or approximate Riemann solver) such as the local Lax–Friedrichs flux (9.5).

The numerical flux resolves the discontinuity at the element edges and again provides the only mechanism by which adjacent elements interact. The finite-volume component of the DG method is the boundary flux integral, which in fact bridges the discontinuous elements together. The flux exchange at the boundaries is responsible for "communicating" physical information across the domain, and it preserves the local conservation properties. Thus the flux integration procedure is extremely important and its accurate evaluation is pivotal to maintaining the overall accuracy of the DG scheme. Following is a simplified version of (9.33) with the numerical flux $\hat{\mathbf{F}} = (\hat{F}_1, \hat{F}_2)$ (for brevity dependencies on (x, y) and t are omitted).

$$\frac{d}{dt}\int_{\Omega_e} U_h\, \varphi_h\, d\Omega - \int_{\Omega_e} \mathbf{F}(U_h) \cdot \nabla\varphi_h\, d\Omega + \int_{\Gamma_e} \hat{\mathbf{F}} \cdot \mathbf{n}\,\varphi_h\, d\Gamma = \int_{\Omega_e} S(U_h)\,\varphi_h d\Omega. \tag{9.34}$$

9.3.1 Space Discretization

The space discretization consists of simplifying the integrals in (9.34) by choosing an appropriate set of local orthogonal basis functions. As introduced in the 1D problem, the basis set can be either a set of Legendre polynomials for the *modal* case or a set of Lagrange–Legendre polynomials for the *nodal* case. In either case, the 2D basis set can be constructed with a tensor product of 1D basis functions. This approach significantly simplifies the computational procedure. In order to exploit this option, we introduce the local independent variables (ξ, η) such that

$$\xi = \frac{2(x - x_i)}{\Delta x_i}, \quad \eta = \frac{2(y - y_j)}{\Delta y_j}, \tag{9.35}$$

where $x_i = (x_{i+1/2} + x_{i-1/2})/2$ and $y_j = (y_{j+1/2} + y_{j-1/2})/2$. The width of any element Ω_e is defined by $\Delta x_i = (x_{i+1/2} - x_{i-1/2})$ and $\Delta y_j = (y_{j+1/2} - y_{j-1/2})$ along the x- and y-directions, respectively (Fig. 9.8). Irrespective of the physical size of the rectangular element Ω_e, the transformation (9.35) maps Ω_e onto a unique element $Q \equiv [-1, 1] \otimes [-1, 1]$, also known as the reference element. Figure 9.9 shows the mapping between a rectangular element Ω_e and Q. Now the approximate solution, test functions and the basis functions all can be defined in terms of local coordinates on Q. Effectively Q is the computational stencil or *molecule* for the 2D DG discretization, where all the integral and differential operations required in (9.34) are performed.

For the rectangular elements Ω_e the boundary flux integrals in (9.34) along Γ_e can be decomposed in terms of unit vectors **i** and **j**, parallel to the x- and y-axes

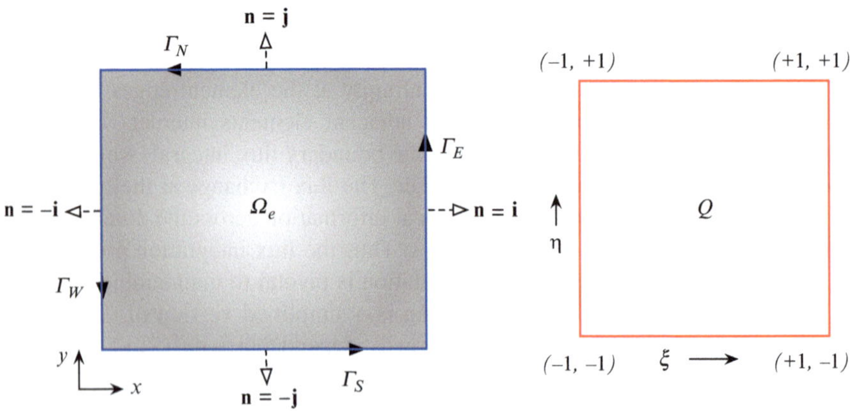

Fig. 9.9 A schematic of the mapping between a rectangular element Ω_e and the reference (*standard*) element Q by (9.35). The local coordinates (ξ, η) on Q are such that $-1 \le \xi, \eta \le 1$. The outward-facing unit normal vector **n** for each wall of Ω_e is marked (*left panel*), and the flux integrals along the boundary Γ_e can be broken into four integrals (9.36) one for each edge as described in the text

respectively:

$$\int_{\Gamma_e} \hat{\mathbf{F}} \cdot \mathbf{n}\,\varphi_h \, d\Gamma = \int_{\Gamma_e} (\hat{F}_1 \mathbf{i} + \hat{F}_2 \mathbf{j}) \cdot \mathbf{n}\,\varphi_h \, d\Gamma,$$

where the outward-facing unit normal vector $\mathbf{n}$ takes the values $\mathbf{i}, \mathbf{j}, -\mathbf{i}$ and $-\mathbf{j}$ along the east (Γ_E), north (Γ_N), west (Γ_W), and the south (Γ_S) walls, respectively, as shown in Fig. 9.9. The boundary integrals can then be written as

$$\int_{\Gamma_e} \hat{\mathbf{F}} \cdot \mathbf{n}\,\varphi_h \, d\Gamma = \int_{\Gamma_E} \hat{F}_1 \varphi_h \, d\Gamma + \int_{\Gamma_N} \hat{F}_2 \varphi_h \, d\Gamma - \int_{\Gamma_W} \hat{F}_1 \varphi_h \, d\Gamma - \int_{\Gamma_S} \hat{F}_2 \varphi_h \, d\Gamma. \tag{9.36}$$

9.3.1.1 2D Modal Form

We first discuss the 2D discretization based on the modal basis set. In the (ξ, η) coordinate system the test function is chosen to be a tensor-product of Legendre polynomials $P_\ell(\xi)\, P_m(\eta)$, which belongs to $\mathbb{P}_N$ in (9.32). The approximate solution $U_h(\xi, \eta, t)$ can be written in terms of the basis functions,

$$U_h(\xi, \eta, t) = \sum_{\ell=0}^{N} \sum_{m=0}^{N} U_h^{\ell m}(t)\, P_\ell(\xi)\, P_m(\eta) \quad \text{for} \quad -1 \le \xi, \eta \le 1 \tag{9.37}$$

where $U_h^{\ell m}(t)$ are the time dependent 2D moments (*dofs*) and defined to be

$$U_h^{\ell m}(t) = \frac{(2\,\ell + 1)(2\,m + 1)}{4} \int_{-1}^{1} \int_{-1}^{1} U(\xi, \eta, t)\, P_\ell(\xi)\, P_m(\eta)\, d\xi\, d\eta. \tag{9.38}$$

The weak formulation (9.33) can be further simplified by mapping the integrals onto Q using the transformation (9.35), and the properties of Legendre polynomials (basis functions). The mass matrix (9.15) associated with the 2D discretization is also diagonal and can be easily inverted. The final computational form can be written as a decoupled system of time-dependent ODEs for every element Ω_e,

$$\frac{d}{dt} U_h^{\ell m}(t) = \frac{(2\,\ell + 1)(2\,m + 1)}{2\, \Delta x_i\, \Delta y_j} \left[I_G + I_{F_1} + I_{F_2} + I_S \right], \tag{9.39}$$

where $0 \le \ell, m \le N$. Note that the source term $S(U) = 0$ in (9.30) for the pure advection problem; for generality we consider a non-zero source term. The integrals appearing in the right-side of (9.39) can be defined on Q as below,

$$I_G = \int_{-1}^{1} \int_{-1}^{1} \left[\Delta y_j\, F_1(U_h)\, P'_\ell(\xi)\, P_m(\eta) + \Delta x_i\, F_2(U_h)\, P_\ell(\xi)\, P'_m(\eta) \right] d\xi\, d\eta \tag{9.40}$$

$$I_{F_1} = -\Delta y_j \int_{-1}^{1} \left[\hat{F}_1(U(1,\eta,t)) - (-1)^{\ell} \hat{F}_1(U(-1,\eta,t) \right] P_m(\eta)\, d\eta \tag{9.41}$$

$$I_{F_2} = -\Delta x_i \int_{-1}^{1} \left[\hat{F}_2(U(\xi,1,t) - (-1)^{m} \hat{F}_2(U(\xi,-1,t) \right] P_\ell(\xi)\, d\xi \tag{9.42}$$

$$I_S = \frac{\Delta x_i \Delta y_j}{2} \int_{-1}^{1} \int_{-1}^{1} S(U_h(\xi,\eta,t))\, P_\ell(\xi)\, P_m(\eta) d\xi\, d\eta, \tag{9.43}$$

where I_G and I_S are the surface integrals corresponding to the gradient and the source terms in (9.33), respectively, and I_{F_1} and I_{F_2} are boundary flux integrals (9.36) along the η and ξ-directions, respectively. $\hat{F}_1$ and $\hat{F}_2$ are the numerical fluxes at the element interfaces, which can be computed by using (9.5).

The integrals appearing in (9.39) are evaluated using high-order accurate Gaussian quadrature rules and will be discussed in the following section. The modes $U_h^{\ell m}$ are predicted at a new time level by (9.39), then the corresponding approximate solution $U_h(\xi,\eta)$ is computed from (9.37). However, this process involves transformations from the spectral to the physical space as discussed in 1D case. The ODE (9.39) can be solved by the SSP-RK procedure given in (9.29).

9.3.1.2 2D Nodal Form

The basic difference between the modal and the nodal form is the choice of basis set. The mapping between the element Ω_e and standard element Q remains the same as in the modal case. In the 2D nodal case, the test function φ_h as well as the approximate solution U_h are expanded in terms of the tensor-product of 1D functions from the nodal basis set. In the (ξ,η) coordinate system the test function is chosen to be $h_\ell(\xi)\, h_m(\eta)$, a tensor-product of Lagrange–Legendre polynomials (9.21) with roots at GLL quadrature points; $h_\ell(\xi)\, h_m(\eta)$ belongs to $\mathbb{P}_N$ in (9.32). Thus the approximate solution $U_h(\xi,\eta,t)$ can be expanded as

$$U_h(\xi,\eta,t) = \sum_{\ell=0}^{N} \sum_{m=0}^{N} U_{\ell m}(t)\, h_\ell(\xi)\, h_m(\eta), \quad \text{for} \quad -1 \le \xi,\eta \le 1, \tag{9.44}$$

where $U_{\ell m}(t)$ are the grid-point values (*dofs*) of the approximate solution at the 2D GLL points. The weak formulation (9.33) is simplified by mapping the elements onto the reference element Q, and the procedure is quite analogous to the modal case. The final approximation of (9.30) for an element Ω_e takes the form

$$\frac{d}{dt} U_{\ell m}(t) = \frac{4}{\Delta x_i\, \Delta y_j\, w_\ell w_m} \left[I_G + I_{F_1} + I_{F_2} + I_S \right], \tag{9.45}$$

where w_ℓ and w_m are the weights associated with the GLL quadrature rule and I_G is the surface integral corresponding to the gradient term. I_{F_1} and I_{F_2} are the line integrals along the η- and ξ-directions, respectively, and they are grouped according to (9.36). The simplification (9.45) is possible because the mass matrix associated with discretization is diagonal and easily invertible.

The explicit forms of these integrals are quite similar to those for the modal case (9.40)–(9.43), however, we take an additional step and discretize them using the GLL quadrature rule. The surface (2D) integrals are approximated by a tensor-product of 1D integrals based on the Nth-order GLL quadrature rule. Thus on Q there are $(N+1)^2$ GLL quadrature points with coordinates (ξ_l, η_n); $l, n \in \{0, 1, \ldots, N\}$. In this particular case we have the following approximations by using the discrete orthogonality relation (9.22) and the property (9.23).

$$I_G \approx \frac{\Delta y_j}{2}\, w_m \sum_{l=0}^{N} F_{1,lm}(t)\, h'_\ell(\xi_l) w_l + \frac{\Delta x_i}{2}\, w_\ell \sum_{n=0}^{N} F_{2,\ell n}(t)\, h'_m(\xi_n) w_n, \tag{9.46}$$

$$I_{F_1} \approx -\frac{\Delta y_j}{2} w_m \left[\hat{F}_1(U(1,\eta_m,t))\,\delta_{\ell N} - \hat{F}_1(U(-1,\eta_m,t))\,\delta_{\ell 0}\right], \tag{9.47}$$

$$I_{F_2} \approx -\frac{\Delta x_i}{2} w_\ell \left[\hat{F}_2(U(\xi_\ell,1,t))\,\delta_{Nm} - \hat{F}_2(U(\xi_\ell,-1,t))\,\delta_{0m}\right], \text{ and} \tag{9.48}$$

$$I_S \approx \frac{\Delta x_i \Delta y_j}{4}\, S_{\ell m}(t)\, w_\ell\, w_m, \tag{9.49}$$

where h'_ℓ and h'_m are the derivatives of the Lagrange polynomial as defined in (9.27) and $\delta_{\ell m}$ is the Kronecker delta function defined in (9.22).

9.3.1.3 Approximating the Integrals

The integrals appearing in the ODEs (9.39) and (9.45) are surface integrals for the internal points and line integrals for the boundaries. Approximation of these integrals has a major role in maintaining the accuracy and computational efficiency of the 2D space discretization. As we saw in the 1D case, Gaussian quadrature rules are the most accurate and efficient means for evaluating integrals. Quadrature formulas such as the Gauss–Legendre (GL) or GLL are widely used for this purpose.

The GL quadrature rule employing $N+1$ quadrature points is *exact* for polynomials of degree $2N+1$ while the GLL quadrature rule with the same number of quadrature points is exact for polynomials of degree $2N-1$. If the integrand is a polynomial of degree $2N$, as in the case of flux integrals, then the integration resulting from the GLL quadrature is *inexact*. In the analysis by Cockburn et al. (1990), it is shown that, for a $(N+1)$-th order DG scheme using polynomials of degree N, the quadrature rule used for the surface (internal) integrals should be exact for polynomials of degree $2N$ and the quadrature rule used for boundary flux integrals should be exact for polynomials of degree $2N+1$. In a strict sense, this indicates that there is no single set of $N+1$ quadrature points that can be used to evaluate all the integrals to the required accuracy (Atkins and Shu 1996). In order to meet the requirements for the exact internal integration and consistent boundary (flux) integration, these integrals are usually treated with different orders of quadrature formulas.

Utilizing the same type of high-order quadrature rule for both internal and boundary integrals is certainly an option. This is very convenient for practical applications and leads to computationally efficient code development. Nevertheless, it is reported that, for some applications, *over-integration* resulting from keeping the boundary and flux integrals of the same order may lead to instabilities (Lomtev et al. 2000). On the other hand, computational domains with complex geometries that consist of strong curvature or curved boundaries may require more quadrature points than simple Cartesian cases; this is necessary to maintain a specific order of accuracy in the discretization. In other words, the choice of a particular quadrature rule is application dependent, and is also based on the practical consideration of computational efficiency and ease of implementation.

We now review the GL and GLL quadrature rules for the integrals. A tensor-product of 1D quadratures is usually employed to efficiently evaluate the 2D integrals (Deville et al. 2002). Figure 9.10a is a GLL grid with 4×4 quadrature points. Figures 9.10b and c are the GL grids with 3×3 quadrature points associated with the 2D GLL and GL quadrature rules, respectively; the internal (solution) points are marked as filled circles. The filled-squares along the boundaries in Fig. 9.10b and c indicate flux points which are interpolated from the solution. Technically both of the quadratures are exact for polynomials of degree up to $k = 5$, and sufficient for a third-order or P^2 DG method. The GLL grid has more points (*dofs*) than the GL case, but the internal integral is still inexact for a P^3 method.

The GLL quadrature must employ more points than the GL quadrature to guarantee the same order of accuracy. However, the GLL grid has some inherent computational advantages. The GLL quadrature points include points along the boundary lines and corners of the square domain $[-1, 1]^2$ – computing the flux

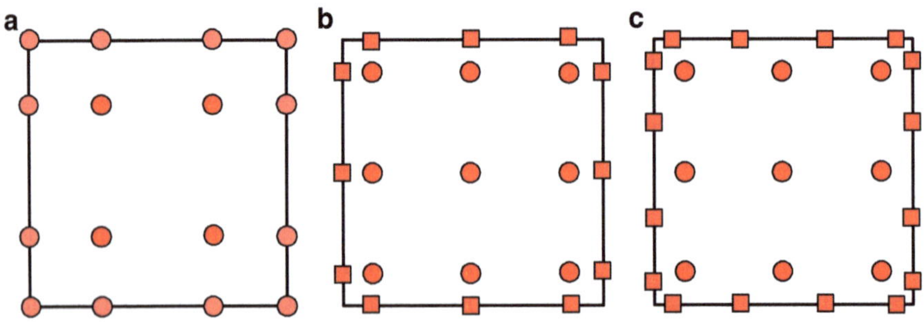

Fig. 9.10 Different types of 2D grid configurations based on Gauss–Legendre–Lobatto (GLL) and Gauss–Legendre (GL) quadrature rules on a square domain $[-1, 1]^2$. The solution points are marked by *filled circles* and flux points along the boundaries are marked by *filled squares*. (**a**) GLL grid with 4×4 quadrature points where the flux points on the boundary coincide with the solution points. (**b**) GL grid with 3×3 points for internal integrals and three flux points on each boundary. (**c**) Same as in case (**b**) but with four flux points on each boundary

integrals (I_{F_1} and I_{F_2} in (9.39)) along the boundaries is trivial in this case, because the solution and flux points coincide at the quadrature points. This avoids the interpolations required for the flux evaluation, which is a significant computational savings. However, a caveat for this GLL grid configuration is that the boundary flux integral reduces to the same order of accuracy as the internal integral and leads to inexact integration. This may be an issue when the degree of the polynomial is low ($k \leq 3$) because losing an order of accuracy is not affordable, but for higher values of k the loss of an order of accuracy is often outweighed by the computational efficiency and ease of implementation (Nair 2009). In practice, this type of GLL grid is used for many high-order nodal DG implementations (Hesthaven and Warburton 2008).

The GL grid as shown in Fig. 9.10b is exact for the DG P^2 scheme but the boundary flux integrals have the same order of accuracy as the internal integrals. In Fig. 9.10c, the order of accuracy of the flux integrals exceeds that of the internal integral as per the theoretical requirement pointed out by Cockburn et al. (1990). In order to compute the fluxes along the boundaries, interpolations are required to transfer the solution to the boundary quadrature points – the basis functions may be used for the accurate interpolation of solution (9.37). This will, of course, increase the computational expense. As previously noted, the GL quadrature rule does not use the end points ± 1 in $[-1, 1]$, which means that in 2D the corner points are excluded. For rectangular domains, the problematic corner singularities may be avoided by the GL grids. So the GL quadrature may be beneficial for domains with isolated singularities such as the latitude-longitude sphere. An interesting discussion about the choice of quadrature rules can be found in a recent paper by Kopriva and Gassner (2010). In the following section we consider several examples with both GL and GLL grids.

9.3.2 Computational Examples: Advection Tests

Two standard tests for advection problems are the solid-body rotation test and deformational flow test. We examine these non-divergent test cases individually.

9.3.2.1 Solid-Body Rotation Test

To test the DG schemes discussed above we first consider a solid-body rotation problem with a smooth function on a square domain. The domain D in (9.30) is chosen to be $[-\pi, \pi]^2$ with periodic boundary conditions and the initial condition is the Gaussian hill $U(x, y, t = 0) = \exp[-5((x - x_c)^2 + (y - y_c)^2))]$ centered at (x_c, y_c). The velocity is prescribed as $(u, v) = (-\pi y, \pi x)$ and the flux function is $\mathbf{F}(U) = (uU, vU)$. The Gaussian hill is placed at the center of the domain ($x_c = 0$,

$y_c = 0$) for the convergence study so that U is continuous at the (periodic) boundaries.

The tests are conducted with both modal and nodal versions of the DG discretization and for different spatial resolution. We vary both the total number of elements ($N_{elm} \in \{20^2, 40^2, 80^2, 160^2\}$) and the polynomial degree ($k \in \{1, 2, 3, 4\}$). The normalized standard l_2 error is computed after one complete rotation, and Fig. 9.11a shows the results with the modal version employing GLL quadrature (the nodal version gives visibly indistinguishable results). Two types of errors, h-error and p-error, are used for the convergence tests of element-based high-order Galerkin methods such as DG. The h-error measures the error computed by varying number of elements and keeping the polynomial degree (k) constant, while the p-error measures the error when the polynomial degree is varied but the number of elements is kept fixed. For a given N_{elm} the p-error is reduced as the polynomial degree increases, in Fig. 9.11a it is shown as black dots aligned in the vertical direction. The measures of the p-error vary more rapidly (at an exponential rate) than that of the h-error. The exponential (spectral) convergence is also reported for similar tests in Levy et al. (2007).

Figure 9.11b shows the *strong scaling* results on a parallel computer architecture, a measure of parallel efficiency when the problem size is held constant. Ideally, the total work would be split evenly among processors so that doubling the number of processors would halve the runtime. This is measured by 'speed-up,' the ratio of the runtime on one processor to the runtime on a given number of processors. In this sense Fig. 9.11b shows almost perfect scaling for the nodal DG scheme (run with $N_{elm} = 80^2$ elements, each with 6×6 GLL nodes). This simulation consisted of 40,000 time steps on a 1,024 dual-node BlueGene/L cluster. Spectral convergence (for smooth problems) and excellent scaling are two remarkable properties of DG algorithms.

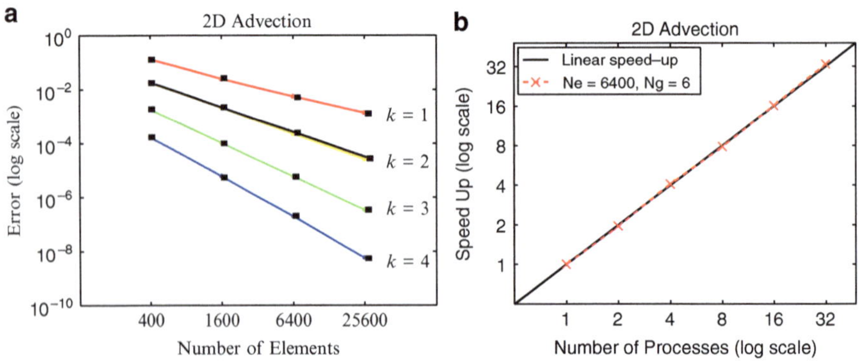

Fig. 9.11 (**a**) Convergence results (l_2 error) for the solid-body rotation test at different resolutions and varying polynomial degree (k). (**b**) The strong scaling results as measured with a resolution of $N_{elm} = 80^2$ and each element containing 6×6 GLL points

9.3.2.2 Deformational Flow Test

For the deformational flow test case we consider the test proposed by Smolarkiewicz (1982). This problem is relevant to meteorology because it simulates the effect of closed vortices on warm air parcels. The test describes the advection of a scalar field (i.e., U in (9.30)), which is initially defined to be a cone of height 1 and radius 15 units located at the center of a square domain of side $L = 100$ units. The non-divergent flow field is defined by the stream function,

$$\psi(x, y) = 8 \sin(4\pi\, x/L) \cos(4\pi\, y/L), \quad u = -\frac{\partial \psi}{\partial y}, \ v = \frac{\partial \psi}{\partial x},$$

where u and v are the components of the wind field. Staniforth et al. (1987) provide an analytical solution for this test in terms of elliptic functions and showed that there is a breaking time $T_b = 2637.6$, beyond which the length scale of the exact solution diminishes as a function of time. We examine the DG solutions at time $t = T_b/50$, when the solution exhibits very fine structures of deformation. This test is very challenging because of the severe deformation of the fields and sharp gradients which evolve in time.

The numerical results are presented in Fig. 9.12 on a 40×40 element domain employing the GLL and GL quadratures (grids) as shown in Fig. 9.10, plotted on the native computational grid to avoid interpolation errors. Figure 9.12a shows the results for the nodal DG scheme with a 4×4 GLL grid, where the boundary integrals use the same order GLL quadrature. This choice of quadrature exhibits spurious overshoots and undershoots, and the modal DG scheme with the GLL quadrature produces a similar result. Changing the spatial order of accuracy (up to 7×7 quadrature points) with the GLL nodes does not improve the results, and similar results are reported by Crowell et al. (2009). Figure 9.12b shows the results with a modal version of the DG P^2 method employing 3×3 GL points. With GL grids, the solution is significantly smoother. Again, the nodal version produces similar results.

This indicates that, irrespective of the modal or nodal variant of the DG method, the GL quadrature has some qualitative advantage over the GLL quadrature; especially when the flow field is very complex. The DG schemes employing GL quadrature are more robust than those with the GLL quadrature. On the other hand, for a fixed order of accuracy, we noticed that the DG/GLL combination has a more lenient CFL stability restriction than the DG/GL combination. This is mainly due to the distribution of the internal quadrature points in the reference element (see Fig. 9.10). In the case of the GL quadrature points, the shortest distance between the internal points and the boundary is smaller than that of the GLL points, leading to relatively smaller grid spacing (Δx). In other words, the CFL estimate discussed in Sect. 9.2.5, $1/(2N + 1)$ for the DG P^N method (Cockburn and Shu 2001), appears to be an overestimate when the DG/GLL combination is used.

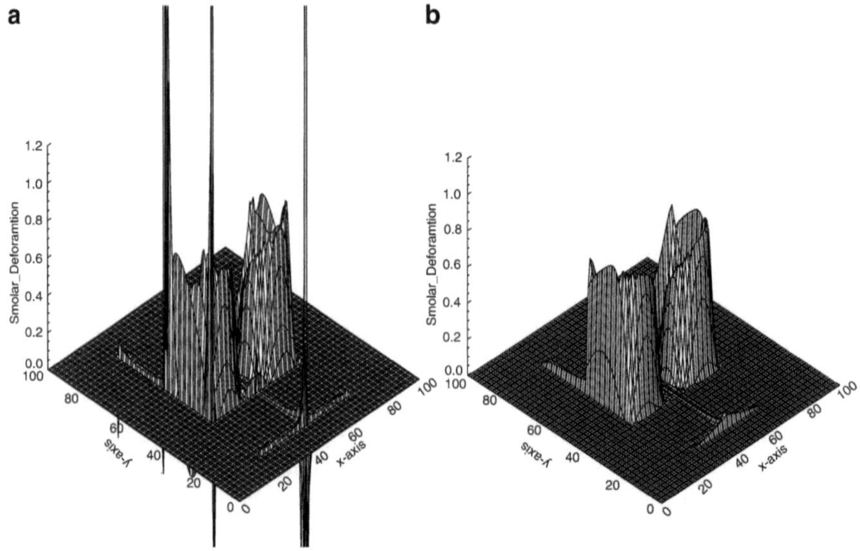

Fig. 9.12 Numerical solutions for the deformational flow test at time $t = T_b/50$ with the DG advection scheme on a 2D Cartesian domain with 40×40 elements. (**a**) Solution with the nodal DG scheme employing 4×4 GLL points (as shown in Fig. 9.10a) on each element. The boundary flux integrals are approximated with the same order 1D GLL quadrature rule. (**b**) Solution with the modal DG scheme employing 3×3 GL points (as shown in Fig. 9.10b or c) on each element. The flux integrals are performed with the same order GL quadrature

9.3.2.3 Barotropic Vorticity Equation

We now discuss a general form of (9.30) with a non-zero source term, a simple non-divergent barotropic model based on the classical barotropic vorticity equation (BVE). A barotropic atmosphere is a single-layered fluid; under this assumption there is no vertical component, so the equation to be solved is 2D. The BVE has special importance in meteorology and a historical perspective of the BVE can be found in Lynch (2008). The BVE is useful for modeling the (idealized) evolution of tropical cyclones (DeMaria 1985), and also for the theoretical study of the interactions of vortices in close proximity. Recently, Levy et al. (2009) have developed an element-based Galerkin method for solving the BVE using the DG discretization; we review this model in the present context.

The BVE can be cast in the following form (Levy et al. 2009):

$$\frac{\partial \zeta}{\partial t} + \frac{\partial}{\partial x}(u\zeta) + \frac{\partial}{\partial y}(v\zeta) = -\beta v, \tag{9.50}$$

where u and v are the horizontal components of the wind vector $\mathbf{v}$ such that $\mathbf{v} = (u, v)$, $\zeta = (\nabla \times \mathbf{v}) \cdot \hat{\mathbf{k}}$ is the relative vorticity and $\hat{\mathbf{k}}$ is a unit normal vector in the vertical direction. In (9.50), $\beta = \partial f/\partial y$ is based on the beta-plane approximation (Vallis 2006) where f is the Coriolis parameter. The solution process involves predicting ζ at every time step, however, the (u, v) field also evolves in time and therefore needs to be computed at every new time step. Since the wind field is non-divergent it can be prescribed in terms of the stream function ψ such that $u = -\psi_y$ and $v = \psi_x$, where the suffixes denote partial differentiation. The relation $\zeta = v_x - u_y$ leads to the following Poisson equation for ψ:

$$\nabla^2 \psi = \zeta. \tag{9.51}$$

Usually the initial conditions for (9.50) are prescribed in terms of the tangential velocity, from which the initial values for $\mathbf{v}$ and ζ can be derived. At every time step ζ is predicted and the corresponding stream function at the new time-level is computed by solving the Poisson problem (9.51). This is required because the wind field (u, v) must be available for the new prediction cycle; as mentioned, it can be computed directly from ψ using the relation $(u, v) = (-\psi_y, \psi_x)$.

Thus the solution process for the BVE involves solving the advection equation (9.50) and the Poisson equation (9.51) as a system. The elliptic type equation (9.51) may be solved using the DG method as described in Rivière (2008), the high-order spectral method (Kopriva 2009), or any number of other methods. Since our focus is primarily on hyperbolic problems, we do not consider the solution procedure for (9.51) here, except to say that we adopt a spectral-element based Poisson solver (Levy 2009) for the BVE model.

The initial wind profile for the vortex centered at (x_c, y_c) can be expressed in terms of tangential velocity $V(r)$ where $r = [(x - x_c)^2 + (y - y_c)^2]^{1/2}$ is the radial distance from the center. The wind field and $V(r)$ are given by

$$u = -V(r)(y - y_c)/r, \quad v = V(r)(x - x_c)/r, \quad V(r) = \frac{2V_m r \, \exp[-a(r/r_m)^b]}{r_m[1 + (r/r_m)^2]}. \tag{9.52}$$

The initial relative vorticity can be derived as

$$\zeta(r) = \begin{cases} V'(r) + V(r)/r & \text{if } r \neq 0, \\ 2V'(0) & \text{if } r = 0. \end{cases} \tag{9.53}$$

The physical dimension of the domain D is a 4,000 × 4,000 km square, and D is periodic in both directions. The other parameters used in (9.52) are $V_m = 30$ m/s, $r_m = 80$ km, $a = 10^{-6}$, $b = 6$, the vortex center (x_c, y_c) positioned at $(2000, 2000)$ km, and β is computed at the latitude 20°N.

The formulation for the BVE (9.50) may be considered as a special case of the flux-form transport equation (9.30) with a non-zero source term $S(U)$. Therefore it is clear that the "conservative" variable is $U = \zeta$, the flux function is $\mathbf{F}(U) = (u\zeta, v\zeta)$ and the source is $S(U) = -\beta v$. For the DG discretization of (9.50), we employ the nodal scheme as described in Sect. 9.3.1.2. The computational domain consists of 100×100 elements each with 4×4 GLL points (Fig. 9.10a) so the average horizontal resolution is approximately 13.3 km. A third-order Runge–Kutta scheme (9.29) is used to solve the ODE (9.45) corresponding to (9.50), with a (sub-optimal) time step of $\Delta t = 90$ s.

In the nodal formulation the relative vorticity ζ and stream function ψ (from (9.51)) are approximated at the GLL quadrature points (ξ_l, η_n) using the summation (9.44). To find the non-divergent wind at any time-level from the stream function fields at the GLL points the following collocation differentiation can be employed,

$$u(\xi,\eta) = -\psi_\eta \approx -\sum_{\ell=0}^{N}\sum_{m=0}^{N} \psi_{\ell m}\, h_\ell(\xi)\, h'_m(\eta),$$

$$v(\xi,\eta) = \psi_\xi \approx \sum_{\ell=0}^{N}\sum_{m=0}^{N} \psi_{\ell m}\, h'_\ell(\xi)\, h_m(\eta).$$

The numerical results are shown in Fig. 9.13. The leftmost panel shows the initial relative vorticity fields, and simulated results after 24 and 72 h are shown in the central and right panels, respectively. As expected, the center of cyclonic vortex is well resolved and the cyclonic motion has drifted in the northwestward direction (DeMaria 1985). Realistic hurricane simulation needs high-resolution complex 3D models capable of fast simulations. The DG methods are well-suited to address this problem because DG algorithms are known for their high parallel efficiency and adaptive mesh refinement capabilities.

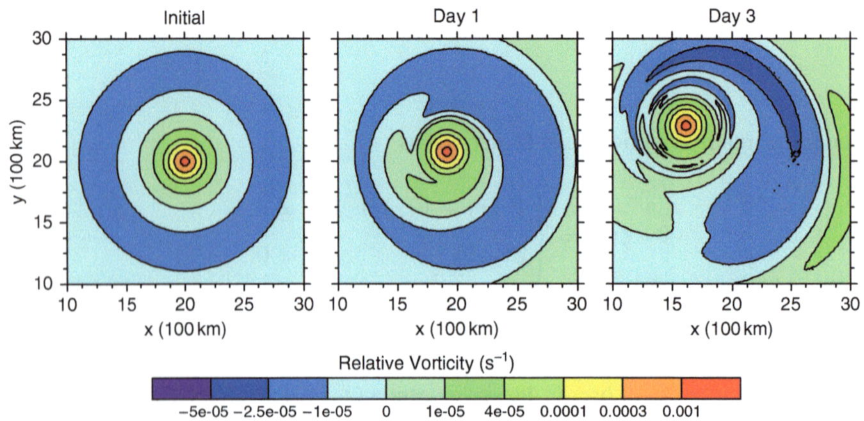

Fig. 9.13 Contours of the vorticity field (ζ) in the tropical cyclone simulation, shown after 1 and 3 days. The *left panel* shows the initial fields; the simulated results after 1 and 3 days are shown in the central and the *right panels*. Calculations are done on a square domain consisting of 100×100 elements each with 4×4 GLL points

9.4 Limiters for DG Methods

High-order numerical schemes will produce spurious oscillations in the vicinity of discontinuities or shocks and near under-resolved solution gradients. The unphysical oscillations not only pollute the solution but may lead to numerical instabilities. Preservation of physically realizable properties of the solution such as monotonicity (shape-preservation) or the less restrictive positivity is of great importance in atmospheric transport modeling (Chap. 8). For instance, the mixing ratio (e.g., relative humidity) or density simulated by an atmospheric model should always preserve its positive sign (positive-definite). Even oscillations with small amplitudes can create negative density which in turn produce physically unacceptable negative mass – this might arise even if a minute negative density is multiplied by the volume (or integrated over a region). The process of controlling or completely eliminating the spurious oscillations in the numerical solution is often referred to as limiting. A limiter also provides nonlinear stability to the solution.

The Godunov theorem (Godunov 1959) asserts that the "monotone linear schemes are at most first-order accurate." For high-order methods this implies that designing a monotone scheme is a daunting task because the coexistence of monotonicity and the high-order nature of the solution is difficult if not impossible. The monotonic limiting is a non-linear process that removes the oscillations from the solution at regions (points) where monotonicity is violated, and when activated the limiter reduces the oscillatory (high-order) solution to first-order. It is required that a limiter does not violate the mass conservation property (i.e., preservation of the cell-average) of the underlying conservative numerical scheme and, to the greatest extent possible, it should retain the high-order accuracy of the solution. Therefore, a limiter should be applied to the high-order scheme in a surgical manner and it should not be activated in smooth regions of the solution. Thus it is very important to have a criterion for limiting that guides *when* and *where* to limit the solution.

Another potential venue for controlling numerical noise due to under-resolved solution gradients is the application of so-called h-p adaptivity. Here h stands for number of elements in the domain and p is the polynomial order within in the element (Karniadakis and Sherwin 2005). Since shocks are not really present in atmospheric model, the requirement is to prevent the generation of under-resolved gradients on the grid. The problem here is optimize the h-p *dofs* to the local structure of the solution. For example, high-order elements where high-gradients are developing can be divided into two elements of order $p/2$ to prevent the growth of oscillations. Ultimately one could end-up with p first order elements that are guaranteed to preserve the extrema of the solution. This approach may be more intensive on software engineering and grid refinement based on error estimators, but could be an alternative to the brute force approach of slope limiters. The DG methods are amenable to adaptive mesh refinement (AMR) strategy based on h-p adaptivity. Development of models based on AMR is an active area of research in geosciences (St-Cyr and Neckels 2009; Kubatko et al. 2009).

The second-order finite-volume (FV) schemes can successfully incorporate limiters such as the slope limiters (van Leer 1974) or flux limiters (Boris and Book 1973). This is done either by designing a scheme which inherently prohibits oscillatory solution (Smolarkiewicz 1984) or by applying the limiter in the reconstruction or the post-processing stage. As the order of the numerical scheme increases the limiting procedure becomes more complex and computationally expensive. A class of high-order finite-volume schemes known as essentially non-oscillatory (ENO) developed by Harten et al. (1987) and its advanced variant weighted essentially non-oscillatory (WENO) by Liu et al. (1994) can successfully control spurious oscillations in the solution. As the name suggests the ENO or WENO solutions are not strictly monotonic. The solution may still have oscillations of small amplitude but they do not grow with time. These schemes use adaptive stencils in the reconstruction procedure which are based on local smoothness of the numerical solution, and automatically achieve high-order accuracy and non-oscillatory properties near discontinuities.

Since the DG method has a strong FV-like connection, it may be technically possible to extend the limiters developed for FV methods to at least low-order DG methods. However, for DG schemes the direct application of a FV-based limiter such as the flux limiter is not trivial because the *dof* evolved in time per element (cell) is higher than that of the FV method. Limiting high-order DG methods on general meshes is still an open question. Here we consider the basic slope limiter (Cockburn and Shu 1989) and the WENO-based limiting proposed by Qui and Shu (2005b) for relatively low-order DG methods.

9.4.1 The 1D Limiters for DG Methods

The basic limiter developed for the DG scheme (Cockburn and Shu 1989) relies on the MUSCL (Monotonic Upstream Centered Schemes for Conservation Laws) slope limiting technique (van Leer 1977). The MUSCL approach employs the piecewise linear reconstruction for the subgrid-cell distributions resulting in a second-order accurate scheme. The reconstruction process in this case is constrained to be free from spurious oscillations (monotonic) by applying the *minmod* limiter. To understand how the minmod limiter works, we consider the piecewise linear reconstruction for the 1D grid used in Sect. 9.2.3.

9.4.1.1 The Minmod Limiter

Let $U_j(x)$ be the density distribution in a cell of width $\Delta x_j = x_{j+1/2} - x_{j-1/2}$. The piecewise linear representation of $U_j(x)$ can be expressed in terms of the slope $U_{x,j}$ and the cell-averaged density $\overline{U}_j$ (see Fig. 9.14),

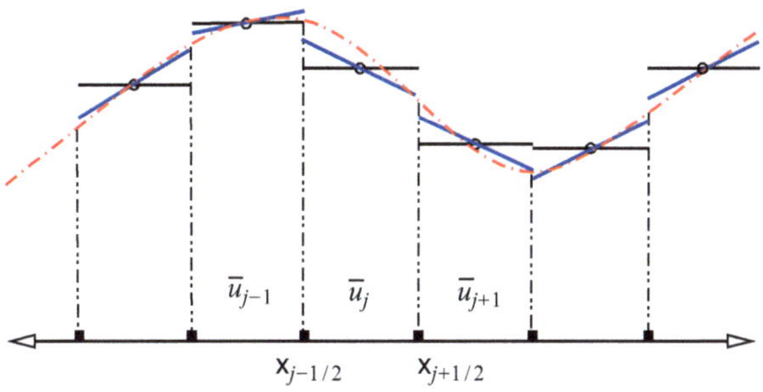

Fig. 9.14 A schematic illustration of the piecewise linear reconstruction. The cell averages $\overline{U}_j$ are shown as *horizontal lines* and the cell boundaries are labeled by $x_{j\pm1/2}$. The *smooth dashed-line* indicates the actual solution $U(x)$ which is approximated by piecewise linear distributions (*broken thick lines*) on each cell

$$U_j(x) = \overline{U}_j + (x - x_j)U_{x,j}, \quad \overline{U}_j = \frac{1}{\Delta x_j}\int_{x_{j-1/2}}^{x_{j+1/2}} U_j(x)dx, \tag{9.54}$$

where $x_j = (x_{j-1/2} + x_{j+1/2})/2$. There are an infinite number of possibilities to choose the value of $U_{x,j}$ in (9.54) without violating mass conservation (preserving $\overline{U}_j$), nonetheless, we choose the limited slope $\tilde{U}_{x,j}$ based on the minmod approach. A minmod function has three arguments. The first argument is the slope of the cell in question and remaining arguments are the slopes of the neighboring cells. If the *left* and the *right* slopes preserve the same sign, then the minmod function returns the minimum of the absolute value of the slopes with the same sign; otherwise, if the signs are opposite, it sets the slope to zero. This can be written as follows:

$$U(x)_j = \overline{U}_j + (x - x_j)\tilde{U}_{x,j}, \quad \tilde{U}_{x,j} \Leftarrow \text{minmod}(U_{x,j}, U_{x,j-1/2}, U_{x,j+1/2}), \tag{9.55}$$

where the arrow indicates the replacement of the slope $U_{x,j}$ by the limited slope $\tilde{U}_{x,j}$; the minmod function is formally defined to be

$$\text{minmod}(a,b,c) = \begin{cases} s\ \min(|a|,|b|,|c|) & \text{if } s = \text{sign}(a) = \text{sign}(b) = \text{sign}(c), \\ 0 & \text{otherwise}. \end{cases} \tag{9.56}$$

The slopes of the neighboring cells (on a non-uniform grid) are given by

$$U_{x,j-1/2} = \frac{\overline{U}_j - \overline{U}_{j-1}}{(\Delta x_j + \Delta x_{j-1})/2}, \quad U_{x,j+1/2} = \frac{\overline{U}_{j+1} - \overline{U}_j}{(\Delta x_{j+1} + \Delta x_j)/2}.$$

This limiter falls under the class of the total variation diminishing (TVD) limiters (Toro 1999). The minmod limiter is strictly non-oscillatory, but unfortunately it clips the legitimate extrema of smooth solutions and degrades high-order accuracy. However, the excessive limiting of the minmod function at smooth regions can be controlled to some extent by modifying (relaxing) the limiting criteria in (9.56). The resulting *modified minmod limiter* has the total variation bounded (TVB) property, which preserves high-order accuracy at smooth extrema at the cost of allowing minor oscillations in the solution. Let 'Minmod' be the modified minmod function which is defined to be

$$\text{Minmod}(a,b,c) = \begin{cases} a & \text{if } |a| \le M_l, \\ \text{minmod}(a,b,c), & \text{otherwise}, \end{cases} \tag{9.57}$$

where M_l is a problem-dependent positive number. This parameter is more or less a *magic* number which works quite well for a few sets of problems (see, Cockburn and Shu 2001). Smaller values of M_l introduce greater local dissipation, but larger values produce oscillations in the solution. Although there are efforts to make M_l problem independent (Ghostine et al. 2009), a generalized approach for various applications particularly in multi-dimensional systems has yet to be established.

9.4.1.2 Generalized Slope Limiter

The modal expansion (9.10) for the approximate solution $U_j(\xi)$ can be rearranged as follows (with the time dependency omitted for brevity):

$$U_j(\xi) = U_j^0 + U_j^1\, \xi + \sum_{k=2}^{N} U_j^k\, P_k(\xi), \tag{9.58}$$

where the expansion coefficients (or moments) U_j^k are defined in (9.11). If the solution $U_j(\xi)$ is approximated as element-wise linear functions, then $U_j(\xi) = U_j^0 + U_j^1\xi$, where the coefficients $U_j^0 = \overline{U}_j$ is the average value and $U_j^1 = U_j'(\xi)$ is the slope. This is simply the P^1 part of the solution (9.58), which is analogous to the piecewise linear reconstruction (9.54). Therefore the limited solution for the P^1 case, in terms of ξ, can be written as

$$U_j(\xi) = \overline{U}_j + \xi\, \tilde{U}_j^1, \quad \tilde{U}_j^1 \Leftarrow \text{minmod}\left(U_j^1, \frac{\overline{U}_j - \overline{U}_{j-1}}{\Delta\xi}, \frac{\overline{U}_{j+1} - \overline{U}_j}{\Delta\xi}\right), \tag{9.59}$$

where $\tilde{U}_j^1$ is the limited slope by the minmod function and $\Delta\xi = 2$. If $\tilde{U}_j^1 \neq U_j^1$ then it indicates that minmod limiter is in action; otherwise, if $\tilde{U}_j^1 = U_j^1$ then the indication is that the element is non-oscillatory and does not need limiting. In other words the minmod function may also be used to detect elements which require limiting.

Note that the *left* and *right* slopes used in the minmod function in (9.59) may be replaced with the less restrictive slopes $2(\overline{U}_j - \overline{U}_{j-1})/\Delta\xi$ and $2(\overline{U}_{j+1} - \overline{U}_j)/\Delta\xi$, respectively, (Cockburn and Shu 2001). This leads to a simplified slope estimate at the element edges in the ξ-coordinate as employed in (9.59).

$$U_j(\xi) = \overline{U}_j + \xi\, \tilde{U}_j^1, \quad \tilde{U}_j^1 \Leftarrow \text{minmod}\,(U_j^1, \overline{U}_j - \overline{U}_{j-1}, \overline{U}_{j+1} - \overline{U}_j). \quad (9.60)$$

In the context of the high-order DG method, Cockburn and Shu (1989) further extended the minmod limiter to the *generalized slope limiter*. This is achieved by selectively applying the limiter (9.60) to the high-order solution (9.58) where the solution is not smooth. The selection procedure (i.e., detecting the elements which require limiting) involves finding the left and right edge values $U_{j\pm1/2} = U_j(\xi = \pm1)$ from (9.58), and checking for oscillation using the minmod function:

$$\tilde{U}^-_{j+1/2} = \overline{U}_j + \text{minmod}\,(U^-_{j+1/2} - \overline{U}_j, \overline{U}_j - \overline{U}_{j-1}, \overline{U}_{j+1} - \overline{U}_j), \quad (9.61)$$
$$\tilde{U}^+_{j-1/2} = \overline{U}_j - \text{minmod}\,(\overline{U}_j - U^+_{j+1/2}, \overline{U}_j - \overline{U}_{j-1}, \overline{U}_{j+1} - \overline{U}_j), \quad (9.62)$$

where $U^-_{j+1/2}$ and $U^+_{j-1/2}$ denote the left and right limits (see Fig. 9.1) of the edge values $U_{j+1/2}$ and $U_{j-1/2}$, respectively.

Now the generalized slope limiter algorithm for a high-order solution (9.58) can be summarized as follows:

- First, compute the limited edge values $\tilde{U}^-_{j+1/2}$ and $\tilde{U}^-_{j+1/2}$ using (9.61) and (9.62).
- If $\tilde{U}^-_{j+1/2} = U^-_{j+1/2}$ and $\tilde{U}^+_{j-1/2} = U^+_{j-1/2}$, then it indicates that there is no spurious oscillation (or no need for limiting) in the element in question, and the solution (9.58) is acceptable as is.
- If $\tilde{U}^-_{j+1/2} \neq U^-_{j+1/2}$ and/or $\tilde{U}^+_{j-1/2} \neq U^+_{j-1/2}$, then it indicates there is oscillation is in the element and the solution should be limited by using (9.60).
- In the limited case only the limited P^1-part of solution is considered, all the high-order coefficients in (9.58) $U_j^k = 0$ for $k \geq 2$.

As discussed above the minmod limiters are dissipative, and may not be suitable for some applications. In such cases, if a solution with oscillations of small amplitude is acceptable, then it is appropriate to use the more relaxed Minmod function (9.57) instead of the regular minmod function.

9.4.1.3 The Moment Limiter

Biswas et al. (1994) generalized the minmod limiter to a *moment limiter* suitable for limiting high-order DG methods. The moment limiter limits the derivative of the solution starting with the highest order, and it is given by

$$\tilde{U}_j^k = \frac{1}{2k-1}\text{minmod}\,((2k-1)U_j^k,\, U_j^{k-1} - U_{j-1}^{k-1},\, U_{j+1}^{k-1} - U_j^{k-1}). \quad (9.63)$$

When $k = 1$, clearly the limiter (9.63) reduces to the minmod limiter in (9.60). The limiter is applied in an adaptive manner starting with the highest-order coefficient (moment) U_j^k. If $\tilde{U}_j^k = U_j^k$ then it indicates limiting is not required; if not, limiting is required and (9.63) is applied to the next lower- level coefficients U_j^{k-1}. The process stops when no modification of the coefficient occurs by applying (9.63); otherwise, the next highest order coefficient is limited. The moment limiter performs better than the generalized slope limiter at least in the 1D case; however, extending the algorithm to multi-dimension (Krivodonova 2007) is computationally expensive.

9.4.1.4 The WENO-Based Limiter

There is a novel class of limiters for DG methods recently introduced by Qui and Shu (2005b) based on the WENO method. A major advantage of this approach is its ability to retain high-order accuracy for the DG scheme while suppressing spurious oscillations. The WENO based limiting strategy for DG methods consists of two crucial steps. These are the identification of so-called *troubled cells* or the cells (elements) that need limiting, followed by a reconstruction step for the non-oscillatory solution in the troubled cells using the neighboring cell-averages. To identify the troubled cells one may use any of the slope limiting techniques described above. If, for example, the slope in a cell changes when using the minmod limiter, then that particular cell is declared a troubled cell and limiting is performed by using the WENO approach. Although the WENO limiter does not adversely affect the order of accuracy of the solution in a smooth cell, a judicious identification of troubled cells is required to avoid unnecessary computations in smooth regions.

The details of the WENO limiter implementation is given in Qui and Shu (2005b), and we do not discuss it herein. A DG P^N method is formally $(N+1)$th order accurate if the quadrature rule is exact for polynomials of degree at least $2N+1$. In order to match the same order of accuracy, a WENO reconstruction should be at least $(2N+1)$th order accurate as well. A WENO-based limiter of this order requires $2N+1$ neighboring elements $\Omega_{j-N}, \ldots, \Omega_{j+N}$ to limit an element Ω_j located at the center of the stencil. Unfortunately, this requirement necessitates a wider computational stencil when $N > 2$, which impedes the local nature (and, therefore, the parallel efficiency) of the combined DG-WENO scheme.

9.4.1.5 Computational Examples with Limiters

We repeat the numerical examples used in Sect. 9.2.6 to demonstrate the effectiveness of the limiters as discussed above. First, the simple linear advection problem $U_t + U_x = 0$ is solved with initial conditions representing two extreme cases, a Gaussian hill (smooth case) and a rectangular wave (non-smooth case). A modal version of the DG discretization is employed with 50 elements, each with 4 GLL quadrature points, in the domain $[-1, 1]$, and 400 time steps are used for a complete revolution. Ideally, the challenge for a limiter is to preserve high-order accuracy in smooth regions of the solution while eliminating spurious oscillations *only* from the non-smooth regions.

Figure 9.15 shows the numerical solutions with the basic minmod limiter (9.55) and the generalized slope limiter combined with the modified Minmod limiter

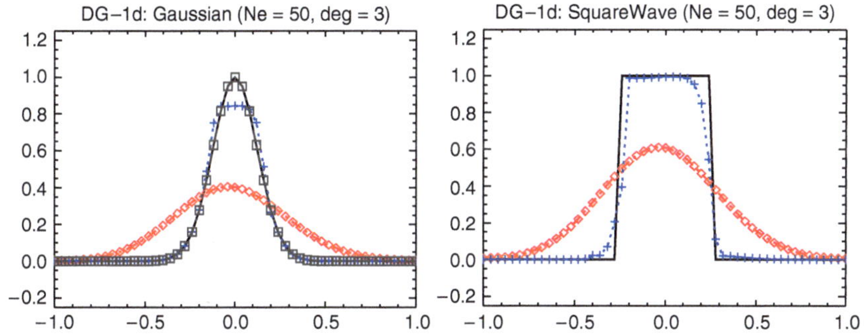

Fig. 9.15 Numerical solution after one revolution with the modal DG scheme combined with various limiters for the linear advection problem (9.2). The *left and right panels* show solutions for the smooth case (*Gaussian hill*) and non-smooth case (*rectangular wave*) as initial conditions, respectively, where the solid line indicates the exact solution. The *diamond points* (*diamond*) show the solution with the basic minmod limiter, and '+' *points* indicate the solution with the generalized slope limiter employing the Minmod function with the parameter $M_l = 0.02$. On the *left panel*, square points show the solution with the generalized slope limiter but the parameter $M_l = 0.06$

(9.57). Only one point per element is plotted for clarity. The Minmod function employs a problem-dependent parameter M_l which controls the limiting operation. The exact solution (initial condition) is shown as solid lines in Fig. 9.15, and the solution with the basic minmod limiter is very diffused in both cases ('$\diamond$' *points*). For the non-smooth case (Fig. 9.15 *right panel*), relatively *better* solutions are obtained with the generalized slope limiter ('+' *points*) for which the parameter value is $M_l = 0.02$; however, for $M_l > 0.02$ the limiter reintroduces oscillations. For the same value $M_l = 0.02$, the generalized slope limiter clips the peak smooth regions of the Gaussian hill as seen in the left panel of Fig. 9.15 ('+' *points*). Nevertheless, when M_l is increased to 0.06 the limiter further relaxes without destroying the legitimate extrema of the Gaussian hill (square points). Although these limiters are simple and easy to implement, a major drawback is that they have a strong dependence on the parameter M_l. Moreover, the basic minmod (P^1) limiter is unacceptably diffusive for high-order DG methods for practical applications (Iskandarani et al. 2005).

Now we consider the same experiment with the moment limiter (9.63) and a third-order WENO- based limiter. For the computational examples considered here we employ a DG P^2 case combined with a WENO limiter employing the GLL quadrature rule with 4 points. Figure 9.16 shows the limited solution with the WENO limiter ('$\diamond$' *points*) and moment limiter ('+' *points*). Both the limiters perform very well for the two extreme cases. However, the WENO based limiter is very robust and performs slightly better than the moment limiter in terms of the symmetry of the solution (shape preservation). The WENO limiter unfortunately comes with a higher computational cost because for the third-order (P^2) case a 5-element wide stencil is required for the reconstructions.

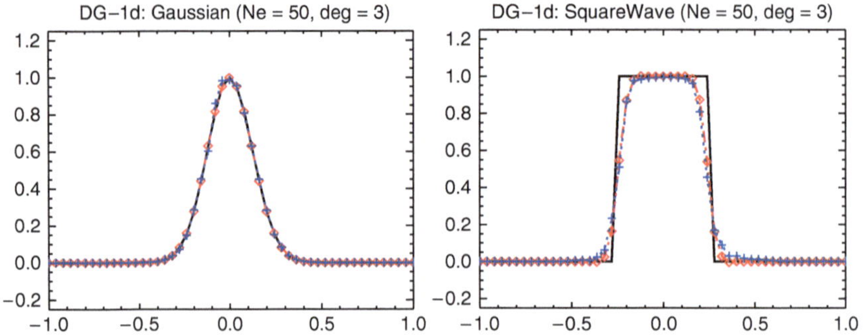

Fig. 9.16 Same as in Fig. 9.15 but with the third-order WENO limiter (*'diamond' points*) and the moment limiter ('+' *points*). The computational domain $[-1, 1]$ consists of 50 elements each with 4 GLL quadrature points

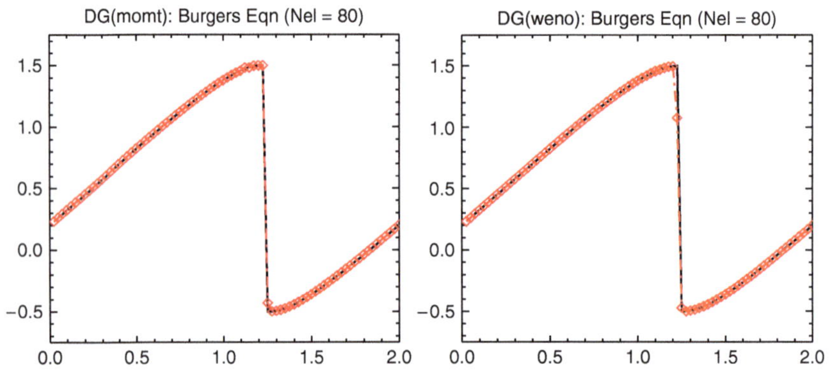

Fig. 9.17 Limited numerical solution for the inviscid Burgers equation at time $t = 3/(2\pi)$ with the modal DG scheme. The *solid-line* indicates the exact solution and *'diamond' points* show the limited DG solution. The *left panel* shows solution by DG scheme combined with the moment limiter and *right panel* shows DG solutions combined with the WENO limiter

The moment limiter (9.63) and the third-order WENO limiter are applied to the P^2 DG scheme for solving the inviscid Burgers equation $U_t + (U^2/2)_x = 0$, with the initial condition $U_0(x) = 1/2 + \sin(\pi x)$. As mentioned in Sect. 9.2.6, this is a simple non-linear case where a shock wave develops during the integration, but the analytic solution is known at any time. The computational domain $[0, 2]$ consists of 80 elements, each containing 4 GLL points. The limited numerical solution at time $t = 3/(2\pi)$ (1,000 time steps) is shown in Fig. 9.17, where the left and right panels show solutions with the moment and WENO limiters, respectively. For clarity, only one point per element is sampled for displaying the numerical results. Both limiters successfully eliminate spurious oscillations near the shock (as seen in Fig. 9.7), and the computed solutions are very similar to the reference solutions. Note that in Fig. 9.7, for which no limiting is employed, shocks develop during the integration and oscillations appear at $t = 9/(8\pi)$ (750 time steps). Eventually the growing spurious oscillations contaminate the numerical solution in this case.

9.4.2 2D Limiters for the DG Method

The 1D limiters used for the high-order DG method are quite successful in eliminating spurious oscillations. Unfortunately, extending these limiters to 2D problems is not trivial. In addition to the slopes (derivatives), the high-order derivatives and cross-derivative terms are also subject to limiting, making the limiting process algorithmically complex and computationally expensive. The development of limiting techniques for high-order DG methods is an active area of research, and two promising approaches in this direction are based on the moment limiter (Biswas et al. 1994) and the WENO limiter (Qui and Shu 2005b). Recently, the moment limiter has been rigorously extended to 2D problems with high computational expense (Krivodonova 2007). A major advantage of this limiter is that it only needs information from the nearest neighbors of the element which is to be limited. The 1D WENO limiter can be extended to 2D problems in a tensor-product form as demonstrated in Levy et al. (2007).

However, recently a compact limiter based on the Hermite WENO (or H-WENO) method has been proposed by Qui and Shu (2005a). This new limiter has been successfully implemented in applications involving system of conservation laws (Balsara et al. 2007; Luo et al. 2007). The H-WENO limiter not only exploits the cell-averages but also the readily available derivative information (high-order moments) from the nearest neighboring cells. This enables the WENO reconstruction process to rely on narrow stencils, and as a result the limiter is computationally attractive. However, for the 2D case we only consider the moment limiter combined with a positivity-preserving slope limiter.

9.4.2.1 A Limiter for the DG P^2 Method

We consider a third-order (P^2) modal DG scheme with the expansion (9.37) employing the basis set $\mathscr{B} = \{1, \xi, \eta, \xi\eta, (3\xi^2 - 1)/2, (3\eta^2 - 1)/2\}$. The approximate solution $U_{ij}(\xi, \eta)$ corresponding to element Ω_{ij} is then given as

$$U_{ij}(\xi,\eta) = U_{ij}^{0,0} + U_{ij}^{1,0}\,\xi + U_{ij}^{0,1}\,\eta + U_{ij}^{1,1}\,\xi\,\eta + U_{ij}^{2,0}\,(3\xi^2 - 1)/2 + U_{ij}^{0,2}\,(3\eta^2 - 1)/2, \tag{9.64}$$

where the coefficients $U_{ij}^{\ell,m}$ correspond to the moments (9.38), and $U_{ij}^{0,0}$ is the average value over Ω_{ij}. In the tensor-product expansion (9.37) for the P^2 case, the basis set employs additional basis functions $P_2(\xi)P_1(\eta)$, $P_1(\xi)P_2(\eta)$ and $P_2(\xi)P_2(\eta)$ in $\mathscr{B}$. However, for the sake of simplicity we exclude additional basis functions in (9.64).

The moment limiter (9.63) introduced for the 1D case can be extended for the 2D case (Biswas et al. 1994; Krivodonova 2007). We denote the limited coefficients in (9.64) as $\tilde{U}_{ij}^{\ell,m}$ which are modified by a generalized version of the minmod limiter (9.63):

$$\tilde{U}_{ij}^{2,0} = \text{minmod}\left[U_{ij}^{2,0},\, \alpha_l(U_{ij}^{1,0} - U_{i-1,j}^{1,0}),\, \alpha_l(U_{i+1,j}^{1,0} - U_{ij}^{1,0})\right],$$

$$\tilde{U}_{ij}^{0,2} = \text{minmod}\left[U_{ij}^{0,2},\, \alpha_l(U_{ij}^{0,1} - U_{i,j-1}^{0,1}),\, \alpha_l(U_{i,j+1}^{0,1} - U_{ij}^{0,1})\right],$$

$$\tilde{U}_{ij}^{1,1} = \text{minmod}\left[U_{ij}^{1,1},\, \alpha_l(U_{ij}^{1,0} - U_{i,j-1}^{1,0}),\, \alpha_l(U_{i,j+1}^{1,0} - U_{ij}^{1,0}),\right.$$
$$\left.\alpha_l(U_{ij}^{0,1} - U_{i-1,j}^{0,1}),\, \alpha_l(U_{i+1,j}^{0,1} - U_{ij}^{0,1})\right], \qquad (9.65)$$

$$\tilde{U}_{ij}^{1,0} = \text{minmod}\left[U_{ij}^{1,0},\, \alpha_l(U_{ij}^{0,0} - U_{i-1,j}^{0,0}),\, \alpha_l(U_{i+1,j}^{0,0} - U_{ij}^{0,0})\right],$$

$$\tilde{U}_{ij}^{0,1} = \text{minmod}\left[U_{ij}^{0,1},\, \alpha_l(U_{ij}^{0,0} - U_{i,j-1}^{0,0}),\, \alpha_l(U_{i,j+1}^{0,0} - U_{ij}^{0,0})\right],$$

where α_l is a parameter in $[0, 1]$ which controls the effect (dissipation) of limiting. Smaller α_l values reduce the effect of limiting. Note that the minmod function used in (9.65) has five arguments, but it acts as the standard minmod function defined in (9.56): it returns the minimum of the absolute value of arguments if all of the arguments have the same sign, otherwise it returns zero. The limiting algorithm for (9.64) can be summarized in the following steps:

- If $\tilde{U}_{ij}^{2,0} = U_{ij}^{2,0}$ and $\tilde{U}_{ij}^{0,2} = U_{ij}^{0,2}$ then there is no need for limiting and the limiting process can be stopped. If not, replace the coefficients $U_{ij}^{2,0}$ and $U_{ij}^{0,2}$ by the corresponding limited coefficients and move to the next step.
- If $\tilde{U}_{ij}^{1,1} = U_{ij}^{1,1}$ then stop limiting, otherwise replace the coefficient $U_{ij}^{1,1}$ by the limited coefficient $\tilde{U}_{ij}^{1,1}$ and move to the last step.
- If $\tilde{U}_{ij}^{1,0} = U_{ij}^{1,0}$ and $\tilde{U}_{ij}^{0,1} = U_{ij}^{0,1}$ then stop limiting. If not, replace the coefficient by the corresponding limited coefficients (i.e., slopes).

Limiting an element Ω_{ij} using the above algorithm requires information from the nearest-neighboring four elements ($\Omega_{i\pm1,j}$, $\Omega_{i,j\pm1}$). The most influential factor controlling the quality of the limited solution is the set of coefficients corresponding to the slopes ($U_{ij}^{0,1}$ and $U_{ij}^{1,0}$) used in the last step. As shown in the 1D case, excessive use of the minmod slope limiter (MUSCL) makes the solution very dissipative. For the moment limiter described above, the limiting hierarchy starts with the highest order coefficients and prevents excessive slope limiting at the last step. We also examine a positivity-preserving limiter (which is less restrictive than the minmod slope limiter) in the following section as an alternative to the slope limiter at the last step of the limiting algorithm.

9.4.2.2 A Positivity-Preserving Slope Limiter

A positivity-preserving (PP) scheme guarantees that the cell-averages which evolve in time will lie in a certain range governed by the initial conditions. Although the solution may contain minor oscillations within this range, it is less dissipative than the rigorous monotonic case. Recently, Zhang and Shu (2010) introduced a uniformly high-order accurate PP scheme for the compressible Euler equations. This scheme avoids creating negative pressure and density in the solution at a reasonable computational cost.

The PP solution is acceptable for practical applications such as atmospheric tracer transport modeling, where positivity preservation is a highly desirable property. The minmod limiter introduced in the MUSCL scheme is strictly monotonic; unfortunately, it is very diffusive too. However, the 2D PP limiter introduced by Suresh (2000) for FV methods is less restrictive than the basic minmod limiter (9.56), and, unlike the modified Minmod limiter (9.57), does not have a problem-dependent parameter.

We adapt this PP limiter as a replacement for the minmod slope limiter used in the above-mentioned limiting process for the coefficients $U_{ij}^{0,1}$ and $U_{ij}^{1,0}$. The 2D PP limiter requires information from the nearest neighbors as well as the corner elements ($\Omega_{i\pm1,j\pm1}$), which create a 3×3 halo region with Ω_{ij} at the center. The average value of the solution on Ω_{ij} is denoted $\overline{U}_{ij}$. To understand how the PP scheme works we use the linear part of (9.64), which can be written as

$$U_{ij}(\xi, \eta) = \overline{U}_{ij} + U_{ij}^{1,0}\, \xi + U_{ij}^{0,1}\, \eta. \tag{9.66}$$

In order to advance in time, the MUSCL scheme requires a reconstruction step (9.66) which involves computing new slopes $U_{ij}^{1,0}$ and $U_{ij}^{0,1}$ from the neighboring cell averages $\overline{U}_{i\pm1,j}$ and $\overline{U}_{i,j\pm1}$. The minmod slope limiter is constrained in such a way that the U_{ij} in (9.66) lie in the range of $\overline{U}_{ij}$ and four independent cell averages $\overline{U}_{i\pm1,j}$ and $\overline{U}_{i,j\pm1}$. The PP limiter essentially extends this range by adding the corner cell-averages. The slopes are then restricted so that the reconstructed values at the corner points also lie within the new ranges (Suresh 2000). The modification of the slopes is done so that they continuously depend on the neighboring data. The following procedure briefly outlines the process of modifying the slopes.

We first construct a 3×3 matrix $\mathbf{D}^{(\epsilon)}$ consisting of the differences between the averages of each element in the halo region and the average value of the element Ω_{ij} that require limiting:

$$\mathbf{D}^{(\epsilon)} = \begin{bmatrix} \overline{U}_{i-1,j+1} - \overline{U}_{ij} & \overline{U}_{i,j+1} - \overline{U}_{ij} & \overline{U}_{i+1,j+1} - \overline{U}_{ij} \\ \overline{U}_{i-1,j} - \overline{U}_{ij} & \epsilon & \overline{U}_{i+1,j} - \overline{U}_{ij} \\ \overline{U}_{i-1,j-1} - \overline{U}_{ij} & \overline{U}_{i,j-1} - \overline{U}_{ij} & \overline{U}_{i+1,j-1} - \overline{U}_{ij} \end{bmatrix}, \tag{9.67}$$

where ϵ is a small positive number ($O(10^{-20})$) in order to make the algorithm robust. The extreme values of $\mathbf{D}^{(\epsilon)}$ are computed as

$$V_{\min} = \min[\mathbf{D}^{(-\epsilon)}], \quad V_{\max} = \max[\mathbf{D}^{(+\epsilon)}].$$

The corner values of the reconstructed solution (9.66) can be effectively bounded within in the interval $[V_{\min}, V_{\max}]$ by restricting the slopes $|U_{ij}^{0,1}| + |U_{ij}^{1,0}|$. In other words we rescale the slopes using the ratio

$$V_s = \frac{\min(|V_{\min}|, |V_{\max}|)}{|U_{ij}^{0,1}| + |U_{ij}^{1,0}|}.$$

The final PP limited slopes are given by

$$\tilde{U}_{ij}^{0,1} = \min(1, V_s)\, U_{ij}^{0,1}, \quad \tilde{U}_{ij}^{1,0} = \min(1, V_s)\, U_{ij}^{1,0} \tag{9.68}$$

The modified slopes in (9.68) may be used as a substitute for the slopes computed by the minmod in the moment limiter.

9.4.2.3 2D Numerical Experiments

To test the limiter we use the 2D advection problem (9.30) for a solid-body rotation test. The test consists of quasi-continuous data and provides an excellent test for the monotonicity of the advecting field (LeVeque 2002; Cheruvu et al. 2007). The velocity field is given by $(u, v) = (y, -x)$ on a square domain D where $x, y \in [-1, 1]$, and the initial condition is defined in a piecewise fashion: $U(x, y, t = 0) = U_0 = 0$ except in a square region where $U_0 = 1$ and a circular region where U_0 is cone-shaped, growing to the maximum value 1 at the center. Formally,

$$U_0(x, y) = \begin{cases} 1 & \text{if } 0.1 < x < 0.6 \quad \text{and} \quad -0.25 < y < 0.25, \\ 1 - \rho_c/0.35 & \text{if } \rho_c = \sqrt{(x + 0.45)^2 + y^2} \ < 0.35, \\ 0 & \text{otherwise.} \end{cases} \tag{9.69}$$

The initial conditions are shown in Fig. 9.18. The domain consists of 80^2 elements and the time step is $\Delta t = 2\pi/1000$ so 1,000 iterations are required for one complete revolution.

Figure 9.19 shows the solution after one revolution with and without the moment limiter. The left panel of Fig. 9.19 shows the DG P^2 numerical solution without any limiting, and the dashed lines indicate oscillations. The right panel shows the limited solution with the moment limiter where the slopes $U_{ij}^{0,1}$ and $U_{ij}^{1,0}$ are limited with a minmod limiter. The solution is very diffusive, the cone height has been reduced to about 60% of its initial height, and the square-block has been smoothly deformed. In Fig. 9.20 the numerical solution with the moment limiter combined with the PP slope limiter is shown. The PP limiter (9.68) is only used as a substitute for the minmod limiter in the last step of the limiting algorithm. There is a significant improvement in the solution as compared to Fig. 9.19: the cone and square-block both preserve their maximum height, although the numerical solution still suffers from slight diffusion.

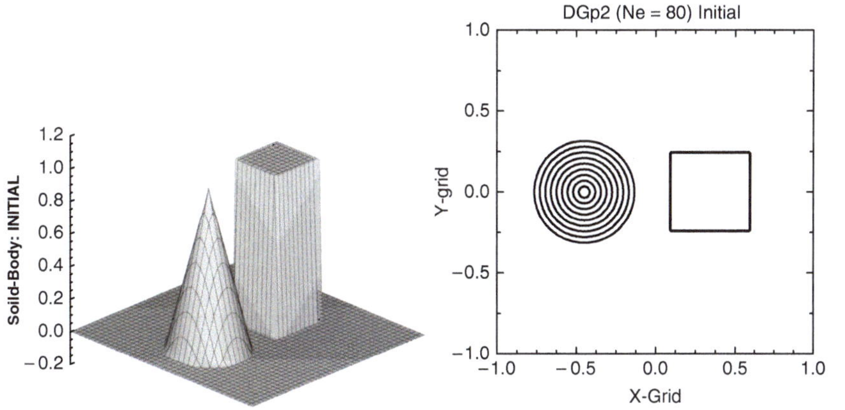

Fig. 9.18 Initial conditions for the solid-body rotation test. The initial scalar field consists of a quasi-smooth cone and a non-smooth square block whose height range form 0 to 1. The domain is $[-1, 1]^2$ with 80 elements in each direction. Only one value per element is sampled in the plots for clarity

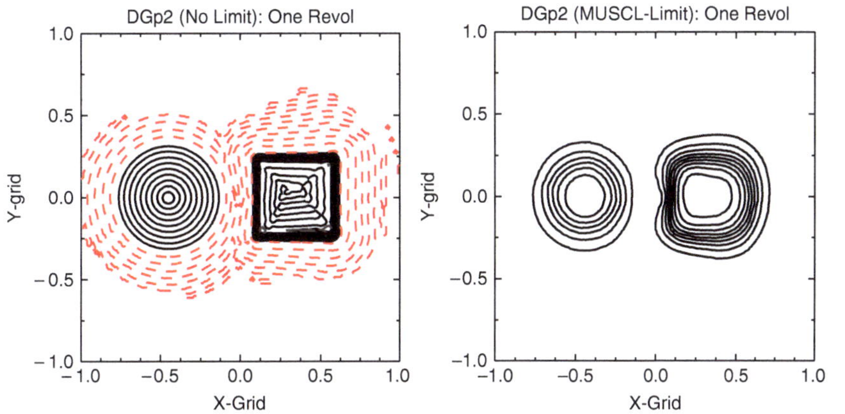

Fig. 9.19 Numerical solution with a third-order DG scheme after one revolution. The *left panel* shows the solution without limiting where the dashed lines correspond to the zero-contours, indicating spurious undershoots. The *right panel* shows limited monotonic solution with a moment limiter, where a MUSCL type minmod limiter is employed for limiting the coefficients $U^{0,1}$ and $U^{1,0}$ corresponding to the slopes

9.5 The DG Methods on the Sphere

There are several geometrical options for discretizing a sphere for global modeling. The choice of a particular spherical grid system is based on various factors including the numerical method being considered (Williamson 2007). For element-based Galerkin approaches such as the spectral element or DG method, the cubed-sphere geometry provides an excellent choice. The cubed-sphere topology introduced by

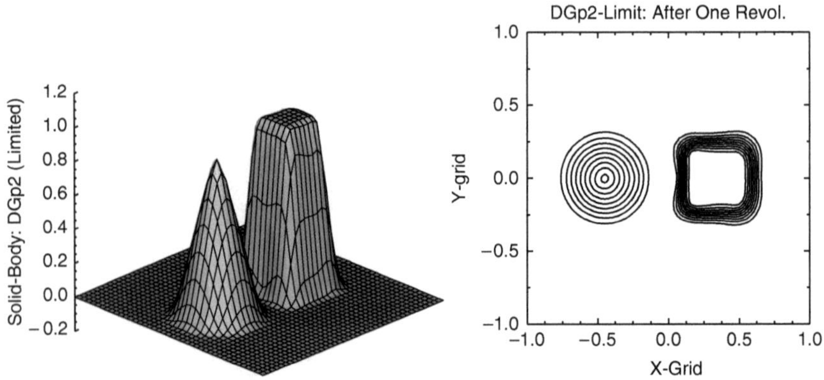

Fig. 9.20 Numerical solution with a third-order DG scheme combined with the moment limiter after one revolution. The coefficients $U^{0,1}$ and $U^{1,0}$ (*corresponding to the slopes*) are limited using the positivity-preserving limiter

Sadourny (1972) consists of a rectangular (quasi-uniform) tiling of the sphere $\mathscr{S}$, representing the planet Earth, which facilitates an efficient implementation of the DG method on the sphere. As an application of the DG method on the sphere, we consider the global shallow water model as reviewed below.

9.5.1 The Shallow Water Model on the Sphere

The shallow water (SW) equations are a system of hyperbolic PDEs. They are widely used for studying horizontal aspects of atmospheric dynamics (Vallis 2006), and also serve as a testbed to evaluate various discretization techniques (Williamson et al. 1992). The flux-form (or conservative form) SW equations on a rotating sphere can be written as

$$\frac{\partial h\mathbf{v}}{\partial t} + \nabla \cdot (\mathbf{v}\, h\mathbf{v}) = -f\hat{\mathbf{k}} \times h\mathbf{v} - gh\nabla(h + h_s) \tag{9.70}$$

$$\frac{\partial h}{\partial t} + \nabla \cdot (h\mathbf{v}) = 0 \tag{9.71}$$

Here, h is the depth of the fluid above the solid surface and is related to the free surface geopotential height (above sea level) $\Phi = g\,(h_s + h)$, where h_s denotes the height of the underlying topography and g is the gravitational acceleration. $\mathbf{v}$ is the horizontal wind vector, f is the Coriolis parameter, and $\hat{\mathbf{k}}$ is the unit vector along the outward radial direction. The 2D divergence ($\nabla\cdot$) and gradient (∇) operators are general and not specific to a particular spherical grid system. Note that $\mathbf{v}\, h\mathbf{v}$ is a *dyadic* (or a second-order tensor) term and can also be written in the tensor-product notation $h\mathbf{v}\otimes\mathbf{v}$. Although (9.70) is widely used in computational fluid dynamics, for meteorological modeling application a simplified version of the momentum equations, the so-called "vector invariant form" is popular and is given by (Sadourny 1972; Arakawa and Lamb 1977),

$$\frac{\partial \mathbf{v}}{\partial t} + \nabla(\Phi + \frac{1}{2}\mathbf{v}\cdot\mathbf{v}) = -(\zeta + f)\hat{\mathbf{k}} \times \mathbf{v}, \tag{9.72}$$

where $\zeta = \hat{\mathbf{k}} \cdot (\nabla \times \mathbf{v})$ is the relative vorticity. The vector invariant form (9.72), as the name suggests, preserves its formal form under coordinate transformations. In a rigorous sense (9.72) is not in momentum conserving form, and when combined with (9.71) it leads to a weakly hyperbolic SW system (Toro 2001). Nevertheless, (9.72) is still in flux-form, although the fluxes being addressed are the energy fluxes $\Phi + \mathbf{v}\cdot\mathbf{v}/2$, rather than the momentum fluxes $h\mathbf{v}$ as used in (9.70).

9.5.2 The Cubed-Sphere Geometry

Here we consider the cubed-sphere geometry employing the equiangular central (gnomonic) projection as described in Nair et al. (2005b). The physical domain $\mathcal{S}$ is partitioned into six identical regions (sub-domains), which are obtained by the central projection of the faces of the inscribed cube onto the surface of $\mathcal{S}$, (see Fig. 9.21a). Each of the local coordinate systems is free of singularities, employs identical metric terms, and creates a non-orthogonal curvilinear coordinate system on $\mathcal{S}$. However, the edges of the six faces are discontinuous.

Because of the non-orthogonal nature of the grid system on $\mathcal{S}$, a tensorial form is convenient for describing the local vectors and the fluid motion in general. In order to be consistent with tensor notations, we choose (x^1, x^2) as the independent variables, which are the central angles of the gnomonic projection (Nair et al. 2005b). Thus the local coordinates for each face are $x^1 = x^1(\lambda, \theta)$, $x^2 = x^2(\lambda, \theta)$ such that $x^1, x^2 \in [-\pi/4, \pi/4]$, where λ and θ are the longitude and latitude, respectively, of a sphere with radius R. The metric tensor, G_{ij}, associated with the transformation is

$$G_{ij} = \frac{R^2}{\rho^4 \cos^2 x^1 \cos^2 x^2} \begin{bmatrix} 1 + \tan^2 x^1 & -\tan x^1 \tan x^2 \\ -\tan x^1 \tan x^2 & 1 + \tan^2 x^2 \end{bmatrix}, \tag{9.73}$$

where $i, j \in \{1, 2\}$ and $\rho^2 = 1 + \tan^2 x^1 + \tan^2 x^2$. The Jacobian of the transformation (the metric or curvature term) is $\sqrt{G} = [\det(G_{ij})]^{1/2}$, which is identical for each face of the cubed-sphere. For a unit sphere the curvature term has a maximum value of 1 at the center of each panel and a minimum of $1/\sqrt{2}$ at the center of the edges (see Fig. 9.21b). Although the cells are uniform on the cube, the quadrilateral cell on the sphere is most deformed at the corners of the cubed-sphere and the ratio between the maximum and minimum grid width for the gnomonic cubed-sphere has an upper bound approximately 1.3 at any resolution (Rančić et al. 1996).

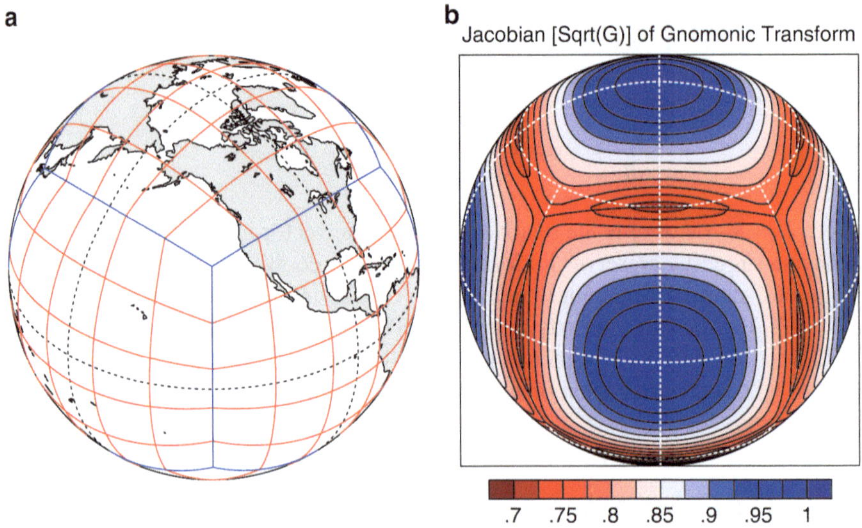

Fig. 9.21 (**a**) A cubed-sphere with 5×5 elements on each face, so 150 elements span the entire surface of the sphere. (**b**) The Jacobian $\sqrt{G}$ (also referred to as the metric or curvature term) associated with the gnomonic transformation from a cube onto a sphere. For a unit sphere $\sqrt{G}$ has a maximum value of 1 at the center of each face, and has a minimum value $1/\sqrt{2}$ at the center of the edges. The cubed-sphere gridlines are great-circle arcs and they are orthogonal only at the center of each panel

9.5.3 The Shallow Water Model on the Cubed-Sphere

On the cubed-sphere the SW equations are treated in tensor form with covariant (u_1, u_2) and contravariant (u^1, u^2) wind vectors. These vectors are related through the matrix equations:

$$\begin{bmatrix} u_1 \\ u_2 \end{bmatrix} = \begin{bmatrix} G_{11} \; G_{12} \\ G_{21} \; G_{22} \end{bmatrix} \begin{bmatrix} u^1 \\ u^2 \end{bmatrix}, \quad \begin{bmatrix} u^1 \\ u^2 \end{bmatrix} = \begin{bmatrix} G^{11} \; G^{12} \\ G^{21} \; G^{22} \end{bmatrix} \begin{bmatrix} u_1 \\ u_2 \end{bmatrix}, \tag{9.74}$$

where $G^{ij} = G_{ij}^{-1}$ and can be computed from (9.73).

The orthogonal components of the spherical wind vector $\mathbf{v}(\lambda, \theta) = (u, v)$ – i.e., the physical zonal and meridional components of the horizontal wind – can be expressed in terms of contravariant vectors (u^1, u^2) as follows:

$$\begin{bmatrix} u \\ v \end{bmatrix} = \mathbf{A} \begin{bmatrix} u^1 \\ u^2 \end{bmatrix}, \; \mathbf{A} = \begin{bmatrix} R\cos\theta \, \partial\lambda/\partial x^1 & R\cos\theta \, \partial\lambda/\partial x^2 \\ R\,\partial\theta/\partial x^1 & R\,\partial\theta/\partial x^2 \end{bmatrix}; \; \mathbf{A}^T\mathbf{A} = G_{ij}. \tag{9.75}$$

The details of the local transformation laws and the transformation matrix $\mathbf{A}$ for each face of the cubed-sphere can be found in Nair et al. (2005b).

The SW equations of a thin layer of fluid in 2D are the horizontal momentum equations and the continuity equation for the height h. The momentum equations are cast in terms of covariant (u_1, u_2) vectors, which leads to a flux-form formulation suitable for methods based on hyperbolic conservation laws (Nair et al. 2005a). Note that this particular formulation preserves the vector invariant form of momentum equations (9.72). Thus the prognostic variables are u_1, u_2 and h, and the shallow water equations on $\mathscr{S}$ can be written in a compact form following the inviscid formulation described in Nair (2009):

$$\frac{\partial}{\partial t}\mathbf{U} + \frac{\partial}{\partial x^1}\mathbf{F}_1(\mathbf{U}) + \frac{\partial}{\partial x^2}\mathbf{F}_2(\mathbf{U}) = \mathbf{S}(\mathbf{U}), \tag{9.76}$$

where the state vector $\mathbf{U}$ and the flux vectors $\mathbf{F}_1, \mathbf{F}_2$ are defined by

$$\mathbf{U} = \left[u_1, u_2, \sqrt{G}h\right]^T, \quad \mathbf{F}_1 = \left[E, 0, \sqrt{G}hu^1\right]^T, \quad \mathbf{F}_2 = \left[0, E, \sqrt{G}hu^2\right]^T,$$

and $E = \Phi + \frac{1}{2}(u_1 u^1 + u_2 u^2)$ is the energy term. The divergence δ and relative vorticity ζ on $\mathscr{S}$ are defined as

$$\delta = \frac{1}{\sqrt{G}}\left[\frac{\partial\sqrt{G}u^1}{\partial x^1} + \frac{\partial\sqrt{G}u^2}{\partial x^2}\right], \quad \zeta = \frac{1}{\sqrt{G}}\left[\frac{\partial u_2}{\partial x^1} - \frac{\partial u_1}{\partial x^2}\right] \tag{9.77}$$

The source term, $\mathbf{S}$, is a function of the relative vorticity ζ, the Coriolis parameter f, and the contravariant wind vector (u^1, u^2), and is defined as

$$\mathbf{S}(\mathbf{U}) = \left[\sqrt{G}u^2(f + \zeta), -\sqrt{G}\,u^1(f + \zeta), 0\right]^T.$$

9.5.4 The Computational Domain

The spherical SW equations can be discretized either in physical space or in the computational (transformed) space. Since the SW equations (9.76) are already in the computational (x^1, x^2) space (due to the central projection), it makes sense to discretize the system in the same space. The computational domain may be considered as the surface of a logical cube $\mathscr{C}$ such that each face of $\mathscr{C}$ is defined in terms of local orthogonal Cartesian coordinates $x^1, x^2 \in [-\pi/4, \pi/4]$, as shown in Fig. 9.22. Thus $\mathscr{C}$ is essentially a union of six non-overlapping sub-domains (faces) and any point on $\mathscr{C}$ can be uniquely represented by the ordered triple (x^1, x^2, ν) where $\nu = 1, \ldots, 6$, is the cube-face or panel index. The projections and the logical orientation of the cube panels are described in Nair et al. (2005b) and Lauritzen et al. (2010).

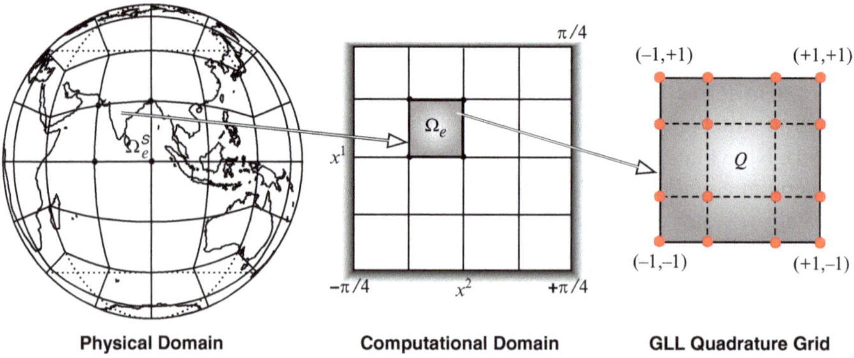

Fig. 9.22 A schematic diagram showing the mapping between each spherical tile (*element*) Ω_e^S of the physical domain (*cubed-sphere*) $\mathscr{S}$ onto a planar element Ω_e on the computational domain $\mathscr{C}$ (*cube*). For a DG discretization each element on the cube is further mapped onto a unique reference element Q, which is defined by the Gauss–Lobatto–Legendre (GLL) quadrature points. The horizontal discretization of the HOMME dynamical cores relies on this grid system

The equiangular central projection results in a uniform element width ($\Delta x^1 = \Delta x^2$) on $\mathscr{C}$, which is an advantage for practical implementation. Figure 9.22 provides a schematic diagram of the mapping between the physical domain $\mathscr{S}$ (*cubed-sphere*) and the computational domain $\mathscr{C}$ (*cube*).

The cubed-sphere has the attractive feature that the domain $\mathscr{S}$ is naturally decomposed into non-overlapping quadrilateral elements (tiles) Ω_e^S. This topology is well-suited for high-order element-based methods such as spectral element or DG methods, and amenable to efficient parallel implementation. Each face of the cubed-sphere has $N_e \times N_e$ elements, thus $N_{elm} = 6\,N_e^2$ elements span the entire spherical domain such that $\mathscr{S} = \cup_{e=1}^{N_{elm}} \Omega_e^S$; in Fig. 9.22 N_e is 4. There exists a one-to-one correspondence between the spherical element Ω_e^S on $\mathscr{S}$ and the planar element Ω_e on $\mathscr{C}$ as depicted in Fig. 9.22. The element-wise continuous mapping allows us to perform integrations on the sphere in a mapped (local) Cartesian geometry rather than on the surface of the sphere. The High-Order Method Modeling Environment (HOMME) developed at NCAR relies on this grid system (Dennis et al. 2005).

9.5.5 The DG Discretization of the SW Equations

The SW model developed in Nair et al. (2005a) is based on a modal DG discretization, however, here we consider the nodal inviscid version of the SW model as implemented in HOMME (Nair 2009). The discretization process for a multi-dimensional system of equations (9.76) is quite similar to the 2D case considered in Sect. 9.3. However, as we discuss in Sect. 9.5.5.1, the flux operations (Riemann solvers) along the cubed-sphere edges are not trivial to implement. For notational simplicity, we consider a generic component of the system (9.76) as follows,

$$\frac{\partial \psi}{\partial t} + \nabla_c \cdot \mathbf{F}(\psi) = S(\psi), \quad \text{in} \quad \mathscr{C} \times (0, T], \tag{9.78}$$

where $\mathbf{F} = (F_1, F_2)$ is the flux function and T is the prescribed time of integration. The Cartesian gradient operator ∇_c on $\mathscr{C}$ is defined to be

$$\nabla_c \equiv \left(\frac{\partial}{\partial x^1}, \frac{\partial}{\partial x^2} \right) \quad \Rightarrow \quad \nabla_c \cdot \mathbf{F} = \frac{\partial F_1}{\partial x^1} + \frac{\partial F_2}{\partial x^2}$$

For example, (9.78) may be considered the continuity equation (or the flux-form transport equation) for the SW system (9.76); in this case $\psi = \sqrt{G}h$, $\mathbf{F} = (\psi u^1, \psi u^2)$ and the source term is $S = 0$. Similarly, the components of the momentum equation in (9.76) can be cast in the Cartesian form (9.78).

Analogous to the 2D case considered earlier, the weak Galerkin form corresponding to (9.78) on any element Ω_e with boundary Γ_e on $\mathscr{C}$ can be written as follows:

$$\begin{aligned} \frac{d}{dt} \int_{\Omega_e} \psi_h \varphi_h \, d\Omega - \int_{\Omega_e} \mathbf{F}(\psi_h) \cdot \nabla_c \varphi_h \, d\Omega + \int_{\Gamma_e} \hat{\mathbf{F}} \cdot \mathbf{n} \varphi_h \, d\Gamma \\ = \int_{\Omega_e} S(\psi_h) \varphi_h \, d\Omega, \end{aligned} \tag{9.79}$$

where ψ_h is the approximate solution and φ_h is a test function in $\mathscr{V}_h$. $\hat{\mathbf{F}}$ is the numerical flux, $\mathbf{n}$ is the outward-facing unit normal vector on the element boundary Γ_e and the element of integration is $d\Omega = dx^1 dx^2$. For the numerical flux we employ the local Lax-Friedrichs flux formula as follows:

$$\hat{\mathbf{F}}(\psi_h) = \frac{1}{2} \left[(\mathbf{F}(\psi_h^-) + \mathbf{F}(\psi_h^+)) - \alpha^i_{\max} (\psi_h^+ - \psi_h^-) \right], \tag{9.80}$$

where $\alpha^i_{\max}$ is the absolute maximum of the eigenvalues of the flux Jacobian; ψ_h^- and ψ_h^+, respectively, are the left and right limits of ψ_h along the boundary Γ_e.

Recall that for each component of the system (9.76) the weak formulation (9.79) is valid, however, $\alpha^i_{\max}$ must be computed for the entire system. Nair et al. (2005a) derived the flux Jacobian for the SW system on the cubed-sphere, which is a 3×3 matrix, and its maximum eigenvalues along the x^1 and x^2-directions are,

$$\alpha_1 = |u^1| + \sqrt{G^{11}\Phi}, \quad \alpha_2 = |u^2| + \sqrt{G^{22}\Phi}. \tag{9.81}$$

These values are nothing but the maximum phase speed of the SW system in the curvilinear coordinate directions. From (9.81) the local maximum values computed from both sides along the element wall (Γ_e), are $\alpha^1_{\max} = \max(\alpha_1^-, \alpha_1^+)$ and $\alpha^2_{\max} = \max(\alpha_2^-, \alpha_2^+)$, as required in (9.80).

9.5.5.1 Flux Exchanges at the Cubed-Sphere Edges

For DG methods, the flux exchanges at the element edges are managed by the numerical flux formulas such as (9.5), and this is the only mechanism by which the adjacent elements *communicate*. Because local coordinates are discontinuous at the cubed-sphere edges, the flux exchange across the edges require special attention. The local transformation of vectors using (9.75) at the cubed-sphere edges can be used for exchanging vector quantities including fluxes. For example, consider a point on the cubed-sphere edge separated by two neighboring faces 'm' and 'n'. The local vector on the point $(u^1, u^2)_m$ belonging to a face m can be transformed into the global spherical components $(u, v)_s$ using (9.75), and then transformed back to the local vector $(u^1, u^2)_n$ of the adjacent edges on the face n.

The flux operations on the cubed-sphere edges also follow a similar procedure. To compute the flux on an edge (or interface) using (9.5), both the left, $\mathbf{F}^-$, and the right, $\mathbf{F}^+$, contributions of $\mathbf{F} = (F_{u^1}, F_{u^2})$ are required. For instance, if $\mathbf{F}^-$ on the panel m is available then the corresponding $\mathbf{F}^+$ belongs to the adjacent panel n, and can be transformed in terms of the local vectors in the panel m by employing the following dual transformation,

$$\begin{bmatrix} F_{u^1} \\ F_{u^2} \end{bmatrix}_m^+ = \mathbf{A}_m^{-1}\mathbf{A}_n \begin{bmatrix} F_{u^1} \\ F_{u^2} \end{bmatrix}_n^+ , \tag{9.82}$$

where the suffixes m, n indicate the adjacent panel indices such that $m, n \in \{1, 2, \ldots, 6\}$. $\mathbf{A}_m$, $\mathbf{A}_n$ are transformation matrices defined in (9.75), and for the sake of computational efficiency the dual transformation matrices $\mathbf{A}_m^{-1}\mathbf{A}_n$ in (9.82) as well as the metric terms can be pre-computed.

9.5.5.2 Numerical Integration of the SW Model

The integral and the differential operators required in the DG discretization (9.79) of the SW system can be approximated on each Ω_e with boundary Γ_e. The element-wise discretization is quite similar to the 2D case considered earlier, therefore, we just outline the procedure in terms of the weak form (9.79) and the SW system (9.76).

Here we adopt the nodal basis set used for the HOMME dynamical core (Nair 2009). In order to take advantage of efficient quadrature rules, new independent variables $\xi^i = \xi^i(x^i), i \in \{1, 2\}$ are introduced such that $\xi^i \in [-1, 1]$. This leads to a mapping of each element $\Omega_e \in \mathscr{C}$ to a unique reference element $Q = [-1, 1] \otimes [-1, 1]$, as illustrated schematically in Fig. 9.22. The nodal basis functions are the Lagrange polynomials $h_\ell(\xi^i)$, with roots at the GLL quadrature points. The nodal basis set is chosen to be a tensor-product of polynomials $h_k(\xi^1)h_\ell(\xi^2)$. Now the approximate solution ψ_h and test function φ_h in $\mathscr{V}_h$ can be expanded in terms of a tensor-product of the Lagrange basis functions, and, in the case of ψ_h, such that

$$\psi_h(\xi^1, \xi^2) = \sum_{k=0}^{N} \sum_{\ell=0}^{N} \psi_h(\xi_k^1, \xi_\ell^2)\, h_k(\xi^1) h_\ell(\xi^2), \tag{9.83}$$

where $\{\xi_\ell^i\}_{\ell=0}^N$ are the GLL quadrature points on the reference element Q. In other words, there are $N_v \times N_v$ GLL points on Q (where $N_v = N + 1$), therefore the total degrees of freedom on $\mathscr{C}$ is $6N_e^2 N_v^2$. The equivalent resolution of the cubed-sphere with respect to the regular latitude-longitude sphere at the equator is approximately $90°/(N_e \times N)$. However, a latitude-longitude spherical grid with the same resolution at the equator will have approximately 30% more grid points. For the sake of computational efficiency we use the same order GLL quadrature rule for the internal integrals in Ω_e and the boundary flux integrals along Γ_e, at the cost of nominal loss of accuracy due to inexact integration (see Sect. 9.3.1.3).

Substitution of the expansion (9.83) for ψ_h and φ_h in the weak formulations and further simplification leads to a system of ODEs in time corresponding to the continuous problem (9.76),

$$\frac{d\mathbf{U}}{dt} = L(\mathbf{U}) \;\text{ in } (0, T], \tag{9.84}$$

where $\mathbf{U}$ are the time dependent nodal gridpoint values for the SW system (9.76). In the present study we use the third-order accurate explicit strong stability-preserving (SSP) Runge–Kutta as discussed in Sect. 9.2.5.

9.5.6 Numerical Experiments

Discussion of the solutions to the SW equations on the cubed-sphere based on the DG method with the Williamson et al. (1992) test suite can be found in Nair et al. (2005a,b) or, with a viscous SW model, in Nair (2009). In this section we consider a new deformational test and the barotropic instability test case proposed by Galewsky et al. (2004).

9.5.6.1 Advection Test

The flux-form advection equation (9.71) on the cubed-sphere can be written as

$$\frac{\partial}{\partial t}\sqrt{G}\phi + \frac{\partial}{\partial x^1}(\sqrt{G}\phi u^1) + \frac{\partial}{\partial x^2}(\sqrt{G}\phi u^2) = 0, \tag{9.85}$$

where ϕ is the scalar field and the advecting wind is given by the contravariant vector field (u^1, u^2). In fact, this is the continuity equation in the SW system (9.76). If we introduce $\psi = \sqrt{G}\phi$ and the fluxes $F_1 = \psi u^1$ and $F_2 = \psi u^2$ then (9.85) can be written in a form analogous to the 2D Cartesian case.

9.5.6.2 Deformational Flow Test

We consider a new deformational flow test introduced in Nair and Lauritzen (2010). For this problem, the initial distributions undergo severe deformation for a prescribed time and then the flow reverses its course, returning the deforming fields to their initial states (the "boomerang effect"). A special feature of this test is that the trajectories of the flow are non-trivial (not along a circle or straight line) and consequently the deformation is severe, making the test very challenging.

This test is prescribed on a unit sphere and quasi-smooth cosine-bell patterns (a C^1 function) are used as the initial scalar fields. Two symmetrically located cosine bells are defined by

$$\phi(\lambda, \theta) = \frac{1}{2}[1 + \cos(\pi r_i / r)] \quad \text{if} \quad r_i < r, \tag{9.86}$$

where $r = 1/2$ is base radius of the bells, $r_i = r_i(\lambda, \theta)$ is the great-circle distance between (λ, θ) and a specified center (λ_i, θ_i) of the cosine bell, which is given by

$$r_i(\lambda, \theta) = \cos^{-1}[\sin\theta_i \sin\theta + \cos\theta_i \cos\theta \cos(\lambda - \lambda_i)]. \tag{9.87}$$

The scalar values are initially set to zero ($\phi(\lambda, \theta) = 0$), and then two cosine bells (cones) are generated using (9.86) at known points $(\lambda_1, \theta_1) = (5\pi/6, 0)$ and $(\lambda_2, \theta_2) = (7\pi/6, 0)$ as the bell centers. The flow field is non-divergent and the time dependent velocity fields $\mathbf{v}(\lambda, \theta, t)$ are prescribed in longitude-latitude coordinates,

$$u(\lambda, \theta, t) = \kappa \sin^2(\lambda) \sin(2\theta) \cos(\pi t / T) \tag{9.88}$$

$$v(\lambda, \theta, t) = \kappa \sin(2\lambda) \cos(\theta) \cos(\pi t / T), \tag{9.89}$$

where the parameter $\kappa = 2$ and the final time of the simulation is $T = 5$ non-dimensional.

The DG transport scheme employs a 4×4 GLL grid with $N_e = 20$. This corresponds to an approximate resolution of 1.5° at the equator. The third-order SSP RK scheme (9.29) is used with a time step $\Delta t = 5/1200$ for the simulations (1,200 time steps are required for the total simulation). Figure 9.23 shows the initial conditions and simulated results for the deformational test with the DG scheme. The cosine bells move away from the initial positions (Fig. 9.23a) and deform into thin spiral shapes at time $t = T/2$ (Fig. 9.23b). The trajectories for the non-divergent flow are complex and the cosine bells pass along the edges and corners, covering the six faces of the cubed-sphere. The DG scheme successfully simulates the deformations and retains the initial position as well as shape of the distribution at the end of the simulation ($t = T$), as shown in Fig. 9.23c. Since the final solution is identical to the initial conditions by design, the global standard errors norms l_1, l_2 and l_∞ (Williamson et al. 1992) can be computed.

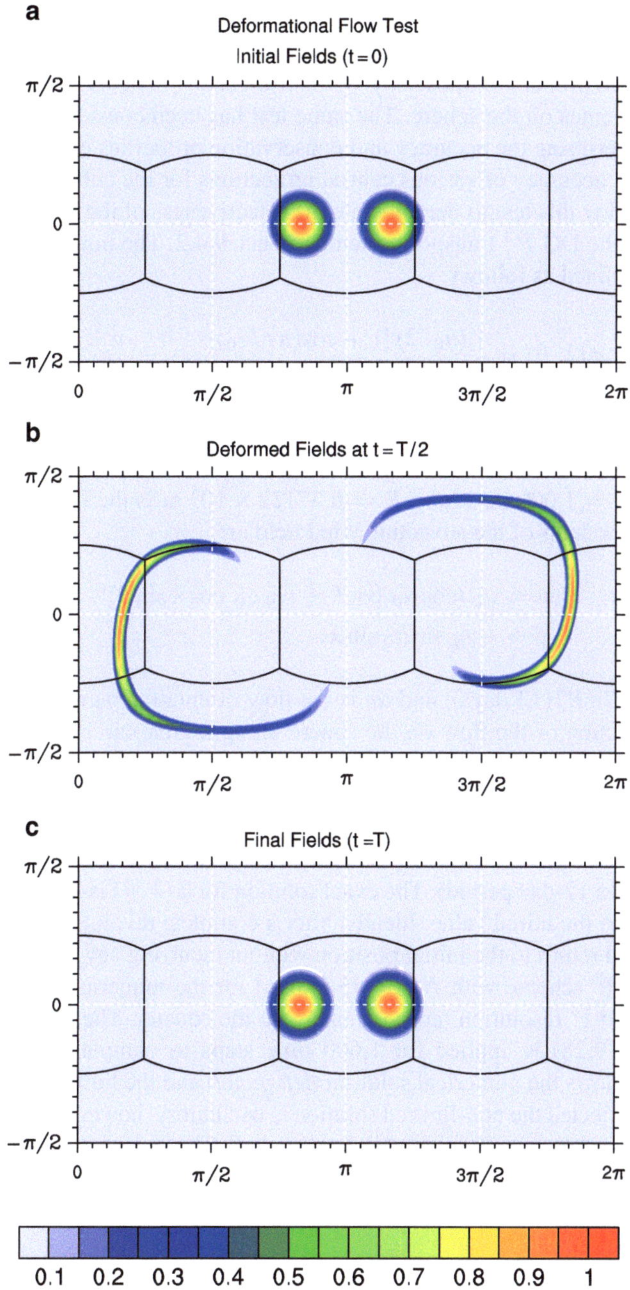

Fig. 9.23 Deformational flow test with the DG transport scheme on the cubed-sphere. The equivalent resolution at the equator (with $N_e = 20$) is approximately 1.5°. (**a**) The initial positions of the scalar field (*cosine bells*) centered at $(\lambda_i, \theta_i) = (5\pi/6, 0)$ and $(7\pi/6, 0)$. (**b**) Deformed scalar fields at half-time ($t = T/2$) of the simulation. (**c**) The scalar fields (numerical solution) return back to the initial positions at the final time ($t = T$)

9.5.6.3 Solid-Body Rotation Test

The cosine-bell problem proposed by Williamson et al. (1992) is widely used to test advection schemes on the sphere. The same test has been considered in Nair et al. (2005b) for verifying the accuracy and conservation properties of the DG schemes as well as the accuracy of various central projections for the cubed-sphere system. Here we employ this test to demonstrate the effectiveness of the monotonic limiter designed for the DG P^2 transport scheme in Sect. 9.4.2. The initial scalar field is a cosine bell defined as follows,

$$\phi(\lambda,\theta) = \begin{cases} (h_0/2)\,[1+\cos(\pi r/r_0)] & \text{if} \quad r < r_0 \\ 0 & \text{if} \quad r \geq r_0, \end{cases} \tag{9.90}$$

where r is the great-circle distance between (λ,θ) and the bell center $(3\pi/2, 0)$ as given in (9.87). The cosine-bell radius is $r_0 = R/3$ and the maximum height of the bell is $h_0 = 1{,}000$ m, where $R = 6.37122 \times 10^6$ m is the Earth's radius. The velocity components of the advecting wind field are

$$\begin{aligned} u &= u_0\,(\cos\alpha_0\, \cos\theta + \sin\alpha_0\, \cos\lambda\, \sin\theta), \\ v &= -u_0\, \sin\alpha_0\, \sin\lambda, \end{aligned}$$

where $u_0 = 2\pi R/(12 \text{ days})$, and α_0 is the flow orientation parameter which controls the direction of the flow on the sphere along a great-circle trajectory. When the value of α_0 is equal to zero or $\pi/2$, the flow direction is along the equator or in the north-south (meridional) direction, respectively. For the cubed-sphere, flow along the north-east direction ($\alpha_0 = \pi/4$) is more challenging because the cosine-bell pattern passes over four vertices and two edges of the cube during a complete revolution (in a 12-day period). The exact solution for $\phi(\lambda,\theta)$ is known for this test and is equal to the initial value. Ideally, after a complete revolution the cosine-bell pattern should return to the initial position without incurring any deformation.

The DG P^2 scheme with $N_e = 45$ is used for the numerical simulation, this corresponds to 1° resolution (approximately) at the equator. The second-order SSP RK scheme (9.28) is applied for 1,600 time steps to complete one revolution. Figure 9.24 shows the numerical solution (*left panel*) and the limited solution (*right panel*). As expected the non-limited solution is oscillatory, however, oscillations are confined to a smaller region around the cosine-bell. The monotonic limiter removes spurious oscillations but slightly deforms the shape of the bell. The additional computational expense required for the limiter is nominal, for the cosine-bell advection test it is found to be less than 5%.

9.5.6.4 Barotropic Instability Test

The barotropic instability test proposed by Galewsky et al. (2004) is an interesting test for the SW models developed on the cubed-sphere grids. The test describes

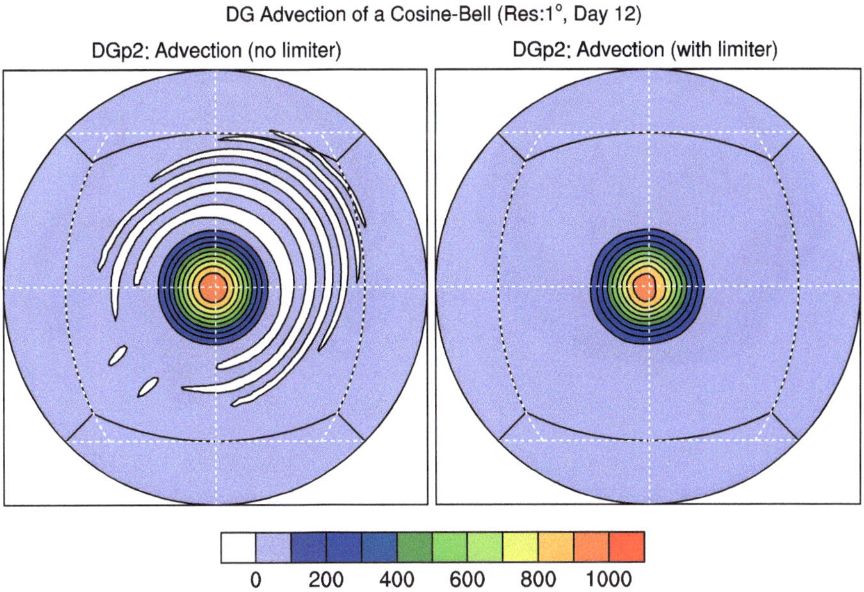

Fig. 9.24 The cosine-bell advection test on the sphere. The *left panel* shows the DG (P^2) numerical solution after a complete revolution along northeast direction, where spurious oscillations in the solution can be seen for the zero contour value. The *right panel* shows the limited solution by applying a monotonic limiter that completely removes the oscillations

the evolution of a barotropic wave in the northern hemisphere and exhibits continuous nonlinear transfer of energy at the midlatitudes from large to small scales. The test is particularly challenging on the cubed-sphere because the vigorous barotropic instability activities are located at the discontinuous edges of the top panel of the cubed-sphere grid. This test *exposes* artifacts from wave number four due to the cube-edge discontinuities at low resolutions for various SW models (St-Cyr et al. 2008; Chen and Xiao 2008; Levy 2009).

The initial conditions are zonally symmetric, and nearly in balance but physically unstable. This introduces a strong zonal jet along the midlatitudes; details can be found in Galewsky et al. (2004). The test recommends a simulation time of 6 days with and without diffusion. Fine features of the vorticity fields can be captured at a resolution of about 1.25° or higher (St-Cyr et al. 2008), and the DG results agree with this observation. Figure 9.25 shows a high-resolution DG simulation of relative vorticity (ζ) at days 4 and 6, respectively. The approximate equatorial resolution is 0.64° ($N_e = 20$, $N_v = 8$) and a time step $\Delta t = 6$s is used for these simulations. The fine features of the vortex are well captured by the DG SW model and comparable to the reference solution given in Galewsky et al. (2004). Small-scale noise at the sharp gradients in Fig. 9.25 can be effectively controlled by using a diffusion scheme.

We briefly outline the diffusion process as used for DG methods in the context of the barotropic vorticity evolution. Diffusion and dissipation mechanisms are

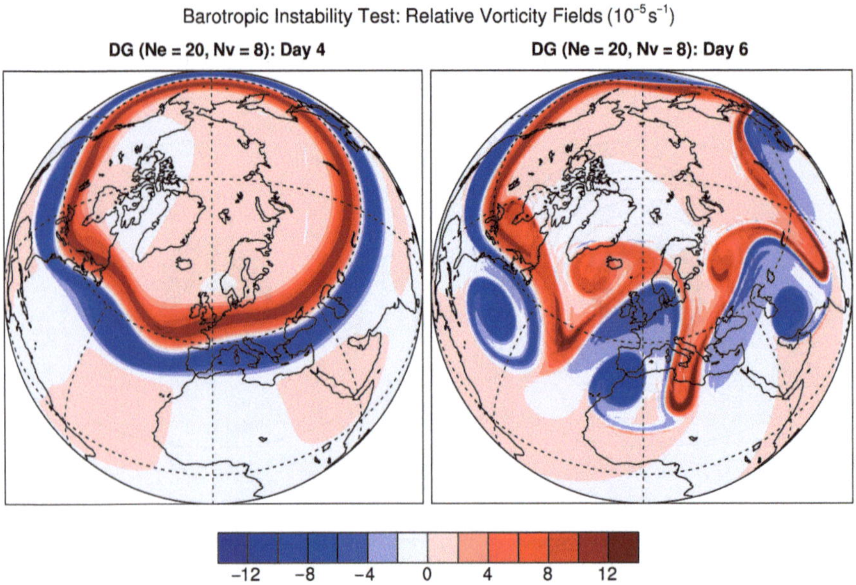

Fig. 9.25 The simulated relative vorticity fields (ζ) for the barotropic instability test at a high-resolution. The *left panel* shows ζ at day 4 and the *right panel*, ζ at day 6

inevitable for practical atmospheric models. For example, momentum diffusion transfers energy from the resolved scales into the unresolved scales. However, in a discrete climate model, diffusion tries to mimic the effects of unresolved scales on the resolved fluid flow (Chap. 13). Moreover, the diffusion process prevents spurious accumulation of energy and enstrophy at the model grid scale. The DG method is amenable to efficient implementation of robust diffusion schemes. This is based on the so-called Local DG or LDG method by Cockburn and Shu (1998), which is a generalization of the explicit diffusion scheme proposed originally by Bassi and Rebay (1997). Recently, Nair (2009) developed a second-order LDG diffusion scheme for the viscous SW model on the cubed-sphere. The vorticity evolution results shown in Nair (2009) confirm that the LDG based diffusion mechanism removes small-scale noise such that the solution converges monotonically to a diffused state. The convergence is dependent on the coefficient of diffusion.

9.6 Concluding Remarks

The DG method combined with explicit strong stability-preserving Runge-Kutta time-stepping is particularly attractive for wave propagation problems because of the ability to use local high-order polynomial approximations for the solution, providing an efficient way to control phase and dissipation errors. The DG method is becoming popular in geophysical fluid dynamics modeling, with several efforts to develop global SW models based on DG methods (Giraldo et al. 2002; Nair et al. 2005a; Läuter et al. 2008; Nair 2009). Very recently, DG methods have been further extended to hydrostatic (Nair et al. 2009) and non-hydrostatic (Giraldo and Restelli 2008; St-Cyr and Neckels 2009) atmospheric models. Currently there are new efforts by various research groups to develop sophisticated DG-based atmospheric models, including some with adaptive meshes. Motivations for choosing the DG method as the primary numerical technique for these model developments are based on various factors such as the high-order accuracy, conservation, geometric flexibility and parallel efficiency. Nevertheless, there are some computational issues associated with the explicit DG discretization.

A major drawback of the DG algorithm is the severe CFL stability restriction associated with explicit time-stepping. For practical climate models and high resolution non-hydrostatic NWP models, overall computational efficiency is very much contingent on the model's ability to take larger time steps. A moderate order DG scheme employing third- or fourth-order spatial discretization (i.e., a P^2 or P^3 method) can address the stringent stability requirement to some extent. Implicit time integration approaches are also popular for DG methods in CFD applications (Diosady and Darmofal 2009; Bassi et al. 2009). The numerical algorithms for such methods are far more complex and require considerably more computational resources than explicit schemes. If such techniques permit at least 3-fold longer time steps for unsteady problems as compared to the explicit method, then they may be worth considering for atmospheric modeling applications.

Development of efficient time integration methods for DG methods is an active area of research. The semi-implicit time integration method for a DG non-hydrostatic model introduced by Restelli and Giraldo (2009) appears to be promising. The recent novel time integration approaches such as the ADER (Arbitrary high order DERivatives) by Käser et al. (2007) and IMEX (implicit explicit) RK methods by Kanevsky et al. (2007) have been shown to be efficient time integration options for DG methods. These new time integration techniques could be extended to DG atmospheric models.

Acknowledgments The authors are thankful to IMAGe (NCAR) colleagues, particularly Dr. Duane Rosenberg for an internal review of the manuscript. The authors would also like to thank two anonymous reviewers for several helpful suggestions. This project is partially supported by the U.S. Department of Energy under the awards DE-FG02-07ER64464 and DE-SC0001658. The National Center for Atmospheric Research is sponsored by the National Science Foundation.

References

Arakawa A, Lamb VR (1977) Computational design of the basic dynamical process of the UCLA general circulation model. In: Chang J (ed) Methods in Computational Physics, Academic Press, pp 173–265

Atkins HL, Shu CW (1996) Quadrature-free implementation of the discontinuous Galerkin method for hyperbolic equations. In: 2nd AIAA/CEAS Aeroacoustic Conference, Paper 96-1683.

Balsara DS, Altman C, Munz CD, Dumbser M (2007) Sub-cell based indicator for troubled zones in RKDG schemes and a novel class of hybrid RKDG+HWENO schemes. J Comput Phys 226(1):586–620

Bassi F, Rebay S (1997) A high-order accurate discontinuous finite element method for the numerical solution of the compressible Navier-Stokes equations. J Comput Phys 131:267–279

Bassi F, Ghidoni A, Rebay S, Tesini P (2009) High-order accurate p-multigrid discontinuous Galerkin solution of the Euler equations. Int J Numer Meth Fluids 60:847–865, doi:10.1002/fld.1917

Biswas R, Devine K, Flaherty J (1994) Parallel adaptive finite-element methods for conservation laws. Appl Num Math 14:255–283

Boris JP, Book DL (1973) Flux-Corrected Transport. I. SASHSTA, a fluid transport algorithm that works. J Comput Phys 11(1):38–69

Butcher JC (2008) Numerical Methods for Ordinary Differential equations, Second edn. Wiley, ISBN 978-0-470-72335-7 463 pp.

Canuto C, Hussaini MY, Quarteroni A, Zang TA (2007) Spectral Methods: Evolution of Complex Geometries and Application to Fluid Dynamics. Springer, ISBN 978-3-540-30727-3, 596 pp.

Chen C, Xiao F (2008) Shallow water model on cubed-sphere by multi-moment finite volume method. J Comput Phys 227(10):5019–5044

Cheruvu V, Nair RD, Tufo HM (2007) A spectral finite volume transport scheme on the cubed-sphere. Appl Num Math 57:1021–1032

Cockburn B, Shu CW (1989) TVB Runge-Kutta local projection discontinuous Galerkin finite element method for conservative laws II. Math Comp 52:411–435

Cockburn B, Shu CW (1998) The local discontinuous Galerkin method for time-dependent convection-diffusion schemes. SIAM J Numer Anal 35:2440–2463

Cockburn B, Shu CW (2001) The Runge-Kutta discontinuous Galerkin method for convection-dominated problems. J Sci Computing 16:173–261

Cockburn B, Hou S, Shu CW (1990) TVB Runge-Kutta local projection discontinuous Galerkin finite element method for conservation laws IV. Math Comp 54:545–581

Cockburn B, Johnson C, Shu CW, Tadmor E (1997) Advanced Numerical Approximation of Nonlinear Hyperbolic Equations. Springer, LNM 1697

Cockburn B, Karniadakis GE, Shu CW (2000) The development of discontinuous Galerkin methods. In: Cockburn B, Karniadakis GE, Shu CW (eds) Discontinuous Galerkin Methods: Theory, Computation, and Applications. Lecture Notes in Computational Science and Engineering, vol 11, Springer, 470 pp.

Colella P, Woodward PR (1984) The Piecewise Parabolic Method (PPM) for gas-dynamical simulations. J Comput Phys 54:174–201

Crowell S, Williams D, Marviplis C, Wicker L (2009) Comparison of traditional and novel discretization methods for advection models in numerical weather prediction. In: Lecture Notes in Computer Science, vol 5545, Springer-Verlag, pp 263–272, ICCS 2009, Part II.

DeMaria M (1985) Tropical cyclone motion in a nondivergent barotropic model. Mon Wea Rev 113:119–1210

Dennis J, Fournier A, Spotz WF, St-Cyr A, Taylor MA, Thomas SJ, Tufo H (2005) High-resolution mesh convergence properties and parallel efficiency of a spectral element atmospheric dynamical core. Int J High Perf Computing Appl 19(3):225–235

Deville MO, Fisher PF, Mund EM (2002) High-Order Methods for Incompressible Fluid Flow. Cambridge University Press, ISBN 0-521-45309-7, 499 pp.

Diosady LT, Darmofal DL (2009) Preconditioning methods for discontinuous Galerkin solutions of the compressible Navier-Stokes equations. J Comput Phys 228:3917–3835

Galewsky J, Polvani LM, Scott RK (2004) An initial-value problem to test numerical models of the shallow water equations. Tellus 56A:429–440

Ghostine R, Kessewani G, Mosé R, Vazquez J, Ghenaim A (2009) An improvement of classical slope limiters for high-order discontinuous Galerkin method. Int J Numer Meth Fluids 59: 423–442

Giraldo FX, Restelli M (2008) A study of spectral element and discontinuous Galerkin methods for the Navier-Stokes equations in nonhydrostatic mesoscale atmospheric modeling: Equation sets and test cases. J Comput Phys 227:3849–3877
Giraldo FX, Hesthaven JS, Wartburton T (2002) Nodal high-order discontinous Galerkin methods for the shallow water equations. J Comput Phys 181:499–525
Godunov SK (1959) A difference method for numerical calculation of discontinuous solutions of the equations of hydrodynamics. Mat Sb 47:271–306
Gottlieb S, Shu CW, Tadmor E (2001) Strong stability preserving high-order time discretization methods. SIAM Rev 43:89–112
Harten A, Engquist B, Osher S, Chakravarthy S (1987) Uniformly high order essentially non-oscillatory schemes, III. J Comput Phys 71:231–303
Hesthaven JS, Warburton T (2008) Nodal Discontinuous Galerkin Methods: Algorithms, Analysis, and Applications. Springer, ISBN 978-0-387-72065-4, 500 pp.
Iskandarani M, Levin J, Choi BJ, Haidgovel D (2005) Comparison of advection schemes for high-order h-p finite element and finite volume methods. Ocean Modeling 10:233–252
Kanevsky A, Carpenter MH, Gottlieb D, Hasthevan JS (2007) Application of implicit-explicit high order Runge-Kutta methods to discontinuous-Galerkin schemes. J Comput Phys 225: 1753–1781
Karniadakis GE, Sherwin S (2005) Spectral/hp Element Methods for Computational Fluid Dynamics. Oxford University Press, ISBN 0-19-852869-8, 657 pp.
Käser M, Dumbser M, Puente J, Igel H (2007) An arbitrary high-order discontinuous Galerkin method for elastic waves on unstructured meshes-III. Geophys J Int 168:224–242
Kopriva DA (2009) Implementing Spectral Methods for Partial Differential Equations. Springer, ISBN 978-90-481-2260-8, 394 pp.
Kopriva DA, Gassner G (2010) On the quadrature and weak form choices in collocation type discontinuous Galerkin spectral element methods. J Sci Comput 44:136–155
Krivodonova L (2007) Limiters for high-order dicontinuous Galerkin methods. J Comput Phys 226:879–896
Kubatko EJ, Bunya S, Dawson C, Westerink JJ (2009) Dynamic p-adaptive Runge–Kutta discontinuous Galerkin methods for the shallow water equations. Comput Methods Appl Mech Engrg 198:1766–1774
Lauritzen PH, Nair RD, Ullrich PA (2010) A conservative semi-Lagrangian multi-tracer transport scheme (CSLAM) on the cubed-sphere grid. J Comput Phys 229:1401–1424
Läuter M, Giraldo FX, Handorf D, Dethloff K (2008) A discontinuous Galerkin method for shallow water equations in spherical traingular coordinates. J Comput Phys 227:10, 226–10, 242
van Leer B (1974) Towards the ultimate conservative difference scheme. II. Monotonicity and conservation combined in a second-order scheme. J Comput Phys 14:361–370
van Leer B (1977) Towards the ultimate conservative difference scheme. IV. A new approach to numerical convection. J Comput Phys 23:276–299
Lesaint P, Raviart P (1974) Mathematical Aspects of Finite Elements in Partial Differential Equations, Academic Press, New York, chap On a finite element method for solving neutron transport eqaution, pp 89–123
LeVeque RJ (2002) Finite Volume Methods for Hyperbolic Problems. Cambridge University Press, ISBN 0-19-00924-3, 558 pp.
Levy MN (2009) A high-order element-based Galerkin method for the global shallow water equations. PhD thesis, University of Colorado at Boulder, Department of Applied Mathematics, 108 pp.
Levy MN, Nair RD, Tufo HM (2007) High-order Galerkin methods for scalable global atmospheric models. Computers & Geosciences 33(8):1022–1035
Levy MN, Nair RD, Tufo HM (2009) A high-order element-based Galerkin method for the barotropic vorticity equation. Int J Numer Meth Fluids 59(12):1369–1387
Liu XD, Osher S, Chan T (1994) Weighted essentially non-oscillatory schemes. J Comput Phys 115:200–212

Lomtev I, Kirby RM, Karniadakis GE (2000) A discontinuous Galerkin method in moving domains. In: Cockburn B, Karniadakis GE, Shu CW (eds) Discontinuous Galerkin Methods: Theory, Computation, and Applications. Lecture Notes in Computational Science and Engineering, vol 11, Springer, 470 pp.
Luo H, Baum JD, Löhner R (2007) A Hermite WENO-based limiter for discontinuous Galerkin method on unstructured grids. J Comput Phys 225(1):686–713
Lynch P (2008) The origins of computer weather prediction and climate modeling. J Comput Phys 227:3431–3444
Nair RD (2009) Diffusion experiments with a global discontinuous Galerkin shallow-water model. Mon Wea Rev 137:3339–3350
Nair RD, Lauritzen PH (2010) A class of deformational flow test cases for linear transport problems on the sphere. J Comput Phys 229:8868–8887
Nair RD, Thomas SJ, Loft RD (2005a) A discontinuous Galerkin global shallow water model. Mon Wea Rev 133:876–888
Nair RD, Thomas SJ, Loft RD (2005b) A discontinuous Galerkin transport scheme on the cubed-sphere. Mon Wea Rev 133:814–828
Nair RD, Choi HW, Tufo HM (2009) Computational aspects of a scalable high-order discontinuous Galerkin atmospheric dynamical core. Computers and Fluids 38:309–319
Prather MJ (1986) Numerical advection by conservation of second-order moments. J Geophys Res 91:6671–6681
Qiu J, Khoo BC, Shu CW (2006) A numerical study for the performance of the Runge-Kutta discontinuous Galerkin method based on different numerical fluxes. J Comput Phys 212(2):540–565
Qui J, Shu CW (2005a) Hermite WENO schemes and their application as limiters for Runge-Kutta discontinuous Galerkin methods II: Two dimesional case. Computers & Fluids 34:642–663
Qui J, Shu CW (2005b) Runge-Kutta discontinuous Galerkin methods using WENO limiters. SIAM J Sci Computing 26:907–927
Rančić M, Purser R, Mesinger F (1996) A global shallow water model using an expanded spherical cube. Q J R Meteorol Soc 122:959–982
Reed WH, Hill TR (1973) Triangular mesh method for neutron transport equation. Technical Report LA-UR-73-479, Los Alamos Scientific Laboratory
Remacle JF, Flaherty JE, Sheppard MS (2003) An adaptive discontinuous Galerkin technique with an orthogonal basis applied to compressible flow problems. SIAM Review 45:53–72
Restelli M, Giraldo FX (2009) A conservative discontinuous Galerkin semi-implicit formulation for the Navier-Stokes equations in nonhydrostatic mesoscale modeling. SIAM J Sci Comput 31:2231–2257
Rivière B (2008) Discontinuous Galerkin Method for Solving Elliptic and Parabolic Equations: Theory and Implementation. SIAM, ISBN 978-0-898716-56-6, 187 pp.
Sadourny R (1972) Conservative finite-difference approximations of the primitive equations on quasi-uniform spherical grids. Mon Wea Rev 100:136–144
Shu CW (1997) Advanced Numerical Approximation of Nonlinear Hyperbolic Equations, Springer, chap Essentially non-oscillatory and weighted essentially non-oscillatory schemes for conservation laws, pp 324–432. LNM 1697
Simmons AJ, Burridge DM (1981) An energy and angular-momentum conserving vertical finite-difference scheme and hybrid vertical coordinates. Mon Wea Rev 109:758–766
Smolarkiewicz PK (1982) The multi-dimensional Crowley advection scheme. Mon Wea Rev 110:1968–1983
Smolarkiewicz PK (1984) A fully muliti-dimensional positive definite advection transport algorithm with small implicit diffusion. J Comput Phys 54:325–362
St-Cyr A, Neckels D (2009) A fully implicit Jacobian-free high-order discontinuous Galerkin mesoscale flow solver. In: Lecture Notes in Computer Science, vol 5545, Springer-Verlag, pp 243–252, ICCS 2009, Part II.
St-Cyr A, Jablonowski C, Dennis JM, Tufo HM, Thomas SJ (2008) A comparison of two shallow water models with nonconforming advaptive grids. Mon Wea Rev 136:1898–1922

Staniforth A, Coté J, Pudickiewicz J (1987) Comments on "Smolarkiewicz's deformational flow". Mon Wea Rev 115:894–900

Suresh A (2000) Positivity preserving schemes in multidimensions. SIAM J Sci Comput 22:1184–1198

Toro EF (1999) Riemann Solvers and Numerical Methods for Fluid Dynamics. A Practical Introduction (2nd Ed.). Springer-Verlag, New York

Toro EF (2001) Shock-Capturing Methods for Free-Surface Shallow Flows. John Wiley & Sons, England, ISBN 0-471-98766-2, 305 pp.

Vallis GK (2006) Atmopsheric and Ocenaic Fluid Dynamics. Cambridge University Press, ISBN 978-0-521-84969-2, 745 pp.

Williamson DL (2007) The evolution of dynamical cores for global atmospheric models. J Meteo Soc of Japan 85:241–269

Williamson DL, Drake JB, Hack JJ, Jakob R, Swarztrauber PN (1992) A standard test set for numerical approximations to the shallow water equations in spherical geometry. J Comput Phys 102:211–224

Zhang X, Shu CW (2010) On positivity-preserving high-order discontinuous Galerkin schemes for compressible Euler equations on rectangular meshes. J Comput Phys 229(23):8918–8934, doi:10.1016/j.jcp.2010.08.016

Chapter 10
Voronoi Tessellations and Their Application to Climate and Global Modeling

Lili Ju, Todd Ringler, and Max Gunzburger

Abstract We review the use of Voronoi tessellations for grid generation, especially on the whole sphere or in regions on the sphere. Voronoi tessellations and the corresponding Delaunay tessellations in regions and surfaces on Euclidean space are defined and properties they possess that make them well-suited for grid generation purposes are discussed, as are algorithms for their construction. This is followed by a more detailed look at one very special type of Voronoi tessellation, the centroidal Voronoi tessellation (CVT). After defining them, discussing some of their properties, and presenting algorithms for their construction, we illustrate the use of CVTs for producing both quasi-uniform and variable resolution meshes in the plane and on the sphere. Finally, we briefly discuss the computational solution of model equations based on CVTs on the sphere.

10.1 Introduction

Given two sets A and B and a distance metric $d(a,b)$ defined for $a \in A$ and $b \in B$, a Voronoi diagram or tessellation is a subdivision of A into subsets, each of which contains the objects in A that are closer, with respect to the distance metric, to one object in B than to any other object in B. Although Voronoi tessellations can be defined for a wide variety of sets and metrics, of interest here is the situation

L. Ju (✉)
Department of Mathematics, University of South Carolina, Columbia, SC 29208, USA
e-mail: ju@math.sc.edu

T. Ringler
T-3 Fluid Dynamics Group, Theoretical Division, Los Alamos National Laboratory, Los Alamos, NM 87545, USA
e-mail: ringler@lanl.gov

M. Gunzburger
Department of Scientific Computing, Florida State University, Tallahassee, FL 32306-4120, USA
e-mail: gunzburg@fsu.edu

P.H. Lauritzen et al. (eds.), *Numerical Techniques for Global Atmospheric Models*, Lecture Notes in Computational Science and Engineering 80,
DOI 10.1007/978-3-642-11640-7_10,

for which the set A is a region or surface in Euclidean space, B is a finite set of points also in Euclidean space, and the metric is the Euclidean distance.

Voronoi tessellation have a long history, probably because Voronoi-like arrangements often appear in nature. Voronoi-like tessellations appeared in 1644 in the work of Decartes on the distribution of matter in the cosmic region near our sun. The first systematic treatment of what we now call Voronoi tessellations was given by Dirichlet (1850) in his study of two- and three-dimensional quadratic forms, i.e., homogeneous, multivariate polynomials of degree two; hence, Voronoi regions are often referred to as Dirichlet cells. Voronoi (1907) generalized the work of Dirichlet to arbitrary dimensions, again using what are now referred to as Voronoi tessellations or diagrams.

The first documented application of Voronoi tessellations appeared in the classic treatise of Snow (1855) on the 1854 cholera epidemic in London in which he demonstrated that proximity to a particular well was strongly correlated to deaths due to the disease. Voronoi tessellations have continued to be very useful in the social sciences, e.g., in the study of dialect variations, demographics, territorial systems, economics, and markets. Starting in the late nineteenth century and continuing to this day, Voronoi tessellations have also been used in crystallography, especially in the study of space-filling polyhedra, although, in this setting, various other names have been used to denote Voronoi regions, e.g., stereohedra, fundamental area, sphere of influence, domain of action, and plesiohedra.

It is not surprising, due to their ubiquity and usefulness, that throughout the twentieth century, Voronoi tessellations were rediscovered many times. As a result, Voronoi regions have been called by many different names. *Thiessen polygons* refer to the work of Theissen on developing more accurate estimates for the average rainfall in a region. *Area of influence polygons* was a term coined in connection with the processing of data about ore distributions obtained from boreholes. *Wigner-Seitz regions*, *domain of an atom*, and *Meijering cells* were terms that arose from work on crystal lattices and the Voronoi cell of the reciprocal crystal lattice is referred to as the *Brillouin zone* (Kittel 2004; Ziman 1979). In the study of codes by, e.g., Shannon, Voronoi cells are called *maximum likelihood regions* (Weaver and Shannon 1963). The field of ecology gave rise to two more alternate labels: *area potentially available* and *plant polygons* for a Voronoi region associated with a particular tree or plant. *Capillary domains* refers to Voronoi regions in a tissue based on the centers of capillaries.

For a long time, the routine use of Voronoi tessellations in applications was hindered by the lack of efficient means for their construction. This situation has now been remedied, at least in two and three dimensions. Voronoi tessellations also became closely intertwined with computational geometry. For example, Shamos and Hoey (1975) not only provided an algorithm for constructing Voronoi tessellations, but also showed how they could be used to answer several fundamental questions in computational geometry.

Delaunay tessellations,[1] the dual concept to Voronoi tessellations, also have a long history and have been called by other names. They originated with Voronoi (1908) who called them *the ensemble (L) of simplices*. Delaunay (1928, 1934) was the first to define the tessellations bearing his name[2] in terms of empty spheres; he referred to them in terminology similar to that of Voronoi and, even today, some refer to Delaunay tessellations as *L-partitions*. The name Delaunay was first associated with Delaunay triangulations by Rogers (1964). Delaunay tessellations have also proven to be very useful, especially for grid generation.

The first applications of Voronoi tessellations to global atmospheric modeling were made by Williamson (1968) and Sadourny et al. (1968) wherein the barotropic vorticity equation was integrated forward in time. Neither Williamson nor Sadourny referred to their meshes as Voronoi tessellations; Williamson referred to the underlying tessellation as a "geodesic grid," a colloquialism that is used in much of the literature discussing the use of Voronoi tessellations in global climate modeling.[3] Both of these efforts produced promising results as compared to other models available at the time. The reason for their success was really due to not having a longitudinal polar filter which distorted the earlier solutions on latitude-longitude grids, or to not using a reduced grid which also distorted the solutions. In addition, it helped as well that the discrete formation of the Jacobian put forth by Arakawa (1966) could be readily translated to their respective "geodesic grids." Williamson (1970) continued this line of research with the integration of the shallow-water system in primitive variable form. While Williamson's tessellation was extremely uniform, in a global sense, as compared to the latitude-longitude meshes being using in other model development efforts (Kasahara and Washington 1967), the truncation error analysis by Williamson clearly reflected the fact that the Voronoi tessellation

[1] Delaunay tessellations are often referred to as Delaunay triangulations because, in two dimensions, they consist of a triangulation of the points that generate the Voronoi tessellation. We choose to refer to them as Delaunay tessellations to emphasize the fact that the concept of a dual to Voronoi tessellations is quite general and not limited to two dimensions. When dealing with two-dimensional settings, we will however, call them Delaunay triangulations.

[2] Delone was a Russian number theorist who used the spelling Delaunay when writing papers in French or German. He was also the first to coin both the descriptors "Dirichlet domains" and "Voronoi regions."

[3] Adjectives such as "geodesic," "bisection," and "icosahedral" are often used to describe grids on the sphere. However, there seems to be a lack of consistency about what these qualifiers mean. In this paper, we use the following terminology.

Geodesic grids refer to any grid on the sphere such that the edges of the grid cells are geodesic arcs, i.e., arcs of great circles. According to this definition, all Voronoi grids on the sphere are, by construction, geodesic grids.

Bisection grids refer to any grid constructed through repeated bisection of a platonic solid having vertices on the sphere and edges projected onto the sphere. Bisection grids are by construction geodesic grids. One may also define a bisection grid by repeated bisection of the Delaunay triangulation corresponding to the platonic solid.

Octahedral-bisection grids refer to bisection grids that are based on the platonic octahedra having 12 pentagonal faces. Note that this grid is often referred to as a "geodesic grid" or a "bisection grid" but here we make a finder distinction between these terminologies.

was less uniform in a local sense; the discrete operators used in (Williamson 1970) resulted in first-order truncation errors that could quickly corrupt the solution. Williamson's barotropic primitive equation model was discretized using a collocated grid where thickness and velocity reside at the same location.[4] Since a collocated grid makes little (if any) use of the dual mesh, it is not clear if this geodesic grid was or was not a Voronoi tessellation.

Following Williamson (1970), the idea of solving the barotropic primitive equations based on a Voronoi tessellation was essentially abandoned for 15 years. It appears that this idea did not gain traction for two reasons. First, global spectral models emerged as a superior choice to their finite-volume or finite-difference counterparts because they are based on the natural polar filter so to speak and have no pole problem. Their spectral accuracy and the reintroduction of the fast Fourier transform (Cooley and Tukey 1965) also contributed significantly. Second, while numerical schemes situated on latitude-longitude meshes were burdened with truncation errors comparable to those found by Williamson (1970), progress toward methods to mitigate the impact of these errors on the long-term stability of simulations was much more rapid for quadrilateral meshes; see, e.g., (Arakawa and Lamb 1977). Unfortunately, the numerical methods developed for the solution of the barotropic primitive equations on quadrilateral meshes did not readily translate to Voronoi tessellations. For example, while C-grid staggered quadrilateral meshes were essentially operational by the mid 1970s, a comparable C-grid scheme for general Voronoi tessellations was not derived until Thuburn et al. (2009) in 2009.

Because it appeared, at least at the time, that Voronoi tessellations were not well suited for the integration of the primitive equations, when this idea was revisited by Masuda and Ohnishi (1986) they chose a different system of equations to discretize. Masuda and Ohnishi formulated the shallow-water system in vorticity and divergence variables, instead of primitive variables. In this approach, the thickness, vorticity, and divergence are collocated at the center of each Voronoi cell. Other similar work on solving shallow water equations based on Voronoi mesh was done by Augenbaum (1984) and Augenbaum and Peskin (1985). Randall (1994) would later show that the collocation of variables in the vorticity-divergence system, termed the Z-grid, leads to a simulation of geostrophic adjustment that is better than any of the other staggerings based on primitive variables. The superior simulation of geostropic adjustment along with the direct control over the evolution of vorticity led to robust simulations of the shallow-water system. Heikes and Randall (1995a,b) continued this line of research with the implementation of a geometric multigrid solver to mitigate the cost associated with solving the vorticity-divergence system. In turn, this work led to the creation, by Ringler et al. (2000) in 2000, of the first global atmosphere dynamical core situated on a Voronoi tessellation.

The demonstration that Voronoi tessellations could be used to successfully model global atmosphere dynamics created considerable interest. By and large, all global atmosphere models using finite-volume methods were based on latitude-longitude

[4] The collocated grid was later named the "A-Grid" in (Arakawa and Lamb 1977).

grids. With no satisfactory solutions to overcome the grid singularities present at the poles of latitude-longitude grids, the quasi-uniform grid offered by Voronoi tessellations was a compelling alternative. This stimulated research toward finding numerical schemes based on primitive variables that would essentially translate the A-, B- and C-grid staggerings from quadrilateral meshes to Voronoi tessellations. The collocated, A-grid staggering, first proposed by Williamson (1970), was successfully implemented by Tomita et al. (2001). That effort resulted in the first ever global cloud resolving simulation by Tomita et al. (2005). It is important to note that (Tomita et al. 2001, 2005) do not employ a Voronoi tessellation since the location of the cell vertices are placed at the barycenter[5] of the Delaunay triangulation, instead of the circumcenter of the Delaunay triangulation. As a result, the powerful results that follows from a Voronoi tessellation are not immediately applicable to their mesh. The B-grid staggering was successfully developed for Voronoi tessellations by Ringler and Randall (2002). It is only at this point, fully two decades after the energy and potential enstrophy conserving schemes for quadrilateral grids were derived (Arakawa and Lamb 1981), that the numerical methods on Voronoi tessellations are comparable to their quadrilateral counterparts.

With the successful implementation of both the discrete vorticity-divergence system and various discrete forms of the primitive equation system on quasi-uniform Voronoi tessellations, attention is now turning toward the use of variable resolution Voronoi tessellations. During this process we are essentially revisiting the truncation error problems that Williamson (1970) identified four decades ago when using quasi-uniform Voronoi tessellations. When pairing low-order, finite-volume methods with variable resolution Voronoi tessellations, truncation error will be increased, at least locally, in the regions of mesh transition. To overcome the challenge presented by this truncation error behavior, we see three routes forward. First, increase the accuracy of the underlying finite-volume method to reduce truncation error to acceptable levels; this approach was successfully employed in (Du et al. 2003b; Du and Ju 2005; Weller 2009; Weller and Weller 2008). Second, develop numerical schemes that respect both geostrophic adjustment and the need for nonlinear stability, even when the mesh is highly distorted; this approach has been developed by (Thuburn et al. 2009; Ringler et al. 2010). And finally, we can attempt to optimize the quality of the variable resolution meshes in order to limit the extent of the problem. In the end, some combination of these three approaches will likely lead to the creation of a variable-resolution global climate system model.

This chapter is focused on the mesh generation aspect of Voronoi tessellations and, more importantly, the inherent properties that these meshes are guaranteed to possess. We first provide a mathematical description of Voronoi tessellations and their Delaunay triangulation counterparts. This is followed by a detailed analysis of one very special type of Voronoi tessellation, the *centroidal Voronoi tessellation*. We then explore the properties of centroidal Voronoi tessellations when producing both

[5] The barycenter is the center of mass; thus, for a triangle, the barycenter is at the intersection of the three lines joining the vertices and the centers of the opposite sides whereas the circumcenter is at the intersection of the perpendicular bisectors of the three sides.

quasi-uniform and variable resolution meshes. Finally, we briefly discuss the numerical implementation of models using centroidal Voronoi tessellations. We defer until Sect. 10.3 a discussion about why centroidal Voronoi tessellations are especially well suited as a basis for grid generation.

10.2 Voronoi and Delaunay Tessellations

10.2.1 Definitions and Properties

We are given an open bounded domain $\Omega \in \mathbb{R}^d$ and a set of distinct points $\{\mathbf{x}_i\}_{i=1}^n \subset \Omega$. For each point $\mathbf{x}_i$, $i = 1, \ldots, n$, the corresponding *Voronoi region* V_i, $i = 1, \ldots, n$, is defined by

$$V_i = \big\{\mathbf{x} \in \Omega \mid \|\mathbf{x} - \mathbf{x}_i\| < \|\mathbf{x} - \mathbf{x}_j\| \quad \text{for} \quad j = 1, \cdots, n \quad \text{and} \quad j \neq i\big\}, \tag{10.1}$$

where $\|\cdot\|$ denotes the Euclidean distance (the L^2 metric) in $\mathbb{R}^d$. Clearly $V_i \cap V_j = \emptyset$ for $i \neq j$, and[6] $\cup_{i=1}^n \overline{V}_i = \overline{\Omega}$ so that $\{V_i\}_{i=1}^n$ is a *tessellation* of Ω. We refer to $\{V_i\}_{i=1}^n$ as the *Voronoi tessellation* or *Voronoi diagram* of Ω (Okabe et al. 2000) associated with the point set $\{\mathbf{x}_i\}_{i=1}^n$. A point $\mathbf{x}_i$ is called a *generator*; a subdomain $V_i \subset \Omega$ is referred to as the *Voronoi region* or *Voronoi cell* corresponding to the generator $\mathbf{x}_i$.

It is clear that, except for "sides" that are part of the boundary of Ω, Voronoi regions $\{V_i\}_{i=1}^n$ are polygons in two dimensions and polyhedra in three dimensions. Figure 10.1 (*upper left*) presents a Voronoi tessellation of the unit square in two dimensions corresponding to ten randomly selected generators. It is guaranteed that the line segment connecting two neighbor generators is orthogonal to the shared edge/face and is bisected by that edge/face.

The dual of a Voronoi tessellation in the graph-theoretical sense (i.e., by connecting all pair of neighbor generators) is called a *Delaunay tessellation* or, in two dimensions, a *Delaunay triangulation* (Okabe et al. 2000) associated with the point set $\{\mathbf{x}_i\}_{i=1}^n$. Elements of a Delaunay tessellation consist of triangles in two dimensions and tetrahedra in three dimensions. The Delaunay triangulation corresponding to the above ten generators is shown in Fig. 10.1 (*top right*). Note that each triangle of the Delaunay triangulation is associated with a single vertex of its dual Voronoi tessellation. That Voronoi vertex is located at the center of the circumscribed circle of the triangle; see an illustration in Fig. 10.1 (*bottom*). Each cell edge of the Voronoi tessellations is uniquely associated with one cell edge of the dual Delaunay triangulation; each pair of edges are orthogonal, but do not necessarily intersect. If the pair of edges do intersect (or if the lines segments are extended to a point where

[6] For the open region Ω, $\overline{\Omega}$ denotes its closure, i.e., Ω together with its boundary points.

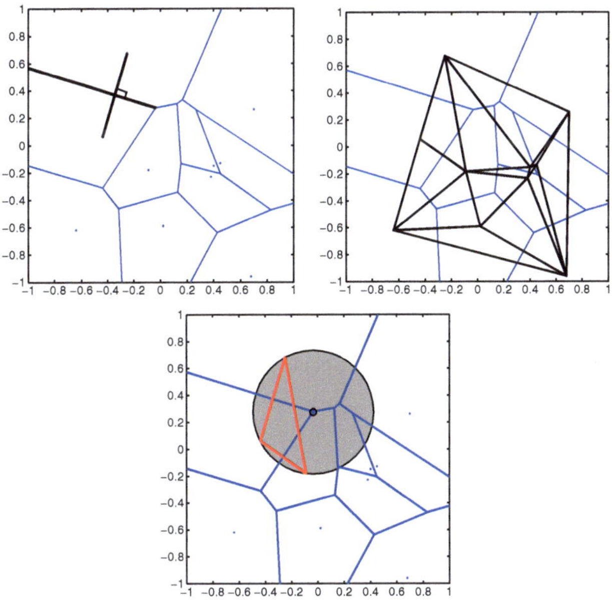

Fig. 10.1 The Voronoi tessellation of the unit square corresponding to ten randomly selected generators. *Top-left*: the bisection property; *top-right*: the corresponding Delaunay triangulation; *bottom*: the circumcircle property

they intersect), then the intersection point will bisect the line segment connecting generators.

In two dimensions, the Delaunay triangulation maximizes the minimum angle, i.e., compared to any other triangulation of the points, the smallest angle in the Delaunay triangulation is at least as large as the smallest angle in any other. This property does not hold in higher dimensions. Note also that for a given set of generators, the Voronoi tessellation is always unique; however, the Delaunay tessellation may not be unique in certain special situations, e.g., when four generators in two dimensions form a rectangle that does not contain any other generator.

Voronoi and Delaunay tessellations of a general *surface* or *manifold* also have been widely studied in the field of computer graphics; see, e.g., (Boissonnat and Oudot 2005). In particular, spherical Voronoi tessellation and Delaunay triangulation and related algorithms are developed in (Renka 1997).

10.2.2 Construction Algorithms

For a given set of distinct points $\{\mathbf{x}_i\}_{i=1}^n \subset \Omega$, the construction of the corresponding Voronoi tessellation and Delaunay triangulation in Euclidean space has been well studied in past decades; see (Okabe et al. 2000). Note that some algorithms directly compute the Delaunay tessellation whereas others compute the Voronoi tessellation.

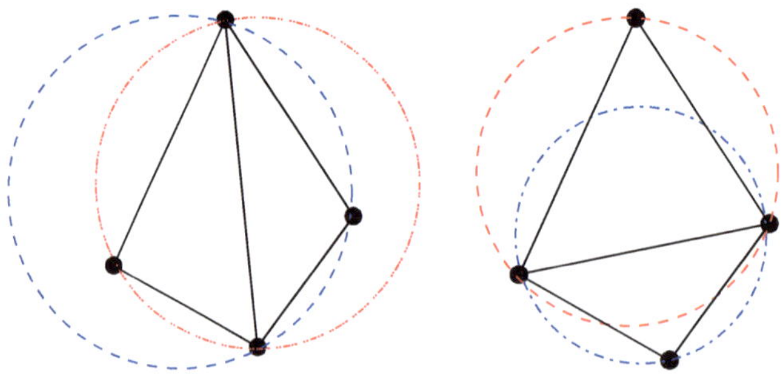

Fig. 10.2 The flipping operation. *Left*: the triangulation does not meet the Delaunay condition, i.e., the circumcircles contain more than three points; *right*: flipping the common edge produces a Delaunay triangulation for the four points

As illustrated in Fig. 10.1, a property of the Delaunay triangulation is that the circle circumscribing any Delaunay triangle does not contain any other generators in its interior. This is an important property because it allows the use of a *flipping* technique. If a triangle is non-Delaunay, we can flip one of its edges; see Fig. 10.2 for an illustration. This leads to the simple *flip algorithm*: construct any triangulation of the points, and then flip edges until no triangle is non-Delaunay. Unfortunately, this can take $O(n^2)$ edge flips. It is worth noting that this edge-flipping technique does not directly extend to three or higher dimensions; on the other hand, the circumcircle property itself does generalize, e.g., to circumspheres of the Delaunay tetrahedra in three dimensions, and some topological operations analogous to flipping have been proposed and discussed in three dimensions (Freitag and Ollivier-Gooch 1997; Du and Wang 2003; Alliez et al. 2005).

A usually more efficient way to construct the Delaunay triangulation is to repeatedly add one vertex at a time and then re-triangulate the affected parts of the graph. When a point $\mathbf{x}_i$ is added, the triangle containing $\mathbf{x}_i$ is split into three triangles and then the flip algorithm is applied. This procedure is called the *incremental algorithm*. It takes $O(n)$ time to search through all the triangles to find the one that contains $\mathbf{x}_i$, after which we potentially flip in every triangle. The overall runtime is theoretically $O(n^2)$ (Guiba et al. 1992), but often in practice this algorithm has better than expected performance (Bentley et al. 1980). While the technique extends to higher dimension, the complexity could grow exponentially in the dimension, even if the final Delaunay triangulation is small (Edelsbrunner and Shah 1996).

An efficient *divide and conquer algorithm* (Lee and Schachter 1980; Guibas and Stolfi 1985) for constructing a Voronoi tessellation of a given set of generators in the plane is defined as follows. One recursively draws a line to split the generators into two sets having roughly the same number of points. Voronoi tessellations of the two subsets are separately constructed. Then, a piecewise linear dividing line between the two subsets is determined. Each segment of this line is itself a segment of the perpendicular bisector corresponding to two generators belonging to different subsets. Then, all edges or part of edges from the Voronoi tessellations of each subset that lie on the opposite side of the dividing line are deleted. Finally, the Voronoi

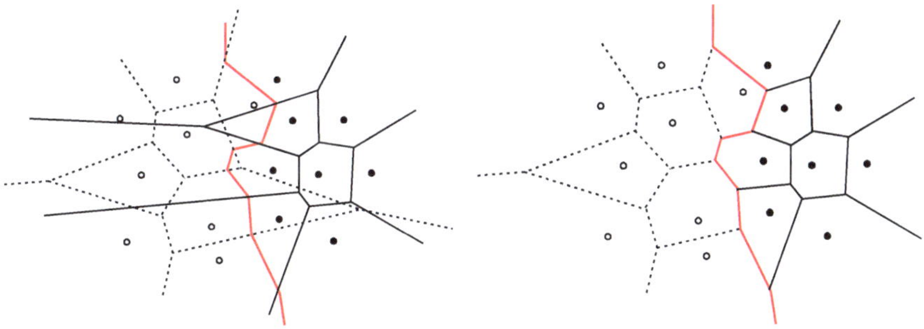

Fig. 10.3 *Left*: the divide-and-conquer algorithm after the given generators are divided into two subsets (*the open and filled circles*), the two Voronoi tessellations of the subsets have been constructed (*the dashed lines and thin solid lines*), and the piecewise linear dividing line has been determined (*the thick, red lines*). *Right*: the Voronoi tessellation of all the generators found by deleting appropriate portions of the Voronoi tessellations of the two subsets

tessellation of the original set of generators is given by the union of the remaining edge segments of the Voronoi tessellations of the two subsets and the piecewise linear dividing line. See Fig. 10.3 for illustrative sketches of the divide and conquer algorithm. Carefully implemented, this divide and conquer method for constructing a Voronoi tessellation of a given set of generators has complexity $O(n \log n)$. A divide and conquer paradigm for constructing a triangulation in d-dimensions was developed in Cignoni et al. (1998).

Another efficient algorithm, *Fortune's sweep line algorithm* (Fortune 1986), is based on the sweep line technique (Sedgewick 1983) and involves not only a sweep line, but also a beach line that actually consists of parabolic arcs. Without loss of generality, one can assume that the sweep line is vertical and that it moves from left to right. Generators to the right of the sweep line have yet to be considered. The beach line is to the left of the sweep line. It is defined as follows: first, for each generator to the left of the sweep line whose Voronoi region has yet to be completely determined, one defines the parabola that separates the points that are closer to the sweep line from those that are closer the generator; then, the beach line is determined as the right-most points in the union of the parabolas. Clearly, a vertex of the beach line is equidistant from the two generators corresponding to the parabolas meeting at that vertex. Thus, as the sweep line moves from left to right, the vertices of the beach line move along the edges of the Voronoi tessellation. A parabolic arc is added to the beach line whenever the sweep line passes a new generator; an arc is removed from the beach line whenever the Voronoi cell for the corresponding generator has been completely determined. The latter situation occurs whenever the sweep line is tangent to a circle passing through three generators whose parabolas form consecutive arcs of the beach line. See Fig. 10.4 for an illustrative sketch for Fortune's algorithm. Carefully implemented, Fortune's algorithm for constructing a Voronoi tessellation of a given set of generators has complexity $O(n \log n)$.

Finally, we mention *"convex hull" algorithms* (Chynoweth and Sewell 1990) for, e.g., Delaunay tessellation construction in Euclidean regions. For example, in

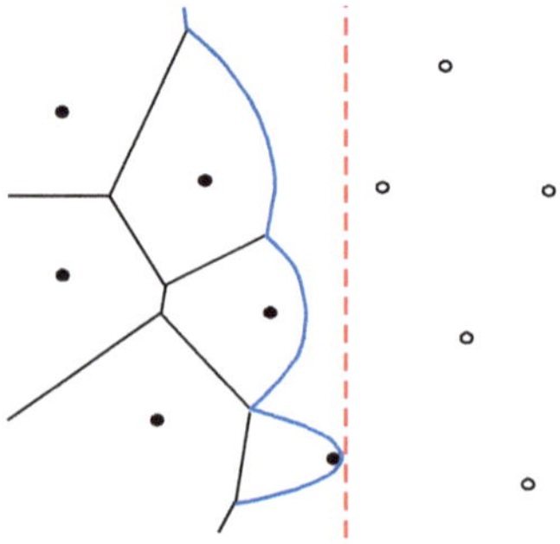

Fig. 10.4 Set-up in Fortune's algorithm. *The red, dashed straight line* is the sweep line that is moving from *left to right*; *the blue, piecewise parabolic curve* is the *beach line*. *The filled circles* are the generators already visited by the sweep line whereas the *open circles* are generators yet to be visited. *The thin black lines* are edges or edge segments of Voronoi regions already constructed

the two-dimensional case, one can vertically project the generators from their plane onto a paraboloidal surface whose axis is perpendicular to that plane. The lower boundary of the convex hull of the points on the paraboloid is generally a triangulated shell whose vertical projection back onto the original plane gives the Delaunay triangulation. This geometrical characterization also explains the circumcircle property mentioned above. The plane of any triangular facet of the assumed convex shell intersects the paraboloid on a closed curve whose projection is that projected triangle's circumcircle. Thus, other generators lying strictly inside that circle would have to correspond to points of the paraboloid that necessarily lie outside the putative convex hull, in violation of the original assumption that a convex hull was constructed. See (Chynoweth and Sewell 1990; Sewell 2002) for detailed discussions on this characterization.

10.3 Centroidal Voronoi Tessellations

Centroidal Voronoi tessellations (CVTs) are special Voronoi tessellations having the property that the *generators* of the Voronoi tessellation are also the *centers of mass* (or *centroids* or *barycenters*), with respect to a given density function, of the corresponding Voronoi regions. CVT methodologies produce high-quality point distributions in regions and surfaces in $\mathbb{R}^d$ or within sets of discrete data. In the latter context and in its simplest form, CVT reduces to the well-known k-means clustering algorithm (Gersho and Gray 1992; Hartigan 1975; Kanungo et al. 2002). The dual tessellation corresponding to a centroidal Voronoi tessellation is referred to as a *centroidal Voronoi Delaunay tessellation* (CVDT).

CVTs and CVDTs possess certain properties, which we discuss below, that make them very well suited for grid generation which is a focus of this paper. In addition, in (Nguyen et al. 2009), several quality measures were used to effect a quantitative comparison of uniform triangular mesh generators in convex and non-convex planar regions; it was found that CVDTs result in higher quality meshes compared to those constructed using most other algorithms, with only the method given in (Persson and Strang 2004) that uses spring dynamics being somewhat competitive.

Although uniform CVT-based grids have been shown to be competitive with (or even better than) other uniform mesh generators in planar, three-dimensional, and spherical regions, perhaps they have even greater utility for the construction of nonuniform meshes. For one thing, through a point density function, CVT grid generation methodologies allow for a simple means of controlling the local grid size; moreover, the density function can easily be connected to error estimators, resulting in effective adaptive refinement strategies (Ju et al. 2002b). For another thing, CVT-based grids feature smooth transitions from coarse to fine grids; see Sect. 10.4.1.2 for an illustration. Smooth grid transitions can greatly reduce deleterious effects, e.g., non-physical wave reflections, that can occur if grid sizes change abruptly.

10.3.1 Definitions and Properties

Given a density function $\rho(\mathbf{x}) \geq 0$ defined on Ω, for any region $V \subset \Omega$, the standard *mass center* (or *centroid*) $\mathbf{x}^*$ of V is given by

$$\mathbf{x}^* = \frac{\displaystyle\int_V \mathbf{x}\rho(\mathbf{x})\,d\mathbf{x}}{\displaystyle\int_V \rho(\mathbf{x})\,d\mathbf{x}}. \tag{10.2}$$

Note that it is often required that ρ is integrable with respect to Ω and the volume of the set $\{\mathbf{x} \mid \rho(\mathbf{x}) = 0\}$ is zero in order to make sure (10.2) is well defined in practice. A special family of Voronoi tessellations are defined as follows.

Definition 1. *(*Du et al. 1999*) Given a density function $\rho(\mathbf{x})$ defined on Ω, we refer to a Voronoi tessellation $\{(\mathbf{x}_i, V_i)\}_{i=1}^n$ of Ω as a* centroidal Voronoi tessellation (CVT) *if and only if the points $\{\mathbf{x}_i\}_{i=1}^n$ which serve as the generators of the associated Voronoi regions $\{V_i\}_{i=1}^n$ are also the centroids, with respect to $\rho(\mathbf{x})$, of those regions, i.e., if and only if we have that $\mathbf{x}_i = \mathbf{x}_i^*$ for $i = 1, \ldots, n$. The corresponding dual triangulation is called a* centroidal Voronoi Delaunay tessellation (CVDT).

A generic Voronoi tessellation does not in general satisfy the CVT property; see Fig. 10.5 for an illustration as well as for an illustration of CVT. On the other hand, given a density function ρ and the number n of generators, the CVT of a domain always exists, although it may not be unique.

CVTs possess an optimization property that can be used as a basis for various extensions. Given any set of points $\widetilde{\mathbf{X}} = \{\widetilde{\mathbf{x}}_i\}_{i=1}^n$ in Ω and any tessellation $\widetilde{\mathbf{V}} = \{\widetilde{V}_i\}_{i=1}^n$ of Ω, define a *clustering energy* by[7]

[7] Note that, a priori, $\mathbf{V}$ need not be a Voronoi tessellation and $\mathbf{x}_i$ need not be in V_i.

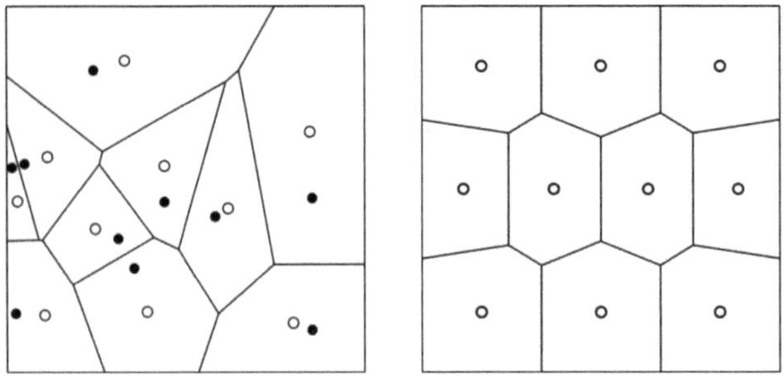

Fig. 10.5 (Du et al. 1999) *Left*: a Voronoi tessellation of the unit square with ten randomly selected generators (*the filled circles*); the open circles denote the centroids of the Voronoi polygons with respect to a uniform density; the centroids do not coincide with the generators. *Right*: a ten-generator centroidal Voronoi tessellation of the square for a uniform density; the generators and centroids coincide

$$\mathscr{K}\big(\widetilde{\mathbf{X}},\widetilde{\mathbf{V}}\big) = \sum_{i=1}^{n} \int_{\widetilde{V}_i} \rho(\mathbf{x})\|\mathbf{x}-\widetilde{\mathbf{x}}_i\|^2\,\mathrm{d}\mathbf{x}\,. \tag{10.3}$$

Then, it can be shown that $\mathscr{K}$ is minimized only if $\{(\widetilde{\mathbf{x}}_i, \widetilde{V}_i)\}_{i=1}^n$ forms a CVT[8] of Ω. Note that if $\{(\mathbf{x}_i, V_i)\}_{i=1}^n$ forms a CVT, it does not necessarily minimize $\mathscr{K}$, e.g., it may define a saddle point (Du et al. 1999) of (10.3). In many applications, the *clustering energy* functional $\mathscr{K}$ is often naturally associated with quantities such as *quantization error*, *variance*, and *cost*.

Asymptotically, as the number of generators becomes larger and larger, Gersho's conjecture (Gersho 1979) states that, locally, the optimal CVT (in the sense of minimizing the clustering energy) under the Euclidean metric forms a regular tessellation consisting of the replication of a single polytope whose shape depends only on the spatial dimension.[9] The regular hexagon provides a confirmation of the conjecture in two dimensions for the constant density case (Newman 1982). For the three-dimensional case and a constant density function, it has been proved (Barnes and Sloane 1983; Du and Wang 2005) that among all lattice-based CVTs,[10] the CVT corresponding to the body-centered cubic lattice for which the Voronoi regions are the space-filling *truncated octahedra* is the optimal one. For more general, non-lattice cases and for non-constant densities, the question remains open, although extensive numerical simulations given in (Du and Wang 2005) demonstrated that

[8] In fact, this can be used as an *analytical* definition of CVTs alternate to the geometric definition given in Definition 1.

[9] In other words, Gersho's conjecture states that, at least for smooth density functions, if the number of generators n is large enough and one focuses on a small enough region, then a CVT appears to be a uniform tessellation involving congruent polytopes.

[10] A lattice-based CVT is one whose generators are located on a lattice so that the Voronoi regions form congruent polytopes.

the truncated octahedra remains the likely candidate. It is interesting to note that, in two dimensions, Gersho's conjecture implies that the dual Delaunay triangulation asymptotically consists of a replications of a single polygon, namely congruent equilateral triangles. In three dimensions, the dual Delaunay tessellation cannot consist of congruent equilateral tetrahedra because the latter cannot cover three space.

10.3.1.1 Centroidal Voronoi Tessellations of Surfaces

Extensions of the VT and CVT concept to *surfaces* (or *manifolds*) are possible; for example, tessellations of surfaces under the Euclidean metric are considered in (Du et al. 2003a). Suppose that Ω is a compact and continuous hypersurface in $\mathbb{R}^{d+1}$. Then, for any subregion $V \subset \Omega$, we call $\mathbf{x}^c$ a *constrained mass center* of V if it is a solution of the problem:

$$\text{find } \mathbf{x}^c \text{ such that } \quad \int_V \rho(\mathbf{y}) \|\mathbf{y} - \mathbf{x}\|^2 \, d\mathbf{y} \quad \text{is minimized over } \mathbf{x} \in V. \tag{10.4}$$

Existence of minimizers of the problem (10.4) can be easily demonstrated using the continuity and compactness of the objective function; however, solutions may not be unique. It is worth noting that if Ω is a flat surface, then $\mathbf{x}^c$ coincides with $\mathbf{x}^*$, the standard center of mass center of V. If we replace $\mathbf{x}_i^*$ in Definition 1 by $\mathbf{x}_i^c$, then the resulting Voronoi tessellation $\{(\mathbf{x}_i, V_i)\}_{i=1}^n$ of the surface Ω is called a *constrained centroidal Voronoi tessellation* (CCVT) (Du et al. 2003a), and its dual tessellation is called a *constrained centroidal Voronoi Delaunay triangulation* (CCVDT). In particular, when Ω is the surface of a sphere, we call $\{(\mathbf{x}_i, V_i)\}_{i=1}^n$ a *spherical centroidal Voronoi tessellation* (SCVT). Figure 10.6 presents an illustration of non-centroidal and centroidal Voronoi tessellations of the sphere.

The calculation of the constrained centroid $\mathbf{x}^c$ for any given subregion V of a smooth surface Ω can be effected using Newton's method or a damped Newton's method (Ju 2007). However, a more direct and less costly approach may be used

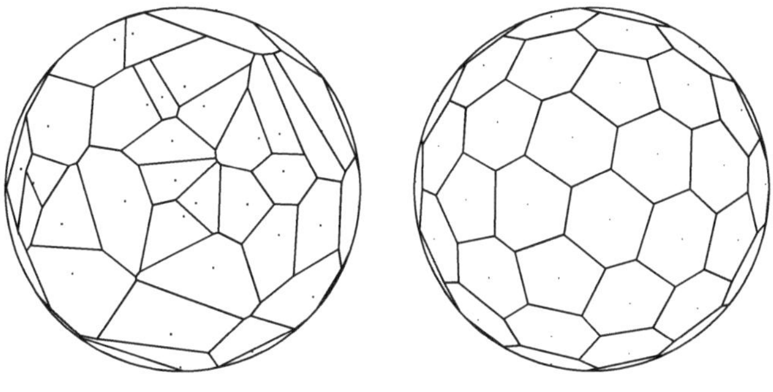

Fig. 10.6 (Du et al. 2003a) *Left*: A spherical Voronoi tessellation with 64 randomly selected generators. *Right*: a 64-generator spherical centroidal Voronoi tessellation for the uniform density

instead. One can first compute the standard centroid $\mathbf{x}^*$ of the subregion V as defined in (10.2). Note that, in general, the standard centroid $\mathbf{x}^*$ of V does not lie on the surface Ω; for example, for a region on the sphere, $\mathbf{x}^*$ is inside the sphere. Then, as is shown in (Du et al. 2003a), the constrained centroid $\mathbf{x}^c$ of $V \in \Omega$ can be found by projecting $\mathbf{x}^*$ onto Ω along the normal direction at $\mathbf{x}^c$. In particular, if V is a subset of the surface of a sphere of radius r, we have that its constrained center of mass is given by $\mathbf{x}^c = r\mathbf{x}^*/\|\mathbf{x}^*\|$.

10.3.2 Algorithms for Constructing CVTs

CVTs can be constructed either using probabilistic methods typified by MacQueen's random algorithm (MacQueen 1967) (which simply alternates between sampling and averaging points) or deterministic methods typified by Lloyd's method (Lloyd 1982) (which simply alternates between constructing Voronoi tessellations and mass centroids). Due to its effectiveness and simplicity, much attention has been focused on Lloyd's method.

Algorithm 1. (**Lloyd's Method**) *Given a domain Ω, a density function ρ defined on Ω, and a positive integer n (the number of generators).*

1. *Select an initial set of n points $\{\mathbf{x}_i\}_{i=1}^n$ on Ω.*
2. *Construct the Voronoi regions $\{V_i\}_{i=1}^n$ of Ω associated with $\{\mathbf{x}_i\}_{i=1}^n$.*
3. *Determine the centroids (or constrained centroids), with respect to the given density function, of the Voronoi regions $\{V_i\}_{i=1}^n$; these centroids form the new set of points $\{\mathbf{x}_i\}_{i=1}^n$; if Ω is a hypersurface, then $\mathbf{x}_i$ must be projected onto Ω.*
4. *If the new points meet some convergence criterion, return $\{(\mathbf{x}_i, V_i)\}_{i=1}^n$ and terminate; otherwise, go to Step* 2.

It has been shown (Du et al. 1999) that the energy $\mathscr{K}$ associated with the Voronoi tessellation *decreases monotonically* during the Lloyd iteration until a CVT is reached. Some convergence analyses of the Lloyd's method are given in (Du et al. 2006; Emelianenko et al. 2008).

In Step 1 of Algorithm 1, the initial set of points can be selected at random. However, because Lloyd's method only finds local minima of the clustering energy $\mathscr{K}$, the generator positions of the final CVT produced is affected by the initial distribution of generators.[11] Therefore, in some situations, one may want to use less noisy starting conditions; an example is given in Sect. 10.4.1.1.

For the second step, the methods described in Sect. 10.2.2 can be applied. There also exist software packages that may be used for Voronoi tessellation construction. For example, on the sphere, there is the STRIPACK package (Renka 1997).

[11] This is true for other CVT construction methods because, invariably, they only find local minima of the clustering energy.

The computation of centroids of the Voronoi regions in the third step of Algorithm 1 can be effected by first decomposing each Voronoi region into a set of triangles/tetrahedra and then using a high-order quadrature rule for triangles/tetrahedra to approximate integrals appearing in (10.2). Note that if the region of interest is a surface, e.g., part of the sphere or the whole sphere, then this step also includes the projection of the Euclidean centroid onto the surface; see Sect. 10.3.1.1.

For the fourth step, an example of a stopping criterion is if some measure, e.g., the root mean square, of the movement of the generators from one iteration to the next is smaller than a prescribed tolerance; alternately, one can stop if the change in the (computable) clustering energy is smaller than a prescribed tolerance.

A probabilistic version of a generalized Lloyd's method was proposed in (Ju et al. 2002a) together with its parallel implementation.[12]

Algorithm 2. **(Probabilistic Generalized Lloyd's Method)** *Given a domain Ω, a density function ρ defined on Ω, and a positive integer n.*

1. *Choose a positive integer q (the number of sampling points per iteration) and constants $\{\alpha_i, \beta_i\}_{i=1}^2$ such that $\alpha_2 > 0$, $\beta_2 > 0$, $\alpha_1 + \alpha_2 = 1$, and $\beta_1 + \beta_2 = 1$; choose an initial set of n points $\{\mathbf{x}_i\}_{i=1}^n$; set $j_i = 1$ for $i = 1, \ldots, n$.*
2. *Choose q sample points $\{\mathbf{y}_r\}_{r=1}^q$ in Ω at random, e.g., by a Monte Carlo method, with the density function $\rho(\mathbf{x})$ acting as the probability density function.*
3. *For $i = 1, \ldots, n$, gather together in the set W_i all sampled points $\mathbf{y}_r$ closest to $\mathbf{x}_i$ among $\{\mathbf{x}_i\}_{i=1}^n$, i.e., all sampled points in the Voronoi region of $\mathbf{x}_i$; if the set W_i is empty, do nothing; otherwise, compute the average $\mathbf{u}_i$ of the set W_i and set*

$$\mathbf{x}_i \leftarrow \frac{(\alpha_1 j_i + \beta_1)\mathbf{x}_i + (\alpha_2 j_i + \beta_2)\mathbf{u}_i}{j_i + 1} \quad \text{and} \quad j_i \leftarrow j_i + 1; \qquad (10.5)$$

   the new set of $\{\mathbf{x}_i\}$, along with the unchanged $\{\mathbf{x}_j\}$ corresponding to empty W_j, form the new set of points $\{\mathbf{x}_i\}_{i=1}^n$; if Ω is a hypersurface, then $\mathbf{x}_i$ must be projected onto Ω.
4. *If the new points meet some convergence criterion, terminate; otherwise, return to Step 2.*

In Steps 1 and 2 of Algorithm 2 as well as in Step 1 of Algorithm 1, points need to be sampled according to a given density function ρ. Such sampling steps may be accomplished by a rejection method (Du et al. 2003a; Ju et al. 2002a; Ross 1998) which we now describe. Given a general domain Ω in the plane or on the sphere, determine an enclosing rectangle D whose sides are parallel to the coordinate axes or, on the sphere, are latitude and longitude lines, and which contains all points in $\overline{\Omega}$. Set $\rho_{max} = \max_{\mathbf{x}\in\Omega} \rho(\mathbf{x})$. Then, there are two rejection tests applied. First,

[12] This algorithm can also be viewed as a generalization of MacQueen's method (MacQueen 1967); see (Ju et al. 2002a). In fact, if in (10.5), $q = 1$, $\alpha_2 = \beta_1 = 0$, and $\alpha_1 = \beta_2 = 1$, Algorithm 2 reduces to MacQueen's method.

a point $\mathbf{y}$ in D is sampled according to a uniform distribution;[13] this is done by uniformly sampling each coordinate; all computer systems have a built-in uniform random sampling method. If the sampled point $\mathbf{y}$ is not in $\overline{\Omega}$, it is rejected and one samples again. If the sampled point is in $\overline{\Omega}$, a scalar ϕ is sampled uniformly in the interval [0, 1]. If $\phi < \rho(\mathbf{y})/\rho_{max}$, then the sample point is accepted; otherwise, it is rejected.[14]

In Step 3 of Algorithm 2, the average $\mathbf{u}_i$ of the sampled points in W_i is given by

$$\mathbf{u}_i = \frac{\sum_{\mathbf{y} \in W_i} \mathbf{y}}{\# W_i},$$

where $\# W_i$ denotes the number of elements in W_i. Since the points in W_i are randomly selected points in the Voronoi region corresponding to $\mathbf{x}_i$, one may view $\mathbf{u}_i$ as a probabilistic approximation to the centroid (or constrained centroid) of V_i; the larger is q, the better the centroid approximations.[15] Note that j_i keeps track of the number of times that $\mathbf{x}_i$ has been previously updated. Some over-relaxation updating methods can be defined by appropriately choosing $\{\alpha_1, \alpha_2, \beta_1, \beta_2\}$; see (Ju et al. 2002a).

Algorithm 2 is much easier to implement and code than Algorithm 1. For Algorithm 1, one has to explicitly construct Voronoi tessellations and determine centers of mass of Voronoi regions. These steps are doable in two-dimensional settings such as planar regions and regions on the sphere and in three-dimensional volumes, but involve considerable coding. On general surfaces in three-dimensions, algorithms for Voronoi tessellations are not generally available and in regions in four and higher dimensions, the calculation of centers of mass become impractical. On the other hand, to find the generators of a CVT, Algorithm 2 does not require the construction of Voronoi tessellations or of centers of mass; both are approximated via sampling. Thus Algorithm 2 can be applied to regions and hypersurfaces in arbitrary dimensions.

The accuracy of Algorithm 1 is limited only by machine precision, although, in practice, one would not want to iterate to that level of accuracy. On the other hand, for Algorithm 2, accuracy is limited by the sampling errors made in Step 2. The q sampled points are divided among the generators so that, say, in a uniform density setting, each generator would only be assigned roughly q/n points, where n denotes the number of generators. Thus, if, say, Monte Carlo sampling is used, the errors in the probabilistic approximations of the centroids of the Voronoi regions

---

[13] Instead of random, i.e., Monte Carlo, sampling, one can, in conjunction with the rejection steps, use quasi-Monte Carlo, Latin hypercube, etc. sampling methods (McKay and Beckman 1979; Niederreiter 1992; Saltelli et al. 2004) appropriate for hypercubes.

[14] Note that both rejection tests can be incorporated into a single test because an alternate means for rejecting points that are outside of $\overline{\Omega}$ is to simply set $\rho(\mathbf{x}) = 0$ outside of $\overline{\Omega}$.

[15] If $\alpha_1 = \beta_1 = 0$, and $\alpha_2 = \beta_2 = 1$, we have in (10.5) that $\mathbf{x}_i \leftarrow \mathbf{u}_i$, i.e., the new generators are probabilistic approximations of the centroid of the Voronoi regions; this justifies saying that Algorithm 2 is a probabilistic generalized Lloyd's method.

would be proportional to $\sqrt{n/q}$ so that this is the best accuracy one can expect from Algorithm 2. Note that, for fixed q, the accuracy degrades as we increase the number of generators n and that, for fixed n, greater accuracy can be achieved by increasing the number of sample points q. Also, note that it is useless to set a tolerance in whatever stopping criterion is used in Step 4 of Algorithm 2 to be smaller than $O(\sqrt{n/q})$.

Because accuracy control is better served by Algorithm 1, it is usually the algorithm of choice for regions in the plane and on the sphere and for three-dimensional regions. For other cases, e.g., higher-dimensional regions and general surfaces in three dimensions, Algorithm 2 becomes more practical.

We close this section on algorithms for CVT construction by noting that several other schemes for computing CVTs such as Newton-type algorithms and multi-level methods are studied in (Du and Emelianenko 2006, 2008; Liu et al. 2009).

10.3.3 The Relation Between the Density Function and the Local Mesh Size

An interesting problem about the asymptotic behavior CVTs is the distribution of the energy $\mathscr{K}$ defined in (10.3). It was shown in (Du et al. 1999), that in the one-dimensional case, for the CVT of n generators $\{(\mathbf{x}_i, V_i)\}_{i=1}^n$ with a smooth density function ρ, we have

$$\mathscr{K}_i \approx \frac{1}{12}\rho(\mathbf{x}_i)h_i^3 \approx \frac{\mathscr{K}}{n}, \quad \forall\, 1 \le i \le n, \tag{10.6}$$

where h_i denotes the diameter of V_i, $\mathscr{K}_i = \int_{V_i} \rho(\mathbf{x})\|\mathbf{x} - \mathbf{x}_i\|^2 \, d\mathbf{x}$, and $\mathscr{K} = \sum_{i=1}^n \mathscr{K}_i$, i.e., under some assumptions on the density function, asymptotically speaking, the energy is equally distributed in the Voronoi intervals and the diameter of Voronoi intervals are inversely proportional to the one-third power of the underlying density. Based on (10.6) and the fact $\sum_{i=1}^n h_i$ = length of Ω, we then obtain an approximation of total clustering energy of the CVT in one dimension given by, for n large,

$$\mathscr{K} \approx \frac{1}{12}\frac{\left(\int_\Omega \rho^{1/3}\, d\mathbf{x}\right)^3}{n^2}.$$

Let d denote the space dimension and set $d' = d - 1$ if Ω is a hypersurface and $d' = d$ otherwise. For higher dimensions, a similar conjecture about CVTs or CCVTs can be stated as follows:

$$\mathscr{K}_i \approx c_1\rho(\mathbf{x}_i)h_i^{d'+2} \approx \frac{\mathscr{K}}{n}, \quad \forall\, 1 \le i \le n, \tag{10.7}$$

$$\mathcal{K} \approx \frac{c_2}{n^{2/d'}} \left(\int_{\Omega} \rho^{d'/(d'+2)} \, \mathrm{d}\mathbf{x} \right)^{(d'+2)/d'}, \tag{10.8}$$

where c_1, c_2 are constants depending only on d'. This conjecture still remains open for $d \geq 2$ although its validity has been supported through many numerical studies and widely used for applications in vector quantizations (Gersho and Gray 1992) and image processing.

A direct consequence of (10.7) is

$$\frac{h_i}{h_j} \approx \left(\frac{\rho(\mathbf{x}_j)}{\rho(\mathbf{x}_i)} \right)^{1/(d'+2)}. \tag{10.9}$$

The relation (10.9) between the density function and the local mesh sizes is also very useful in CVT-based adaptive mesh generation and optimization (Ju 2007; Ju et al. 2002b).

10.4 Application to Climate and Global Modeling

10.4.1 Global SCVT Meshes

We define quantitative measures of grid quality that we can use to assess the quality of meshing schemes on the sphere.

Given a Voronoi mesh $\{(\mathbf{x}_i, V_i)\}_{i=1}^n$, set $Q = \{(i, j) \mid \mathbf{x}_i \text{ and } \mathbf{x}_j \text{ are neighbors}\}$ and let

$$h_{min} = \min_{(i,j)\in Q} \|\mathbf{x}_i - \mathbf{x}_j\| \qquad \text{and} \qquad h_{max} = \max_{(i,j)\in Q} \|\mathbf{x}_i - \mathbf{x}_j\|.$$

Clearly, the ratio (Du et al. 2003b)

$$\mu = \frac{h_{max}}{h_{min}} \tag{10.10}$$

is a natural measurement of the *global uniformity* of the Voronoi mesh $\{\mathbf{x}_i, V_i\}_{i=1}^n$. It is clear that $\mu \geq 1$ and the smaller is μ, the more globally uniform is the Voronoi mesh.

Letting χ_i denote the set of neighbor generators of $\mathbf{x}_i$, a measure of the *local quality* or *local uniformity* of the Voronoi mesh at $\mathbf{x}_i$ is given by

$$\sigma_i = \frac{\min_{j\in\chi_i} \|\mathbf{x}_i - \mathbf{x}_j\|}{\max_{j\in\chi_i} \|\mathbf{x}_i - \mathbf{x}_j\|}.$$

Clearly $0 < \sigma_i \leq 1$ and the larger is σ, the better the local uniformity.

We apply the commonly used q-measure (Field 2000) to evaluate the quality of dual Delaunay triangular meshes, where, for any triangle T_i, q_i is defined to be twice the ratio of the radius R_{T_i} of the largest inscribed circle and the radius r_{T_i} of the smallest circumscribed circle, i.e.,

$$q_i = 2\frac{R_{T_i}}{r_{T_i}} = \frac{(b+c-a)(c+a-b)(a+b-c)}{abc}, \tag{10.11}$$

where a, b, and c denote the side lengths of T_i. Clearly, $0 < q_i \le 1$ and $q_i = 1$ corresponds to the equilateral triangle.

We then define the mesh quality measures

$$\sigma_{min} = \min_{i=1,\dots,m_D} \sigma_i, \quad \sigma_{avg} = \frac{1}{n}\sum_{i=1}^{m_d} \sigma_i, \quad q_{min} = \min_{i=1,\dots,m_D} q_i,$$
$$q_{avg} = \frac{1}{m_D}\sum_{i=1}^{m_d} q_i,$$

where m_D denotes the number of dual Delaunay triangles. The closer these measures are to unity, the better the mesh.

10.4.1.1 Uniform SCVT Meshes vs. Icosahedral-Bisection Meshes

Icosahedral-bisection meshes on the sphere have been widely used in the climate and global modeling communities; icosahedral-bisection meshes from a family of hierarchical meshes with $10 \times 4^{\ell-1} + 2$ nodes at level ℓ, in which there are 12 pentagons and all others cells are hexagons. The level $\ell = 1$ and $\ell = 2$ meshes having 12 and 42 nodes, respectively, are SCVT meshes with respect to the uniform density, but all other members of the family with levels $l > 2$ are not SCVTs, although they are quite uniform. We use the centroids of the Voronoi cells of each icosahedral-bisection mesh as the initial guess and apply Lloyd's method with a uniform density to generate a sequence of SCVT meshes; see Fig. 10.7. The quality measures of Sect. 10.4 for the icosahedral-bisection and uniform SCVT meshes are given in Table 10.1. The SCVT meshes do better with respect to the local mesh quality measures, i.e., with respect to local mesh uniformity, although they get worse with respect to global mesh uniformity due to the shrinking relative size of the pentagonal cells as the mesh size decreases.

10.4.1.2 Locally Refined SCVT Meshes

Let a point $\mathbf{x}$ on the sphere be represented by its spherical coordinate $\mathbf{x} = (lat, lon)$ with $-\pi/2 \le lat \le \pi/2$ and $0 \le lon < 2\pi$. Set $\mathbf{x}_c = (\pi/6, 3\pi/2)$ and define

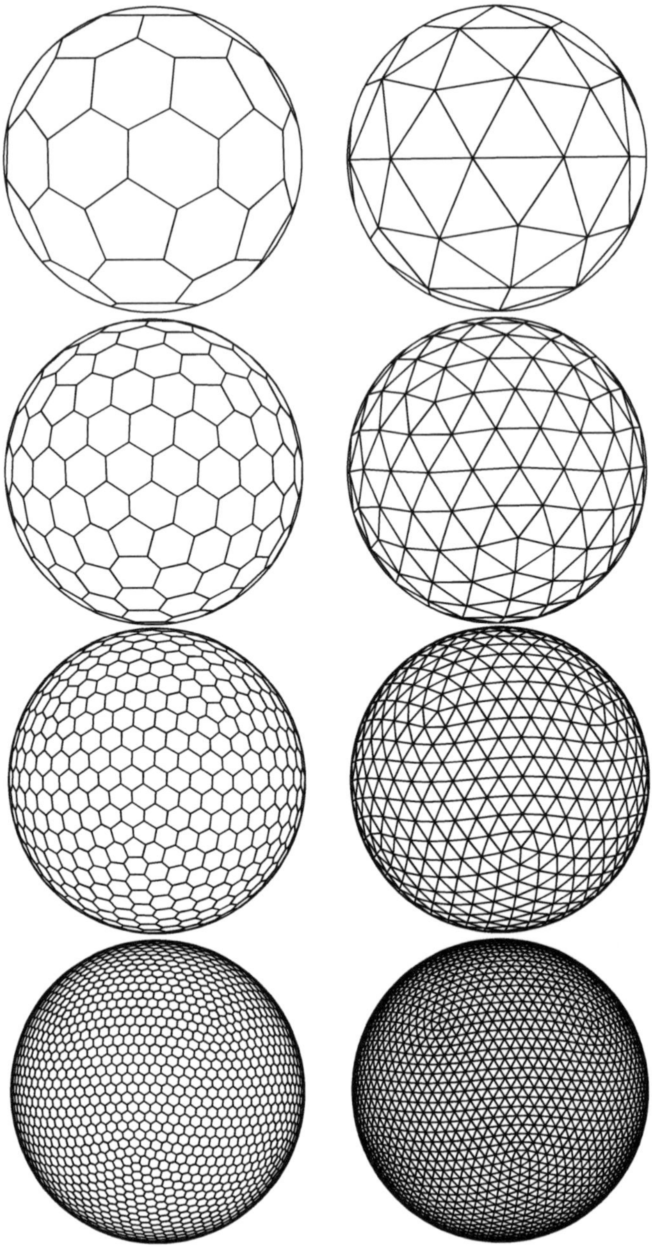

Fig. 10.7 *From top to bottom*: spherical centroidal Voronoi tessellations (*left column*) with 42, 162, 642, 2,562 generators for a uniform density and the corresponding spherical Delaunay triangulations (*right column*)

Table 10.1 Comparisons of quality of icosahedral-bisection (I-B) and uniform spherical centroidal Voronoi tessellation (SCVT) meshes

Level ℓ	# of generators	Mesh types	μ	σ_{avg}	σ_{min}	q_{avg}	q_{min}
2	42	I-B	1.1308	0.9174	0.8843	0.9872	0.9829
		SCVT	1.1308	0.9174	0.8843	0.9872	0.9829
3	162	I-B	1.1777	0.9111	0.8586	0.9904	0.9729
		SCVT	1.1647	0.9174	0.8843	0.9872	0.9829
4	642	I-B	1.1907	0.8737	0.8482	0.9865	0.9701
		SCVT	1.1592	0.9121	0.8525	0.9923	0.9701
5	2,562	I-B	1.1940	0.8803	0.8405	0.9866	0.9694
		SCVT	1.2335	0.9141	0.8511	0.9931	0.9694
6	10,242	I-B	1.1948	0.8879	0.8386	0.9866	0.9692
		SCVT	1.2710	0.9157	0.8507	0.9934	0.9692
7	40,962	I-B	1.1951	0.8932	0.8380	0.9866	0.9692
		SCVT	1.3107	0.9168	0.8504	0.9935	0.9692
8	163,842	I-B	1.1951	0.8966	0.8379	0.9870	0.9692
		SCVT	1.3526	0.9173	0.8494	0.9952	0.9687
9	655,362	I-B	1.1952	0.8970	0.8378	0.9952	0.9691
		SCVT	1.4080	0.9167	0.8465	0.9987	0.9675

$$d_s(\mathbf{x}, \mathbf{x}_c) = \sqrt{(lat - \pi/6)^2 + (lon - 3\pi/2)^2}.$$

Define the subregion of the sphere

$$S_{mt} = \{\mathbf{x} = (lat, lon) \mid d_s(\mathbf{x}, \mathbf{x}_c) \leq \pi/6\}.$$

In the subregion, we want a high-quality mesh having a local mesh size that is γ_s times smaller than that outside the subregion. We also want a smooth transition between the coarse and fine grid regions.

Using the density-mesh size relation (10.9), the density function is set to

$$\rho(\mathbf{x}) = \begin{cases} \gamma_s^4 & \text{if } d_s(\mathbf{x}, \mathbf{x}_c) \leq \pi/6 \\ ((1 - s_{\mathbf{x}})\gamma_s + s_{\mathbf{x}})^4 & \text{if } \pi/6 < d_s(\mathbf{x}, \mathbf{x}_c) \leq \pi/6 + \epsilon_s \\ 1 & \text{otherwise,} \end{cases} \quad (10.12)$$

where ϵ_s denotes the width of the transition layer and $s_{\mathbf{x}} = \frac{d_s(\mathbf{x},\mathbf{x}_c) - \pi/6}{\epsilon_s}$; we set $\gamma_s = 3$ and $\epsilon_s = \pi/12$ here. The resulting SCVT with 2,562 generators produced by Lloyd's method and the corresponding dual Delaunay triangulation are presented in Fig. 10.8 (*top row*). Variations in the Voronoi cell sizes and areas are plotted in Fig. 10.8 (*bottom row*). The histogram of the size distribution clearly indicates that there are two dominant mesh sizes; cells 1–1,250 have one size, cells 1,500–2,500 have another size, and these two cells sizes differ by a factor of three as predicted by (10.9). For this example, we have $\mu = 5.4018$, $\sigma_{avg} = 0.8712$, $\sigma_{min} = 0.4533$, $q_{avg} = 0.9854$, and $q_{min} = 0.6886$.

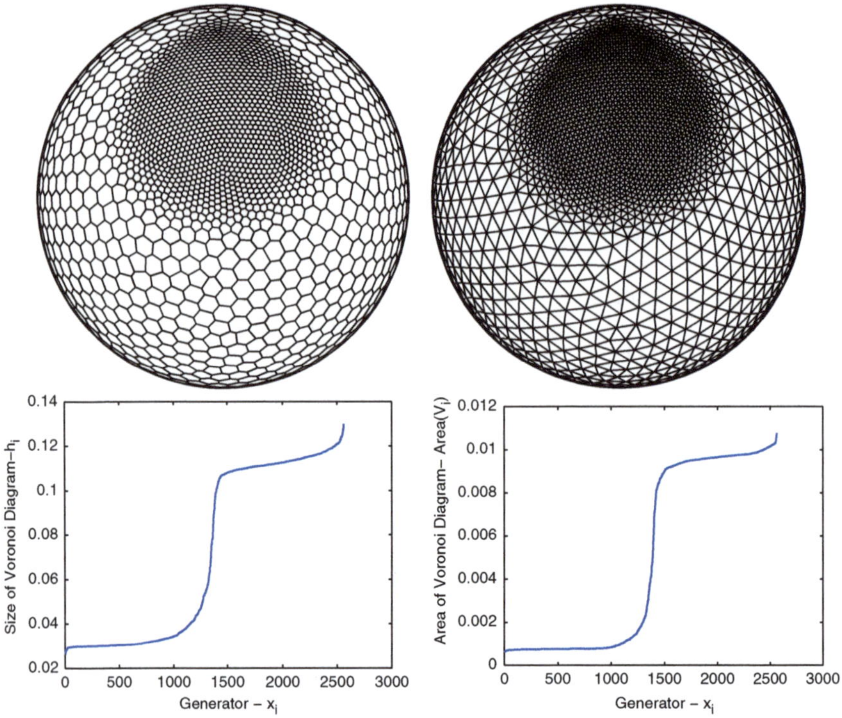

Fig. 10.8 *Top row*: a spherical centroidal Voronoi tessellation (*left*) and its dual spherical Delaunay triangulation (*right*) with 2,562 generators and the density (10.12); *bottom row*: plot of Voronoi cell sizes (*left*) and areas (*right*)

Figure 10.8 as well as Fig. 10.9 below illustrate an important feature of nonuniform CVT and SCVT grids, namely smooth transitions from coarse to fine grids. This can always be effected within the CVT/SCVT framework through the use of smooth density functions so that, if a given density function is not smooth, it is often beneficial to smooth it before using it to generate CVT/SCVT grids; see Sect. 10.4.2.

10.4.1.3 Nested SCVT Meshes

For this example, the region of interest covers most of North America, i.e.,

$$S_{n\ell} = \{\mathbf{x} = (lat, lon) \mid -5^\circ \le lat \le 60^\circ, \quad 225^\circ \le lat \le 310^\circ\}.$$

Again, we want a high-quality mesh with local mesh size in $S_{n\ell}$ being approximately $\gamma_s = 3$ times smaller that in outside that region. This time we use a different means to generate a locally refined SCVT mesh because we wish to make use of global uniform SCVT meshes.

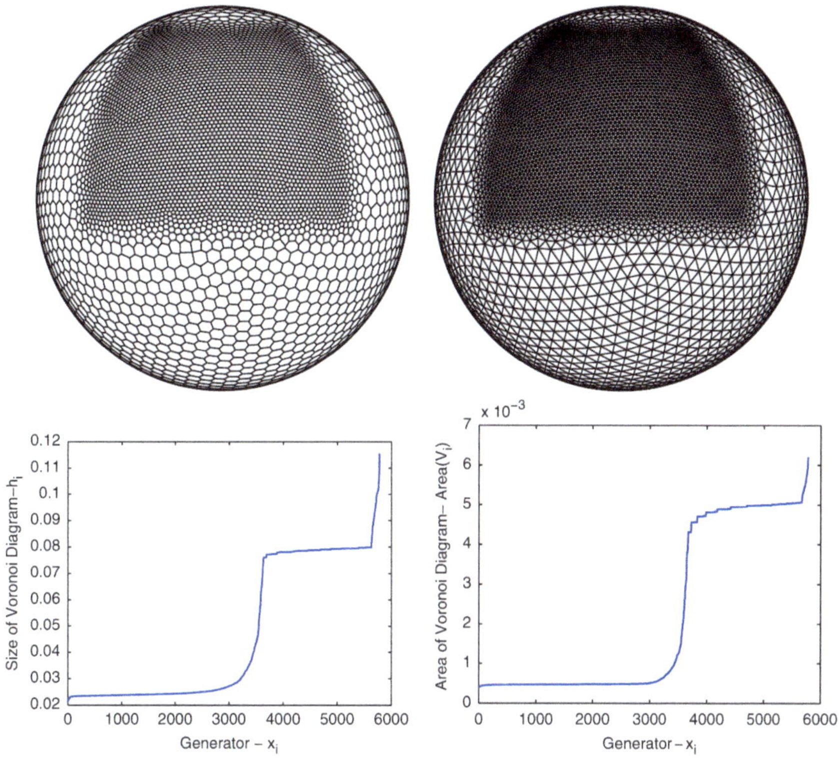

Fig. 10.9 *Top row*: a spherical centroidal Voronoi tessellation (*left*) with 5,781 generators and its dual spherical Delaunay triangulation (*right*) produced by the nested method; *bottom row*: plot of Voronoi cell sizes (*left*) and areas (*right*)

We begin with the global uniform SCVT with 2,562 nodes shown in Sect. 10.4.1.1. The submesh falling inside $S_{n\ell}$ has about 355 nodes. We refine this submesh to get a new mesh of $S_{n\ell}$ with 3,574 nodes (about ten times more nodes). We then merge the refined submesh with the remaining generators of the original unform SCVT outside of $S_{n\ell}$ and produce a new global nonuniform Voronoi mesh with 5,781 generators; the result is clearly not a SCVT but we use it as an initial guess for Lloyd's method. We choose a Similar to (10.12), we choose the density function

$$\rho(\mathbf{x}) = \begin{cases} \gamma_s^4 & \text{if } \mathbf{x} \in S_{n\ell} \\ ((1 - s_{\mathbf{x}})\gamma_s + s_{\mathbf{x}})^4 & \text{if } 0 < d(\mathbf{x}, S_{n\ell}) \leq \epsilon_s \\ 1 & \text{otherwise,} \end{cases} \qquad (10.13)$$

where $s_{\mathbf{x}} = \frac{d(\mathbf{x}, S_{n\ell})}{\epsilon_s}$ and the width of the transition layer $\epsilon_s = 0.24$. Then, we apply Lloyd's method with this density, adding one more restriction: all generators $\mathbf{x}_i$ are fixed during the iterations if $d(\mathbf{x}_i, S_{n\ell}) > \epsilon_s$.

The resulting SCVT with 5,781 generators and its dual Delaunay triangulation are presented in Fig. 10.9 (*top row*). Variations of Voronoi cell sizes and areas are plotted in the bottom row. For this example, we have $\mu = 5.7079$, $\sigma_{avg} = 0.9006$, $\sigma_{min} = 0.4012$, $q_{avg} = 0.9904$ and $q_{min} = 0.7114$.

10.4.2 CVT-Based Regional Meshes of the North Atlantic Ocean

Figure 10.10 (*top left*) shows the time-mean kinetic energy from a global 0.1° simulation of the North Atlantic Ocean (Smith et al. 2000). We use this data set to determine both the boundary of the North Atlantic ocean and an appropriate density function, and then construct the CVT mesh based on this information; see (Ringler et al. 2008) for details.[16]

Based on the kinetic energy, *KE*, we defined the density function

$$\rho = \max\left[0.1, \frac{KE}{KE_{\max}}\right]^4,$$

where KE_{max} is the maximum kinetic energy in the domain. The lower bound 0.1 insures that the grid in quiescent regions is not overly coarse. We also raise the value of the density function as we approach the boundary of the ocean so that the boundary is resolved to a desired resolution; this is accomplished by making the density in regions near a land boundary also depend in an inverse manner on the distance to the boundary. The resulting mesh has a grid spacing that varies by a factor of 10.

In order to allow for a smooth transition between regions of high and low resolution, we apply a substantial amount, e.g., approximately 20 passes, of Laplacian smoothing[17] to our density function. Figure 10.10 shows some of the resulting CVT meshes. Whereas the two examples given above produce a mesh with two dominant resolutions, in this example a wide spectrum of resolutions are present. Note that this type of mesh will lead to additional complications related to parameter settings of sub-grid closures but that is also offers the opportunity to adaptively select multiple closure models whose efficacy depends on the local grid size. All in all, variational resolution meshes such as the one illustrated in Fig. 10.10 are significantly more ambitious than those considered in Sects. 10.4.1.2 and 10.4.1.3.

[16] In practice, we would not use such a proxy to determine a variable resolution CVT grid, but instead would adaptively determine the grid from the simulation model output.

[17] In the current context, Laplacian smoothing is a process of smoothing the a function defined on a grid. One replaces the value of a function at a point by first averaging its value at neighboring points and then averaging that result with its own value at the point.

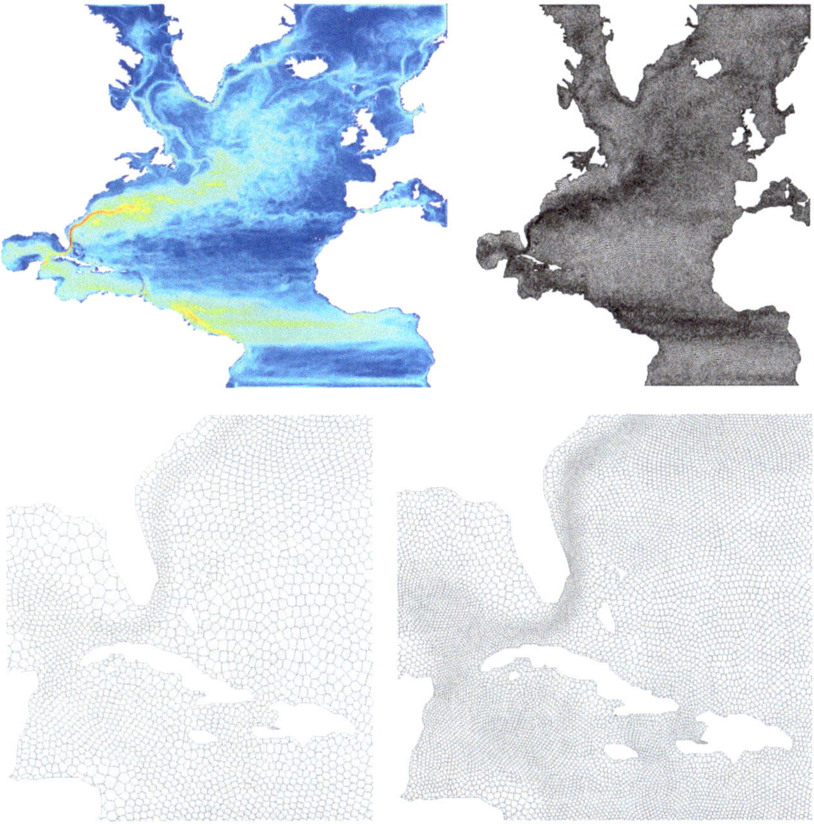

Fig. 10.10 *Top-left*: time-mean kinetic energy of the North Atlantic Ocean; *top-right*: a CVT mesh with 47,305 generators of the North Atlantic; *bottom-left*: a zoom-in of the CVT mesh; *bottom-right*: a zoom-in of the same region of a CVT mesh with 183,907 generators

10.4.3 Numerical Simulations with SCVT Meshes

10.4.3.1 Mesh Decomposition for Parallel Computing

We take a global SCVT mesh with 40,962 generators (about 120 km resolution) and separate it into 642 blocks; see Fig. 10.11. These blocks are created so as to balance the work-per-block and to minimize the amount of information that must be communicated between blocks; the software package "METIS" (Karypis and Kumar 1998) in which a family of multilevel partitioning algorithms is implemented is used for this purpose. We can assign an arbitrary number of blocks per processor so that two types of parallelism within are supported within this framework, i.e., distributed memory across nodes and shared memory within a node.

Fig. 10.11 Decomposition of a global SCVT mesh of 40,962 generators into 642 blocks. The blocks can be distributed across computational nodes for implementation on high-performance architectures

10.4.3.2 Example Numerical Methods

As discussed in Sect. 10.1, all typical finite-volume grid staggerings used for quadrilateral meshes, i.e., A-, B-, C- and Z-grid staggerings, have been successfully applied to Voronoi tessellations. C-grid staggering has shown promising results, particularly when applied to variable resolution meshes. See Check: Ringler Dyncore Chapter for a broad discussion of C-grid staggerings and see (Thuburn et al. 2009; Ringler et al. 2010) for an in-depth discussion of C-grid staggering applied to the nonlinear shallow-water equations.

We apply the methods developed in (Thuburn et al. 2009; Ringler et al. 2010) to test case 5 of the standard shallow-water test cases developed in (Williamson et al. 2001). A flow in geostrophic balance is confronted with a large-scale orographic feature at the start of the simulation, $t = 0$. The transient forcing at $t = 0$ leads to the generation of large-amplitude gravity waves and Rossby waves. The sole forcing mechanism is the presence of the orographic forcing. While no analytical solution is known, results from high-resolution global spectral models (Lipscomb and Ringler 2005) are adequate reference solutions for the simulations conducted here.

Figure 10.12 shows the potential vorticity and kinetic energy at day 50 when using a SCVT with 40,962 cells based on a uniform density function. Shallow-water test case 5 is shown to breakdown into 2D turbulence after day 25, so Fig. 10.12 shows a snapshot of this turbulent behavior. Even in the presence of fully-developed 2D turbulence, the simulation is stable and robust while conserving total energy to within time truncation error. Simulations of this same test case, but using the variable resolution meshes shown in Figs. 10.8 and 10.9, produce equally robust results.

10.5 Summary

Voronoi tessellations and, in particular, centroidal Voronoi tessellations, offer a robust approach to tiling the surface of the sphere. The Delaunay triangulation is the dual of the Voronoi tessellations, so whether hexagons or triangles are of

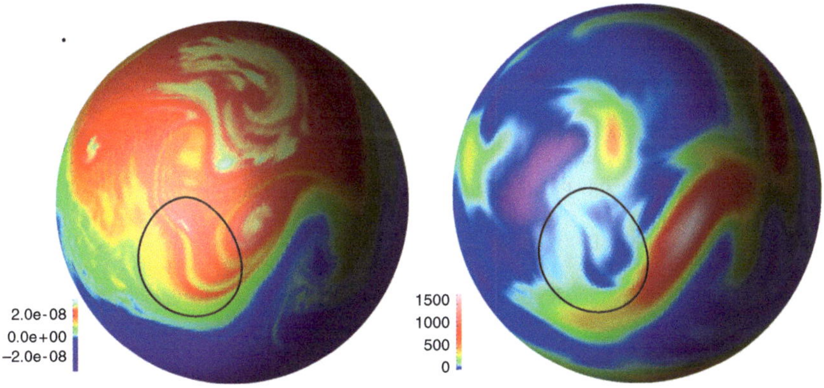

Fig. 10.12 Simulation results at day 50 using a uniform SCVT mesh with the method outlined in (Ringler et al. 2010). The figure depicts the potential vorticity field (*left*) and the kinetic energy field (*right*). The simulation conserves potential vorticity to machine precision and total energy to within time-truncation error

interest, this approach will result in high-quality uniform and nonuniform meshes. Centroidal Voronoi tessellations are particularly well-suited for the generation of smoothly varying meshes, thus providing a possible alternative to traditional nesting approaches. With the recent discovery of a class of finite-volume methods that are directly applicable to variable resolution meshes (Thuburn et al. 2009; Ringler et al. 2010), it appears that the creation of variable resolution, global climate system models is now possible.

Acknowledgments This work was supported by the US Department of Energy Office of Science Climate Change Prediction Program through grant numbers DE-FG02-07ER64431 and DE-FG02-07ER64432, and the US National Science Foundation under grant numbers DMS-0609575 and DMS-0913491.

The authors thank the reviewers and editors for the many helpful comments that resulted in substantial improvements to this chapter.

References

Alliez P, Cohen-Steiner D, Yvinec M, Desbrun M (2005) Variational tetrahedral meshing. In: Proceedings of SIGGRAPH, pp 617–625

Arakawa A (1966) Computational design for long-term numerical integration of the equations of fluid motion: Two-dimensional incompressible flow. J Comput Phys 1:119–143

Arakawa A, Lamb V (1977) Compuational design of the basic dynamical processes in the UCLA general circulation model. J Comput Phys 17:173–265

Arakawa A, Lamb V (1981) A potential enstrophy and energy conserving scheme for the shallow water equations. Mon Wea Rev 109:18–36

Augenbaum J (1984) A lagrangian method for the shallow water equations based on a voronoi mesh - one dimensional results. J Comput Phys 53:240–265

Augenbaum J, Peskin C (1985) On the construction of the Voronoi mesh on the sphere. J Comput Phys 59:177–192
Barnes E, Sloane N (1983) The optimal lattice quantizer in three dimensions. SIAM J Algeb Disc Meth 4:31–40
Bentley J, Weide B, Yao A (1980) Optimal expected-time algorithms for cloest point problems. ACM Trans Math Soft 6:563–580
Boissonnat J, Oudot S (2005) Provably good sampling and meshing of surfaces. Graph Models 67:405–451
Chynoweth S, Sewell M (1990) Mesh duality and Legendre duality. Proc R Soc London A 428:351–377
Cignoni P, Montani C, Scopigno R (1998) Dewall: A fast divide and conquer delaunay triangulation algorithm in e^d. Computer-Aided Design 30:333–341
Cooley J, Tukey J (1965) An algorithm for the machine calculation of complex Fourier series. Math Comp 19:197–301
Delaunay B (1928) Sur la sphère vide. In: Proceedings of the International Mathematical Congress, pp 695–700
Delaunay B (1934) Sur la sphère vide. Izvestia Akademii Nauk SSSR, Otdelenie Matematicheskikh i Estestvennykh Nauk 7:793–800
Dirichlet G (1850) Uber die Reduktion der positiven quadratischen Formen mit drei unbestimmten ganzen Zahlen. J Reine Angew Math 40:209–227
Du Q, Emelianenko M (2006) Acceleration schemes for computing the centroidal Voronoi tessellations. Numer Linear Alg Appl 13:173–192
Du Q, Emelianenko M (2008) Uniform convergence of a nonlinear energy-based multilevel quantization scheme via centroidal Voronoi tessellations. SIAM J Numer Anal 46:1483–1502
Du Q, Ju L (2005) Finite volume methods on spheres and spherical centroidal Voronoi meshes. SIAM J Numer Anal 43:1673–1692
Du Q, Wang D (2003) Tetrahedral mesh generation and optimization based on centroidal Voronoi tessellations. Int J Numer Methods Engrg 56:1355–1373
Du Q, Wang D (2005) On the optimal centroidal Voronoi tessellations and Gersho's conjecture in the three dimensional space. Comput Math Appl 49:1355–1373
Du Q, Faber V, Gunzburger M (1999) Centroidal Voronoi tessellations: Applications and algorithms. SIAM Review 41:637–676
Du Q, Gunzburger M, Ju L (2003a) Constrained centroidal Voronoi tessellations for surfaces. SIAM J Sci Comput 24:1488–1506
Du Q, Gunzburger M, Ju L (2003b) Voronoi-based finite volume methods, optimal Voronoi meshes and PDEs on the sphere. Comput Meth Appl Mech Engrg 192:3933–3957
Du Q, Emelianenko M, Ju L (2006) Convergence of the lloyd algorithm for computing centroidal Voronoi tessellations. SIAM J Numer Anal 44:102–119
Edelsbrunner H, Shah N (1996) Incremental topological flipping works for regular triangulations. Algorithmica 15:223–241
Emelianenko M, Ju L, Rand A (2008) Nondegeneracy and weak global convergence of the Lloyd algorithm in $\mathbb{R}^d$. SIAM J Numer Anal 46:1423–1441
Field D (2000) Quantitative measures for initial meshes. Int J Numer Meth Engrg 47:887–906
Fortune S (1986) A sweepline algorithm for Voronoi diagrams. In: Proceedings of the Second Ann. Symp. Comput. Geom., pp 313–322
Freitag L, Ollivier-Gooch C (1997) Tetrahedral mesh improvement using swapping and smoothing. Int J Numer Methods Engrg 40:3979–4002
Gersho A (1979) Asymptotically optimal block quantization. IEEE Trans Inform Theory 25:373–380
Gersho A, Gray R (1992) Vector Quantization and Signal Compression. Kluwer, Boston
Guiba L, Knuth D, Sharir M (1992) Randomized incremental construction of Delaunay and Voronoi diagrams. Algorithmica 7:381–413
Guibas L, Stolfi J (1985) Primitives for the manipulation of general subdivisions and the computation of Voronoi diagrams. ACM Trans Graphics 4:74–123

Hartigan J (1975) Clustering Algorithms. Wiley, New York
Heikes R, Randall D (1995a) Numerical integration of the shallow-water equations on a twisted icosahedral grid. Part i: basic design and results of tests. Mon Wea Rev 123:1862–1880
Heikes R, Randall D (1995b) Numerical integration of the shallow-water equations on a twisted icosahedral grid. Part ii. a detailed description of the grid and an analysis of numerical accuracy. Mon Wea Rev 123:1881–1887
Ju L (2007) Conforming centroidal Voronoi Delaunay triangulation for quality mesh generation. Inter J Numer Anal Model 4:531–547
Ju L, Du Q, Gunzburger M (2002a) Probabilistic methods for centroidal Voronoi tessellations and their parallel implementations. Para Comput 28:1477–1500
Ju L, Gunzburger M, Zhao W (2002b) Adaptive finite element methods for elliptic PDEs based on conforming centroidal Voronoi-Delaunay triangulations. SIAM J Sci Comput 28:2023–2053
Kanungo T, Mount D, Netanyahu N, Piatko C, Silverman R, Wu A (2002) An efficient k-means clustering algorithm: Analysis and implementation. IEEE Trans Pattern Anal Mach Intel 24:881–892
Karypis G, Kumar V (1998) A fast and high quality multilevel scheme for partitioning irregular graphs. SIAM J Sci Comput 20:359–392
Kasahara A, Washington W (1967) NCAR global general circulation model of the atmosphere. Mon Wea Rev 95:389–402
Kittel C (2004) Introduction to Solid State Pgysics. Wiley, New York
Lee D, Schachter B (1980) Two algorithms for constructing a Delaunay triangulation. Int J Inform Sci 9:219–242
Lipscomb W, Ringler T (2005) An incremental remapping transport scheme on a spherical geodesic grid. Mon Wea Rev 133:2335–2350
Liu Y, Wang W, Levy B, Sung F, Yan DM (2009) On centroidal Voronoi tessellation - Energy smoothness and fast computation. ACM Trans Graphics 28:1–17
Lloyd S (1982) Least squares quantization in PCM. IEEE Trans Inform Theory 28:129–137
MacQueen J (1967) Some methods for classification and analysis of multivariate observations. In: Proceedings of the Fifth Berkeley Symp. Math. Stat. and Prob., Ed. by L. LeCam and J. Neyman, vol I, pp 181–197
Masuda Y, Ohnishi H (1986) An integration scheme of the primitive equations model with an icosahedral-hexagonal grid system and its application to the shallow water equations. In: Short- and Medium-Range Numerical Weather Prediction, Ed. by T. Matsuno, Universal Academy Press, pp 317–326
McKay M, Beckman WCR (1979) A comparison of three methods for selecting values of input variables in the analysis of output from a computer code. Technometrics 21:239–245
Newman D (1982) The hexagon theorem. IEEE Trans Inform Theo 28:137–139
Nguyen H, Burkardt J, Gunzburger M, Ju L, Saka Y (2009) Constrained CVT meshes and a comparison of triangular mesh generators. Comp Geom: Theory Appl 42:1–19
Niederreiter H (1992) Random Number Generation and Quasi-Monte Carlo Methods. SIAM, Philadelphia
Okabe A, Boots B, Sugihara K, Chiu S (2000) Spatial Tessellations: Concepts and Applications of Voronoi Diagrams. Second Edition, Wiley, Chichester
Persson PO, Strang G (2004) A simple mesh generator in Matlab. SIAM Rev 46:329–345
Randall D (1994) Geostrophic adjustment and the finite-difference shallow water equations. Mon Wea Rev 122:1371–1377
Renka R (1997) Algorithm 772. STRIPACK: Delaunay triangulation and Voronoi diagrams on the surface of a sphere. ACM Trans Math Soft 23:416–434
Ringler T, Randall D (2002) A potential enstrophy and energy conserving numerical scheme for solution of the shallow-water equations on a geodesic grid. Mon Wea Rev 130:1397–1410
Ringler T, Heikes R, Randall D (2000) Modeling the atmospheric general circulation using a spherical geodesic grid: A new class of dynamical cores. Mon Wea Rev 128:2471–2490
Ringler T, Ju L, Gunzburger M (2008) A multi-resolution method for climate system modeling. Ocean Dynamics 58:475–498

Ringler T, Thuburn J, Klemp J, Skamarock W (2010) A unified approach to energy conservation and potential vorticity dynamics on arbitrarily structured c-grids. J Comput Phys 229:3065–3090

Rogers C (1964) Packing and Covering. Cambridge University Press, Cambridge

Ross S (1998) A First Course in Probability. Prentice Hall, Englewood Cliffs

Sadourny R, Arakawa A, Mintz Y (1968) Integration of the nondivergent barotropic vorticity equation with an icosahedral-hexagonal grid for the sphere. Mon Wea Rev 96:351–356

Saltelli A, Chan K, Scott E (2004) Sensitivity Analysis. Wiley, Chichester

Sedgewick R (1983) Algorithms. Addison Wesley

Sewell M (2002) Some applications of transformation theory in mechanics. In: Large-Scale Atmosphere-Ocean Dynamics, Ed. by J. Norbury and I. Roulstone, vol II, Chapter 5

Shamos M, Hoey D (1975) Closest-point problems. In: Proceedings ot the Sixteenth IEEE Annual Symposium on Foundations of Computer Science, pp 151–162

Smith R, Maltrud M, Bryan F, Hecht M (2000) Numerical simulation of the north atlantic ocean at $1/10°$. J Phys Ocean 20:1532–1561

Snow J (1855) On the Mode of Communication of Cholera. Churchill Livingstone, London

Thuburn J, Ringler T, Klemp J, Skamarock W (2009) Numerical representation of geostrophic modes on arbitrarily structured c-grids. J Comput Phys 228:8321–8335

Tomita H, Tsugawa M, Satoh M, Goto K (2001) Shallow water model on a modified icosahedral geodesic grid by using spring dynamics. J Comput Phys 174:579–613

Tomita H, Miura H, Iga S, Nasuno T, Satoh M (2005) A global cloud-resolving simulation: Preliminary results from an aqua planet experiment. Geophys Res Lett 32:L08,805

Voronoi G (1907) Nouvelles applications des paramètres continus à la théorie des formes quadratiques. Primiere Mémoire: Sur quelques prepriétés des formes quadratiques positives parfaites. J Reine Angew Math 133:97–178

Voronoi G (1908) Nouvelles applications des paramètres continus à la théorie des formes quadratiques. Deuxième Mémoire: Recherches usr les parallélloèdres primitifs,. J Reine Angew Math 134:198–287

Weaver W, Shannon C (1963) The Mathematical Theory of Communication. University of Illinois Press, Champaign

Weller H (2009) Predicting mesh density for adaptive modelling of the global atmosphere. Philosophical Transactions of the Royal Society A: Mathematical, Physical and Engineering Sciences 367:4523–4542

Weller H, Weller H (2008) A high-order arbitrarily unstructured finite-volume model of the global atmosphere: Tests solving the shallow-water equations. Inter J Numer Meth Fluids 56: 1589–1596

Williamson D (1968) Integration of the barotropic vorticity equation on a spherical geodesic grid. Tellus 20:642–653

Williamson D (1970) Integration of the primitive barotropic model over a spherical geodesic grid. Mon Wea Rev 98:512–520

Williamson D, Drake J, Hack J, Jakob R, Swarztrauber P (2001) A standard test set for numerical approximations to the shallow water equations in spherical geometry. J Comput Phys 102: 211–224

Ziman J (1979) Principles of the Theory of Solids. Second Ed., Cambridge University Press, Cambridge

Part III
Practical Considerations for Dynamical Cores in Weather and Climate Models

Chapter 11
Conservation in Dynamical Cores: What, How and Why?

John Thuburn

Abstract The conservation properties of the continuous, adiabatic and frictionless equations governing atmospheric flow are summarized. It is often considered desirable for atmospheric models to possess analogues of these conservation properties; some of the techniques for obtaining such analogues are noted. However, there is no widespread agreement in the literature on which conservation properties are most important and why. Here we suggest some ways of thinking about these questions, taking into account the atmospheric flow regimes that global numerical models are intended to represent.

11.1 Introduction

It is usually considered desirable for an atmospheric model dynamical core to have analogous conservation properties to those of the adiabatic and frictionless continuous governing equations. Although apparently obvious at first glance, this idea turns out to involve a number of subtle issues. In this lecture we touch on several of those issues. A fuller discussion is given by Thuburn (2008).

Section 11.2 summarizes the conservation properties of the continuous governing equations; in fact there are infinitely many such properties. This, then, raises the questions of which of these properties *can* we obtain in our numerical models (Sect. 11.3), and which of these properties *ought* we to try and obtain (Sect. 11.4)?

J. Thuburn
School of Engineering, Computing and Mathematics, University of Exeter, North Park Road, Exeter, EX4 4QF, UK
e-mail: j.thuburn@ex.ac.uk

P.H. Lauritzen et al. (eds.), *Numerical Techniques for Global Atmospheric Models*, Lecture Notes in Computational Science and Engineering 80,
DOI 10.1007/978-3-642-11640-7_11,

11.2 Conservation Properties of the Continuous Adiabatic Frictionless Governing Equations

First we review the conservation properties of the continuous adiabatic frictionless governing equations. It is convenient to classify them in four categories: flux-form conservation laws, Lagrangian conservation laws, conserved integral quantities, and kinematic identities. The definitions of the conserved quantities given in this section assume we are working with the unapproximated compressible Euler equations. For approximated equation sets of practical interest, such as hydrostatic and/or shallow atmosphere, all of the listed conservation properties continue to hold, but the definition of the conserved quantity may need to be modified. For example, under the hydrostatic approximation the contribution of the vertical velocity to the kinetic energy must be neglected (e.g. White et al. 2005).

11.2.1 Flux-Form Conservation Laws

A number of quantities satisfy conservation laws of the form

$$\frac{\partial A}{\partial t} + \nabla.\mathbf{F} = 0, \tag{11.1}$$

where A is the density of the conserved quantity and $\mathbf{F}$ is the flux. Table 11.1 lists three such quantities and gives expression for A and $\mathbf{F}$.

Equation (11.1) implies that the global integral of A is conserved; however, the local conservation property described by (11.1) is also considered important.

11.2.2 Lagrangian Conservation Laws

Certain quantities χ are materially conserved, that is, they satisfy

$$\frac{D\chi}{Dt} = 0 \tag{11.2}$$

Table 11.1 Some quantities satisfying flux-form conservation laws

Quantity	A	$\mathbf{F}$
Mass	ρ	$\rho\mathbf{u}$
Angular momentum	$\rho\hat{\mathbf{z}}.[\mathbf{r}\times(\mathbf{u}+\boldsymbol{\Omega}\times\mathbf{r})]$	$\mathbf{u}A + p\hat{\mathbf{z}}\times\mathbf{r}$
Energy	$\rho\left(\frac{1}{2}\mathbf{u}^2 + c_v T + \Phi\right)$	$\mathbf{u}(A+p)$

Here $\mathbf{r}$ is the position vector relative to the Earth's centre, $\boldsymbol{\Omega}$ is the Earth's rotation vector, $\hat{\mathbf{z}}$ is the unit vector in the direction of $\boldsymbol{\Omega}$, $\mathbf{u}$ is the velocity vector, ρ is density, p is pressure, T is temperature, c_v is the specific heat capacity at constant volume, and Φ is geopotential

Table 11.2 Some quantities satisfying Lagrangian conservation laws

Potential temperature	$\chi = \theta$
Potential vorticity	$\chi = Q = \boldsymbol{\zeta}.\nabla\theta/\rho$
Specific tracer	$\chi = q$
or tracer mixing ratio	$\chi = \eta$

Table 11.3 Some conserved integral quantities

Mass per unit θ in an isentropic layer	$\mathscr{F}(\theta) = \int \rho / \lvert\nabla\theta\rvert \, dA$
Mass per unit θ in an isentropic layer within a material contour	$\mathscr{M} = \int_D \rho / \lvert\nabla\theta\rvert \, dA$
Absolute circulation around an isentropic material contour	$\mathscr{C} = \oint_\Gamma \mathbf{v}_a.d\mathbf{r} = \int_D \rho Q / \lvert\nabla\theta\rvert \, dA$

where D/Dt is the material derivative, i.e. the time derivative following the flow. This immediately implies

$$\frac{Df(\chi)}{Dt} = 0 \tag{11.3}$$

for an arbitrary function $f(\chi)$. Table 11.2 lists some examples.

Each Lagrangian conservation law can be combined with the flux form conservation law for ρ to generate an infinite family of flux form conservation laws

$$\frac{\partial \rho f(\chi)}{\partial t} + \nabla.(\rho \mathbf{u} f(\chi)) = 0. \tag{11.4}$$

11.2.3 Conserved Integrals

The Lagrangian conservation laws for potential temperature and potential vorticity, along with conservation of mass, imply that certain integral quantities are conserved. Table 11.3 lists some of them. Here, the integral that appears in the definition of $\mathscr{F}$ is over the global extent of an isentropic surface, the domain D that appears in the definition of $\mathscr{M}$ and $\mathscr{C}$ is a region of an isentropic surface bounded by a material contour, and Γ is that material contour. Γ may be a potential vorticity contour. The quantity $\mathbf{v}_a$ is the absolute velocity, i.e. the velocity viewed in an inertial frame rather than one rotating with the Earth.

11.2.4 Kinematic Identities

The global integrals of horizontal divergence

$$\int_D \delta \, dA \tag{11.5}$$

and the vertical component of vorticity

$$\int_D \zeta \, dA \tag{11.6}$$

must vanish on any isosurface of the vertical coordinate that wraps the sphere. Although these properties are kinematic identities, rather than any consequence of the governing dynamical equations, they may nevertheless be of practical importance for models that predict δ and ζ. For example, if the numerical scheme that predicts ζ cannot maintain a global integral of zero then, unless an ad hoc fixer is applied, it will not be possible to solve $\nabla^2\psi = \zeta$ to obtain the stream function ψ and hence the rotational part of the horizontal velocity.

11.3 What Conservation Properties can we Obtain in Numerical Models?

The simplest technique for obtaining a discrete analogue of a flux-form conservation law is to use the conserved quantity as one of the predicted variables and discretize the conservation law itself in a conservative way, for example,

$$\frac{A_j^{n+1} - A_j^n}{\Delta t} + \frac{1}{V_j}\sum_k F_{j,k} S_{j,k} = 0. \tag{11.7}$$

Here, A_j^n is the average over cell j of the density of the conserved quantity at time step n, $F_{j,k}$ is the flux per unit area of the conserved quantity across face k of cell j, averaged over the time step, V_j is the volume of cell j, and $S_{j,k}$ is the area of face k of cell j. By making sure that F is uniquely defined at each face, so that the flux out of one cell across a particular face is equal to the flux into a neighbouring cell across the same face, we ensure that the predicted quantity is indeed conserved. Although local conservation of mass might be regarded as a fundamental requirement, historically it has often been sacrificed, for example to improve efficiency through the use of a (non-conservative) semi-Lagrangian advection scheme, or to reduce noise by predicting log of surface pressure rather than surface pressure itself in spectral models; global (but not local) conservation of mass can then be restored through an *ad hoc* fixer (e.g. Williamson and Olson 1994). More detailed and up-to-date discussion on the discretization of flux-form conservation laws is given in Chaps. 7 and 8.

This approach can only work for up to n_p conserved quantities, where n_p is the number of predicted variables. In particular, for a materially conserved quantity like θ, we can predict and conserve $\rho\theta$, but we would not automatically conserve higher moments unless we also predict those higher moments; but this would be an expensive (and very unusual) thing to do. Moreover, it is debatable whether we should attempt to conserve higher moments – see Sect. 11.4.

Some conservation properties are derived for the continuous equations through manipulation and making certain cancellations. Analogous conservation properties can sometimes be obtained for numerical schemes by designing them so that they respect analogous cancellations. The energy and angular momentum conservation properties of the Simmons and Burridge (1981) scheme, discussed in Chap. 4 are achieved in this way. Other well-known examples include the cancellation of Coriolis terms in the energy budget on a C-grid (Chap. 3, Arakawa and Lamb 1977), and the Arakawa Jacobian (Arakawa 1966). Historically, the development of schemes that could conserve quadratic quantities such as energy or enstrophy was important for controlling nonlinear instability, enabling long model integrations to be carried out (e.g. Arakawa 1966; Sadourny 1975). In some cases there are systematic ways of deriving such schemes using Poisson bracket and Nambu bracket ideas (e.g. Salmon 2004; Gassmann and Herzog 2008).

Lagrangian conservation properties can most obviously be obtained by using a Lagrangian solution technique. However, fully Lagrangian solution techniques are not yet developed to the point where they can be used for operational atmospheric model dynamical cores. On the other hand, the use of a Lagrangian vertical coordinate, or an entropy-based quasi-Lagrangian vertical coordinate, can improve Lagrangian conservation properties (Chap. 4, Johnson et al. 2000).

Lagrangian conservation implies, among other things, that extrema are not amplified. Schemes that prevent the spurious amplification of extrema ('overshoots' and 'undershoots') therefore respect this aspect of Lagrangian conservation. A variety of techniques exist for constructing non-oscillatory advection schemes. These may solve the flux form conservation law (ensuring conservation in that sense) while carefully constraining the fluxes to eliminate or minimize overshoots and undershoots. Semi-Lagrangian advection schemes can also be constructed to prevent overshoots and undershoots. Non-oscillatory advection schemes have often been applied to the prediction of tracers such as water vapour and chemical constituents, as well as potential temperature. Non-oscillatory advection schemes have also been applied to improve the Lagrangian conservation of potential vorticity, but much less often because this approach requires some non-trivial calculations to recover wind information from the predicted potential vorticity.

A particular problem with many atmospheric model dynamical cores is excessive dissipation of energy. The typical forms of dissipation included in most dynamical cores, e.g. a ∇^{2m} term added to the prognostic equations or the inherent dissipation due to interpolation in semi-Lagrangian schemes, are unable to dissipate the required amount of potential enstrophy without excessively dissipating energy. This may be understood heuristically as follows. Suppose the dissipation mechanism removes energy and enstrophy at wavenumber k_{diss} (and let us use enstrophy as a proxy for potential enstrophy here); the rate of dissipation of enstrophy will then be of the order k_{diss}^2 times the rate of dissipation of energy. Now k_{diss} is bounded above by the maximum resolvable wavenumber k_{max}, and so, provided the dissipation rate is positive at all scales, the ratio of energy dissipation to enstrophy dissipation is bounded below by k_{max}^{-2}. For currently affordable resolutions this bound is greater than the observed dissipation ratio. See Thuburn (2008) for more detailed discussion. This situation has led researchers to propose alternative forms of dissipation that can dissipate potential enstrophy while conserving energy (the Anticipated Potential Vorticity Method, Sadourny and Basdevant 1985) or that return some energy to the larger scales while dissipating it from the smaller resolved scales (Koshyk and Boer 1995). This latter idea is closely related to the idea of energy 'backscatter', which can also be parameterized in a stochastic way (e.g. Shutts 2005). For further discussion of dissipation mechanisms in atmospheric models see Chaps. 13 and 14.

11.4 Which Conservation Properties are the Most Relevant or Important?

Given that the continuous governing equations have infinitely many conserved quantities, and it is impossible to have analogues of all of these in a numerical model, it is natural to ask which of these conservation properties are the most relevant or important. This section suggests some arguments for helping to decide the answer to this question.

11.4.1 Finite Resolution Effects

In this subsection it is argued that the finite difference or finite volume analogous of higher moments of some conservable quantity only include resolved contributions and neglect unresolved contributions. For simplicity, let the fluid density $\rho \equiv 1$, and let χ be the mass mixing ratio of some materially conserved tracer. Define

$$V_j = \int_{\text{cell } j} dV \tag{11.8}$$

to be the volume of grid cell j,

$$m_j V_j = \int_{\text{cell } j} \chi \, dV \tag{11.9}$$

to be the mass of tracer in cell j, and

$$r_j V_j = \int_{\text{cell } j} \chi^2 \, dV \tag{11.10}$$

to be the contribution to the second moment of the tracer from cell j. Then the total mass of tracer is indeed exactly equal to its discrete analogue:

$$\int \chi dV = \sum_j m_j V_j. \tag{11.11}$$

However, the total second moment of the tracer is underestimated by a discrete analogue expressed in terms of m_j:

$$\int \chi^2 dV = \sum_j r_j V_j \geq \sum_j m_j^2 V_j. \tag{11.12}$$

Consequently, there is not a very strong argument for demanding conservation of analogues of second and higher order moments. This is particularly so for quantities such as tracer variance or potential enstrophy that have a systematic downscale cascade and therefore a systematic transfer from resolved to unresolved scales.

11.4.2 The Adiabatic Frictionless Limit

A dynamical core is usually thought of as a discretization of the adiabatic, frictionless governing equations. However, it may be more natural to think of it as a discretization of the governing equations in the adiabatic, frictionless limit. Quantities like tracer variance and potential enstrophy that cascade downscale are dissipated even in the limit of vanishing viscosity and thermal diffusivity: they are *non-Robust* invariants. Any numerical model, when applied to flow of realistic complexity, must therefore be able to dissipate such quantities.

Figure 11.1 shows an example result from a numerical integration of the barotropic vorticity equation. The initial condition is the same as that used in Chap. 1, Fig. 1.8, and the same spectral integration scheme is used, with the exception that no scale-selective dissipation term is included so that the model conserves both energy and enstrophy *on resolved scales*. After just a few vortex turnover times

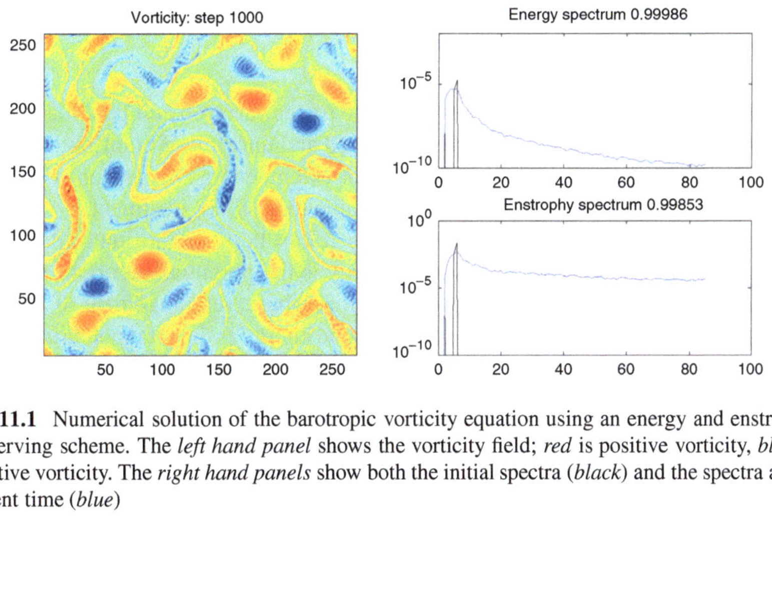

Fig. 11.1 Numerical solution of the barotropic vorticity equation using an energy and enstrophy conserving scheme. The *left hand panel* shows the vorticity field; *red* is positive vorticity, *blue* is negative vorticity. The *right hand panels* show both the initial spectra (*black*) and the spectra at the current time (*blue*)

the solution has become very noisy. Enstrophy has cascaded downscale towards the resolution limit at wavenumber 85, but is unable to cascade any further (or to be dissipated), resulting in a build up of small-scale enstrophy. This phenomenon is sometimes called ‘spectral blocking’. This example clearly shows that it is not always appropriate to conserve discrete analogues of quantities conserved by the continuous adiabatic, frictionless equations.

Accepting that it is in fact essential to dissipate non-robust invariants in numerical models for realistically complex flows, there are then two philosophies the model designer can adopt. One is to attempt to design the dynamical core to conserve the non-robust invariants, then supplement the model with a *sub-grid model*, i.e. additional terms in the equations intended to represent the effect of the unresolved scales on the resolved scales. Sub-grid models can vary from a simple ∇^{2m} scale-selective dissipation to a range of more sophisticated schemes (e.g. Smagorinsky 1963; Sadourny and Basdevant 1985). (Boundary layer parameterizations and convection parameterizations are also examples of sub-grid models designed to capture particular processes.)

An alternative philosophy is to use an inherently dissipative discretization of the resolved flow. This approach often uses high order schemes supplemented by some form of flux limiter, and is referred to as Implicit Large Eddy Simulation (ILES). There is some theoretical justification for this approach in terms of modified equation analysis, and some empirical evidence for its effectiveness in neutrally stratified three-dimensional turbulent flows. The book by Grinstein et al. (2007) provides a broad introduction and a route into the literature. However, despite the success claimed for the approach, for the flow regimes relevant to modelling the global atmosphere there has been relatively little analysis of how well the idea works, even though practical models based on semi-Lagrangian or non-oscillatory advection schemes are, in effect, using the ILES approach.

11.4.3 Energy

Energy is a nonlinear quantity and so, like the tracer variance and potential enstrophy discussed in Sect. 11.4.1, will have resolved and unresolved contributions. This then raises the question of whether, or to what extent, energy too is non-robust and therefore may require dissipation or be amenable to an ILES treatment.

Energy is particularly interesting because it can be split into unavailable and available contributions. The unavailable energy is the potential energy of the minimum energy state that can be obtained by an adiabatic rearrangement of the fluid parcels in the atmosphere. The available potential energy is the total potential energy minus the unavailable energy; it gives an upper bound on the amount of potential energy that is available for conversion into kinetic energy by adiabatic dynamics. The available potential energy plus the kinetic energy gives the available energy. For Earth’s atmosphere the unavailable energy is about 2,000 times as large as the kinetic energy and the available potential energy is about four times as large as the kinetic energy (Peixoto and Oort 1992).

The unavailable energy is a function of the $\mathscr{F}(\theta)$ defined in Table 11.3, and is therefore conserved for adiabatic frictionless flow separately from the total energy. Moreover, the $\mathscr{F}(\theta)$ are almost robust invariants because the strong restoring force due to stratification inhibits vertical overturning and mixing. (However some mixing, and hence non-conservation of unavailable energy, is unavoidable because of vertical propagation and eventual breaking of gravity waves.) This near-robustness suggests that it may be desirable to conserve a discrete analogue of unavailable energy, and indeed a family of analogues of the $\mathscr{F}(\theta)$. Interestingly, an isentropic-coordinate dynamical core that conserved mass in each isentropic layer would do just that.

The available energy is much smaller than the unavailable energy. However, precisely because the available energy is involved in atmospheric motions, its conservation (or non-conservation) remains important. Idealized two dimensional turbulence theory suggests that, in an inertial range, energy cascades predominantly upscale (e.g. Salmon 1998). Although real atmospheric flows are far from satisfying the assumptions of this theory, several pieces of evidence (see Thuburn 2008 for a discussion) suggest that about 5–10% of the throughput of available energy cascades downscale, while the rest goes upscale before being dissipated primarily by the planetary boundary layer. This implies that the available energy is not analogous to the downscale cascading non-robust invariants such as tracer variance and potential enstrophy, and therefore that its budget will not be adequately captured by using a simple scale-selective dissipation or the most straightforward ILES treatment, as already suggested in Sect. 11.3.

One final point to note is that a scheme that conserves the total energy while allowing spurious conversions between the unavailable and available components could lead to poor behaviour in an atmospheric model.

11.4.4 Spurious Sources vs Physical Sources

It may be argued that our numerical solutions should be accurate provided any spurious numerical sources of conservable quantities are much weaker than the true physical sources of those quantities. The strengths of the physical sources may be conveniently expressed in terms of time scales.

The physical sources and sinks of mass of dry air are completely negligible for modelling the atmosphere. For all practical purposes its timescale is infinite.

Next consider momentum. Locally the adjustment towards hydrostatic and geostrophic balance is fast, with typical timescale ranging from a few tens of seconds to a few tens of hours. This suggests that, for accurate modelling of the momentum equation, the most essential factor is an ability to capture balance accurately. However, in a zonal mean, the terms in the zonal momentum equation are not in geostropic balance (because the zonal mean of the zonal derivative of pressure must be identically zero). Therefore, the above argument about adjustment to balance does not apply, and conservation of angular momentum becomes more important. Comparison of a typical global mean angular momentum ($\pm 0.4 \times 10^{26}\,\mathrm{kg\,m^2 s^{-1}}$) with a typical surface torque ($\pm 0.5 \times 10^{20}\,\mathrm{kg\,m^2 s^{-2}}$) implies an angular momentum timescale of around 10 days (Peixoto and Oort 1992). Locally the angular momentum timescale can be much longer. For example, in the tropical lower stratosphere it is of the order of years. Thus, successful simulation of the quasi-biennial oscillation of the zonal winds in the tropical lower stratosphere will require an accurate treatment of angular momentum conservation.

Potential enstrophy budgets for the atmosphere have not been computed, but enstrophy budgets (e.g. Koshyk and Boer 1995 suggest that the physical timescale is of the order of 10 days. Variance budgets for long-lived tracers have also not been calculated, but estimates of "mixdown time" (e.g. Thuburn and Tan 1997) suggest a timescale of the order of 10–20 days. Thus, these non-robust invariants have comparable timescales, as might have been anticipated.

As suggested in Sect. 11.4.3, the unavailable and available contributions to the energy should be considered separately. Comparing a global mean value of the unavailable energy ($3 \times 10^9\,\mathrm{Jm^{-2}}$) with the total energy throughput of the climate system ($240\,\mathrm{Wm^{-2}}$) implies a timescale of about 150 days for the unavailable energy. Comparing a global mean value of the available energy ($6 \times 10^6\,\mathrm{Jm^{-2}}$) with a typical available energy throughput ($\sim\, 2\,\mathrm{Wm^{-2}}$) implies a timescale of about 30 days for available energy. According to this argument, there is a stronger case for attempting to conserve the unavailable energy than the available energy.

Table 11.4 summarizes the timescales for these quantities, along with entropy, which is closely related to the unavailable energy and the $\mathscr{F}(\theta)$. The arguments presented in this section suggest the following:

- Most benefit will be obtained by conserving quantities with long physical timescales
- Most benefit will be obtained by conserving robust invariants
- Quantities that cascade to small scales need to be dissipated, and may be amenable to an ILES treatment

Table 11.4 Summary of physical source timescales and other properties of some conservable quantities

Quantity	Robust	Cascade	Approx. timescale
Mass	Yes		Infinite
Momentum			Minutes to hours
Angular momentum			10 days (locally longer)
Potential enstrophy		Yes	10 days
Tracer variance		Yes	10 days
Unavailable energy	Almost		150 days
Available energy		Yes (5–10%)	20–30 days
Entropy	Almost		Variable

11.5 Conclusion

There is considerable discussion in the atmospheric model development literature of techniques for obtaining one conservation property or another. At the same time there is no widespread agreement in the literature on which conservation properties are most important and why. In this chapter we have suggested some ways of thinking about these questions, bearing in mind the particular fluid dynamical regimes that global atmospheric models are intended to capture. These arguments suggest that model developers should give prime consideration to conservation of quantities that are robustly conserved, do not systematically cascade downscale, or have long physical source timescales.

References

Arakawa A (1966) Computational design for long-term numerical integration of the equations of fluid motion: Two-dimensional incompressible flow. J Comput Phys 1:119–143

Arakawa A, Lamb VR (1977) Computational design and the basic dynamical processes of the UCLA general circulation model. Methods in Computational Physics 17:172–265

Gassmann A, Herzog HJ (2008) Towards a consistent numerical compressible non-hydrostatic model using generalized Hamiltonian tools. Q J R Meteorol Soc 134:1597–1613

Grinstein FF, Margolin LG, Rider WJ (eds) (2007) Implicit Large Eddy Simulation: Computing turbulent fluid dynamics. Cambridge University Press

Johnson DR, Lenzen AJ, Zapotocny TH, Schaak TK (2000) Numerical uncertainties in the simulation of reversible isentropic processes and entropy conservation. J Clim 13:3860–3884

Koshyk JN, Boer GJ (1995) Parametrization of dynamical subgrid-scale processes in a spectral GCM. J Atmos Sci 52:965–976

Peixoto JP, Oort M (1992) Physics of Climate. American Institute of Physics

Sadourny R (1975) The dynamics of finite-difference models of the shallow-water equations. J Atmos Sci 32:680–989

Sadourny R, Basdevant C (1985) Parameterization of subgrid scale barotropic and baroclinic eddies in quasi-geostrophic models: Anticipated potential vorticity method. J Atmos Sci 42:1353–1363

Salmon R (2004) Poisson-bracket approach to the construction of energy- and potential-enstrophy-conserving algorithms for the shallow-water equations. J Atmos Sci 61:2016–2036

Salmon R (1998) Lectures on Geophysical Fluid Dynamics. Oxford University Press

Shutts G (2005) A kinetic energy backscatter algorithm for use in ensemble prediction systems. Quart J Roy Meteorol Soc 131:3079–3102

Simmons AJ, Burridge DM (1981) An energy and angular-momentum conserving vertical finite-difference scheme and hybrid vertical coordinates. Mon Wea Rev 109(4):758–766

Smagorinsky J (1963) General circulation experiments with the primitive equations: I. the basic experiment. Mon Wea Rev 101:99–164

Thuburn J (2008) Some conservation issues for the dynamical cores of NWP and climate models. J Comput Phys 227:3715 – 3730

Thuburn J, Tan DGH (1997) A parameterization of mixdown time for atmospheric chemicals. J Geophys Res 102:13,037–13,049

White AA, Hoskins BJ, Roulstone I, Staniforth A (2005) Consistent approximate models of the global atmosphere: Shallow, deep, hydrostatic, quasi-hydrostatic and non-hydrostatic. Quart J Roy Meteorol Soc 131:2081–2107

Williamson DL, Olson J (1994) Climate simulations with a semi-Lagrangian version of the NCAR Community Climate Model. Mon Wea Rev 122(7):1594–1610

Chapter 12
Conservation of Mass and Energy for the Moist Atmospheric Primitive Equations on Unstructured Grids

Mark A. Taylor

Abstract The primitive variable formulation of the moist hydrostatic equations conserves mass and moist total energy due to the property that the divergence and gradient operators are adjoints. Any *compatible* numerical method, which has a discrete analog of this property will conserve a discrete mass and total energy. We demonstrate this using aqua-planet simulations performed with CAM-HOMME (NCAR's Community Atmospheric Model with the High-Order Method Modeling Environment dynamical core). CAM-HOMME uses a compatible numerical method on arbitrary unstructured quadrilateral grids. The equations described here are the full set of dynamical equations used by CAM. Aqua-planet simulations use the full suite of physics parametrizations as well. The only simplification is the use of idealized surface conditions. We report on the magnitude of the total energy budget in the dynamical core including estimates for the non-adiabatic processes. The practice of *fixing* dry total energy as opposed to the conserved total moist energy is shown to generate a forcing of $-0.56\,\mathrm{W/m^2}$.

12.1 Introduction

Today's petascale computers have hundreds of thousands of processors and the next generation machines could have millions of processors. As we no longer see much increase in single processor performance, these machines are relying almost entirely on increasing performance through increased parallelism. Translating this to application performance is thus only possible with very scalable applications. Achieving the required level of scalability in modern climate models remains a challenge due to several scalability bottlenecks. The largest bottleneck in these models is created by the numerical methods used in the dynamical core of the atmospheric model component. The dynamical core solves the partial differential equations governing the

M.A. Taylor
Sandia National Laboratories, Albuquerque, NM 87185, USA
e-mail: mataylo@sandia.gov

P.H. Lauritzen et al. (eds.), *Numerical Techniques for Global Atmospheric Models*, Lecture Notes in Computational Science and Engineering 80,
DOI 10.1007/978-3-642-11640-7_12,

Fig. 12.1 A latitude–longitude grid showing the clustering of *grid lines* and reduced grid spacing at the poles

fluid dynamical aspects of the atmosphere, but does not include the suite of subgrid parametrizations used for unresolved physical processes such as convection, precipitation and radiative forcings. Currently, most dynamical cores use latitude–longitude based grids (Fig. 12.1) for the horizontal directions (surface of the sphere) coupled with finite differences for the vertical (radial) direction. The latitude–longitude grid is a logically Cartesian orthogonal grid suitable for a wide array of numerical methods, including finite differences, finite volumes and spherical harmonic based spectral methods. The grid lines cluster at the pole, creating a severe Courant–Friedrichs–Lewy (CFL) restriction on the time-step. There are many successful techniques to handle this pole problem, however most of them substantially degrade parallel scalability by requiring too much inter-processor communication.

Thus there is a renewed interest in highly scalable dynamical cores based on more uniform grids for the sphere which avoid the pole problem. There are many approaches that have been extensively studied and recently surveyed in Williamson (2007). One can generate a quasi-uniform grid by patching together a few large regions, where each region uses methods developed for logically Cartesian grids and the challenge is how to couple the different patches together. Early examples of this approach include (Phillips 1957; Sadourny 1972; Browning et al. 1989). As an alternative, one can use numerical methods developed for general meshes that do not require grid lines to be aligned with a coordinate system and that instead can make use of unstructured (or less structured) grids constructed by tiling the sphere with polygons: typically triangles, quadrilaterals, or a combination of pentagons and hexagons are used. Early examples include Williamson (1968) and Sadourny et al. (1968).

These quasi-uniform grids present new challenges for the development of numerical methods, such as numerically conserving a suitable subset of the quantities conserved by the continuum equations being discretized (see Chap. 11). In this chapter, we focus on the issue of developing mass and energy conserving discretizations for unstructured grids. Energy conservation in particular has not received enough attention (e.g., Williamson 2007, Sect. 4.4). Operational models at typical resolutions appear to require about 2 W/m^2 of horizontal kinetic energy diffusion, which is conjectured to be unphysically large (Thuburn 2008). To conserve total energy, this diffusion must be added back to the internal energy as heating. Energy conserving methods are needed to access the impacts of various approaches to dissipating kinetic energy and the associated heating.

The use of finite volume methods is a popular approach to obtaining conservation in atmospheric models. It treats the equations in flux form with control volumes, obtaining conservation through careful discretization of the control volume fluxes. Obtaining conservation with finite difference methods is more complex, requiring the construction of intricate stencils (Sadourny 1972; Arakawa and Lamb 1977). Galerkin finite element methods have long been recognized as providing a natural way to obtain conservation based on the fact that a Galerkin discretization will preserve integral properties of the derivative operator (Cliffe 1981; Yakimiw and Girard 1987; Laprise and Girard 1990). Here we discuss a formalization of this approach, based on *compatible* (or *mimetic*) discretizations (Samarskiĭ et al. 1981; Margolin and Tarwater 1987; Nicolaides 1992; Shashkov and Steinberg 1995; Shashkov 1996; Hyman and Shashkov 1997a,b). Compatible discretizations obtain conservation by mimicking key integral properties of the divergence, gradient and curl operators. The compatibility property most connected to conservation is the requirement that the discrete divergence and gradient operators are adjoints with respect to the discrete integral used to define the conserved quantity. If this property holds locally, with suitably defined control volumes (where the adjoint relation includes a discrete boundary integral) then the conservation will also be local, meaning that the flux of the conserved quantity out of the control volume will be equal to the flux into the adjacent control volumes.

There is no formal definition of a compatible numerical method. We will only make use of the divergence/gradient adjoint relation, but other properties often considered are that the curl operator is self-adjoint, $\nabla\times\nabla() = 0$ and $\nabla\cdot\nabla\times() = 0$. The later two identities mean that the range of the gradient operator (or curl operator) be contained in the null space of the curl operator (or divergence operator). For some applications it is also required that the range be equal to the null space.

Compatible discretizations are suitable for finite difference, finite volumes and finite element methods and are considered in a common framework in Bochev and Hyman (2006). The finite element method in particular has only recently been associated with *local* conservation (Hughes et al. 2000). Here we are interested in finite elements because of its long history of successfully dealing with unstructured grids. Examples of compatible finite element methods include (Arnold et al. 2006; Bochev and Ridzal 2008). In the finite element method, instead of developing discrete approximations to derivative operators, one develops a discrete functional space, and then finds the function in this space which solves the equations of interest in a minimum residual sense. As compared to finite volume methods, there is less choice in how one constructs the discrete derivative operators in this setting, since functions in the discrete space are represented in terms of known basis functions whose derivatives are known, often analytically.

In the case of energy conservation, compatible methods are of interest because they allow conservation of energy without utilizing a total energy equation (Margolin and Tarwater 1987; Margolin and Shashkov 2008). In one-dimension, this property was used earlier to obtain energy and angular momentum conservation (Simmons and Burridge 1981). Energy is conserved by the careful mimicking of the energy balance in the original equation: the conversion between kinetic, internal and potential energy terms will exactly balance and the advection operator will not dissipate any kinetic energy. Kinetic energy dissipation, if needed, must be explicitly added in a compatible method via the introduction of limiters, hyper-viscosity or large-eddy-simulation based approaches.

Recent work on compatible methods for global atmospheric models with finite element methods on the cubed-sphere grid includes (Taylor et al. 2007; Taylor and Fournier 2010) and with finite volume methods on geodesic grids includes Gassmann and Herzog (2008) (using a closely related approach based on discretizations which preserve properties of the Hamiltonian (Salmon 2004, 2005, 2007)). In this chapter we show that an atmospheric model will conserve the moist total energy if one uses a combination of:

- A discretization for the surface of the sphere that has a discrete version of the Gauss divergence theorem. The divergence theorem for any two-dimensional surface without a boundary (such as the surface of the sphere) can be written

$$\int h\nabla\cdot\mathbf{u} + \int \mathbf{u}\cdot\nabla h = 0$$

  for any smooth scalar h and vector $\mathbf{u}$. It expresses the adjoint relationship between the divergence and gradient operators.
- The Simmons and Burridge (1981) compatible vertical discretization.
- A standard formulation of the moist hydrostatic equations.

The conservation is *semi-discrete*, meaning exact with exact time-stepping. We will use CAM-HOMME aqua-planet simulations to demonstrate the conservation and to measure the energy dissipation introduced by the Robert filtered leapfrog time-stepping method. CAM-HOMME (Taylor et al. 2008) is an experimental version of the Community Atmospheric Model (Collins et al. 2004) with the High-Order Method Modeling Environment (Dennis et al. 2005) cubed-sphere dynamical core framework running the compatible finite element discretization.

12.2 Quadrilateral Tilings of the Sphere

We first give an example as to how tilings of the sphere force us into a non-uniform geometry. Consider the case of a grid for the sphere consisting of only quadrilaterals, such as the cubed-sphere (Fig. 12.2), first used for global circulation modeling in Sadourny (1972). It is a conforming quadrilateral grid, meaning a tiling of the sphere where all tiles can be mapped to quadrilaterals and contain exactly four vertex points.

Fig. 12.2 Tiling the surface of the sphere with quadrilaterals. An inscribed cube is projected on the surface of the sphere. The faces of the cubed-sphere are further subdivided to form a quadrilateral grid of the desired resolution. Coordinate lines from the Gnomonic equal angle projection are shown

In the cubed-sphere grid, the eight corner points of the cube create eight vertices belonging to only three edges, while all other vertices belong to four edges. This non-uniformity is unavoidable. To see this, consider the vertices as defining a convex polyhedron. From Euler's formula for polyhedra, we have that

$$V - E + F = 2$$

where V is the number of vertices, E is the number of edges and F is the number of faces (quadrilaterals). For a conforming quadrilateral grid every face contains four edges, and every edge is shared by two faces, so $E = 2F$. We define the degree d of each vertex to be the number of edges that vertex belongs, and we let V_d, $d \geq 3$ be the number of vertices in our polyhedron that are of degree d, so that $V = \sum_d V_d$. In the cubed-sphere grid, every vertex is of degree 3 or 4, but more general grids may have vertices of higher degree. Since each point of degree d belongs to dV_d edges, and all these edges are shared by exactly two such points, summing dV_d will count every edge twice, and thus we have $\sum_d dV_d = 2E$. Combining these results, we see that

$$\sum_d (4 - d)V_d = 8$$

or

$$V_3 = 8 + V_5 + 2V_6 + 3V_7 + \cdots$$

This relation places no restrictions on V_4, the only type of vertex which appears in a Cartesian grid. But it does place a restriction on V_3, showing that any pure quadrilateral grid on the sphere must have at least eight vertices of degree d. The most uniform pure quadrilateral grid for the sphere, with no nodes of degree $d > 4$ must thus contain exactly 8 nodes of degree 3. This explains the popularity of the cubed-sphere grid, with eight vertices of degree 3 and all remaining vertices are of degree 4. Another such grid is based on stitching together two octagons (Purser and

Rančić 1997). This grid is topologically distinct from the cubed-sphere grid and its eight vertices of degree 3 all lie on the equator.

We note an interesting recently developed grid for the sphere for which every vertex is of degree 4 (Calhoun et al. 2008), but the grid contains both quadrilaterals and triangles. This grid has the property that it can be mapped to a single Cartesian block, and is thus logically rectangular if the triangles are treated as degenerate quadrilaterals. It is made up almost entirely of quadrilaterals, but the constraint derived above does not apply because of the presence of one or more triangles.

12.3 Continuum Formulation of the Equations

We now give a formulation of the moist primitive equations which will conserve energy when discretized by a compatible method in the horizontal directions coupled to a conservative discretization in the vertical. For the vertical discretization we use the hybrid η pressure vertical coordinate system modeled after CAM and based on (Kasahara 1974; Simmons and Burridge 1981; Simmons and Strüfing 1981) and also described in Chap. 4. This formulation differs mainly in that we use surface pressure as opposed to its logarithm as a prognostic variable, and we consider the moist total energy as opposed to the dry total energy.

In the η coordinate system, the pressure is given by

$$p(\eta) = A(\eta)p_0 + B(\eta)p_s$$

with $\eta = A(\eta) + B(\eta)$ and a constant reference pressure $p_0 \sim 1{,}000$ hPa. The functions A and B are prescribed to control the spacing of the model surfaces. They are chosen to allow the coordinate system to transition from a pure pressure coordinate system near the top of the atmosphere ($\eta = \eta_{\text{top}}$) to a terrain following coordinate system near the surface ($\eta = 1$), as shown in Fig. 12.3. At the surface, we require $A(1) = 0$ and $B(1) = 1$. We require $B(\eta_{\text{top}}) = 0$ so that the model top is at a constant pressure p_{top}. The value of $A(\eta_{\text{top}})$ is chosen to achieve the desired p_{top} (usually $\sim$ 1 hPa). In η-coordinates, the hydrostatic approximation $\partial p/\partial z = -g\rho$ can be used to replace the mass density ρ by an η-coordinate pseudo-density $\partial p/\partial \eta$. The material derivative in η coordinates can be written (e.g., Satoh 2004, Sect.3.3),

$$\frac{DX}{Dt} = \frac{\partial X}{\partial t} + \mathbf{u} \cdot \nabla X + \dot{\eta}\frac{\partial X}{\partial \eta}$$

where the $\nabla()$ operator (as well as $\nabla \cdot ()$ and $\nabla\times()$ below) is the two-dimensional gradient on constant η-surfaces, $\partial/\partial\eta$ is the vertical derivative, $\dot{\eta} = D\eta/Dt$ is a vertical flow velocity and $\mathbf{u}$ is the horizontal velocity component (tangent to constant z-surfaces, not η-surfaces).

The η-coordinate atmospheric primitive equations, neglecting dissipation and forcing terms can then be written as

$$\frac{\partial \mathbf{u}}{\partial t} + (\zeta + f)\hat{k}\times\mathbf{u} + \nabla\left(\frac{1}{2}\mathbf{u}^2 + \Phi\right) + \dot{\eta}\frac{\partial \mathbf{u}}{\partial \eta} + \frac{RT_v}{p}\nabla p = 0 \tag{12.1}$$

$$\frac{\partial T}{\partial t} + \mathbf{u}\cdot\nabla T + \dot{\eta}\frac{\partial T}{\partial \eta} - \frac{RT_v}{c_p^* p}\omega = 0 \tag{12.2}$$

$$\frac{\partial}{\partial t}\left(\frac{\partial p}{\partial \eta}\right) + \nabla\cdot\left(\frac{\partial p}{\partial \eta}\mathbf{u}\right) + \frac{\partial}{\partial \eta}\left(\dot{\eta}\frac{\partial p}{\partial \eta}\right) = 0 \tag{12.3}$$

$$\frac{\partial}{\partial t}\left(\frac{\partial p}{\partial \eta}q\right) + \nabla\cdot\left(\frac{\partial p}{\partial \eta}q\mathbf{u}\right) + \frac{\partial}{\partial \eta}\left(\dot{\eta}\frac{\partial p}{\partial \eta}q\right) = 0. \tag{12.4}$$

These are prognostic equations for $\mathbf{u}$, the temperature T, density $\frac{\partial p}{\partial \eta}$, and $\frac{\partial p}{\partial \eta}q$ where q is the specific humidity (the ratio of water vapor to air). The prognostic variables are functions of time t, vertical coordinate η and two coordinates describing the surface of the sphere. The unit vector normal to the surface of the sphere is denoted by $\hat{k}$. This formulation has already incorporated the hydrostatic equation and the ideal gas law, $p = \rho R T_v$. There is a no-flux ($\dot{\eta} = 0$) boundary condition at $\eta = 1$ and $\eta = \eta_{\text{top}}$. The vorticity is denoted by $\zeta = \hat{k}\cdot\nabla\times\mathbf{u}$, f is a Coriolis term and $\omega = Dp/Dt$ is the pressure vertical velocity. The virtual temperature T_v and variable-of-convenience c_p^* are defined by

$$RT_v = RT + (R_v - R)\,qT \qquad c_p^* = c_p + \left(c_{pv} - c_p\right)q$$

where R and R_v the ideal gas constants for dry air and water vapor, respectively and c_p, c_{pv} the specific heat at constant pressure for dry air and water vapor, respectively. Later we will also make use of c_v and c_{vv}, the corresponding specific heats defined at constant volume.

The diagnostic equations for the geopotential height field Φ is

$$\Phi = \Phi_s + \int_\eta^1 \frac{RT_v}{p}\frac{\partial p}{\partial \eta}\,d\eta \tag{12.5}$$

where Φ_s is the prescribed surface geopotential height (given at $\eta = 1$). To complete the system, we need diagnostic equations for $\dot{\eta}$ and ω, which come from integrating (12.3) with respect to η. In fact, (12.3) can be replaced by a diagnostic equation for $\dot{\eta}\frac{\partial p}{\partial \eta}$ and a prognostic equation for surface pressure p_s

$$\frac{\partial}{\partial t}p_s + \int_{\eta_{\text{top}}}^1 \nabla\cdot\left(\frac{\partial p}{\partial \eta}\mathbf{u}\right)d\eta = 0 \tag{12.6}$$

$$\dot{\eta}\frac{\partial p}{\partial \eta} = -\frac{\partial p}{\partial t} - \int_{\eta_{\text{top}}}^{\eta} \nabla\cdot\left(\frac{\partial p}{\partial \eta'}\mathbf{u}\right)d\eta', \tag{12.7}$$

where (12.6) is (12.7) evaluated at the model bottom ($\eta = 1$) after using that $\partial p/\partial t = B(\eta)\partial p_s/\partial t$ and $\dot{\eta}(1) = 0, B(1) = 1$. Using (12.7), we can derive a diagnostic equation for the pressure vertical velocity $\omega = Dp/Dt$,

$$\omega = \frac{\partial p}{\partial t} + \mathbf{u} \cdot \nabla p + \dot{\eta} \frac{\partial p}{\partial \eta} = \mathbf{u} \cdot \nabla p - \int_{\eta_{\text{top}}}^{\eta} \nabla \cdot \left(\frac{\partial p}{\partial \eta} \mathbf{u} \right) d\eta'$$

The equations have infinitely many conserved quantities (Chap. 11). Here we will only consider the total mass, tracer mass, potential temperature defined by

$$M_X = \iint \frac{\partial p}{\partial \eta} X \, d\eta dA$$

with ($X = 1, q$ or $(p/p_0)^{-\kappa} T$) and the total moist energy E defined by

$$E = \iint \frac{\partial p}{\partial \eta} \left(\frac{1}{2} \mathbf{u}^2 + c_p^* T \right) d\eta dA + \int p_s \Phi_s \, dA \tag{12.8}$$

where dA is the spherical area measure. To compute these quantities in their traditional units they should be divided by the constant of gravity g. We have omitted this scaling since g has also been scaled out from (12.1) to (12.4). We note that in this formulation of the primitive equations, the pressure p is a moist pressure, representing the effects of both dry air and water vapor. The unforced equations conserve both the moist air mass ($X = 1$ above) and the dry air mass ($X = 1 - q$). However, in the presence of a forcing term in (12.4) (representing sources and sinks of water vapor as would be present in a full model) a corresponding forcing term must be added to (12.3) to ensure that dry air mass is conserved.

The energy (12.8) is specific to the hydrostatic equations. We have omitted terms from the physical total energy which are constant under the evolution of the unforced hydrostatic equations (Staniforth et al. 2003). It can be converted into a more universal form involving $\frac{1}{2}\mathbf{u}^2 + c_v^* T + \Phi$, with c_v^* defined similarly to c_p^*, so that $c_v^* = c_v + (c_{vv} - c_v)q$. We note that $c_p = R + c_v$ and $c_{pv} = R_v + c_{vv}$ so that $c_p^* T = c_v^* T + R T_v$. Expanding $c_p^* T$ with this expression, integrating by parts with respect to η and making use of the fact that the model top is at a constant pressure

$$\int \frac{\partial p}{\partial \eta} R T_v \, d\eta = - \int p \frac{\partial \Phi}{\partial \eta} \, d\eta = \int \frac{\partial p}{\partial \eta} \Phi \, d\eta - (p\Phi) \Big|_{\eta=\eta_{\text{top}}}^{\eta=1}$$

and thus

$$E = \iint \frac{\partial p}{\partial \eta} \left(\frac{1}{2} \mathbf{u}^2 + c_v^* T + \Phi \right) d\eta dA + \int p_{\text{top}} \Phi(\eta_{\text{top}}) \, dA. \tag{12.9}$$

The model top boundary term in (12.9) vanishes if $p_{\text{top}} = 0$. Otherwise it must be included to be consistent with the hydrostatic equations. It is related to the form drag, which is the transfer of momentum between the atmosphere and the solid earth due to topography (e.g., Vallis 2006, Sect. 3.5).

12.4 Discrete Formulation of the Equations

We discretize the equations exactly in the form shown in (12.1), (12.2), (12.4), (12.6) and (12.7). The equations are written with $\mathbf{u}$ and T as the prognostic variables as opposed to conservative variables so as to minimize the number of worse-than-quadratic non-linearities. We use $\nabla_h \cdot$ and ∇_h to denote the discrete divergence and gradient operators and δ_η to denote the discrete $\partial/\partial\eta$ operator. We also replace the η-integrals by sums. The Simmons and Burridge (1981) coordinate system uses a Lorenz staggering of the variables (Chap. 4) as shown in Fig. 12.3. Let L be the total number of layers, with variables $\mathbf{u}, T, q, \omega, \Phi$ at layer mid points denoted by $i = 1, 2, \ldots, L$. We denote layer interfaces by $i + \frac{1}{2}, i = 0, 1, \ldots, L$, so that $\eta_{1/2} = \eta_{\text{top}}$ and $\eta_{L+1/2} = 1$. The δ_η operator uses centered differences to compute derivatives with respect to η at layer mid point from layer interface values, $\delta_\eta(X)_i = (X_{i+1/2} - X_{i-1/2})/(\eta_{i+1/2} - \eta_{i-1/2})$. We will use the over-bar notation for vertical averaging, $\overline{q}_{i+1/2} = (q_{i+1} + q_i)/2$. We also introduce the symbol π to denote the discrete pseudo-density $\frac{\partial p}{\partial \eta}$ given by $\pi_i = \delta_\eta(p)$.

We will use $\overline{\dot{\eta}\delta_\eta}$ to denote the discrete form of the $\dot{\eta}\partial/\partial\eta$ operator. This operator acts on quantities defined at layer mid-points and returns a result also at layer mid-points. It is defined in terms of δ_η and π by

$$\overline{\dot{\eta}\delta_\eta}(X)_i = \frac{1}{2\pi_i \Delta\eta_i} \left[(\dot{\eta}\pi)_{i+1/2} (X_{i+1} - X_i) + (\dot{\eta}\pi)_{i-1/2} (X_i - X_{i-1}) \right] \quad (12.10)$$

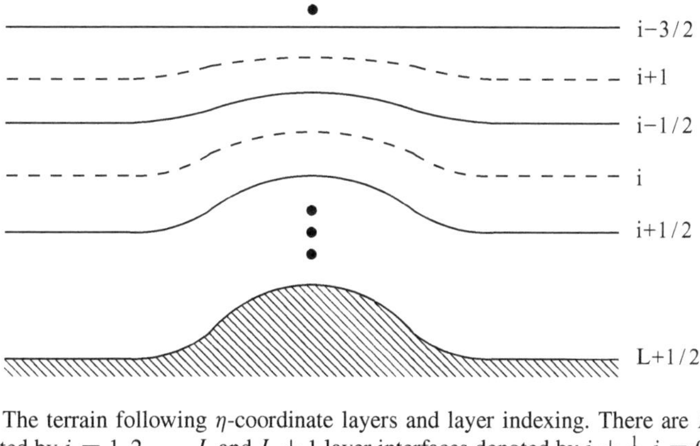

Fig. 12.3 The terrain following η-coordinate layers and layer indexing. There are L layer mid points denoted by $i = 1, 2, \ldots, L$ and $L + 1$ layer interfaces denoted by $i + \frac{1}{2}, i = 0, 1, \ldots, L$

where $\Delta\eta_i = \eta_{i+1/2} - \eta_{i-1/2}$. We use the over-bar notation since the formula can be seen as a π-weighted average of a layer interface centered difference approximation to $\dot{\eta}\partial/\partial\eta$. This formulation was constructed in Simmons and Burridge (1981) in order to ensure mass and energy conservation. Here we will use an equivalent expression that can be written in terms of δ_η,

$$\overline{\dot{\eta}\delta_\eta}(X)_i = \frac{1}{\pi_i}\Big[\delta_\eta\left(\dot{\eta}\pi\overline{X}\right)_i - X\,\delta_\eta\left(\dot{\eta}\pi\right)_i\Big]. \tag{12.11}$$

The discrete equations can now be written as

$$\frac{\partial \mathbf{u}}{\partial t} = -(\zeta + f)\,\hat{k}\times\mathbf{u} - \nabla_h\left(\frac{1}{2}\mathbf{u}^2 + \Phi\right) - \overline{\dot{\eta}\delta_\eta}(\mathbf{u}) - \frac{RT_v}{p}\nabla_h(p) \tag{12.12}$$

$$\frac{\partial T}{\partial t} = -\mathbf{u}\cdot\nabla_h(T) - \overline{\dot{\eta}\delta_\eta}(T) + \frac{RT_v}{c_p^* p}\omega \tag{12.13}$$

$$\frac{\partial}{\partial t}(\pi q) = -\nabla_h\cdot(\pi q\mathbf{u}) - \delta_\eta\left((\dot{\eta}\pi)\overline{q}\right) \tag{12.14}$$

$$\frac{\partial p_s}{\partial t} = -\sum_{j=1}^{L}\nabla_h\cdot(\pi\mathbf{u})_j\,\Delta\eta_j \tag{12.15}$$

$$(\dot{\eta}\pi)_{i+1/2} = -B(\eta_{i+1/2})\frac{\partial p_s}{\partial t} - \sum_{j=1}^{i}\nabla_h\cdot(\pi\mathbf{u})_j\,\Delta\eta_j. \tag{12.16}$$

We consider $(\dot{\eta}\pi)$ a single quantity given at layer interfaces and defined by (12.16). The no-flux boundary condition is $(\dot{\eta}\pi)_{1/2} = (\dot{\eta}\pi)_{L+1/2} = 0$. In (12.16), we used a midpoint quadrature rule to evaluate the indefinite integral from (12.7). In practice $\Delta\eta$ can be eliminated from the discrete equations by scaling π, but here we retain them so as to have a direct correspondence with the continuum form of the equations written in terms of $\frac{\partial p}{\partial\eta}$.

Finally we give the approximations for the diagnostic equations. We first integrate to layer interface $i - \frac{1}{2}$ using the same mid-point rule as used to derive (12.16), and then add an additional term representing the integral from $i - \frac{1}{2}$ to i:

$$\omega_i = (\mathbf{u}\cdot\nabla_h p)_i - \sum_{j=1}^{i-1}\nabla_h\cdot(\pi\mathbf{u})_j\,\Delta\eta_j + \nabla_h\cdot(\pi\mathbf{u})_i\,\frac{\Delta\eta_i}{2} \tag{12.17}$$

$$= (\mathbf{u}\cdot\nabla_h p)_i - \sum_{j=1}^{L} C_{ij}\nabla_h\cdot(\pi\mathbf{u})_j \tag{12.18}$$

where

$$C_{ij} = \begin{cases} \Delta\eta_j & i > j \\ \Delta\eta_j/2 & i = j \\ 0 & i < j \end{cases}$$

and similar for Φ,

$$(\Phi - \Phi_s)_i = \left(\frac{RT_v}{p}\pi\right)_i \frac{\Delta\eta_i}{2} + \sum_{j=i+1}^{L} \left(\frac{RT_v}{p}\pi\right)_j \Delta\eta_j \tag{12.19}$$

$$= \sum_{j=1}^{L} H_{ij} \left(\frac{RT_v}{p}\pi\right)_j \tag{12.20}$$

where

$$H_{ij} = \begin{cases} \Delta\eta_j & i < j \\ \Delta\eta_j/2 & i = j \\ 0 & i > j \end{cases}$$

We note for later use that

$$\Delta\eta_i \, C_{ij} = \Delta\eta_j \, H_{ji} \tag{12.21}$$

12.4.1 Consistency

It is important that the discrete equations be as consistent as possible. In particular, we need a discrete version of (12.3), the non-vertically averaged continuity equation. Equation (12.16) implicitly implies such an equation. To see this, apply δ_η to (12.16) and then we can derive, at layer mid-points,

$$\frac{\partial}{\partial t}\pi = -\nabla_h \cdot (\pi \mathbf{u}) - \delta_\eta \left(\dot{\eta}\pi\right). \tag{12.22}$$

A second type of consistency that has been identified as important is that (12.17), the discrete equation for ω, be consistent with (12.16), the discrete continuity equation (Williamson and Olson 1994). The two discrete equations should imply a reasonable discretization of $\omega = Dp/Dt$. To show this, we take the average of (12.16) at layers $i - 1/2$ and $i + 1/2$ and combine this with (12.17) (at layer mid-points i) and assuming that $B(\eta_i) = B(\eta_{i-1/2}) + B(\eta_{i+1/2})$ we obtain

$$\omega_i = B(\eta_i)\frac{\partial p_s}{\partial t} + (\mathbf{u} \cdot \nabla_h p)_i + \frac{1}{2}\left((\dot{\eta}\,\delta_\eta)_{i-1/2} + (\dot{\eta}\,\delta_\eta)_{i+1/2}\right),$$

which, since $\mathbf{u} \cdot \nabla_h p$ is given at layer mid-points and $\dot{\eta}\pi$ at layer interfaces, is the natural discretization of $\omega = \partial p/\partial t + \mathbf{u} \cdot \nabla_h p + \dot{\eta}\pi$.

12.4.2 Discrete Global Integral

Depending on the characteristics of a compatible method, it is often possible to define a control volume and show that the change in energy in the control volume is given by the flux of energy through the control volume boundary. These calculations are beyond the scope of this chapter and we instead focus on the conservation of the total energy. We denote the discrete global integral by

$$\langle X \rangle = \sum_{mn} w_{mn} \sum_{i=1}^{L} \Delta\eta_i X_{i,m,n} \approx \iint X \, dA d\eta$$

where we use our previously defined quadrature formula for the integral with respect to η and assume the quadrature formula for the integral over the surface of the sphere with respect to the surface area measure dA is denoted by $\sum w_{mn}$. The quadrature weights w_{mn} will be specific to the numerical method.

12.4.3 Compatibility Identities

For an arbitrary scalar h and vector $\mathbf{u}$ at layer mid-points, our assumption of a compatible method means that we have a discrete version of the divergence/gradient adjoint relation

$$\int h \nabla \cdot \mathbf{u} \, dA + \int \mathbf{u} \nabla h \, dA = 0$$

which we write as

$$\sum_{mn} w_{mn} \, h \nabla_h \cdot \mathbf{u} + \sum_{mn} w_{mn} \, \mathbf{u} \cdot \nabla_h h = 0 \tag{12.23}$$

This is the key property of the horizontal discretization that is needed to show conservation. In the vertical, Simmons and Burridge (1981) showed that the δ_η and $\overline{\dot{\eta}\delta_\eta}$ operators needed to satisfy two integral identities to ensure conservation. Let $\dot{\eta}$ be any layer interface variable which satisfies $\dot{\eta}_{1/2} = \dot{\eta}_{L+1/2} = 0$ and f, g arbitrary functions of layer mid points. The first identity is the adjoint property (compatibility) for δ_η and π,

$$\sum_{i=1}^{L} \Delta\eta_i \, \pi_i \, \overline{\dot{\eta}\delta_\eta}(f) + \sum_{i=1}^{L} \Delta\eta_i \, f_i \, \delta_\eta(\dot{\eta}\pi) = 0 \tag{12.24}$$

which follows directly from the definition of the $\overline{\dot{\eta}\delta_\eta}$ difference operator given in (12.11). The second identity we write in terms of δ_η,

$$\sum_{i=1}^{L} \Delta\eta_i \, fg \, \delta_\eta(\dot{\eta}\pi) = \sum_{i=1}^{L} \Delta\eta_i \, f \, \delta_\eta(\dot{\eta}\pi\overline{g}) + \sum_{i=1}^{L} \Delta\eta_i \, g \, \delta_\eta(\dot{\eta}\pi\overline{f}) \tag{12.25}$$

which is a discrete integrated-by-parts analog of $\partial(fg) = f\partial g + g\partial f$. Construction of methods with both properties on a staggered unequally spaced grid is the reason behind the complex definition for $\overline{\dot{\eta}\delta_\eta}$ in (12.11).

12.4.4 Discrete Conservation of Mass and Tracer Mass

Conservation of quantities advected in conservation form, such as mass and tracer mass in (12.16) and (12.14) are trivially conserved due to the compatibility properties. Considering (12.14), we see that

$$\frac{\partial}{\partial t}\langle \pi q\rangle = \left\langle \frac{\partial}{\partial t}(\pi q)\right\rangle = -\langle -\nabla_h \cdot (\pi q \mathbf{u})\rangle - \left\langle \delta_\eta \; (\dot{\eta}\pi\overline{q})\right\rangle = 0 \tag{12.26}$$

after applying (12.23) and (12.24) and using the fact differentiating a constant is zero ($\nabla_h(1) = 0$ and $\overline{\dot{\eta}\delta_\eta}(1) = 0$). We note that this equation will hold for any reasonable time-stepping method (one that can preserve the constant solution to $\partial q/\partial t = 0$) and thus in practice these quantities will be conserved to machine precision.

Assuming exact time integration, a compatible method can also conserve tracer mass if one advects concentration instead. Consider

$$\frac{\partial q}{\partial t} + \mathbf{u} \cdot \nabla_h q + \overline{\dot{\eta}\delta_\eta}(q) = 0. \tag{12.27}$$

Multiplying this equation by π, summing with the product of (12.22) and q, and then applying (12.23) and (12.24), we have

$$\begin{aligned} \frac{\partial}{\partial t}\langle \pi q\rangle &= \left\langle \pi\frac{\partial q}{\partial t}\right\rangle + \left\langle q\frac{\partial \pi}{\partial t}\right\rangle = \\ &- \langle -\pi \mathbf{u} \cdot \nabla_h q\rangle - \left\langle \pi \; \overline{\dot{\eta}\delta_\eta}(q)\right\rangle - \langle q \nabla_h \cdot (\pi \mathbf{u})\rangle - \left\langle q \; \delta_\eta \; (\dot{\eta}\pi)\right\rangle = 0 \end{aligned} \tag{12.28}$$

With inexact time stepping, to conserve this quantity to machine precision would require a time-stepping scheme which has a discrete analog of the product rule, $\partial(q\pi)/\partial t = q\partial\pi/\partial t + \pi\partial q/\partial t$. This is not common, but we note that the leapfrog method satisfies this identity if we let $Q = \pi q$ and consider the discrete tracer mass at half time levels defined by

$$Q(t + \tfrac{1}{2}\Delta t) = \tfrac{1}{2}(q(t)\pi(t + \Delta t) + q(t + \Delta t)\pi(t)).$$

12.4.5 Discrete Conservation of Energy

A compatible method obtains global energy conservation by mimicking the behavior of the continuum energy dynamics on a term-by-term basis. The discrete form of the terms in the energy equation which are responsible for the transfer between kinetic, internal and potential will be in exact balance, while the advection terms will vanish as in the continuum form of the equations.

We decompose the conserved total energy into kinetic, internal and potential, $< E >=< K > + < I > + < P >$ where

$$K = \frac{1}{2}\pi \mathbf{u}^2 \qquad I = \pi c_p^* T \qquad P = \pi \Phi_s.$$

Starting with (12.12), (12.13), (12.14) and (12.22) and the identities (12.23), (12.24) and (12.25) we show

$$\frac{\partial}{\partial t}\langle K\rangle = \langle T_1\rangle + \langle T_2\rangle + \langle T_3\rangle \tag{12.29}$$

$$\frac{\partial}{\partial t}\langle I\rangle = -\langle T_2\rangle - \langle T_3\rangle \tag{12.30}$$

$$\frac{\partial}{\partial t}\langle P\rangle = -\langle T_1\rangle \tag{12.31}$$

which implies $d/dt < E >= 0$. Here $\langle T_1\rangle$ is the transfer of potential energy to kinetic energy defined by

$$T_1 = \Phi_s \nabla_h \cdot (\pi \mathbf{u})$$

and $\langle T_2\rangle + \langle T_3\rangle$ is the transfer of internal energy to kinetic energy defined by

$$T_2 = -\pi \mathbf{u} \cdot \frac{RT_v}{p}\nabla_h(p) \qquad T_3 = (\Phi - \Phi_s)\nabla_h \cdot (\pi \mathbf{u}).$$

To derive (12.29), we sum the product of (12.12) with $\pi\mathbf{u}$ and the product of (12.22) with $\frac{1}{2}\mathbf{u}^2$ to obtain (assuming exact time integration)

$$\frac{\partial}{\partial t}K = -\pi\mathbf{u}\cdot\nabla_h\left(\frac{1}{2}\mathbf{u}^2\right) - \frac{1}{2}\mathbf{u}^2\nabla_h\cdot(\pi\mathbf{u}) - \pi\mathbf{u}\cdot\overline{\dot{\eta}\delta_\eta(\mathbf{u})} - \frac{1}{2}\mathbf{u}^2\,\delta_\eta(\dot{\eta}\pi) - \pi\mathbf{u}\cdot\nabla_h(\Phi) - \pi\mathbf{u}\cdot\frac{RT_v}{p}\nabla_h p. \tag{12.32}$$

When the discrete integral $\langle\cdot\rangle$ is applied, the first two terms on the RHS will vanish by (12.23). The next two terms will vanish by (12.24) and (12.25) with f and g replaced by $\mathbf{u}$. Applying (12.23) to the fifth term we establish (12.29).

To derive (12.30), we start with

$$\frac{\partial}{\partial t}I = c_p\frac{\partial}{\partial t}(\pi T) + (c_{pv} - c_p)\frac{\partial}{\partial t}(q\pi T). \tag{12.33}$$

The first term is the dry internal energy. To derive the discrete equation for it we sum the product of (12.13) with $c_p\pi$ and the product of (12.22) with c_pT. The second term is the moist contribution, for which we obtain an equation by summing the product of (12.13) with $(c_{pv} - c_p)q\pi$ and the product of (12.14) with $(c_{pv} - c_p)T$. The result (assuming exact time integration) gives

$$\begin{aligned}\frac{\partial}{\partial t}I &= c_p\pi\mathbf{v}\cdot\nabla_h T + c_pT\nabla_h\cdot(\pi\mathbf{v}) + (c_{pv}-c_p)q\pi\mathbf{v}\cdot\nabla_h T + (c_{pv}-c_p)T\nabla_h\cdot q\pi\mathbf{v}\\ &+ c_p\pi\,\overline{\dot\eta\delta_\eta}(T) + c_pT\,\delta_\eta\,(\pi\dot\eta) + (c_{pv}-c_p)q\pi\,\overline{\dot\eta\delta_\eta}(T) + (c_{pv}-c_p)T\,\delta_\eta\,q\pi\dot\eta\\ &+ \frac{RT_v}{p}\pi\omega\end{aligned} \tag{12.34}$$

After applying $\langle\cdot\rangle$, the first four terms on the RHS will vanish due to (12.23). The next four terms will vanish due to (12.24). Expanding ω with (12.18), we see that

$$\left\langle\frac{RT_v}{p}\pi\omega\right\rangle = -\langle T_2\rangle - \left\langle\frac{RT_v}{p}\pi\sum_{j=1}^{L}C_{ij}\nabla_h\cdot(\pi\mathbf{u})_j\right\rangle$$

Using (12.21), we have that

$$\begin{aligned}\sum_{i=1}^{L}\Delta\eta_i\frac{RT_v}{p}\pi\sum_{j=1}^{L}C_{ij}\nabla_h\cdot(\pi\mathbf{u})_j &= \sum_{j=1}^{L}\Delta\eta_j\nabla_h\cdot(\pi\mathbf{u})_j\sum_{i=1}^{L}H_{ji}\left(\frac{RT_v}{p}\pi\right)_i\\ &= \sum_{j=1}^{L}\Delta\eta_j\nabla_h\cdot(\pi\mathbf{u})_j\,(\Phi-\Phi_s)_j = T_3\end{aligned} \tag{12.35}$$

and thus

$$\frac{\partial}{\partial t}\langle I\rangle = \left\langle\frac{RT_v}{p}\pi\omega\right\rangle = -\langle T_2\rangle - \langle T_3\rangle. \tag{12.36}$$

Finally, to derive (12.31), we take the discrete integral of the product of (12.22) and Φ_s, then apply (12.23) and note that $\overline{\dot\eta\delta_\eta}\,\Phi_s = 0$.

We have thus shown that a compatible discretization of (12.12)–(12.16) will also satisfy the energy balance equations, (12.29)–(12.31), to within time-truncation error.

12.4.6 Potential Temperature Formulation

If one prefers to advect potential temperature $\theta = T(p/p_0)^{-\kappa}$ with $\kappa = R/c_p$ instead of temperature, as motivated in Chap. 4, it is still possible to obtain energy conservation in the dry case ($q = 0$). We reformulate (12.12), (12.13) and (12.20) with

$$\frac{\partial \mathbf{u}}{\partial t} = -(\zeta + f)\hat{k}\times\mathbf{u} + \nabla_h\left(\frac{1}{2}\mathbf{u}^2 + \Phi\right) - \overline{\dot{\eta}\delta_\eta(\mathbf{u})} - c_p\theta\nabla_h\left((p/p_0)^k\right) \tag{12.37}$$

$$\frac{\partial}{\partial t}(\pi\theta) = -\nabla_h\cdot(\pi\theta\mathbf{u}) - \delta_\eta\left(\dot{\eta}\pi\overline{\theta}\right) \tag{12.38}$$

$$(\Phi - \Phi_s)_i = c_p\sum_{j=1}^{L} H_{ij}\theta\,\delta_\eta\left((p/p_o)^k\right) \tag{12.39}$$

and solve this system in conjunction with (12.15) and (12.16). The energy balance equations (12.29), (12.30) and (12.31) are unchanged, but with

$$T_2 = -\pi\mathbf{u}\cdot c_p\theta\nabla_h\left((p/p_0)^\kappa\right).$$

The calculations needed to show (12.29) and (12.31) are identical to those used in Sect. 12.4.5. To show (12.30), we need to make liberal use of the exact time integration assumption and consider

$$\frac{\partial}{\partial t}I = \frac{\partial}{\partial t}\left((p/p_0)^\kappa\pi\theta\right) = \theta\,\delta_\eta\left((p/p_0)^\kappa\right)\frac{\partial p}{\partial t} + (p/p_0)^\kappa\frac{\partial}{\partial t}(\pi\theta) \tag{12.40}$$

and then apply the same algebra used in Sect. 12.4.5.

12.5 Example Computations

We now present results using the CAM-HOMME global atmospheric model in an aqua-planet configuration (Neale and Hoskins 2000a,b). In these experiments, CAM-HOMME is run with the full CAM atmospheric physics parametrizations, but the surface boundary conditions are greatly simplified by prescribing a planet covered with water with a fixed zonally symmetric sea surface temperature. A perpetual March equinox diurnal cycle is used.

HOMME uses a continuous Galerkin hp finite element discretization. It solves the equations of interest in integral form. The discrete inner product, denoted by $\sum_{mn} w_{mn}()$ in Sect. 12.4.2, is defined by decomposing the integral over the surface of the sphere into a sum of integrals within each element, and then approximating each element integral with the $p \times p$ tensor-product Gauss-Lobatto quadrature rule. The global basis and test functions span the space of C_0 functions which are polynomials of order p (up to degree $p - 1$) within each element. With a nodal basis defined at the Gauss-Lobatto quadrature nodes, the finite element mass matrix will be diagonal. This makes the method a very efficient way to obtain a high-order explicit method on unstructured grids for time dependent equations. For details of the Gauss-Lobatto quadrature and the nodal basis, see Chap. 9.

For our aqua-planet simulations, we use CAM 3.5.1 physics (Gent et al., 2009). This version advects three tracers: specific humidity q, cloud ice and cloud water, each using (12.14). The forcing terms computed by the CAM physics are applied with a time-split coupling (Williamson 2002), meaning that the forcings due to the physics are applied as an adjustment to the prognostic variables and then the flow is evolved by the HOMME dynamical core without a forcing term. The forcing is applied every 30 and 15 min in the 3.75 and 0.5 degree simulations, respectively. In these aqua-planet simulations, $\Phi_s = 0$, so $P = T_1 = 0$ and there is no potential energy term in the total energy budget. A typical snapshot showing fully developed turbulent flow and the realistic nature of the aqua-planet atmosphere is shown in Fig. 12.4.

In all cases, mass and tracer mass is conserved to machine precision. For energy conservation, we saw in Sect. 12.4.5 that a compatible method will exactly mimic all adiabatic processes in the dynamics. However there are several non-adiabatic terms not considered in Sect. 12.4.5 which will impact total energy conservation of a model in practice. The largest term in the CAM-HOMME dynamical core is the horizontal dissipation of kinetic energy via a hyper-viscosity term. For this term a corresponding heating term is added to the temperature equation so that total energy

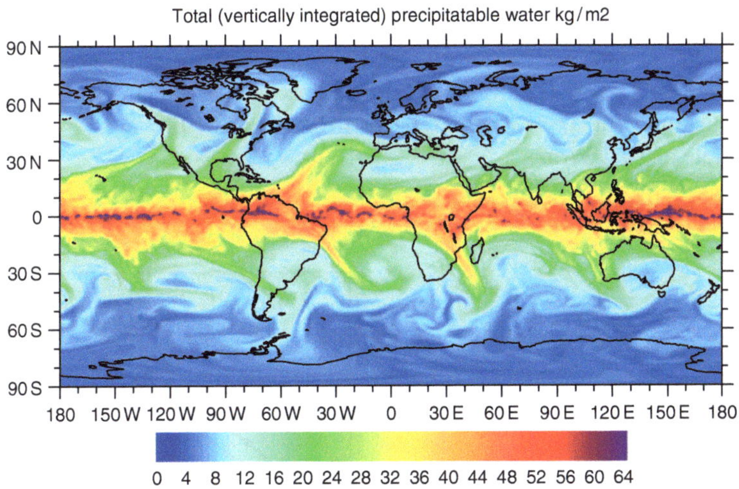

Fig. 12.4 A snapshot of the vertically summed atmospheric water content (liquid, ice and vapor) over the surface of an Aqua-planet, simulated with CAM-HOMME

remains conserved. That is, if the hyper-viscosity term on the right-hand-side of the momentum equation is represented by $\mathbf{D}$, the amount of energy dissipated is $< \frac{\partial p}{\partial \eta}\mathbf{u} \cdot \mathbf{D} >$. To add this back to the system as internal energy, one can add the term $-(1/c_p)\mathbf{u} \cdot \mathbf{D}$ to the right-hand-side of the temperature equation (Williamson et al. 2009). For the other non-adiabatic processes in CAM-HOMME we do not write down corresponding heating terms. These include the leapfrog-Robert filter, moisture variance dissipation and temperature variance dissipation. In this section we first disable these non-adiabatic terms and demonstrate that CAM-HOMME conserves total energy to machine precision. We then measure the level of conservation with these terms enabled.

12.5.1 Adiabatic Results

We first verify that the CAM-HOMME leapfrog time discretization of (12.12)–(12.16) will satisfy the energy balance equations, (12.29)–(12.30), to within time-truncation error. We use CAM's standard 26 vertical levels and coarse resolution in the horizontal (3.75° average grid spacing at the equator). We use coarse spatial resolution to verify that conservation is obtained in the presence of large truncation error levels. For the initial condition, we use a fully spun-up state generated by a regular run of CAM-HOMME with all dissipation terms needed in the full atmospheric model. Starting with this initial condition, we then run CAM-HOMME without the Robert filter and with all dissipation terms disabled. When run in this manor, the flow will soon become unrealistic as enstrophy will accumulate at the small scales and the leapfrog scheme has a computational mode. But this inviscid configuration can be run for short simulations. It is of interest because the only errors in total moist energy conservation is from the second-order accurate leapfrog time-stepping, and thus this error will decrease to machine precision at a second-order rate, as shown in Fig. 12.5. We plot the relative error in total moist energy conservation after 30 min, $|E(t+30\text{m}) - E(t)|/E(t)$, from eight simulations with Δt ranging from 300 to 1 s. The simulation with $\Delta t = 1$ has a relative error of 5.6×10^{-15} corresponding to a heating rate of 10^{-8} W/m^2.

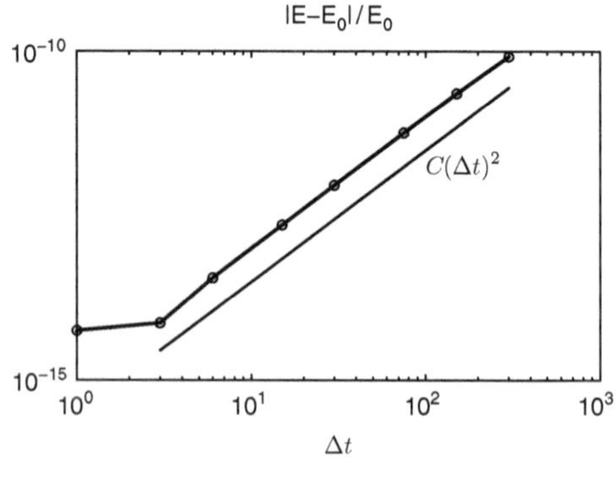

Fig. 12.5 Relative error in total moist energy conservation from low resolution CAM-HOMME aqua-planet simulations. The error converges to machine precision at a second-order rate

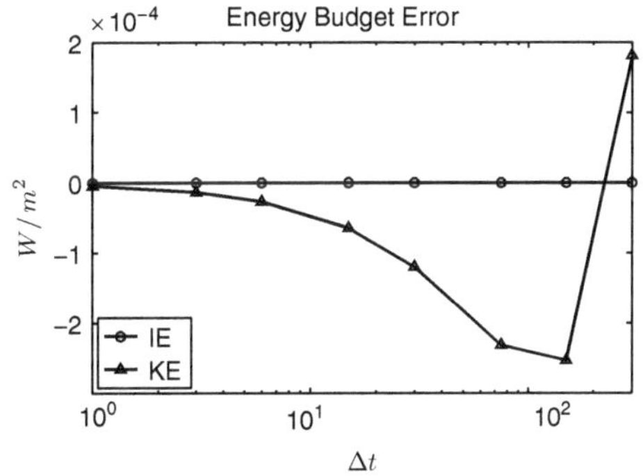

Fig. 12.6 Error in the kinetic and internal energy budget equations from low resolution CAM-HOMME aqua-planet simulations. The internal energy budget error is at machine precision for any time-step. The kinetic energy budget error converges to machine precision with decreasing time-step

One could achieve exact energy conservation by advecting the total energy instead of temperature or potential temperature. However, such an approach does not mean that one is accurately solving the energy budget equations, (12.29) and (12.30). A method which advects total energy can have large errors in these balance equations which are then effectively lumped into the temperature when it is recovered from the total energy. To show that the conservation here is in fact due to the correct representation of these budgets, we plot the error in (12.29) and (12.30) in Fig. 12.6. From the simulations used in Fig. 12.5, we compute the terms on the RHS of (12.29) and (12.30) from the flow snapshot at $t = 30$ min. The terms on the LHS are computed with centered-in-time differencing of K and I at $t = 30$ from their values defined at $t \pm \Delta t/2$ using the half-time level definition given at the end of Sect. 12.4.4. With this definition, (12.30) holds to machine precision when using leapfrog without the Robert filter. The error in (12.29) converges to machine precision with decreasing time-step.

12.5.2 Non-Adiabatic Results

We now consider the equations including all dissipation terms needed in the full atmospheric model:

- A hyper-viscosity term added to (12.12) with a corresponding heating term added to (12.13)
- A hyper-viscosity term added to (12.13) to dissipate temperature variance
- A sign-preserving reconstruction (Taylor et al. 2009) is used with the horizontal advection operator and the vertical advection operator is replaced by a Lagrange-remap approach (Lin 2004) with monotone reconstruction (Zerroukat et al. 2005).

The hyper-viscosity operators mentioned above are modeled after the terms used in the Eulerian dynamical core in CAM. In this non-adiabatic case, the energy budget in the dynamical core can be written as

$$\frac{\partial}{\partial t}\langle K\rangle = \langle T_2\rangle + \langle T_3\rangle + D_1 \tag{12.41}$$

$$\frac{\partial}{\partial t}\langle I\rangle = -\langle T_2\rangle - \langle T_3\rangle - D_1 + D_2. \tag{12.42}$$

where D_1 serves to represent the kinetic energy dissipation from the hyper-viscosity operator and the (much smaller) effects of the Robert filter and time-truncation errors in the kinetic energy budget. We included it in (12.42) with the opposite sign to represent the contribution of the compensating heating term used in the model. In CAM-HOMME, the heating term is not exactly equal to the kinetic energy dissipation, so this is only an approximation. The remaining term, D_2, thus contains the errors in the approximation, time-truncation errors and effects of all other dissipation mechanisms in the model, and $d/dt\,\langle E\rangle = D_2$.

In these simulations, we compute $d/dt < K >, d/dt < I >, T_2$ and T_3 as in the adiabatic case, and then use (12.41) and (12.42) to solve for D_1 and D_2. In Table 12.1 we present results from the high resolution aqua planet simulation pictured in Fig. 12.4. The simulation used the standard CAM 26 vertical levels, a high horizontal resolution (0.5 degree average grid spacing at the equator) and a time-step of 40 s for both tracers and dynamics. We use high resolution for these runs so that the data reported will be typical of modern simulations. The data is computed from the instantaneous values, sampled hourly, over a one month simulation time starting with a spun-up initial condition. For completeness, we also include the impacts of the velocity and temperature forcing terms applied by the CAM physics routines which do not appear in the energy budget for the dynamics. For all quantities except the forcings, there is little variation. That and the lack of seasons in aqua-planet suggests these global means will be typical for the whole simulation.

The data shows that the moist total energy dissipation in the full model, $d/dt\,\langle E\rangle = D_2 = -0.013\text{W/m}^2$, is quite small relative to the other terms. This value corresponds to a relative change, $|E - E_0|/E_0$, of 2×10^{-10} per time-step. We also followed the methodology used in the adiabatic case above and made short runs with only selected dissipation mechanisms turned on. These runs verify that the contribution to D_2 from the Robert filter, the hyper-viscosity term acting on T and the various types of dissipation on q are all individually less than 0.013 W/m^2 (not shown). The only significant diffusive term in the model is the horizontal dissipation of kinetic energy (0.6 W/m^2), and thus this is the only term for which a compensating heating term must be included in order to obtain conservation to a level of 0.013 W/m^2.

Finally, we show that the common practice of running a moist primitive equation model with a dry energy fixer, as in CAM (Williamson et al. 2009), results in a not insignificant amount of cooling. The dry total energy is defined as $E_{\text{dry}} = K + I_{\text{dry}} + P$, with $I_{\text{dry}} = c_p \pi T$. It differs from E by only 0.2%. Running CAM-HOMME

Table 12.1 Averages and standard deviations for some of the terms in the energy budget (12.41)–(12.42), from a high-resolution CAM-HOMME aqua-planet simulation. The CAM physics forcings of K and I are given as F_K and F_I

Variable	Average	Standard deviation
$\langle K \rangle$	2.653×10^6 J/m^2	9.5×10^4
$\langle I \rangle$	2.574×10^9 J/m^2	1.8×10^6
$\langle I_{\text{dry}} \rangle$	2.570×10^9 J/m^2	1.8×10^6
$\langle F_K \rangle$	-2.53 W/m^2	0.23
$\langle F_I \rangle$	2.26 W/m^2	6.1
$\langle T_2 + T_3 \rangle$	3.20 W/m^2	0.48
D_1	-0.59 W/m^2	0.055
D_2	-0.013 W/m^2	0.0014

with CAM's dry energy fixer, we measure the forcing introduced by the fixer as $-0.56 \pm 0.05\text{W/m}^2$. This is not the result of non-adiabatic moist processes, but is due entirely to adiabatic terms in the energy budget. Neglecting dissipative terms,

$$\frac{\partial}{\partial t}\left\langle E_{\text{dry}} - E\right\rangle = \left\langle \frac{c_p R T_v}{c_p^* p}\pi\omega \right\rangle - \left\langle \frac{R T_v}{p}\pi\omega \right\rangle.$$

and this term in our simulations is $0.56 \pm 0.04\,\text{W/m}^2$. If $d/dt < E >= 0$, we have that $d/dt < E_{\text{dry}} >= 0.56$. Thus if one wishes to maintain a constant total dry energy in a conservative moist hydrostatic model, one must compensate this level of heating via some type of fixer.

12.6 Conclusions

Compatible numerical methods are an effective way to obtain conservative methods on unstructured grids. Here we showed that a compatible method will conserve mass and moist total energy when used to discretize a standard primitive variable formulation of the hydrostatic equations. In one dimension, the approach is well known, an early example includes Simmons and Burridge (1981). For two and three dimensional unstructured quadrilateral grids a recent example is the finite element method which has been implemented in CAM-HOMME. Using CAM-HOMME, we confirmed that without dissipative processes the method conserves moist total energy to within a second-order time truncation error, which can be reduced to machine precision by reducing the time step. In the full model, at 0.5 degree resolution, the dissipative processes in the dynamics are dominated by the horizontal diffusion of kinetic energy at $-0.6\,\text{W/m}^2$. When this diffusion is implemented via hyper-viscosity, a heating term can be added which compensates to better than $0.013\,\text{W/m}^2$. The remaining terms (diffusion of temperature variance, monotone and sign preserving limiters on moisture and the Robert filter) are individually less than 0.013 W/m^2 in magnitude. The common practice of fixing the dry total energy introduces an additional forcing of 0.5 W/m^2.

Acknowledgments We thank the reviewers for many useful and detailed comments and P.H. Lauritzen for the vertical coordinate figure. This work supported by DOE/BER grant 06-13194.

References

Arakawa A, Lamb V (1977) Computational design of the basic dynamical processes in the UCLA general circulation model. In: Chang J (ed) Methods in computational physics. Vol. 17: General circulation models of the atmosphere, Academic Press, pp 174–264

Arnold DN, Falk RS, Winther R (2006) Finite element exterior calculus, homological techniques, and applications. Acta Numerica 15:1–155, DOI 10.1017/S0962492906210018

Bochev P, Hyman M (2006) Principles of compatible discretizations. In: Arnold DN, Bochev P, Lehoucq R, Nicolaides R, Shashkov M (eds) Compatible Discretizations, Proceedings of IMA Hot Topics Workshop on Compatible Discretizations, Springer Verlag, vol IMA 142, pp 89–120

Bochev P, Ridzal D (2008) Rehabilitation of the lowest-order Raviart-Thomas element on quadrilateral grids. SIAM J Numer Anal 47(1):487–507

Browning GL, Hack JJ, Swarztrauber PN (1989) A Comparison of Three Numerical Methods for Solving Differential Equations on the Sphere. Mon Wea Rev 117(5):1058–1075

Calhoun DA, Helzel C, LeVeque RJ (2008) Logically rectangular finite volume grids and methods for 'circular' and 'spherical' domains. SIAM Rev 50:723–752

Cliffe KA (1981) On conservative finite element formulations of the inviscid Boussinesq equations. International Journal for Numerical Methods in Fluids 1:117–127

Collins WD, Rasch PJ, Boville BA, Hack JJ, McCaa JR, Williamson DL, Kiehl JT, Briegleb BP, Bitz CM, Lin SJ, Zhang M, Dai Y (2004) Description of the NCAR Community Atmosphere Model (CAM3.0). NCAR Technical Note NCAR/TN-464+STR, National Center for Atmospheric Research, Boulder, Colorado, 214 pp., available from http://www.ucar.edu/library/collections/technotes/technotes.jsp

Dennis J, Fournier A, Spotz WF, St-Cyr A, Taylor MA, Thomas SJ, Tufo H (2005) High resolution mesh convergence properties and parallel efficiency of a spectral element atmospheric dynamical core. Int J High Perf Comput Appl 19:225–235

Gassmann A, Herzog HJ (2008) Towards a consistent numerical compressible non-hydrostatic model using generalized Hamiltonian tools. Quart J Roy Meteor Soc 134(635):1597–1613, DOI 10.1002/qj.297

Gent PR, Yeager SG, Neale RB, Levis S, Bailey DA (2009) Improvements in a half degree atmosphere/land version of the ccsm. Clim Dyn DOI 10.1007/s00382-009-0614-8

Hughes TJR, Engel G, Mazzei L, Larson MG (2000) The continuous Galerkin method is locally conservative. J Comput Phys 163:467–488, DOI 10.1006/jcph.2000.6577

Hyman JM, Shashkov M (1997a) Adjoint operators for the natural discretizations of the divergence, gradient, and curl on logically rectangular grids. Appl Numer Math 25:413–442

Hyman JM, Shashkov M (1997b) Natural discretizations for the divergence, gradient and curl on logically rectangular grids. International Journal of Applied Numerical Mathematics 33:81–104

Kasahara A (1974) Various vertical coordinate systems used for numerical weather prediction. Mon Wea Rev 102:509–522

Laprise R, Girard C (1990) A spectral general circulation model using a piecewise-constant finite-element representation on a hybrid vertical coordinate system. J Climate 3:32–52

Lin SJ (2004) A vertically Lagrangian finite-volume dynamical core for global models. Mon Wea Rev 132:2293–2397

Margolin LG, Shashkov M (2008) Finite volume methods and the equations of finite scale: A mimetic approach. Int J Numer Meth Fluids 56(8):991–1002, DOI 10.1002/fld.1592
Margolin LG, Tarwater AE (1987) A diffusion operator for Lagrangian codes. In: Lewis R, Morgan K, Habashi W (eds) Numerical Methods for Heat Transfer, Pineridge Press, Swansea, pp 1252–1260
Neale RB, Hoskins BJ (2000a) A standard test case for AGCMs including their physical parametrizations: I: The proposal. Atmos Sci Lett 1:101–107, DOI 10.1006/asle.2000.0022
Neale RB, Hoskins BJ (2000b) A standard test case for AGCMs including their physical parametrizations: II: Results for the Met Office Model. Atmos Sci Lett 1:108–114, DOI 10.1006/asle.2000.0024
Nicolaides R (1992) Direct discretization of planar div-curl problems. SIAM J Numer Anal pp 32–56
Phillips NA (1957) A map projection system suitable for large-scale numerical weather prediction. J Meteor Soc Japan, 75th Anniversary Volume pp 262–267
Purser RJ, Rančić M (1997) Conformal octagon: An attractive framework for global models offering quasi-uniform regional enhancement of resolution. Meteorology and Atmospheric Physics 62:33–48, DOI 10.1007/BF01037478
Sadourny R (1972) Conservative finite-difference approximations of the primitive equations on quasi-uniform spherical grids. Mon Wea Rev 100(2):136–144
Sadourny RA, Arakawa A, Mintz Y (1968) Integration of the nondivergent barotropic vorticity equation with an icosahedral-hexagonal grid for the sphere. Mon Wea Rev 96:351–356
Salmon R (2004) Poisson-bracket approach to the construction of energy- and potential-enstrophy-conserving algorithms for the shallow-water equations. J Atmos Sci 61:2016–2036
Salmon R (2005) Poisson-bracket approach to the construction of energy- and potential-enstrophy-conserving algorithms for the shallow-water equations. Nonlinearity 18:R1–R16
Salmon R (2007) A general method for conserving energy and potential enstrophy in shallow-water models. J Atmos Sci 64:515–531
Samarskiĭ AA, Tishkin VF, Favorskiĭ AP, Shashkov MY (1981) Operator-difference schemes. Differentsial′nye Uravneniya 17(7):1317–1327, 1344
Satoh M (2004) Atmospheric circulation dynamics and general circulation models, 1st edn. Springer (Praxis), 643 pp.
Shashkov M (1996) Conservative Finite Difference Methods on General Grids. CRC-Press, Boca Raton, FL, 384 pages
Shashkov M, Steinberg S (1995) Support-operator finite-difference algorithms for general elliptic problems. J Comput Phys 118:131–151
Simmons AJ, Burridge DM (1981) An energy and angular momentum conserving vertical finite-difference scheme and hybrid vertical coordinates. Mon Wea Rev 109:758–766
Simmons AJ, Strüfing R (1981) An energy and angular-momentum conserving finite-difference scheme, hybrid coordinates and medium-range weather prediction. Tech. Rep. 28, European Centre for Medium-Range Weather Forecasts, Reading, U.K., 68 pages
Staniforth A, Wood N, Girard C (2003) Energy and energy-like invariants for deep non-hydrostatic atmospheres. Quart J Roy Meteor Soc 129:3495–3499
Taylor MA, Fournier A (2010) A compatible and conservative spectral element method on unstructured grids. J Comput Phys 229:5879–5895, DOI 10.1016/j.jcp.2010.04.008
Taylor MA, Edwards J, Thomas S, Nair R (2007) A mass and energy conserving spectral element atmospheric dynamical core on the cubed-sphere grid. J Phys Conf Ser 78(012074), DOI 10.1088/1742-6596/78/1/012074
Taylor MA, Edwards J, St-Cyr A (2008) Petascale atmospheric models for the community climate system model: New developments and evaluation of scalable dynamical cores. J Phys Conf Ser 125(012023), DOI 10.1088/1742-6596/125/1/012023
Taylor MA, St-Cyr A, Fournier A (2009) A non-oscillatory advection operator for the compatible spectral element method. In: Computational Science ICCS 2009 Part II, Lecture Notes in Computer Science 5545, Springer, Berlin / Heidelberg, pp 273–282

Thuburn J (2008) Some conservation issues for the dynamical cores of NWP and climate models. J Comput Phys 227:3715–3730

Vallis, G. K. (2006) Atmospheric and Oceanic Fluid Dynamics, Cambridge University Press, Cambridge, U.K., 745 pages

Williamson DL (1968) Integration of the barotropic vorticity equation on a spherical geodesic grid. Tellus 20:642–653

Williamson DL (2002) Time-split versus process-split coupling of parameterization and dynamical core. Mon Wea Rev 130:2024–2041

Williamson DL (2007) The evolution of dynamical cores for global atmospheric models. J Meteor Soc Japan 85B:242–269

Williamson DL, Olson JG (1994) Climate simulations with a semi-Lagrangian version of the NCAR community climate model. Mon Wea Rev 122:1594–1610

Williamson DL, Olson J, Jablonowski C (2009) Two dynamical core formulation flaws exposed by a baroclinic instability test case. Mon Wea Rev 137:790–796

Yakimiw E, Girard C (1987) Experimental results on the accuracy of a global forecast spectral mode1 with different vertical discretization schemes. Atmosphere-Ocean 25(3):304–325

Zerroukat M, Wood N, Staniforth A (2005) A monotonic and positive-definite filter for a semi-Lagrangian inherently conserving and efficient (SLICE) scheme. Quart J Roy Meteor Soc 131(611):2923–2936, DOI 10.1256/qj.04.97

Chapter 13
The Pros and Cons of Diffusion, Filters and Fixers in Atmospheric General Circulation Models

Christiane Jablonowski and David L. Williamson

Abstract All atmospheric General Circulation Models (GCMs) need some form of dissipation, either explicitly specified or inherent in the chosen numerical schemes for the spatial and temporal discretizations. This dissipation may serve many purposes, including cleaning up numerical noise generated by dispersion errors or computational modes, and the Gibbs ringing in spectral models. Damping processes might also be used to crudely represent subgrid Reynolds stresses, eliminate undesirable noise due to poor initialization or grid-scale forcing from the physics parameterizations, cover up weak computational stability, damp tracer variance, and prevent the accumulation of potential enstrophy or energy at the smallest grid scales. This chapter critically reviews the wide selection of dissipative processes in GCMs. They are the explicitly added diffusion and hyper-diffusion mechanisms, divergence damping, vorticity damping, external mode damping, sponge layers, spatial and temporal filters, inherent diffusion properties of the numerical schemes, and a posteriori fixers used to restore lost conservation properties. All theoretical considerations are supported by many practical examples from a wide selection of GCMs. The examples utilize idealized test cases to isolate causes and effects, and thereby highlight the pros and cons of the diffusion, filters and fixers in GCMs.

13.1 Introduction

There are many design aspects that need to be considered when building the fluid dynamics component, the so-called dynamical core, for atmospheric General Circulation Models (GCMs). Among them are the choice of the equation set

C. Jablonowski (✉)
Department of Atmospheric, Oceanic and Space Sciences, University of Michigan, 2455 Hayward Street, Ann Arbor, MI 48109, USA
e-mail: cjablono@umich.edu

D.L. Williamson
National Center for Atmospheric Research, 1850 Table Mesa Drive, Boulder, CO 80305, USA
e-mail: wmson@ucar.edu

P.H. Lauritzen et al. (eds.), *Numerical Techniques for Global Atmospheric Models*, Lecture Notes in Computational Science and Engineering 80,
DOI 10.1007/978-3-642-11640-7_13,

and prognostic variables, the computational grid and grid staggering options, the spatial and temporal numerical discretizations, built-in conservation properties, and the choice of dissipative processes that (1) might be needed to keep a model simulation numerically stable and (2) might truthfully mimic the cumulative effects of unresolved subgrid-scale processes on the resolved fluid flow. The latter aspect is at least a "hope" in the GCM modeling community. Here, the phrase *subgrid-scale* denotes the dry adiabatic unresolved processes in the dynamical core. This is in contrast to all other unresolved processes that lead to physical parameterizations such as radiation, convection, cloud processes and planetary boundary layer phenomena. These are not considered here, even though they are intimately coupled to the equations of motion. This chapter sheds light on the pros and cons of the most popular processes to handle both "physical" or "unphysical" subgrid-scale flow and mixing, and reviews the use of explicitly added and inherent diffusion, filters and fixers in GCMs. These are rarely documented in the refereed GCM literature but might be detailed in technical model descriptions.

It is common practice in GCMs to include a parameterization of the effects of subgrid-scale motions in the horizontal momentum and thermodynamic equations that is formulated as a local diffusive mixing. In fact, all numerical models need some form of dissipation, either explicitly specified or inherent in the chosen numerical schemes for the spatial and temporal discretizations. This dissipation may serve many purposes, including cleaning up numerical noise generated by dispersion errors, computational modes, or the Gibbs ringing, crudely representing subgrid Reynolds stresses, eliminating undesirable noise due to poor initialization or grid-scale forcing from the physics parameterizations, covering up weak computational stability, damping tracer variance, and preventing the accumulation of potential enstrophy or energy at the grid scale (Wood et al. 2007; Thuburn 2008a). Such an accumulation of energy is due to the physical downscale cascade and can result in excessive small-scale noise. It is known as spectral blocking and leads to an upturn (hook) or flattening in the kinetic energy spectrum at the smallest scales. Furthermore, physical "noise" in GCMs might originate from parameterized grid-scale forcings or from surface boundary conditions such as orography, the land-sea or land-use mask.

An accumulation of energy and enstrophy at the smallest scales may also arise due to a numerical misrepresentation of nonlinear interactions, the so-called aliasing effect. Nonlinear interactions and aliasing mostly originate from the quadratic or higher-order terms in the equations of motion. In essence, products of waves can create new waves that are shorter than $2\Delta x$ where Δx is the physical grid spacing. These waves cannot be represented on a model grid and are aliased into longer waves. Aliasing, if left unchecked, can lead to a blow up of the solution. This phenomenon is characterized as nonlinear computational instability as first discussed by Phillips (1959). Note that almost all GCMs suffer to some degree from aliasing. Exceptions are spectral transform models with quadratic transform grids which eliminate the aliasing of quadratic advection terms, the most problematic form, but do not completely eliminate aliasing from higher-order terms. Nonlinear computational instability does not occur in models that conserve quadratic quantities like enstrophy and kinetic energy (Arakawa 1966; Arakawa and Lamb 1981). Aliasing errors are not necessarily fatal. Whether an amplification of the waves occurs depends on the phase relation between the misrepresented and original waves in the model. More information on nonlinear computational instability and aliasing is provided in textbooks like Durran (1999, 2010), Kalnay (2003) or Lin (2007).

All mixing processes remove energy and enstrophy from the simulation which would otherwise build up to unrealistic proportions. Frequently, the included dissipation is restricted to be in the horizontal, as there is usually sufficient vertical mixing or diffusion in the physical boundary layer turbulence and convective parameterizations in full GCMs or sufficient inherent numerical diffusion to control noise in the vertical direction. Sometimes, vertical diffusion is also explicitly included in the dynamical core and applied throughout the whole troposphere. An example is the model by Tomita and Satoh (2004) which is discussed later.

A common expectation might be that dissipative formulations based on turbulence theory or observations provide a physical foundation for the subgrid-scale mixing. However, such physical motivation is not guaranteed and each ad hoc mixing process in a GCM must be critically reviewed. As pointed out by Mellor (1985) the horizontal diffusivities in use by GCMs are typically many orders of magnitude larger than those which would be appropriate for turbulence closures. Thus, horizontal diffusion used by most models cannot be considered a representation of turbulence but should be viewed as a substitute mechanism for unresolved horizontal advective processes. Awareness of this might offer some guidance in choosing an adequate subgrid-scale mixing scheme.

Mixing in GCMs generally serves as a numerical filter and neither reflects the mathematical representation of the energy or enstrophy transfer to small scales nor the representation of physical molecular diffusion (Koshyk and Boer 1995). Subgrid-scale processes, although small, can have a profound impact on the large-scale circulation. For example, diffusive mechanisms affect the propagation of waves and thereby the mean flow through wave-mean flow interactions. In addition, both inherent and explicitly added dissipation processes smear out sharp gradients in the tracer fields, and may lead to unphysical and overly strong mixing. Such mixing might provide feedbacks to the physical parameterizations. For example, the precipitation field might be highly influenced by the diffusive characteristics of the moisture transport algorithm in the dynamical core. The notion of overly diffusive GCMs was discussed by Shutts (2005). He argued that numerical advection errors, horizontal diffusion and parameterization schemes like the gravity wave drag or convection, act as energy sinks and lead to excessive energy dissipation in GCMs. However, such a conclusion might be highly model dependent.

In summary, some mixing processes are used for purely numerical reasons to prevent the model from becoming unstable. Others are meant to mimic subgrid-scale turbulence processes that are unsolved on the chosen model grid. In practice, many filters and mixing processes are used at once, which makes it more difficult to evaluate their individual effects. The form of the diffusion processes in atmospheric dynamical cores varies widely and is somewhat arbitrary. There are explicit dissipation processes and filters, inherent numerical dissipation, and fixers in GCMs.

Throughout the chapter we associate the phrase *explicit diffusion* with processes explicity added to the equations of motion. The phrase *implicit diffusion* characterizes the inherent dissipation of numerical schemes. These phrases are intended to make a distinction from explicit and implicit numerical approximations to diffusion operators. Note that the words "diffusion", "dissipation" and "viscosity" are often used interchangeably in the literature. Other characterizations of damping are smoothing, filtering and mixing.

13.1.1 Model Equations and the Representation of Explicit Diffusion

Mixing processes in GCMs can appear in many forms. A very dominant form is based on explicit dissipation mechanisms that get appended to the equations of motion shown in Chap. 15. In the continuous equations this mixing symbolizes molecular diffusion. However, GCMs are not capable of representing molecular diffusion at the nm or mm scale since they are typically applied with horizontal grid spacings between 20 and 300 km. Nonhydrostatic GCMs (Tomita et al. 2005; Fudeyasu et al. 2008) and mesoscale limited-area models like the Weather Research and Forecasting Model WRF (Skamarock et al. 2008) are also run with finer grid spacings of a few kilometers. Other atmospheric models with even finer scales might utilize the Large Eddy Simulation (LES) approach. LES is a mathematical model for turbulence that is based upon the Navier–Stokes equations with built-in low-pass filter. The underlying idea was initially proposed by Smagorinsky (1963) and further developed by Deardorff (1970). LES has been extensively used to study small-scale physical processes and mixing in the atmosphere. But in any case, models truncate the multi-scale spectrum of atmospheric motions well above the molecular diffusion scales. The unresolved part is typically modeled as dissipation and one might hope that it adequately captures the adiabatic subgrid-scale processes in some poorly understood way.

Explicit dissipation can be added to the momentum and thermodynamic equations in the symbolic form

$$\frac{\partial \psi}{\partial t} = Dyn(\psi) + Phys(\psi) + F_{\psi} \tag{13.1}$$

where $Dyn(\psi)$ denotes the time tendencies of the prognostic variable ψ according to the resolved adiabatic dynamics, $Phys(\psi)$ symbolizes the time tendencies from the subgrid-scale diabatic physical parameterizations, and F_{ψ} is the dissipation. The actual form of this dissipation is model dependent. For example, models in momentum form, that utilize the zonal and meridional velocities u, v and temperature T, might append the diffusive terms F_u, F_v, F_T. Models in vorticity-divergence (ζ, δ) form add the diffusion $F_{\zeta}, F_{\delta}, F_T$, or even replace F_{ζ} with a diffusion of the absolute vorticity $F_{\zeta+f}$ where f symbolizes the Coriolis parameter. Alternatively, if the potential temperature Θ is selected in the thermodynamic equation a diffusive term F_Θ might be chosen. Dissipation might also be applied to the tracer transport equations, and in case of nonhydrostatic models to the vertical velocity. Whether explicit diffusion is *needed* for computational stability is model dependent. Some models prefer to control the smallest scales via inherent numerical dissipation and select $F_\psi = 0$. However, the form of F_ψ is one of the main foci in Sects. 13.3–13.5, and therefore we introduce the generic form of the forcing here.

13.1.2 Overview of the Chapter

This chapter presents a comprehensive review of dissipative processes and fixers in general circulation models. Many pointers to references are given, and we illustrate the practical implications of the diffusion, filters and fixers on the fluid flow in atmospheric dynamical cores. In particular, we review the principles behind the different dissipative formulations, isolate causes and effects, provide many examples from today's GCMs and utilize idealized dynamical core test cases and so-called aqua-planet simulations to demonstrate the concepts. These test cases are briefly outlined in Sect. 13.2. Overall, we quote or show examples from over 20 different dynamical cores to highlight the broad spectrum of the dissipative processes in GCMs. The models are listed in Sect. 13.2. We characterize fourteen of them in greater detail in the Appendix since they are used as examples throughout the chapter.

The chapter is organized as follows. Sections 13.3 and 13.4 discuss the most popular explicit diffusion and damping mechanisms in the dynamical cores of GCMs. Section 13.3 includes the classical linear and nonlinear horizontal diffusion and hyper-diffusion, their diffusion coefficients and stability constraints, and physical consistency arguments. Section 13.4 discusses the 2D and 3D divergence damping, vorticity damping, Rayleigh friction and diffusive sponges near the model top, and external mode damping. In general, it is debatable whether filters are considered explicit dissipation or just a computational technique to keep a model numerically stable. Here, we choose to present them in their own category in Sect. 13.5 where both temporal and spatial filter are assessed. Section 13.6 captures the basic ideas behind inherent numerical dissipation which is nonlinear and sometimes is interpreted as physically motivated diffusion. Section 13.7 sheds light on the conservation properties of atmospheric GCMs and introduces a posteriori fixers. They include the dry air mass fixer, fixers for tracer masses and total energy fixers. Some final thoughts are presented in Sect. 13.8.

13.2 Selected Dynamical Cores and Test Cases

We illustrate many of the effects of the diffusion, filters and fixers in GCMs with the help of 2D shallow water or 3D hydrostatic model runs to discuss the practical implications of the theoretical concepts. Throughout this chapter, we point to the specific implementations of the dissipative processes, and quote typical values for the empirical coefficients from a variety of models to shows their spread in the GCMs. The intention is to give hands-on guidance and present the design options.

In particular, this chapter features examples from the dynamical cores CAM Eulerian, CAM semi-Lagrangian, COSMO, ECHAM5, FV, FVcubed, GEOS, GME, HOMME, ICON, IFS, NICAM, UM and WRF. The Appendix explains the acronyms and briefly characterizes the numerical schemes. Most models are global hydrostatic GCMs and use the shallow-atmosphere approximation. The only exceptions are the nonhydrostatic models COSMO, NICAM, UM and WRF that are built upon the deep atmosphere equation set (see White et al. 2005 for a review of the equations). We also briefly refer to other models such as NASA's ModelE by the Goddard Institute for Space Studies, the Atmospheric GCM for the Earth Simulator (AFES) developed by the Center for Climate System Research at the University of Tokyo and the National Institute for Environmental Studies (Japan), the Global Environmental Multiscale (GEM) model from the Canadian Meteorological Centre, the Global Forecast System (GFS) and the Eta model developed by the National Centers for Environmental Prediction (NCEP). The references for these models are given later.

The model simulations utilize a variety of idealized test cases. They include the steady-state and baroclinic wave test cases for dynamical cores suggested by Jablonowski and Williamson (2006a,b), selected shallow water test cases from the Williamson et al. (1992) test suite, the Held–Suarez climate forcing (Held and Suarez 1994), a variant of the Held–Suarez test with modified equilibrium temperatures in the stratosphere (Williamson et al. 1998), and the aqua-planet configuration as proposed by Neale and Hoskins (2000). The adiabatic dynamical core and shallow water test cases generally assess the properties of the numerical schemes in short deterministic model runs of up to 30 days. The idealized Held-Suarez-type simulations utilize a prescribed Newtonian temperature relaxation and boundary layer friction that replace the physical parameterization package. These model runs are typically integrated for multiple years to assess the climate statistics on a dry and spherical earth with no mountains. The aqua-planet assessments are the most complex simulations discussed in this chapter. They represent moist GCM runs that include the full physical parameterization suite but utilize a simplified lower boundary condition. In essence, the lower boundary is replaced by a water covered earth with prescribed sea surface temperatures. In addition, the settings of the physical constants and a symmetric ozone data set are prescribed in aqua-planet simulations.

13.3 Explicit Horizontal Diffusion

This section discusses the ideas behind explicit horizontal diffusion mechanisms in GCMs. In particular, we assess the linear second-order diffusion, the higher-order and thereby more scale-selective hyper-diffusion, reveal the selection criteria for the diffusion coefficients in both spectral transform and grid point models, discuss the concept of spectral viscosity, and review the stability constraints for the diffusion equation. In addition, we introduce the principles behind nonlinear horizontal diffusion and briefly survey the physical consistency of explicit diffusion schemes.

13.3.1 Generic Form of the Explicit Diffusion Mechanism

The generic form of the explicit linear diffusion is given by

$$F_{\psi} = (-1)^{q+1} K_{2q} \nabla^{2q} \psi \tag{13.2}$$

where $q \geq 1$ is a positive integer, $2q$ denotes the order of the diffusion, K_{2q} stands for the diffusion coefficient and ∇ is the gradient operator. Both the horizontal 2D gradient operator or an approximated 3D gradient operator have been used for the horizontal diffusion as further explained below. Setting $q = 1$ yields a second-order diffusion that emerges from physical principles such as the heat diffusion, molecular diffusion and Brownian motion. However, molecular diffusion acts on the nanometer to millimeter scale, and is therefore unresolved on a GCM model grid.

In practice, second-order diffusion is often applied as an artificial sponge near the top boundary, and has very little resemblance with its physical counterpart. In general, more scale-selective hyper-diffusion schemes with $q = 2, 3, 4$ are selected in the majority of the model domain. The most popular choice is the fourth-order hyper-diffusion with $q = 2$ that is also called bi-harmonic diffusion or superviscosity. The use of hyper-diffusion is often motivated by the need to maximize the ratio of enstrophy to energy dissipation since 2D turbulence theory predicts a vanishing energy dissipation rate at increasing Reynolds numbers (Sadourny and Maynard 1997). The higher the order of the hyper-diffusion, the higher the ratio of enstrophy to energy dissipation becomes. Farge and Sadourny (1989) even suggested using a 16th-order hyper-diffusion.

13.3.2 Particular Forms of Explicit Diffusion in GCMs

The exact form of the damping varies widely in GCMs. Typically, for convenience the horizontal operators are applied along model levels with the possible exception of the formulation for the scalar temperature diffusion. We now list several examples to illustrate the variety of the diffusion mechanisms. Our first example is taken from the weather forecast model GME which has been developed at the German Weather Service. It applies the fourth-order hyper-diffusion

$$F_u = -K_4 \nabla^4 u \tag{13.3}$$
$$F_v = -K_4 \nabla^4 v \tag{13.4}$$
$$F_T = -K_4 \nabla^4 (T - T_{ref}) \tag{13.5}$$

where ∇ is the horizontal gradient operator and T_{ref} is a reference temperature that depends only on height (Majewski et al. 2002). At upper levels near the model top, a second-order diffusion is applied. GME utilizes local basis functions that are orthogonal and conform perfectly to the spherical surface. They are locally anchored in each triangle of GME's icosahedral grid. Within the local neighborhood of a grid point the coordinate system is therefore nearly Cartesian. Note that Cartesian coordinates simplify the representation of the ∇^{2q} operator since the metric terms are equal to unity.

If the diffusion is expressed in spherical coordinates many metric terms are present. Here, we first show the operators for three spatial dimensions before simplifying them. The scalar 3D Laplacian $\nabla^2_{(3D)}$ operator in spherical coordinates for a prognostic variable ψ has the form

$$\nabla^2_{(3D)}\psi = \frac{1}{r^2 \cos^2\phi}\partial_{\lambda\lambda}\psi + \frac{1}{r^2\cos\phi}\partial_\phi(\cos\phi\,\partial_\phi\psi) + \frac{1}{r^2}\partial_r(r^2\partial_r\psi) \tag{13.6}$$

where r denotes the radial distance in the local vertical direction from the center of the earth, λ and ϕ are the longitude and latitude, and the notation ∂_x symbolizes a partial derivative in the x direction where x is a placeholder for λ, ϕ and r. In addition, the 3D vector Laplacian for the three-dimensional wind vector $\mathbf{v_3} = (u, v, w)$ is given by

$$\nabla^2_{(3D)}\mathbf{v_3} = \begin{pmatrix} \nabla^2_{(3D)}u - \frac{2\sin\phi}{r^2\cos^2\phi}\partial_\lambda v + \frac{2}{r^2\cos\phi}\partial_\lambda w - \frac{1}{r^2\cos^2\phi}u \\ \nabla^2_{(3D)}v - \frac{1}{r^2\cos^2\phi}v + \frac{2}{r^2}\partial_\phi w + \frac{2\sin\phi}{r^2\cos^2\phi}\partial_\lambda u \\ \nabla^2_{(3D)}w - \frac{2}{r^2}w - \frac{2}{r^2\cos\phi}\partial_\lambda u - \frac{2}{r^2\cos\phi}\partial_\phi(\cos\phi\, v) \end{pmatrix}. \tag{13.7}$$

These expressions are e.g., shown in Appendix A in Satoh (2004). The extra terms besides the scalar $\nabla^2_{(3D)}$ operator arise due to the spatial variation of the unit vectors in spherical coordinates. With the exception of the undifferentiated term in each of the components the extra terms are not necessarily negligible in comparison with those of the scalar diffusion operator. In fact, some of them are crucial in ensuring that the diffusion operator conserves angular momentum as outlined by Staniforth et al. (2006).

Generally, approximated forms of (13.6) and (13.7) are chosen to express the horizontal diffusion. The distance r is often approximated by the constant radius of the earth a, the terms containing the vertical derivative ∂_r and vertical velocity w are dropped, and the vertical component of the vector Laplacian is neglected to create a 2D diffusion operator. The replacement of the distance r by a is in fact a necessity in hydrostatic models based on the primitive equations (White et al. 2005). The scalar 2D Laplacian then simplifies in the following way

$$\nabla^2 \psi = \frac{1}{a^2 \cos^2 \phi} \partial_{\lambda\lambda} \psi + \frac{1}{a^2 \cos \phi} \partial_\phi (\cos \phi \partial_\phi \psi). \tag{13.8}$$

The 2D vector Laplacian for the two-dimensional wind vector $\mathbf{v} = (u, v)$ is given by

$$\nabla^2 \mathbf{v} = \begin{pmatrix} \nabla^2 u - \dfrac{2 \sin \phi}{a^2 \cos^2 \phi} \partial_\lambda v - \dfrac{1}{a^2 \cos^2 \phi} u \\ \nabla^2 v - \dfrac{1}{a^2 \cos^2 \phi} v + \dfrac{2 \sin \phi}{a^2 \cos^2 \phi} \partial_\lambda u \end{pmatrix}. \tag{13.9}$$

This form of the vector Laplacian leads to the "conventional form" of the horizontal momentum diffusion as characterized by Becker (2001). Unfortunately, this form does not conserve angular momentum as further discussed in Sect. 13.3.7. Some models also drop the extra terms and only apply the scalar 2D Laplacian operator to the vector wind (u, v) in spherical geometry. Such a simplified form is e.g., provided as an optional sponge layer damping mechanism near the model top in the finite-volume (FV) dynamical core in the Community Atmosphere Model CAM (version 5) (Neale et al. 2010). The model CAM FV is used at the National Center for Atmospheric Research (NCAR).

There is another caveat. Both formulations of the Laplacian in (13.7) or (13.9) would lead to an undesired damping of a solid body rotation as thoroughly analyzed by Staniforth et al. (2006) for the Unified Model (UM) developed at the UK Met Office. Therefore in practice, a more complicated form of the momentum diffusion is chosen in the model UM that is applied to the velocity components u, v and w (see Staniforth et al. (2006) for the derivation). The NCAR CAM spectral transform Eulerian (EUL) and semi-Lagrangian (SLD) dynamical cores (Collins et al. 2004) also include such a correction for solid body rotation as explained later.

Concerning the scalar diffusion in the model UM, a form similar to (13.8) is selected for the diffusion of potential temperature. It is applied twice (including a sign reversal) to resemble a fourth-order hyper-diffusion mechanism. The main differences to (13.8) are that (1) the model UM does not utilize a shallow-atmosphere approximation and retains the radial distance r, (2) a slope correction is utilized over steep terrain to lessen the spurious mixing along UM's deformed orography-following vertical coordinate, and (3) the diffusion coefficient is different in the two horizontal directions. The coefficient is constant in the meridional direction, but the strength of the diffusion in longitudinal direction is allowed to vary with latitude. This leads to non-isotropic diffusion and is further explained in Sect. 13.3.5. Note that the model UM does not need horizontal diffusion for computational stability reasons due to the inherent numerical dissipation in the interpolations of its semi-Lagrangian scheme. In practice, the explicit diffusion is therefore optional and not used by default. For example, it is never utilized in short weather prediction simulations (Terry Davies, personal communication).

Scalar diffusion of type (13.8) is also applied in other models such as the spectral transform Integrated Forecasting System (IFS) at the European Centre for Medium-Range Weather Forecast (Ritchie et al. 1995; ECMWF 2010). The model utilizes a second-order ($q = 1$) diffusion scheme close to the model top and a fourth-order ($q = 2$) hyper-diffusion of the prognostic scalar variables relative vorticity ζ, horizontal divergence δ and temperature. It yields the explicit diffusion

$$F_\zeta = (-1)^{q+1} K_{2q} \nabla^{2q} \zeta \tag{13.10}$$
$$F_\delta = (-1)^{q+1} K_{2q} \nabla^{2q} \delta \tag{13.11}$$
$$F_T = (-1)^{q+1} K_{2q} \nabla^{2q} T. \tag{13.12}$$

This form of the diffusion is furthermore utilized by the spectral transform model ECHAM5 developed at the Max–Planck Institute for Meteorology, where even higher-order diffusion operators are chosen below the sponge layer at the model top (Roeckner et al. 2003).

The application of the diffusion along sloping general vertical coordinates, like the hybrid pressure-based η-coordinate (Simmons and Burridge 1981), is straightforward to implement, but as mentioned before can cause spurious mixing over mountains, especially in the neighborhood of steep terrain. This is largely due to the presence of large vertical temperature variations along the sloping surfaces that overlay the horizontal gradients. Such spurious mixing triggered by the vertical variations is undesirable and may grow to significant proportions. Therefore in practice, the diffusion of the temperature in the model IFS is modified to approximate the horizontal diffusion on surfaces of constant pressure rather than on the sloping η-coordinate surfaces. This is further explained in the technical model documentation of the CAM EUL and SLD dynamical cores (Collins et al. 2004). CAM EUL and SLD apply the fourth-order temperature diffusion

$$F_T = -K_4 \left[\nabla^4 T - p_s \frac{\partial T}{\partial p} \frac{\partial p}{\partial p_s} \nabla^4 \ln p_s \right] \tag{13.13}$$

where p is the pressure and p_s symbolizes the surface pressure. The second term in F_T consists of the leading term in the transformation of the ∇^4 operator from η surfaces to pressure surfaces. CAM also applies a second-order sponge-layer diffusion at upper levels. But since the upper levels in CAM coincide with pure pressure levels the correction is not needed there. In general, it is unclear whether diffusion should be applied along constant model levels, constant pressure levels or even along constant height or isentropic levels. If the diffusion primarily counteracts numerical artifacts, arguments can be found that it should be applied along model levels. However, if the primary motivation is to characterize physical mixing, height, pressure or isentropic levels are advantageous as explained in detail by Staniforth et al. (2006).

Note that NCAR's EUL and SLD dynamical cores actually apply a variant of the diffusion shown in (13.10) and (13.11) to the relative vorticity and horizontal divergence fields which generalizes the approach by Bourke et al. (1977). It is given by

$$F_{\zeta} = (-1)^{q+1} K_{2q} \left[\nabla^{2q} (\zeta + f) + (-1)^{q+1} (\zeta + f) \left(\frac{2}{a^2} \right)^q \right] \quad (13.14)$$

$$F_{\delta} = (-1)^{q+1} K_{2q} \left[\nabla^{2q} \delta + (-1)^{q+1} \delta \left(\frac{2}{a^2} \right)^q \right] \quad (13.15)$$

which diffuses the absolute vorticity $(\zeta + f)$ instead of ζ in (13.14). The undifferentiated correction terms are added to the vorticity and divergence diffusion to prevent the damping of uniform solid-body rotations. For example, a solid body wind distribution like

$$u = u_0 \cos \phi \quad (13.16)$$

$$v = 0 \quad (13.17)$$

with a constant velocity u_0 does not experience any damping by the diffusion shown in (13.14) and (13.15). The models ECHAM5 and IFS with the diffusion terms (13.10) and (13.11) do not apply this correction and thereby damp such a solid-body rotation.

The derivation of (13.14) and (13.15) can be understood when taking a second look at the scalar 3D Laplacian operator (13.6) as explained by Satoh (2004) (his Chap. 17). Substitute $\mathbf{v}_3$ with $\mathbf{v}_3 = (u, v, 0)$ and assume that the horizontal wind components u and v are proportional to the distance r to help simplify the Laplacian. This could be envisioned by assuming an idealized profile like $u(\lambda, \phi, r) \sim r k_u u_H(\lambda, \phi)$ with the constant k_u and velocity u_H that only varies in the horizontal direction. The scalar 3D Laplacian (13.6) for the velocity $\psi = u$ then yields

$$\nabla^2_{(3D)} u = \frac{1}{r^2 \cos^2 \phi} \partial_{\lambda\lambda} u + \frac{1}{r^2 \cos \phi} \partial_{\phi} (\cos \phi \partial_{\phi} u) + \frac{2}{r^2} u. \quad (13.18)$$

After replacing the distance r by the radius a we obtain

$$\nabla^2_{(3D)} u = \frac{1}{a^2 \cos^2 \phi} \partial_{\lambda\lambda} u + \frac{1}{a^2 \cos \phi} \partial_{\phi} (\cos \phi \partial_{\phi} u) + \frac{2}{a^2} u. \quad (13.19)$$

A similar expression holds for v. With these approximations, the horizontal diffusion expressed in (13.14) and (13.15) can be derived as shown by Satoh (2004).

The explicit diffusion mechanisms for the horizontal divergence in spectral models (13.11) or (13.15) readily controls gravity waves as also pointed out by Randall (1994). This suggests an interesting analogy to another type of explicit damping called horizontal divergence damping. The latter is sometimes used in GCMs written in momentum (u, v) form and is explained in detail in Sect. 13.4.1. Although both mechanisms are characterized with different names they accomplish a similar or even identical physical effect, namely they damp the divergent motions with either a second-order or higher order diffusion. We return to this analogy later.

The diffusion discussed so far is linear. It can therefore easily be calculated in spectral space in the spectral models ECHAM5, IFS, EUL or SLD since the spherical harmonics basis functions are eigenfunctions of the Laplacian operator on the sphere. These models utilize a constant diffusion coefficient and apply the diffusion with an implicit temporal discretization to increase the numerical stability of the diffusion mechanism. This is discussed below in Sects. 13.3.4.1 and 13.3.5.

13.3.3 Practical Considerations: Linear High-Order Diffusion Operators

Second-order diffusion schemes are not very scale selective and can therefore impact the well-resolved scales in a negative way. In practice, higher-order hyper-diffusion formulations are generally preferred to improve the scale selectivity. This is despite the fact that higher-order diffusion does not posses a physical foundation. High-order diffusion such as a fourth-order hyper-diffusion based on the ∇^4 operator is most often chosen. Even sixth- or eighth-order hyper-diffusion schemes are applied in GCMs, e.g., in ECHAM5 (Roeckner et al. 2003). The higher order can either be achieved via multiple applications of the ∇^2 operator with sign reversals as in Staniforth et al. (2006) or via the direct discretization of the higher-order operators.

As a sneak preview of the practical aspects of the diffusion discussion in Sects. 13.3, 13.4 and 13.6, we start the assessment by isolating the effects of the fourth-order hyper-diffusion and second-order sponge-layer diffusion (applied in the top three levels $< 14\,\text{hPa}$) in an idealized dynamical core simulation. In particular, we choose the CAM (version 4) SLD dynamical core at the triangular (T) truncation T85 ($\approx 156\,\text{km}$) with 26 levels. A steady-state test case, described in Jablonowski and Williamson (2006a), is used and run for 30 days with (a) no explicit diffusion and (b) the default second-order and fourth-order diffusion using the default coefficients $K_2 = 2.5 \times 10^5\ \text{m}^2\ \text{s}^{-1}$ and $K_4 = 1 \times 10^{15}\ \text{m}^4\ \text{s}^{-1}$. Note that this K_2 value is the base value at the third level from the top. It is doubled at the second level and doubled again at the top level (Neale et al. 2010). This adds a vertical dependency to the formulation of K_2 that is purely based on the level number without taking the actual pressure or height position into account. The second-order diffusion serves as a sponge near the model top. Figure 13.1 shows the corresponding zonal-mean zonal wind fields at day 30 and the difference plot between the run with default diffusion and no diffusion. The run without diffusion (Fig. 13.1a) keeps an almost perfect steady state that is visually indistinguishable from the initial state. In the default diffusion simulation (Fig. 13.1b) the fourth-order hyper-diffusion acts throughout the entire atmosphere except in the uppermost three levels above 14 hPa. Its influence on this very smooth steady state solution is negligible and the absolute differences in the lower layers are on the order of $\pm 0.025\ \text{m s}^{-1}$. The difference plot in Fig. 13.1c is clearly dominated by the effects of the second-order diffusion in the sponge layer. The sponge layer diffusion changes the shape of the midlatitudinal

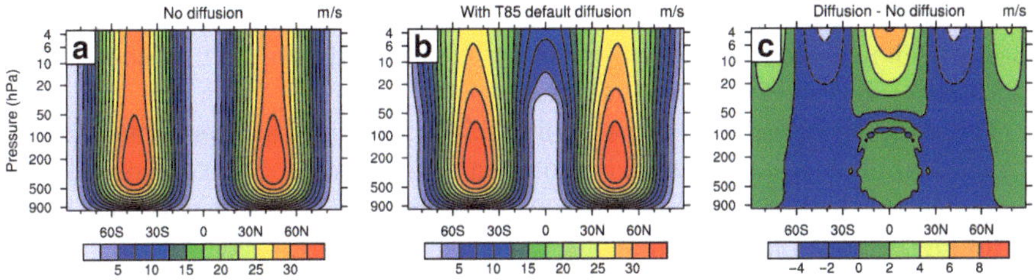

Fig. 13.1 Zonal-mean zonal wind (m/s) at day 30 of the steady-state test case of Jablonowski and Williamson (2006a) in the CAM 4 SLD dynamical core at the resolution T85L26 with (**a**) no diffusion, (**b**) the default second-order and fourth-order diffusion (see text), (**c**) zonal wind difference between the simulation with diffusion and no diffusion. A logarithmic pressure coordinate is used. No decentering is applied. The time step is $\Delta t = 1{,}800$ s

zonal jets considerably which leads to a decrease in the peak wind speeds at the model top by about $4\,\mathrm{m\ s^{-1}}$. The diffusion also causes significant increases in the wind speeds in the formerly calm equatorial and polar regions by up to $8\,\mathrm{m\ s^{-1}}$. It thereby smoothes out the sharp gradients in the zonal wind field. Note that the influence of the second-order diffusive sponge is not just limited to the top three layers. It clearly modulates the wind profile in the uppermost six levels which lie above 54 hPa. These sponge layer effects are discussed in greater detail later in Sect. 13.4.5.

As an aside for completeness, no decentering of the trajectories was used ($\epsilon = 0$), as will be explained later in Sect. 13.6.3, and the trajectory calculation utilizes only spherical coordinates to suppress any signal from non-zonal geodesic trajectory calculations in polar regions (typically poleward of 70°). The local geodesic coordinate is essentially a rotated spherical coordinate system whose equator goes through the arrival point of the trajectory (see details in Williamson and Rasch 1989). Of course, omitting decentering is only reasonable in the absence of mountains, and the exclusive use of spherical coordinates is only reasonable in the case of zonal advection as considered in the special case here. These deviations from the default CAM SLD configuration are selected to truly isolate the damping effects from the linear horizontal diffusion. In practice, the damping of all explicit and implicit dissipation mechanisms as well as filters and fixers act in concert, and they are generally difficult to isolate individually.

13.3.4 Choice of the Diffusion Coefficient: Damping Time Scales

The choice of the ∇^2, ∇^4 or even higher-order diffusion coefficient is most often motivated by empirical arguments and chosen in a somewhat arbitrary manner. It is sometimes even considered a model *tuning* parameter. However as seen in Fig. 13.1 and also shown later, the diffusion can have a profound impact on the global circulation, and must be chosen with care. This was also noted by Stephenson (1994) who points out that relatively few systematic horizontal diffusion studies have been performed with realistic GCMs. Among the few are the studies by Williamson (1978), MacVean (1983) and Laursen and Eliasen (1989).

This raises questions about the scaling of the subgrid-scale horizontal mixing parameterizations with horizontal resolution. For example, Smagorinsky (1963) suggested a second-order diffusion based on turbulence concepts in which the Eddy diffusivity depends on the square of the model horizontal grid spacing and the deformation of the flow field (see Sect. 13.3.6). Higher degree hyper-diffusion operators, such as ∇^4, ∇^6 or ∇^8, are commonly used in spectral transform models. Takahashi et al. (2006) showed with the help of an Eulerian spectral transform model with a truncation limit of about wavenumber $n_0 \leq 100$ that the coefficients for the ∇^4 form may be chosen to yield a straight kinetic energy spectrum with a slope of n^{-3} for spherical wavenumbers n between $[15, n_0]$. For significantly higher truncations, coefficients can be found which yield a slope of $\mathrm{n}^{-5/3}$ beyond $n = 100$. Appendix B of Jakob et al. (1993) provides details about the calculation of such kinetic energy spectra. As a physical motivation, Skamarock (2004) (see also Chap. 14), Takahashi et al. (2006) and Hamilton et al. (2008) discussed the desirability of modeling such slopes based on observational evidence (Nastrom and Gage 1985) and the theoretical reasons why they may or may not be expected. The latter is also addressed in Chen and Wiin-Nielsen (1978) and Boer and Shepherd (1983). Examples of kinetic energy spectra in the resolution range from 224 km down to 3.5 km are shown in Terasaki et al. (2009) for the nonhydrostatic global model NICAM (Satoh et al. 2008).

In practice, coefficients for different resolutions are found experimentally with the model configured for earth-like simulations so that in the mid- to upper-troposphere the kinetic energy spectra have the desired straight tails for each resolution. For example, Boville (1991) empirically determined diffusion coefficients via trial and error with the NCAR Community Climate Model, Version 1 (CCM1). Boville tested coefficients in short model integrations and adjusted them until he obtained kinetic energy spectrum at 250 hPa which did not change shape near the truncation limit. Using the same approach, diffusion coefficients have also been found for CAM 3.1 which provide kinetic energy spectra similar to those of Takahashi et al. (2006). The dynamical component of CAM 3.1 is the Eulerian spectral transform scheme as in the model used by Takahashi et al. (2006), although CAM 3.1 has a different subgrid-scale physics parameterization package and different water vapor transport. The model of Takahashi et al. (2006) used the Eulerian spectral transform method for the water vapor transport and applied the diffusion to water vapor as well as temperature, vorticity and divergence. In contrast, CAM 3.1 uses shape preserving semi-Lagrangian approximations for water vapor transport. Diffusion is applied to temperature, vorticity and divergence, but not to water vapor.

The choice of the diffusion coefficient needs to obey physical and numerical constraints. From a physical viewpoint, the coefficient influences the damping time scales for all waves and should be as small as possible for the resolved large scales to avoid overly strong dissipation of the physically relevant signals while still providing enough damping to prevent the build-up of energy and enstrophy at the smallest scale. From a numerical viewpoint, the coefficient needs to be sufficiently large to guarantee stable computations if the numerical approximation requires such damping. The stability aspect deserves special attention in explicit time-stepping schemes that place upper stability limits on the strength of the coefficient or impose restrictive time step sizes. An inadequately chosen coefficient can even act as the source of grid-scale noise and numerical instability. Both aspects are further explained below.

Ideally the rate of the energy dissipation near the truncation limit of a model should mimic the true energy transfer rates of the atmosphere at these scales. But unfortunately the knowledge from atmospheric observations of such transfer or dissipation rates is relatively poor as discussed by MacVean (1983). However, it helps to associate the value of the diffusion coefficient with a damping time scale at the smallest spatial scale in the model, since time scales can readily be understood from a physical viewpoint.

13.3.4.1 Diffusion Coefficients in Spectral Transform Models

Recall that $2q$ is the order of the diffusion where $q \geq 1$ is a positive integer. In spectral transform models the diffusion coefficient K_{2q} is typically represented by

$$K_{2q} = \frac{1}{\tau}\left(\frac{a^2}{n_0(n_0+1)}\right)^q \tag{13.20}$$

as for example shown by MacVean (1983), Sardeshmukh and Hoskins (1984) and Roeckner et al. (2003). τ is the e-folding time scale for the diffusion at the smallest wavelength, a denotes the Earth's radius and n_0 symbolizes the maximum wavenumber corresponding to the smallest wavelength. The wavenumber is e.g., specified by a triangular truncation limit like T85 with $n_0 = 85$. Equation (13.20) means that the (n, m)-th spectral component of the diffused quantity in the time-continuous case will be damped by the response function

$$E_n = \exp(-\Delta t\, d_n) = \exp\left\{ -\Delta t \left[\frac{1}{\tau}\left(\frac{n(n+1)}{n_0(n_0+1)}\right)^q \right] \right\} \tag{13.21}$$

$$= \exp\left\{ -\Delta t\, K_{2q} \left(\frac{n(n+1)}{a^2}\right)^q \right\} \tag{13.22}$$

where Δt symbolizes the length of the time step and d_n determines the strength of the damping (Sardeshmukh and Hoskins 1984; von Storch 2004). Here, n denotes the total (also called spherical) wavenumber and m stands for the zonal wavenumber as discussed in textbooks like Kalnay (2003) or Durran (1999, 2010). Note that the damping is independent of m. The response function is equivalent to an "amplification factor" represented by

$$E_n = \frac{\Psi_n^{t+\Delta t}}{\Psi_n^t} \tag{13.23}$$

that expresses the ratio of the wave amplitudes Ψ_n for each wavenumber n at the discrete future $t + \Delta t$ and current t time levels. The response function provides a damping mechanism for all $E_n < 1$.

The time-discretized form of the response function (13.21) yields

$$E_n \approx 1 - \Delta t \left[\frac{1}{\tau} \left(\frac{n(n+1)}{n_0(n_0+1)} \right)^q \right] \tag{13.24}$$

which can be transformed into the approximate form

$$E_n \approx \left\{ 1 + \Delta t \left[\frac{1}{\tau} \left(\frac{n(n+1)}{n_0(n_0+1)} \right)^q \right] \right\}^{-1} . \tag{13.25}$$

An almost identical response function in comparison to (13.25) is also shown in Collins et al. (2004) for the temperature diffusion in the CAM EUL spectral transform dynamical core. However, this Eulerian dynamical core applies the damping over a duration of $2\Delta t$ due to the chosen leapfrog time-stepping scheme. In fact, CAM EUL uses time splitting and applies the temperature diffusion with an implicit temporal discretization in spectral space after some other temporal updates

$$T_n^m = \tilde{T}_n^m - 2\Delta t \; K_{2q} \left(\frac{n(n+1)}{a^2} \right)^q T_n^m . \tag{13.26}$$

T_n^m and $\tilde{T}_n^m$ symbolize the spectral coefficients for the temperature at the future and partially updated past time, respectively. Note that (13.26) can also be rewritten as

$$T_n^m = E_n \, \tilde{T}_n^m \tag{13.27}$$

$$E_n = \left\{ 1 + 2\Delta t \; K_{2q} \left(\frac{n(n+1)}{a^2} \right)^q \right\}^{-1} \tag{13.28}$$

which confirms that the response function E_n plays the role of a damping mechanism as stated above in (13.23). The only difference between (13.28) and (13.25) is the duration of the time interval which depends on the time-stepping scheme in the dynamical core. More details on the application of the diffusion in the model CAM EUL are provided in Collins et al. (2004) (their Chap. 3.1.14) that also explains how the correction to pressure levels is computed (13.13). The diffusion mechanism in (13.27) and (13.28) can also be compared to the concept of "spectral viscosity" which is discussed below in Sect. 13.3.4.2.

Equations (13.20) and (13.21) reflect the strength of the damping for the diffusion operators shown in (13.10)–(13.12). If the variant of the diffusion is used that does not diffuse a solid body rotation (13.14)–(13.15) the relationship between the diffusion coefficient and the time scale becomes

$$K_{2q} = \frac{1}{\tau}\left(\frac{a^{2q}}{[n_0(n_0+1)]^q - 2^q}\right). \tag{13.29}$$

This leads to the time-continuous response function

$$E_n = \exp\left\{ -\Delta t\left[\frac{1}{\tau}\left(\frac{[n(n+1)]^q - 2^q}{[n_0(n_0+1)]^q - 2^q}\right)\right]\right\} \tag{13.30}$$

for the vorticity and divergence diffusion. Analogous to (13.25) the time-discretized form of (13.30) can be approximated by

$$E_n \approx \left\{1 + \Delta t\ K_{2q}\left(\frac{[n(n+1)]^q - 2^q}{a^{2q}}\right)\right\}^{-1} \tag{13.31}$$

that reintroduces the corresponding diffusion coefficient K_{2q} (13.29) into the equation.

The strength of the horizontal diffusion, as explicitly stated in (13.25), (13.28) and (13.31), is scale-dependent. This is confirmed for (13.28) in Fig. 13.2. The figure is drawn for a T85 ($n_0 = 85$) triangular truncation with an assumed time step of $2\Delta t = 1{,}200$ s and a damping time scale of $\tau = 8$ h. These settings are close to the default values for the fourth-order hyper-diffusion in CAM EUL (see Sect. 13.3.4.4) that utilizes the leapfrog time-stepping scheme. The response function (13.28) for the second-, fourth-, sixth- and eighth-order horizontal diffusion clearly shows that the higher-order diffusion provides less damping at long spatial scales (low wavenumbers) and rapidly increases in strength towards the high

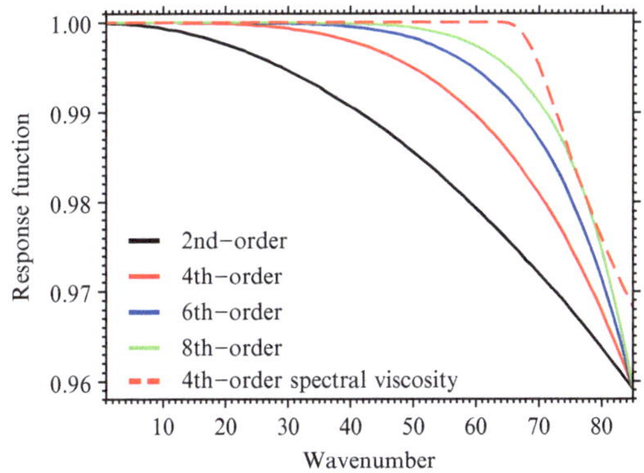

Fig. 13.2 Scale selectivity of the response function E_n shown in (13.28) for a spectral T85 triangular truncation and second-, fourth-, sixth- and eighth-order horizontal diffusion. In addition, the filter function σ_n (13.39) of the spectral viscosity is shown. A time step of $2\Delta t = 1{,}200$ s and damping time scale of $\tau = 8$ h are assumed

wavenumbers. The highest spherical wavenumber $n = 85$ is damped at an identical rate in this example. The figure also shows the damping characteristics of the spectral viscosity method (13.39) that is discussed in the next subsection.

As an aside, Leith (1971) suggested an alternative formulation of the response function. His derivation is based on an analysis of the numerical dissipation function that arises from the energy transfer between wavenumbers in a two-dimensional Cartesian turbulence closure model known as the Eddy-damped Markovian approximation. Gelb and Gleeson (2001) showed the corresponding response function for the Leith (1971) fourth-order hyper-diffusion when utilizing a leapfrog time-stepping scheme. It yields

$$E_n = \left\{ 1 + 2\Delta t \; K_L \; \frac{(n - n_L)^2 (n - n_L + 1)^2}{a^4} \right\}^{-1} \tag{13.32}$$

where the parameter $n_L = 0.55 n_0$ is the cutoff wavenumber and the diffusion coefficient K_L is defined as

$$K_L = \begin{cases} 0 & n \le n_c \,, \\ \dfrac{K_4}{(0.45)^4} & n_c < n \le n_0 \quad . \end{cases} \tag{13.33}$$

K_4 is given in (13.20) for $q = 2$. A second-order version of the Leith (1971) diffusion has been implemented in the weather prediction model "Global Forecast System" (GFS 2003) which is an operational spectral transform model at the National Centers for Environmental Prediction (NCEP). A comparison of the fourth-order traditional diffusion (13.28), the Leith (1971) diffusion (13.32) and spectral viscosity is provided in Gelb and Gleeson (2001).

13.3.4.2 The Concept of Spectral Viscosity

The horizontal diffusion in spectral transform models can also be replaced with a spherical "spectral viscosity" operator as proposed by Gelb and Gleeson (2001). They suggested a spectral viscosity method which is built upon rigorous mathematical principles for nonlinear conservation laws. It yields a viscosity term that depends on the spatial scale of the waves and is equal to zero for low wavenumbers. It is thereby highly scale-selective and does not damp well-resolved wave modes.

Note that the traditional fourth-order hyper-diffusion of the form $F_\Psi = -K_4 \nabla^4 \Psi$ (13.2) has the analytic spectral representation

$$\{F_\Psi\}_n^m = - \, K_4 \; \frac{n^2 (n+1)^2}{a^4} \; \Psi_n^m . \tag{13.34}$$

The spectral viscosity (SV) approach then translates (13.34) into the new form

$$\{F_\Psi^{SV}\}_n^m = - \, \epsilon \; \hat{q}_n^2 \; \frac{n^2 (n+1)^2}{a^4} \; \Psi_n^m \tag{13.35}$$

where ϵ is a tuning parameter that, according to the mathematical theory behind spectral viscosity, scales like $\epsilon \sim n_0^{-3}$. $\hat{q}_n$ is given by

$$\hat{q}_n = \begin{cases} 0 & n \leq n_c, \\ \exp\left(-\dfrac{(n-n_0)^2}{2(n-n_c)^2}\right) & n_c < n \leq n_0. \end{cases} \tag{13.36}$$

As before, n_0 is the maximum spherical wavenumber, a is the earth's radius, and n_c is a cutoff wavenumber. The spectral viscosity is zero for all waves with wavelengths that are larger than the cutoff wavelength. Gelb and Gleeson (2001) recommended the scaling parameters

$$\epsilon = \frac{c\,a^3}{n_0^3} \qquad \text{and} \qquad n_c = 2\,n_0^{\frac{3}{4}} \tag{13.37}$$

with $c = 2\,\text{m s}^{-1}$ and tested these in an Eulerian spectral transform shallow water model with a leapfrog time-stepping scheme. The parameter c needs to carry velocity units to match the physical dimensions, which were originally omitted. Gelb and Gleeson (2001) noted that these scalings might not be universal since they were tuned for a single shallow water test case.

The fourth-order spectral viscosity yields

$$\Psi_n^m = \sigma_n\,\tilde{\Psi}_n^m \tag{13.38}$$

$$\sigma_n = \left\{1 + 2\Delta t\;[\epsilon\;\hat{q}_n^2]\;\frac{n^2(n+1)^2}{a^4}\right\}^{-1} \tag{13.39}$$

where Ψ_n^m stands for the spectral coefficients of a prognostic variable, such as temperature, at the future time step, and $\tilde{\Psi}_n^m$ represents the spectral coefficients at a partially updated past time level, as discussed earlier in Sect. 13.3.4.1. The spectral viscosity operation specified in (13.38) is equivalent to applying a spectral filter of form σ_n at each time step (Canuto et al. 1987). Equation (13.38) is formally identical to the equation for the fourth-order ($q = 2$) hyper-diffusion (13.27). The most important difference is the definition of the response function σ_n (13.39) in comparison to E_n given in (13.28). The comparison reveals that the wavenumber-dependent viscosity parameter $[\epsilon\;\hat{q}_n^2]$ replaces the constant diffusion coefficient K_4. The different damping characteristics of σ_n and E_n can clearly be seen in Fig. 13.2. The dashed red curve depicts the spectral viscosity response function that can readily be compared to the traditional fourth-order hyper-diffusion (solid red curve). It confirms that the spectral viscosity does not damp the low wavenumbers (e.g., up to spherical wavenumber $n = 56$ in this example) and then quickly increases in strength. However, at this particular resolution and time step the damping effect of the spectral viscosity with parameter set (13.37) is always weaker than the damping of the traditional diffusion.

Gelb and Gleeson (2001) showed via 2D shallow water tests that the spectral viscosity method gives appreciably superior results if the flow field is underresolved.

In a well-resolved flow field the accuracy of SV and hyper-diffusion simulations was comparable. However, they observed that the use of spectral viscosity improved the conservation of invariants by the numerical scheme and led to more accurate energy spectra. Spectral viscosity has not been used in operational 3D spectral transform GCMs so far.

13.3.4.3 Diffusion Coefficients in Grid-Point Models

It is less straightforward to represent the relation between the damping time and the K coefficient in grid point models since the exact relation depends on the type of spatial discretization and the model grid. We therefore pick an example that illustrates the relationship and then generalize it.

The example reflects the UK Met Office's Unified Model on a latitude–longitude grid that utilizes a finite-difference approach in spherical coordinates. Following the arguments by Staniforth et al. (2006) in their Chap. 12 the response function for a centered finite-difference approximation of the second-order diffusion yields

$$E = 1 - \Delta t \left(\frac{K_\lambda \sin^2(k_\lambda \Delta\lambda/2)}{a^2 \cos^2 \phi (\Delta\lambda/2)^2} + \frac{K_\phi \sin^2(k_\phi \Delta\phi/2)}{a^2 (\Delta\phi/2)^2} \right) \tag{13.40}$$

where K_λ and K_ϕ are the second-order diffusion coefficients, k_λ and k_ϕ stand for the longitudinal and latitudinal wavenumbers, and $\Delta\lambda$ and $\Delta\phi$ are the grid spacings (in radians) in the longitudinal and meridional direction, respectively. Selecting an isotropic diffusion with coefficients $K_\lambda = K_\phi$ reveals a stability concern in (13.40), as further discussed in Sect. 13.3.5. In practice, $K_\lambda \cos^2 \phi = K_\phi = constant \equiv K$ is therefore chosen (Staniforth et al. 2006) which leads to the damping for each pair of wavenumbers

$$E = 1 - \Delta t \, K \left(\frac{\sin^2(k_\lambda \Delta\lambda/2)}{a^2 (\Delta\lambda/2)^2} + \frac{\sin^2(k_\phi \Delta\phi/2)}{a^2 (\Delta\phi/2)^2} \right). \tag{13.41}$$

Assuming that the highest wavenumbers $k_\lambda = 2\pi/L_x$ and $k_\phi = 2\pi/L_y$ are represented by the smallest resolvable wavelengths $L_x = 2\Delta\lambda$ and $L_y = 2\Delta\phi$ the relation between the diffusion coefficient and the time scale τ for the shortest waves becomes

$$\frac{1}{\tau} = K \left(\frac{1}{a^2 (\Delta\lambda/2)^2} + \frac{1}{a^2 (\Delta\phi/2)^2} \right) \tag{13.42}$$

or equivalently

$$K = \frac{1}{\tau} \left(\frac{1}{(a\Delta\lambda/2)^2} + \frac{1}{(a\Delta\phi/2)^2} \right)^{-1}. \tag{13.43}$$

Note that $\Delta x = a\Delta\lambda$ and $\Delta y = a\Delta\phi$ express the physical grid spacings at the equator.

A generalization of this approach for grid point models with approximately uniform physical grid spacings $\Delta x = \Delta y$ and $2q$-th-order hyper-diffusion yields

$$K_{2q} = \frac{1}{2\tau}\left(\frac{\Delta x}{2}\right)^{2q}. \tag{13.44}$$

Examples of such models are GME on an icosahedral grid (Majewski et al. 2002) or the dynamical core HOMME on a cubed-sphere mesh (see Taylor et al. (2007) and Chap. 12). Other models like the nonhydrostatic icosahedral dynamical core NICAM (Tomita and Satoh 2004) define the horizontal and vertical diffusion coefficients as

$$K_{2q\,H} = \gamma_H \, \frac{\overline{\Delta x}^{\,2q}}{\Delta t} \tag{13.45}$$

$$K_{2q\,V} = \gamma_V \, \frac{(\Delta \xi)^{2q}}{\Delta t} \tag{13.46}$$

where γ_H and γ_V are non-dimensional empirical factors and $\Delta\xi$ is the vertical grid spacing in the generalized terrain-following height coordinate ξ. Here $\overline{\Delta x}$ symbolizes the average grid spacing in their quasi-uniform triangular grid given by

$$\overline{\Delta x} = \sqrt{\frac{4\pi a^2}{N}} \tag{13.47}$$

with a total of N grid points per model level. $4\pi a^2$ denotes the total surface area of the sphere. After an empirical factor γ and thereby K_{2q} is chosen, its damping time scale is

$$\tau = \frac{\Delta t}{2^{2q+1}\,\gamma} \tag{13.48}$$

according to (13.44). The symbol γ is used as a placeholder for either γ_H or γ_V. Concrete values for γ are presented below in Sect. 13.3.4.5. As an aside, explicitly added vertical diffusion is often considered part of the physical parameterization suite, and is rarely included in a dynamical core.

The details of the assessments above will differ somewhat based on the numerical scheme and the degree of non-uniformity of the computational grid. Nevertheless, (13.44) gives guidance when selecting the appropriate damping time scales. Note again, that models with explicit time discretizations enforce stability limits on the strength of the coefficient (Jakimow et al. 1992; Staniforth et al. 2006) for a given time step.

13.3.4.4 Examples of Diffusion Coefficients in Spectral Models

As a practical example, we briefly discuss the heterogeneous horizontal diffusion mechanism in the spectral model ECHAM5. First, the order of the hyper-diffusion scheme varies depending on the model level with highly scale-selective sixth or eighth-order diffusion at low levels and increased fourth- and second-order diffusion in higher regions. This increased diffusion near the model top serves as a sponge to lessen the spurious reflection of planetary and gravity waves at the upper boundary. Second, the strength of the diffusion depends on the type of prognostic variable. Assume that τ_{vor} is the e-folding damping time of the highest resolvable wavenumber for the horizontal vorticity diffusion. In ECHAM5, τ is independent of the vertical position and order of the diffusion but changes with horizontal resolution (Roeckner et al. 2003; Wan et al. 2008). The e-folding times for the horizontal divergence and temperature diffusion are then chosen as

$$\tau_{div} = 0.2\,\tau_{vor} \tag{13.49}$$

$$\tau_T = 2.5\,\tau_{vor}. \tag{13.50}$$

For low resolution T42, T63 and T85 ECHAM5 simulations, the vorticity e-folding times τ_{vor} are typically set to 9, 7 and 5 h, respectively. These resolutions correspond to equatorial grid spacings of about 313, 208 and 156 km.

In contrast, other models such as the semi-Lagrangian and Eulerian spectral transform dynamical cores in CAM utilize the same diffusion coefficient for all prognostic variables. Boville (1991) recommended damping time scales of 14.4 and 7.2 h at the resolutions T42 and T63 for NCAR's Eulerian model with fourth-order hyper-diffusion. Similar damping time scales were suggested by Takahashi et al. (2006) who also investigated the higher-resolution spectral truncations T159, T319 and T639 with the spectral transform Eulerian model AFES (Enomoto et al. 2008). More concretely, Hamilton et al. (2008) reported AFES damping time scales around 9.6, 8.4, 4.8, 2.4 h for the spectral truncations T79, T159, T319 and T639. They yield the K_4 diffusion coefficients $1.19 \times 10^{15}, 8.41 \times 10^{13}, 9.14 \times 10^{12}$ and 1.14×10^{12} m^4 s^{-1}, respectively, according to (13.20). The corresponding grid spacings are approximately 167, 83, 42 and 21 km.

As mentioned before, the diffusivity parameter is generally empirically adjusted in each simulation to produce results in which the ends of the kinetic energy spectra follow a power law and do not change with the model resolution. Both Boville (1991) and Takahashi et al. (2006) found that the diffusivity coefficient needs to be scaled at about the inverse third power of the spectral truncation. As a consequence, the diffusion time scale of the smallest resolved scale drops with increased resolution as also suggested by Williamson (2008a) for CAM 3.1. In particular, Williamson (2008a) proposed damping time scales of 14.0, 8.6, 3.6 and 1.5 h for the T42, T85, T170 and T340 spectral truncations in the Eulerian dynamical core which yield the K_4 coefficients $1 \times 10^{16}, 1 \times 10^{15}, 1.5 \times 10^{14}$ and 2.25×10^{13} m^4 s^{-1}, respectively. In addition, a second-order diffusive sponge was employed in the three uppermost levels. The base diffusion coefficient at the third level from the top was held constant with $K_2 = 2.5 \times 10^5$ m^2 s^{-1} regardless of resolution which corresponds to damping time scales of about 25, 6.2, 1.6 and 0.4 h at the resolutions T42, T85, T170 and T340. The two highest resolutions T170 and T340 correspond to equatorial grid spacings of about 78 and 39 km. Note again that the K_2 coefficient was doubled at the second level and doubled again at the first level from the top. The second-order sponge layer diffusion was introduced to damp the vertically propagating resolved waves and prevent them from being reflected back down. Therefore, the coefficient is not reduced with resolution. It was chosen by trial and error to yield reasonable stratospheric polar night jet speeds and reduce the polar cold bias in the stratosphere. As resolutions increase further in the future, this choice of the diffusion coefficient K_2 might need to be revisited since a sudden onset of a strong sponge layer with very short time scales can act as a wave reflector by itself. This would defeat the purpose of the sponge layer.

As a concrete example, we now evaluate the effects of horizontal dissipation on the development of baroclinic waves. This was also suggested by MacVean (1983) who conducted a systematic study on the effects of hyper-diffusion and energy transfers in idealized GCM experiments. The effects of different types of hyper-diffusion are also discussed later in Sect. 13.3.8. Here, we start by isolating the effects of a varying fourth-order horizontal diffusion coefficient on growing baroclinic waves, and utilize the idealized dynamical core test case by Jablonowski and Williamson (2006a).

Figure 13.3 shows the surface pressure, 850 hPa temperature and vertical pressure velocity fields at day 9 of two simulations with the CAM Eulerian spectral transform dynamical core at the triangular truncation T85 with 26 levels. Both simulations applied the fourth-order hyper-diffusion with (left column) the default coefficient $K_4 = 1 \times 10^{15}\,\mathrm{m}^4\,\mathrm{s}^{-1}$, and (right column) an increased coefficient by the factor

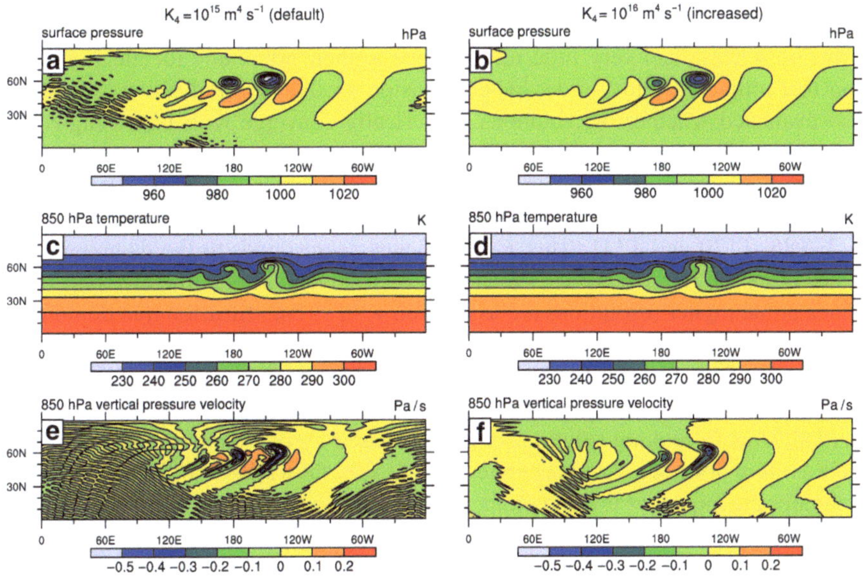

Fig. 13.3 (**a**,**b**) Surface pressure (hPa), (**c**,**d**) 850 hPa temperature (K) and (**e**,**f**) vertical pressure velocity (Pa/s) at day 9 of the growing baroclinic wave test case of Jablonowski and Williamson (2006a) in the CAM T85L26 Eulerian spectral dynamical core with ∇^4 diffusion. *Left*: default diffusion coefficient $K_4 = 1 \times 10^{15}\,\mathrm{m}^4\,\mathrm{s}^{-1}$, *Right*: increased diffusion coefficient by a factor of 10. A time step of $\Delta t = 600$ s is used

of 10. The corresponding damping time scales are 8.6 and 0.86 h, respectively. A second-order diffusive sponge near the model top is also applied, but is irrelevant for the discussion here. The figure shows that the baroclinic wave grows in both simulations, but that the circulation is relatively damped in the increased diffusion run. The peak magnitudes of the surface pressure and vertical pressure velocity fields are clearly reduced (Fig. 13.3b,f) and the otherwise sharp gradients in the temperature field are smeared out in the highly diffusive run (Fig. 13.3d). However, the increased diffusion diminished (but not eliminated) the numerical noise, the so-called Gibbs ringing, which is very dominant in the default configuration. The Gibbs phenomena are introduced by the need to represent fields with discontinuities or sharp gradients by smooth global basis functions. The noise does not grow unstable in the default EUL run, but such noise in the vertical velocity field can have detrimental effects on other quantities like precipitation in case of full GCMs with physical parameterizations. Spurious rainfall with such a noisy signature is sometimes called "spectral rain", and is undesirable. Therefore, a delicate balance needs to be found between sufficient scale-selective damping and too diffusive simulations. As an aside, except for the spectral ringing the evolution of the baroclinic wave in the default EUL configuration resembles high-resolution reference solutions of other models quite closely as shown in Jablonowski and Williamson (2006a,b) and Lauritzen et al. (2010a).

The impact of explicitly added diffusion on the evolution of baroclinic waves was also investigated by Polvani et al. (2004). They demonstrated that the choice of the diffusion mechanism can fundamentally change the characteristics of the flow field. In particular, their spectral transform simulation with a second-order diffusion scheme had very little resemblance to their nominally identical simulation that utilized a fourth-order hyper-diffusion mechanism. Polvani et al. (2004) also evaluated whether the solutions numerically converge with increasing resolution when keeping a constant diffusion coefficient. For example, they tested a fourth-order hyper-diffusion with the constant coefficient $K_4 = 2.5 \times 10^{16}\,\mathrm{m}^4\,\mathrm{s}^{-1}$ and simulated the evolution of a baroclinic wave at the triangular truncations T21, T42, T85, T170 and T341. This K_4 coefficient corresponds to the damping time scales around 85.8, 5.6, 0.34 h, 78 and 5 s, respectively. The extremely short time scales at the higher resolutions are associated with strong diffusion that dominates the flow and thereby allows the numerical solutions to converge. It means that diffusion can effectively reduce the spatial resolution by suppressing the generation of finer-scale structures that are normally resolved at higher resolutions. As a note of caution, smooth-looking solutions could be caused by overly strong diffusion and are therefore not necessarily accurate. This is also shown later in Sect. 13.6.1 that compares the impact of inherent numerical dissipation.

13.3.4.5 Examples of Diffusion Coefficients in Grid Point Models

All aforementioned diffusion coefficients for spectral models are quite comparable and lie within a factor of 2–3 at a given resolution. But note that the amount of explicit diffusion needed for a numerical scheme is highly dependent on the inherent diffusive characteristics of the numerical scheme, and all other filters or fixers in the GCM as further explained in this chapter. For example, a second-order grid point model might have different diffusion needs than spectral models. The diffusion coefficients for such a finite-difference-type grid point model on an icosahedral grid are listed in Majewski et al. (2002). They show the grid-size dependencies of the fourth-order linear diffusion coefficients in the weather forecast model GME for various horizontal grid spacings between 10 and 160 km. A comparison with Williamson (2008a) reveals that the K_4 diffusion coefficients in the model GME are higher by factors between 3.5 and 5. A comparison of the damping time scales in GME with the K_4 damping coefficients 5.25×10^{15} and 6.5×10^{14} m^4 s^{-1} at the approximate $\Delta x = \Delta y$ grid spacings 160 and 80 km yields damping time scales of about 1.1 and 0.55 h according to (13.44). The result confirms that GME employs a stronger hyper-diffusion mechanism in comparison to spectral models at similar resolutions.

Lastly, we comment on the empirically tuned diffusions as shown above in (13.45) and (13.46). In particular, Tomita and Satoh (2004) chose a ($q = 2$) fourth-order diffusion in the horizontal direction and a ($q = 3$) sixth-order diffusion in the vertical. At the $\Delta x \approx 240$ km horizontal resolution with a time step of $\Delta t = 1{,}800$ s the empirical factor γ_H was most often set to 6.25×10^{-3}. In Tomita and Satoh (2004) this led to the diffusion coefficient $K_{4H} = 1.152 \times 10^{16}$ m^4 s^{-1} which corresponded to the damping time scale of 2.5 h according to (13.48).

13.3.4.6 Caveats

Unfortunately and as a word of caution, the choice of the diffusion coefficients often remains undocumented in the refereed literature. Even the defaults are difficult to find, and the resolution-dependencies are only rarely mentioned. Official model documentation often lacks the specific information.

13.3.5 Choice of the Diffusion Coefficient: Stability

As pointed out by Mesinger and Arakawa (1976) and Wood et al. (2007) explicit time approximations of diffusion equations, such as those presented in Staniforth et al. (2006)

$$\frac{\psi^{j+1} - \psi^j}{\Delta t} = \frac{1}{a^2}\left[\frac{\partial}{\partial \lambda}\left(\frac{K_\lambda}{\cos^2\phi}\frac{\partial \psi^j}{\partial \lambda}\right) + \frac{\partial}{\partial \phi}\left(K_\phi \frac{\partial \psi^j}{\partial \phi}\right)\right], \tag{13.51}$$

are only conditionally stable. Here, the index j symbolizes a discrete time level. This is especially problematic on latitude–longitude grids since the time step can easily violate the condition for stability close to the poles. This is mainly due to the convergence of the meridians in polar regions that leads to shrinking longitudinal spacings and higher Courant–Friedrichs–Lewy (CFL) numbers.

If the horizontal diffusion operator is applied with an explicit time discretization, which is for example the case in the UK Met Office model UM, the diffusion coefficient needs to obey strict stability constraints. A comprehensive stability analysis for a finite-difference representation of the diffusion operator on a latitude–longitude grid is shown in Staniforth et al. (2006) (their Chap. 12). In two dimensions, the stability constraint for the second-order diffusion with $q = 1$ in the model UM is given by

$$\frac{\Delta t}{r^2}\left(\frac{K_\lambda}{\cos^2\phi\,\Delta\lambda^2} + \frac{K_\phi}{\Delta\phi^2}\right) \leq \frac{1}{4} \tag{13.52}$$

where the radial distance r from the center of the earth can also be approximated by the radius a, and K_λ and K_ϕ are the diffusion coefficients in the longitudinal and latitudinal directions. This restriction guarantees that the corresponding response function (13.41) lies between [0, 1] and does not change sign on alternate time steps. From a physical viewpoint, an isotropic choice of the diffusion coefficient $K_\lambda = K_\phi$ is advantageous to damp physical scales at the same rate. However, such a choice would enforce very stringent stability conditions on the maximum allowable time step due to the dependence on the $\cos^2\phi$ term. Therefore, $K_\lambda/\cos^2\phi = K_\phi =$ *constant* is generally selected in the model UM that leads to the less restrictive condition

$$\frac{K_\phi \Delta t}{a^2}\left(\frac{1}{\Delta\lambda^2} + \frac{1}{\Delta\phi^2}\right) \leq \frac{1}{4}. \tag{13.53}$$

An undesirable side effect is that the diffusion becomes highly anisotropic, particularly in polar regions where diffusion is probably most needed, and noise is much less controlled in the east–west direction than in the north–south direction (Staniforth et al. 2006). Note that the second-order response function (13.41) can also be generalized for higher-order diffusion schemes

$$E = 1 - \left\{\frac{K_\phi \Delta t}{a^2}\left[\frac{\sin^2(k_\lambda \Delta\lambda/2)}{(\Delta\lambda/2)^2} + \frac{\sin^2(k_\phi \Delta\phi/2)}{(\Delta\phi/2)^2}\right]\right\}^q \tag{13.54}$$

where $2q$ denotes the order of the hyper-diffusion.

The non-isotropy of the diffusion scheme in the model UM has a caveat. The damping for a particular physical scale decreases polewards because the physical scale represented by any given wavenumber decreases by a factor $\cos^{-1}\phi$ as the poles are approached and the response function of the operator is the same everywhere. As a result, a feature which moves equatorwards from the poles experiences increased damping which has the effect of creating a boundary effect for the propagation of that feature (Staniforth et al. 2006). In addition, the shrinking longitudinal spacing near the poles supports smaller and smaller physical scales. If these small scales are not damped enough it increases the risk of developing noise near the poles. In practice, the application of an explicit diffusion scheme on a latitude–longitude grid is often paired with the application of a polar filter that removes linear and nonlinear instabilities if any. The applications of an additional filter also allows violations of the CFL stability condition at high latitudes as further explained below in Sect. 13.5.

To avoid the stability limitations of the horizontal diffusion mechanism an implicit temporal approximation of the diffusion is generally desirable as suggested by Jakimow et al. (1992) and Li et al. (1994). On the other hand, this can adversely affect the computational efficiency of the scheme since implicit calculations require the solution of a Helmholtz equation. As an example, an implicit representation of the horizontal diffusion shown in (13.51) yields

$$\frac{\psi^{j+1} - \psi^{j}}{\Delta t} = \frac{1}{a^2}\left[\frac{\partial}{\partial\lambda}\left(\frac{K_\lambda}{\cos^2\phi}\frac{\partial\psi^{j+1}}{\partial\lambda}\right) + \frac{\partial}{\partial\phi}\left(K_\phi\frac{\partial\psi^{j+1}}{\partial\phi}\right)\right] \tag{13.55}$$

which can also be symbolically written in the (Helmholtz) matrix form

$$[I - \Delta t(D_{\lambda\lambda} + D_{\phi\phi})]\psi_{\lambda,\phi}^{j+1} = \psi_{\lambda,\phi}^{j}. \tag{13.56}$$

I stands for the identity matrix, and $D_{\lambda\lambda}$ and $D_{\phi\phi}$ symbolize discretized diffusion operators in matrix form. However, inverting the three-dimensional matrix $[I - \Delta t(D_{\lambda\lambda} + D_{\phi\phi})]$ at every time step would be computationally expensive, although approximations of (13.56) can be formulated to make the implicit formulation more attractive (Staniforth et al. 2006).

As mentioned before and shown in (13.26), (13.27) and (13.28), spectral transform models like ECHAM5, IFS or CAM EUL or SLD always compute the horizontal diffusion implicitly in spectral space. This can be done in a straightforward way since the Laplacian operator has an an analytic spectral representation as e.g., presented for the operator ∇^4 in (13.34). The implicit calculation does not need to obey stability constraints and remains stable even with a high isotropic diffusion coefficient in both directions (Collins et al. 2004; ECMWF 2010). A thorough stability analysis of linear diffusion can also be found in Williamson and Laprise (2000).

13.3.6 Nonlinear Horizontal Diffusion

The choice of the horizontal diffusion mechanism is sometimes linked to turbulence concepts (Boer and Shepherd 1983) which e.g., can be based on nonlinear horizontal mixing coefficients. However, as stated earlier the horizontal diffusivities used in GCMs are typically many orders of magnitude larger than those which would be appropriate for turbulence closures (Mellor 1985). Thus, this association needs to be made with care. It even could offer some guidance in choosing a suitable subgrid-scale mixing scheme. For example, in NCEP's Eta model (Black 1994) this awareness led to the use of only second-order horizontal diffusion since the choice was not based on the desire to select "scale selective" dissipation. Instead, the intention was that the diffusion scheme should mimic the impact of grid box filamentation due to deformation-dependent stretching (Fedor Mesinger, personal communication, Janjić (1990)).

Nonlinear diffusion typically defines the diffusion coefficient in terms of a nonlinear function of the horizontal wind. Nonlinear second-order diffusion was originally proposed by Smagorinsky (1963) who used a Cartesian coordinate system to derive deformation-based Eddy viscosity coefficients. The basic design of this diffusion mechanism is connected to mixing-length concepts. The latter can be motivated if the ideas of Prandtl are applied to the dissipation of enstrophy in 2D turbulence (Becker and Burkhardt 2007). As shown below Smagorinsky's nonlinear parameterization might appear to be "more physical" than any linear diffusion scheme. However, there is little theoretical basis for such a nonlinear formulation at large geophysical scales.

Nonlinear harmonic (second-order) diffusion depends on the flow field and damps only at times and places of strong horizontal shear. The generic form of this second-order diffusion needs to be written in flux form

$$F_\psi = +\nabla \cdot (K_H \nabla \psi) \tag{13.57}$$

where $\nabla\cdot$ symbolizes the divergence operator. The Eddy viscosity coefficient K_H can be symbolically associated with a length (L) and time (T) scale

$$K_H = \frac{L^2}{T} \tag{13.58}$$

and is proportional to the norm of the strain tensor and the quadratic grid spacing. The inverse time scale $T^{-1} = |D|$ is determined by the deformation rate $|D|$ which is the norm of the strain tensor. Smagorinsky (1963) defines the nonlinear coefficient as

$$K_H = (k_0 \Delta)^2 \sqrt{\left(\partial_x u - \partial_y v\right)^2 + \left(\partial_x v + \partial_y u\right)^2} \tag{13.59}$$

where $\partial_x = (a \cos\phi)^{-1} \partial\lambda$ and $\partial_y = a^{-1}\partial\phi$ denote the partial derivatives in the longitudinal and latitudinal directions in spherical coordinates. k_0 is a unitless empirical constant and Δ is a measure of the physical grid spacing, such as $\Delta = \sqrt{\Delta x \Delta y}$ (Skamarock et al. 2008), $\Delta = \Delta y$ (Koshyk and Hamilton 2001) or $\Delta = \min(\Delta x, \Delta y)$ (Griffies and Hallberg 2000). k_0 is typically set to a value between $[0.1, 0.3]$ as suggested by Smagorinsky (1963, 1993), Andrews et al. (1983), Koshyk and Hamilton (2001) or Skamarock et al. (2008), but both smaller and larger k_0 values have been tried in GCMs. Regardless of the application, the formulation of Smagorinsky's viscosity coefficient utilizes the horizontal tension $D_T = (\partial_x u - \partial_y v)$ and horizontal shearing strain $D_S = (\partial_x v + \partial_y u)$ and is thereby linked to the local deformation rate via $|D| = \sqrt{D_T^2 + D_S^2}$.

Other variants of the nonlinear diffusion coefficient exist as shown by Gordon and Stern (1982) and Andrews et al. (1983). These are summarized in Becker and Burkhardt (2007) who also list the later modification by Smagorinsky (1993)

$$K_H = (k_0 \Delta)^2 \sqrt{\left(\partial_x u - \partial_y v - v \tan\phi/a\right)^2 + \left(\partial_x v + \partial_y u + u \tan\phi/a\right)^2}. \tag{13.60}$$

This revised version can be applied consistently in spherical geometry. Note that nonlinear viscosity is not widely used in atmospheric GCMs today. However, it is commonly used in large-scale ocean models as outlined in Griffies and Hallberg (2000). Griffies and Hallberg (2000) even extended the definition of K_H to incorporate biharmonic (fourth-order) diffusion that enhances the scale selectivity of the Smagorinsky scheme.

A Smagorinsky-type Eddy viscosity is also often chosen as a subgrid-scale (SGS) model in large-eddy simulations to represent the effects of small-scale turbulence in the inertial range. LES has been widely used to study atmospheric boundary layer dynamics and vertical mixing processes. LES explicitly resolves the dynamics of large-scale eddies. They contain most of the energy and are the primary transport mechanism, while small-scale eddies in LES are parameterized by the SGS model as e.g., outlined by Huang et al. (2008). The limited-area mesoscale model WRF also includes a Smagorinsky-type option for its horizontal and vertical diffusion scheme. WRF can base the Eddy viscosities either on a 3D Smagorinsky turbulence closure or on the flow deformation (Skamarock et al. 2008).

13.3.7 Physical Consistency

Explicitly adding linear horizontal diffusion to the equations of motion is popular in GCMs, but most often the implementations are physically inconsistent (Burkhardt and Becker 2006). For example, if the horizontal momentum diffusion is applied in the form of hyper-diffusion the conservation of angular momentum is generally violated. This is e.g., discussed in Becker (2001) who argued that consistent friction in GCMs must be formulated as the divergence of a symmetric Reynolds stress tensor. In contrast, conventional formulations of the horizontal diffusion correspond to a nonsymmetric stress tensor.

In general, viscous dissipation is the conversion of mechanical energy to thermal energy by the flow working against viscous stresses. Diffusion therefore removes kinetic energy from the flow field which could be interpreted as the transfer of kinetic energy from the resolved scales to subgrid scales and finally to a turbulent microscale, where it needs to be converted into heat. A very important aspect is that the conversion is irreversible and always needs to appear as a positive frictional heating on the right-hand side of the thermodynamic equation. However, frictional heating is commonly ignored in GCMs since it is small in comparison to other contributions to the heat budget such as radiative forcing or latent heating (Fiedler 2000; Burkhardt and Becker 2006).

Nevertheless, the kinetic energy loss due to dissipation cannot be neglected over long time periods since it violates the conservation of total energy. This is especially important for climate simulations to prevent artificial climate drifts. According to the estimates by Becker (2003) using a simple GCM the total dissipation averages to approximately 2 W m^{-2} with horizontal dissipation making up about a third of the overall dissipation. Neglecting this heating would cause a spurious thermal forcing of about 0.6 W m^{-2} which is on the order of climate change signals. The conversion of the dissipated energy into heat must therefore be explicitly added to the thermodynamic equation. Similar energy losses around 2 W m^{-2} due to the explicit and inherent damping mechanisms in NCAR's CAM model were also reported by Williamson (2007).

Frictional heating due to horizontal diffusion is explicitly included in the EUL and HOMME dynamical cores of NCAR's CAM 5 model (Neale et al. 2010) and in a variant of the ECHAM4 model (Burkhardt and Becker 2006). But as noted in Becker (2001) CAM does not utilize a symmetric stress tensor formulation. The dissipative heating can therefore only be approximated as outlined in Boville and Bretherton (2003) and thereby allows (at least theoretically) dissipative "cooling" which is unphysical. Furthermore, Boville and Bretherton (2003) and Becker (2003) discussed how to include frictional heating due to vertical diffusion processes in the physical parameterizations. Alternatively or even in addition, the total energy budget can also be simply "fixed" via an a posteriori energy fixer as explained later in Sect. 13.7.3. This is common practice in dynamical cores since other types of damping, such as filters or inherent nonlinear numerical dissipation in semi-Lagrangian or finite-volume schemes, cannot be analytically quantified. Instead, their effects collectively appear in form of a residual in the total energy equation.

Two other aspects need to be considered. As pointed out by Wood et al. (2007) some diffusion schemes do not properly maintain steady-state solutions, but instead wrongly distort them. These errors in steady-state solutions may lead to systematic biases and climate drift as discussed for physical parameterization by Dubal et al. (2004, 2006). In addition, only the second-order diffusion operator guarantees the preservation of the monotonicity of the diffused field. The more scale-selective and thereby desirable higher order operators have the potential to introduce spurious new extrema and violate monotonicity constraints. The amplitude of unphysical over- and undershoots can even increase with increasing order of the diffusion (Sardeshmukh and Hoskins 1984). Over- and undershoots can cause nonlinear interactions between the physics and dynamics and result in undesired side effects such as spurious rainfall. Higher-order diffusion is especially discouraged for positive definite moisture and other tracer fields since it can lead to supersaturation or even negative tracer values. Negative values would then need to be artificially "filled" as discussed later in Sect. 13.7.2. Note that horizontal diffusion operators do not necessarily preserve the global volume integral of the diffused quantity. Most often, they are therefore not conservative as outlined in Staniforth et al. (2006).

In order to avoid over- and undershoots that are triggered by linear high-order diffusion Xue (2000) suggested a monotonic diffusion scheme with simple flux limiters. The basic idea is to interpret the generic form of the diffusion (13.2) as a flux divergence such as

$$F_{\psi} = -\nabla \cdot \left[(-1)^q K_{2q} \nabla^{2q-1} \psi\right]. \tag{13.61}$$

The diffusive fluxes inside the outermost bracket can then be limited via flux limiters to prevent numerically triggered over- and undershoots as for example shown by Knievel et al. (2007) for a sixth-order scheme in the model WRF. The form (13.61) of the diffusion was also utilized for the model COSMO by Doms and Schättler (2002) who further improved the monotonic characteristics of the Xue (2000) scheme.

13.3.8 Diffusion Properties in Practice: The Model CAM 3.1

We now illustrate the practical aspects of the linear horizontal hyper-diffusion in GCM simulations with NCAR's CAM Eulerian spectral transform model that is utilized in an aqua-planet mode (Neale and Hoskins 2000) and with the idealized baroclinic wave test case by Jablonowski and Williamson (2006a,b). In addition, we compare the EUL simulations to the CAM Finite Volume (FV) model to gain an appreciation for the subtle differences between explicit hyper-diffusion and inherent numerical dissipation. Selected aspects of the inherent numerical dissipation are discussed below in Sect. 13.6.

The left panel of Fig. 13.4 shows 250 hPa kinetic energy spectra from CAM 3.1 simulations with the Eulerian spectral dynamical core for a variety of spectral truncations. The spectra are calculated from aqua-planet simulations (Williamson 2008a) and except for the highest resolution, are averaged over 100 samples separated by 30 h. The spectrum for the highest resolution is averaged every 6 h for the last 3 days of a 10-day run which started from a lower resolution aqua-planet state. Such sampling is adequate to determine the spectra. As listed above

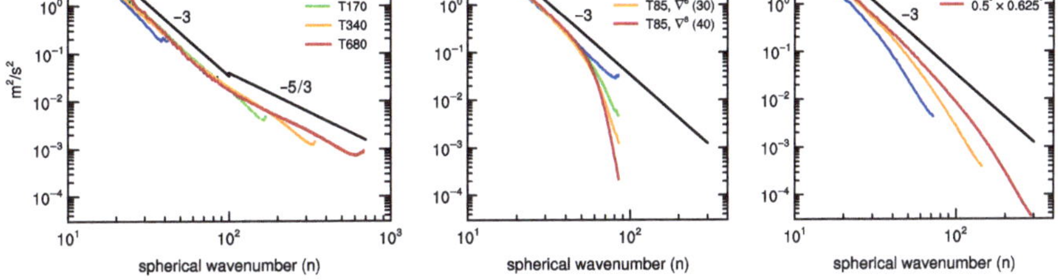

Fig. 13.4 250 hPa kinetic energy spectra as a function of the spherical wavenumber (n) in aqua-planet simulations from (*left*) CAM Eulerian spectral dynamical core with ∇^4 diffusion for different resolutions, (*center*) T85L26 Eulerian spectral dynamical with ∇^4, ∇^6 and ∇^8 diffusion, and (*right*) CAM Finite Volume (FV) dynamical core for different *lat* × *lon* resolutions in degrees and 26 levels

in Sect. 13.3.4.4, the diffusion coefficients used here are 1.0×10^{16}, 1.0×10^{15}, 1.5×10^{14}, 1.5×10^{13} and $1.0 \times 10^{12}\,\mathrm{m^4\,s^{-1}}$ for T42, T85, T170, T340 and T680 truncations, respectively. The corresponding quadratic unaliased transform grids are approximately 2.8°, 1.4°, 0.7°, 0.35° and 0.17°, respectively. For resolutions below T340, the spectra all have a slope close to n^{-3}, and lie very close to each other. There is a small upturn at the truncation limits also found and discussed by Takahashi et al. (2006) and Hamilton et al. (2008). These spectra are also similar to those presented in Williamson (2008a). The T340 spectrum starts to deviate from n^{-3} around $n = 100$. This is the region where the observed atmospheric spectra transition from an n^{-3} slope to an $n^{-5/3}$ slope (Nastrom and Gage 1985). The model starts to make a similar transition but is then overwhelmed at the highest wavenumbers by the diffusion. Perhaps a smaller coefficient would allow the transition to form. The T680 simulation makes the transition and exhibits an $n^{-5/3}$ slope above $n = 100$ in agreement with spectra estimates from observations. The ∇^4 diffusion coefficient for T680 was chosen to yield this $n^{-5/3}$ slope. As an aside, at the 2008 Community Climate System Model (CCSM) Annual Workshop in Breckenridge, Colorado, it was reported that the spectral element model HOMME also exhibits a transition from n^{-3} to $n^{-5/3}$ (Taylor 2008). The transition of the slopes is furthermore evident in high-resolution simulations with the model NICAM when grid spacings below $\approx$ 10 km are employed (Terasaki et al. 2009).

It could be argued that the modeled spectra should follow n^{-3} or $n^{-5/3}$ slope to the truncation limit, depending on resolution as illustrated above, because that is what observations indicate they should do. It could also be argued that that is not a good discrete modeling strategy. For example, the University of Reading spectral model and the model ECHAM5 (Roeckner et al. 2003) have always used ∇^6 or ∇^8 forms of diffusion which lead to steeper kinetic energy spectra approaching the truncation limit. They argue that the scales near the truncation limit are not calculated accurately and cannot be trusted and that a discrete model should treat the end of the spectrum smoothly, including the transition to zero energy at the truncation limit (personal communication, Mike Blackburn). This requires a steeper spectrum which minimizes ringing or noise arising from a sudden transition. The ringing is a non-physical artifact arising from sharp truncation in the discrete system.

MacVean (1983) studied the effect of higher degree hyper-diffusions on baroclinic development in a spectral model truncated at T42. His simplest, but most subjective criterion, was a visual assessment of the degree to which synoptic scale detail was retained and the level of small-scale noise present. He also used other, more objective measures. He concluded that ∇^4 formulation is not scale-selective enough with T42 truncation, which, by the way, is still used today in climate models. The ∇^6 and ∇^8 forms, with appropriate coefficients, appear to be better and both are equally satisfactory.

The concept of applying stronger damping to the smaller scales than needed to maintain the observed spectra to the truncations limit is taken one step further by Lander and Hoskins (1997) with their introduction of the concept of "believable scales". They argue that since the shortest scales are not calculated accurately they should be filtered out before calculating the parameterizations, i.e., the parameterizations should be calculated on scales calculated accurately by the dynamics.

Higher degree hyper-diffusions such as ∇^6 or ∇^8 lead to steeper slopes approaching the truncation limit as seen in the center panel of Fig. 13.4 for the CAM EUL. The T85, ∇^4 line duplicates that in the left panel, the ∇^6 diffusion coefficients are chosen to damp wavenumber 40 or 30 at the same rate as the ∇^4 case, and the ∇^8 coefficient is chosen to damp wavenumber 40 the same rate as the ∇^4. The effect of diffusion on spectral ringing or noise in the solution is seen in Fig. 13.5 which shows the surface pressure from day 9 of calculations of the growing baroclinic wave test case of Jablonowski and Williamson (2006a,b). The top panel replots Fig. 13.3a, which was shown earlier for the default T85L26 CAM EUL simulation, but now we choose rather unconventional contour values which accentuate the noise. The second and third panels show T85L26 solutions with ∇^6 diffusion with two different diffusion coefficients. The coefficient for the second panel is chosen so that wavenumber 40 is damped at the same rate as in the ∇^4 calculations. The coefficient for the third panel is chosen so that wavenumber 30 is damped at the same rate as in the ∇^4 calculations. The noise is reduced but still present with ∇^6 diffusion. The larger coefficient (third panel) reduces the noise more than the smaller one (second panel). The noise is largely eliminated with ∇^8 diffusion (fourth panel). The coefficient is chosen so that wave 40 is damped at the same rate as it is in the ∇^4 case. Close examination of the figure shows that minimal noise is still present in the ∇^8 solution as indicated by the 999.98 hPa contour line in the left half of the plot. Quite possibly this noise could be reduced further with other choices for the coefficient values, but the whole process of minimizing the noise via diffusion operators is rather arbitrary and perhaps case dependent.

In contrast to the Eulerian spectral solutions, Fig. 13.5, bottom panel, shows that the solution from the CAM FV dynamical core in CAM 3.1 on a 1° latitude–longitude grid has no indication of noise in the surface pressure field. The contours are smoother than any of the other examples. The FV numerical approximations are shape preserving and thus do not generate small scale noisy structures. This is achieved via monotonicity constraints which are discussed later in Sect. 13.6.

Figure 13.4, right panel, shows the 250 hPa kinetic energy spectra from the CAM FV model for 2°, 1° and 0.5° grids. The spectra tail-off faster than in the Eulerian spectral model with ∇^4 diffusion. They behave more like the Eulerian spectral ∇^6 and ∇^8 diffusion cases, except the departure from an n^{-3} slope begins at lower wavenumbers relative to the truncation limit with the Finite Volume. These FV spectra are dominated by the rotational component of the flow, for which the numerical approximations are shape preserving. The divergent component is not similarly approximated and can become relatively large near the tail of the spectrum if no extra damping of the divergence is included. In these experiments the divergent component is controlled by a divergence damping mechanism (Neale et al. 2010; Whitehead et al. 2011) so that the divergent kinetic energy remains smaller than the rotational kinetic energy. A thorough discussion of horizontal divergence damping is provided in Sect. 13.4.1.

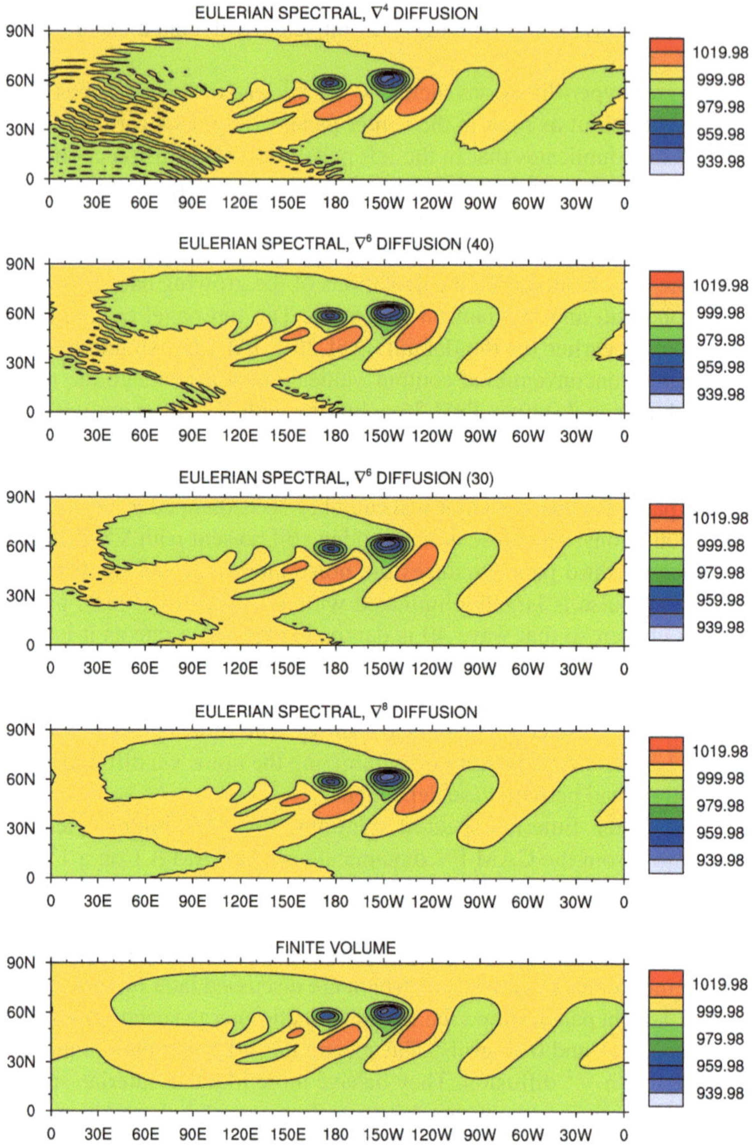

Fig. 13.5 Surface pressure (hPa) at day 9 of the growing baroclinic wave test case of Jablonowski and Williamson (2006a) from T85L26 CAM Eulerian spectral dynamical core with ∇^4, ∇^6 and ∇^8 diffusion, and $1^\circ \times 1.25^\circ$ CAM FV model. A time step of $\Delta t = 600$ s is used

The Eulerian spectral model can obtain smooth fields through the application of an arbitrary higher degree hyper-diffusion term. The FV model obtains them through the application of a physical condition, namely shape preservation. Shape preservation provides a more physically based method to obtain smooth solutions.

Perhaps the physical realism in the FV discrete system indicates that the spectrum should fall off faster than n^{-3} in the discrete system, and that it is a better discrete modeling strategy.

In Fig. 13.4 the 1° FV spectrum begins to depart from the n^{-3} slope around $n = 40$, although this evaluation is somewhat subjective. The 2° departs at a much lower wavenumber and generally has lower amplitude at the lower wavenumbers. The T85 spectral model with ∇^6 and ∇^8 diffusions also begins to depart from the n^{-3} slope around $n = 40$. The wavenumber of the departure from n^{-3} might imply something about the resolution of the models. For example, Skamarock (2004) argues that such a departure defines the *effective resolution* of a model (see also Chap. 14). This would imply that the CAM 1° FV and the T85 Eulerian spectral models with ∇^6 and ∇^8 diffusions have the same effective resolution. On the other hand, the T85 spectral model with ∇^4 diffusion maintains the n^{-3} slope to the truncation limit, yet most modelers would not argue that its effective resolution was T85. Even though the spectral method is accurate to the truncation limit for linear problems, the nonlinear interactions cannot be accurate there since they would involve unresolved scales. In addition the arbitrary diffusion is a dominant component in the equations at the smallest scales. Of more practical interest is the question of equivalent resolution of different schemes applied to a problem of interest. By examining a variety of simulated climate statistics in aqua-planet simulations Williamson (2008b) concluded that the T85 Eulerian spectral model with ∇^4 diffusion and 1° FV model reflect equivalent resolutions when applied to the aqua-planet problem. He also mentioned that experiments with the Eulerian spectral model with ∇^6 and ∇^8 diffusions produced results similar to those from the spectral Eulerian model with ∇^4 diffusion.

In summary, it could be argued that the modeled kinetic energy spectra should follow an n^{-3} or $n^{-5/3}$ slope to the truncation limit, depending on resolution, since that is what is observed in the atmosphere. However, this approach can lead to noise in the smallest scales of the solutions as seen in the Eulerian spectral model. It could also be argued that a drop off in the spectra as seen here in the FV model simulations indicates excessive damping by the numerics. Other modelers, however, argue that the modeled kinetic energy spectra should be steeper approaching the truncation limit so that the discrete, truncated spectra transitions to zero energy at the truncation limit more smoothly. In this interpretation shape preserving approximations such as those used in the FV model are not necessarily overly diffusive. They provide a physical condition (smoothness) which determines the shape of the spectra. This is compared to adding arbitrary higher degree hyper-diffusion terms to the equations which require subjective evaluations to determine the diffusion coefficient. The superiority of these two approaches remains a matter of discussion. In general, we recommend using kinetic energy spectra in combination with other quality measures, but not as the sole criterion, to judge the diffusive characteristics of GCMs.

13.4 Divergence and Vorticity Damping, External Mode Damping and Sponge Layers

Besides the very popular explicitly added horizontal diffusion and hyper-diffusion techniques discussed above, GCMs might also apply other explicit damping mechanisms. They include the 2D and 3D divergence damping, vorticity damping, an external mode damping approach or sponge layers near the model top. These are discussed in the next subsections.

13.4.1 2D Divergence Damping

Adding a horizontal divergence damping term to the horizontal momentum equations is a simple way of reducing high-frequency gravity wave noise. In this approach the time rates of change of the zonal and meridional velocities u and v are forced by a damping term. Recall the generic prognostic equation (13.1) for variable ψ where ψ now stands for the horizontal velocity vector $\mathbf{v} = (u, v)$ and $\mathbf{F_v}$ symbolizes the vector of the horizontal divergence damping. The divergence damping mechanism of order $2q$ is then given by

$$\mathbf{F_v} = (-1)^{q+1} \nabla \left(\nu_{2q} \, \nabla^{2q-1} \cdot \mathbf{v} \right) \tag{13.62}$$

where $q \geq 1$ is a positive integer value analogous to the discussion of the horizontal diffusion in Sect. 13.3. ν_{2q} stands for the divergence damping coefficient. Applying the horizontal divergence operator $\nabla \cdot$ to (13.1) and utilizing (13.62) yields an evolution equation for the divergence

$$\frac{\partial D}{\partial t} = \cdots + (-1)^{q+1} \nu_{2q} \nabla^{2q} \, D \tag{13.63}$$

where $D = \nabla \cdot \mathbf{v}$ denotes the horizontal divergence defined by

$$D = \frac{1}{a \cos \phi} \left(\frac{\partial u}{\partial \lambda} + \frac{\partial (v \cos \phi)}{\partial \phi} \right) \tag{13.64}$$

in spherical coordinates. As before, a symbolizes the Earth's radius and λ,ϕ are the longitude and latitude, respectively. Equation (13.63) assumes that the coefficient is constant in the horizontal direction. The equation demonstrates that the divergence damping represents a horizontal diffusion of the divergent part of the flow that is generally closely associated with inertia-gravity waves. Divergence damping can easily be explicitly added to models written in (u, v) form that do not utilize a prognostic equation for D.

As a concrete example, the second-order divergence damping mechanism (with $q = 1$) in component form yields

$$\frac{\partial u}{\partial t} = \cdots + \frac{1}{a \cos\phi} \frac{\partial}{\partial \lambda} \big(\nu_2 D\big) \tag{13.65}$$

$$\frac{\partial v}{\partial t} = \cdots + \frac{1}{a} \frac{\partial}{\partial \phi} \big(\nu_2 D\big). \tag{13.66}$$

This type of divergence damping was e.g., suggested by Shuman and Stackpole (1969), Sadourny (1975), Dey (1978), Haltiner and Williams (1980), Bates et al. (1990) and Lin (2004), mainly for numerical stability reasons. Note that Dey (1978) called it *divergence diffusion*. It leaves the rotational motion unaffected, selectively damps inertia-gravity waves, controls numerical noise and prevents the spurious build-up of energy near the cut-off grid scale. Sadourny and Maynard (1997) argued that 2D divergence damping can be viewed as a model of nonlinear interactions between inertia-gravity waves and rotational motion. The use of divergence damping was also explored by McDonald and Haugen (1992) and Gravel et al. (1993) to control gravity wave noise in their two-time-level semi-Lagrangian schemes.

The horizontal Laplacian-type ∇^{2q} hyper-diffusive term shown in (13.63) damps all scales, but is strongest at higher wavenumbers. Generally, divergence damping becomes more scale-selective at higher orders. This is e.g., shown by Whitehead et al. (2011) who also provide a linear von Neumann stability analysis of the divergence damping technique for explicit time-stepping schemes. In particular, the fourth-order divergence damping (with $q = 2$) can be expressed as

$$\frac{\partial u}{\partial t} = \cdots - \frac{1}{a \cos\phi} \frac{\partial}{\partial \lambda} \big(\nu_4 \nabla^2 D\big) \tag{13.67}$$

$$\frac{\partial v}{\partial t} = \cdots - \frac{1}{a} \frac{\partial}{\partial \phi} \big(\nu_4 \nabla^2 D\big) \tag{13.68}$$

where ν_4 is the fourth-order damping coefficient. This leads to the evolution equation for the divergence

$$\frac{\partial D}{\partial t} = \cdots - \nu_4 \nabla^4 D \tag{13.69}$$

in case of a horizontally constant coefficient. Fourth-order damping is an option in NCAR's model CAM 5 (Neale et al. 2010) which utilizes the FV dynamical core on a latitude–longitude grid. Even high-order divergence damping mechanisms like a sixth or eighth-order scheme are under investigation in the cubed-sphere version of the FV model (FVcubed) at the NOAA Geophysical Fluid Dynamics Laboratory (S.-J. Lin and William Putman, personal communication). FVcubed (Putman and Lin 2007, 2009) is also part of the most recent internal version of NASA's GEOS model which will be officially named GEOS6 upon its public release.

As pointed out earlier in Sect. 13.3 the explicit diffusion mechanisms for the horizontal divergence in spectral models (13.11) and (13.15) resemble the form of the divergence damping in (13.63) and (13.69). Again, both mechanisms are characterized with different names, but they accomplish a similar or even identical physical effect, namely they damp the divergent motions with either a second-order or higher order diffusion.

13.4.1.1 Selection of the 2D Divergence Damping Coefficient

As an example, we present the formulation of the damping coefficients in NCAR's CAM 5 model with the FV dynamical core (Lin 2004). The default second-order diffusion coefficient ν_2 in CAM is given by

$$\nu_2 = C_2(i)\frac{a^2 \Delta\lambda \Delta\phi}{\Delta t} \tag{13.70}$$

where the parameter $C_2(i)$ is a function of pressure to provide a sponge at the model top. Let p_{top} be the pressure at the model top and $[p_{ref}(i) = (a(i) + b(i))p_0]$ the reference pressure at a given model level with index i. Here, $a(i)$ and $b(i)$ are the hybrid coefficients of the vertical η coordinate (Simmons and Burridge 1981; Jablonowski and Williamson 2006b) and the surface pressure is assumed to be $p_0 =$ 1000 hPa. Then, as implemented in CAM, the parameter C_2 is given by

$$C_2(i) = \frac{1}{128}\max\left\{1, 8\left[1 + \tanh\left(\ln\left(\frac{p_{top}}{p_{ref}(i)}\right)\right)\right]\right\} \tag{13.71}$$

which lets the divergence damping coefficient ν_2 increase by up to a factor of 8 close to the model top. Generally, the model top in CAM lies around $p_{top} \sim 3$ hPa or even at a lower pressure so that the troposphere and lower stratosphere are unaffected by this increase in the strength. C_2 is then constant with $C_2 = 1/128 \approx 0.0078$ at almost all levels except the uppermost two or three.

The formulation of ν_2 (13.70) is proportional to the area of a grid cell at the equator, and inversely proportional to the time step. Dimensionally this is an appropriate choice of the damping coefficient, but reliance on the area of the grid cell at the equator, and not the true area of the grid cell (with appropriate latitudinal dependence) places the same damping effect on a given physical wavelength, regardless of discretization or latitudinal location. Alternatively, a latitude-dependent coefficient could also be selected that takes the variation of the grid cell areas on a latitude–longitude grid into account

$$\nu_2 = C_2(i)\,\frac{a^2 \cos\phi \Delta\lambda \Delta\phi}{\Delta t}. \tag{13.72}$$

Such an area-dependent coefficient is selected for the optional fourth-order divergence damping mechanism in CAM 5 (Neale et al. 2010). It is given by

$$\nu_4 = C_4\,\frac{a^4 \cos^2\phi\,(\Delta\lambda)^2\,(\Delta\phi)^2}{\Delta t} \tag{13.73}$$

where the parameter C_4 is set to a default value of 0.01. The alternative formulation without the variation of the grid cell area yields

$$\nu_4 = C_4 \frac{a^4 (\Delta\lambda)^2 (\Delta\phi)^2}{\Delta t}. \tag{13.74}$$

Again, the stability aspects of both coefficients ν_2 and ν_4 are discussed in Whitehead et al. (2011) who determine upper limits for the parameters C_2 and C_4 in case of explicit time-stepping schemes.

Alternatively, NASA's FVcubed model on the almost uniform-resolution cubed-sphere grid defines the second-order and fourth-order divergence damping coefficients as

$$\nu_2 = C_2 \frac{A_{min}}{\Delta t} \tag{13.75}$$

$$\nu_4 = C_4 \frac{A_{min}^2}{\Delta t} \tag{13.76}$$

where $A_{min} \sim \Delta x_{min} \Delta y_{min}$ symbolizes the minimum grid area of the cubed sphere grid cells at a given resolution. C_2 and C_4 are often set to 0.005 and 0.0025, respectively, as documented in Lauritzen et al. (2010a). At coarser resolutions $\geq 1°$ these coefficients are typically lowered (William Putman, personal communication). Second- and fourth-order divergence damping can be used concurrently in the model FVcubed.

As the last point of comparison Bates et al. (1990) used the constant second-order damping coefficient $\nu_2 = 9 \times 10^7$ m^2 s^{-1} for their semi-implicit semi-Lagrangian shallow water simulations at the grid resolution $\Delta\lambda = \Delta\phi = 3.75°$ with both a $\Delta t = 600$ s and $\Delta t = 3{,}600$ s time step. According to (13.70) these correspond to the parameters $C_2 \approx 0.31$ and $C_2 \approx 1.86$, respectively, and are thereby considerably higher than the values in CAM FV. Such high values are unstable for explicit time-stepping schemes on latitude–longitude grids and necessitate an implicit treatment.

13.4.1.2 Example: The Effects of Divergence Damping

As a concrete example, we now illustrate the effects of the second-order and fourth-order divergence damping mechanisms in the Finite Volume dynamical core on the latitude–longitude grid. Both techniques can be selected in NCAR's model version CAM 5 at run time, with the second-order divergence damping technique being the default. In addition, we present FV simulations without divergence damping. We again utilize the growing baroclinic wave dynamical core test case by Jablonowski and Williamson (2006a) as depicted before in Sect. 13.3 (Figs. 13.3 and 13.5).

Figure 13.6 shows the surface pressure and 850 hPa meridional velocity field at day 9 for three CAM 5 FV simulations at the resolution $1° \times 1°$ with 26 levels. The only difference between the simulations is the selected divergence damping technique with their specific default coefficient. The upper row displays the simulation with the default second-order damping using the coefficient shown in (13.70)

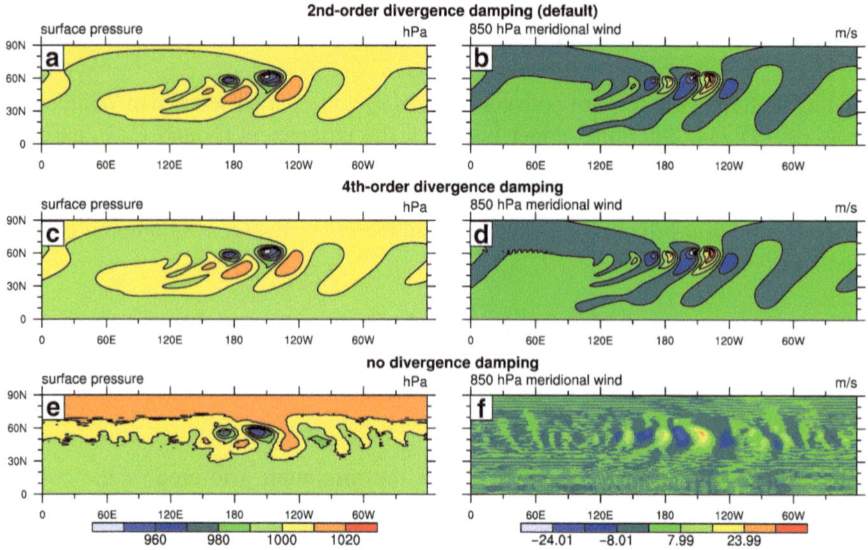

Fig. 13.6 Surface pressure (hPa) and 850 hPa meridional velocity (m/s) at day 9 of the growing baroclinic wave test case of Jablonowski and Williamson (2006a) from the CAM FV dynamical core at the resolution $1^\circ \times 1^\circ$ L26. (**a**,**b**): default second-order divergence damping, (**c**,**d**) fourth-order divergence damping, (**e**,**f**) no divergence damping. The unusual contour interval for the v wind emphasizes the very weak oscillations in (**d**). A dynamics time step of $\Delta t = 180$ s is used

with the default base value $C_2 = 1/128$, the middle row depicts the simulation with the optional fourth-order damping and the coefficient from (13.73) with the default $C_4 = 0.01$, and the bottom row reflects the simulation without divergence damping. Slightly unusual contour values for the meridional velocity are chosen to highlight the differences between the second-order and fourth-order divergence damping mechanisms. The figure shows that the simulation without divergence damping is corrupted by small-scale noise which suppresses the evolution of the baroclinic wave. This solution has little resemblance with Fig. 13.6a–d or additional high-resolution reference solutions from other models (Jablonowski and Williamson 2006a).

The differences between the simulation with second-order and fourth-order divergence damping are more subtle. As argued in Whitehead et al. (2011) fourth-order divergence damping is more scale-selective and introduces very strong damping near the grid scale (2-4 Δx) whereas longer scales are damped slightly less in comparison to the second-order damping scheme. The exact break-even point of the damping and the corresponding wavelength depends on the grid aspect ratio $\Delta\lambda/\Delta\phi$, the damping coefficients and the latitudinal position ϕ. However, it lies around 4-5 Δx at 60° in the current simulation. Presumably, this is the reason why there are still some very minor oscillations in the otherwise smooth 850 hPa meridional wind field in Fig. 13.6d. These oscillations have an approximate wavelength of 4-5 Δx. They are not obvious though in the surface pressure field or if the

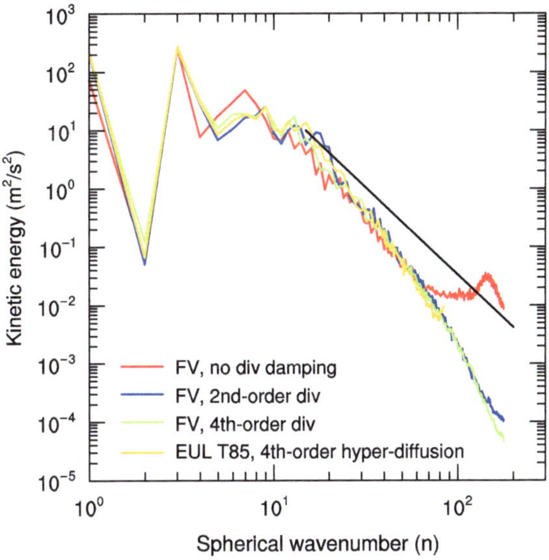

Fig. 13.7 700 hPa kinetic energy spectra at day 30 as a function of the spherical wavenumber n using the baroclinic wave test case of Jablonowski and Williamson (2006a). The spectra of the CAM FV model at the resolution $1^\circ \times 1^\circ$ L26 without divergence damping, the default second-order divergence damping and the fourth-order divergence damping are depicted. The dynamics time step is $\Delta t = 180$ s. A CAM Eulerian T85L26 run (with $\Delta t = 600$ s) with fourth-order hyper-diffusion is shown for comparison. The *black line* indicates an n^{-3} slope

contour values are integer multiples of 8 m s^{-1} (not shown). These oscillations do not grow over time. The visual comparison of the surface pressure fields suggests that the fourth-order divergence damping provides indeed less damping since the low pressure systems have deepened slightly more as indicated by the contour lines.

A more quantitative comparison of the divergence damping is provided in Fig. 13.7. The figure shows the 700 hPa kinetic energy spectra at day 30 of the simulations. The slightly rugged tails of the spectra could be smoothed via time-averaging, but is not of importance here. The kinetic energy spectra present a single snapshot in time. The figure displays that there is insufficient damping near the tail of the kinetic energy spectrum in the simulation without divergence damping. The upturn in the spectrum is a sign of numerical grid-scale noise and small-scale gravity waves which are connected to unbalanced (divergent) ageostrophic motions (O'Sullivan and Dunkerton 1995). The presence of too much energy in the divergent part of the spectrum is confirmed in the right figure of Fig. 14.6 in Chap. 14. The hook in the spectrum in Fig. 14.6 is solely triggered by the divergent motions (dotted red line) in this CAM FV aqua-planet run without divergence damping. In our example in Fig. 13.7 the divergent part of the spectrum causes a similar build-up of energy at the smallest spatial scales which is diffused by the divergence damping in the two additional FV simulations. The two FV simulations with divergence damping almost overlay each other, but show small differences in the steepness of the kinetic energy spectra at the very end of the tail. It again suggests that fourth-order divergence damping damps the shortest scales more aggressively than second-order damping since the tail falls off more quickly. As a point of comparison, the figure also depicts the kinetic energy spectrum of the CAM Eulerian T85L26 simulation with default fourth-order hyper-diffusion (see Sect. 13.3.4.4). The EUL curve is very similar to the Finite Volume simulations. Recall that CAM EUL applies fourth-order hyper-diffusion to both the divergence and vorticity fields.

Similar results were also found by Becker and Burkhardt (2007) who observed a hook at the end of the kinetic energy spectrum in a simple general circulation model. The hook was eliminated via hyper-diffusion of the horizontal divergence in their spectral transform model. Note that the spectra in Fig. 13.7 fall off faster than the spectra shown earlier in Fig. 13.4. This is mainly due to the nature of the circulations in the different test cases and the fact that they are plotted at different pressure levels. The low 700 hPa level is selected here since the idealized baroclinic wave simulation is most pronounced at lower levels in this deterministic dynamical core assessment.

Divergence damping provides the major source of the diffusion for the divergent part of the motion in CAM FV. The rotational motions are damped via inherent numerical diffusion as shown later (see also Lin (2004)). The divergence damping stabilizes the FV dynamical core by smoothing the small-scale noise and preventing the hook in the tail of the kinetic energy spectrum. However, there are other physical effects that need to be considered. For example, it has been observed that divergence damping impacts the precipitation field in aqua-planet simulations at very coarse resolutions (Peter H. Lauritzen, personal communication). This is depicted in Fig. 13.8 that shows the annual average of the frequency distribution for tropical precipitation between 10°S and 10°N. The yellow curve represents a CAM 3.5 FV simulation with standard second-order divergence damping (13.70) at the coarse $2.7° \times 3.3°$ L26 resolution, the blue curve denotes an identical FV simulation, but with a doubled divergence damping coefficient. The red curve depicts a CAM EUL aqua-planet simulation at the resolution T31L26. The variation of the divergence damping coefficient has profound impact on the likelihood of heavy precipitation events in FV. The increase in the diffusion (blue curve) leads to a sharp decrease in the likelihood of heavy precipitation events and lets the FV simulation resemble the CAM EUL run. However, there is no "true solution" so the physical realism cannot be judged from these experiments alone. The figure only demonstrates the high sensitivity of the precipitation to the choice of the coefficient. The plots can also be compared to similar figures shown in Williamson (2008a,b).

13.4.1.3 2D Divergence Damping: Avoiding Confusion

As a word of caution, the spectral Eulerian dynamical core in NCAR's CAM model is formulated in vorticity-divergence form (Collins et al. 2004) and defines a Rayleigh friction technique (see Sect. 13.4.5.1 below) with the term *divergence damping*. However, the two mechanisms are very different. CAM EUL's definition is

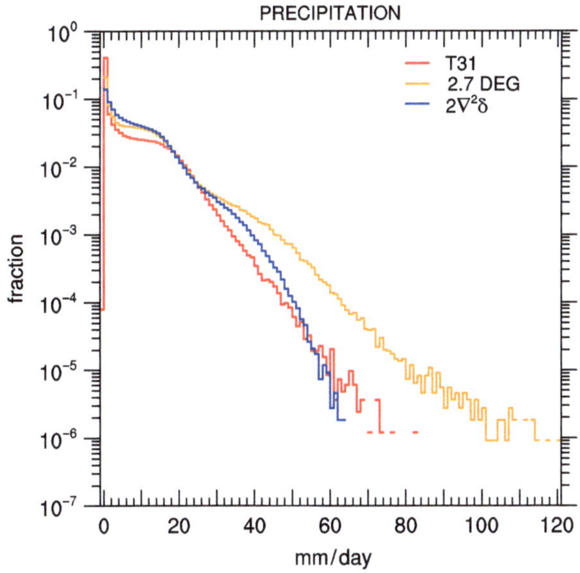

Fig. 13.8 Fraction of the time the tropical precipitation is in 1 mm day^{-1} bins ranging from 0 to 120 mm day^{-1}, calculated from 6-h averages for all grid points between $\pm 10°$. This frequency distribution is an annual average. The aqua-planet simulations are (*blue*, *yellow*) CAM FV at the coarse *lat* × *lon* resolution 2.7° × 3.3° L26 and (*red*) CAM EUL at the resolution T31L26 (with time step $\Delta t = 1{,}800$ s). *Yellow FV curve*: standard second-order divergence damping (13.70). *Blue curve*: FV simulation with a doubled coefficient. The figure is courtesy of Peter H. Lauritzen, NCAR

$$\frac{\partial D}{\partial t} = -rD \tag{13.77}$$

where r symbolizes an inverse damping time scale like $1/T$. The damping time scale determines the strength of the friction and is user-defined. In particular, the damping has an initial e-folding time of T and linearly decreases to zero over a time period of T_d, usually set to 2 days. It yields

$$r = \max\left[\frac{1}{T}\frac{T_d - t}{T_d}, 0\right] \tag{13.78}$$

where t stands for the elapsed time after the start of the model. In the CAM Eulerian or semi-Lagrangian dynamical core the damping is computed implicitly in spectral space via time splitting after the horizontal diffusion. If activated by the user it is only applied at the beginning of a model climate simulation to damp the gravity wave propagation arising from poorly balanced initial states. They usually result from interpolating a model simulated state to a different resolution with no attempt to maintain geostrophic balance. The initial behavior of a climate simulation is generally of no interest. This damping should never be used for short-term forecasts when the initial behavior is of interest. After day T_d this damping mechanism is no longer active. Dey (1978) uses the phrase *divergence dissipation* for this type of damping.

13.4.2 3D Divergence Damping (or Acoustic Mode Filtering)

3D divergence damping is a smoothing mechanism for nonhydrostatic models and is, from a design perspective, very similar to the 2D divergence damping presented above in Sect. 13.4.1. However, there is a principal difference concerning its impact on the circulation. 2D divergence damping mainly affects internal gravity waves, whereas gravity waves are not noticeably impacted by 3D divergence damping. This is due to the fact that their velocity fields are almost non-divergent in three dimensions.

3D divergence damping serves two purposes. First, it is an effective damping mechanism for acoustic modes in nonhydrostatic models. Second, it eliminates spurious high-wavenumber modes caused by the instabilities in a partially-split (split-explicit) time-stepping scheme (Tatsumi 1983; Skamarock and Klemp 1992). Time-split schemes separate the terms in the equations of motion into slow and fast processes, and integrate them with large and multiple small timesteps, respectively. This technique is sometimes used to increase the computational efficiency of mesoscale models since fast, but meteorologically less important, sound waves can then be treated with a lower order and cheaper numerical scheme. Partially split numerical schemes are for example used in the models WRF (Skamarock et al. 2008) and COSMO (Doms and Schättler 2002; Gassmann and Herzog 2007; Baldauf 2010) and both models apply a 3D divergence damper to stabilize the schemes. Skamarock and Klemp (1992) showed that 3D divergence damping only caused very minor reductions in the amplitudes of gravity waves whereas it effectively damped both acoustic waves and the spurious noise associated with the time-split discretization. They also used the phrase *acoustic mode filtering* to describe 3D divergence damping. As an aside, acoustic mode filtering can also be accomplished by forward biasing an implicit time-stepping scheme of Crank-Nicolson or trapezoidal type (see textbooks like Durran (1999, 2010) or Kalnay (2003). This is briefly discussed in Sect. 13.6.3 that characterizes such an off-centering approach as inherent numerical diffusion.

The second-order 3D divergence damping formulation in vector form is given by

$$\mathbf{F_v} = \nabla_{2/3}\big(\nu\,\nabla_3 \cdot \mathbf{v}\big) \tag{13.79}$$

where ν is the divergence damping coefficient, $\mathbf{v} = (u, v, w)$ is the 3D velocity vector, $\nabla_{2/3}$ symbolizes either the two- or three-dimensional gradient operator and $(\nabla_3\cdot)$ denotes the 3D divergence operator along constant coordinate surfaces. The term $\mathbf{F_v}$ can be appended exclusively to the 2D horizontal momentum equations as in Dudhia (1993), Doms and Schättler (2002) and Skamarock et al. (2008), or to the 3D momentum equations as in Skamarock and Klemp (1992), Xue et al. (2000), Tomita and Satoh (2004) and Gassmann and Herzog (2007). Adding (13.79) to the momentum equations effectively introduces a second-order diffusion of the 3D divergence. This is highly specific to filtering acoustic modes.

The analysis in Gassmann and Herzog (2007) showed that an isotropic (constant ν) application to all three momentum equations should be the preferred choice. However, if the horizontal grid spacing is much larger than the vertical spacing as in typical GCMs, it is not possible to use the same value of ν in all directions. This was argued in Tomita and Satoh (2004) who also give guidance on the choice of the divergence damping coefficient for the model NICAM. In particular, Tomita and Satoh (2004) selected

$$\nu_x = \alpha'_x \, c_{s0}^2 \, \Delta\tau \tag{13.80}$$

where x serves a placeholder for either the horizontal (H) or vertical (V) direction, c_{s0} is the speed of sound at a temperature of $T = 273$ K, and $\Delta\tau$ is the length of the small time step in their split-explicit time-stepping scheme ($\sim \Delta t/4$). Typically, horizontal values around $\alpha'_H \in [0.05, 0.2]$ were chosen for horizontal grid spacings between 120 km and 240 km, and α'_V was set to zero (nonisotropic case) in their selected dynamical core experiments.

13.4.3 Vorticity Damping

A second-order vorticity damping formulation was suggested by Shuman (1969) and McPherson and Stackpole (1973) for models written in momentum (u, v) form. It is represented by

$$\frac{\partial u}{\partial t} = \cdots - \nu_\zeta \, \frac{1}{a} \frac{\partial \zeta}{\partial \phi} \tag{13.81}$$

$$\frac{\partial v}{\partial t} = \cdots + \nu_\zeta \, \frac{1}{a \cos\phi} \frac{\partial \zeta}{\partial \lambda} \tag{13.82}$$

where ν_ζ symbolizes the vorticity damping coefficient. In spherical coordinates the relative vorticity ζ is defined as

$$\zeta = \frac{1}{a \cos\phi} \left(\frac{\partial v}{\partial \lambda} - \frac{\partial (u \cos\phi)}{\partial \phi} \right) \tag{13.83}$$

and expresses the vertical component of the 3D vorticity vector. Formulating the evolution equation for the vorticity based on (13.81) and (13.82) yields

$$\frac{\partial \zeta}{\partial t} = \cdots + \nu_\zeta \, \nabla^2 \zeta. \tag{13.84}$$

It shows that the vorticity damping selectively diffuses the rotational part of the motion. Divergent motions remain unaffected.

An interesting analogy can again be drawn between the second-order vorticity damping presented here and the vorticity diffusion described earlier in the context of spectral transform models (13.10) and (13.14). Despite the different names, the physical effects of both damping mechanisms are similar. In practice though, none of today's GCMs in (u, v) form use such a vorticity damping. They employ alternatives such as the hyper-diffusion of u and v in the model GME (Majewski et al. 2002) that damps both the rotational and divergent motions. Other alternatives include the use of inherent numerical dissipation. For example, the model FV (Lin 2004) damps rotational motions via monotonicity constraints that are built into in its finite volume scheme (see also Sect. 13.6.2).

13.4.4 External Mode Damping

Noise in a numerical model can also manifest itself in form of pressure oscillations that are almost independent of the vertical level. These can be identified as the Lamb wave that is also called the "external inertia-gravity wave" mode. Lamb waves are fast moving horizontal acoustic modes with imaginary vertical wavenumbers that do not propagate in the vertical (and are therefore described as *external*). As shown in Kalnay (2003) Lamb waves are equivalent to gravity waves in a shallow water model.

Lamb waves are associated with fluctuations of the mean divergence in an atmospheric column. Recall that the change of pressure at a point is determined by the vertical integral of the mass divergence at this location. Therefore, damping the mass-weighted vertical integral of the divergence controls spurious pressure oscillations. Washington and Baumhefner (1975) explored this type of damping mechanism for model initialization purposes. In particular, they *Lamb filtered* the initial velocity data by modifying the divergence and successfully suppressed the external mode and high-speed oscillations. This connection was also pointed out in Dey (1978).

The so-called *external mode damping* for models in the native (u, v) momentum form is given by

$$\frac{\partial \mathbf{v}}{\partial t} = \cdots + K_e \nabla \left[\frac{1}{p_s - p_{top}} \int_{p_{top}}^{p_s} D \, dp \right] \tag{13.85}$$

where the term in the bracket is the mass-weighted vertical integral of the horizontal divergence D, and K_e is the damping coefficient. p_s stands for the surface pressure, and p_{top} is the pressure at the model top. Applying the horizontal divergence operator to (13.85) and integrating this equation again in the vertical then yields

$$\frac{\partial \overline{D}}{\partial t} = \cdots + K_e \nabla^2 \overline{D} \tag{13.86}$$

with the vertically integrated divergence $\overline{D}$. It shows that external mode damping indeed acts as a second-order diffusion mechanism and affects the vertically averaged divergence of the column.

All that is left is the definition of the external mode damping coefficient. Analogous to the discussion of the horizontal second-order diffusion it carries the physical dimensions $m^2\ s^{-1}$. Typically, it is defined by

$$K_e = \beta \frac{A}{\Delta t} = \beta \frac{\Delta x \Delta y}{\Delta t} \tag{13.87}$$

where the area A of a grid cell can be expressed by the physical grid spacings Δx and Δy in the two horizontal directions. A similar form of the external mode filter was also shown in Klemp et al. (2007) who discussed its definition in the weather forecast model WFR (note that there is a minus sign missing in Klemp et al.'s definition of D_h after their (46)).

In the model WRF the dimensionless and positive coefficient β is typically set to $\beta = 0.01$ (Klemp et al. 2007). In NASA's FVcubed dynamical core on a cubed sphere grid (Putman and Lin 2007, 2009) β is often set to 0.02 (William Putman, personal communication). In addition, the area measure A for the individual grid cells is replaced by the minimum cell area A_{min} within the FVcubed model. Equation (13.87) therefore defines a constant coefficient for all grid points regardless of their actual size. For latitude–longitude grids this choice might need to be reconsidered. External mode damping is only rarely used in GCMs today.

13.4.5 Sponge Layer Mechanisms at the Model Top

Setting appropriate upper boundary conditions in atmospheric models has been proven difficult for many years. The choices include radiation boundary conditions that allow energy to radiate outward at some finite height, the choice of a zero vertical velocity at the model top, or absorbing boundary conditions that absorb most or all incoming energy (Rasch 1986). Radiation boundary condition are popular in research models as e.g., suggested by Klemp and Durran (1983) but they cannot easily be implemented in GCMs. Therefore, operational models generally do not apply a radiation boundary condition but impose the condition that the vertical velocity is zero at the top. However, the presence of such a rigid top can lead to spurious wave reflections and even trigger instabilities at the top. Extra diffusion is then often utilized near the model top to absorb the reflections and slow down the wind speeds. This is common practice in almost all GCMs.

The type of extra dissipation in these sponge layers varies widely though. For example, the models CAM EUL, CAM SLD, ECHAM5 and GME switch from a linear hyper-diffusion to a second-order diffusion which is applied in a few (around three) levels near the top. Sometimes the diffusion coefficient also increases upward. The model FV increases the strength of the divergence damping (see Sect. 13.4.1) and furthermore, utilizes a lower order numerical scheme to provide inherent diffusion as explained later in Sect. 13.6.1. The ECMWF model IFS (until cycle Cy35r3, September 2009), the spectral transform model AFES (Enomoto et al. 2008), NASA's ModelE (Schmidt et al. 2006) developed at the Goddard Institute for Space Studies and the model COSMO generally apply Rayleigh friction to the horizontal wind field. Rayleigh friction is also optional in the model WRF (Skamarock et al. 2008). These sponge layer mechanisms are outlined in more detail below.

All sponge layer mechanisms have one feature in common. As pointed out by Rasch (1986) sponge layers need to be thick enough and have adequate resolution to capture the waves reasonably well that they are supposed to damp. Sponge layers should ideally resolve the vertical wavelength with 8–10 vertical levels in order to damp waves effectively. They also need to guarantee a smooth transition region since a sudden onset of a strong sponge can cause wave reflections by itself. Sponge layers are simple to use, but can become computationally expensive in case extra vertical levels are needed for the sole purpose of providing a damping layer. An analysis of the properties of some dissipative sponge layer mechanisms can be found in Klemp and Lilly (1978).

13.4.5.1 Rayleigh Friction

A Rayleigh friction sponge is based on a linear relaxation which can be appended to the prognostic equations in the generic form

$$\frac{\partial \psi}{\partial t} = \cdots - k_R\,(\psi - \psi_r). \qquad (13.88)$$

Here, k_R symbolizes a possibly spatially varying Rayleigh friction coefficient that expresses the inverse of a friction time scale, ψ is a placeholder for the velocities u or v, and ψ_r is a prescribed reference profile that might vary in space and time. Most often, the reference profile for the wind components u_r and v_r is set to zero. If "Rayleigh friction" is applied to the temperature field, it is called a *temperature relaxation* or *Newtonian heating or cooling*. Such a temperature relaxation always utilizes the reference profile since a temperature relaxation towards zero would be overly strong and unrealistic. This type of Rayleigh forcing was for example suggested by Held and Suarez (1994) and Boer and Denis (1997) for idealized dynamical core assessments.

In practice, Rayleigh friction might be used for two reasons at the upper levels in GCMs. First, it is considered a very crude approximation for gravity wave breaking in models with a very high model top in the upper stratosphere or mesosphere (Boville 1986; Boville and Randel 1986). Basically, it then replaces missing physics mechanisms that would otherwise capture the momentum deposition by breaking gravity waves that travel upwards from the troposphere. These gravity waves induce a force on the large-scale flow, but note that Rayleigh friction is too simplistic to drive the observed reversal of the jets between the upper mesosphere and the lower thermosphere, or the tropical zonal wind oscillations in the stratosphere. Second, Rayleigh friction might be applied as a pure numerical sponge to prevent wave reflections at the model top. However, this requires some care in the formulation of the Rayleigh damping coefficient since as argued above, a sudden onset of a strong Rayleigh friction can also act as a reflector. Most vertical profiles of the Rayleigh damping coefficient therefore provide a smooth transition with increasing strength towards the model top. Typical damping time scales are 50 days or longer in the middle stratosphere and 1–2 days at the model top in the mesosphere.

We now present several damping profiles which have been suggested in the literature. A commonly used profile for the damping coefficient k_R is

$$k_R = \frac{1}{\tau}\left[1 + \tanh\left(\frac{z - z_1}{h}\right)\right] \tag{13.89}$$

where the height z is given by

$$z = H \ln\left(\frac{p_0}{p}\right). \tag{13.90}$$

$H = R_d \overline{T} g^{-1}$ symbolizes a constant scale height of the atmosphere with the gas constant for dry air R_d, a constant temperature $\overline{T}$ and the gravity g. p_0 is a reference pressure set to 1,000 hPa, p denotes the pressure at the model level, τ is a damping time scale, z_1 presents the approximate height of the model top, and h is a scaling parameter with height units. Such a profile has for example been defined in Boville and Randel (1986) and was the default in the ECMWF model IFS until September 2009 (Orr and Wedi 2009; Orr et al. 2010). Boville and Randel (1986) suggested the parameters $\tau = 3$ days, $z_1 = 63$ km and $h = 7.5$ km for a middle atmosphere GCM with a model top around 63 km. The model IFS (cycle 18R3, November 1997) set the parameters to $\tau = 3$ days, $z_1 = 61$ km and $h = 7.7$ km for a 50-level version with a model top at 0.1 hPa (about 61 km). IFS only applied the Rayleigh friction to the zonal wind field at model levels above 10 hPa. We refer to these IFS settings as "strong Rayleigh friction" in the examples below. Alternatively, Boville and Baumhefner (1990) used $\tau = 3$ days, $z_1 = 75$ km and $h = 7.5$ km which we characterize as "weak Rayleigh friction" (note that there is a sign error in their original definition). These latter two profiles of the Rayleigh damping coefficient are shown in Fig. 13.9. The figure also depicts the additional profile

$$k_R = \begin{cases} 0 & \text{if } p \geq p_{sponge} \\ k_{max}\left[(p_{sponge} - p)/p_{sponge}\right]^2 & \text{if } p < p_{sponge} \end{cases} \tag{13.91}$$

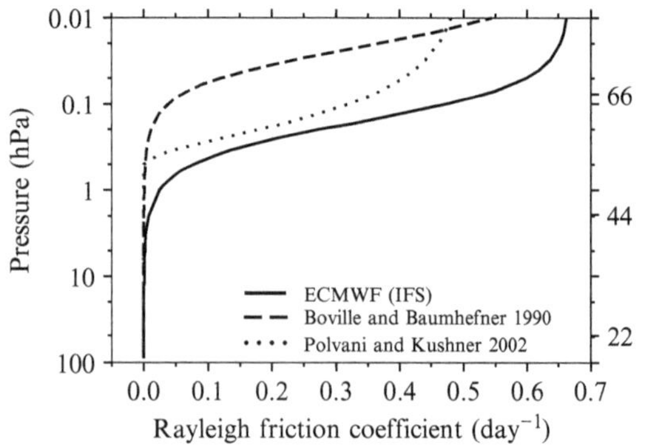

Fig. 13.9 Vertical profiles of three Rayleigh friction coefficients k_R in units day^{-1}

as suggested by Polvani and Kushner (2002) with the parameters $k_{max} = 0.5$ day^{-1} and $p_{sponge} = 0.5$ hPa. The onset of this Rayleigh friction at 0.5 hPa is more sudden. Rayleigh friction is sometimes only added to the zonal momentum equation. If the numerical stability of this damping process is a concern, it can easily be applied implicitly via a time-splitting approach.

As an aside, some meso-scale models such as WRF (Skamarock et al. 2008) offer optional ultra-strong Rayleigh friction sponges with profiles like

$$k_R = \begin{cases} 0 & \text{if } z < (z_{top} - z_d) \\ k_{max} \sin^2\left[\frac{\pi}{2}\left(1 - \frac{z_{top}-z}{z_d}\right)\right] & \text{if } z \geq (z_{top} - z_d) \end{cases} \tag{13.92}$$

where z_{top} denotes the height of the model top, z_d stands for the thickness of the damping layer as measured from the model top, and k_{max} is set to 0.2 s^{-1}. The latter corresponds to the damping time scale of $\tau = 5$ s at the model top which is very short in comparison to typical GCM settings of 1–2 days at upper levels. This damping is not used operationally in WRF. However, in idealized mountain wave test cases it has been found that about a third of the vertical domain must be classified as a sponge layer to suppress gravity wave reflections. Doms and Schättler (2002) also suggested using one third of the total domain height or at least one vertical wavelength as a sponge in the regional nonhydrostatic weather forecast model COSMO. They picked the formulation

$$k_R = \begin{cases} 0 & \text{if } z < z_{damp} \\ k_{max}\left[1 - \cos\left(\pi\left(\frac{z-z_{damp}}{z_{top}-z_{damp}}\right)\right)\right] & \text{if } z \geq z_{damp} \end{cases} \tag{13.93}$$

where z_{damp} symbolizes the starting position of the damping layer. The default values in COSMO are $z_{damp} = 11$ km and a coefficient of $k_{max} = (20\Delta t)^{-1}$ where Δt denotes the model time step. Similar Rayleigh friction profiles for meso-scale mountain wave simulations are also presented in Durran and Klemp (1983).

We now illustrate the effects of "weak" and "strong" Rayleigh friction in dynamical core simulations with CAM SLD at the triangular truncation T63 ($\approx$ 210 km)

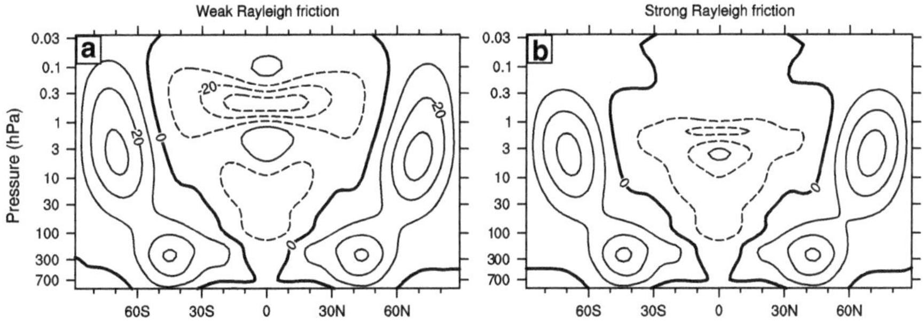

Fig. 13.10 Zonal-mean 1110-day-mean zonal wind in the CAM semi-Lagrangian dynamical core run with the Williamson et al. (1998) forcing at the resolution T63L55 with a model top at 0.015 hPa with (**a**) weak Rayleigh friction corresponding to the Boville and Baumhefner (1990) profile, (**b**) strong Rayleigh friction (ECMWF IFS). The friction is applied above 1 hPa in both simulations. The contour interval is 10 m/s. Negative contours are *dashed*. The time step is $\Delta t = 1{,}800$ s

with 55 vertical levels. The model top is placed at 0.015 hPa which lies around 75 km. The sponge-layer Rayleigh friction is only appended to the zonal momentum equation above 1 hPa and applied at every time step with $\Delta t = 1{,}800$ s. The experiments utilize the Held and Suarez (1994) forcing with the modification by Williamson et al. (1998) that provides strong meridional gradients in the stratospheric reference temperature field. This reference field is used for a temperature relaxation and causes zonal jets in the middle atmosphere according to the thermal wind relationship.

Figure 13.10 depicts the zonal-mean 1100-day-mean zonal wind fields of the two SLD simulations with (a) the weak Rayleigh friction according to Boville and Baumhefner (1990) and (b) the strong Rayleigh friction used in previous versions of IFS. Both model simulations were spun up with identical initial conditions and run for 1,440 days. The time average is over days 360–1,440. The figure shows that the flow field at the upper levels is very different in the two simulations. The "strong" Rayleigh friction almost damps out all motion above 1 hPa which is quite drastic. In contrast, the effect in the "weak" Rayleigh friction simulation is less pronounced in the region between 1 and 0.1 hPa. However, it is clearly visible near the model top where the wind speed has slowed down considerably. These figures can also be directly compared to a gentler diffusive sponge discussed in the following subsection (Fig. 13.11).

As a word of caution, Rayleigh drag changes the upper atmospheric angular momentum which should be balanced by a correction in the troposphere in order to conserve angular momentum. From a physical viewpoint, such a compensation mimics in some way the transfer of momentum by the unresolved gravity waves and is included in the Rayleigh friction implementation in NASA's ModelE (Schmidt et al. 2006). However, angular momentum conservation is not generally considered in GCMs. In case Rayleigh drag violates the angular momentum conservation it leads to a physically-spurious large-scale meridional circulation extending to the

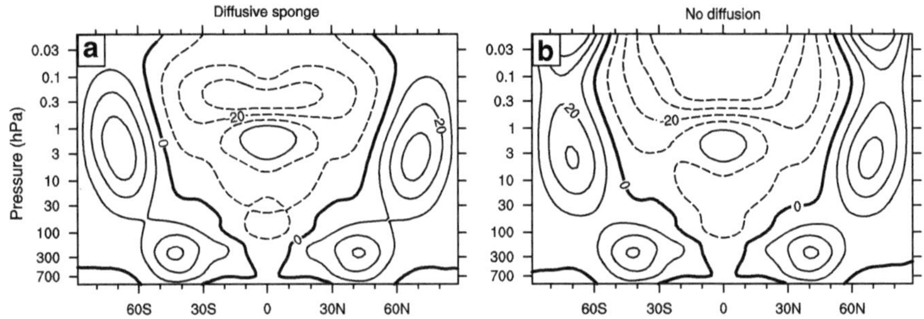

Fig. 13.11 Zonal-mean 1110-day-mean zonal wind in the CAM semi-Lagrangian dynamical core run with the Williamson et al. (1998) forcing at the resolution T63L55 with a model top at 0.015 hPa with (**a**) second-order diffusive sponge near the model top, (**b**) no ∇^2 and ∇^4 diffusion. The contour interval is 10 m/s. Negative contours are *dashed*. The time step is $\Delta t = 1{,}800$ s

surface of the earth as shown in Shepherd et al. (1996) and Shepherd and Shaw (2004). Therefore, Shepherd and Shaw (2004) recommended using Rayleigh friction very selectively, if at all, and only applying it to the wind perturbations from the mean state. A Rayleigh friction sponge is rarely used in operational GCMs today. Newtonian heating or cooling for tropospheric models is also often frowned upon because it forces the simulation toward a prescribed state rather than letting the state evolve freely.

13.4.5.2 Diffusive Sponges

The most popular sponge layer mechanism in GCMs is an increase in the horizontal diffusion, either via an increase in the diffusion coefficient, as e.g., discussed in Klemp and Lilly (1978), or a decrease in the order of the diffusion. The latter or even a combination of the two are most often chosen. Then a second-order diffusion replaces the usual hyper-diffusion in a few layers near the model top. For example, the model ECHAM5 decreases the order of the diffusion from a sixth-order to a fourth-order and finally to a second-order diffusion at the model top while using a constant time scale for all K_6, K_4 and K_2 diffusion coefficients (Roeckner et al. 2003). The effects of this diffusive sponge in ECHAM5 are clearly visible in the idealized dynamical core simulations by Wan et al. (2008) who utilized the Held and Suarez (1994) forcing. The sponge leads to a strong damping of the zonal wind field at upper levels in the equatorial region which can be compared to the simulations with the model CAM in Sect. 13.7.1 (Fig. 13.29).

A diffusive sponge is quite effective in reducing reflections and slowing down the wind speeds near the model top. The latter is illustrated in dynamical core simulations with CAM SLD at the triangular truncation T63 (≈ 210 km) with 55 vertical levels. As explained above in Sect. 13.4.5.1 the experiments utilize the Held-Suarez forcing with the modifications of the equilibrium temperature in the stratosphere according to Williamson et al. (1998). The model top is set to 0.015 hPa and no

Rayleigh friction is applied. Instead, we compare a simulation with a linear second-order sponge in the uppermost three layers (between 0.015 and 0.05 hPa) and very weak hyper-diffusion in the rest of the domain to a simulation without any (neither a K_2 nor K_4) explicit diffusion. In particular, the base diffusion coefficient for the sponge layer starting at the third level from the top is $K_2 = 2.5 \times 10^5\,\mathrm{m}^2\,\mathrm{s}^{-1}$ which corresponds to a damping time scale of 11.2 h, and the hyper-diffusion parameter K_4 is set to $1 \times 10^{15}\,\mathrm{m}^4\,\mathrm{s}^{-1}$ which corresponds to the long damping time scale of 28.1 h according to (13.20).

Figure 13.11 shows the zonal-mean 1100-day-mean zonal wind fields of the two SLD simulations with (a) the diffusive sponge and (b) neither ∇^2 nor ∇^4 diffusion. As before, both model simulations were spun up with identical initial conditions and run for 1440 days. The time average is over days 360–1440. The wind speeds near the model top are strongly reduced in the simulation with the sponge (Fig. 13.11a) whereas the winds almost reach their maxima at the model top in the model run without diffusion (Fig. 13.11b). This emphasizes the very strong influence of the additional sponge-layer dissipation on the circulation in the upper atmosphere. The impact of the sponge is not just confined to the top three layers but extends further down into the domain to about 1 hPa which incorporates 14 vertical levels. A similar effect was also shown earlier for a 26-level setup in Fig. 13.1. The impact of the ∇^4 diffusion at lower levels below the sponge is harder to evaluate since the stratospheric polar jets show some variability even in these long time averages. However, the impact of the ∇^4 diffusion seems to be minor in Fig. 13.11a. As a word of caution, if sponge layers are applied to model simulations the upper layers cannot be used for scientific evaluations. But unfortunately, it remains unclear whether the simulation without diffusion is any more reliable at upper levels due to the potential impact of artificial wave reflections. As seen before in the Rayleigh friction case, the sponge-layer dissipation dominates the upper level flow field. Recall that is has no physical foundation.

13.4.5.3 Vertical Velocity Damping

If Rayleigh damping is explicitly applied to the prognostic vertical velocity field in a nonhydrostatic model it is sometimes called *vertical velocity damping*. Rayleigh friction can also be implicitly applied within the implicit solution technique for vertically propagating acoustic modes. The latter has been found to be a very effective and robust absorbing sponge layer mechanism in nonhydrostatic models as suggested by Klemp et al. (2008).

As an example, the model WRF offers both an implicit Rayleigh damping of the vertical velocity and optional explicit vertical velocity damping to foster the robustness of the numerical scheme (Skamarock et al. 2008). Here, we briefly describe the explicit vertical velocity damping which is not just a sponge-layer technique. It generally damps the vertical motion whenever a violation of the CFL condition in the vertical direction is imminent, and thereby prevents the model from becoming unstable. The damping coefficient is locally determined and utilizes a critical

Courant number C_r defined by

$$C_r = \left| \frac{\dot{\eta} \Delta t}{\Delta \eta} \right| \tag{13.94}$$

where $\dot{\eta}$ is the vertical velocity in the generalized vertical coordinate η with grid spacing $\Delta\eta$. If C_r exceeds an activation Courant number C_a a Rayleigh friction is switched on. Assuming a forecast equation for the vertical velocity w, it yields

$$\frac{\partial w}{\partial t} = \cdots - sign(w) \; \gamma_w \; (C_r - C_a) \tag{13.95}$$

where γ_w is a damping coefficient and $sign(w)$ symbolizes the sign of w. The typical coefficients in WRF are $C_a = 1$ and $\gamma = 0.3$ m s^{-2}. This process does not possess a physical foundation and needs to be used with care. The regional nonhydrostatic weather forecast model COSMO (Doms and Schättler 2002) even includes a similar CFL-dependent Rayleigh friction as a Courant number limiter in the forecast equations for the horizontal wind speeds.

13.4.5.4 Courant Number Limiter

Occasionally in global GCM simulations the polar night jet becomes very strong at the top of the model and then the CFL stability condition is violated for the shortest longitudinal waves. Without further action the model would blow up. Rather than damp the jet speed further, some models simply remove the short waves that are unstable. This is generally only done at the top few levels of the model and only while the jet speed remains overly strong. The elimination is particularly simple in spectral transform models. If the maximum wind speed is sufficiently large, then the amplitudes of waves with wavenumber $n > n_c$ are set to zero, where the cutoff wave length is $n_c = a\Delta t / \max|V|$. a symbolizes the radius of the earth. This condition is applied whenever the maximum wind speed $\max|V|$ is large enough that n_c is less than the truncation limit and temporarily reduces the effective resolution of the model at the affected levels, but it does not affect the remaining scales.

To avoid adding code to sweep through spectral space, the dynamical core CAM EUL includes this process in the solution of the horizontal diffusion. Recall (13.12) that expresses the $2q$-th order temperature diffusion and for simplicity, let us focus the discussion on the diffusion that is applied along model levels. The time-discretized response function (13.28) for CAM EUL then becomes

$$E_n = \left\{ 1 + 2\Delta t \; D_n \; K_{2q} \left(\frac{n(n+1)}{a^2} \right)^q \right\}^{-1} \tag{13.96}$$

where the so-called "Courant number limiter" D_n has been added. The response function (13.96) can also be viewed as a wavenumber-dependent damping function that is denoted by $K_n^{(2)}$ and $K_n^{(4)}$ in the technical CAM documentation (Collins et al. 2004).

Generally, the factor D_n is set to 1. However, the diffusion coefficient K_{2q} is increased by a factor of $D_n = 1000$ for those waves which are to be eliminated. This is possible because the diffusion is approximated implicitly in spectral space in the model. Since the diffusion is linear, the solution for each wavenumber is independent of all other wavenumbers. Note that this should not be thought of as increased overall diffusion since that is a process in physical space and affects all scales. The diffusion code is simply a convenient place to eliminate the selected waves. This Courant number limiter should be thought of as simply removing the shortest waves that would otherwise be unstable and thereby temporarily reducing the horizontal resolution. Therefore it is generally only applied near the model top where the solution is already contaminated by the sponge layer. Here, we only list the principle design of this limiter and refer to Collins et al. (2004) for the exact application in spectral space. Note this type of limiter can only be safely used if the diffusion is approximated implicitly as it is the case in CAM EUL. Otherwise the diffusion process will likely be unstable. The stability limitations for explicit time stepping schemes are discussed above in Sect. 13.3.5.

13.5 Explicit Filtering Techniques

Filtering is a fairly common smoothing technique in GCMs. Two types of filters need to be distinguished as explained below. The first category encompasses the temporal filters. The second type includes spatial filters such as digital grid point filters, or spectral filters like the Fast Fourier Transform (FFT) or Boyd-Vandeven filter. Spatial filters are popular in grid point models, especially on latitude–longitude grids at high latitudes, where they are often called *polar filters*. In addition, temporal digital filters are sometimes used as model initialization schemes to damp out high-frequency noise in analyzed data as discussed in Lynch and Huang (1992).

Spatial filters damp both linear and nonlinear computational instabilities. Linear instability arises if the CFL stability condition is violated by e.g., fast moving inertia-gravity waves which can easily occur on longitude-latitude grids with converging meridians near the poles. Filtering then allows the use of longer CFL-violating time steps which would otherwise grow unstable. Nonlinear computational instability is associated with quadratic terms in the equations of motion (Phillips 1959; Orszag 1971). In particular, products of waves can create new waves with wavelengths shorter than $2\Delta x$ as discussed in Durran (1999, 2010). Since these waves cannot be resolved on a model grid, they are aliased into longer wavelengths. This tends to accumulate energy at the smallest scales in a nonlinear model which can be removed by a filter to maintain computational stability. High-frequency noise might also be introduced by truncation or observational errors in the initial data. Whether filtering is necessary for stable computations depends on the characteristics of the numerical scheme and the choice of the model grid. In general, no attempt is made to associate filters with physical processes. Filtering is an ad-hoc smoothing process in atmospheric models.

13.5.1 Time Filters

13.5.1.1 Robert-Asselin Filter

Time filters are commonly applied to multi-step time-stepping schemes, such as the three-time-level Adam-Bashforth scheme or the popular leapfrog method. These time-stepping schemes are e.g., described in Haltiner and Williams (1980), Durran (1999, 2010) or Kalnay (2003). In particular, the leapfrog scheme for the variable ψ is given by

$$\psi^{j+1} = \psi^{j-1} + 2\Delta t\, F(\psi^{j}) \tag{13.97}$$

where $j+1$, j and $j-1$ represent the future, current and previous time level of the three-time-level scheme, and $F(\psi^{j})$ is the forcing by the dynamical and physical processes. A major problem with the leapfrog scheme is that every wavenumber is associated with two frequencies (Durran 1999, 2010). One is a physical mode, the second is an undamped computational mode that is associated with the handling of the even and odd time steps. It manifests itself as a spurious oscillation between even and odd time steps that amplifies during nonlinear simulations. It ultimately grows explosively and causes the model to blow up.

The decoupling of the solutions at odd and even time steps can be avoided when applying a recursive time filter at each time step as suggested by Robert (1966) and Asselin (1972). This filter is referred to as the Robert–Asselin, Robert or Asselin filter. It is defined by

$$\overline{\psi}^{j} = \psi^{j} + \alpha\left(\overline{\psi}^{j-1} - 2\psi^{j} + \psi^{j+1}\right) \tag{13.98}$$

where the overbar denotes the time-filtered variable, and the unitless positive coefficient α determines the strength of the filter. It leads to the Asselin-leapfrog scheme

$$\psi^{j+1} = \overline{\psi}^{j-1} + 2\Delta t\, F(\psi^{j}). \tag{13.99}$$

In general, the Asselin filter can be viewed as a second-order diffusion mechanism in time. It effectively damps the computational mode of the leapfrog time-stepping scheme, but unfortunately also affects the physical mode by slightly changing the phase and amplitude of the resolved waves. This reduces the formal order of accuracy of the leapfrog scheme to first order and can degrade the accuracy of model simulations. For example, physical quantities such as energy that are conserved by the time-continuous equations are not necessarily conserved by the time-discretized equations which is even true with or without the Asselin filter. The damping and non-conservation may be acceptable for short model simulations, but could become more prevalent and questionable in long GCM runs. The Asselin filter is generally applied at each time step to all prognostic variables along model levels. The accuracy and stability aspects of the Asselin filter were investigated by Tandon (1987), Robert and Lépine (1997), Cordero and Staniforth (2004) and Williams (2009). In particular, Williams (2009) proposed a modification of the Asselin-filter that greatly reduces its negative impact on the physical mode and increases the numerical accuracy, but does so at the expense of a very slight instability.

A comprehensive list of atmospheric models with the Asselin filter is documented in Williams (2009). However, the strength of the damping coefficient α is highly model-dependent and rarely mentioned in the refereed literature. The typical values in GCMs vary between 0.02 and 0.2. More specifically, the spectral element dynamical core HOMME utilizes a default coefficient of $\alpha = 0.05$ as listed in Lauritzen et al. (2010a), the spectral element model by Giraldo and Rosmond (2004) uses $\alpha = 0.02$, the NCAR Eulerian spectral dynamical core in CAM 3.1 applies $\alpha = 0.06$ (Collins et al. 2004), Skamarock and Klemp (1992) and McDonald and Haugen (1992) set α to 0.1, the model ICON defines $\alpha = 0.1$, and the model GME specifies $\alpha = 0.15$ (Majewski et al. 2008). Schlesinger et al. (1983) even suggested using values in the range 0.25–0.3 for certain advection-diffusion problems. A systematic sensitivity analysis to the Asselin filter coefficient has been conducted by Rípodas et al. (2009) with the shallow-water version of the ICON model. They found that $\alpha \geq 0.05$ was required to keep their model stable. As an aside, there are other methods for controlling the computational mode of the leapfrog scheme, which do not involve the application of a time filter. These methods are briefly outlined in Williams (2009).

Figure 13.12 assesses the effect of the Asselin filter on a short dynamical core simulation with the CAM Eulerian spectral transform model at the triangular truncation T85 and 26 levels (L26). The figure shows the surface pressure and 850 hPa temperature field at day 9 of the growing baroclinic wave test case of Jablonowski and Williamson (2006a) with and without strong Asselin time filtering. The quantitative comparison suggests that the simulation of the baroclinic wave without Asselin filtering develops slightly stronger high and low pressure systems by day 9. There might also be a very minor shift in the position of the wave as indicated by the difference plots (Fig. 13.12e,f). However, the difference fields are mostly dominated by amplitude errors provided the unfiltered simulation is considered more accurate and closer to the high-resolution reference solutions shown in Jablonowski and Williamson (2006a,b). The computational mode in the unfiltered simulation is not obvious at day 9, but grows unstable by day 14. The application of the Asselin filter is therefore paramount. Recall that the default in CAM EUL is generally set to $\alpha = 0.06$ which provides significantly weaker filtering.

As a word of caution, Déqué and Cariolle (1986) stated that despite the demonstrated ability of the Asselin filter to stabilize numerical solutions to the equations of motion for certain combinations of temporal differencing and physical forcings, even a very weak Asselin filter may have the potential to trigger an instability. Déqué and Cariolle (1986) suppressed this instability by a severe reduction of the time step in their model runs. Some unexpected anomalous behavior of the time filter was also highlighted by Robert and Lépine (1997).

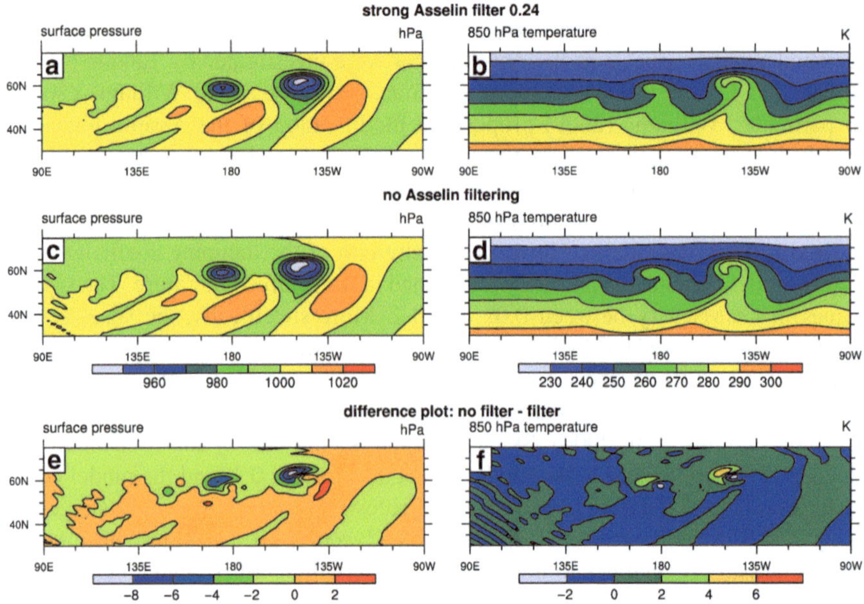

Fig. 13.12 Surface pressure (hPa) and 850 hPa temperature (K) at day 9 of the growing baroclinic wave test case of Jablonowski and Williamson (2006a) in the CAM T85L26 Eulerian spectral dynamical core with and without Asselin time filtering. (**a**,**b**) strong Asselin filter with $\alpha = 0.24$, (**c**,**d**): no Asselin filter, (**e**,**f**) difference between the unfiltered and filtered simulation. A time step of $\Delta t = 600$ s is used

13.5.1.2 Time Filter for Extrapolated Values

Semi-implicit semi-Lagrangian models need information about future parcel trajectories and therefore require information about the wind velocities at the future half-time level $t^{j+1/2}$ (Staniforth and Côté 1991). A popular method for estimating these time-centered wind speeds is time extrapolation. Two different time extrapolators are commonly used that either utilize two or three time levels. For the wind vector $\mathbf{v}$ the two-term extrapolator with times t^j and t^{j-1} yields

$$\mathbf{v}^{j+1/2} = \frac{3\mathbf{v}^j - \mathbf{v}^{j-1}}{2}. \tag{13.100}$$

which has been widely used in two-time level semi-implicit semi-Lagrangian schemes by e.g., Temperton and Staniforth (1987), McDonald and Haugen (1992, 1993) or ECMWF (2010). The three-term extrapolator is defined as

$$\mathbf{v}^{j+1/2} = \frac{15\mathbf{v}^j - 10\mathbf{v}^{j-1} + 3\mathbf{v}^{j-2}}{8} \tag{13.101}$$

which includes the additional time level t^{j-2}.

However, time extrapolation is a potential source of instability. For example, McDonald and Haugen (1992) showed that the noise introduced by the extrapolations in a σ-level model could be efficiently controlled by an *implicit divergence damper* that damps gravity waves. Here, the phrase *divergence damper* referred to an inherent numerical dissipation via a decentering scheme as explained later (Sect. 13.6.3). But the noise, that mainly originates from discretized nonlinear terms, was no longer controlled by decentering when switching to hybrid vertical coordinates (McDonald and Haugen 1993). Therefore, they introduced the following time filtered equations for both the extrapolated wind velocities and nonlinear terms

$$\psi^{j+1/2} = \frac{3\psi^j - \overline{\psi}^{j-1}}{2} \tag{13.102}$$

$$\psi^{j+1/2} = \frac{15\psi^j - 10\overline{\psi}^{j-1} + 3\overline{\psi}^{j-2}}{8}. \tag{13.103}$$

The time-filtered quantity is defined by

$$\overline{\psi}^j = \psi^j + \epsilon\left(\overline{\psi}^{j-1} - 2\psi^j + \psi^{j+1}\right) \tag{13.104}$$

where ψ serves as a placeholder variable. This time filter with the unitless and positive filter coefficient ϵ is formally the same as the Asselin filter (13.98). However, it is only selectively applied to nonlinear terms and the centering of the trajectory departure points in the semi-Lagrangian method. The linear terms are untouched which causes minimal decreases in accuracy according to McDonald and Haugen (1993). Their recommended ϵ values, in combination with a decentering scheme, ranged from $[0.05, 0.3]$.

13.5.2 Spatial Filters

Spatial filtering techniques have long been used for global atmospheric modeling. As mentioned above, spatial filtering suppresses linear and nonlinear instabilities, but conservation properties can get lost and might necessitate the use of a-posteriori restoration mechanisms (Takacs 1988). For example the conservation of mass gets lost if the mass variable needs to be filtered for numerical stability reasons. Similar difficulties emerge with respect to the conservation of total energy which is most often lost even without filtering. These aspects are assessed in Sect. 13.7. Here, we focus on the discussion of digital and spectral filters.

13.5.2.1 Digital Grid Point Filters

Digital filters are local grid point filters that only take neighboring grid points in the horizontal direction into account. They can be applied in one or two dimensions.

There are many digital filtering techniques that have been proposed in the literature. Examples are the filters by Shuman (1957), Shapiro (1970, 1971, 1975), Nelson and Weible (1980), Raymond and Garder (1988), Raymond (1988), Purser (1987) and Staniforth et al. (2006).

The most popular digital filter still used in some models today is the Shapiro filter which is based on constant-coefficient grid point operators of order n. The order determines the width of the numerical stencil. In general, the higher the order the higher the computational cost due to the wider stencil, which leads to more scale-selective filters. These minimize the damping of long and thereby flow-relevant waves.

In this section we focus on the one-dimensional Shapiro operator of order n to illustrate the general characteristics of digital filtering and show examples from shallow water simulations with a finite-volume model (Jablonowski 2004). The smoothing operation of the so-called *optimal* or *ideal* second-order ($n = 2$) Shapiro filter (Shapiro 1975) is defined by

$$\bar{\psi}_i = \frac{1}{16}\big(- \psi_{i-2} + 4\psi_{i-1} + 10\psi_i + 4\psi_{i+1} - \psi_{i+2}\big) \tag{13.105}$$

where the overbar symbolizes the smoothed variable ψ at grid point index i. The width of the stencil in one dimension is $2n + 1$. The coefficients for higher-order filters are listed in Shapiro (1975). As shown by Purser (1987) the Shapiro filter can also be generalized to describe a family of symmetric-stencil filters. This was first discussed in Hamming (1977) but has received very little attention in the atmospheric sciences.

The $n = 2$ Shapiro filter completely eliminates the unwanted $2\Delta x$ waves and significantly reduces the amplitudes of other poorly-resolved short waves, especially the $3, 4\Delta x$ waves that also tend to accumulate energy during model integrations. Each application of the Shapiro filter reduces the amplitude of a Fourier wave component $\exp(\imath k_x \Delta x)$ by the factor

$$\begin{aligned} \rho_n(k_x) &= 1 - sin^{2n}\left(\frac{k_x\,\Delta x}{2}\right) \\ &= 1 - sin^{2n}\left(\frac{\pi\,\Delta x}{L}\right) \end{aligned} \tag{13.106}$$

where $\imath = \sqrt{-1}$ is the imaginary unit, $k_x = 2\pi/L_x$ is the wavenumber in the x-direction and L_x is the wavelength, expressed as multiples of the grid spacing Δx. $\rho_n(k_x)$ is the response or damping function of the Shapiro filter (Shapiro 1971). In two dimensions the response function becomes

$$\rho_n(k_x, k_y) = \left[1 - sin^{2n}\left(\frac{k_x\,\Delta x}{2}\right)\right]\left[1 - sin^{2n}\left(\frac{k_y\,\Delta y}{2}\right)\right] \tag{13.107}$$

where k_y symbolizes the wavenumber in the y-direction with grid spacing Δy. This form expresses the response function for subsequent applications of the 1D filter operators.

Note that there is an inconsistency in the literature how to denote the order of the Shapiro filter. Here, we choose to follow the notation n as suggested in Shapiro (1971). Other modelers like Fox-Rabinovitz et al. (1997) or Ruge et al. (1995) use the notation $2n$ which corresponds to the order 4 in (13.105). The differences in the notations might have been motivated by Shapiro's observation that the one-dimensional ideal operator of order n is equivalent to the incorporation of a one-dimensional linear diffusion of order $2n$ with a coefficient $K = (-1)^{n-1}(\Delta x/2)^{2n}/\Delta t$. This draws an interesting analogy to the 1D linear diffusion mechanism. However, this result does not entirely generalize in two dimensions as discussed in Shapiro (1971). In 2D, the ideal n-th-order Shapiro operators only resemble the $2n$-th-order linear horizontal diffusion with the addition of a $4n$-th-order mixed damping term. This renders the Shapiro filter more scale selective than linear horizontal diffusion and fully eliminates $2\Delta x$ waves after each application.

The response of different 1D Shapiro operators (13.106) with respect to the wave spectrum is illustrated in Fig. 13.13. The figure shows the filter responses after one and 1,000 applications, and clearly depicts the cumulative character of the smoothing operation, especially for low-order filters. In practice, this is not a concern for overresolved waves close to the pole points on latitude–longitude grids, but must be taken into consideration in case filtering is to be applied at lower latitudes. For example, if filtering in midlatitudes or even in the tropics becomes necessary due to stability reasons, a low-order filter like the second-order Shapiro filter should be avoided and replaced by either a highly scale-selective FFT or higher order digital filters. A commonly used higher order filter is the eighth- or sixteenth-order Shapiro filter. An $n = 8$ Shapiro operator effectively eliminates all components

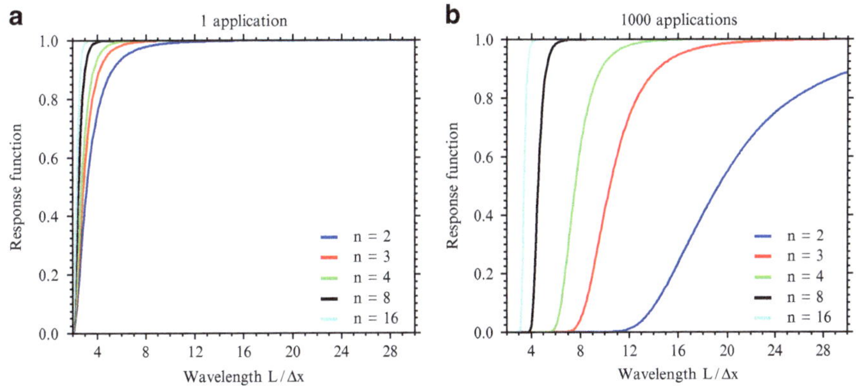

Fig. 13.13 Response function of different 1D Shapiro filters after (**a**) one application and (**b**) 1,000 applications. n indicates the order of the ideal Shapiro operator

with wavelengths less than four grid intervals but does not damp the waves with wavelengths greater than six grid intervals (Shapiro 1975). This filter has been successfully utilized by Ruge et al. (1995) who applied the $n = 8$ Shapiro filter every twelve time steps to eliminate the nonlinear computational instability in a shallow water model.

13.5.2.2 Examples: Applications of the Shapiro Filter

In the following shallow water examples, the Shapiro filter is solely applied in the longitudinal direction poleward of 60° N/S. The shallow water model is described in Jablonowski (2004) and applied at the constant resolution $2.5° \times 2.5°$. It utilizes the Lin and Rood (1997) finite-volume approach and is built upon a latitude–longitude grid with converging meridians near the pole points. The examples are chosen to briefly outline the advantages and disadvantages of the Shapiro filtering approach.

Filtering techniques must be carefully chosen. As mentioned before, strong low-order filters, like the second-, third- or fourth-order Shapiro filters, applied at low- or midlatitudes can significantly diffuse and degenerate the flow pattern and should therefore only be rarely used. An example that illustrates the detrimental effects of rather aggressive and unnecessary filtering is presented in Fig. 13.14. The example depicts a Rossby-Haurwitz wave which corresponds to test case 6 of the standard shallow water test suite proposed by Williamson et al. (1992). This flow field comprises a wavenumber four pattern that moves from west to east with only minor changes in shape. We assess two model runs that both apply an FFT filter to the horizontal wind components poleward of 75°N/S. Additionally, the strong second-order Shapiro filter is applied to the wind speeds at every time step in (a) the limited model domain 60°–75°N/S and (b) the whole remaining 75°S–75°N model area. Figure 13.14 shows the geopotential height field at model day 14. It can clearly be seen in Fig. 13.14b that the additional digital filtering in midlatitudes and the tropics leads to a very diffusive and inaccurate solution in comparison to the (a) less-filtered simulation which resembles reference solutions. The filter effects accumulate significantly during the course of the 3,360 time step simulation which confirms the cumulative effect in Fig. 13.13b. The errors are pure amplitude errors without changes in the phase speed of the wave.

The crucial need for a filtering mechanism in the polar 60°–75°N/S range in the finite-volume shallow water model is further depicted in Fig. 13.15 which displays a geostrophically balanced flow field in the Northern Hemisphere (test case 3 in Williamson et al. (1992)). The test results are shown for the geopotential height field after a 23-h simulation with and without the digital second-order Shapiro filtering technique. Again, an FFT filter is applied poleward of 75°N/S. Here, the chosen time step $\Delta t = 600$ s purposely violates the CFL condition for gravity waves in the polar region and as a consequence, a numerical instability equatorwards of 75°N develops in Fig. 13.15a. This position clearly marks the edge of the FFT filtering mechanism. In Fig. 13.15b the instability is removed by an $n = 2$ Shapiro filter. Of

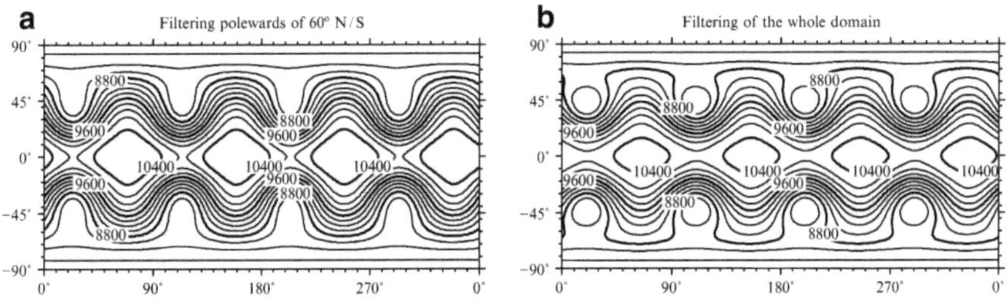

Fig. 13.14 Latitude–longitude plot of the geopotential height at day 14 for the Rossby–Haurwitz wave (test case 6) in the shallow water version of the FV model (Jablonowski 2004). (**a**) Second-order Shapiro filter is applied to the wind fields u and v between 60°–75°N/S. (**b**) Same filter is applied in the whole 75°S–75°N domain. Contour interval is 200 m. Resolution is 2.5° × 2.5° with $\Delta t = 360$ s

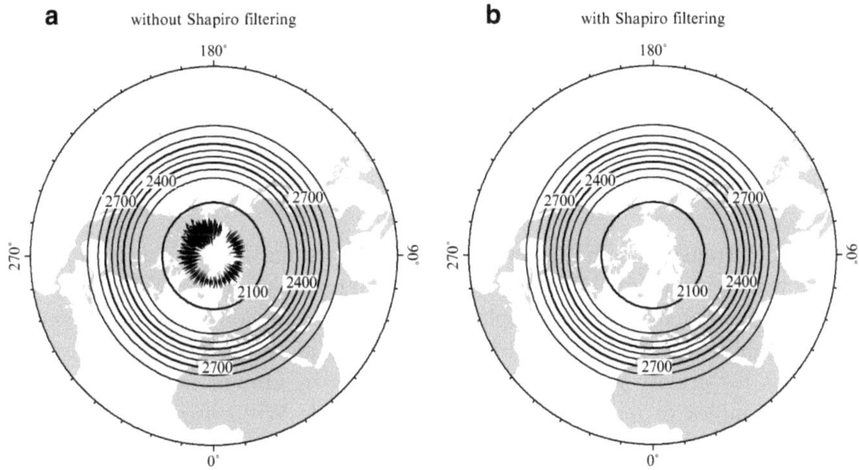

Fig. 13.15 Geopotential height field after 23 model hours (test case 3) in the shallow water version of the FV model (Jablonowski 2004). A north-polar stereographic projection is shown (*outer circle* is the equator). (**a**) No Shapiro filtering poleward of 60°, (**b**) Second order Shapiro filter is applied between 60°–75°. Contour interval is 100 m. Resolution is 2.5° × 2.5° with $\Delta t = 600$ s

course, the Fourier filter could have also been used to eliminate this linear instability at lower latitudes.

In summary, digital filtering promotes computational stability by eliminating or severely dampening the CFL unstable waves, especially at high latitudes. The filtering must be selectively applied in order to avoid a detrimental damping effect in the midlatitudes or tropical regions. In the finite-volume shallow water example, digital filtering techniques complement the even more effective FFT filtering technique used near the pole points. However, attempts to entirely replace the FFT filter with a digital Shapiro (1975) or Purser (1987) algorithm did not prevent numerical instabilities close to the poles. As an aside, an FFT-Shapiro filter mix was also promoted by Kalnay-Rivas et al. (1977), Ruge et al. (1995), Fox-Rabinovitz et al. (1997) and Takacs et al. (1999).

The main motivation behind such a mix is that local digital filtering has computationally advantages on parallel computing architectures. Shapiro filtering only requires the use of $2n + 1$ neighboring points whereas Fourier filtering incorporates all grid cells along a latitude circle. The latter becomes problematic on parallel machines that distribute the entire GCM domain over many processors. It necessitates parallel communication and triggers computational overhead. A comprehensive discussion of such computational aspects in GCMs is also provided in Chap. 16.

Many choices are still left open. Digital filters can be applied at each time step, sporadically or only with a fraction of their full strength. The latter was suggested by Fox-Rabinovitz et al. (1997). In addition, decisions need to be made about the filtered variables. For example, Fox-Rabinovitz et al. (1997) do not apply their 2D high-order Shapiro filter to the surface pressure field so that the conservation of mass is not affected. Instead, the time tendencies of all other forecast variables are filtered.

13.5.2.3 Spectral Filters: Fourier Filtering

Fast Fourier Transform (FFT) filters are spectral filters that are popular in grid point models with latitude–longitude grids, especially if explicit time-stepping is used. In general, a Fourier filter is only applied in the zonal direction to promote computational stability at mid- and high latitudes, and to allow a violation of the CFL conditions for gravity waves in the filtered region by eliminating short unstable waves. Examples of models with FFT filters include Williamson and Browning (1973), Williamson (1976), Purser (1987, 1988), Fox-Rabinovitz et al. (1997) and Lin (2004). In particular, Purser (1988) examined different filtering strategies and highlighted their advantageous and detrimental effects. Takacs and Balgovind (1983) compared the spectral filtering of tendencies to the spectral filtering of the prognostic variables and assessed the side effects of polar filtering. This fuels the discussion of conservation properties in Sect. 13.7. Generally, Fourier filtering is solely applied along coordinate surfaces without adjustments to constant pressure or height levels as was discussed in Sect. 13.3.2 for the divergence and vorticity diffusion.

The polar FFT filtering of a variable ψ is accomplished by first applying a 1D Fourier (forward) transformation along an entire latitude circle with constant longitudinal grid spacing $\Delta\lambda = 2\pi/n_x$. Here, n_x symbolizes the total number of grid points in the zonal direction. The Fourier coefficients for wavenumbers that exceed a prescribed threshold are then modified which corresponds to a damping mechanism in spectral space. The filtered coefficients are finally transformed back to physical space which completes the filter step. This filter application can be written for all dimensionless wavenumbers k as

$$\hat{\psi}(k)_{filtered} = f(k)\,\hat{\psi}(k) \tag{13.108}$$

where $\hat{\psi}(k)$ and $\hat{\psi}(k)_{filtered}$ are the Fourier coefficients before and after the filtering step. The filter coefficients $f(k)$ are generally defined as a function of k and the latitudinal position ϕ. Most commonly, they are defined as

$$f(k) = \min\left[1., \left(\frac{\cos\phi}{\cos\phi_c}\right)^{2q} \frac{1}{\sin^2(k\,\Delta\lambda/2)}\right] \tag{13.109}$$

with the cutoff latitude ϕ_c. The polar filter is then only active polewards of ϕ_c where $|\phi| > \phi_c$. Its strength gradually increases toward the pole by increasing the number of affected wavenumbers and decreasing the response function $f(k)$ by which they are damped. The positive integer coefficient q can be used to adjust the filter strength. Models like WRF (Skamarock et al. 2008), CAM FV (Collins et al. 2004) or the stretched-grid finite-difference model by Fox-Rabinovitz et al. (1997) set the default to $q = 1$. Older versions of NASA's GEOS model (version 2, Suarez and Takacs (1995)) and its stretched-grid variant (Takacs et al. 1999) utilized a weaker Fourier filter with the coefficients

$$f(k) = \min\left[1., \left(\frac{\cos\phi}{\cos\phi_c}\right) \frac{1}{\sin(k\,\Delta\lambda/2)}\right]. \tag{13.110}$$

For an even number of grid points n_x the mesh supports the dimensionless wavenumbers $k = 1, 2, 3, \cdots, n_x/2$ that can travel in both the west or east direction. Note that the last entry $k = n_x/2$ corresponds to a $2\Delta\lambda$ wave which is the shortest resolvable wavelength. The $2\Delta\lambda$ mode is stationary and its phase cannot be resolved. For odd n_x, the unitless wavenumber range is $k = 1, 2, 3, \cdots, (n_x - 1)/2$.

The cutoff latitude is model-dependent. In CAM FV and NASA's GEOS5 model with the FV dynamical core on a latitude–longitude grid the cutoff ϕ_c is determined by

$$\phi_c = \arccos\Big[\min\big(0.81, \Delta\phi/\Delta\lambda\big)\Big]. \tag{13.111}$$

For equal grid spacings in both directions this cutoff lies around $\phi_c \sim 36°$ which is the minimum latitude. For default CAM FV grid spacings with $\Delta\phi/\Delta\lambda \approx 0.754$ though, the cutoff is placed near $\phi_c \sim 41°$. Fox-Rabinovitz et al. (1997) set their cutoff latitude to $\phi_c = 45°$. In general, the cutoff is most often placed in the midlatitudes which leaves the tropical region unfiltered. The midlatitudes are often empirically chosen since the grid spacing does not change much equatorwards. An example of the response function $f(x)$ for the cutoff latitude $\phi_c = 40°$ is presented in Fig. 13.16. The figure shows the wavenumber dependency of the filter for both the strong ((13.109) with $q = 1$) and weaker (13.110) Fourier filtering at the latitudes $\phi = 45°, 60°$ and $85°$. The strong filter response is depicted by the solid lines. The figure confirms that the filter is scale-selective and primarily damps the higher wavenumbers (shorter scales) which depend on the latitudinal position. The filtering

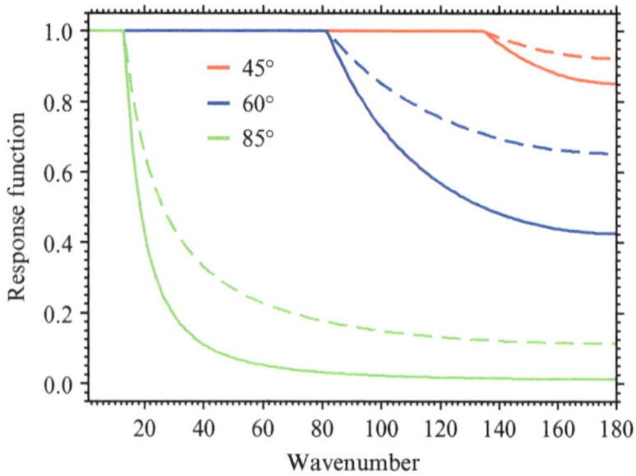

Fig. 13.16 Response function $f(x)$ for the Fourier filter at the latitudes $\phi = 45°, 60°$ and $85°$. The *solid line* denotes the strong filter (13.109), the *dashed line* shows the weaker filter (13.110). The cutoff latitude is $\phi_c = 40°$

gets stronger at higher latitudes. Close to the poles almost all wavenumbers are heavily damped.

The Fourier filtering can be made mass-conservative as discussed in Skamarock et al. (2008), but it is neither monotonic nor positive definite. The latter becomes especially important if tracer variables are filtered since the filter can create negative tracer mass.

13.5.2.4 Local Spectral Filters

Local spectral filters are popular in GCMs that utilize local spectral methods like the spectral element method or the discontinous Galerkin (DG) approach. Despite their high accuracy at high orders the spectral element or DG methods are susceptible to nonlinear aliasing errors which introduce nonphysical high-frequency oscillations. A local spectral filter is then often employed as an alternative to hyper-diffusion to prevent this noise from contaminating the solution.

The most common local spectral filter is the Boyd-Vandeven filter which is a variant of the Vandeven (1991) filter that was developed by Boyd (1996, 1998). The Boyd-Vandeven filter of order p is given by

$$\sigma(x) = \frac{1}{2}\mathrm{erfc}\left(2\sqrt{p}\Omega\sqrt{-\frac{\log(1-4\Omega^2)}{4\Omega^2}}\right) \qquad \text{with } \Omega = |x| - \frac{1}{2} \tag{13.112}$$

where $\mathrm{erfc}(x) = 1 - \mathrm{erf}(x)$ symbolizes the complementary error function and $\mathrm{erf}(x)$ is the error function. Taylor et al. (1997) explain how this filter is applied to a 1D function $f(x)$ in the x-direction that can be written as the sum of the first N Legendre polynomials $P_k(x)$ with coefficients f_k

$$f(x) = \sum_{k=0}^{N-1} f_k P_k(x). \tag{13.113}$$

The filtered function $f'(x)$ is then given by

$$f'(x) = (1 - \mu) f(x) + \mu \sum_{k=0}^{N-1} w_k(x) f_k P_k(x) \qquad (13.114)$$

where the weights w_k are defined as

$$w_k(x) = \begin{cases} 1 & \text{if } k < s \\ \sigma(x)\left(\frac{k-s}{N-s}\right) & \text{if } s \leq k \leq N. \end{cases} \qquad (13.115)$$

The parameter μ represents the filter viscosity which can range from 0 (no filtering) to 1 (full filtering). The coefficient s specifies the filter lag, e.g., setting $s = 2N/3$ determines that the filter is only applied to the last one third of the spectrum of f. Taylor et al. (1997) showed that this filter is very scale-selective, especially at high order. It can be applied without sacrificing the spectral accuracy of the spectral element or DG scheme.

The Boyd-Vandeven filter has for example been used by Taylor et al. (1997), Giraldo et al. (2002), Giraldo and Rosmond (2004), Thomas and Loft (2005) and St-Cyr et al. (2008) for idealized dynamical core assessments. These references also discuss the specifics of the 2D filter implementation in the particular model. A common choice is to apply a weak twelfth-order ($p = 12$) filter every few time steps with the parameters $\mu = 0.2$ and $s = 2N/3$. These parameters are generally chosen through experimentation. Note that Thomas and Loft (2005) observed that a much larger $\mu = 0.4$ was often required for stable integrations when switching from a pure vertical σ-coordinate to hybrid η-coordinates. They also needed to apply the filter at every time step.

In practice though, the filter might not be strong enough for full GCMs simulations with parameterized physics (Mark Taylor, personal communication). For example, it has been found in the spectral element model HOMME that spectral filtering alone did not prevent the so-called grid imprinting of a cubed-sphere computational mesh in the numerical solution. The cubed-sphere structure of the mesh was mirrored e.g., in the precipitation field. Such artificial effects were avoided when switching to a fourth-order hyper-diffusion in HOMME.

13.6 Inherent Numerical Damping

Inherent numerical dissipation comes in many forms and is a source of nonlinear flow-dependent damping in model simulations. For example, it is embedded in semi-Lagrangian advection schemes due to the necessary interpolations at every time step. In addition, dissipation is inherent in finite volume methods that are upwind-biased or utilize monotonicity constraints to avoid unphyscial over- and undershoots in the solution. The inherent damping is not necessarily a weakness of the numerical scheme. It can be turned into a useful property as e.g., demonstrated by Váňa et al. (2008) who used the damping abilities of interpolations in a semi-Lagrangian scheme as targeted diffusive filtering. They named their technique "semi-Lagrangian horizontal diffusion" which brought beneficial new skills to their forecast model. Skamarock and Klemp (2008) pointed out that their upwind-biased advection schemes in the model WRF provides significant filtering of the small scales. They estimated that the effective hyperviscosity coefficient is proportional to the Courant number, and thereby most active at higher Courant numbers where phase errors are most likely to produce noise. Skamarock and Klemp (2008) also found that the horizontal mixing provided by the fifth-order upwind-biased advection scheme in WRF is sufficient to control small-scale noise in weather prediction applications for grid spacings larger than 10 km.

The topic of inherent numerical dissipation is rather broad and cannot be exhaustively covered in this chapter. Therefore, we only present selected aspects to highlight the principal design considerations and characteristics of this type of nonlinear dissipation. In particular, we discuss the inherent dissipation that is embedded in the nominal order of a finite volume scheme, assess the diffusive properties of monotonicity constraints, briefly review the use of the decentering technique as e.g., used in semi-implicit semi-Lagrangian schemes, and shed light on the damping characteristics of semi-Lagrangian methods.

13.6.1 Order of the Numerical Scheme

As a specific example of the inherent dissipation in a finite volume scheme we discuss the properties of a first-, second- and third-order approximation in the model FV. The advection method implemented in the FV dynamical core can be viewed as a multi-dimensional extension of higher-order Godunov-type schemes like the van Leer scheme (van Leer 1974, 1977) or the Piecewise Parabolic Method (PPM, Colella and Woodward (1984)). Finite volume schemes are based on the reconstruct-evolve-average approach as e.g., explained by LeVeque (2002) and in Chap. 8, and use constant, piecewise linear (van Leer), piecewise parabolic (PPM) or even piecewise cubic subgrid distributions for the piecewise continuous reconstruction of the flow field. The transport problem is then solved exactly and new initial data at the future time step are obtained by averaging the transported quantity over each control volume.

The first-order upwind method is based on a constant subgrid distribution and is thereby very diffusive by design. It is for example explained by Lin and Rood (1996) who introduced the FV advection scheme. The second-order van Leer advection scheme (van Leer 1977) is based on the reconstruction of linear subgrid distributions in each finite volume cell. We now briefly review the design of such subgrid distributions to motivate the subsequent discussion. The linear subgrid distribution $h(x)$ of a model variable h is given by

$$h(x, y) = \bar{h} + \Delta a^x\, x + \Delta a^y\, y \tag{13.116}$$

where $\bar{h} = \int_{-1/2}^{1/2} \int_{-1/2}^{1/2} h(x, y)\, dx\, dy$ is the volume-averaged value with normalized local coordinates $x, y \in [\frac{-1}{2}, \frac{1}{2}]$. Δa^x and Δa^y denote the slopes in the x and y direction at a grid point (i, j) which in the model FV are defined via van Leer's scheme I

$$\Delta a^x = \frac{1}{2}\left(h_{i+1,j} - h_{i-1,j}\right) \tag{13.117}$$

$$\Delta a^y = \frac{1}{2}\left(h_{i,j+1} - h_{i,j-1}\right). \tag{13.118}$$

This assessment uses centered finite differences. The slopes can be further manipulated if monotonicity constraints are required. Then the monotonized central-difference (MC) slope limiter (van Leer 1977) can e.g., be used

$$\begin{aligned} \Delta a^x &= min\left(|\Delta a^x|, 2|h_{i+1,j} - h_{i,j}|, 2|h_{i,j} - h_{i-1,j}|\right)\, sgn(\Delta a^x) \\ &\qquad \text{if } (h_{i+1,j} - h_{i,j})(h_{i,j} - h_{i-1,j}) > 0 \\ &= 0 \quad \text{otherwise} \end{aligned} \tag{13.119}$$

that picks out the smallest magnitude among three slopes which are the centered difference and the two one-sided differences. However, if the discrete value of $h_{i,j}$ represents an extreme value, the slope is set to zero. The same principle applies to the MC slope limiter in the y direction. The $sgn(\Delta a^x)$ function extracts the sign of the argument.

Alternatively, the third-order PPM scheme can be applied. The corresponding 2D biparabolic subgrid distribution is then given by

$$h(x, y) = \bar{h} + \delta a^x\, x + b^x \left(\frac{1}{12} - x^2\right) + \delta a^y\, y + b^y \left(\frac{1}{12} - y^2\right) \tag{13.120}$$

where the coefficients of the parabola $\delta a^x, b^x, \delta a^y$ and b^y are defined by Colella and Woodward (1984) or Carpenter et al. (1990). The coefficients can again be modified in order to enforce monotonicity constraints as explained in Lin and Rood (1996) and Lin (2004). Both the order and the choice of the monotonicity constraint (see Sect. 13.6.2) determine the inherent diffusion in the FV advection scheme.

Figure 13.17 visualizes the effects of the inherent diffusion in the dynamical core CAM FV with the grid spacing $1^\circ \times 1^\circ$ and 26 vertical levels. CAM FV has runtime options to run with the aforementioned first-order upwind advection scheme, the second-order van Leer algorithm or with the nominally third-order PPM method. Note however, that the 2D implementation of the PPM algorithm in the model FV does not exhibit a third-order convergence and is in practice closer to a second-order scheme (Lin and Rood 1996; Jablonowski et al. 2006). Both the van Leer scheme and PPM apply monotonicity constraints which are the MC limiter in case of van Leer, and the "relaxed" monotonicity constraint in case of PPM (Lin 2004).

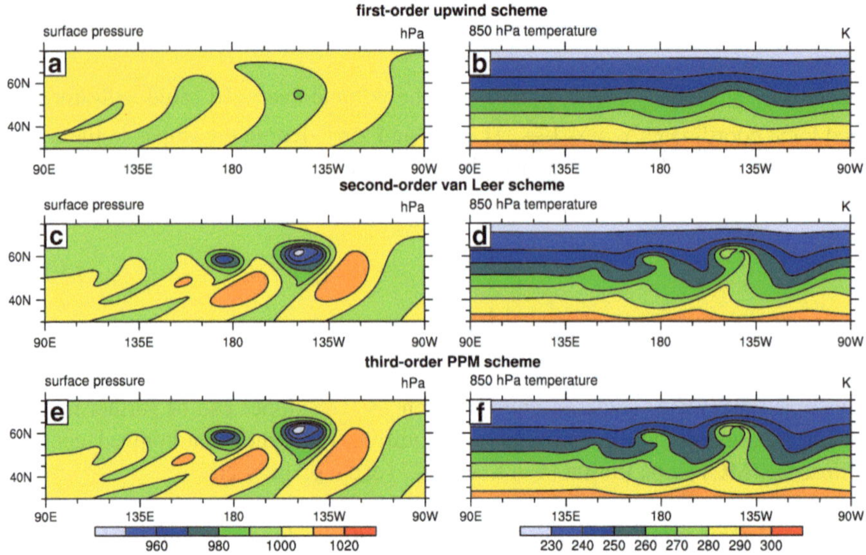

Fig. 13.17 Surface pressure (hPa) and 850 hPa temperature (K) at day 9 of the growing baroclinic wave test case of Jablonowski and Williamson (2006a) in the CAM FV dynamical core at the resolution $1^\circ \times 1^\circ$ L26. (**a**,**b**) first-order upwind scheme, (**c**,**d**) second-order van Leer scheme, (**e**,**f**) default third-order PPM scheme with the relaxed monotonicity constraint (Lin 2004). The dynamics time step is $\Delta t = 180$ s

The figure shows the surface pressure and 850 hPa temperature fields at day 9 of the growing baroclinic wave described in Jablonowski and Williamson (2006a). The impact of the inherent numerical dissipation can clearly be seen in all fields. The first-order method (Fig. 13.17a,b) hardly captures the evolving baroclinic instability and only shows hints of a wave. The second-order van Leer scheme shows a clear evolution of the baroclinic wave and exhibits a slightly early wave breaking event in the temperature field (Fig. 13.17d). The peak amplitudes of the surface pressure field at day 9 are $(p_s)_{min} = 948.19$ hPa and $(p_s)_{max} = 1018.78$ hPa with the second-order van Leer (Fig. 13.17c) algorithm. The evolution of the wave simulated with the PPM scheme is very similar to the van Leer simulation. However, the peak surface pressure amplitudes are slightly intensified and the values read $(p_s)_{min} = 947.04$ hPa and $(p_s)_{max} = 1018.74$ hPa in Fig. 13.17e. In addition, the temperature field in the PPM simulation (Fig. 13.17f) shows slightly sharper frontal zones without wave breaking.

A more quantitative comparison of the baroclinic wave simulation is provided in Fig. 13.18. The figure depicts the 700 hPa kinetic energy spectra at day 30 of the baroclinic wave simulation for the first-order upwind method, the second-order van Leer scheme and two PPM simulations. The only difference between the PPM simulations is the selection of the monotonicity constraint. Here, we compare the "relaxed" constraint by Lin (2004) and "positive definite" constraint described in Lin and Rood (1996). The latter only prevents negative undershoots and has

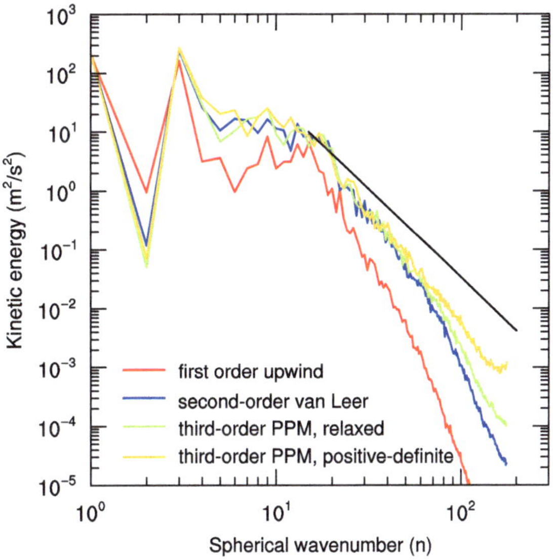

Fig. 13.18 700 hPa kinetic energy spectra at day 30 of the growing baroclinic wave test case of Jablonowski and Williamson (2006a) in the CAM FV dynamical core at the resolution $1^\circ \times 1^\circ$ L26: first-order upwind, second-order van Leer, third-order PPM scheme with the *relaxed* and *positive definite* monotonicity constraints (Lin 2004). The *black line* indicates an n^{-3} slope. The dynamics time step is $\Delta t = 180$ s

originally been designed for pure tracer advection. Therefore, its application in the dynamical core is generally not recommended, but we only use it here to demonstrate different diffusion properties. The curves confirm that the first-order method is the most diffusive as indicated by the sharp drop off of the spectrum and the significant damping of the longer wavenumbers 4–10. The differences between the van Leer and PPM simulations are more subtle. The van Leer and relaxed PPM curve almost overlay each other until about wavenumber 60 or so before the van Leer curve exhibits a slightly faster drop off than the relaxed PPM run. In contrast, the positive definite PPM simulation is less diffusive than the relaxed PPM scheme and almost runs parallel to the n^{-3} slope. More analysis on the monotonicity constraints in provided in the next Sect. 13.6.2.

Increased inherent dissipation can also be used as a sponge at the model top. This is for example accomplished by lowering of the order of the numerical scheme in the uppermost levels. Such a technique is used in the CAM FV dynamical core that switches the numerical scheme from the PPM algorithm to the more diffusive van-Leer scheme in the uppermost $nlev/8$ layers where $nlev$ is the number of total levels.

Inherently dissipative schemes can often be run without explicit diffusion which is for example true for the semi-Lagrangian models CAM SLD and UM. However, additional explicit diffusion might still be applied in long climate simulations. Inherent numerical dissipation can also be viewed as an application of a symmetric low-pass sine filter as suggested by Raymond and Garder (1991). These similarities between inherent dissipation in finite-difference models and numerical filters were also pointed out in Purser and Leslie (1994).

13.6.2 Monotonicity Constraints and Shape Preservation

The advection algorithm in the CAM Finite Volume model is upstream-biased and monotonic if limiters are applied to the subgrid distributions. As mentioned in the previous subsection such monotonicity constraints are used in case of the default PPM scheme which leads to a total variation diminishing (TVD) method. A short review of limiters for finite volume schemes is given in Chap. 8 or in the textbook by Durran (2010). Note that Chap. 9 introduces limiters for discontinuous Galerkin methods which are an active research topic (see also Nair (2009)). In general, limiters can be grouped into slope/curvature limiters or flux limiters. Here we only briefly assess the impact of slope/curvature limiters that a-priori limit the bi-parabolic subgrid distribution used in the PPM scheme.

The limiting of the subgrid distribution clips extreme values and thereby introduces inherent nonlinear dissipation into the finite volume scheme. From a design perspective the clipping suppresses over- and undershoots in the advection step. It should be as strict as necessary to prevent unphysical oscillations but as nonintrusive as possible to minimize the associated dissipation. Figure 13.19 schematically illustrates how the clipping by a monotonicity constraint affects a flow field (Jablonowski 2004). In this particular example, an interpolation of a zonal wind field from a $2.5° \times 2.5°$ to a $1.25° \times 1.25°$ latitude–longitude grid is performed. The analytically prescribed zonal wind field at the $2.5° \times 2.5°$ resolution is depicted in

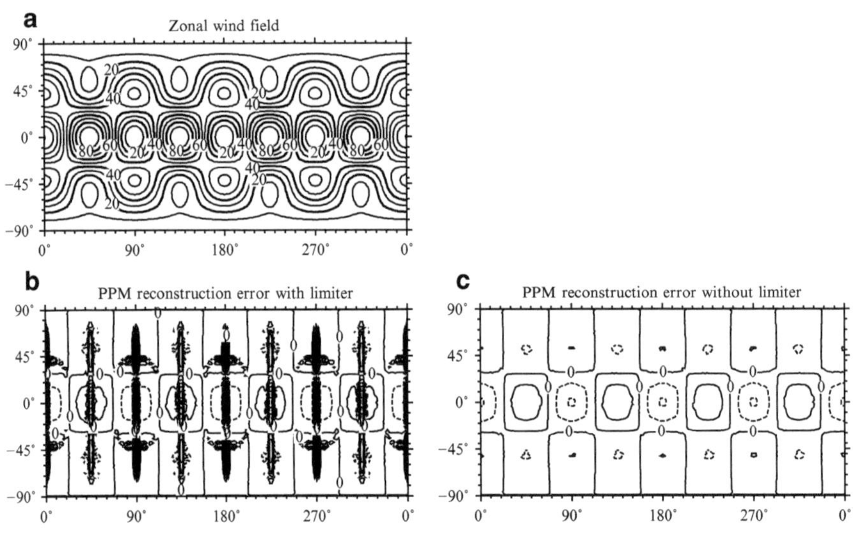

Fig. 13.19 Latitude–longitude plot of the zonal wind field in the Rossby–Haurwitz wave test case. (**a**) Zonal wind field at a $2.5° \times 2.5°$ resolution, (**b**) absolute error of the zonal wind field after a PPM-based interpolation with limiters to a $1.25° \times 1.25°$ grid, (**c**) absolute error of the zonal wind field after a PPM-based interpolation without limiters to a $1.25° \times 1.25°$ grid. The contour intervals are (**a**) $10\,m\,s^{-1}$ and (**b**,**c**) $0.025\,m\,s^{-1}$. Negative contours are *dashed*

Fig. 13.19a. The field corresponds to the Rossby–Haurwitz wave with wavenumber 4 that is part of the standard shallow-water test suite by Williamson et al. (1992). The interpolation uses the bi-parabolic subgrid-scale reconstruction defined by the PPM algorithm (Colella and Woodward 1984) which are then integrated over the new $1.25^\circ \times 1.25^\circ$ domain.

Figure 13.19b,c assess the absolute errors of these interpolations with and without monotonicity constraints. In particular, the original PPM monotonicity constraint without steepening (Colella and Woodward 1984) was selected. The distribution of the errors in Fig. 13.19b confirms that the monotonicity constraint mostly affects the areas with extreme values. This can be seen when comparing the error pattern to the original zonal wind field (Fig. 13.19a). In contrast, the interpolation errors are clearly diminished without the limiter (Fig. 13.19c). In fact, small-magnitude over- and undershoots are present in Fig. 13.19c with peak values around ± 0.05 m s^{-1}. The overshoots appear in the regions of the wind maxima, the undershoots are concentrated near the wind minima. These are eliminated by the monotonicity constraint (Fig. 13.19) that, on the downside, increases the overall errors to ± 0.22 m s^{-1}. Note that these errors are assessed with the help of the analytic solution at the $1.25^\circ \times 1.25^\circ$ resolution.

Limiting can enforce a strictly monotonic advection algorithm in the 1D case as also discussed by Thuburn (1993, 1997) and Mesinger and Jovic (2002). However, very minor violations of the monotonicity constraint in two-dimensional flows are possible and have indeed been observed for the horizontal FV advection algorithm by Lin and Rood (1996). Limiters can also be designed to only avoid negative undershoots for tracer transport application, and allow an overestimation of the transported quantity. The limiters in finite volume schemes or shape-preservation constraints in semi-Lagrangian models are physically motivated and provide a smoothness constraint. They are a form of nonlinear inherent numerical dissipation that is guided by the flow field. Shape preservation constraints for semi-Lagrangian dynamical cores and advection schemes are for example discussed in Williamson and Rasch (1989), Williamson (1990), Rasch and Williamson (1990b) and Bermejo and Staniforth (1992).

A detailed documentation of the many limiter options in the model FV is beyond the scope of this section and we refer to the associated literature for the exact explanations of the algorithms (Colella and Woodward 1984; Carpenter et al. 1990; Lin and Rood 1996; Huynh 1996; Lin 2004; Putman and Lin 2007). The main focus here is to qualitatively demonstrate that limiters determine the amount of inherent dissipation in finite volume scheme and need to be used with care. Figure 13.20 visualizes the influence of different PPM monotonicity constraints on the evolution of the baroclinic wave in the FV dynamical core. Here, two models are depicted which both utilize almost identical versions of the FV dynamics described in Lin (2004). They are the CAM 5 FV implementation with the grid spacing $1^\circ \times 1^\circ$ and the GFDL FV model with the *lat* $\times$ *lon* grid spacing $1^\circ \times 1.25^\circ$. The figure can also be directly compared to the surface pressure fields in Fig. 13.17.

The figure demonstrates the effects of the positive-definite (Fig. 13.20a) and quasi- (or semi-) monotone constraints (Fig. 13.20b) described in Lin and Rood

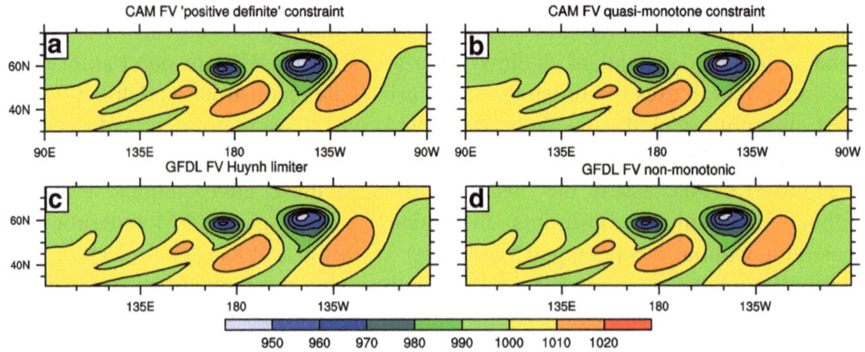

Fig. 13.20 Surface pressure (hPa) at day 9 of the growing baroclinic wave test case of Jablonowski and Williamson (2006a) in the (**a**,**b**) CAM FV dynamical core at the resolution 1° × 1° L26 and (**c**,**d**) GFDL FV model at the resolution M90 1° × 1.25° with 26 levels. (**a**) positive definite constraint that only eliminates negative undershoots, (**b**) quasi-monotone constraint (Lin and Rood 1996) that is the default in CAM FV, (**c**) second constraint discussed in Huynh (1996), (**d**) non-monotonic simulation with slope- but no curvature-limiter in the PPM scheme

(1996), the second monotonicity constraint by Huynh (1996) (Fig. 13.20c) and a non-monotonic simulation (Fig. 13.20d) that only utilized the MC slope limiter but no curvature constraint. Option (b) is the default in CAM FV. Overall, the strengths of the pressure systems in all four simulations are comparable and the differences are subtle. The quasi-monotone simulation appears to be slightly more diffusive than the other three model runs as indicated by the fewer contour lines in the low pressure system located at about 175°E. The "positive definite" simulation, that only prevents negative undershoots, is the least diffusive and has developed the deepest low pressure systems. This has also been demonstrated by the kinetic energy spectra in Fig. 13.18. However as mentioned before, this monotonicity option has been specifically designed for positive definite tracer transports and should not be used as the basis for the dynamical core. Other even less stringent limiters are available for PPM-type algorithms as e.g., documented by Colella and Sekora (2008) and McCorquodale and Colella (2010).

The choice of the limiter should be motivated by the design criteria of the model as argued in Chap. 15. Note that the cumulative damping effect of the limiters cannot be quantified analytically. Therefore, the model CAM 5 FV applies a total energy fixer that provides dissipative heating (Neale et al. 2010).

13.6.3 Decentering Mechanisms

Decentering adds inherent dissipation to the numerical scheme and is tightly linked to semi-implicit time discretizations. Here we briefly review the decentering used in semi-implit semi-Lagrangian models and in other semi-implicit approaches.

13.6.3.1 Decentering in Semi-implicit Semi-Lagrangian Models

The decentering mechanism, sometimes also called uncentering or off-centering technique, is usually applied in semi-implicit semi-Lagrangian (SISL) models. Its primary purpose is to suppress computational noise and orographic resonance in regions of steep orography and high Courant numbers, and maintain stability, especially at high resolutions (Bates et al. 1993; Rivest et al. 1994). A comprehensive discussion of the orographic resonance problem in semi-Lagrangian models is provided in Rivest et al. (1994), Côté et al. (1995) and Lindberg and Alexeev (2000), and is not repeated here. Tanguay et al. (1992) suggested a first-order decentering of the semi-implicit terms along the trajectory. Rivest et al. (1994) discussed both first and second-order decentering schemes in a 1D SISL shallow water model. A thorough stability analysis of the decentering method is presented in Tanguay et al. (1992) and Gravel et al. (1993).

To illustrate the basic idea behind the first-order decentering technique consider the prognostic equation

$$\frac{D\psi}{Dt} = S \tag{13.121}$$

where D/Dt is the total time derivative, ψ is a scalar variable and S is a source term which may incorporate ψ. In a two-time-level semi-Lagrangian scheme a conventional discretization of the trajectory calculation leads to

$$\psi^{j+1} - \psi_d^j = \int_{t^j}^{t^{j+\Delta t}} S\,dt = \bar{S}\,\Delta t \tag{13.122}$$

where ψ^{j+1} is the value of the prognostic variable at the arrival point at time t^{j+1} and ψ_d^j is the value at the departure point of the trajectory at time t^j. $\bar{S}$ denotes the time-averaged source term along the trajectory that can be replaced by

$$\psi^{j+1} - \psi_d^j = \Delta t\left(\frac{1+\epsilon}{2}\,S^{j+1} + \frac{1-\epsilon}{2}\,S_d^j\right) \tag{13.123}$$

with the decentering (time-weighting) parameter ϵ. The averaged source term represents both a temporal and spatial average. A centered two-time-level scheme with $\epsilon = 0$ is second-order accurate. For a decentered scheme with $0 < \epsilon < 1$ the truncation error is first-order. A decentered SISL scheme is generally more accurate and less damping the closer ϵ is to 0, and less accurate and more damping the closer ϵ is to unity. In practice though, some decentering is desirable or even necessary in SISL schemes to suppress the spurious orographic resonance.

Decentering is for example used in the operational Global Environmental Multiscale (GEM) model developed at the Canadian Meteorological Centre (Côté et al. 1998a,b), in the spectral transform model CAM SLD (Collins et al. 2004) and the grid point model UM (Staniforth et al. 2006). Typical decentering parameters in GCMs are $\epsilon = 0.1$ in the model GEM, $\epsilon = 0.2$ in CAM SLD and $\epsilon \in [0.2, 0.4]$ in the model UM (Davies et al. 2005). Davies et al. (2005) reported that the smaller

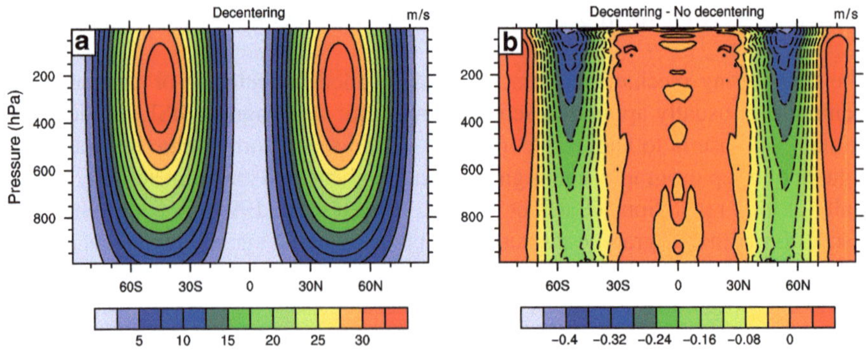

Fig. 13.21 (**a**) Zonal-mean zonal wind (m/s) at day 30 of the steady-state test case of Jablonowski and Williamson (2006a) in the CAM SLD dynamical core at the resolution T85L26. The default decentering parameter $\epsilon = 0.2$ is used. (**b**) Difference of the zonal-mean zonal wind at day 30 between the run with decentering and a run without decentering ($\epsilon = 0$). No explicit diffusion is used, the time step is $\Delta t = 1{,}800$ s

value $\epsilon = 0.2$ is sufficient at low resolutions, but it needed to be replaced with $\epsilon = 0.4$ at high weather forecast resolutions to suppress noise near strong jets. A discussion of the impact of decentering and its stability properties can also be found in Chap. 14.

As a practical example, we now isolate the effect of the inherent dissipation from the decentering mechanism in idealized dynamical core simulations. As before in Sect. 13.3.3, we choose the CAM 4 semi-Lagrangian dynamical core at the triangular truncation T85 with 26 levels. A steady-state test case, described in Jablonowski and Williamson (2006a), is used and run for 30 days with varying decentering parameter ϵ. Figure 13.21 shows the zonal-mean zonal wind field at day 30 with the default decentering parameter $\epsilon = 0.2$ and the zonal-mean zonal wind difference between the run with decentering and no decentering. No explicitly added diffusion was used. The influence of the decentering can clearly be seen in the difference plot (Fig. 13.21b) throughout the entire atmosphere but the impact is strongest in the midlatitudes in this test case, especially near the model top. The decentering damps the zonal wind speed with magnitudes of up to 0.5 m s^{-1} during this 30-day simulation. Note that Fig. 13.21 can also be readily compared to Fig. 13.1 that isolates the effects of the fourth-order hyper-diffusion and second-order sponge layer diffusion (without decentering) with the help of the same test case.

A quantitative comparison of the damping due to decentering is depicted in Fig. 13.22 that shows the time evolution of the global root-mean square l_2 zonal wind error during the 30-day steady-state simulation. For this analysis the zonal-mean zonal wind field $\overline{u}$ is compared to the analytic solution at time $t = 0$ (see Jablonowski and Williamson (2006a) for the definition of the error norm). The decentering parameter ϵ is set to 0, 0.1, 0.2 and 0.3, respectively. Again, no explicitly added diffusion was used and, as briefly discussed before for Fig. 13.1, the semi-Lagrangian trajectory calculation utilized only spherical coordinates to suppress

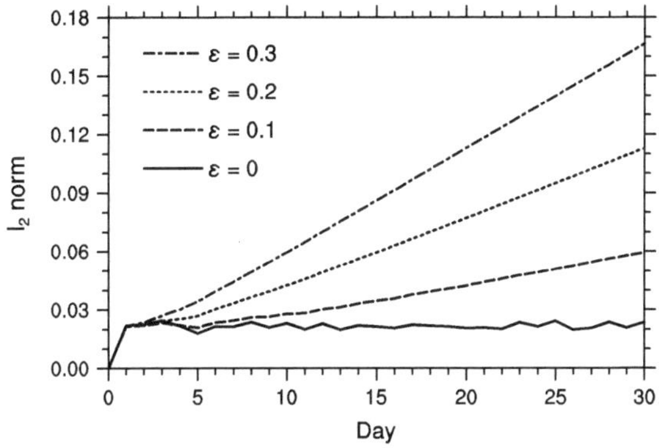

Fig. 13.22 Time evolution of the $l_2(\overline{u}(t) - \overline{u}(t=0))$ error norms (in m/s) of the zonal-mean zonal wind field $\overline{u}$ in the steady-state test case of Jablonowski and Williamson (2006a). The CAM SLD dynamical core simulations at the resolution T85L26 with decentering parameters between $\epsilon = 0$ and $\epsilon = 0.3$ are shown. No explicit diffusion is used, the time step is $\Delta t = 1{,}800$ s

any additional damping from non-zonal geodesic trajectory calculations in polar regions (Williamson and Rasch 1989). The latter is only reasonable in the case of zonal advection as considered here. These deviations from the default CAM SLD configuration are selected to truly isolate the damping effects from the decentering parameter. Of course, in practice the damping of all explicit and inherent dissipation mechanisms as well as filters and fixers act in concert, and they are generally difficult to isolate individually. Figure 13.22 confirms that the inherent dissipation in these steady-state simulations strongly depends on the decentering parameter. The l_2 zonal wind errors grow steadily over time, and there is an almost linear relationship between the magnitude of the error at day 30 and the magnitude of the decentering parameter. Recall that CAM SLD selects $\epsilon = 0.2$ by default which poses a compromise between accuracy and the suppression of orographic noise in practice.

13.6.3.2 Forward-Biasing of Trapezoidal Time Integrations

Of similar spirit as the SISL decentering approach is the forward-biasing technique for the implicit trapezoidal time integration method that provides damping of high-frequency modes. A short discussion can be found in Durran (1999) (his Chap. 7.3). The principal difference between the SISL decentering and the forward-biasing is that the SISL decentering represents a mix of a spatial and temporal average since both the departure and arrival point information are involved in the estimate of the decentered trajectory. In contrast, the forward-biasing technique only represents a temporal average at a single location. However, forward-biasing is sometimes called off-centering, uncentering or decentering, but despite the same nominal names the differences should be kept in mind.

Forward biasing is accomplished by replacing time-centered source terms of the form $(S^{j+1} + S^j)/2$ with the off-centered expression

$$\left(\frac{1+\epsilon}{2}\right)S^{j+1} + \left(\frac{1-\epsilon}{2}\right)S^j \tag{13.124}$$

where $0 \leq \epsilon \leq 1$. Choosing $\epsilon = 0$ recovers a second-order in time centered discretization that e.g., represents the implicit Crank–Nicolson scheme

$$\frac{\psi^{j+1} - \psi^j}{\Delta t} = \frac{1}{2}(S^{j+1} + S^j). \tag{13.125}$$

Any $\epsilon > 0$ formally reduces the order of accuracy of the temporal discretization. Off-centering the time discretization with $\epsilon > 0$ adds inherent numerical dissipation and numerically stabilizes the solution. Values in the range of $[0.2, 0.4]$ are quite common for models with two-time-level semi-implicit schemes. Note that ϵ is also often called *implicitness parameter*.

As an example, Durran and Klemp (1983) used forward-biasing of the trapezoidal time-differencing scheme for vertical derivatives. They found that a value of $\epsilon = 0.2$ provided sufficient damping that did not noticeably modify the gravity waves. Bonaventura and Ringler (2005) also used $\epsilon = 0.2$ and argued that such an inherently dissipative scheme can often be used without adding further explicit diffusion. As discussed by Skamarock et al. (2008) forward-in-time weighting of the vertically-implicit acoustic-time-step terms damps instabilities associated with vertically-propagating sound waves and the partially-split temporal discretization. The forward weighting also damps instabilities associated with sloping model levels and horizontally propagating sound waves as shown in Durran and Klemp (1983). A value of $\epsilon = 0.1$ is used as the default in the nonhydrostatic limited-area model WRF. The regional model COSMO sets the default parameter to $\epsilon = 0.4$ (Gassmann and Herzog 2007). Recently, Baldauf (2010) assessed suitable limits for the off-centering parameter for both buoyancy terms and sound wave terms in the regional weather forecast model COSMO.

13.6.4 Damping by Semi-Lagrangian Interpolation

Semi-Lagrangian schemes require spatial interpolations at every time step to determine the transported variables at the departure points. These interpolations provide a source of uncontrolled damping in model simulations. Conventional wisdom says that semi-Lagrangian approximations damp more over a given length integration when run with short time steps than when run with long time steps. The argument is that more interpolations are performed with the shorter time step, thus the net damping will be larger. The damping from interpolation generally increases as the wavelength decreases. Thus spectra of, for example, the kinetic energy or vertical

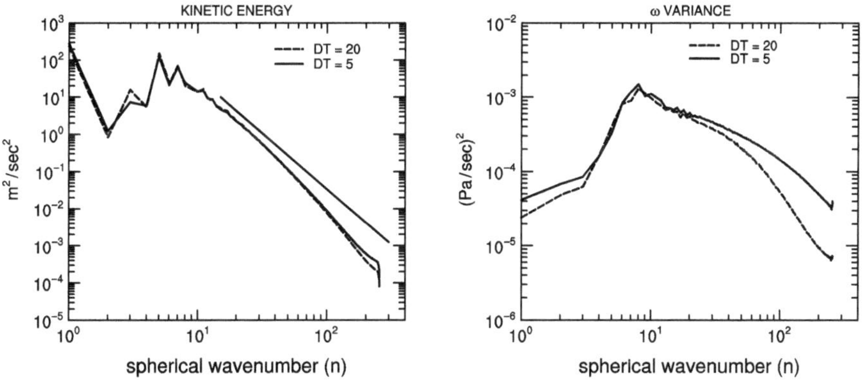

Fig. 13.23 Spectra as function of the spherical wavenumber (n) from TL255 CAM 3.1 semi-Lagrangian integrations with linear transform grid using 5 and 20 min time steps for (*left*) 250 hPa kinetic energy and (*right*) 500 hPa pressure vertical velocity (ω) variance

velocity variance, will be steeper approaching the truncation limit in an experiment using a short time step than in one with a long time step.

Figure 13.23, however, shows that this is not necessarily the case. This figure plots the 250 hPa kinetic energy and 500 hPa pressure vertical velocity variance spectra as a function of spherical wavenumber (n) from integrations with the semi-Lagrangian version of CAM 3.1 at TL255 truncation with 5 and 20 min time steps. This semi-Lagrangian model uses an optional linear (TL) unaliased grid of approximately 0.7° (or 78 km at the equator) which is the same as the quadratically unaliased grid used by the T170 Eulerian spectral model. The linear grid is defined to be the minimum grid required for transformations of a field from spectral space to grid point space and back again to spectral without loss of information. With such a grid, only linear terms are unaliased (Williamson 1997). The numerical algorithms are detailed in Collins et al. (2004). The simulations presented in this subsection are for an aqua-planet (Williamson 2008a) but in our experience, except possibly for the long waves, the shape of spectra in aqua-planet simulations is the same as in earth-like simulations. For both variables the spectra fall off faster for the long time step than for the short time step.

So where does conventional wisdom go wrong? It does not take into account that the short and long time step departure or interpolation points are not at the same relative location in a grid interval and thus the damping rate for a single interpolation is not the same. Figure 13.24a shows the response function of selected wavelengths for cubic Lagrange polynomial interpolation as used in CAM 3.1 SLD as a function of relative position in the grid interval (β) following Williamson and Laprise (2000). For all waves, the amplitude damping increases from the edge of the grid interval ($\beta = 0$ or 1), where it is zero, to the center of the interval ($\beta = 0.5$) where the damping of each wave is greatest, with the $2\Delta x$ wave annihilated there. As a specific example of relative damping assume the long time step is four times the short time step, and that the short time step yields a departure point location with $\beta = 0.125$.

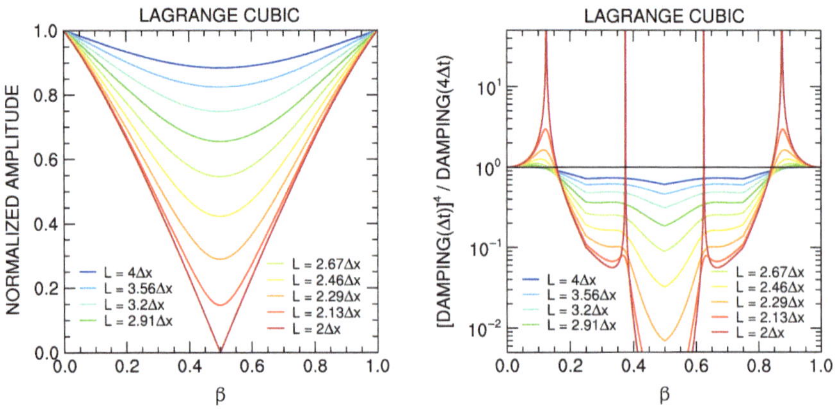

Fig. 13.24 (*Left*) Response function for Lagrange cubic interpolation as function of location within grid interval for selected wavelengths. (*Right*) Ratio of response function for short time step to the power 4 to the response function for long time step as a function of the location in the grid interval of the short time step. The long time step is four times the short time step

The long time step then has a departure point with $\beta = 0.5$ and the amplitude of the $2\Delta x$ wave is zero after interpolation. The amplitude after one short time step ($\beta = 0.125$) is 0.805. However four such interpolations are required to reach the same forecast time as one long time step, therefore the total damping with the short time step is $0.805^4 = 0.42$. Clearly, in this extreme example, the damping is less with the short time step than with the long one.

The general situation is shown in Fig. 13.24b. Here and in the following we consider the situation with the long time step being four times the short one. Later, specific results from the semi-Lagrangian CAM 3.1 at TL255 truncation will use 5 and 20-min time steps. Figure 13.24b plots the ratio of the short time step damping to the fourth power to the damping of the long time step as a function of β which is the location in the grid interval for the short time step. There is a region with $\beta <\sim 0.15$ and a mirror one $\beta >\sim 0.85$ in which the net damping from the short time step is less than that from the long. Elsewhere ($\sim 0.15 < \beta <\sim 0.85$) the net damping from the short time step is greater. The only exception being the $2\Delta x$ wave at $\beta = 0.375$ and 0.625 where the corresponding long time step is $\beta = 1.5$ and 2.5, both of which are equivalent to $\beta = 0.5$ with zero response function. With smaller time step ratios the zero crossings shift inward toward $\beta = 0.5$ with a structure similar to Fig. 13.24b except the secondary interior $2\Delta x$ ratio of ∞ does not occur (not shown). With a time step ratio of 3, the damping ratio for the $2\Delta x$ wave crosses 1 at β around 0.2 and 0.8, and with a time step ratio of 2, the ramping ratio for the $2\Delta x$ wave crosses 1 at β around 0.3 and 0.7.

Therefore, the overall damping in a semi-Lagrangian integration will depend on the population of departure points. Figure 13.25 shows the frequency distribution of the departure point locations from the 5 and 20 min TL255 integrations. The frequency distribution is calculated over all grid points at the 250 hPa model level

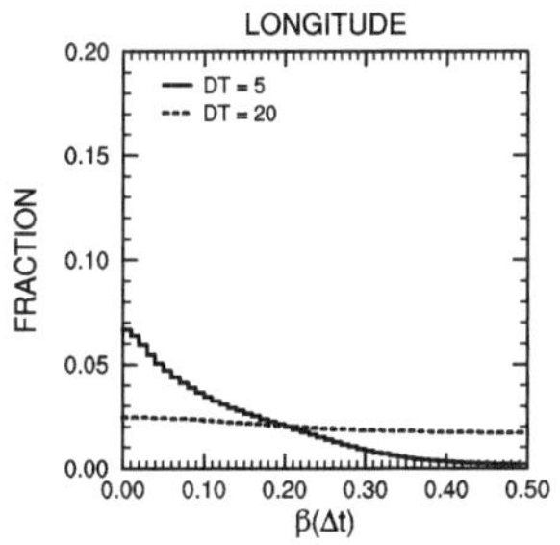

Fig. 13.25 Fraction of departure points falling in 0.01 bins ranging from 0.0 to 0.5 of the grid interval from semi-Lagrangian integrations with $\Delta t = 5$ and 20 min for (*left*) longitude, (*center*) latitude and (*right*) vertical directions

(corresponding to the spectra in Fig. 13.23) every 6 h for 125 days. Because of natural symmetries only β in the range 0–0.5 need be considered; other values fold back into this range. CAM 3.1 SLD uses a tensor product interpolation in which interpolations are done sequentially in longitude, in latitude and then in the vertical coordinate. The frequency distributions are shown for each coordinate. The plots show the fraction of the time the departure point falls in each 0.01 interval from 0.0 to 0.5. We consider the 5 min case first. In the vertical almost all departure points have $\beta < 0.05$. In latitude almost all departure points have $\beta < 0.25$, with only a small subset between 0.15 and 0.25. In longitude, on the other hand, departure points have values as large as $\beta = 0.5$, although the fraction is small approaching 0.5. Five minutes is not a particularly long time step for TL255 resolution. In fact it is the time step normally used for the Eulerian version of CAM 3.1 at T170 truncation. The longitudinal frequency distribution for integrations with 20 min is relatively flat decreasing from 0.022 for the first bin $\beta = [0.0, 0.01)$ to 0.019 for the last bin $\beta = (0.49, 0.05]$. (Recall there are 50 bins, $50 \times 0.01 = 0.5$) The 20 min latitudinal frequency distribution is much less steep than the 5 min one, starting at 0.05 for the first bin, decreasing to 0.02 at $\beta = 0.2$ and continuing to around 0.005 for the last bin. The 20 min vertical frequency distribution starts at 0.55 at the first bin and is close to zero by $\beta = 0.1$

The frequency distribution for the ratio of the damping of the short time step to that of the long time step from the CAM 3.1 SLD integrations is shown in Fig. 13.26 for the $2\Delta x$ wave. To calculate the frequency distribution, equivalent 20-min departure points were calculated based on the 5-min ones captured from the model integration. The ratio of the damping to the fourth power of each 5-min departure point to the damping from the matching 20-min departure point was calculated. The frequency distribution for the log of the ratio is plotted in Fig. 13.26 for the $2\Delta x$ wave. A value of 0 is neutral, positive values imply less damping for the short time step and negative values imply more damping for the short time step. To avoid including the neutral 0 values in either the positive or negative bin, they are given a special bin of their own. This bin contains the fraction of values within rounding of zero and is indicated by the dot plotted at zero abscissa. Before discussing Fig. 13.26

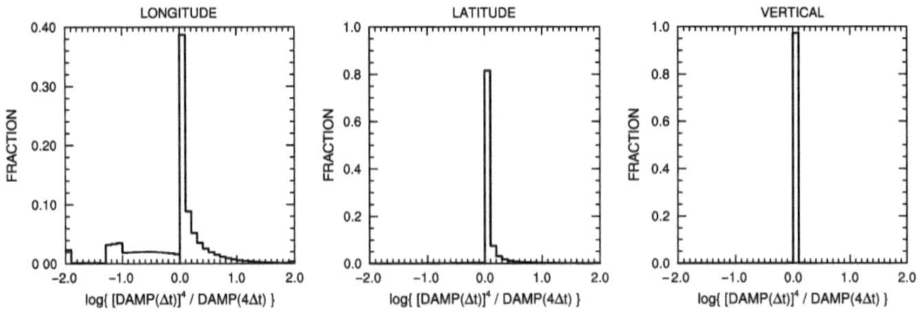

Fig. 13.26 Fraction of log of ratio of short time step damping to fourth power to long time step damping for $2\Delta x$ wave in 0.1 bins from −2.0 to 2.0 for (*left*) longitude, (*center*) latitude and (*right*) vertical interpolations

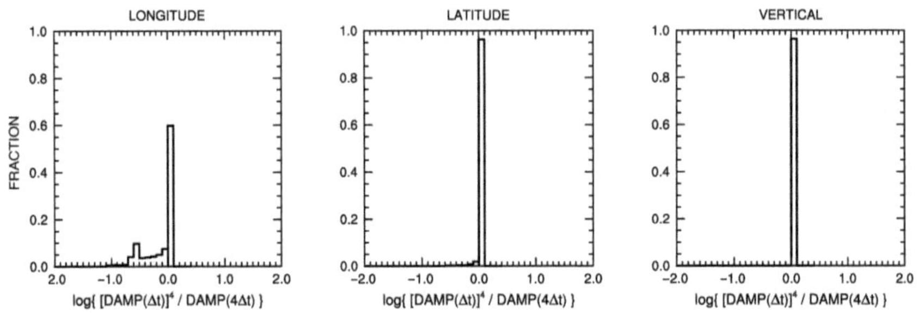

Fig. 13.27 Fraction of log of ratio of short time step damping to fourth power to long time step damping for $2.7\Delta x$ wave in 0.1 bins from −2.0 to 2.0 for (*left*) longitude, (*center*) latitude and (*right*) vertical interpolations

we note that when we base the calculation on the sampled departure points from the 20-min CAM 3.1 SLD experiment and calculate the equivalent 5-min departure points, the plot of the damping ratio is virtually indistinguishable from Fig. 13.26.

The frequency distribution for longitude in Fig. 13.26 is non-zero for both positive and negative log of ratios, indicating there are points where the short time step damps less and points where it damps more. The frequency distribution for latitude however indicates the short time step always damps less, there are no negative values. The frequency distribution for the vertical trajectory also indicates the short time step always damps less; however almost all values are in the first positive bin, and the remainder have the log of the ratio within rounding of zero (the dot in the figure.)

Figure 13.27 shows the same frequency distribution but now for the $2.7\Delta x$ wave. In longitude, at 60% of the points the short time step damps less, but the damping ratio is between 1 and 1.26 (the first positive bin with log ranging from 0 to 0.1). At the remainder of the points the long time step damps less, and the ratio of damping is primarily distributed over bins with ratios 0.2 to 1 (log ranging from −0.7 to 0) with some ratios from 0.08 to 0.2 (log ranging from −1.1 to −0.7). In latitude, at over 95% of the points the small time step damps less. As in longitude, the positive log damping ratio is in the first positive bin, where the damping ratio is between 1 and 1.26. In the vertical, the small time step damps less at 97% of the points. The ratio at the remaining points is within rounding of 1 (i.e., log is 0).

In summary, semi-Lagrangian integrations with short time steps do not necessarily damp shorter waves more than integrations with long time steps do. The different time steps yield different departure points and therefore different damping from the interpolations. The overall damping depends on the population of departure points which in turn depends on the atmospheric flow and model time step.

13.7 Fixers and Thoughts About Conservation Properties

It is generally desirable for a dynamical core to possess discrete analogues of the conservation properties of the adiabatic and frictionless continuous equations of motion as e.g., laid out in Chap. 11. However, the continuous equations possess an infinite number of invariants, such as mass, tracer mass, total energy, enstrophy and angular momentum just to name a few, whereas a numerical model can only conserve very few quantities. A straightforward way to ensure the conservation of an invariant is to choose it as a prognostic variable and utilize a flux-form finite volume discretization. Such a built-in conservation law is then a design feature of a dynamical core and its numerical scheme. This design decision needs to be carefully weighted against other beneficial properties like the computational efficiency or accurate wave dispersion characteristics.

Conservation can also be obtained through special mathematical properties of the numerical discretization. For example, spatial discretizations can be formulated so that they enforce the conservation of global integrals, such as mass, total energy and potential enstrophy (Arakawa 1966; Arakawa and Lamb 1981; Arakawa and Hsu 1990). This is also discussed in Chap. 12. The basic question is how accurately a dynamical core *needs* to capture various conservation properties and whether global conservation is sufficient or local conservation needs to be enforced. These issues are addressed in Thuburn (2008b) who gives guidance concerning the desirable conservation properties of GCMs.

In practice, there are many reasons why numerical models might lose even the most basic conservation properties of the continuous equations like the conservation of dry air mass, tracer mass or total energy. A prominent reason is that the equations are often not formulated in conservation form. But even if a conservation form is chosen the inevitable dissipation, either explicitly specified or inherent in the numerical schemes, and use of filters can violate the conservation in the discretized case (Takacs 1988). For example the conservation of mass principle is violated if the mass variable needs to be time-filtered or spatially filtered for numerical stability reasons. In addition, full GCMs contain physical parameterizations that represent the unresolved often dissipative processes at the subgrid-scale such as boundary layer turbulence. Kinetic energy is therefore generally lost due to dissipation which might translate into a loss of total energy. Conservation must therefore be addressed not only in the dynamical core and its numerical discretization but also in full GCMs with physics packages.

If a conservation property is violated in a GCM, the global conservation can still be artificially recovered. This can be done through the use of so-called *fixers*. Fixers are modeling paradigms that allow an ad-hoc and a posteriori restoration of conserved quantities at each time step. There is no physical basis for such restorations other than that the conservation is a necessary or desirable characteristic of the GCM. A general expectation might be that the GCM simulations become more trustworthy if conservation properties are obeyed. This is especially true for the conservation of the dry air mass and the total energy which prevents the model climate from drifting into unrealistic states. However, the use of fixers does not imply that the physical processes and scale interactions are better represented. In addition, it is often imperative to fix unphysical negative tracer masses to prevent the model from "exploding" in the physical parameterizations.

This section discusses three types of a posteriori fixers that are broadly used in GCMs today. They are the mass fixers for dry air, filling algorithms for tracers and total energy fixers. Most often, the application of a fixer is an undocumented design feature of a GCM.

13.7.1 Dry Air Mass Fixer

In nature, dry air mass has no true physical sources and sinks, and is conserved regardless of diabatic or frictional processes. Conservation of dry air mass is probably the most fundamental conservation property that should be enforced in GCMs. In fact, the conservation of mass is paramount for long climate simulations where any drift in the total mass translates into a drift of the pressure distribution through the equation of state. This leads to spurious motions and artificial drifts of the simulated climate. For short weather predictions though, GCMs have put less emphasis on the conservation of mass. This design decision is probably justified since the drifts in the mass over short 10-day forecasts are generally negligible for practical purposes.

In the absence of sources and sinks, the mass of water vapor is conserved just as the mass of the dry atmosphere is. Since total air is a mixture of water vapor and dry air the conservation of both mass of water vapor and of dry air are often considered together. This is especially true if the moist surface pressure is the prognostic forecast variable. If a model prognoses the dry air pressure, a sole dry air mass fixer of course suffices.

The most popular dry air mass fixers are built upon an ad-hoc correction of the global surface pressure field regardless of the origin of the pressure drift which is often unknown. In models that predict $\ln p_s$ like CAM EUL or SLD this is done by adjusting the surface pressure at all grid points so that the gradient of the logarithm of the surface pressure field $\nabla \ln p_s$ is unaffected. The fixer thereby preserves the gradients of the pressure gradient force. The latter is an important driver in the momentum equations that should not be arbitrarily changed. Such a mass fixer is for example documented in Williamson and Olson (1994). For brevity, we only present the design of the fixer for dry air masses. The extensions for moist air is shown in Williamson and Olson (1994), Collins et al. (2004) and Rasch et al. (1995).

Conceptually, the global dry air mass in hydrostatic models is represented by the global integral P of the dry surface pressure as given by

$$P(t) = \frac{1}{4\pi} \int_{-\pi/2}^{\pi/2} \int_{0}^{2\pi} p_{s\,dry}(\lambda, \phi, t) \cos\phi \, d\lambda d\phi \tag{13.126}$$

in spherical coordinates. The total mass, that needs to be conserved after each time step, is determined by the global mean surface pressure of the initial state $P(t = 0)$. Let t^+ denote the future time after completion of a time step but before the application of the mass fixer. The values of the surface pressure field at t^+ are therefore provisional. The surface pressure at the final future time step t^{j+1} is then fixed in the following way

$$p_s(\lambda, \phi, t^{j+1}) = M p_s(\lambda, \phi, t^+) \tag{13.127}$$

with the correction factor

$$M = \frac{P(t = 0)}{P(t^+)}. \tag{13.128}$$

Such a fixer is applied by default at each time step in NCAR's CAM Eulerian and semi-Lagrangian spectral transform dynamical cores. This formulation is only valid for hydrostatic dynamical cores.

To highlight the effects of the mass fixer on idealized dry dynamical core simulations we present results from both CAM EUL and CAM SLD (version 4) at the triangular truncation T85 ($\approx$ 156 km) with 26 levels (L26). In particular, the models were run with and without the mass fixer for 1,800 days and utilized the Held and Suarez (1994) forcing. All simulations start from identical initial conditions that contain a global mean dry surface pressure of 1,000 hPa. After 1,800 days, the Eulerian simulation without the mass fixer exhibits a global mean surface pressure of 999.9992 hPa which is a quite accurate. On the other hand, the SLD simulation without the mass fixer shows a steady, almost linear, increase in the amount of total mass. After 900 days the global mean surface pressure is 1005.29 hPa and increases to 1011.11 hPa by day 1,800. This is a substantial increase that would prevent credible climate simulations unless a mass fixer is employed. The mass is perfectly conserved in both dynamical core simulations with the mass fixer, as expected.

It is informative to evaluate the changes in the model climate due to the mass fixer. Such an assessment can reveal whether there are any systematic differences in the circulation when modeled with and without the mass fixer. Figures 13.28a-d and 13.29 show the zonal-mean 1200-day-mean temperature and zonal wind fields from the simulations with the mass fixer, and the differences between the runs with and without the mass fixer for both EUL and SLD. No total energy fixer is applied.

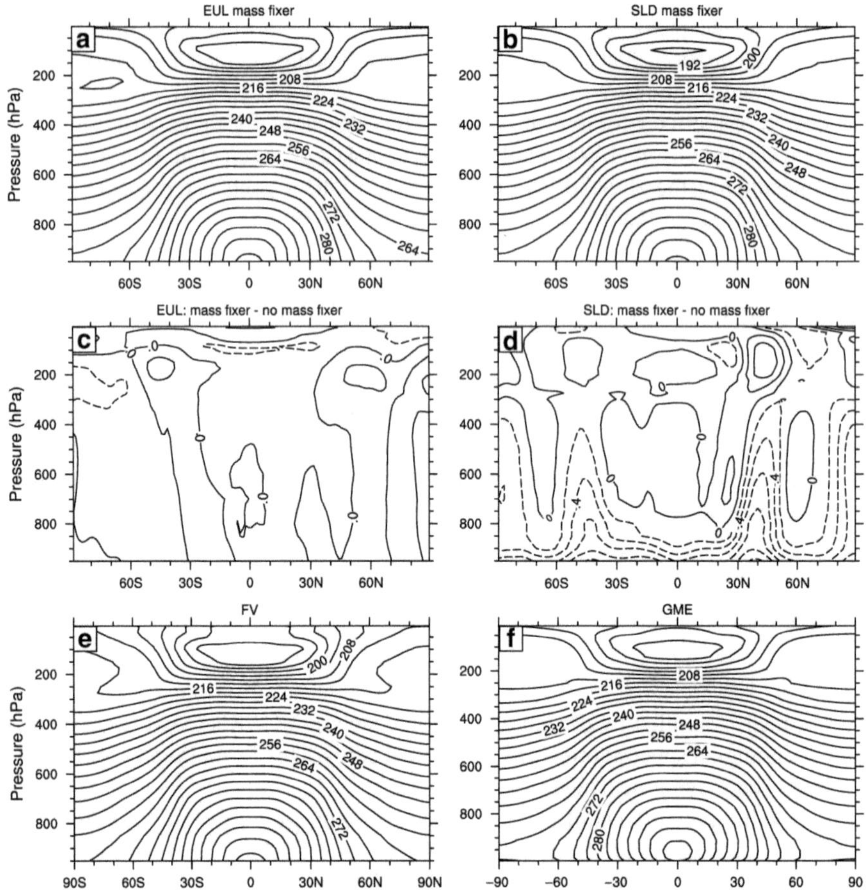

Fig. 13.28 Zonal-mean time-mean temperature field (K) forced with the Held-Suarez forcing: (**a**,**b**) CAM 4 EUL and CAM 4 SLD simulations at the resolution T85L26 with mass fixer, (**c**,**d**) difference between the EUL and SLD simulations with and without mass fixer, and comparisons to (**e**) CAM 4 FV at the resolution 1° × 1° L26 and (**f**) GME at the resolution $ni = 64$ L19 (≈ 120 km). (**a–d**) are 1200-day means, (**e**) is a 450-day mean, (**f**) 900-day mean. The contour intervals in (**a**,**b**,**e**,**f**) are 4 K and in (**c**,**d**) 0.2 K. Negative contours are *dashed*. The time step is $\Delta t = 600$ s (EUL) and $\Delta t = 1{,}800$ s (SLD)

The EUL simulation used the T85 default horizontal diffusion with the coefficients $K_4 = 1 \times 10^{15}$ m^4 s^{-1} and $K_2 = 2.5 \times 10^5$ m^2 s^{-1} whereas no explicit diffusion was utilized in the SLD run. Note again that the K_2 value is the base value at the third level from the top. It is doubled at the second level and doubled again at the top level. The 1200-day time averages incorporate day 600–1,800 of the model simulations. Furthermore, we show Held-Suarez temperature results from two additional models CAM 4 FV and GME without a mass fixer (Fig. 13.28e,f) to demonstrate that the EUL and SLD simulations are visually very similar to other models (Jablonowski

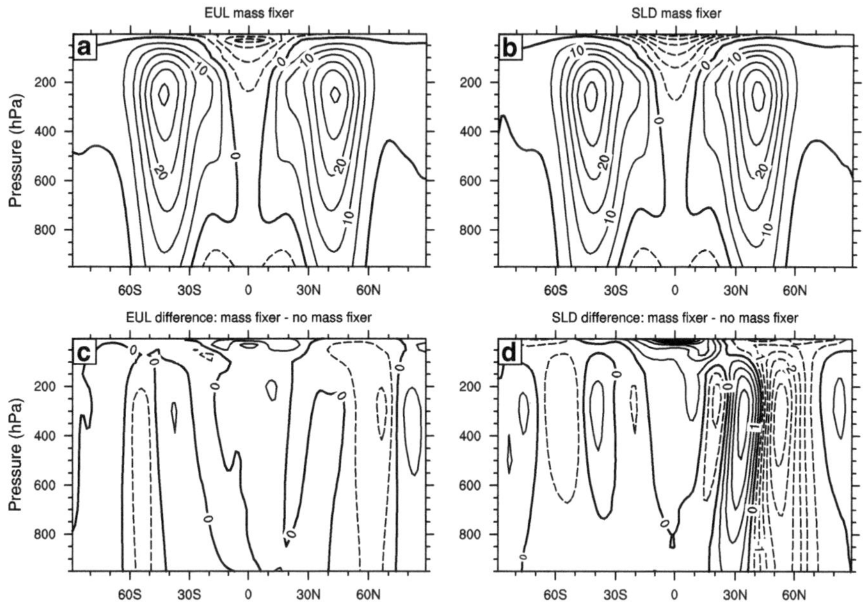

Fig. 13.29 Zonal-mean 1200-day-mean zonal wind (m/s) in CAM 4 EUL and CAM 4 SLD at the resolution T85L26 forced with the Held-Suarez forcing: (**a**,**b**) simulations with mass fixer, (**c**,**d**) difference between a simulation with and without mass fixer. The contour intervals are (**a**,**b**) 5 m/s and (**c**,**d**) 0.5 m/s. Negative contours are *dashed*. The time step is $\Delta t = 600$ s (EUL) and $\Delta t = 1{,}800$ s (SLD)

1998; Wan et al. 2008). FV is mass-conserving by design, and the small mass gain in GME over its 1440-day simulation period was on the order of 0.5 hPa.

The temperature and zonal wind difference plots for the EUL simulation shown in Figs. 13.28c and 13.29c suggest that the mass fixer has negligible impact on the mean EUL circulation. The temperature distributions and positions of the midlatitudinal jets in both hemispheres are almost identical in both runs. This is somewhat expected since the Eulerian simulation hardly lost mass over the 1800-day simulation so the impact of the mass fixer should be minimal. A clearer modulation of the mean circulation can be seen in the corresponding SLD runs (Figs. 13.28d and 13.29d). The simulation with the mass fixer appears to be slightly warmer in the upper atmosphere whereas the lower troposphere and the midlatitudes up to 400 hPa show systematically colder temperatures. In particular, the SLD mass fixer run is about 0.4–0.8 K colder in the levels below 800 hPa. The jets in the zonal wind field are slightly shifted equatorwards in the SLD mass fixer run (Fig. 13.29d). However, it is unclear whether these shifts are statistically significant without further investigations. As an aside, the differences in the zonal wind fields between the EUL and SLD runs (Figs. 13.29a,b) close to the model top are caused by the sponge-layer diffusion in the Eulerian model which is discussed in detail in Sect. 13.4.5.

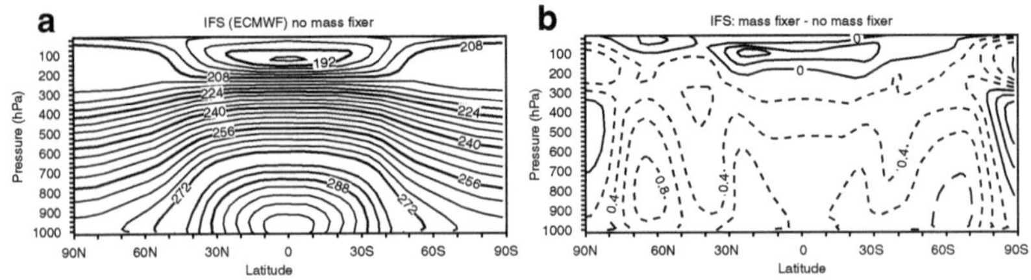

Fig. 13.30 Zonal-mean 1000-day-mean temperature field (K) in ECMWF's IFS model cycle 18R3 at the resolution T63L31 forced with the Held-Suarez forcing: (**a**) simulations without the mass fixer, (**b**) difference between a simulation with and without mass fixer. The contour intervals are (**a**) 4 K and (**b**) 0.2 K. Negative contours are *dashed*

To put the CAM 4 SLD changes due to the mass fixer into perspective we also show results from an older version of ECMWF's model IFS (cycle 18R3, November 1997). The dynamical core of IFS is a two-time-level semi-implicit semi-Lagrangian spectral transform model and therefore very similar to CAM SLD. The Held-Suarez test was run for 1,200 days with and without a mass fixer at the triangular truncation T63 ($\approx$ 210 km) on a reduced Gaussian grid with 31 vertical levels (L31). The specific design of the reduced grid and semi-Lagrangian model are explained in Hortal and Simmons (1991) and Hortal (2002), and are not of interest for the following discussion. Here, we solely concentrate on the effect of the mass fixer on the model climate. The IFS mass fixer follows the identical design principle as CAM's EUL/SLD mass fixer. Similar to CAM SLD, the *unfixed* IFS shows a systematic, almost linear, increase in mass over the 1200-day forecast period. The rate is +0.012% per 10 days. Assuming an initial global surface pressure of 1000 hPa this amounts to about 1014.4 hPa after 1,200 days. This increase in mass is resolution dependent. At the higher triangular truncation T106 ($\approx$ 125 km) the rate is reduced to +0.0079% per 10 days which yields a global surface pressure of 1009.5 hPa by day 1200. The changes in IFS's mass are thereby slightly higher but comparable to the changes of the mass in CAM 4 SLD.

Figure 13.30 shows the zonal-mean 1000-day-mean temperature distribution of the IFS run without the mass fixer and the temperature differences between the runs with and without the mass fixer. The overall temperature distribution in Fig. 13.30a resembles the CAM EUL and SLD runs with similar peak temperatures. The temperature difference in Fig. 13.30b also exhibits some structural resemblance to the SLD difference plot (Fig. 13.28d). The IFS mass fixer run is systematically colder throughout the lower and middle troposphere, and warmer near the poles and near the tropopause. The cold temperature difference peaks in midlatitudes with a magnitude of 0.8 K. Note, that the use of the mass fixer in SLD and IFS would be paramount for long climate simulations. Most likely, the warming and cooling signatures in the runs without the mass fixer are entirely spurious and related to the unphysical gain in mass. As an aside, the design of IFS's dynamical core has only slightly changed in comparison to more current hydrostatic versions of IFS (ECMWF 2010).

13.7.2 Filling Algorithms for Tracers

Mass conservation is one of the most important design aspects of tracer transport algorithms. If a scheme is nonconservative, it can significantly underestimate or overestimate the concentration of trace gases in long time integrations. This is particularly true if the transported quantity has a large residence time in the atmosphere like methane or nitrous oxide.

An additional desirable characteristic of transport schemes is monotonicity and thereby the prevention of non-physical under- and overshoots in the solution. They can lead to negative trace constituents or even the supersaturation in the humidity field. In particular, negative mixing ratios are undesirable since physical parameterizations cannot deal with e.g., negative moisture quantities. Ideally, both monotonic (also called shape-preserving or non-oscillatory) and mass-conserving advection schemes should therefore be used to assure the physical consistency of the advection process, and prevent negative tracer constituents from occurring in the first place. Examples are the flux-corrected advection scheme of Zalesak (1979) or the mass-conservative and monotonic semi-Lagrangian advection scheme by Lauritzen et al. (2010b). Alternatively, positive definite advection algorithms can be employed that, at least, prevent negative undershoots. A comprehensive overview of possible advection algorithms and their characteristics is given by Rood (1987) and Chap. 8.

In case negative tracer constituents occur during a model integration an a posteriori borrowing and filling algorithm is most often employed to *fix* the negative tracer mass. The basic idea is that a grid box with negative tracer values is filled to a minimum small positive value, and an equivalent amount is subtracted (borrowed) from other grid cells. This ensures that the total constituent mass remains the same. However, it does not eliminate overshoots or undershoots that are associated with non-negative parts of the field. Such a fixer is characterized as a *conservative* fixer. Fixers might also be used as *positivity* fixers that only obey a positive-definite constraint but neglect global conservation.

The mixing induced by both types of fixers can trigger nonlinear interactions. For example, Rasch and Williamson (1990a) showed that positivity fixers can greatly influence the water vapor budget in a spectral transport scheme due to the interactions of the fixed specific humidity field with physical parameterizations. This was especially true in the polar regions which are rather dry. The fixer operated as a moisture transport algorithm, yielding a local moisture source, and although it only brought the moisture up to a positive minimum value at negative points, it increased the overall moisture in the polar regions with strong impact on the clouds and precipitation.

Examples of tracer filling algorithms for the use in GCMs are presented in Mahlmann and Sinclair (1977), Royer (1986), Rasch and Williamson (1990a) and Rasch and Williamson (1991). They can be either local or global, and resemble a nonlinear diffusive process. Both, monotonicity constraints and filling algorithms are designed to control numerical dispersion errors, and could also be viewed as an implicit or explicit numerical filter, respectively. Note that even tracer advection algorithms that are strictly monotonic in one dimension might lead to violations of the monotonicity when applied in multiple dimensions. Therefore, even schemes with monotonicity constraints such as the finite volume advection algorithm in the CAM FV dynamical core can trigger small under- and overshoots in multiple dimensions as reported in Lin and Rood (1996). Therefore, CAM FV applies a local filling algorithms to eliminate negative tracer masses as documented in Neale et al. (2010).

Another highly desirable property of a tracer advection scheme is that it should be consistent with the mass continuity equation. This is for example outlined in Lin and Rood (1996), Jöckel et al. (2001), Gross et al. (2002), Satoh et al. (2008) and Chap. 8, and is not discussed in detail here. As an aside, the expression "borrow and fill" is somewhat misleading since the amount taken from a neighboring grid cell is never given back to the cell it is taken from. A more appropriate description is "take and fill" or even "steal and fill" (Fedor Mesinger, personal communication). However, we stick with the expression "borrow and fill" in this chapter since it is widely used throughout the literature.

13.7.2.1 Local Filling Algorithms

Local filling algorithms are "borrow and fill" fixers that try to borrow mass primarily from the four nearest cells in the east, west, north, and south directions at the same model layer. However, if there is insufficient mass they might also borrow from a level below or above. Most often, the filling algorithm starts downward from the model top as described in Rasch and Williamson (1990a). It is not straightforward to write down a concise set of equations to describe a local borrower scheme. Therefore, we only describe the underlying ideas and present two variants documented in the literature. Note that there are many additional variants in practice, but borrowing schemes are not necessarily documented and should ideally be avoided with the help of improved tracer advection algorithms.

Variant 1 After identifying a grid cell with negative tracer mass the surrounding neighbors with positive tracer masses are determined and an equal percentage is borrowed from each. The negative mass is set to small minimum value. If there is insufficient mass in the neighboring cells, no action is taken and the next point is evaluated. The borrowing might be limited to one-third of the total mass available in the neighboring cells as suggested by Reames and Zapotocny (1999).

Variant 2 Reames and Zapotocny (1999) also tested a borrowing scheme that weighted the borrowed mass by the amount available in the neighboring boxes and by the velocity components toward these grid cells. This extends a suggestion by Mahlmann and Sinclair (1977) who argued that borrowing should first come from a neighboring grid cell that is downstream. Only if there is not enough mass in the downstream direction, mass is borrowed from an upstream point or even a more distant point.

The local filling algorithm can still leave residual negative values in case immediate neighbours do not have enough mass to fill a point. An additional global borrowing scheme can then be used to remove this residual as explained next. Of course, a global borrowing scheme can also be used by itself without a local filling algorithm. As pointed out by Rood (1987) a characteristic of local filling algorithms is that many decisions have to be made. Therefore, a local filling scheme is quite expensive from a computational viewpoint. It can also violate the monotonicity of the tracer field.

13.7.2.2 Global Filling Algorithms

Global "borrow and fill" algorithms are less time-consuming but require global communication on parallel computing architectures. There are two classes of global borrowing schemes which can be characterized as a subtractive and multiplicative method as outlined by Rood (1987). First, the total area- or volume-weighted negative (N) and positive (P) tracer masses are computed and negative tracer values are set to a small minimum value. In the subtractive method, a fraction of the extra mass N is then subtracted from all grid points from which mass can be subtracted without creating new negative values. This subtraction needs to obey the constraint that the total mass is constant after the correction which could require some further searching. In the multiplicative method, each positive tracer value q^+ is replaced by $(1 - N/P)q^+$ which automatically conserves the global integral of the constituent mass and guarantees that the new tracer distribution is positive definite. The correction can either be applied level-by-level or globally in case the volume-weights are computed (Rasch and Williamson 1990b). Such a multiplicative fixer was for example used by Zubov et al. (1999) and Reames and Zapotocny (1999) .

Rasch and Williamson (1991) found that this variational adjustment of the tracer mass does neither improve nor degrade the accuracy of an unadjusted tracer transport scheme, but that it merely enforces the conservation and positivity constraints. However, Jöckel et al. (2001) argued that simple mass fixer algorithms for tracers have severe disadvantages since they either violate the monotonicity requirement or introduce non-physical transport. It is therefore best to select a conservative tracer transport scheme that is also consistent with the mass continuity equation as outlined in Chap. 8. It means that the discretized tracer transport scheme should reduce to the discretized continuity equation for air if the tracer mixing ratio is unity. The discretization then mimics the characteristics of the continuous equations.

13.7.3 Total Energy Fixers

Global invariants like the total energy provide useful constraints on the design of numerical schemes which makes a built-in conservation of total energy a desirable choice for GCMs. As suggested by Thuburn (2008b) the conservation of total energy and enstrophy in full GCMs might even play a major role in accurately capturing nonlinear transfers to small scales. However in practice, many aspects of today's GCMs and in particular their dynamical cores are not energy-conserving. This includes the horizontal diffusion in case of neglected dissipative heating, the spatial and time filters, the time differencing, inherent numerical dissipation, or the discretization technique for the energy conversion term. The latter aspect is emphasized in Simmons and Burridge (1981) who developed an energy and angular-momentum conserving vertical finite-difference scheme.

Total energy conservation is paramount for long climate runs to prevent drifts in the mean circulation (Boville 2000). In early GCMs though, energy conservation was not a significant design consideration. For example the energy imbalance in an early version of NCAR's Community Climate Model CCM0 was 10 W m^{-2} as reported in Williamson (1988). The loss in energy originated primarily from inconsistencies in the vertical numerical approximations. In later versions, energy conservation received more attention, as e.g., version 1 of the Community System Model (CSM1). It conserved energy to about 0.4 W m^{-2} as analyzed by Boville and Gent (1998) but this level of conservation was partly due to a cancellation of errors. Today, total energy conservation has become a serious concern since even small imbalances of order 0.4 W m^{-2} can cause spurious long-term trends in multicentury coupled ocean-atmosphere simulations.

In general, the variation of energy during a simulation can manifest itself as either a gain or loss of total energy. Most often though, energy is lost which is mainly attributable to the kinetic energy dissipation. Kinetic energy dissipation is due to explicit horizontal diffusion, inherent dissipation in the numerical approximations and filters. In practice, it averages to an energy loss of about 2 W m^{-2} in the three CAM dynamical cores EUL, SLD and FV when applied at typical climate resolutions (Williamson 2007). This amount is clearly not negligible and only the contribution from the explicit horizontal diffusion can be analytically quantified. Some models therefore include a frictional heating term associated with the explicit horizontal momentum diffusion (Collins et al. 2004) as also outlined in Sect.13.3.7. A thorough review of the kinetic energy dissipation in NCAR's CAM model and the required compensating heating is provided in Boville and Bretherton (2003). As an aside, Bowler et al. (2009) estimated that the energy dissipation due to the interpolation error alone in the semi-Lagrangian advection scheme in the UK Met Office model is about 0.75 W m^{-2}. They suggested using a stochastic kinetic energy backscatter scheme to reintroduce the missing energy from the explicit horizontal diffusion and semi-Lagrangian interpolations, partly into the resolved scales of the flow. An alternative energy backscattering method was also presented by Shutts (2005).

In order to maintain the energy balance, the globally averaged inherent dissipative heating can be determined via a residual calculation. The heating is then added to the temperature field in the thermodynamic equation. This can be done as either a globally uniform heating or cooling, or another ad hoc function. The choice of such ad hoc functions is arbitrary but there are adequate and inadequate choices. As revealed in Williamson et al. (2009) with the help of an idealized dynamical core test, an inadequate "bad" energy fixer has detrimental effects on the circulation. This was not obvious by a pure inspection of the ad hoc correction algorithm and not immediately obvious in full GCM runs with physical parameterizations.

To remedy the effect, Williamson et al. (2009) recommended using only very simple corrections like uniform adjustments at all grid points in the global domain. We discuss the "wrong" and "right" choices below, but first start with a brief review of the total energy equation.

13.7.3.1 Different Forms of the Total Energy Equation

The total energy equation for dry air can be obtained by adding the kinetic and potential energy equations to the first law of thermodynamics. The derivations of these equations may be found in Gill (1982). Here, we present the form of the total energy equation for adiabatic and hydrostatic dynamical cores that utilize the primitive equations.

As pointed out by Staniforth et al. (2003) total energy is formally only conserved if the model employs a rigid lid as the upper boundary condition. Such a rigid lid needs to be fixed in time and space but is allowed to vary with latitude. Models with elastic isobaric lids, like the popular choice of the upper boundary at constant pressure, are formally non-energy-conserving. But an energy-like invariant exists that gives these approaches merit (Kasahara 1974; Laprise and Girard 1990). The specific form of the total energy equation is tightly linked to the choice of the vertical coordinate. This is outlined in Arakawa and Suarez (1983) and briefly shown for pressure- and height-based vertical coordinates below.

In the continuous primitive equations with the hybrid pressure-based vertical coordinate η (Simmons and Burridge 1981) total energy is conserved if the following relationship holds (Laprise and Girard 1990)

$$\frac{\partial}{\partial t}\int_A\int_{\eta_{top}}^{\eta_s}\left(\frac{\mathbf{v}^2}{2}+c_pT\right)\frac{\partial p}{\partial \eta}\,d\eta\,dA=-\int_A\left(\Phi_s\frac{\partial p_s}{\partial t}-\Phi_{top}\frac{\partial p_{top}}{\partial t}\right)dA. \quad (13.129)$$

This equation is valid in the absence of diabatic and frictional effects. Φ_s, p_s and Φ_{top}, p_{top} are the geopotential and pressure at the surface and the model top, c_p is the specific heat of dry air at constant pressure and $\mathbf{v}=(u,v)$ stands for the horizontal velocity vector with the zonal and meridional wind components u and v. Furthermore, T symbolizes the temperature, p is the pressure, and t denotes the time. The integrals span the 3D and 2D domains where A symbolizes the horizontal area of the sphere. The vertical integral is bounded by the value η_s at the surface and η_{top} at the model top. Here, η_s is identical to unity and η_{top} is equivalent to p_{top}/p_0 with reference pressure $p_0 = 1000\,\text{hPa}$. Note that η_{top} is non-zero for constant $p_{top} > 0\,\text{hPa}$. A constant pressure at the model top ensures the global conservation of total energy in the continuous equations and simplifies the 2D integral. Equation (13.129) then becomes

$$\frac{\partial}{\partial t}\left\{\int_A\frac{1}{g}\left[\left(\int_{\eta_{top}}^{\eta_s}\left(\frac{\mathbf{v}^2}{2}+c_pT\right)\frac{\partial p}{\partial \eta}\,d\eta\right)+\Phi_s p_s\right]dA\right\}=0. \quad (13.130)$$

Here we divided (13.129) by the gravity g to recover energy units (Kasahara 1974). This expression is equivalent to $\partial(TE)/\partial t = 0$ where TE symbolizes the global integral of the total energy as shown by the term in the curly bracket in (13.130). In the semi-discrete system with $\partial p/\partial\eta \approx \Delta p/\Delta\eta$ and $d\eta \approx \Delta\eta$, the domain-integrated total energy TE is given by

$$TE = \int_A \frac{1}{g}\left[\left(\sum_{k=1}^{K_{max}} \left(\frac{u_k^2 + v_k^2}{2} + c_p T_k\right)\Delta p_k\right) + \Phi_s p_s\right] dA. \qquad (13.131)$$

The summation index k indicates the vertical index of a full model level with the maximum level number K_{max} near the surface. The pressure difference Δp_k is defined as

$$\Delta p_k = p_{k+1/2} - p_{k-1/2} = p_0\,\Delta A_k + p_s\,\Delta B_k \qquad (13.132)$$

with $\Delta A_k = A_{k+1/2} - A_{k-1/2}$ and $\Delta B_k = B_{k+1/2} - B_{k-1/2}$. As an example, the discrete positions of the hybrid coefficients $A_{k+1/2}$ and $B_{k+1/2}$ at the model interface levels for the CAM EUL, SLD and FV dynamical cores (versions 3.1 and 4) are listed in Jablonowski and Williamson (2006b). $\Delta\eta_k$ is given by $\Delta\eta_k = \eta_{k+1/2} - \eta_{k-1/2} = \Delta A_k + \Delta B_k$. Note that the form of the domain-integrated total energy equation TE in the optional CAM 5 dynamical core HOMME (see (12.8) in Chap. 12) differs from (13.130). The main difference is that HOMME utilizes the parameter $[c_p^* = c_p + (c_{pv} - c_p)q]$ instead of c_p where c_p^* symbolizes the specific heat of moist air at constant pressure, c_{pv} denotes the specific heat of water vapor at constant pressure and q stands for the specific humidity.

If hydrostatic models with pressure-based σ coordinates like

$$\sigma = \frac{p - p_{top}}{p_s - p_{top}} \qquad (13.133)$$

are considered (Phillips 1957; Kasahara 1974) the global integral of the dry total energy becomes

$$TE = \frac{1}{g}\int_A \int_{\sigma_{top}}^{\sigma_s} \left(\frac{\mathbf{v}^2}{2} + c_p T + \Phi_s\right)\frac{\partial p}{\partial\sigma}\, d\sigma\, dA \qquad (13.134)$$

$$\approx \frac{1}{g}\int_A \left[\sum_{k=1}^{K_{max}} \left(\frac{u_k^2 + v_k^2}{2} + c_p T_k + \Phi_s\right)\Delta p_k\right] dA. \qquad (13.135)$$

The pressure differences Δp_k are determined by

$$\Delta p_k = p_{k+1/2} - p_{k-1/2} = (\sigma_{k+1/2} - \sigma_{k-1/2})\,(p_s - p_{top}). \qquad (13.136)$$

Note that the conservation of energy requires $\sigma_{top} = \sigma(p = p_{top}) = 0$ which is guaranteed in (13.133). The lower boundary at the surface is $\sigma_s = \sigma(p = p_s) = 1$.

As noted in Kasahara (1974) the global integral of the total energy for hydrostatic models with a pure height coordinate z in the vertical direction is represented by

$$TE = \int_A \int_{z_s}^{z_{top}} \left(\frac{\mathbf{v}^2}{2} + c_v T + gz\right) \rho \, dz \, dA \tag{13.137}$$

$$\approx \int_A \left[\sum_{k=1}^{K_{max}} \left(\frac{u_k^2 + v_k^2}{2} + c_v T_k + g z_k\right) \rho_k \, \Delta z_k \right] dA \tag{13.138}$$

where Δz_k symbolizes the height thickness of a layer with $\Delta z_k = z_{k-1/2} - z_{k+1/2}$ and z_{top} stands for the height of the model top. The quantities $\rho \mathbf{v}^2/2$, $\rho c_v T$ and $\rho g z$ are the kinetic, internal and potential energy per unit volume, respectively. c_v is the specific heat of dry air at constant volume and defined by $c_v = c_p - R_d$. R_d is the gas constant for dry air and ρ denotes the density which is defined by the ideal gas law $\rho = p/(R_d \, T)$. z_s is the height of the orography. If height-based orography-following coordinates like

$$\xi = \frac{z_{top} - z}{z_{top} - z_s} \tag{13.139}$$

are used the domain integral of the dry total energy transforms to

$$TE = \int_A \int_{\xi_{top}}^{\xi_s} \left(\frac{\mathbf{v}^2}{2} + c_v T + gz\right) \rho \, d\xi \, dA \tag{13.140}$$

$$\approx \int_A \left[\sum_{k=1}^{K_{max}} \left(\frac{u_k^2 + v_k^2}{2} + c_v T_k + g z_k\right) \rho_k \, \Delta \xi_k \right] dA \tag{13.141}$$

with the lower and upper integration limits $\xi_s = \xi(z = z_s)$ and $\xi_{top} = \xi(z = z_{top})$. $\Delta \xi_k = \xi_{k+1/2} - \xi_{k-1/2}$ is the thickness of a layer in the transformed ξ-coordinate.

As an aside, the full 3D velocity vector $\mathbf{v}_{3D} = (u, v, w)$ needs to be used for the computation of the kinetic energy in nonhydrostatic dynamical cores. A discussion of the total energy equation for dry shallow- and deep-atmosphere nonhydrostatic equation sets is provided in Staniforth et al. (2003). For moist dynamical cores the assessment of the total energy needs to be adjusted. Then p represents the pressure of the moist air and $\rho = p/(R_d \, T_v)$ is the moist density which utilizes the virtual temperature T_v. Other modifications might include the use of moist physical constants such as c_p^* as indicated above for the model HOMME. Satoh et al. (2008) discussed a particular form of the moist total energy equation for the nonhydrostatic dynamical core NICAM (their appendix B). Unfortunately, GCMs use many different and often undocumented approaches to calculating the total energy of moist air. For example, some models include the latent heat contributions Lq in *TE* where L symbolizes the latent heat of vaporization. Some models also add the energy contributions from cloud liquid water or even cloud ice. On the other hand, some models do not include either of these. In this chapter we avoid this confusion by focusing on the dry dynamical core. This allows us to assess the impact of the total energy fixer for dry air in isolation.

13.7.3.2 Ad Hoc Corrections of Total Energy

We now present three ad hoc functions for the total energy correction that we label FIXER 1–3, and demonstrate that FIXER 1 leads to bad results. This is shown via dry idealized baroclinic wave simulations (Jablonowski and Williamson 2006a) with NCAR's CAM semi-Lagrangian spectral transform dynamical core at the resolution T170 L26. First, we assess the spirit of the three total energy fixers and review how they are applied. This is also discussed in Williamson et al. (2009).

Let $\left(\hat{T}^+, \hat{\mathbf{v}}^+, \hat{p}_s^+\right)$ symbolize the temperature, horizontal wind vector and surface pressure at the end of a time step and $\left(T^-, \mathbf{v}^-, p_s^-\right)$ denote the values at the beginning of the time step. In total energy-conserving model formulations the residual

$$RES = \widehat{TE}^+ - TE^- \tag{13.142}$$

would be zero if there are no diabatic sources and sinks. Note that Collins et al. (2004) describe how to include sources and sinks which is not discussed for the dynamical cores considered here. However, if GCMs with physical parameterizations or idealized forcings like the Held and Suarez (1994) forcing are utilized these sources and sinks need to be added. According to the total energy equation for hybrid coordinates (13.131) $\widehat{TE}^+$ is given by

$$\widehat{TE}^+ = \int_A \frac{1}{g} \left\{ \left[\sum_{k=1}^{K_{max}} \left(\frac{(\hat{\mathbf{v}}_k^+)^2}{2} + c_p \hat{T}_k^+ \right) (p_0 \, \Delta A_k + \hat{p}_s^+ \, \Delta B_k) \right] + \Phi_s \hat{p}_s^+ \right\} dA. \tag{13.143}$$

The equation for TE^- is identical to (13.143) except the superscript + is replaced by superscript −, and the hat (ˆ) is removed from TE, $\mathbf{v}$, T and p_s.

In general, RES is not zero due to the inherent and explicitly imposed diffusion processes in the dynamical cores. Therefore, modifications can be made to the provisional forecast values $\left(\hat{T}^+, \hat{\mathbf{v}}^+, \hat{p}_s^+\right)$. This adjustment yields updated values $\left(T^+, \mathbf{v}^+, p_s^+\right)$ which, if substituted for the provisional values in (13.142), yield a zero residual. This is the underlying concept of the energy fixer.

The form of the energy fixer used with the CAM semi-Lagrangian model only modifies the temperature. This could be interpreted as diffusive heating in case of a positive temperature adjustment. However, in case of cooling no such physical analogy can be drawn. The future wind and surface pressure fields are set to $\mathbf{v}^+ = \hat{\mathbf{v}}^+$ and $p_s^+ = M\,\hat{p}_s^+$ where M symbolizes the mass fixer if applied (see also Sect. 13.7.1).

We now define FIXER 1. Its temperature modifications are proportional to the magnitude of the local change in T at that time step and are given by

$$T^+(\lambda,\phi,\eta) = \hat{T}^+(\lambda,\phi,\eta) + \beta_1 |\hat{T}^+(\lambda,\phi,\eta) - T^-(\lambda,\phi,\eta)|. \qquad (13.144)$$

This temperature adjustment follows the spirit of the water vapor fixer developed by Rasch and Williamson (1991) and Williamson and Rasch (1994) for predecessors of CAM 3. Its physical motivation follows the argument that if the change in water vapor is small, then it is most likely not responsible for a lack of conservation and therefore the effect of the fixer should be small. The constant β_1 is determined by replacing $\hat{T}^+$ with T^+ in (13.143) and setting $RES = 0$ in (13.142). Equation (13.144) is then substituted for T^+ in (13.143). Lastly, (13.143) and the equation for TE^- are plugged into (13.142) which is solved for β_1.

Alternatively, two other total energy fixers are suggested. FIXER 2 is given by

$$T^+(\lambda,\phi,\eta) = \hat{T}^+(\lambda,\phi,\eta) + \beta_2 \qquad (13.145)$$

and FIXER 3 is formulated as

$$T^+(\lambda,\phi,\eta) = (1+\beta_3)\,\hat{T}^+(\lambda,\phi,\eta). \qquad (13.146)$$

The energy fixer FIXER 2 changes the provisional temperature by a constant, whereas FIXER 3 changes it proportionally. Both constants β_2 and β_3 are determined as described above for FIXER 1 and β_2 is shown in Williamson et al. (2009). FIXER 2 is adopted in the EUL and SLD dynamical cores of NCAR's CAM 3.1 GCM and was used operationally for climate applications. The FV dynamical core in CAM 4 and CAM 5 applies a slightly different total energy fixer that is described in Neale et al. (2010).

Despite the pseudo-physical justification for FIXER 1, it has detrimental effects on the flow fields in the baroclinic wave test case which which had not been obvious in long Earth-like simulations. Figure 13.31 shows the surface pressure field at day 10 from the SLD baroclinic wave simulation with FIXER 1 and FIXER 2. The two are clearly different, whereas the FIXER 2 simulation is visually indistinguishable

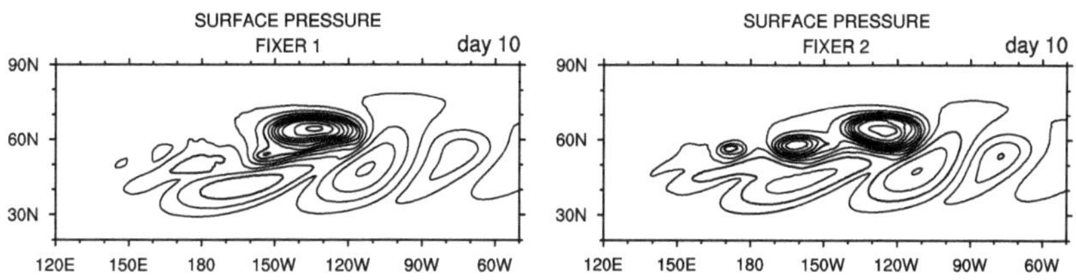

Fig. 13.31 Surface pressure at day 10 from the CAM semi-Lagrangian spectral dynamical core at the resolution T170L26 with energy fixer FIXER 1 and FIXER 2. Contour interval is 7.5 hPa with the 980 hPa and 1,010 hPa contours *thicker*. The time step is $\Delta t = 900$ s

from FIXER 3, from simulations without an energy fixer or from CAM EUL and FV reference simulations as shown in Williamson et al. (2009) and Jablonowski and Williamson (2006a,b). Therefore, FIXER 1 should not be used in practice. Here, we list it for demonstration purposes to raise awareness for potential problems with energy fixers.

The negative effect of FIXER 1 can be understood when recognizing that most of the loss of total energy is associated with damping of the wind field and the reduction of kinetic energy. However, with semi-Lagrangian approximations the damping is due in large part to the interpolations fundamental to the numerical method. There is no physical argument to justify making the fixer proportional to the change in temperature as tried in FIXER 1. Although the changes of the energy by the fixers are very small, the energy change due to FIXER 1 is systematic and obviously accumulates to a significant error. The short dynamical core run was able to isolate this problem. In full GCMs with physical parameterizations such detrimental effects of the total energy fixer are masked by many other processes and harder to identify.

Note again that the application of fixers might be unavoidable in long climate simulations to prevent unphysical signals from systematic mass or energy drifts. This is a strong argument for mass-conserving designs of dynamical cores to at least avoid mass-fixing the GCM. The latter is conceptually straightforward and some examples of mass-conserving dynamical cores include FV, FVcubed, HOMME, ICON, NICAM and WRF. NICAM also employs a total energy conservation form (Satoh 2002, 2003). Such a choice of the prognostic equation for total energy written in flux form automatically guarantees the conservation of total energy provided no explicitly added dissipation or filters are utilized. However, built-in total energy conservation in full GCMs with explicit diffusion or filters, and physical parameterizations is harder to accomplish and might not easily be achievable. Today, most climate models use a total energy fixer which is often undocumented and sometimes called with a different name. For example, the Unified Model (UM) has a total energy "correction" which is occasionally, e.g. once a day, applied (Terry Davies, personal communication). It is viewed as a correction that substitutes all missing physics processes.

13.8 Final Thoughts

The intention of this chapter was to remind the GCM modeling community of the many, sometimes hidden, diffusion processes in the dynamical cores of atmospheric general circulation models. There is no universal theory that guides the design of subgrid-scale diffusion, dissipation, mixing, damping, smoothing, filters or fixers, or however we name the many ad hoc mechanisms in GCMs. They are needed to keep the simulation stable or to satisfy important physical properties, and the hope is that they capture and mimic in some unknown way the true processes at the unresolved subgrid scale. There is no physical basis though, that dissipation can accomplish this. Therefore, a lesser goal is that the dissipative processes keep a GCM simulation stable and promote its physical realism, while doing no harm to the resolved fluid flow. The latter is a practical approach, and this chapter highlighted the many considerations that contribute to the practical designs of diffusion, filters and fixers in GCMs. In practice, diffusion processes and filtering in atmospheric models are based on a subjective determination of when the noise and short waves have been sufficiently damped while minimizing the damping of longer wave modes. This might be described as the "Art of Filtering". The selected filters are tuned for the grid resolutions and flow regimes, and this tuning is based on modeling experience.

There is no numerical scheme or diffusive process that incorporates an optimal solution to the filtering problem at all scales. An optimal formulation at a particular resolution or for a specific flow regime might fail if conditions and scales change. Adequate diffusion and filtering mechanisms are scale-dependent. For example, 2D divergence damping is considered adequate for large-scale hydrostatic motions where rotational motions are of main interest. However, small-scale mesoscale regimes are highly influenced by divergent motions, and 2D divergence damping might become detrimental at those scales in nonhydrostatic models. This remains to be seen as we enter the nonhydrostatic modeling era with future global grid resolution near the kilometer scale.

In general, there is no "right" or "wrong" solution to the subgrid-scale filtering problem. All approaches shown in this chapter have been tried in GCMs, but some have more merit or physical motivations than others. There might also be approaches that are clearly adequate or inadequate for a particular model design. In practice, GCMs apply a potpourri of damping mechanisms, either explicitly or implicitly in the numerical scheme. They act in concert, interact in nonlinear ways, and the causes and effects are usually hard to isolate individually. This chapter provided a comprehensive and systematic evaluation of the many dissipative processes in GCMs and evaluated their impact on the flow with the help of idealized test cases. It is difficult to judge how generally applicable the results of our study are with respect to full GCMs with physical parameterizations. We would expect though that many of the conclusions transfer and that new dissipative processes like boundary layer diffusion make the picture even more complex. This will require another round of evaluations, but to say the least, the dynamical core assessments might have given us new guidance and clues for the evaluations of full weather and climate models.

Acknowledgments We would like to thank Jerry Olson (NCAR) for developing the required semi-Lagrangian CAM 3.1 codes and for running the experiments required in Sects. 13.3.8 and 13.6.4. We would also like to thank Mike Blackburn (University of Reading) for discussions on the University of Reading spectral model and for pointing out relevant papers. We thank Fedor Mesinger (NCEP) and the second reviewer for their very insightful suggestions. DLW was partially supported by the Office of Science (BER), U.S. Department of Energy, Cooperative Agreement No. DE-FC02-97ER62402. CJ was supported by the Office of Science (BER), U.S. Department of Energy, Award No. DE-FG02-07ER64446. The National Center for Atmospheric Research is sponsored by the National Science Foundation.

Appendix: Overview of Selected Dynamical Cores

This chapter has featured examples from many GCMs. Here, we briefly review the basic characteristics of their dynamical cores and give pointers to their primary references. Note that the grid staggering options mentioned below are shown in Randall (1994) or Chap. 3.

CAM EUL The Eulerian (EUL) dynamical core is the default in the Community Atmosphere Model CAM version 3.1 at the National Center for Atmospheric Research (NCAR). It is optional in the later versions CAM 4 or 5 (Neale et al. 2010). This hydrostatic shallow-atmosphere dynamical core is formulated in vorticity-divergence form and based on the traditional leapfrog three-time-level, semi-implicit spectral transform approximations (Machenhauer 1979). A quadratically-unaliased Gaussian transform grid with horizontal triangular truncation is utilized (Collins et al. 2004). In the vertical direction, centered finite differences are used. All prognostic variables are co-located.

CAM SLD The semi-Lagrangian (SLD) dynamical core is an optional dynamical core in NCAR's CAM model (versions 3.1, 4 or 5). It utilizes the primitive equations and is based on two-time-level, semi-implicit semi-Lagrangian spectral transform approximations with quasi-cubic Lagrangian polynomial interpolants. A triangular truncation is adopted which can work both on a quadratically-unaliased transform grid or linear Gaussian grid (Williamson and Olson 1994; Collins et al. 2004). In our experiments here, a Gaussian quadratic transform grid was chosen unless noted otherwise. All prognostic variables are co-located.

COSMO The acronym COSMO stands for "Consortium for Small-scale Modeling" and denotes the nonhydrostatic regional weather prediction model at the German Weather Service (former name was LM which is the abbreviation for "Lokal Modell"). COSMO is a deep-atmosphere, finite-difference model on a staggered C-grid with a split-explicit temporal discretization (Doms and Schättler 2002; Gassmann and Herzog 2007; Baldauf 2010). Vertically traveling sound waves are handled implicitly.

ECHAM5 This model has been developed at the Max-Planck Institute for Meteorology (MPI) in Hamburg, Germany (Roeckner et al. 2003). It utilizes a semi-implicit Eulerian spectral transform dynamical core with triangular truncation and a Gaussian quadratic transform grid. The shallow-atmosphere dynamical core is hydrostatic and formulated in vorticity-divergence form. A three-time-level leapfrog time-stepping scheme is employed. All prognostic variables are co-located.

FV (also CAM FV) The mass-conservative Finite-Volume (FV) dynamical core employs the vector-invariant form of the primitive equations. It is written in flux form that is built upon a 2D shallow-water approach in the horizontal plane (Lin and Rood 1997) on a latitude–longitude grid with D-grid staggering. The vertical discretization utilizes a ÔLagrangian control-volumeÕ principle with conservative vertical remapping steps (Lin 2004). An explicit two-time-level time-stepping scheme is employed. The FV dynamical core is the default in

NCAR's model CAM version 4 or 5 (Neale et al. 2010), and is optional in CAM 3.1 (Collins et al. 2004). It is shared with NASA's GEOS5 model (Rienecker et al. 2008) and the NOAA Geophysical Fluid Dynamics Laboratory's (GFDL) atmospheric model AM2.1. A shallow water version and 3D variant of the FV model with an adaptive grid was also implemented by Jablonowski (2004) and Jablonowski et al. (2006, 2009).

FVcubed This variant of the FV model on an almost uniform-resolution cubed sphere grid with D-grid staggering is employed at the NASA Goddard Space Flight Center (GSFC) and GFDL (Putman and Lin 2007, 2009). It is part of the most recent internal version of NASA's GEOS model (version 6) and GFDL's internal release called "Riga". The cubed-sphere version has slightly different inherent dissipation characteristics e.g., due to the use of alternative limiters in the finite-volume algorithm. The FVcubed model also features an optional nonhydrostatic extension.

GEOS The Goddard Earth Observing System (GEOS) model has been under development at the NASA Goddard Space Flight Center. A commonly used version is GEOS5 which is documented in Rienecker et al. (2008). This version utilizes the FV dynamical core on a latitude–longitude grid as described above. Older GEOS versions like GEOS2 were based on the momentum equations in momentum form and utilized a finite-difference method on a latitude–longitude grid with a staggered C-grid arrangement. An explicit time three-time-level time-stepping scheme was used in GEOS2. A comprehensive model description can be found in Suarez and Takacs (1995). GEOS2 has also been used in a stretched-grid variant as documented in Fox-Rabinovitz et al. (1997). The forthcoming GEOS6 version will be based on the FVcubed dynamical core.

GME This primitive equation based dynamical core in vector-invariant form has been developed at the German Weather Service (DWD). It applies a finite-difference approximation with local spherical basis functions at each grid point. The horizontal grid is based on an icosahedral grid. An Arakawa-A grid staggering is chosen that places the prognostic variables at the vertices of the triangles. The semi-implicit numerical scheme is second-order accurate and applies a classical leapfrog three-time-level approach (Majewski et al. 2002, 2008)

HOMME The High Order Method Modeling Environment (HOMME) model is an optional hydrostatic dynamical core in NCAR's model CAM version 5 (Neale et al. 2010). It is a spectral element cubed-sphere dynamical core in momentum form with fourth-order polynomials inside each element (Thomas and Loft 2005; Taylor et al. 2007, 2009). The spectral element method is compatible, making the method elementwise mass-conservative (see Chap. 12 and Taylor and Fournier (2010)). The default time-stepping scheme is explicit and utilizes the three-time-level leapfrog method. Other time-stepping options are also available. All prognostic variables are co-located.

ICON The ICOsahedral Nonhydrostatic general circulation model ICON is a finite-difference model in momentum form and currently under development at MPI and DWD in Germany. It utilizes a dual icosahedral and hexagonal grid with C-grid staggering, is mass-conservative and employs a semi-implicit three-time-level leapfrog time-stepping scheme. The shallow water and hydrostatic version are documented in Rípodas et al. (2009) and Wan (2009).

IFS This Integrated Forecasting System (IFS) is used for weather predictions at the European Centre for Medium-Range Weather Forecasts (ECMWF) in Reading, U.K.. It is a primitive equation based two-time-level semi-implicit semi-Lagrangian spectral transform dynamical core with a linear Gaussian transform grid and triangular truncation (Hortal 2002; ECMWF 2010). A finite-element discretization is employed in the vertical direction (Untch and Hortal 2004). All prognostic variables are co-located.

NICAM The Nonhydrostatic Icosahedral Atmospheric Model (NICAM) is a deep-atmosphere model which has been developed at the Frontier Research Center for Global Change (FRCGC), the Japan Agency for Marine-Earth Science and Technology (JAMSTEC) and the Center for Climate System Research (CCSR) at the University of Tokyo, Japan. It is a finite-difference model written in mass-, momentum- and total energy-conserving form that utilizes a time-splitting scheme (Satoh 2002, 2003; Tomita and Satoh 2004; Satoh et al. 2008). All prognostic variables are co-located (Arakawa A-grid). NICAM's icosahedral grid configuration is optimized for uniformity by using a so-called "spring dynamics" grid.

UM The Unified Model (UM) is a mass-conserving nonhydrostatic GCM for weather and climate assessments that has been developed at the UK Met Office in Exeter, U.K.. It is built upon a finite-difference, deep-atmosphere, two-time-level, semi-implicit, semi-Lagrangian dynamical core in momentum form on a latitude–longitude grid (Davies et al. 2005; Staniforth and Wood 2008). This dynamical core is also called "New Dynamics". The prognostic variables are placed on a staggered C-grid. A comprehensive model description can be found in Staniforth et al. (2006).

WRF The Weather Research and Forecasting (WRF) model developed at NCAR is mostly used as a limited-area model. It is formulated in mass- and momentum-conserving form. WRF is a nonhydrostatic, deep-atmosphere, finite-difference model on a staggered C-grid with a split-explicit temporal discretization (Skamarock and Klemp 2008; Skamarock et al. 2008). Vertically traveling sound waves are handled implicitly. WRF can utilize a variety of map transformations and grids, e.g., the latitude–longitude grid for global applications.

References

Andrews DG, Mahlman JD, Sinclair RW (1983) Eliassen-Palm diagnostics of wave-mean flow interaction in the GFDL "SKYHI" general circulation model. J Atmos Sci 40:2768–2784

Arakawa A (1966) Computational design for long-term numerical integration of the equations of fluid motion: Two-dimensional incompressible flow. Part I. J Comput Phys 1:119–143

Arakawa A, Hsu YJG (1990) Energy conserving and potential-enstrophy dissipating schemes for the shallow water equations. Mon Wea Rev 118:1960–1969

Arakawa A, Lamb VR (1981) A potential enstrophy and energy conserving scheme for the shallow water equations. Mon Wea Rev 109:18–36

Arakawa A, Suarez MJ (1983) Vertical differencing of the primitive equations in sigma coordinates. Mon Wea Rev 111(1):34–45
Asselin R (1972) Frequency filter for time integrations. Mon Wea Rev 100(5):487–490
Baldauf M (2010) Linear stability analysis of Runge-Kutta based partial time-splitting schemes for the Euler equations. Mon Wea Rev 138:4475–4496
Bates JR, Semazzi FHM, Higgins RW, Barros SRM (1990) Integration of the shallow water equations on the sphere using a vector semi-Lagrangian scheme with a multigrid solver. Mon Wea Rev 118:1615–1627
Bates JR, Moorthi S, Higgins RW (1993) A global multilevel atmospheric model using a vector semi-Lagrangian finite difference scheme. Part I: Adiabatic formulation. Mon Wea Rev 121:244–263
Becker E (2001) Symmetric stress tensor formulation of horizontal momentum diffusion in global models of atmospheric circulation. J Atmos Sci 58:269–282
Becker E (2003) Frictional heating in global climate models. Mon Wea Rev 131:508–520
Becker E, Burkhardt U (2007) Nonlinear horizontal diffusion for GCMs. Mon Wea Rev 135: 1439–1454
Bermejo R, Staniforth A (1992) The Conversion of Semi-Lagrangian Advection Schemes to Quasi-Monotone Schemes. Mon Wea Rev 120:2622–2632
Black TL (1994) The new NMC mesoscale Eta model: Description and forecast examples. Weather and Forecasting 9:265–284
Boer GJ, Denis B (1997) Numerical covergence of the dynamics of a GCM. Climate Dynamics 13:359–374
Boer GJ, Shepherd TG (1983) Large-scale two-dimensional turbulence in the atmosphere. J Atmos Sci 40:164–184
Bonaventura L, Ringler T (2005) Analysis of discrete shallow-water models on geodesic Delaunay grids with C-type staggering. Mon Wea Rev 133:2351–2373
Bourke W, McAvaney B, Puri K, Thurling R (1977) Global modeling of atmospheric flow by spectral methods. In: Chang J (ed) Methods in Computational Physics, vol 17, Academic Press, pp 267–324
Boville BA (1986) Wave-mean flow interactions in a general circulation model of the troposphere and stratosphere. J Atmos Sci 43:1711–1725
Boville BA (1991) Sensitivity of simulated climate to model resolution. J Climate 4:469–485
Boville BA (2000) Toward a complete model of the climate system. In: Mote P, O'Neill A (eds) Numerical modeling of the global atmosphere in the climate system, NATO Science Series C: Mathematical and Physical Sciences, vol 550, Kluwer Academic Publishers, pp 419–442
Boville BA, Baumhefner DP (1990) Simulated forecast error and climate drift resulting from the omission of the upper stratosphere in numerical models. Mon Wea Rev 118:1517–1530
Boville BA, Bretherton CS (2003) Heating and kinetic energy dissipation in the NCAR Community Atmosphere Model. J Climate 16:3877–3887
Boville BA, Gent PR (1998) The NCAR Climate System Model, version one. J Climate 11:1115–1130
Boville BA, Randel WJ (1986) Observations and simulation of the variability of the stratosphere and troposphere in January. J Atmos Sci 43:3015–3034
Bowler NE, Arribas A, Beare SE, Mylne KR, Shutts GJ (2009) The local ETKF and SKEB: Upgrades to the MOGREPS short-range ensemble prediction system. Quart J Roy Meteor Soc 135:767–776
Boyd J (1996) The Erfc-Log filter and the asymptotics of the Euler and Vandeven sum accelerations. In: Proc. of the Third International Conference on Spectral and High Order Methods, pp 267–276, ed. by A. V. Ilin and L. R. Scott, Houston J. Mathematics, Houston, Texas
Boyd J (1998) Two comments on filtering (artificial viscosity) for Chebyshev and Legendre spectral and spectral element methods: Preserving boundary conditions and interpretation of the filter as a diffusion. J Comput Phys 143:283–288
Burkhardt U, Becker E (2006) A consistent diffusion dissipation parameterization in the ECHAM climate model. Mon Wea Rev 134:1194–1204

Canuto C, Hussaini MY, Quarteroni A, Zang TA (1987) Spectral Methods in Fluid Dynamics. Springer, 600 pp.

Carpenter RL, Droegemeier KK, Woodward PR, Hane CE (1990) Application of the Piecewise Parabolic Method (PPM) to meteorological modeling. Mon Wea Rev 118:586–612

Chen TC, Wiin-Nielsen A (1978) Nonlinear cascades of atmospheric energy and enstrophy in a two-dimensional spectral index. Tellus 30:313–322

Colella P, Sekora MD (2008) A limiter for PPM that preserves accuracy at smooth extrema. J Comput Phys 227:7069–7076

Colella P, Woodward PR (1984) The Piecewise Parabolic Method (PPM) for gas-dynamical simulations. J Comput Phys 54:174–201

Collins WD, Rasch PJ, Boville BA, Hack JJ, McCaa JR, Williamson DL, Kiehl JT, Briegleb BP, Bitz CM, Lin SJ, Zhang M, Dai Y (2004) Description of the NCAR Community Atmosphere Model (CAM3.0). NCAR Technical Note NCAR/TN-464+STR, National Center for Atmospheric Research, Boulder, Colorado, 214 pp., available from http://www.ucar.edu/library/collections/technotes/technotes.jsp

Cordero E, Staniforth A (2004) A problem with the Robert-Asselin time filter for three-time-level semi-implicit semi-Lagrangian discretizations. Mon Wea Rev 132:600–610

Côté J, Gravel S, Staniforth A (1995) A generalized family of schemes that eliminate the spurious resonant response of semi-Lagrangian schemes to orographic forcing. Mon Wea Rev 123:3605–3613

Côté J, Gravel S, Méthot A, Patoine A, Roch M, Staniforth A (1998a) The operational CMC-MRB Global Environmental Multiscale (GEM) model. Part I: Design considerations and formulation. Mon Wea Rev 126(6):1373–1395

Côté J, Gravel S, Méthot A, Patoine A, Roch M, Staniforth A (1998b) The operational CMC-MRB Global Environmental Multiscale (GEM) model. Part I: Design considerations and formulation. Mon Wea Rev 126(6):1373–1395

Davies T, Cullen MJP, Malcolm AJ, Mawson MH, Staniforth A, White AA, Wood N (2005) A new dynamical core for the Met Office's global and regional modelling of the atmosphere. Quart J Roy Meteor Soc 131(608):1759–1782

Deardorff JW (1970) A numerical study of three-dimensional turbulent channel flow at large Reynolds numbers. J Fluid Mechanics 41:453–480

Déqué M, Cariolle D (1986) Some destabilizing properties of the Asselin time filter. Mon Wea Rev 114:880–884

Dey CH (1978) Noise suppression in a primitive equation prediction model. Mon Wea Rev 106:159–173

Doms G, Schättler U (2002) A description of the nonhydrostatic regional model LM. Part I: Dynamics and numerics. Consortium for Small-Scale Modelling (COSMO) LM F90 2.18, German Weather Service, Offenbach, Germany, 134 pp., available from http://www.cosmo-model.org

Dubal M, Wood N, Staniforth A (2004) Analysis of parallel versus sequential splittings for time-stepping physical parameterizations. Mon Wea Rev 132:121–132

Dubal M, Wood N, Staniforth A (2006) Some numerical properties of approaches to physics dynamics coupling for NWP. Quart J Roy Meteor Soc 132:27–42

Dudhia J (1993) A nonhydrostatic version of the Penn State-NCAR mesoscale model: Validation tests and simulation of an Atlantic cyclone and cold front. Mon Wea Rev 121:1493–1513

Durran DR (1999) Numerical methods for wave equations in geophysical fluid dynamics, First edn. Springer, 465 pp.

Durran DR (2010) Numerical methods for fluid dynamics: with applications in geophysics, Second edn. Springer, 516 pp.

Durran DR, Klemp JB (1983) A compressible model for the simulation of moist mountain waves. Mon Wea Rev 111:2341–2361

ECMWF (2010) Part III: Dynamics and numerical procedures. IFS Documentation - Cy36r1, Operational implementation 26 January 2010, European Centre for Medium-Range Weather Forecasts, Shinfield Park, Reading, England, 29 pp., available from http://www.ecmwf.int/research/ifsdocs/CY36r1/index.html

Enomoto T, Kuwano-Yoshida A, Komori N, Ohfuchi W (2008) Description of AFES 2: Improvements for high-resolution and coupled simulations. In: Hamilton K, Ohfuchi W (eds) High resolution numerical modelling of the atmosphere and ocean, Springer, pp 77–97

Farge M, Sadourny R (1989) Wave-vortex dynamics in rotating shallow water. J Fluid Mech 206:433–462

Fiedler BH (2000) Dissipative heating in climate models. Quart J Roy Meteor Soc 126:925–939

Fox-Rabinovitz MS, Stenchikov GL, Suarez MJ, Takacs LL (1997) A finite-difference GCM dynamical core with a variable-resolution stretched grid. Mon Wea Rev 125:2943–2968

Fudeyasu H, Wang Y, Satoh M, Nasuno T, Miura H, Yanase W (2008) Global cloud-system-resolving model NICAM successfully simulated the lifecycles of two real tropical cyclones. Geophys Res Lett 35:L22,808

Gassmann A, Herzog HJ (2007) A consistent time-split numerical scheme applied to the non-hydrostatic compressible equations. Mon Wea Rev 135:20–36

Gelb A, Gleeson JP (2001) Spectral viscosity for shallow water equations in spherical geometry. Mon Wea Rev 129:2346–2360

GFS (2003) The GFS atmospheric model. NCEP Office Note 442, National Centers for Environmental Prediction (NCEP), Environmental Modeling Center, Camp Springs, Maryland, 14 pp., available from http://www.emc.ncep.noaa.gov/officenotes/index.shtml see also http://wwwt.emc.ncep.noaa.gov/?branch=GFS

Gill AE (1982) Atmosphere-ocean dynamics. Academic Press, 768 pp.

Giraldo FX, Rosmond TE (2004) A scalable Spectral Element Eulerian Atmospheric Model (SEE-AM) for NWP: Dynamical core tests. Mon Wea Rev 132:133–153

Giraldo FX, Hesthaven JS, Wartburton T (2002) Nodal high-order discontinous Galerkin methods for the shallow water equations. J Comput Phys 181:499–525

Gordon CT, Stern WF (1982) A description of the GFDL global spectral model. Mon Wea Rev 110:625–644

Gravel S, Staniforth A, Côté J (1993) A stability analysis of a family of baroclinic semi-Lagrangian forecast models. Mon Wea Rev 121:815–824

Griffies SM, Hallberg RW (2000) Biharmonic friction with a Smagorinsky-like viscosity for use in large-scale Eddy-permitting ocean models. Mon Wea Rev 128:2935–2946

Gross ES, Bonaventura L, Rosatti G (2002) Consistency with continuity in conservative advection schemes for free-surface models. Int J Numer Meth Fluids 38:307–327

Haltiner GJ, Williams RT (1980) Numerical prediction and dynamic meteorology. John Wiley & Sons, 477 pp.

Hamilton K, Takahashi YO, Ohfuchi W (2008) Mesoscale spectrum of atmospheric motions investigated in a very fine resolution global general circulation model. J Geophys Res 113:D18,110

Hamming RW (1977) Digital filters. Prentice Hall, 226 pp.

Held IM, Suarez MJ (1994) A proposal for the intercomparison of the dynamical cores of atmospheric general circulation models. Bull Amer Meteor Soc 75(10):1825–1830

Hortal M (2002) The development and testing of a new two-time-level semi-Lagrangian scheme (SETTLS) in the ECMWF forecast model. Quart J Roy Meteor Soc 128(583):1671–1687

Hortal M, Simmons AJ (1991) Use of reduced grids in spectral models. Mon Wea Rev 119:1057–1074

Huang H, Stevens B, Margulis SA (2008) Application of dynamic subgrid-scale models for large-eddy simulation of the daytime convective boundary layer over heterogeneous surfaces. Boundary-Layer Meteorology 126:327–348

Huynh HY (1996) Schemes and constraints for advection. In: 15th International Conference on Numerical Methods in Fluid Dynamics, Monterey June 24–28, 1996, CA, USA

Jablonowski C (1998) Test der Dynamik zweier globaler Wettervorhersagemodelle des Deutschen Wetterdienstes: Der Held-Suarez Test. Master's thesis, University of Bonn, Germany, Department of Meteorology, 140 pp.

Jablonowski C (2004) Adaptive grids in weather and climate modeling. PhD thesis, University of Michigan, Ann Arbor, MI, Department of Atmospheric, Oceanic and Space Sciences, 292 pp.

Jablonowski C, Williamson DL (2006a) A baroclinic instabilitiy test case for atmospheric model dynamical cores. Quart J Roy Meteor Soc 132(621C):2943–2975

Jablonowski C, Williamson DL (2006b) A baroclinic wave test case for dynamical cores of General Circulation Models: Model intercomparisons. NCAR Tech. Note NCAR/TN-469+STR, National Center for Atmospheric Research, Boulder, Colorado, 89 pp., available from http://www.ucar.edu/library/collections/technotes/technotes.jsp

Jablonowski C, Herzog M, Penner JE, Oehmke RC, Stout QF, van Leer B, Powell KG (2006) Block-structured adaptive grids on the sphere: Advection experiments. Mon Wea Rev 134:3691–3713

Jablonowski C, Oehmke RC, Stout QF (2009) Block-structured adaptive meshes and reduced grids for atmospheric general circulation models. Phil Trans R Soc A 367:4497–4522

Jakimow G, Yakimiw E, Robert A (1992) An implicit formulation for horizontal diffusion in gridpoint models. Mon Wea Rev 120:124–130

Jakob R, Hack JJ, Williamson DL (1993) Solutions to the shallow-water test set using the spectral transform method. NCAR Tech. Note NCAR/TN-388+STR, National Center for Atmospheric Research, Boulder, Colorado, 82 pp., available from http://www.ucar.edu/library/collections/technotes/technotes.jsp

Janjić ZI (1990) The step-mountain coordinate: Physical package. Mon Wea Rev 118:1429–1443

Jöckel P, von Kuhlmann R, Lawrence MG, Steil B, Brenninkmelter CAM, Crutzen PJ, Rasch PJ, Eaton B (2001) On a fundamental problem in implementing flux-form advection schemes for tracer transport in 3-dimensional general circulation and chemistry transport models. Quart J Roy Meteor Soc 127:1035–1052

Kalnay E (2003) Atmospheric modeling, data assimilation and predictability. Cambridge University Press, 341 pp.

Kalnay-Rivas E, Bayliss A, Storch J (1977) The 4th order GISS model of the global atmosphere. Beiträge zur Physik der Atmosphäre 50:299–311

Kasahara A (1974) Various vertical coordinate systems used for numerical weather prediction. Mon Wea Rev 102:509–522

Klemp JB, Durran DR (1983) An upper boundary condition permitting internal gravity wave radiation in numerical mesoscale models. Mon Wea Rev 111:430–444

Klemp JB, Lilly DK (1978) Numerical simulation of hydrostatic mountain waves. J Atmos Sci 35:78–107

Klemp JB, Skamarock WC, Dudhia J (2007) Conservative split-explicit time integration methods for the compressible nonhydrostatic equations. Mon Wea Rev 135:2897–2913

Klemp JB, Dudhia J, Hassiotis AD (2008) An upper gravity-wave absorbing layer for NWP applications. Mon Wea Rev 136:3987–4004

Knievel JC, Bryan GH, Hacker JP (2007) Explicit numerical diffusion in the WRF model. Mon Wea Rev 135:3808–3824

Koshyk JN, Boer GJ (1995) Parameterization of dynamical subgrid-scale processes in a spectral GCM. J Atmos Sci 52(7):965–976

Koshyk JN, Hamilton K (2001) The horizontal kinetic energy spectrum and spectral budget simulated by a high-resolution troposphere-stratosphere-mesosphere GCM. J Atmos Sci 58:329–348

Lander J, Hoskins BJ (1997) Believable scales and parameterizations in a spectral transform model. Mon Wea Rev 125:292–303

Laprise R, Girard C (1990) A spectral General Circulation Model using the piecewise-constant finite-element representation on a hybrid vertical coordinate system. J Climate 3:32–52

Lauritzen PH, Jablonowski C, Taylor MA, Nair RD (2010a) Rotated versions of the Jablonowski steady-state and baroclinic wave test cases: A dynamical core intercomparison. J Adv Model Earth Syst 2:Art. #15, 34 pp.

Lauritzen PH, Nair RD, Ullrich PA (2010b) A conservative semi-Lagrangian multi-tracer transport scheme (CSLAM) on the cubed-sphere grid. J Comput Phys 229:1401–1424

Laursen L, Eliasen E (1989) On the effects of the damping mechanisms in an atmospheric general circulation model. Tellus Series A 41:385–400

van Leer B (1974) Towards the ultimate conservative difference scheme. II. Monotonicity and conservation combined in a second-order scheme. J Comput Phys 14:361–370
van Leer B (1977) Towards the ultimate conservative difference scheme. IV. A new approach to numerical convection. J Comput Phys 23:276–299
Leith CE (1971) Atmospheric predictability and two-dimensional turbulence. J Atmos Sci 28: 145–161
LeVeque RJ (2002) Finite Volume Methods for Hyperbolic Problems. Cambridge University Press, ISBN 0-521-00924-3, 558 pp.
Li Y, Moorthi S, Bates JR (1994) Direct solution of the implicit formulation of fourth order horizontal diffusion for gridpoint models on the sphere. NASA Technical Memorandum 104606, Vol. 2, NASA Goddard Space Flight Center, Greenbelt, Maryland, 42 pp.
Lin SJ (2004) A "vertically Lagrangian" finite-volume dynamical core for global models. Mon Wea Rev 132:2293–2307
Lin SJ, Rood RB (1996) Multidimensional flux-form semi-Lagrangian transport scheme. Mon Wea Rev 124:2046–2070
Lin SJ, Rood RB (1997) An explicit flux-form semi-Lagrangian shallow water model on the sphere. Quart J Roy Meteor Soc 123:2477–2498
Lin YL (2007) Mesoscale dynamics. Cambridge University Press, 630 pp.
Lindberg K, Alexeev VA (2000) A study of the spurious orographic resonance in semi-implicit semi-Lagrangian models. Mon Wea Rev 128:1982–1989
Lynch P, Huang X (1992) Initialization of the HIRLAM model using a digital filter. Mon Wea Rev 120:1019–1034
Machenhauer B (1979) The spectral method. In: Kasahara A (ed) Numerical methods used in atmospheric models, vol 2, GARP Publications Series No 17, WMO and ICSU, Geneva, pp 121–275
MacVean MK (1983) The effects of horizontal diffusion on baroclinic development in a spectral model. Quart J Roy Meteor Soc 109:771–783
Mahlmann JD, Sinclair WW (1977) Tests of various numerical algorithms applied to a simple trace constituent air transport problem. In: Suffett IH (ed) Advances in Environmental Science and Technology, Part 1: The Fate of Pollutants in the Air and Water Environments, vol 8, Wiley, pp 223–252
Majewski D, Liermann D, Prohl P, Ritter B, Buchhold M, Hanisch T, Paul G, Wergen W, Baumgardner J (2002) The operational global icosahedral-hexagonal gridpoint model GME: Description and high-resolution tests. Mon Wea Rev 130:319–338
Majewski D, Frank H, Liermann D (2008) GME User's Guide. Tech. Rep. corresponding to model version gmtri 2.17 and higher, German Weather Service DWD, Frankfurt, Germany, 70 pp.
McCorquodale P, Colella P (2010) A high-order finite-volume method for hyperbolic conservation laws on locally-refined grids. J. Comput. Phys., in review
McDonald A, Haugen J (1992) A two-time-level, three-dimensional semi-Lagrangian, semi-implicit, limited-area gridpoint model of the primitive equations. Mon Wea Rev 120(11):2603–2621
McDonald A, Haugen J (1993) A two time-level, three-dimensional, semi-Lagrangian, semi-implicit, limited-area gridpoint model of the primitive equations. Part II: Extension to hybrid vertical coordinates. Mon Wea Rev 121(7):2077–2087
McPherson RD, Stackpole JD (1973) Noise suppression in the eight-layer global model. Office Note 83, U.S. Department of Commerce, National Oceanic and Atmospheric Administration, National Weather Service, National Meteorological Center, 36 pp., available from http://www.emc.ncep.noaa.gov/officenotes/index.shtml
Mellor GL (1985) Ensemble average, turbulence closure. In: Manabe S (ed) Advances in Geophysics, Issues in Atmos. and Ocean Modeling. Part B: Weather Dynamics, Academic Press, pp 345–357
Mesinger F, Arakawa A (1976) Numerical methods used in atmospheric models. In: GARP Publication Series No. 17, vol 1, World Meteorological Organization, p 64, Geneva, Switzerland

Mesinger F, Jovic D (2002) The Eta slope adjustment: Contender for an optimal steepening in a piecewise-linear advection scheme? Comparison tests. NCEP Office Note 439, National Centers for Environmental Prediction (NCEP), Environmental Modeling Center, Camp Springs, Maryland, 29 pp., available from http://www.emc.ncep.noaa.gov/officenotes/index.shtml

Nair RD (2009) Diffusion experiments with a global discontinuous Galerkin shallow-water model. Mon Wea Rev 137:3339–3350

Nastrom GD, Gage KS (1985) A climatology of atmospheric wavenumber spectra of wind and temperature observed by commercial aircraft. J Atmos Sci 42:950–960

Neale RB, Hoskins BJ (2000) A standard test for AGCMs including their physical parameterizations: I: The proposal. Atmos Sci Lett 1:101–107

Neale RB, Chen CC, Gettelman A, Lauritzen PH, Park S, Williamson DL, Conley AJ, Garcia R, Kinnison D, Lamarque JF, Marsh D, Mills M, Smith AK, Tilmes S, Vitt F, Cameron-Smith P, Collins WD, Iacono MJ, Rasch PJ, Taylor M (2010) Description of the NCAR Community Atmosphere Model (CAM 5.0). NCAR Technical Note NCAR/TN-XXX+STR, National Center for Atmospheric Research, Boulder, Colorado, draft, available from http://www.cesm.ucar.edu/models/cesm1.0/cam/

Nelson SP, Weible ML (1980) Three-dimensional Shuman filter. J Appl Meteor 19:464–469

Orr A, Wedi N (2009) The representation of non-orographic gravity waves in the IFS Part I: Assessment of the middle atmosphere climate with Rayleigh friction. ECMWF Technical Memorandum 592, European Centre for Medium-Range Weather Forecasts, Reading, U.K., 15 pp., available from http://www.ecmwf.int/publications/library/do/references/list/14

Orr A, Bechthold P, Scinocca J, Ern M, Janiskova M (2010) Improved middle atmosphere climate and forecasts in the ECMWF model through a non-orographic gravity wave drag parameterization. J Climate 23:5905–5926

Orszag SA (1971) On the elimination of aliasing in finite-difference schemes by filtering high-wavenumber components. J Atmos Sci 28:1074–1074

O'Sullivan D, Dunkerton TJ (1995) Generation of inertia-gravity waves in a simulated life cycle of baroclinic instability. J Atmos Sci 52:3695–3716

Phillips N (1959) An example of non-linear computational instability. In: Bolin B (ed) The Atmosphere and Sea in Motion, Oxford University Press, pp 501–504

Phillips NA (1957) A coordinate system having some special advantages for numerical forecasting. J Meteor 14:184–185

Polvani LM, Kushner PJ (2002) Tropospheric response to stratospheric perturbations in a relatively simple general circulation model. Geophys Res Lett 29(7):070,000–1

Polvani LM, Scott RK, Thomas SJ (2004) Numerically converged solutions of the global primitive equations for testing the dynamical core of atmospheric GCMs. Mon Wea Rev 132:2539–2552

Purser RJ (1987) The filtering of meteorological fields. J Climate and Appl Meteor 26:1764–1769

Purser RJ (1988) Degradation of numerical differencing caused by Fourier filtering at high latitudes. Mon Wea Rev 116:1057–1066

Purser RJ, Leslie LM (1994) An efficient semi-Lagrangian scheme using third-order semi-implicit time integration and forward trajectories. Mon Wea Rev 122(4):745–756

Putman WM, Lin SJ (2007) Finite-volume transport on various cubed-sphere grids. J Comput Phys 227:55–78

Putman WM, Lin SJ (2009) A finite-volume dynamical core on the cubed-sphere grid. In: Numerical Modeling of Space Plasma Flows: Astronum-2008, Astronomical Society of the Pacific Conference Series, vol 406, pp 268–276

Randall DA (1994) Geostrophic adjustment and the finite-difference shallow-water equations. Mon Wea Rev 122:1371–1377

Rasch PJ (1986) Toward atmospheres without tops: Absorbing upper boundary conditions for numerical models. Quart J Roy Meteor Soc 112:1195–1218

Rasch PJ, Williamson DL (1990a) Computational aspects of moisture transport in global models of the atmosphere. Quart J Roy Meteor Soc 116:1071–1090

Rasch PJ, Williamson DL (1990b) On shape-preserving interpolation and semi-Lagrangian transport. SIAM J Sci Stat Comput 11(4):656–687

Rasch PJ, Williamson DL (1991) The sensitivity of a general circulation model climate to the moisture transport formulation. J Geophy Res 96:13,123–13,137

Rasch PJ, Boville BA, Brasseur GP (1995) A three-dimensional general circulation model with coupled chemistry for the middle atmosphere. J Geophys Res 100:9041–9072

Raymond WH (1988) High-order low-pass implicit tangent filters for use in finite area calculations. Mon Wea Rev 116:2132–2141

Raymond WH, Garder A (1988) A spatial filter for use in finite area calculations. Mon Wea Rev 116:209–222

Raymond WH, Garder A (1991) A review of recursive and implicit filters. Mon Wea Rev 119: 477–495

Reames FM, Zapotocny TH (1999) Inert trace constituent transport in sigma and hybrid isentropic-sigma models. Part I: Nine advection algorithms. Mon Wea Rev 127:173–187

Rienecker MM, Suarez MJ, Todling R, Bacmeister J, Takacs L, Liu HC, Gu W, Sienkiewicz M, Koster RD, Gelaro R, Stajner I, Nielsen E (2008) The GEOS-5 data assimilation system – Documentation of versions 5.0.1 and 5.1.0. Technical Report Series on Global Modeling and Data Assimilation NASA/TM-2007-104606, Vol. 27, NASA Goddard Space Flight Center, Greenbelt, Maryland, 92 pp.

Rípodas P, Gassmann A, Förstner J, Majewski D, Giorgetta M, Korn P, Kornblueh L, Wan H, Zängl G, Bonaventura L, Heinze T (2009) Icosahedral Shallow Water Model (ICOSWM): results of shallow water test cases and sensitivity to model parameters. Geosci Model Dev 2:231–251

Ritchie H, Temperton C, Simmons A, Hortal M, Davies T, Dent D, Hamrud M (1995) Implementation of the Semi-Lagrangian Method in a High-Resolution Version of the ECMWF Forecast Model. Mon Wea Rev 123:489–514

Rivest C, Staniforth A, Robert A (1994) Spurious resonant response of semi-Lagrangian discretizations to orographic forcing: Diagnosis and solution. Mon Wea Rev 122:366–376

Robert A, Lépine M (1997) Anomaly in the behaviour of the time filter used with the leapfrog scheme in atmospheric models. In: Lin CA, Laprise R, Ritchie H (eds) Numerical Methods in Atmospheric and Oceanic Modeling: The André J. Robert Memorial Volume, Canadian Meteorological and Oceanographic Society / NRC Research Press, pp S3–S15

Robert AJ (1966) The Integration of a low order spectral form of the primitive meteorological equations. J Meteor Soc Japan 44(5):237–245

Roeckner E, Bäuml G, Bonaventura L, Brokopf R, Esch M, Hagemann MGS, Kirchner I, Kornblueh L, Manzini E, Rhodin A, Schlese U, Schulzweida U, Tompkins A (2003) The atmospheric general circulation model ECHAM5. Part I model description. Tech. Rep. 349, Max Planck Institute for Meteorology, 127 pp., available from http://www.mpimet.mpg.de/en/wissenschaft/modelle/echam/echam5.html

Rood RB (1987) Numerical advection algorithms and their role in atmospheric transport and chemistry models. Rev Geophys 25:71–100

Royer J (1986) Correction of negative mixing ratios in spectral models by global horizontal borrowing. Mon Wea Rev 114:1406–1410

Ruge JW, McCormick SF, Yee SYK (1995) Multilevel adaptive methods for semi-implicit solution of shallow-water equations on the sphere. Mon Wea Rev 123:2197–2205

Sadourny R (1975) The dynamics of finite-difference models of the shallow-water equations. J Atmos Sci 32:680–689

Sadourny R, Maynard K (1997) Formulations of lateral diffusion in geophysical fluid dynamics models. In: Lin CA, Laprise R, Ritchie H (eds) Numerical Methods in Atmospheric and Oceanic Modeling: The André J. Robert Memorial Volume, Canadian Meteorological and Oceanographic Society / NRC Research Press, pp 547–556

Sardeshmukh PD, Hoskins BI (1984) Spatial smoothing on the sphere. Mon Wea Rev 112:2524–2529

Satoh M (2002) Conservative scheme for the compressible nonhydrostatic models with the horizontally explicit and vertically implicit time integration scheme. Mon Wea Rev 130(5):1227–1245

Satoh M (2003) Conservative scheme for a compressible nonhydrostatic model with moist processes. Mon Wea Rev 131:1033–1050

Satoh M (2004) Atmospheric circulation dynamics and general circulation models. Springer (Praxis), 643 pp.

Satoh M, Matsuno T, Tomita H, Miura H, Nasuno T, Iga S (2008) Nonhydrostatic icosahedral atmospheric model (NICAM) for global cloud resolving simulations. J Comput Phys 227:3486–3514

Schlesinger RE, Uccellini LW, Johnson DR (1983) The effects of the Asselin time filter on numerical solutions to the linearized shallow-water wave equations. Mon, Wea, Rev, 111:455–467

Schmidt GA, Ruedy R, Hansen JE, Aleinov I, Bell N, Bauer M, Bauer S, Cairns B, Canuto V, Cheng Y, Del Genio A, Faluvegi G, Friend AD, Hall TM, Hu Y, Kelley M, Kiang NY, Koch D, Lacis AA, Lerner J, Lo KK, Miller RL, Nazarenko L, Oinas V, Perlwitz J, Perlwitz J, Rind D, Romanou A, Russell GL, Sato M, Shindell DT, Stone PH, Sun S, Tausnev N, Thresher D, Yao MS (2006) Present-day atmospheric simulations using GISS ModelE: Comparison to in situ, satellite, and reanalysis data. J Climate 19:153–192

Shapiro R (1970) Smoothing, filtering, and boundary effects. Rev Geophys and Space Phys 8(2):359–387

Shapiro R (1971) The use of linear filtering as a parameterization of atmospheric diffusion. J Atmos Sci 28:523–531

Shapiro R (1975) Linear filtering. Mathematics of Computation 29(132):1094–1097

Shepherd TG, Shaw TA (2004) The angular momentum constraint on climate sensitivity and downward Influence in the middle atmosphere. J Atmos Sci 61:2899–2908

Shepherd TG, Semeniuk K, Koshyk JN (1996) Sponge layer feedbacks in middle-atmosphere models. J Geophys Res 101:23,447–23,464

Shuman FG (1957) Numerical methods in weather prediction: II. Smoothing and filtering. Mon Wea Rev 85:357–361

Shuman FG (1969) On a special form for viscous terms. Office Note 32, U.S. Department of Commerce, Environmental Science Service Administration, Weather Bureau, 4 pp., available from http://www.emc.ncep.noaa.gov/officenotes/index.shtml

Shuman FG, Stackpole JD (1969) The currently operational NMC model, and results of a recent numerical experiment. In: Proc. WMO/IUGG Symp. Numerical Weather Prediction, Japan Meteorological Agency, Tokyo, Japan, pp II–85–98

Shutts G (2005) A kinetic energy backscatter algorithm for use in ensemble prediction systems. Quart J Roy Meteor Soc 131:3079–3102

Simmons AJ, Burridge DM (1981) An energy and angular-momentum conserving vertical finite-difference scheme and hybrid vertical coordinates. Mon Wea Rev 109:758–766

Skamarock WC (2004) Evaluating mesoscale NWP models using kinetic energy spectra. Mon Wea Rev 132:3019–3032

Skamarock WC, Klemp JB (1992) The stability of time-split numerical methods for the hydrostatic and the nonhydrostatic elastic equations. Mon Wea Rev 120:2109–2127

Skamarock WC, Klemp JB (2008) A time-split nonhydrostatic atmospheric model for weather research and forecasting applications. J Comput Phys 227:3465–3485

Skamarock WC, Klemp JB, Dudhia J, Gill DO, Barker DM, Duda MG, Huang XY, Wang W, Powers JG (2008) A description of the Advanced Research WRF Version 3. NCAR Tech. Note NCAR/TN-475+STR, National Center for Atmospheric Research, Boulder, Colorado, 113 pp., available from http://www.ucar.edu/library/collections/technotes/technotes.jsp

Smagorinsky J (1963) General circulation experiments with the primitive equations. I. The basic experiment. Mon Wea Rev 91:99–164

Smagorinsky J (1993) Some historical remarks on the use of nonlinear viscosities. In: Galperin B, Orszag SA (eds) Large Eddy Simulation of Complex Engineering and Geophysical Flows, Cambridge University Press, pp 3–36

St-Cyr A, Jablonowski C, Dennis J, Thomas S, Tufo H (2008) A comparison of two shallow water models with non-conforming adaptive grids. Mon Wea Rev 136:1898–1922

Staniforth A, Côté J (1991) Semi-Lagrangian integration schemes for atmospheric modeling - A review. Mon Wea Rev 119:2206–2223

Staniforth A, Wood N (2008) Aspects of the dynamical core of a nonhydrostatic, deep-atmosphere, unified weather and climate-prediction model. J Comput Phys 227(7):3445–3464

Staniforth A, White A, Wood N (2003) Analysis of semi-Lagrangian trajectory computations. Quart J Roy Meteor Soc 129(591):2065–2085

Staniforth A, White A, Wood N, Thuburn J, Zerroukat M, Cordero E, Davies T, Diamantakis M (2006) Joy of U.M. 6.3 - model formulation. Unified Model Documentation Paper No 15, UK Meteorological Service, Exeter, England, available from http://research.metoffice.gov.uk/research/nwp/publications/papers/unified_model/

Stephenson DB (1994) The impact of changing the horizontal diffusion scheme on the northern winter climatology of a general circulation model. Quart J Roy Meteor Soc 120:211–226

Suarez MJ, Takacs LL (1995) Documentation of the ARIES/GEOS dynamical core: Version 2. NASA Technical Memorandum 104606, Vol. 5, NASA Goddard Space Flight Center, Greenbelt, Maryland, 45 pp.

Takacs L, Sawyer W, Suarez MJ, Fox-Rabnovitz MS (1999) Filtering techniques on a stretched grid general circulation model. Tech. Rep. NASA/TM-1999-104606, Vol. 16, NASA Goddard Space Flight Center, Greenbelt, Maryland, 50 pp.

Takacs LL (1988) Effect of using a posteriori methods for the conservation of integral invariants. Mon Wea Rev 116:525–545

Takacs LL, Balgovind RC (1983) High-latitude filtering in global grid-point models. Mon Wea Rev 111:2005–2015

Takahashi YO, Hamilton K, Ohfuchi W (2006) Explicit global simulation of the mesoscale spectrum of atmospheric motions. Geophys Res Lett 33:L12,812

Tandon MK (1987) Robert's recursive frequency filter: A reexamination. Meteor Atmos Phys 37:48–59

Tanguay M, Yakimiw E, Ritchie H, Robert A (1992) Advantages of spatial averaging in semi-implicit semi-Lagrangian schemes. Mon Wea Rev 120:113–123

Tatsumi Y (1983) An economical explicit time integration scheme for a primitive model. J Meteorol Soc Japan 61:269–287

Taylor M, Tribbia J, Iskandarani M (1997) The spectral element method for the shallow water equations on the sphere. J Comput Phys 130:92–108

Taylor MA (2008) CAM/HOMME: Parallel scalability and aqua planet results for CAM on the cubed-sphere grid. Presentation at the Community Climate System Model (CCSM) Workshop in Breckenridge, CO, available online at http://www.ccsm.ucar.edu/events/ws.2008/Presentations/Tarn/SEWG/Taylor_ccsm08.pdf

Taylor MA, Fournier A (2010) A compatible and conservative spectral element method on unstructured grids. J Comput Phys 229:5879–5895

Taylor MA, Edwards J, Thomas S, Nair RD (2007) A mass and energy conserving spectral element atmospheric dynamical core on the cubed-sphere grid. Journal of Physics: Conference Series 78:012,074, available online at http://www.iop.org/EJ/toc/1742-6596/78/1

Taylor MA, St-Cyr A, Fournier A (2009) A non-oscillatory advection operator for the compatible spectral element method. In: Allen G (ed) Computational Science Ð ICCS 2009, Part II, Lecture Notes in Computer Science, vol 5545, Springer Berlin/Heidelberg, pp 273–282

Temperton C, Staniforth A (1987) An efficient two-time-level semi-Lagrangian semi-implicit integration scheme. Quart J Roy Meteor Soc 113:1025–1039

Terasaki K, Tanaka HL, Satoh M (2009) Characteristics of the kinetic energy spectrum of NICAM model atmosphere. SOLA 5:180–183

Thomas SJ, Loft RD (2005) The NCAR spectral element climate dynamical core: Semi-implicit Eulerian formulation. J Sci Comput 25:307–322

Thuburn J (1993) Use of a flux-limited scheme for vertical advection in a GCM. Quart J Roy Meteor Soc 119:469–487

Thuburn J (1997) TVD Schemes, Positive Schemes, and the Universal Limiter. Mon Wea Rev 125:1990–1993

Thuburn J (2008a) Numerical wave propagation on the hexagonal C-grid. J Comput Phys 227:5836–5858
Thuburn J (2008b) Some conservation issues for dynamical cores of NWP and climate models. J Comput Phys 227(7):3715–3730
Tomita H, Satoh M (2004) A new dynamical framework of nonhydrostatic global model using the icosahedral grid. Fluid Dyn Res 34:357–400
Tomita H, Miura H, Iga S, Nasuno T, Satoh M (2005) A global cloud-resolving simulation: Preliminary results from an aqua planet experiment. Geophys Res Lett 32:L08,805
Untch A, Hortal M (2004) A finite-element scheme for the vertical discretization of the semi-Lagrangian version of the ECMWF forecast model. Quart J Roy Meteor Soc 130:1505–1530
Vandeven H (1991) Family of spectral filters for discontinuous problems. J Sci Comput 6:159–192
Váňa F, Bénard P, Geleyn J, Simon A, Seity Y (2008) Semi-Lagrangian advection scheme with controlled damping: An alternative to nonlinear horizontal diffusion in a numerical weather prediction model. Quart J Roy Meteor Soc 134:523–537
von Storch J (2004) On statistical dissipation in GCM-climate. Climate Dynamics 23:1–15
Wan H (2009) Developing and testing a hydrostatic atmospheric dynamical core on triangular grids. Reports on Earth System Science 65, Max Planck Institute for Meteorology, Hamburg, Germany, 153 pp., available from http://www.mpimet.mpg.de/en/wissenschaft/publikationen/berichte-erdsystemforschung.html
Wan H, Giorgetta MA, Bonaventura L (2008) Ensemble Held Suarez test with a spectral transform model: Variability, sensitivity, and convergence. Mon Wea Rev 136:1075–1092
Washington WM, Baumhefner DP (1975) A method of removing Lamb waves from initial data for primitive equation models. J Appl Meteor 14:114–119
White AA, B J Hoskins IR, Staniforth A (2005) Consistent approximate models of the global atmosphere: shallow, deep, hydrostatic, quasi-hydrostatic and non-hydrostatic. Quart J Roy Meteor Soc 131:2081–2107
Whitehead J, Jablonowski C, Rood RB, Lauritzen PH (2011) A stability analysis of divergence damping on a latitude-longitude grid. Mon. Wea. Rev., accepted (pending revisions)
Williams PD (2009) A proposed modification to the Robert-Asselin time filter. Mon Wea Rev 137:2538–2546
Williamson DL (1976) Linear stability of finite-difference approximations on a uniform latitude-longitude grid with Fourier filtering. Mon Wea Rev 104:31–41
Williamson DL (1978) The Relative Importance of Resolution, Accuracy and Diffusion in Short-Range Forecasts with the NCAR Global Circulation Model. Mon Wea Rev 106:69–88
Williamson DL (1988) The effect of vertical finite difference approximations on simulations with the NCAR Community Climate Model. J Climate 1(1):40–58
Williamson DL (1990) Semi-Lagrangian moisture transport in the NMC spectral model. Tellus 42A:413–428
Williamson DL (1997) Climate simulations with a spectral, semi-Lagrangian model with linear grids. In: Lin CA, Laprise R, Ritchie H (eds) Numerical Methods in Atmospheric and Oceanic Modeling: The André J. Robert Memorial Volume, Canadian Meteorological and Oceanographic Society / NRC Research Press, pp 280–292
Williamson DL (2007) The evolution of dynamical cores for global atmospheric models. J Meteorol Soc Japan 85B:241–269
Williamson DL (2008a) Convergence of aqua-planet simulations with increasing resolution in the Community Atmospheric Model, Version3. Tellus 60A:848–862
Williamson DL (2008b) Equivalent finite volume and Eulerian spectral transform horizontal resolutions established from aqua-planet simulations. Tellus 60A:839–847
Williamson DL, Browning GL (1973) Comparison of grids and difference approximations for numerical weather prediction over the sphere. J Appl Meteor 12:264–274
Williamson DL, Laprise R (2000) Numerical approximations for global atmospheric general circulation models. In: Mote P, O'Neill A (eds) Numerical modeling of the global atmosphere in the climate system, NATO Science Series C: Mathematical and Physical Sciences, vol 550, Kluwer Academic Publishers, pp 127–219

Williamson DL, Olson JG (1994) Climate simulations with a semi-Lagrangian version of the NCAR Community Climate Model. Mon Wea Rev 122(7):1594–1610

Williamson DL, Rasch PJ (1989) Two-dimensional semi-Lagrangian transport with shape-preserving interpolation. Mon Wea Rev 117:102–129

Williamson DL, Rasch PJ (1994) Water vapor transport in the NCAR CCM2. Tellus A 46:34–51

Williamson DL, Drake JB, Hack JJ, Jakob R, Swarztrauber PN (1992) A standard test set for numerical approximations to the shallow water equations in spherical geometry. J Comput Phys 102:211–224

Williamson DL, Olson JG, Boville BA (1998) A comparison of semi-Lagrangian and Eulerian tropical climate simulations. Mon Wea Rev 126:1001–1012

Williamson DL, Olson J, Jablonowski C (2009) Two dynamical core formulation flaws exposed by a baroclinic instability test case. Mon Wea Rev 137:790–796

Wood N, Diamantakis M, Staniforth A (2007) A monotonically-damping second-order-accurate unconditionally-stable numerical scheme for diffusion. Quart J Roy Meteor Soc 133:1559–1573

Xue M (2000) High-order monotonic numerical diffusion and smoothing. Mon Wea Rev 128:2853–2864

Xue M, Droegemeier KK, Wong V (2000) The Advanced Regional Prediction System (ARPS) - A multi-scale nonhydrostatic atmospheric simulation and prediction model. Part I: Model dynamics and verification. Meteor Atmos Phys 75:161–193

Zalesak ST (1979) Fully multidimensional flux-corrected transport algorithms for fluids. J Comput Phys 31:335–362

Zubov VA, Rozanov EV, Schlesinger ME (1999) Hybrid scheme for three-dimensional advective transport. Mon Wea Rev 127:1335–1346

Chapter 14
Kinetic Energy Spectra and Model Filters

William C. Skamarock

Abstract We wish to maximize efficiency (accuracy/cost) in the design of atmospheric fluid-flow solvers. An important measure of accuracy for weather and climate applications is a model's ability to resolve meteorologically important features at scales approaching the grid-scale. Simulated kinetic energy spectra provide a useful diagnostic for quantifying a model's resolving capability. Using kinetic energy spectra we illustrate some of the issues affecting the resolution capabilities of models arising from the choice of spatial grid staggering, integration schemes and their implicit filters, and explicit filters. In both Eulerian and semi-Lagrangian formulations, C-grid staggering provides the best resolution of divergent modes that are an important part the KE spectrum in the mesoscale which the global models are now beginning to resolve. Other grid staggerings require special filtering that compromise resolution capabilities. The popular semi-Lagrangian semi-implicit formulations are shown to significantly damp resolvable high-frequency modes and adversely affect their resolving capabilities. While less costly at a given grid density, the SLSI models may well be significantly less efficient than Eulerian models.

14.1 Introduction

Clouds and precipitation are among the most important and challenging phenomena that must be accurately treated for climate and Numerical Weather Prediction (NWP) applications. In our existing operational global climate and weather models, clouds and precipitation processes are parameterized, i.e. they are modeled, as opposed to being explicitly represented in the discrete atmospheric fluid-flow

W.C. Skamarock
National Center for Atmospheric Research, Boulder, Colorado, USA
e-mail: skamaroc@ucar.edu
The National Center for Atmospheric Research is supported by the National Science Foundation.

P.H. Lauritzen et al. (eds.), *Numerical Techniques for Global Atmospheric Models*, Lecture Notes in Computational Science and Engineering 80,
DOI 10.1007/978-3-642-11640-7_14,

solvers. The need to resolve clouds and cloud systems, and the availability of more powerful computers, are driving us to apply global atmospheric models at increasingly higher spatial and temporal resolutions. In the research setting, the coming introduction of peta-scale computers will permit us to regularly produce global simulations using horizontal grid spacing of a few kilometers for weather and short-term climate (seasonal) applications. At this grid spacing we remove the deep convective parameterizations that are seen as most problematic, but we are left with the problem of modeling sub-grid entrainment and detrainment in these poorly-resolved clouds (Tao and Moncrieff 2009; Weisman et al. 1997; Bryan et al. 2003). The denser grids will also permit better resolution of topography, land use, land-sea boundaries, and other atmospheric forcing mechanisms as well as provide better resolution of flow dynamics not related to clouds, such as land-sea breezes, fronts, and mountain waves.

When solving the Navier–Stokes equations for atmospheric fluid flow, we increase resolution because we wish to better resolve features that are marginally resolved at current resolutions, and because we wish to explicitly simulate, or resolve, flow phenomena that were absent or parameterized in the less-well-resolved solutions. The former reason is related to the classical numerical analysis understanding regarding the numerical solution of Partial Differential Equations (PDEs) – increasing resolution will lower solution error, with infinite resolution producing a perfect solution. It must be understood, however, that atmospheric flow solutions do not converge in the strict sense. New phenomena appear, and we ultimately strive only for statistical convergence. The explicit simulation of new flow phenomena and forcings is the critical reason to increase resolution in present-day global climate and NWP models. Global model resolutions are now increasing to a level where mesoscale features begin to be resolved (e.g., the larger-scale aspects of convective cloud systems such as hurricanes, etc.), and we are beginning to resolve phenomena that differ dynamically in a fundamental way from that of planetary- and synoptic-scale flows.

The change in dynamical regimes associated with the newly-resolved phenomena, and the lack of strict solution convergence, raise a number of questions concerning global solver design and evaluation. Our primary objective in designing atmospheric flow solvers is to maximize efficiency, that is, we wish to attain a given level of accuracy for the smallest computational cost, or we wish to achieve the highest accuracy for a given cost. In this paper we consider how observations and model simulations of the atmospheric kinetic energy spectra can be used to quantify solution accuracy, thus allowing us to examine solver characteristics involving spatial and temporal discretizations and model filtering, and ultimately model efficiency. Some aspects of existing global models, such as the time integration and spatial interpolation schemes used in semi-Lagrangian models, the choice of horizontal grid staggering, and filter choices used in some large-scale global models, will be shown to adversely affect solver performance for mesoscale and cloudscale phenomena.

14.2 Kinetic Energy Spectra and Atmospheric Dynamics

Nastrom and Gage (1985) used aircraft observations of winds from the Global Atmospheric Sampling Program (GASP) to compute kinetic energy (KE) spectra for horizontal length scales from a few kilometers to several thousand kilometers. Lindborg (1999) similarly used aircraft observations from the Measurement of ozone and water vapor by Airbus in-service aircraft (MOZAIC) program to compute structure functions and a kinetic energy spectrum. Results from both studies, depicted in Fig. 14.1, illustrate the characteristic behavior of the kinetic energy spectrum. At larger scales (horizontal wavelengths greater than approximately several hundred kilometers) the spectrum scales as k^{-3} where k is the horizontal wavenumber. For shorter wavelengths (higher wavenumbers) the spectrum scales as $k^{-5/3}$, and a small transition region exists between the two regimes. While it is widely accepted that the dynamics of the k^{-3} regime correspond to a downscale cascade of enstrophy, there is no consensus concerning the $k^{-5/3}$ regime (Lilly et al. 1998; Lindborg 2006). The characterization of the $k^{-5/3}$ regime represents one of the major unanswered questions in mesoscale atmospheric dynamics.

The KE spectrum can also be computed from model simulations. This spectrum from a high-resolution simulation using the Advanced Research Weather Research and Forecast model (ARW, Skamarock and Klemp 2008; Skamarock et al. 2008) is shown in Fig. 14.2, and this simulated spectrum reproduces the transition. This behavior has also been found in simulations from other models (Lilly et al. 1998; Lindborg and Berthouwer 2007; Hamilton et al. 2008) and, while there are variations in the spectra as a function of pressure, geographical region and weather regime (Skamarock 2004), the transition is always apparent.

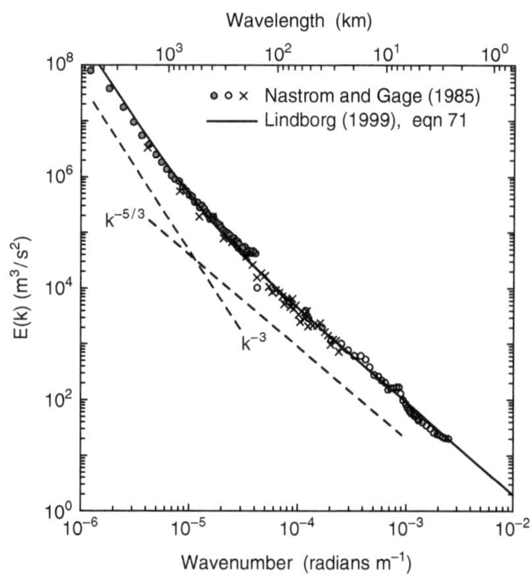

Fig. 14.1 Nastrom and Gage (1985) spectrum derived from the GASP aircraft observations (symbols) and the Lindborg (1999) functional fit to the MOZAIC aircraft observations. The figure is from Skamarock (2004)

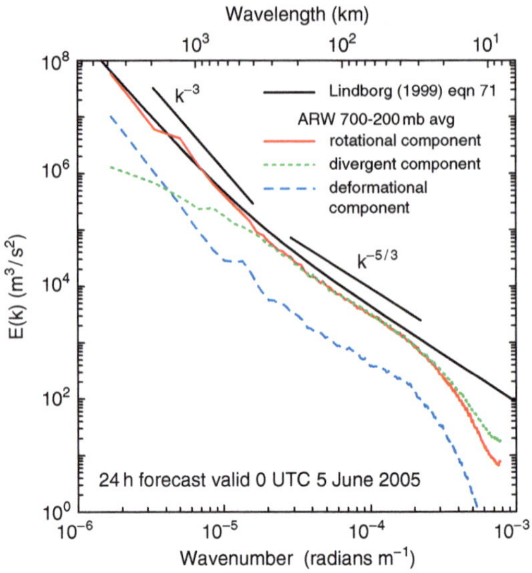

Fig. 14.2 Decomposed kinetic energy spectra for a spring-season forecast over the continental U.S. The forecasts were produced using the ARW with $\Delta x = 4\,\mathrm{km}$. The figure is from Skamarock and Klemp (2008)

The simulated spectrum can be decomposed into a rotational component V_ψ, a divergent component V_χ, and a deformational component V_{def}, where $V = V_\psi + V_\chi + V_{def}$ and the velocities are given as

$$\begin{aligned} V_\psi &= k \times \nabla\psi \\ V_\chi &= \nabla\chi \\ V_{def} &= V - V_\psi - V_\chi . \end{aligned}$$

ψ and χ are defined as

$$\begin{aligned} \nabla^2\psi &= \zeta, \quad \zeta = k \cdot \nabla \times V \\ \nabla^2\chi &= D, \quad D = \nabla \cdot V, \end{aligned}$$

where k is the unit vector normal to the horizontal coordinate surface. The deformational component arises from the presence of lateral boundaries and would be absent in a decomposition of the wind fields over the globe. At large scales the kinetic energy is dominated by rotational energy, and the transition to the $k^{-5/3}$ regime occurs when the energy in the divergent component becomes comparable to that in the rotational component (see Fig. 14.2). Lindborg (2007) argues that it can also be deduced from the aircraft observations. Thus the atmospheric dynamics are seen to be changing in a fundamental way as one moves from large scale to the mesoscale, and one important aspect of this change is the importance of divergent motions to the mesoscale KE spectrum.

Lindborg (1999) also presents results for the kurtosis which is defined as

$$\text{kurtosis: } \frac{<\delta u^4>}{<\delta u^2>^2},$$

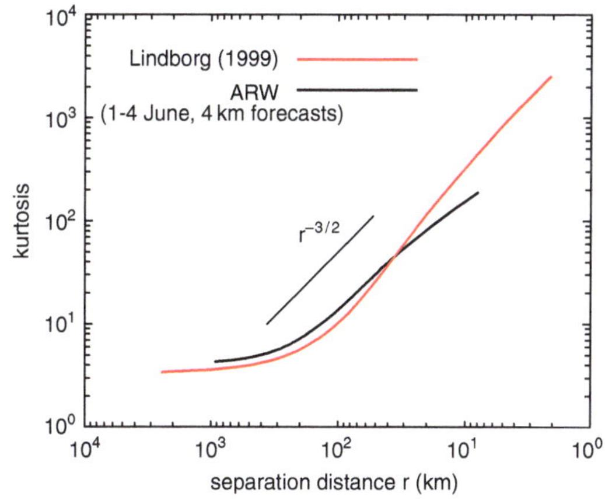

Fig. 14.3 Kurtosis computed using observations from MOZAIC (Lindborg 1999) and from spring forecasts for 1–3 June 2003 over the continental U.S. using the ARW model with $\Delta x = 4$ km (Done et al. 2004)

where $\delta u = u(x+r) - u(x)$ for a distance r along an aircraft track, and u is the along-track velocity. Figure 14.3 shows a plot of the kurtosis computed from the MOZAIC observations and from ARW simulations. For large-scale flow, where the velocity field is observed to possess a Gaussian probability density function (PDF), the kurtosis should equal 3 (Lindborg 1999; Frisch 1995), as is found in the plot. In the mesoscale region the kurtosis increases dramatically and scales as approximately $r^{-3/2}$, indicating that there is significant intermittency at small scales. Both observations and model results exhibit this strong intermittency at the mesoscale and cloudscale where significant energy is in the horizontally divergent component of the flow.

The changes in atmospheric dynamics indicated in the spectrum and the kurtosis have several implications for solver design. The mesoscale and cloud-scale motions that higher-resolution models resolve represent motions that are entirely parameterized in large-scale models. Examples of the parameterizations include those for convection and gravity-wave propagation and breaking (gravity-wave drag). These parameterizations are problematic and are a major impetus for moving to higher resolution where the parameterizations can be removed. Toward this end, the solvers must be able to accurately simulate horizontally divergent motions, which has not been a priority in existing large-scale models. In the following sections we examine some of the issues involved in solver design and ability to accurately simulate divergent motions.

14.3 Model Dissipation and Spectral Damping

In the k^{-3} spectral regime there is a downscale cascade of enstrophy, whereas the $k^{-5/3}$ there appears to be a net downscale cascade of energy (e.g., Lindborg and Cho, 2000). In either case, enstrophy or energy must be removed as it cascades to the highest wavenumbers represented in the model discretization. Failure to provide

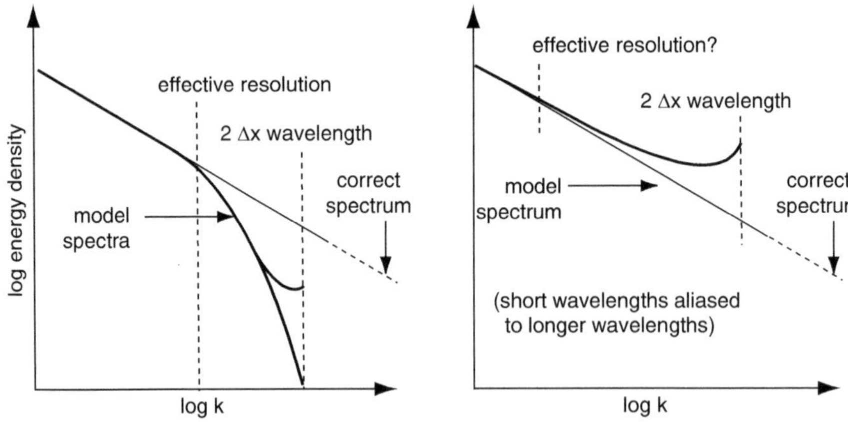

Fig. 14.4 Schematic depicting the possible behavior of spectral tails derived from model forecasts. Using the methodology outlined in the Appendix to compute the spectra, limited area models (including ARW) usually produce the slightly upturned tail shown in the *left panel*. Adapted from Skamarock (2004)

sinks for these cascades results in an unphysical buildup of energy or enstrophy at the smallest scales. In addition, most numerical methods do not accurately simulate these high wavenumber modes, and it is beneficial to remove the energy in these modes so that energy is not erroneously aliased onto the smaller wavenumber (well-resolved) modes. The energy density should drop considerably in the simulated spectrum in the highest wavenumbers as a result of this filtering; this is depicted schematically in Fig. 14.4 and it is apparent in the simulated spectrum in Fig. 14.2. We define the *effective resolution* of a solver as the point at which the slope of the simulated spectrum becomes greater than the slope of the expected (or observed) spectrum, as is indicated in Fig. 14.4. In designing a solver and dissipation mechanisms we wish to maximize the effective resolution (i.e. have an effective resolution at the highest wavenumber possible) while removing energy from the highest wavenumbers thereby minimizing aliasing. In principle, higher-order methods allow a higher effective resolution if they are combined with appropriate energy and enstrophy sinks.

There are many approaches to providing the necessary enstrophy and energy sinks in the solvers (Chap. 13). Many models use explicitly computed horizontal mixing terms of the form

$$\frac{\partial \phi}{\partial t} = \cdots (-1)^{(n+2)/2} \; \nu_n \frac{\partial^n \phi}{\partial x_i^n} \tag{14.1}$$

where n is an even integer and ν_n is referred to as an eddy viscosity when $n = 2$ or a hyperviscosity when $n > 2$. Higher values of n produce filters that are more scale selective (the damping rate as a function of wavenumber drops off more quickly). There is no physical justification for (14.1) for large-scale and mesoscale flows. For atmospheric Large Eddy Simulation (LES) resolutions ($\Delta x < 100\,\mathrm{m}$), theory exists for defining the eddy viscosity ν_2 (Mason 1994; Wyngaard 2004).

A second mechanism for dissipating energy, specifically targeting energy in the divergent motions, is horizontal divergence damping. This filter is usually implemented in the form of a damping term in the horizontal momentum equations:

$$\frac{\partial u_i}{\partial t} = \cdots + \nu_d \frac{\partial}{\partial x_i} (\nabla_h \cdot \mathbf{V}) , \tag{14.2}$$

where $\mathbf{V}$ is the horizontal velocity.

That this damping targets horizontally divergent motions can be seen by taking the horizontal divergence of the momentum equations (14.2) which results in

$$\frac{\partial}{\partial t} (\nabla_h \cdot \mathbf{V}) = \cdots + \nabla_h^2 (\nabla_h \cdot \mathbf{V}) .$$

Horizontal divergence damping has been used in many large-scale models to filter gravity waves, especially those that resulted from imbalances in real-data initializations (Dey 1978; Janjic 1990). There are models today still using this formulation, including the Finite Volume (FV) core in the Community Climate System Model (CCSM) (Collins et al. 2004) and the Nonhydrostatic Mesoscale Model (NMM) used operationally for limited area NWP at the National Centers for Environmental Prediction (NCEP) (Janjic 2003). Horizontal divergence damping can also be extended to higher order ($n > 2$):

$$\frac{\partial u_i}{\partial t} = \cdots + \; (-1)^{(n+2)/2} \; \nu_d \frac{\partial^{(n-1)}}{\partial x_i^{(n-1)}} (\nabla_h \cdot \mathbf{V}) ,$$

or alternatively as in the FV core ($n = 4$; Peter Lauritzen, personal communication)

$$\frac{\partial u_i}{\partial t} = \cdots + \; (-1)^{(n+2)/2} \; \nu_d \frac{\partial}{\partial x_i} \nabla_h^{n-2} (\nabla_h \cdot \mathbf{V}) .$$

Horizontal divergence damping can have a significant impact on a model's ability to produce the $k^{-5/3}$ KE spectra as demonstrated in Skamarock (2004). Given that the KE in the $k^{-5/3}$ region is composed of both rotational and divergent energy, horizontal divergence damping cannot be the sole energy sink employed in a model formulation. Furthermore, convection and convective transport are strongly divergent motions, thus horizontal divergence damping preferentially filters them. Since these processes are becoming increasingly important as we employ higher resolution grids, preferentially damping these modes is counter to our objectives, and no mesoscale or cloudscale resolving models use this form of damping aside from the nonhydrostatic FV core (William Putnam, personal communication) and the NCEP NMM (Janjic 2003). While there may be some computational aspects of model formulations for which the use of horizontal divergence damping may appear beneficial, these problems can be avoided in the model formulation directly or by using other less-deleterious filters.

Another class of filters in numerical models are those that are implicit in the numerical discretization. The filters may damp temporally because the damping is implicit in the time integration, the damping may be part of the transport algorithm or spatial interpolation scheme that is part of the spatial discretization, or it may be intertwined in both the spatial and temporal schemes. These schemes can affect a model's ability to resolve divergent motions and the mesoscale portion of the spectra.

14.4 Grid Staggering and Spatial Discretizations

One dynamical description of the mesoscale is that scale in which the kinetic energy of the horizontally divergent motions approaches the same order as that of the rotational motions, and this occurs where the KE spectrum assumes a $k^{-5/3}$ behavior. Given the critical role of horizontally divergent motions in the mesoscale, it is important to consider how discretizations resolve these motions. One important factor influencing the numerical discretization is the staggering of the variables in the model grid (Chap. 3). Figure 14.5 depicts three different horizontal grid staggerings (commonly referred to as the A, C, and D grids – see Arakawa and Lamb (1977)) used in global and some limited-area mesoscale models.

Consider the simple second-order finite differencing of the height (pressure) gradient, divergence and Coriolis terms for the vector momentum form of the shallow-water equations that is commonly used on these grids.

A grid	C grid	D grid
$h_x \sim \delta_{2\Delta x} h$	$h_x \sim \delta_{\Delta x} h$	$h_x \sim \delta_{2\Delta x} \overline{h}^y$
$u_x \sim \delta_{2\Delta x} u$	$u_x \sim \delta_{\Delta x} u$	$u_x \sim \delta_{2\Delta x} \overline{u}^y$
$fv \sim fv$	$fv \sim f\overline{\overline{v}^x}^y$	$fv \sim f\overline{\overline{v}^x}^y$

where the discrete operators

$$\delta_{\Delta x}\phi = (\phi_{x+\Delta x/2} - \phi_{x-\Delta x/2})/\Delta x$$
$$\overline{\phi}^y = (\phi_{y+\Delta y/2} + \phi_{y-\Delta y/2})/2.$$

Randall (1994) presents a linear analysis of the response of the second-order centered spatial approximations for the linearized shallow water equations (inertia-gravity waves) discretized on these grids using the vector-momentum form of the equations (he also examines the Z-grid discretization that use a vorticity-divergence formulation of the shallow-water equations).

Consider the response for rotationally dominated waves. The A grid response is superior to the response of the C and D grids (see Randall 1994, Fig. 2, $\lambda/d = 0.1$;

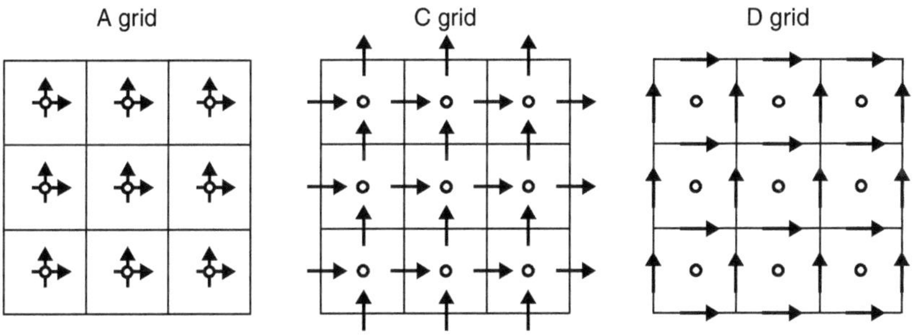

Fig. 14.5 Schematic depicting the Arakawa A, C and D grids

the D-grid response, not shown, is essentially that of the C grid). The superiority of the A grid is especially pronounced in the upper half of the wave spectrum, where the C and D grid frequencies are zero for the $2\Delta x$ and $2\Delta y$ modes. The erroneous zero frequency is a result of the averaging needed for interpolating the tangential velocity ($\overline{\overline{v}^{x}}^{y}$) on these two grids, whereas the A grid needs no averaging for the Coriolis term. The zero frequency for the 2Δ modes on the C and D grids means that the modes are stationary and the grid does not *see* them, and this is referred to as a *null space* in the solution. The poorer response for inertial waves is the primary reason why few large-scale models use the C or D grids.

The frequency response for waves dominated by horizontal divergence (gravity waves) on these grids is given in Randall's figure 1. Here the C grid response is superior to both the A and D grids, and the A and D grids both have erroneous zero-frequencies for the 2Δ modes. The averaging needed on both the A and D grids is responsible for these zero frequencies, and as a result the A and D grids do not *see* these 2Δ modes. Most meso- and cloud-scale models use the C grid because of its superior gravity-wave response, although some modelers use the A grid because it can be advantageous to have all variables defined in the cell center for some discretizations of transport.

To deal with grid-scale modes that have zero frequency, filters must be used to remove the energy in these modes if energy accumulates there. The inertial waves have, however, little energy at scales below the Rossby radius, and for horizontal grid-spacing of $O(100\,\text{km})$ or less, models using the C grid do not appear to have problems with the zero-frequency 2Δ inertial mode. For the gravity-wave modes, at meso- and cloud-scale resolutions there is a downscale cascade of energy that will result in energy accumulating in the 2Δ modes. Filters will be needed on all grids to remove this energy.

To illustrate some of these effects, Fig. 14.6 presents KE spectra computed in the lower stratosphere from aquaplanet simulations for two different CCSM cores – the spectral Eulerian core and the Finite Volume (FV) core (Collins et al. 2004). The KE spectra are decomposed into rotational and divergent components. There is a suggestion of a transition from the large-scale k^{-3} character to a shallower slope occurring around spherical wavenumber 100 in the spectral Eulerian core spectrum, and this

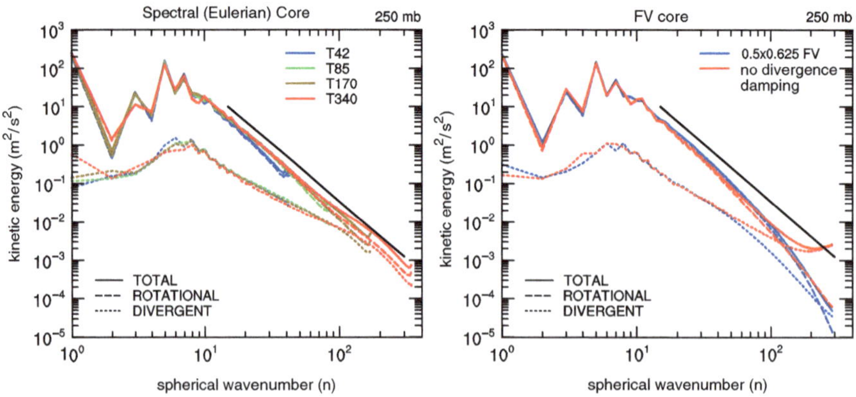

Fig. 14.6 Kinetic energy (*solid lines*) as a function of spherical wavenumber for the CCSM spectral (Eulerian, *left*) core and the CCSM FV core (*right*) derived from aquaplanet simulations. The total KE is broken into divergent and rotational components (*dashed lines*) for both *cores and the solid black lines* shows the k^{-3} slope. The figures are courtesy of David Williamson

is where the energy in the divergent modes, that behaves as $k^{-5/3}$, becomes similar to the rotational mode energy. This behavior is similar to that exhibited by the ARW spectrum (Fig. 14.2) and spectra from other global models (Takahashi et al. 2006), and we would expect that the transition would be better resolved with increasing horizontal resolution. While the spectrum for the spectral Eulerian core does not drop off as rapidly as the ARW spectrum (Fig. 14.2) at the highest wavenumbers, the filters in the model are removing energy as evidenced in the increasing slope beginning around spherical wavenumber 200 in the T340 spectrum.

Two spectra are plotted for the FV dynamical core (Fig. 14.6, *right panel*), one from the standard configuration and a second from a simulation with no horizontal divergence damping. The spectrum from the standard configuration of the FV core depicts a spectral slope that is increasing beyond k^{-3} starting around 15–20 Δx. This evidence of strong filtering appears in both the rotational energy and the divergent energy. There are two numerical filters in the FV core – the monotonicity constraint in the PPM-based advection and interpolation scheme (Lin and Rood 1997), and the horizontal divergence damping. The filtering provided by the horizontal divergence damping is illustrated by comparing the standard-configuration spectrum with that produced with the horizontal divergence damping turned off. There is only a small difference in the rotational component of the spectrum but there is a major buildup of energy at the highest wavenumbers in the divergent component of the spectrum when divergence damping is not used. As discussed in the previous section, the D-grid formulation of the FV core does not *see* the $2\Delta x$ divergent modes. Special filters (horizontal divergence damping in this case) must be used to remove energy for this null space on the grid. C-grid models used in meso- and cloud-scale applications (e.g., ARW Skamarock and Klemp 2008, Coupled Ocean-Atmosphere Prediction System (COAMPS) (Hodur 1997), Advanced Regional Prediction System (ARPS, Xue et al. 1990) and global models using the C-grid discretization (e.g., United Kingdom Meteorological Office (UKMO) model (Staniforth and Wood 2008), global ARW (Skamarock et al. 2008) do not need and do not use these filters, and generally exhibit a much higher effective resolution (typically between 6 and 10 Δx) than evidenced by the FV core spectrum in Fig. 14.6 (15–20 Δx).

14.5 Semi-Implicit Semi-Lagrangian Formulations

Many operational global NWP and climate modeling centers are using semi-implicit formulations in conjunction with semi-Lagrangian dynamics (e.g., UKMO/Hadley Center, Staniforth and Wood 2008; Canadian Meteorological Centre (CMC) Meteorological Research Branch (MRB) Global Environmental Multiscale (GEMS) model, Yeh et al. 2002; European Center for Medium Range Forecasting (ECMWF 2006). These semi-Lagrangian semi-implicit (SLSI) formulations allow for long timesteps because the semi-implicit portion of the formulation removes the timestep restriction associated with propagating gravity (and acoustic) waves while the semi-Lagrangian portion of the scheme largely removes the timestep restriction due to advection. Typically, SLSI models run with timesteps five to ten times that of their Eulerian counterparts. The SLSI cost per timestep is significantly greater than Eulerian models because of the need to compute trajectories and interpolate variables to the departure points, and because of the need to perform a global inversion in the implicit formulation, but this increased cost is offset by using the larger timestep.

The filtering characteristics of SLSI schemes have not been closely examined for meso- and cloud-scale applications. In Shutts (2005), it is shown that the KE spectrum of the ECMWF model does not transition to the $k^{-5/3}$ mesoscale behavior for resolutions where a transition should be resolved. Shutts introduces a backscatter forcing into the system to put energy into these scales, but he does not speculate on why the system does not predict the transition. Palmer (2005, personal communication) found that reducing the timestep in the SLSI model to values used in similarly configured Eulerian models (in this case 1/5 of the SLSI timestep) did not change the KE spectrum – a transition to $k^{-5/3}$ was not observed. Generally, Eulerian models do predict this transition at these resolutions. In this section we examine dissipation mechanisms in SLSI numerics to see if they may be responsible for preferentially damping mesoscale motions.

For the SLSI formulation, consider the linearized 1D shallow water equations with variables U $= U + u(x,t)$ and H $= H + h(x,t)$:

$$\frac{du}{dt} + g\frac{\partial h}{\partial x} = 0,$$

$$\frac{dh}{dt} + H\frac{\partial u}{\partial x} = 0.$$

where g is gravity. The SLSI discretization of these equations can be expressed as

$$u^{t+\Delta t} = \left(u - \frac{1-\epsilon}{2}\Delta t g \delta_x h\right)\Big|_d^t - \left(\frac{1+\epsilon}{2}\Delta t g \delta_x h\right)\Big|^{t+\Delta t}$$

$$h^{t+\Delta t} = \left(h - \frac{1-\epsilon}{2}\Delta t H \delta_x u\right)\Big|_d^t - \left(\frac{1+\epsilon}{2}\Delta t H \delta_x u\right)\Big|^{t+\Delta t},$$

where d refers to the departure point of the trajectory and ϵ is the off-centering parameter for the implicit time integration scheme. Gravel et al. (1993) performed an analysis of this scheme and derived the following amplification factor[1]:

$$\frac{E}{\rho} = \frac{1 - \gamma_3(1-\epsilon^2) \pm 2i\gamma_3^{1/2}}{1 + \gamma_3(1+\epsilon)^2} \tag{14.3}$$

where E is the amplification factor, ρ is the response function for the semi-Lagrangian advection, and $\gamma_3 = gH(k\Delta t)^2/4$ where k is the horizontal wavenumber. It is easily shown that for $|\rho| \leq 1$ and $0 \leq \epsilon \leq 1$ the SLSI scheme is absolutely stable. In most models using the full nonlinear implementation for NWP and climate applications, $0.1 \leq \epsilon \leq 0.2$ is needed for stability. The ECMWF model is run with $\epsilon = 0$ and filtering needed for the two-time-level SETTLS scheme (Stable Extrapolation Two-Time-Level Scheme, Hortal 2002; Durran and Reinecke 2004) is achieved using other mechanisms.

Gravel et al. did not examine the damping properties of the scheme as revealed in the amplification factor (14.3). In order to examine the damping and its effect on the KE spectra, we first need to estimate the spin-up time for motions as a function their horizontal length scale for comparison with the damping timescale. We can expect that damping in a numerical model will be significant for a particular scale when the decay timescale associated with the damping is of the same order or smaller than the spin-up timescale.

Figure 14.7 show the spin-up time for motions as a function of scale as determined using the turbulence theory of Kolmogorov (1990) and Lindborg's functional fit (Lindborg 1999) for the atmospheric KE spectrum. In essence, the spin-up time scale is an eddy turnover time, and this timescale is given by $\tau = L(k)/U(k) = [k^3 E(k)]^{-1/2}$, where L is the eddy length scale, U is a velocity scale, and E is the kinetic energy density. For the energy spectrum depicted in Fig. 14.1, the synoptic scale behaves as k^{-3} hence the spin-up time asymptotes to a constant value ($\tau = .68$ days). This estimate of the spin-up time is consistent with the numerical KE analyses of Skamarock (2004) and Hamilton et al. (2008). The spin-up time decreases dramatically as the spectrum transitions to the $k^{-5/3}$ regime where it behaves as $k^{-2/3}$.

[1] Gravel et al. (1993) analyzed the full primitive equations; here we present its simplified form applicable to the shallow water equations.

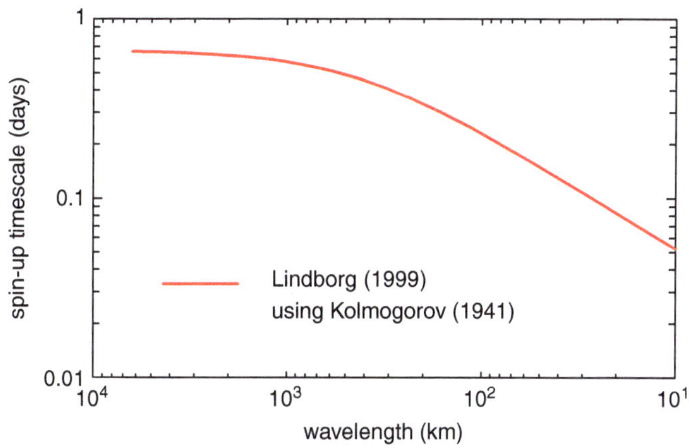

Fig. 14.7 Spin-up timescale using Kolmogorov's theory (Kolmogorov 1990) and atmospheric spectrum result from Lindborg (1999)

In addition to the spin-up timescale, we need estimates of the Courant numbers typically used the SLSI model applications in order to estimate damping from the transport scheme. Figure 14.8 presents a typical distribution of velocities and Courant numbers for two given values of $\Delta t/\Delta x$ typically used in Eulerian and semi-Lagrangian models. The Courant number distribution comes from an ARW forecast on a high-resolution grid ($\Delta x = 5$ km) for the continental US in the winter. For Eulerian timesteps 90% of the Courant numbers on the grid are typically less than 0.2, and only approximately half are less than 0.06. SLSI timesteps are significantly larger than this, and illustrated in the figure is the distribution for an SLSI timestep five times larger than the Eulerian value. For this SLSI timestep 90% of the Courant numbers are less than one, but note that approximately 60% of the Courant numbers are greater than 0.5.

Figure 14.9 presents damping rates normalized by the turnover timescale computed using the amplification factor (14.3) for both Eulerian and SLSI timesteps. We have chosen to examine an $8\Delta x$ wave which is reasonably well resolved in mesoscale models such as ARW (Skamarock 2004). The damping rates for the SLSI formulation are computed using the cubic interpolation from Gravel et al. (1993), and the computation of the damping rates uses the absolute value of E from (14.3).

For the Eulerian timestep (the left panel in Fig. 14.9), the damping of the $8\Delta x$ wave increases dramatically with increasing timestep, and even at a Courant number of 0.05 the decay timescale is only twice the turnover timescale. Most of the Courant numbers will be greater than 0.05 (see Fig. 14.8), hence the damping will be significant. Also notice that the value of the off-centering parameter ϵ in (14.3) has little effect on the damping rate which increases slowly with increasing ϵ. This behavior indicates that it is the damping from the transport scheme that most affects the waves for the Eulerian timesteps. Also shown in the figure is the damping rate for the ARW model using a fifth order advection scheme. Even though the transport

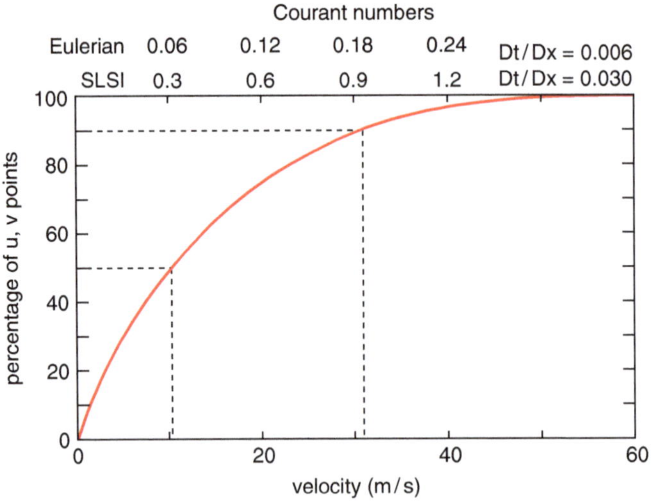

Fig. 14.8 Cumulative Courant number distribution for an ARW simulation over the US continent for a 22 January 2005 forecast. Details of the model configuration are given in Bernardet et al. (2008)

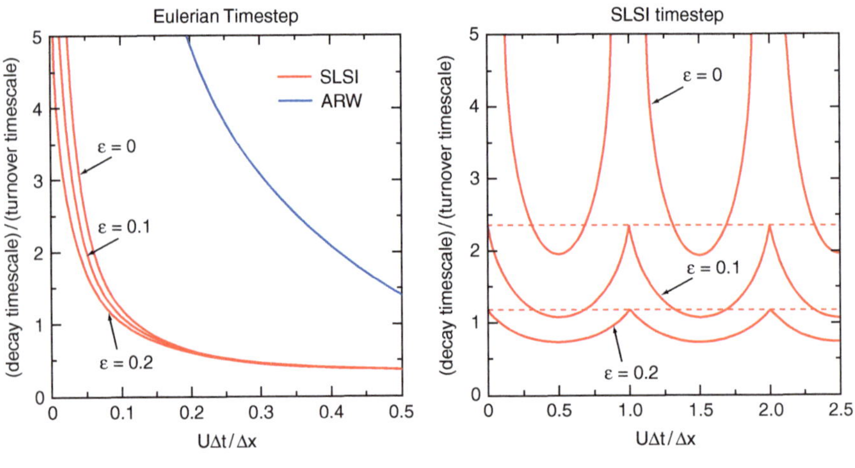

Fig. 14.9 Decay timescale normalized by eddy turnover (spin-up) timescale for a gravity wave with an $8\Delta x$ wavelength with a horizontal grid-length of 10 km and a timestep of 60 s (Eulerian timestep, *left*) and 300 s (SLSI timestep, *right*). The phase speed of the wave ($\sqrt{gH}$ in (14.3); in the absence of a mean wind) is 16.667 m/s, hence the Courant number based on this phase speed is 0.1 and 0.5 for the Eulerian and SLSI timesteps, respectively

scheme in this Eulerian model is upwind biased and dissipative, the overall damping is much less than that shown for the SLSI scheme.

For a more typical SLSI timestep (the right panel in Fig. 14.9), the normalized damping appears dramatically different. The damping rates are plotted for three values of the off-centering parameter $\epsilon = 0, 0.1$, and 0.2, and the damping is seen to increase dramatically with increasing ϵ. Also apparent is the damping associated with the transport scheme. For integer advective Courant numbers $U\Delta t/\Delta x$ the semi-Lagrangian transport scheme does not damp, and for non-integer advective Courant numbers the damping is most pronounced half-way between the integer Courant numbers. Operational SLSI weather and climate models must use the off-centering parameter $0.1 \leq \epsilon \leq 0.2$ for stability in the full nonlinear integrations; hence the damping of the short wavelength high-frequency modes is significant and while there is some damping associated with the semi-Lagrangian interpolation used for transport it does not produce the bulk of the damping for these large timesteps.

These results provide a plausible explanation for the observed behavior of the KE spectrum in SLSI models. Paradoxically, while the damping of the shorter-wavelength high-frequency modes can be attributed to the off-centering of the semi-implicit part of the SLSI formulation when large timesteps are taken, the damping from the interpolations (the response function for the semi-Lagrangian advection, ρ in (14.3)) dominates when the timestep is reduced to Eulerian values. Reduction of this significant damping likely requires more accurate trajectory integrations that will allow smaller off-centering parameters ϵ, but will increase the cost of the integration. The strong damping of the higher-frequency modes also calls into question the efficiency of these schemes for meso- and cloud-scale applications, because, as apparent in the simulated spectra (Shutts 2005) and in this analysis, the long timestep allowable using SLSI formulation comes at a significant cost in accuracy of the small scales. These results are consistent with the theoretical analysis of Bartello and Thomas (1996), who argue that Eulerian timesteps should be used in SLSI schemes at mesoscale resolutions. In addition, Pinty et al. (1995) demonstrate this timestep restriction for accurately simulating vertically propagating gravity waves.

14.6 Conclusions

The goal in the design of atmospheric fluid-flow solvers is to maximize efficiency, where efficiency is defined as solution accuracy divided by cost of a given method that attains that accuracy. Given the lack of convergent solutions for turbulent flow, and the acknowledgement that it is the need to resolve previously sub-gridscale structures that drives increasing resolution, an important and relevant measure of accuracy is the ability of a scheme to resolve scales as close to the limit (the smallest scales) as allowable by the discretization. We have shown that examination of a model's KE spectra provides a way to quantify the resolution limits of a model and to determine a model's effective resolution.

Additionally, as global models push into mesoscale and ultimately the cloud-scale, the horizontally divergent modes become the important modes to resolve at the resolution margin, as opposed to the rotational modes for which most large-scale models were designed. The KE spectra reveal these energetics and we have shown examples where the spectra are used to examine and quantify these regimes and resolution capabilities.

Ultimately, it is the damping characteristics of a model that determine its resolving capabilities. Damping is necessary in atmospheric models because the energy- and enstrophy-cascade dynamics present in the atmosphere demand that there be sinks of energy and enstrophy in the absence of resolved viscous effects. The design of explicit and implicit filters that represent these sinks can impact the effective resolution, but importantly the need for some filters may be dictated by choice of spatial and temporal discretizations, to the detriment of a scheme's resolution capabilities.

For Eulerian discretizations, grid staggering has a significant impact on need for filtering and on the flexibility in filter design and configuration. For example, most mesoscale models use the C-grid staggering which is most accurate for divergent modes. We have shown an example of poor marginal resolution exhibited by the CCSM FV core that uses a D-grid staggering. The FV core needs to use horizontal divergence damping to control grid-scale divergent modes and also uses monotonicity constraints that introduce strong damping into the rotational modes. The effective resolution is only half that of C-grid mesoscale models as revealed in spectra computed from aquaplanet simulations.

Other formulations commonly used in some large scale climate and weather models are also problematic with respect to a scheme's effective resolution. A stability analysis of SLSI schemes that use large timesteps reveals significant damping for high frequency modes because of the need to off-center the semi-implicit time integration in the nonlinear models. The use of Eulerian timesteps in the SLSI models leads to large damping of short wavelengths modes associated with the interpolation scheme in the semi-Lagrangian portion of the algorithm while alleviating the damping associated with the semi-implicit portion of the algorithm.

The SLSI formulations are often stated to be more efficient than Eulerian formulations because the time to solution is less using the SLSI schemes given their comparatively much larger timesteps. However, given their decreased resolution capabilities compared to Eulerian formulations, the SLSI formulations are likely no more efficient and possibly significantly less efficient than Eulerian formulations.

References

Arakawa A, Lamb VR (1977) Computational design of the basic dynamical processes of the UCLA general circulation model. Methods Comput Phys 17:173–265

Bartello P, Thomas SJ (1996) The cost-effectiveness of semi-Lagrangian advection. Mon Wea Rev 124:2883–2897

Bernardet L, Nance L, Demirtas M, Koch S, Szoke E, Fowler T, Loughe A, Mahoney JL, Chuang HY, Pyle M, Gall R (2008) The developmental testbed center and its winter forecast experiment. Bull Amer Meteor Soc 89:611–627

Bryan GH, Wyngaard JC, Fritsch JM (2003) Resolution requirements for the simulation of deep moist convection. Mon Wea Rev 131:2394–2415

Collins WD, Rasch PJ, Boville BA, Hack JJ, McCaa JR, Williamson DL, Kiehl JT, Briegleb B (2004) Description of the NCAR Community Atmosphere Model (CAM 3.0). NCAR technical note NCAR/TN464+STR:226 pp.

Dey CH (1978) Noise suppression in a primitive equations model. Mon Wea Rev 106:159–173
Done J, Davis CA, Weisman ML (2004) The next generation of NWP: Explicit forecasts of convection using the weather research and forecasting model. Atmos Sci Lett 5:110–117
Durran D, Reinecke PA (2004) Instability in a class of explicit two-time-level semi-Lagrangian schemes. Quart J Roy Meteor Soc 130:365–369
ECMWF (2006) Part III: Dynamics and numerical procedures. ECMWF IFS Documentation – Cy31r1 Available at http://www.ecmwf.int/research/ifsdocs
Frisch U (1995) Turbulence. Cambridge University Press
Gravel S, Staniforth A, Côté J (1993) A stability analysis of a family of baroclinic semi-Lagrangian forecast models. Mon Wea Rev 121:815–824
Hamilton K, Takahashi YO, Ohfuchi W (2008) Mesoscale spectrum of atmospheric motions investigated in a very fine resolution global general circulation model. J Geophys Res 113
Hodur R (1997) The Naval Research Laboratory's Coupled Ocean/Atmosphere Mesoscale Prediction System (COAMPS). Mon Wea Rev 125:1414–1430
Hortal M (2002) The development and testing of a new two-time-level semi-Lagrangian scheme (SETTLS) in the ECMWF forecast model. Quart J Roy Meteor Soc 128:1671–1687
Janjic Z (1990) The step-mountain coordinate: Physical package. Mon Wea Rev 118:1429–1443
Janjic Z (2003) A nonhydrostatic model based on a new approach. Meteor Atmos Phys 82:271–285
Kolmogorov A (1990) The local structure of turbulence in incompressible viscous fluid for very large reynolds number, dokl. Akad Nauk SSSR (1941) English translation in Proc R Soc Lond A 434 30:9–13
Lilly D, Bassett G, Droegemeier K, Bartello P (1998) Stratified turbulence in the atmospheric mesoscales. Theor Comput Fluid Dyn 11:139–153
Lin S, Rood RB (1997) An explicit flux-form semi-Lagrangian shallow-water model on the sphere. Quart J Roy Meteor Soc 123:2477–2498
Lindborg E (1999) Can the atmospheric kinetic energy spectrum be explained by two-dimensional turbulence? Fluid Mech 388:259–288
Lindborg E (2006) The energy cascade in a strongly stratified fluid. J Fluid Mech 550:207–242
Lindborg E (2007) Horizontal wavenumber spectra of vertical vorticity and horizontal divergence in the upper troposphere and lower stratosphere. J Atmos Sci 64:1017–1025
Lindborg E, Berthouwer G (2007) Stratified turbulence forced on rotational and divergent modes. J Fluid Mech 586:83–108
Lindborg E, Cho JY (2000) Horizontal velocity structure functions in the upper troposphere and lower stratosphere. 2 Theoretical considerations J Geophys Res D10, 11 106:0233–1024
Mason P (1994) Large-eddy simulation: A critical review of the technique. Quart J Roy Meteor Soc 120:1–26
Nastrom G, Gage KS (1985) A climatology of atmospheric wavenumber spectra of wind and temperature observed by commercial aircraft. J Atmos Sci 42:950–960
Pinty JP, Benoit R, Richard E, Laprise R (1995) Simple tests of a semi-implicit semi-Lagrangian model on 2d mountain wave problems. Mon Wea Rev 123:3042–3058
Randall D (1994) Geostrophic adjustment and the finite difference shallow-water equations. Mon Wea Rev 122:1371–1377
Shutts G (2005) A kinetic energy backscatter algorithm for use in ensemble prediction systems. Q J Roy Meteor Soc 131:3079–3102
Skamarock W (2004) Evaluating mesoscale NWP models using kinetic energy spectra. Mon Wea Rev 132:3019–3032
Skamarock WC, Klemp JB (2008) A time-split nonhydrostatic atmospheric model for weather research and forecasting applications. J Comput Phys 227(7):3465–3485
Skamarock WC, Klemp JB, Dudhia J, Gill DO, Barker DM, Duda MG, Huang XY, Wang W, Powers JG (2008) A description of the Advanced Research WRF version 3. Ncar Technical Journal NCAR/TN-475+STR, 113 pp
Staniforth A, Wood N (2008) Aspects of the dynamical core of a nonhydrostatic, deep-atmosphere, unified weather and climate-prediction model. J Comput Phys 227(7):3445–3464

Takahashi YO, Hamilton K, Ohfuchi W (2006) Explicit global simulation of the mesoscale spectrum of atmospheric motions. Geophys Res Lett 33:L12,812

Tao WK, Moncrieff MW (2009) Multiscale cloud system modeling. Rev Geophys 47: 41pp

Weisman ML, Skamarock WC, Klemp JB (1997) The resolution dependence of explicitly modeled convective systems. Mon Wea Rev 125:527–548

Wyngaard JC (2004) Changing the face of small-scale meteorology. In: Federovich, Rotunno, Stevens (eds) Atmospheric Turbulence and Mesoscale Meteorology, Cambridge University Press, pp 17–34

Xue M, Droegemeier KK, Wong V (1990) The advanced regional prediction system (ARPS) - a multiscale nonhydrostatic atmospheric simulation and prediction tool. Part I Model dynamics and verification Meteor Atmos Phys 75:339–356

Yeh KS, Côté J, Gravel S, Méthot A, Patoine A, Roch M, Staniforth A (2002) The CMC-MRB Global Environmental Multiscale (GEM) model. Part III: Nonhydrostatic formulation. Mon Wea Rev 130:339–356

Chapter 15
A Perspective on the Role of the Dynamical Core in the Development of Weather and Climate Models

Richard B. Rood

Abstract This chapter aims to place the dynamical core of weather and climate models into the context of the model as a system of components. Building from basic definitions that describe models and their applications, the chapter details the component structure of a present-day atmospheric model. This facilitates the categorization of model components into types and the basic description of the dynamical core. An important point in this categorization is that the separation between 'dynamics' and 'physics' is not always clear; there is overlap. This overlap becomes more important as the spatial resolution of models increases, with resolved scales and parameterized processes becoming more conflated. From this categorization an oversimple, intuitive list of the parts of a dynamical core is made. Following this, the equations of motion are analyzed, and the design-based evolution of the dynamical core described in Lin (2004) is discussed. This leads to a more complete description of the dynamical core, which explicitly includes the specification of topography and grids on which the equations of motion are solved. Finally, a set of important problems for future consideration is provided. This set emphasizes the modeling system as a whole and the need to focus on physical consistency, on the scientific investigation of coupling, on the representation of physical and numerical dissipation (sub-scale mixing and filtering), and on the robust representation of divergent flows. This system-based approach of model building stands in contrast to a component-based approach and influences the details of component algorithms.

15.1 Introduction

This is a perspective on the design of physical models for use in the scientific investigation of weather and climate. This perspective follows from a career that involves both model development and the management of the development of institutional models. The point of view is anchored around the role of the dynamical core in

R.B. Rood
Department of Atmospheric, Oceanic and Space Sciences, University of Michigan, 2455 Hayward Street, Ann Arbor, MI 48109, USA
e-mail: rbrood@umich.edu

P.H. Lauritzen et al. (eds.), *Numerical Techniques for Global Atmospheric Models*, Lecture Notes in Computational Science and Engineering 80,
DOI 10.1007/978-3-642-11640-7_15,

atmospheric models. There are numerous books on atmospheric modeling, their history, their construction, and their applications (Trenberth 1992; Randall 2000; Mote and O'Neill 2000; Satoh 2004; Jacobson 2005; Washington and Parkinson 2005). The review paper by Rood (1987) contains many foundational references, and a basic introduction to the problem of numerical advection. The concepts associated with the works of Godunov (1959), Boris and Book (1973), and van Leer (1979) are particularly influential.

The perspective is outlined as follows:

- Definition and Description of the Model
- Construction of Weather and Climate Models
- Analysis of the Atmospheric Equations of Motion
- Numerical Expression of the Atmospheric Equations of Motion
- Synthesis and Future Directions
- Conclusions

15.2 Definition and Description of the Model

Dictionary definitions of model include:

- A work or construction used in testing or perfecting a final product.
- A schematic description of a system, theory, or phenomenon that accounts for its known or inferred properties and may be used for further studies of its characteristics.

In weather and climate modeling a scientist is generally faced with a set of observations of variables, for instance, velocity, temperature, water, ozone, etc., as well as either the knowledge or expectation of correlated behavior between the different variables. A number of types of models could be developed to describe the observations. These include:

- Conceptual or heuristic models which outline in the simplest terms the processes that describe the interrelation between different observed phenomena. These models are often intuitively or theoretically based. An example would be the tropical pipe model of Plumb and Ko (1992), which describes the transport of long-lived tracers in the stratosphere.
- Statistical models which describe the behavior of the observations based on the observations themselves. That is, the observations are described in terms of the mean, the variance, and the correlations of an existing set of observations. Johnson et al. (2000) discuss the use of statistical models in the prediction of tropical sea surface temperatures.
- Physical models which describe the behavior of the observations based on first principle tenets of physics (chemistry, biology, etc.). In general, these principles are expressed as mathematical equations, and these equations are solved using

discrete numerical methods. Detailed discussions of modeling include Trenberth (1992), Randall (2000), Mote and O'Neill (2000), Satoh (2004), Jacobson (2005), and Washington and Parkinson (2005).

In the study of geophysical phenomena there are numerous sub-types of models. These include comprehensive and mechanistic models. Comprehensive models attempt to model all of the relevant couplings or interactions in a system. Mechanistic models have prescribed variables, and the system evolves relative to the prescribed parameters. All of these models have their place in scientific investigation, and it is often the interplay between the different types and sub-types of models that leads to scientific advance.

Models are used in two major roles. The first role is diagnostic, in which the model is used to determine and to test the processes that are thought to describe the observations. In this case, it is determined whether or not the processes are well known and adequately described. In general, since models are an investigative tool, such studies are aimed at determining the nature of unknown or inadequately described processes. The second role is prognostic; that is, the model is used to make a prediction.

In all cases the model represents a management of complexity; that is, a scientist is faced with a complex set of observations and their interactions and is trying to manage those observations in order to develop a quantitative representation. In the case of physical models, which are the focus here, a comprehensive model would represent the cumulative knowledge of the physics (chemistry, biology, etc.) that describe the observations. It is tacit, that an accurate, validated, comprehensive physical model is the most robust way to forecast; that is, to predict the future.

The physical principles represented in an atmospheric model, for example, are a series of conservation laws which quantify the conservation of momentum, mass, and thermodynamic energy. The equation of state describes the relation between the thermodynamic variables. Because of the key roles that phase changes of water play in atmospheric energy exchanges, an equation for the conservation of water is required. Similarly, an equation for salinity is necessary to represent ocean dynamics. Models which include the transport and chemistry of atmosphere trace gases and aerosols require additional conservation equations for these constituents. The conservation equations for mass, trace gases, and aerosols are often called continuity equations.

In general, the conservation equation relates the time rate of change of a quantity to the sum of the quantity's production and loss. The production and loss for momentum follow from the forces described by Newton's Laws of Motion. Since the atmosphere is a fluid, either a Lagrangian or an Eulerian description of the flow can be used (Holton 2004). The Lagrangian description follows a notional fluid parcel, and the Eulerian description relies on spatial and temporal field descriptions of the flow at a particular point in the domain. In this chapter the Eulerian framework will be the primary focus. Holton (2004) provides a thorough introduction to the fundamental equations of motions and their scaling and application to atmospheric dynamics.

Table 15.1 Construction of an atmospheric model (see text for details)

Boundary/Initial conditions	Emissions, topography, sea surface temperature	ϵ
Representative equations	$DA/Dt = P - LA$	ϵ
Discrete/Parameterize	$\left(A_{t+\Delta t} - A_t\right)/\Delta t = \ldots$	ϵ_d,ϵ_p
Theory/Constraints	Geostrophy, Thermal wind	Scale analysis
Primary Products (i.e. A)	$T, u, v, H_2O, O_3, \ldots$	ϵ_b,ϵ_v
Derived Products	Potential Vorticity, Budgets	Consistent

ϵ_d = discretization error, ϵ_p = parameterization error, ϵ_v = variability error, ϵ_b = bias error

In order to provide an overarching background, it is useful to break down the process of the construction of an atmospheric model as shown in Table 15.1. The table lists six major elements (left column), a concrete example of the element (middle column), and a reminder that there are explicit errors, ϵ, at all stages of the construction (right column). The first element points to the boundary and initial conditions. For an atmospheric model, boundary conditions include topography, sea surface temperature, land type, vegetation, etc. Note that boundary conditions are generally prescribed from external sources of information.

The next three items in the table are intimately related. They are the representative equations, the discrete and parameterized equations, and constraints drawn from theory. The representative equations are the continuous forms of the conservation equations. The representative equations used in atmospheric modeling are approximations derived from scaling arguments (see Holton 2004); therefore, even the equations the modeler is trying to solve have *a priori* simplification which can be characterized as errors. Here a conservation equation for an arbitrary quantity, A, is written with an exemplary production, P, and loss, L. The continuous equations are a set of non-linear partial differential equations. The solutions to the representative equations are a balance amongst competing forces and tendencies.

The discrete and parameterized equations arise because it is not possible to solve the representative equations in analytical form. The strategy used by scientists is to develop a numerical representation of the equations. One approach is to define a grid of points which covers the spatial domain of the model. Then a discrete numerical representation of those variables and processes which can be resolved on the grid is written. Processes which take place on spatial scales smaller than the grid are parameterized. These approximate solutions are, at best, discrete estimates to solutions of the analytic equations. The discretization and parameterization of the representative equations introduce a large source of error. This introduces another level of balancing in the model; namely, these errors are generally managed through a subjective balancing process that keeps the numerical solution from producing obviously incorrect estimates.

While all of the terms in the analytic equation are potentially important, there are conditions or times when there is a dominant balance between, for instance, two terms. An example of this is the geostrophic balance and the related thermal wind balance in the middle latitudes of the atmosphere (Holton 2004). It is these balances, generally at the extremes of spatial and temporal scales, which provide the constraints drawn from theory. Such constraints are generally involved in the development of conceptual or heuristic models. If the modeler implements discrete methods which represent the relationship between the analytic equations and the constraints drawn from theory, then the modeler maintains a substantive scientific basis for the interpretation of model results.

The last two items in Table 15.1 represent the products that are drawn from the model. These are divided into two types: primary products and derived products. The primary products are variables such as temperature T, wind (u, v), water (H_2O), and ozone (O_3) – parameters that are, most often, transported by the fluid flow. The primary products might also be called the resolved or prognostic variables. The derived products are of two types. The first type describes those products which are diagnosed from the model's state variables, often in the parameterized physical processes. The second type follows from functional, $F(A)$, relationships between the primary products; for instance, potential vorticity (Holton 2004). A common derived product is the budget – the sum of the different terms of the discretized conservation equations. The budget is studied, explicitly, on how the balance is maintained and how this compares with budgets derived from observations or observations assimilated into predictive models.

In some cases the primary products can be directly evaluated with observations, and errors of bias and variability are estimated. The bias is, for example, the difference between time-averaged model predictions and observations. Variability errors follow from, for example, the representation of the distributions about the temporal mean. If attention has been paid in the discretization of the analytic equations to honor the theoretical constraints, then the derived products will behave consistently with the primary products and theory (see, Table 15.1). In this case consistency is used to state that budgets are balanced, and that the physically based, correlative relationship between variables is represented. In a consistent model, there will be errors of bias and variability, but when a budget is formed from the sum of the terms in the conservation equations, it will balance. That is, the discrete form of the conservation equation is solved.

15.3 Construction of Weather and Climate Models

Weather and climate models are an assembly of components that are composited together to construct integrated functionality. Composites are then composited together, yielding highly complex systems. For example, a physical climate model can be constructed from a sea ice model, a land surface model, an ice sheet model, an ocean model and an atmospheric model. Associated with these composited models are representations of chemical and biological processes important to the physical climate, for example, atmospheric ozone and plant respiration (i.e., carbon dioxide). These models communicate with each other through a coupler.

Big models are made from smaller models, and this concept cascades to increasing granularity. This method of model construction has been described as 'process splitting' or fractional steps and is described in, for instance, Yanenko (1971), Strang (1968), and McCrea et al. (1982). Historically, atmospheric models evolved from efforts focused on specific parts of the atmosphere: thermosphere models (e.g. Dickinson et al. 1981), middle atmosphere models (e.g. Schoeberl and Strobel 1980; Fomichev et al. 2002), and many models of the troposphere – often weather forecast models. The focus on specific parts of the atmosphere was driven by scientific interests, observational and theoretical foundations, and limited computational resources. A primary characteristic of, for example, a model focused on the middle atmosphere (i.e., the stratosphere and mesosphere) is special attention to the physical and chemical parameterizations that are important in the focus region. Connectivity, for example, the influence of the troposphere on the stratosphere is achieved in several ways. Mechoso et al. (1985) coupled the stratosphere to the troposphere with filtered observations at a lower boundary. A natural and comprehensive approach is to extend the domain of a tropospheric model upward or a middle atmosphere model downward with inclusion of appropriate physical algorithms. Only recently, whole atmosphere models have been routinely used for scientific research (e.g. Beres et al. 2005).

Using a specific model type to expose the component structure, a troposphere-stratosphere model might be constructed from a set of components that include, for example, algorithms that represent advection, mixing, the planetary boundary layer, gravity waves, radiation, cumulus convection and clouds. The component of the atmospheric model that represents clouds might then have sub-components that represent the different phases of water, sulfate aerosols (hence, sulfate chemistry), black carbon, etc. Components at all levels need to communicate with each other, and thus, in a generalized sense there is a requirement for coupling of components.

Figure 15.1 shows the Earth System Modeling Framework (ESMF, http://www.esmf.ucar.edu/about_us/) component architecture of the Goddard Earth Observing System, version 5 (GEOS-5) atmospheric model (Rienecker et al. 2008). From the top down, the structure shows the coupling of the atmospheric general circulation model ('agcm'), with the stored, digital 'history' files used in model initialization, diagnostics and application. Below 'agcm' there is a separation of the model components into 'dynamics' and 'physics,' and, again and throughout, the explicit need for coupling.

Those algorithms that are associated with advection and part of the sub-scale mixing (defined below) are often identified as 'the dynamics' and all of the other algorithms are identified as 'the physics'. The dynamical core is identified as 'fvcore' in Fig. 15.1. In this model the physical 'gravity_ wave_ drag' parameterization is counted as part of the dynamics. To be explicit, some algorithms identified with 'the physics' represent adiabatic dynamical processes such as 'turbulence' in the boundary layer. These mixing processes, gravity wave drag and turbulence, are at a resolution smaller than the grid can resolve, but associated with some physical cause not explicitly resolved by the dynamical core. This counting of dynamical processes in both 'dynamics' and 'physics' is a source of ambiguity in the definition of the dynamics of the model and the dynamical core – an ambiguity that becomes more important to address as resolution is increased.

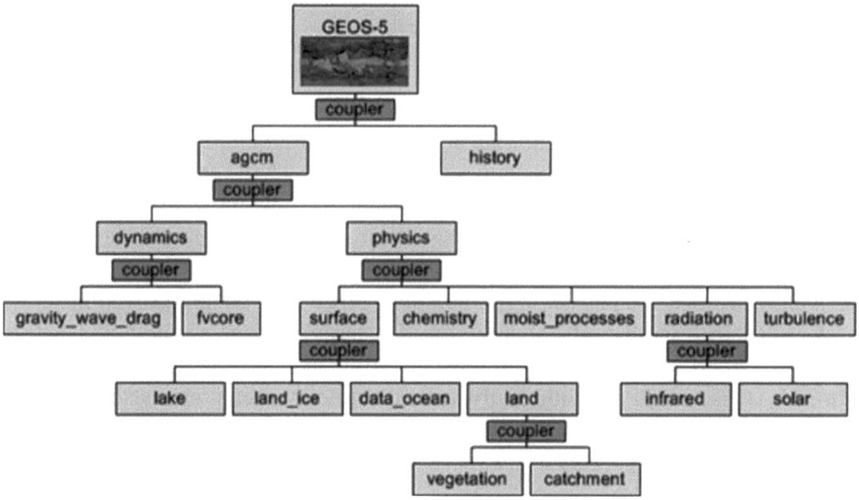

Fig. 15.1 Component architecture of the GEOS-5 atmospheric model

At the 'surface' the atmospheric model is coupled to needed information from other components of the Earth system – or the other components of a climate model. In this case the effects on the atmosphere of lakes ('lake') and ice and snow on the land ('land_ice') are explicitly specified. Data that represent the state of ocean are included in the standard configuration of the model ('data_ocean'). This is where an explicit, interactive ocean model could be coupled. Finally, the interaction of 'vegetation' on 'land' is included. Land-surface hydrology is represented on a spatial discretization based on water 'catchments', rather than the grid used for the atmospheric model.

There is no standard definition of the term 'dynamical core' (in short 'dycore'). Williamson (2007) defines the dynamical core 'to be the resolved fluid flow component of the model'. This definition is one that has been widely shared in model development centers, as is perhaps best represented by the model documentation (e.g. Collins et al. 2004). Within this chapter the following definition from Thuburn (2008b) is used: "The formulation of a numerical model of the atmosphere is usually considered to be made up of a dynamical core, and some parameterizations. Roughly speaking, the dynamical core solves the governing fluid and thermodynamic equations on resolved scales, while the parameterizations represent sub-grid scale processes and other processes not included in the dynamical core such as radiative transfer. Here, no attempt is made to give a precise definition of 'dynamical core' because, as discussed below, there are some open questions concerning exactly which terms and which processes should be included in a dynamical core".

In order to expose the building blocks of a dynamical core and to address the ambiguities and open questions suggested above, (15.1) is used to illustrate the 'dycore' part of the model more concretely. A representative conservation equation for a scalar quantity, A, can be written as

$$\frac{\partial A}{\partial t} = -\nabla \cdot \mathbf{u}A + M + P - LA \qquad (15.1)$$

P represents production and L represents a loss rate. $\mathbf{u}$ is the vector velocity and t is time. M represents dynamical mixing at spatial scales smaller than the grid size. A is, in this example, assumed to be a scalar parameter such as temperature or ozone. Formally in the dynamical core, A would include the velocity components, which yields a nonlinear equation and limits this illustration to only being demonstrative. In analogy with the atmospheric model described in Fig. 15.1, the P and LA terms are identified with 'the physics'. The flux divergence term is the resolved flow and is identified with the dynamical core. The flux term is where a specific advection algorithm (Rood 1987; Williamson 2007) is implemented. The dynamical mixing term, M, is also identified with the dynamical core; however, it may also be related to the mixing associated with the physics and is, hence, not cleanly separated. Part of the purpose of this chapter is to expose this ambiguity and refine the description of the dynamical core.

In their simplest expressions, dynamical cores are generally process split and include the following:

- The resolved advection in the horizontal plane.
- The resolved vertical advection.
- Unresolved sub-scale transport.
- A portfolio of filters and fixers that accommodate errors related to both the numerical technique and the characteristics of the underlying grid.

A more complete description of the dynamical core will be developed below, including discussion of how the dynamical core spans the equations of motion.

As revealed above, models are complex composites of sub-models. These sub-models are, most often, also complex, and they are approximations of varying accuracy that represent physical processes. At the finest levels these models are said to be parameterized, and the algorithms described as parameterizations. The function of the model as a whole is an amalgamation of all of the composites, and is therefore, a function of the errors associated with the components and how those errors are accumulated. For this reason, development of highly accurate sub-models and parameterizations often does not lead directly to an improved function of the model as a whole. The model needs to be rebalanced or tuned, a process that implicitly addresses the balance between both physical processes and error sources.

This description and the representation in Fig. 15.1 explicitly reveal the fact that there are many couplers in a model. Couplers are, *de facto*, yet more model components, and their construction influences the performance of the system as a whole. The robustness and the integrity of the model as a whole are often construed as being based on the construction and the quality of the component algorithms. Ultimately however, it is the function of the system as a whole that is of interest to the discipline scientist, e.g., the climate forecast user. Hence, the physics of the couplers also requires scientific scrutiny.

15.4 Analysis of the Atmospheric Equations of Motion

The equations of motion for the atmosphere in tangential spherical coordinates using the radial distance for the vertical coordinate (λ, ϕ, r) are given by (see also White et al. 2005)

$$\begin{aligned}
\frac{Du}{Dt} - \frac{uv\tan(\phi)}{r} + \frac{uw}{r} &= -\frac{1}{\rho r\cos(\phi)}\frac{\partial p}{\partial \lambda} + 2\Omega v\sin(\phi) - 2\Omega w\cos(\phi) + \nu\nabla^2 u \\
\frac{Dv}{Dt} + \frac{u^2\tan(\phi)}{r} + \frac{vw}{r} &= -\frac{1}{\rho r}\frac{\partial p}{\partial \phi} - 2\Omega u\sin(\phi) + \nu\nabla^2 v \\
\frac{Dw}{Dt} - \frac{u^2+v^2}{r} &= -\frac{1}{\rho}\frac{\partial p}{\partial r} - g + 2\Omega u\cos(\phi) + \nu\nabla^2 w \\
\frac{D\rho}{Dt} &= -\rho\nabla\cdot\mathbf{u} \\
c_v\frac{DT}{Dt} + p\frac{D\alpha}{Dt} &= J \\
p &= \rho R_d T \\
\alpha &= \frac{1}{\rho}
\end{aligned} \tag{15.2}$$

with

$$\frac{D}{Dt} = \frac{\partial}{\partial t} + \mathbf{u}\cdot\nabla \tag{15.3}$$

t denotes the time, λ is longitude, ϕ is latitude, r is the radial distance to the center of the spherical Earth, Ω is the angular velocity of the Earth, g is gravity, ν is a coefficient of viscosity, c_v is specific heat at constant volume, R_d is the gas constant for dry air, ρ is density, T is temperature, p is pressure, $\mathbf{u}$ is the velocity vector $\mathbf{u} = (u, v, w)$, and J stands for the diabatic heating. The first three equations represent the conservation of momentum. The fourth equation is the mass continuity equation, and the fifth equation is the thermodynamic energy equation. The last equation in (15.2) is the equation of state for dry air.

In addition, equations are needed which describe the conservation of trace constituents. The generic form of these continuity equations are:

$$\frac{DQ_i}{Dt} + Q_i\nabla\cdot\mathbf{u} = P_{Q_i} - L_{Q_i} \tag{15.4}$$

Where Q_i is the density of a constituent identified by the subscript i. P_{Q_i} and L_{Q_i} represent the production and loss from phase changes and photochemistry. An equation for water in the atmosphere, $Q_i = Q_{H_2O}$, is required for a comprehensive atmospheric model. For water vapor, the production and loss terms are represented by evaporation and condensation. These are associated with significant consumption and release of heat, which must be accounted for in the heating J, the production and loss term of the thermodynamic energy equation. In the atmosphere below the stratopause, heating due to the chemical reactions of trace constituents are assumed not to impact the heat budget of the atmosphere. It is possible for the spatial distribution of trace constituents, for example ozone, to impact the absorption and emission of radiative energy; hence, there is feedback between the constituent distributions and diabatic processes in the atmosphere.

Dynamical cores are often developed, first, in the two-dimensional shallow-water model (for example Lin and Rood 1997). This brings focus to the momentum equation, with the presumption that if the numerical technique provides a good solution to the nonlinear momentum equations, then spanning the technique across the whole set of the equations of motion is relatively straightforward. The review of Williamson (2007) takes a focus on the 'horizontal aspects of the schemes' and describes the methods used to represent the advective terms in the equations of motion. The extension from two dimensions to three dimensions and consideration of real-world aspects of atmospheric modeling require addressing a set of fundamental issues. These issues lead to a more complete specification of the dynamical core, which will be exposed below.

There are a number of important points to be made directly from the atmospheric equations of motion. In general, the equations are scaled to expose the range of motions that are important to weather and climate models. Consideration of 'large-scale' dynamics, for example motions of spatial scales 1,000 km or greater, leads to a separation of the horizontal and vertical motions in the atmosphere. Similarly, it leads to the conclusion that the flow is dominated by rotational motion, as contrasted with divergent motion. Such scale analysis explicitly impacts the development of dynamical cores in numerical models; for example, the development of different algorithms to treat horizontal and vertical advection. In fact, the consideration of the large-scale characteristics of the atmospheric flow impacts the development of dynamical cores in ways that have such profound influence on the numerical performance that they require algorithmic archeology to expose their impact.

Returning to (15.2), consider the first two terms on the right hand side of the u and v equations. These are the horizontal pressure gradient terms and the Coriolis terms. They are often dominating terms and represent the geostrophic balance. From first principles, the pressure gradient initiates motion. As a large term important to the motion of the atmosphere, it is critical that the pressure gradient term be well represented. Alternatively, if the pressure gradient term is poorly represented, then there will be large negative consequences to the model performance.

Accuracy of the representation of the pressure gradient brings attention to the lower boundary condition and the specification of topography. A common practice in atmospheric modeling is to use a terrain following vertical coordinate (see Holton 2004; Satoh 2004). This eases the specification of the lower boundary. However, it introduces a major challenge in the representation of the horizontal pressure gradient in the presence of steep topography, hence, large pressure gradients, hence, large discretization errors. Though the pressure gradient and the Coriolis force are, abstractly, a momentum source term, these forces are a resolved part of the flow. Therefore, discretization of the pressure gradient term and specification of the Coriolis force are parts of the dynamical core.

There are in the continuous equations of motion, explicit dissipative terms. These include both the viscosity terms as well as the diabatic heating term (J), which includes damping of temperature perturbations. In the continuous equations the viscosity terms are usually very small. In the discrete equations viscosity takes on a far different character. In the estimation of numerical solutions, variability starts to form at the smallest spatial grid scale. This structure comes from a variety of sources ranging from physical advective cascade from large to small scales to numerical dispersion caused by different wavelengths propagating at different speeds. The grid-scale structure can come to dominate the estimated solution; hence, it requires dissipation both to account for a discrete representation of physical mixing and for remediation of unavoidable numerical errors. Therefore, real atmospheric dissipation becomes conflated with many forms of dissipation that are present in dynamical cores for both physical and numerical reasons. Further, this dissipation is not independent of that modeled in the planetary boundary layer parameterization and the gravity wave parameterization – both accounted for as part of the model 'physics'. There is no prescription from first principles on how to address the specification of dissipation, and the modeler is often left with the statement in Farge and Sadourny (1989) that the "validity (of the choice of dissipation) can only be judged on the grounds of numerical results". A thorough review of the dissipative processes in the dynamical cores of general circulation models is provided in Chap. 13.

Finally, consider the constituent continuity equations (15.4). As suggested above, the intuitive focus of the 'dynamical core' is on the algorithm used to solve the momentum equations, or alternatively, the vorticity and divergence equations. An atmospheric model, however, requires the solution of the thermodynamic equation and numerous constituent continuity equations. The mass conservation equation and the equation of state must be tied into the numerical solution. These equations all contain the advection of scalar quantities by the resolved flow, by definition, part of the dynamical core. The thermodynamic and constituent continuity equations can be addressed with different algorithms for the scalar advection than used in the momentum equation (see Rasch and Williamson 1991; Rasch et al. 2006). Without special attention, this explicitly introduces an inconsistency in the formulation of the model as a whole. This inconsistency can be interpreted as using a different vertical velocity for the advection of scalars than is estimated from the solution of the momentum equations (see Lin and Rood 1996; Jöckel et al. 2001; Machenhauer et al. 2008). This will be discussed more fully below.

Compared with the previous sections, the discussion and analysis presented in this section both refines and expands the components that make up the dynamical core. Namely, dynamical cores are generally process split and include algorithms that represent:

- The resolved advection of momentum in the horizontal plane.
- The resolved vertical advection of momentum.
- Unresolved sub-scale transport of momentum.
- A specification of the pressure gradient force.
- A specification of the Coriolis force.
- A specification of topography.

- The resolved advection of scalars in the horizontal plane.
- The resolved vertical advection of scalars.
- Unresolved sub-scale transport of scalars.
- A portfolio of filters and fixers that accommodate errors related to both the numerical technique and the characteristics of the underlying grid.

15.5 Numerical Expression of the Atmospheric Equations of Motion

There are many ways to approach the numerical estimation of solutions to the equations of motion for the atmosphere. A straightforward approach is to develop a discrete representation of variables and derivatives and to estimate, directly, the partial differential equations. It is reasonable to assume that accurate representation of the terms in the equation would lead to a credible numerical solution.

The equations of motions support many scales and types of motion. Some of these motions, such as sound waves, are not of direct relevance to weather and climate models. Unwanted scales are often eliminated either by recasting the continuous equations in such a way as to eliminate the unwanted scales or through numerical techniques such as filtering and scale-selective dissipation. When the discrete equations are formed new types of unwanted, computational motions might be created.

In addition, there are many important relationships that exist in the equations of motion. For example there are energy constraints, such as conservation of total energy for adiabatic, inviscid flows. Scaling arguments reveal strong relationships between, for example, the winds and the thermal structure and the vorticity and the pressure fields (see Holton 2004). Marching through the equation of motions making best estimates of the individual terms in the equations does not assure that these relationships are honored. Such inconsistencies in the discretization can lead to models composed of highly accurate elements that, collectively, do not provide credible simulations.

Therefore, experience suggests an alternative approach to the development of models. In this alternative approach design requirements are specified and numerical algorithms are developed to meet these requirements. Accuracy is sought in the context of integrated design.

This section will investigate the design-based approach of the Lin and Rood (1996) advection scheme and the full dynamical core which has been developed by Lin (2004). Model development by algorithm design is discussed thoroughly by Machenhauer et al. (2008).

The Lin and Rood (1996) advection scheme was motivated by attempts to model the high-quality aircraft observations collected to determine the chemical mechanisms responsible for the Antarctic ozone hole. Of special importance from these observations were the correlations between trace constituents (Fahey et al. 1990). These correlations are conserved in the absence of photochemical losses and sinks; that is, they are conserved in pure advection. Numerical simulations with conventional finite difference and spectral methods showed that correlations were not conserved, and that the lack of conservation was of sufficient magnitude to make comparisons with observations of little scientific value. The inability of these schemes to conserve tracer correlations was directly related to the filtering techniques used to counter the generation of negative tracer concentrations which arise from numerical errors. The strategy for addressing this problem was to adapt piecewise continuous schemes of the sort developed by Bram van Leer to atmospheric problems (see van Leer 1979; Allen et al. 1991). These are finite volume schemes which partition fluid volumes at each time step based on the velocity field. As posed in Lin and Rood (1996) the design criteria were:

- Conservation of mass without *a posteriori* restoration.
- Computation of mass fluxes based on the sub-grid distribution in the upwind direction.
- Generation of no new maxima or minima (ideally, maintain monotonicity).
- Preservation of tracer correlations.
- Computational efficiency in spherical coordinates.

These design criteria in combination with a mixture of higher and lower order numerical techniques led to credible results in a wide variety of chemistry-transport models (Douglass et al. 1997; Bey et al. 2001; Rotman et al. 2001). Implicit in the development was the reduction of numerical diffusion compared with the previously used methods (Allen et al. 1991). Also, implicit in this development is that the advection of well-resolved spatial scales, for example resolved by ten grid cells or more, is well represented. The number of grid cells required to resolve a feature accurately is not a strictly defined quantity. The choice of ten emphasizes that there is an order of magnitude between the number of grid cells and resolved scales; ten is drawn from the discussion of errors in Zalesak (1981). This criterion also directly states that there is a range of scales that are 'resolved,' but not accurately. The advection of accurately resolved waves in modern numerical advection schemes is expected, and therefore, does not serve as a good discriminator between algorithms.

The design features in the Lin and Rood advection scheme can be reframed to state that if a tracer distribution originally has no tracer gradients, then the tracer distribution will not change during the computation of advection. It was often true that chemistry-transport models did not have this feature, which is directly traceable to the underlying mass conservation equation (15.2) not being satisfied. This can be articulated as the vertical velocity that satisfies the momentum and mass conservation equations is not the same as the vertical velocity used in the calculation of the scalar advection. This design criterion is characterized as 'consistency,' where consistency represents the physical relationships that tie together the entire system of the equations of motion and the tracer continuity equations.

Known inadequacies of the Lin and Rood scheme at the time of development included splitting errors that generated negative concentrations in some instances and numerical diffusion related largely to the slope limiters. It was a design decision to take numerical errors in diffusion rather than in dispersion errors. Alternatively, diffusion is used to remedy, not cure, dispersion errors. In practice the scheme conserved constituent correlations in realistic test problems. However, the presence of splitting errors and the nonlinear application of the slope limiters means that there are potentially failures of both monotonicity and the conservation of correlations.

The Lin and Rood (1996) advection scheme was extended to the two-dimensional shallow water equations in Lin and Rood (1997) and to the three-dimensional primitive equations in Lin (2004). Both implementations utilized a regular, equal-angle, latitude-longitude grid. A major goal was to develop a numerical system that treated the momentum equations, the thermodynamic equation, and the tracer continuity equations 'consistently,' as defined above. Also in this development was the specification of quantities on the grid and use of averaging techniques to assure the correlative relationship between geopotential (i.e., a pressure-like variable in a coordinate system that uses pressure as a vertical coordinate) and vorticity. This design decision valued the accurate advection of vorticity. Therefore, the original development of the scheme was implicitly tuned towards the characteristics of large-scale dynamical features in a rotationally dominated flow.

Perhaps more important to model performance than the horizontal advection scheme was the development of methods to represent the horizontal pressure gradient and the treatment of vertical advection. Lin (1997) describes a piecewise continuous, finite volume method to represent the horizontal pressure gradient. This method, which integrates piecewise linear edges of the volume to calculate the balance of pressure forces on a volume, proved to be two orders of magnitude more accurate in the presence of steep topography than finite difference schemes used at the time.

The description of a Lagrangian formulation of the vertical velocity in Lin (2004) completes the development of the dynamical core. This calculation of vertical velocity originally relied on the hydrostatic approximation. It is analogous to the use of isopycnal coordinates in ocean modeling. This approach has a tremendous impact on the fidelity of the model, especially with regard to the representation of the mean meridional circulation important to the general circulation and tracer distributions (Schoeberl et al. 2003).

The design features discussed above suggest another attribute of the Lin (2004) dynamical core that was a desired feature. The net effect of the design is that the scheme is highly localized. The information that is used to calculate the atmospheric dynamics and tracer transport comes from nearby and primarily upstream grid points. This stands in contrast with spectral or pseudospectral methods, which use global basis functions and are formally more accurate (see Rood 1987). The local nature of the scheme has potential positive benefits for the representation of quantities that are derived from the model's physical parameterizations. That is, the locality is relevant to the coupling between the dynamical core and the physics; the physics parameterizations are intrinsically local (see Bala et al. 2008).

The expression of the dynamical core described above in this section addresses the analysis of the equations of motion in the previous section. What has yet to be discussed is the portfolio of filters and fixers that are required for the scheme. The most obvious design feature of the model to address known errors is the slope limiter which is diffusion implemented locally when a new maximum or minimum will be created (see van Leer 1979). The use of slope limiting is a design decision. Slope limiting (van Leer 1979, and references therein) and flux limiters, pioneered by, e.g., Boris and Book (1973), were motivated by consideration of plasma shocks and the prevention of the generation of non-physical ripples at the shock front. This is an error that cannot be overlooked in reactive flow and combustion calculations. A physical analysis of the advective process reveals that advection cannot generate new maxima or minima in the scalar fields. That is, advection is monotonic, and if monotonicity is violated, then the scheme is 'non-physical'.

More generally, there are many errors in the calculation of advection that are non-physical. For example, without special consideration quadratic and higher moments of advected fields are not conserved in numerical algorithms. This is non-physical, and potentially important when considering conservation of energy, the propagation of variance and covariance in data assimilation, or modeling the distribution of droplets and aerosols. Prather (1986) developed a highly accurate advection scheme which conserves moments and vastly reduces numerical diffusion.

In Lin and Rood (1997) it was argued that the nature of the slope limiters, essentially a flow-dependent, nonlinear diffusion, was 'physical'. This argument is not formally true, but it is a statement that the mixing is localized and flow dependent, which is intuitively appealing. The diffusion associated with the limiter is large enough that it was not required to add an additional diffusion to the algorithm to eliminate grid-scale noise in scalar advection. The addition of diffusion is common in atmospheric models (for example Collins et al. 2004; Williamson 2007).

There are a variety of other filters and fixers in the scheme. There is a polar filter, which arises because of the decrease of the grid spacing on the equal-angle grid at high latitudes. More importantly, the scheme generates grid-scale noise, which manifests itself as localized divergent flows. This is countered by damping the divergent part of the flow. There is another digital filter which is used to manage grid-scale noise. All of these filters are ultimately diffusive, essential to the stability and performance of the dynamical core, and have a complex impact on the performance of the model. They are not an unusual portfolio of filters and are conflated with any representation of physical mixing, diffusion, and dissipation (see Chaps. 13 and 14).

15.6 Synthesis and Future Directions

The previous sections provide a high-level view of the structure and construction of weather and climate models. Atmospheric models are used to provide a concrete example. The point of view is from the role of the dynamical core in the model. Adcroft et al. (2004), Adcroft and Hallberg (2006), Adcroft et al. (2008) and White and Adcroft (2008) present a comprehensive representation of a modern oceanic dynamical core with many parallel attributes to what has been presented here for the atmosphere.

15.6.1 Model-Relevant Principles

1. Models are built from components, and the ultimate customers of models are interested in the results of the model as a whole. The application of the model strongly influences the priorities that are given weight in the building of a model. In the example provided in this chapter, correlated behavior of trace constituents and the conservation of advected variables have high priority. Given that model performance as a whole is ultimately required, balanced development of model components is necessary. The benefit of a highly precise algorithm, for say advection, is easily lost because other errors in the model or errors in the coupling of components are large.
2. The model as a whole is explicitly or implicitly optimized, i.e., tuned, towards applications at hand. This tuning includes the balancing of compensating errors. The introduction of a new, better founded algorithm is highly likely to degrade, initially, the performance of the model as a whole. This makes a barrier for the introduction of improved algorithms in models. New tuning is needed.
3. Formally, a validation plan that reflects the expected results of the model as a whole provides a framework for evaluating the impact of algorithms and their coupling. It is within the context of this validation plan that decisions on the potential benefits of improved algorithms should be made.

15.6.2 Lessons Learned about Dynamical Cores

15.6.2.1 Consistency

The enforcement of consistency in the development of dynamical cores has had significant payoff. In the field of tracer advection, the term consistency originally referred to what Machenhauer et al. (2008) call the mass-wind consistency; that is, the potential disconnection that can occur between mass conservation in the fluid and calculation of the transport of trace species (see also Jöckel et al., 2001). In this chapter, consistency is extended to include the theoretical constructs such as the thermal wind, the relation between vorticity and the pressure field obtained from scaling arguments, preservation of constituent correlations, specification of topography, etc. More generally, consistency refers to the correlative behavior that follows from theory, which is of tremendous value in the interpretation of observations and models. Correlative physics is crucial to studies of the attribution of climate change to human's activities (Santer et al. 2000). Attention to consistency improves the robustness of models. Development of consistent numerical schemes is a design decision based on developer's experience (and preference) defined by an application suite.

15.6.2.2 Locality

We have evolved to a state where we need to pay explicit attention to the interaction, that is, the coupling, between the dynamical core and 'the physics'. (see also Williamson 2007). This requires, minimally, presenting to the physics parameterizations physically realizable values of transported quantities with robust relationships to correlated parameters. Given that the physical parameterizations are local, it is intuitive that dynamical cores with localized grid stencils have potential advantage.

15.6.2.3 Horizontal Advection

The credible treatment of resolved horizontal advection is an essential performance criterion that is implicit in all modern dynamical cores. The metrics on which decisions are made are often experiential, and fall within an experiential range. Given that credible performance is realized by many schemes, horizontal advection of resolved scales has progressed to a standard and is not a discriminator of algorithms. All schemes have to balance intrinsic errors of dissipation and dispersion, and tolerance of such errors is a design and application-based decision. Conservation of advected variables without *a posteriori* restoration is, intuitively, a requirement for climate models; however, this, too, is a design decision. The importance of conservation of higher order moments, especially energy, will likely become more important in the future.

15.6.2.4 Vertical Velocity

The vertical velocity is central to the robust representation of weather and climate. Treatment of the vertical velocity is difficult because the vertical velocity is most often much smaller than the horizontal velocity. The vertical velocity is related to horizontal divergence, which is closely related to grid-scale noise and grid-scale forcing by the physical parameterizations. Therefore, the vertical velocity is strongly influenced by the sub-grid mixing, filters, and fixers. It is easy to corrupt the physical consistency of the vertical velocity. Treatment of the vertical velocity requires more attention in the development of dynamical cores.

15.6.2.5 Mixing, Filters and Fixers

The mixing algorithms, filters, and fixers have significant impact on model performance as e.g. discussed in Chaps. 13 and 14. The hydrostatic, geostrophic, and adiabatic balances in the atmosphere are powerful constraints on the flow and offer great theoretical insight. However, it is the difference from these balances that is often most important to weather and climate predictions. Fundamental theory, e.g. Andrews and Mcintyre (1978), shows that difference from balance is due to dissipation, nonlinearity and transience. Mixing algorithms, filters, and fixers are the locations where the artifacts of the discretization and numerical errors are addressed. The specifics of the mixing is important, especially with respect to the dissipation of waves. That mixing processes might be 'small' does not rationalize their being ignored. Far more attention is needed to the formulation and impact of mixing algorithms, filters, and fixers.

15.6.3 Future Directions

15.6.3.1 Divergence

The discussion in this paper reveals a number of facts about the treatment of divergence in atmospheric models. First, in the case of the Lin and Rood (1997) horizontal advection scheme, the development of the scheme is biased towards the advection of vorticity. This bias, implicitly, reflects large-scale, middle-latitude dynamics, and the importance of the conservation of vorticity. This is an acceptable situation for global climate models at resolutions of several hundred kilometers, where the flow is quasi-nondivergent. Second, in the Lin and Rood (1997) scheme, damping is added directly to the divergence in order to manage grid-scale noise and stability (see Collins et al. 2004).

Divergence damping is often used in atmospheric models and warrants more discussion. There are two primary paths of motivation. Bates et al. (1993) formally introduced two-dimensional divergence damping into the equations for the development of their semi-Lagrangian scheme. This damping was subsequently used in development and application (S. Moorthi, personal communication). Divergence damping is routinely used in the North American Model at the National Centers for Environmental Predictions to control noise in both the simulation and assimilation (S. Lord, personal communication). Farge and Sadourny (1989) discuss at length the use of dissipation on both the rotational and divergent parts of the flow to achieve adequate numerical performance. Their discussion is in the context of an investigation using a shallow water model with pseudospectral numerical schemes. They pursue a linear combination of a rotational and divergent form of dissipation (see also Vallis 1992; Gassmann and Herzog 2007).

The second motivational path for divergence damping follows from mesoscale modeling and the development of non-hydrostatic models. In this path the original line of reason was to incorporate three-dimensional divergence damping to remove meteorologically unimportant, computationally demanding acoustic modes (Skamarock and Klemp 1992; Dudhia 1993). Wicker and Skamarock (1998) note that not only are the acoustic modes eliminated, but that the stability of their numerical technique is improved. Therefore, in this path as well, the noise management and stability enhancements of divergence damping have emerged (see also Gassmann and Herzog 2007).

As global models and regional models resolve smaller and smaller scales, the divergent part of the flow becomes important. Furthermore there is forcing at the grid scale, which is formally not resolved, that is a source of physically based divergence. Therefore, these techniques to control noise impact important dynamical features and the interaction between large and small scales. Therefore, increased, direct attention to the physical role and representation of divergent flow is needed.

15.6.3.2 Mixing, Filters, and Fixers (Chap. 13)

The algorithms for mixing, filters, and fixers directly impact both the representation of divergence and the fundamentals of wave dissipation important for climate models. When the consequences of high resolution models are considered, the conflation of these algorithms with the model physics is realized to be even more complex. High resolution models will resolve more and more gravity waves, which are strongly divergent modes and are already 'accounted for' by the gravity wave parameterization. Therefore, the dynamical core and the physics parameterizations will not be as cleanly separated by scales. Similar realizations can be made for the relationship of the dynamical core with the planetary boundary layer parameterization and the convective parameterization. Far more attention is needed to the formulation and impact of mixing algorithms, filters, and fixers.

15.6.3.3 Non-Hydrostatic

As horizontal resolution is increased, the scales of the allowed motion are such that non-hydrostatic motion becomes important. Relaxing the hydrostatic assumption is realized in the vertical momentum equation. This provides a fundamental change in modeling. The strong relation of vertical velocity to small-scale divergence and the complex relationship between small- and large-scale programs again brings attention to the importance of the mixing algorithms, filters, and fixers.

15.6.3.4 Grids

Much attention is currently focused on types of grids (Randall 2000; Ringler et al. 2000; Putman and Lin 2009; Rančić et al. 2008; Walko and Avissar 2008; Thuburn 2008a). The excellent review of Williamson (2007) has a focus on how the development of dynamical cores is strongly influenced by the presence of the polar singularity on regular latitude-longitude grids. The Williamson (2007) review includes a large list of references to grids. Two grids that have received much attention are the cubic sphere (Sadourny 1972; Putman and Lin 2009) and the geodesic grid (Sadourny et al. 1968; Williamson 1968). There are both computational and physical advantages of these grids, and the grid and numerical methods used on the grids will reveal new consistency challenges. The grid has become another element of the dynamical core.

There is currently much discussion about grid artifacts; that is, the underlying grid can be 'seen' in the solutions. These comments also imply that present equal-angle, latitude-longitude grids are free of such artifacts. However, existing grids have a set of filters, especially polar filters, to remedy their artifacts. Indeed, as Williamson (2007) points out, the challenges of the equal-angle, latitude-longitude grid have been a great motivator to develop new techniques. Grid artifacts are currently a fact of modeling, and evaluation of their impact and development of remediation strategies are required; grid artifacts are not, *a priori*, an extraordinary flaw.

15.6.3.5 Coupling

Since climate models are composites of composites of components, there are couplings at many levels. It is easy to lose any advantage of a new numerical method to poor coupling. The coupling of the dynamical core to the physics is especially important because of the conflation of small and large scales at the grid scale and the conflation of numerical and physical mixing at the grid scale. Since model performance relies on the accumulation of the performance of interacting components, the physics of, the consistency of, and the performance of coupling need far more consideration (Staniforth et al. 2002; Williamson 2002, 2007).

15.7 Conclusions

The perspective provided here advocates looking at the function of the model as whole. The model as a whole is a system of interacting components. These components each have their error characteristics. Errors are balanced in the process of optimizing or tuning the model to address specific applications. Therefore, the development of specific components, without regard to the application and the interaction of one component with all components, is likely to have little obvious benefit. Model-building activities should include a formal step of system integration, which should be driven by an application-based validation plan.

With regard to dynamical cores – horizontal advection of resolved scales has evolved to a state of quality that is high compared with other sources of errors in the model. Therefore, in terms of performance it is essentially standardized. The choice of horizontal advection scheme does impact the requirements for filtering and fixers, which are of both theoretical and practical importance. The dynamical core needs to be considered as an integrated module, and the relation of the horizontal advection algorithm to algorithms for the vertical velocity and for mixing, filters, and fixers needs direct attention.

The discussion here brings attention to two items that might be considered values. These are consistency and locality. High value is given to these attributes because of experience in applications and, looking forward, focusing more attention on the coupling of the dynamical core with the physics. Specifically, there is the need to pass to the physics parameterizations physically realizable estimates of transported variables that represent the correlated behavior of the variables.

With these values the following is posed as the suite of elements in the dynamical core. It is implicit that these form an integrated, consistent module, informed by the interface with other components:

- A specification of the grid.
- A specification of topography.
- A specification of the pressure gradient force.
- A specification of the Coriolis force.
- The resolved advection of momentum in the horizontal plane.
- The resolved vertical advection of momentum.
- Unresolved sub-scale transport of momentum.
- The resolved advection of scalars in the horizontal plane.
- The resolved vertical advection of scalars.
- Unresolved sub-scale transport of scalars.
- A portfolio of filters and fixers that accommodate errors related to both the numerical technique and the characteristics of the underlying grid.

The representation of the divergent part of the flow and coupling of model components requires more attention. This demands attention to algorithms that represent mixing, filters, and fixers. High resolution simulations represent divergent circulations explicitly. Such models are poised to better represent the interaction of small and large scales, and ignoring the detritus of the dynamical core undermines efforts focused on the representation of processes from first principles.

The development of weather and climate models does not proceed through a well defined path from first principles. There is a mix of science, engineering, and intuition based upon experience and desired results. In the past decade both atmospheric and oceanic models, whose development has focused on design that gives priority to the underlying correlative physics, have had significant impact. Looking forward, the problems and applications being faced in climate modelers will bring attention to high resolution, the representation of small, divergent scales, and the interaction of small and large scales. This brings direct attention to the difficult and understudied problem of sub-scale mixing and the conflation of physical and numerical processes at the smallest scales. The interaction of the dynamical cores with the physics scheme, coupling in general, needs more rigorous attention and treatment; this is true for both parameterized and cloud-resolving models. Given the intrinsic nature of dissipation and dispersion errors in the numerical representation of advection, new strategies, such as more attention to the conservation of higher order moments, may be required to achieve fidelity between large and small scales.

Acknowledgments I thank Todd Ringler for clarifying the use of the term 'non-physical' to solely refer to violation of monotonicity. I thank Christiane Jablonowski and the editors of this book for their thorough comments, which have significantly improved the quality of the chapter. I thank them for the opportunity to publish this point-of-view analysis on dynamical cores and their role in modeling systems.

I acknowledge and thank, Bram van Leer, Steve Zalesak, Jay Boris, and Elaine Oran for discussions at various points in my career that are at the basis of any contributions I have made to modeling atmospheric advection. I thank Jerry Mahlman for many years of friendship that included discussions and disagreements over the merit of finite difference and finite volume advection. I am grateful to have worked with Shian-Jiann Lin who is most able at visualizing solutions and turning them into substance. I acknowledge contributions from Stephen Cohn about physicality and discrete representation of the equations of motion. And, finally, I want to quote David Burridge who once said of atmospheric modeling, at just the right time, 'There is no magic'. He paused. Then again, 'There is no magic'.

References

Adcroft A, Hallberg R (2006) On methods for solving the oceanic equations of motion in generalized vertical coordinates. Ocean Modeling 11:224–233

Adcroft A, Campin JM, Hill C, Marshall J (2004) Implementation of an atmospheric-ocean general circulation model on the expanded spherical cube. Mon Wea Rev 132:2845–2863

Adcroft A, Hallberg R, Harrison M (2008) A finite volume discretization of the pressure gradient force using analytic integration. Ocean Modeling 22:106–113

Allen DJ, Douglass AR, Rood RB, Guthrie PD (1991) Application of a monotonic upstream-biased transport scheme to three-dimensional constituent transport calculations. Mon Wea Rev 119:2456–2464

Andrews DG, Mcintyre ME (1978) An exact theory of nonlinear waves on a Lagrangian-mean flow. J Fluid Mech 89(4):609–646

Bala G, Rood RB, Bader D, Mirin A, Ivanova D, Drui C (2008) Simulated climate near steep topography: sensitivity to dynamical methods for atmospheric transport. Geophys Res Lett 35, l14807, doi:10.1029/2008GL033204

Bates JR, Moorthi S, Higgins RW (1993) Global multilevel atmospheric model using a vector semi-Lagrangian finite-difference scheme. Part I: Adiabatic formulation. Mon Wea Rev 121(1): 244–263

Beres JH, Garcia RR, Boville BA, Sassi F (2005) Implementation of a gravity wave source spectrum parameterization dependent on the properties of convection in the Whole Atmosphere Community Climate model (WACCM). J Geophys Res 110, d10108,doi:10,1029/2004JD005504

Bey I, Jacob DJ, Yantosca RM, Logan JA, Field B, Fiore AM, Li Q, Liu H, Mickley LJ, Schultz M (2001) Global modeling of tropospheric chemistry with assimilated meteorology: Model description and evaluation. J Geophys Res 106:23,073–23,096

Boris JP, Book DL (1973) Flux corrected transport. I. SHASTA, a fluid transport algorithm that works. J Comput Phys 11:38–69

Collins WD, Rasch PJ, Boville BA, Hack JJ, McCaa JR, Williamson DL, Kiehl JT, Briegleb BP, Bitz CM, Lin SJ, Zhang M, Dai Y (2004) Description of the NCAR Community Atmosphere Model (CAM3.0). NCAR Technical Note NCAR/TN-464+STR, National Center for Atmospheric Research, Boulder, Colorado, 214 pp., available from http://www.ucar.edu/library/collections/technotes/technotes.jsp

Dickinson RE, Ridley EC, Roble RG (1981) A 3-dimensional general-circulation model of the thermosphere. J Geophys Res-Space Physics 86:1499–1512

Douglass AR, Rood RB, Kawa SR, Allen DJ (1997) A three-dimensional simulation of the evolution of the middle latitude winter ozone in the middle stratosphere. J Geophys Res 102:19,217–19,232

Dudhia J (1993) A nonhydrostatic version of the Penn State-NCAR mesoscale model: Validation tests and simulation of an atlantic cyclone and cold front. Mon Wea Rev 121:1493–1513

Fahey DW, Solomon S, Kawa SR, Loewenstein M, Podolske JR, Strahan SE, Chan KR (1990) A diagnostic for denitrification in the winter polar stratospheres. Nature 345:698–702
Farge M, Sadourny R (1989) Wave-vortex dynamics in rotating shallow water. J Fluid Mech 206:433–462
Fomichev VI, Ward WE, Beagley SR, McLandress C, McConnell JC, McFarlane NA, Shepherd TG (2002) Extended Canadian Middle Atmosphere Model: Zonal-mean climatology and physical parameterizations. J Geophys Res - Atmospheres 107(D10):4087
Gassmann A, Herzog HJ (2007) A consistent time-split numerical scheme applied to the non-hydrostatic compressible equations. Mon Wea Rev 135:20–36
Godunov SK (1959) A difference scheme for numerical computation of discontinuous solutions of equations in fluid dynamics. Math Sb 47:271, also: Cornell Aero. Lab. translation
Holton JR (2004) An introduction to dynamic meteorology, Fourth edn. Academic Press, Inc., ISBN 0123540151, 535 pp.
Jacobson MZ (2005) Fundamentals of atmospheric modeling, 2nd edn. Cambridge University Press, 813 pp.
Jöckel P, von Kuhlmann R, Lawrence MG, Steil B, Brenninkmelter CAM, Crutzen PJ, Rasch PJ, Eaton B (2001) On a fundamental problem in implementing flux-form advection schemes for tracer transport in 3-dimensional general circulation and chemistry transport models. Quart J Roy Meteor Soc 127:1035–1052
Johnson SD, Battisti DS, Sarachik ES (2000) Empirically derived Markov models and prediction of tropical Pacific sea surface temperature anomalies. J Climate 13:3–17
van Leer B (1979) Towards the ultimate conservative difference scheme. V. A second-order sequel to Godunov's method. J Comput Phys 32:101–136
Lin SJ (1997) A finite volume integration method for computing pressure gradient forces in general vertical coordinates. Quart J Roy Meteor Soc 123:1749–1762
Lin SJ (2004) A "vertically Lagrangian" finite-volume dynamical core for global models. Mon Wea Rev 132:2293–2307
Lin SJ, Rood RB (1996) Multidimensional flux-form semi-Lagrangian transport scheme. Mon Wea Rev 124:2046–2070
Lin SJ, Rood RB (1997) An explicit flux-form semi-Lagrangian shallow water model on the sphere. Quart J Roy Meteor Soc 123:2477–2498
Machenhauer B, Kaas E, Lauritzen PH (2008) Finite volume techniques in atmospheric models. In: Ciarlet P, Temam R, Tribbia J (eds) Handbook of numerical analysis: Special volume on computational methods for the atmosphere and oceans, vol 14, Elsevier, pp 3–120, 784 pp.
McCrea GJ, Gooden WR, Seinfield JH (1982) Numerical solution of the atmospheric diffusion equation for chemically reacting flows. J Comput Phys 14:1–42
Mechoso CR, Yamazaki K, Kitoh A, Arakawa A (1985) Numerical forecasts of stratospheric warming events during the winter of 1979. Mon Wea Rev 113:1015–1029
Mote P, O'Neill A (eds) (2000) Numerical Modeling of the Global Atmosphere in the Climate System. Kluwer Academic Publishers, NATO Science Series C: Mathematical and Physical Sciences Vol. 550, ISBN 0-7923-6301-9, 517 pp.
Plumb RA, Ko MKW (1992) Interrelationships between mixing ratios of long lived stratospheric constituents. J Geophys Res 97:10,145–10,156
Prather MJ (1986) Numerical advection by conservation of second-order moments. J Geophys Res 91:6671–6681
Putman WM, Lin SJ (2009) A finite-volume dynamical core on the cubed-sphere grid. In: Numerical Modeling of Space Plasma Flows: Astronum-2008, Astronomical Society of the Pacific Conference Series, vol 406, pp 268–276
Rančić M, Zhang H, Savic-Jovcic V (2008) Nonlinear advection schemes on the octagonal grid. Mon Wea Rev 136:4668–4686
Randall DA (2000) General Circulation Model Development: Past, Present, and Future. Academic Press, 807 pp.
Rasch PJ, Williamson DL (1991) The sensitivity of a general circulation model climate to the moisture transport formulation. J Geophy Res 96:13,123–13,137

Rasch PJ, Coleman DB, Mahowald N, Williamson DL, Lin SJ, Boville BA, Hess P (2006) Characteristics of atmospheric transport using three numerical formulations for atmospheric dynamics in a single gcm framework. J Climate 19:2243–2266

Rienecker MM, Suarez MJ, Todling R, Bacmeister J, Takacs L, Liu HC, Gu W, Sienkiewicz M, Koster RD, Gelaro R, Stajner I, Nielsen E (2008) The GEOS-5 data assimilation system – Documentation of versions 5.0.1 and 5.1.0. Technical Report Series on Global Modeling and Data Assimilation NASA/TM-2007-104606, Vol. 27, NASA Goddard Space Flight Center, 92 pp.

Ringler TD, Heikes RP, Randall DA (2000) Modeling the atmospheric general circulation using a spherical geodesic grid: A new class of dynamical cores. Mon Wea Rev 128:2471–2489

Rood RB (1987) Numerical advection algorithms and their role in atmospheric transport and chemistry models. Rev Geophys 25:71–100

Rotman D, Tannahill JR, Kinnison DE, Connell PS, Bergmann D, Proctor D, Rodriguez JM, Lin SJ, Rood RB, Prather MJ, Rasch PJ, Considine DB, Ramaroson R, Kawa SR (2001) Global modeling initiative assessment model: Model description, integration, and testing of the transport shell. J Geophys Res 106(D2)(10.1029/2000JD900463):1669–1692

Sadourny R (1972) Conservative finite-difference approximations of the primitive equations on quasi-uniform spherical grids. Mon Wea Rev 100:136–144

Sadourny R, Arakawa A, Mintz Y (1968) Integration of the nondivergent barotropic vorticity equation with an icosahedral-hexagonal grid for the sphere. Mon Wea Rev 96:351–356

Santer BD, Wigley TML, Gaffen DJ, Bengtsson L, Doutriaux C, Boyle JS, Esch M, Hnilo JJ, Jones PD, Meehl GA, Roeckner E, Taylor KE, Wehner MF (2000) Interpreting differential temperature trends at the surface and in the lower troposphere. Science 287:1227–1232

Satoh M (2004) Atmospheric circulation dynamics and general circulation models. Springer (Praxis), 643 pp.

Schoeberl MR, Strobel DF (1980) Numerical-simulation of sudden stratospheric warmings. J Atmos Sci 37:214–236

Schoeberl MR, Douglass AR, Zhu Z, Pawson S (2003) A comparison of the lower stratospheric age-spectra derived from a general circulation model and two data assimilation systems. J Geophys Res 108(D3)(10.1029/2002JD002652):4113

Skamarock WC, Klemp JB (1992) The stability of time-split numerical methods for the hydrostatic and the nonhydrostatic elastic equations. Mon Wea Rev 120:2109–2127

Staniforth A, Wood N, Cole J (2002) Analysis of the numerics of physics-dynamics coupling. Quart J R Meteor Soc 128(586):2779–2799

Strang G (1968) On the construction and comparison of difference schemes. SIAM J Numer Anal 5:506–517

Thuburn J (2008a) Numerical wave propagation on the hexagonal C-grid. J Comput Phys 227: 5836–5858

Thuburn J (2008b) Some conservation issues for dynamical cores of NWP and climate models. J Comput Phys 227(7):3715–3730

Trenberth KE (ed) (1992) Climate System Modeling. Cambridge University Press, 788 pp.

Vallis GK (1992) Mechanism and parameterizations of geostrophic adjustment and a variational approach to balanced flow. J Atmos Sci 49:1144–1160

Walko RL, Avissar R (2008) The Ocean-Land-Atmosphere Model (OLAM). Part I: Shallow-water tests. Mon Wea Rev 136:4033–4044

Washington WM, Parkinson CL (12005) An introduction to three-dimensional climate modeling, 2nd edn. University Science Books, ISBN: 1-891389-35-1, 353 pp.

White AA, B J Hoskins IR, Staniforth A (2005) Consistent approximate models of the global atmosphere: shallow, deep, hydrostatic, quasi-hydrostatic and non-hydrostatic. Quart J Roy Meteor Soc 131:2081–2107

White L, Adcroft A (2008) A high-order finite volume remapping scheme for nonuniform grids: The piecewise quartic method (PQM). J Comput Phys 227:7394–7422

Wicker LJ, Skamarock WC (1998) A time-splitting scheme for the elastic equations incorporating second-order Runge-Kutta time differencing. Mon Wea Rev 126:1992–1999

Williamson DL (1968) Integration of the barotropic vorticity equations on a spherical geodesic grid. Tellus 20:642–653
Williamson DL (2002) Time-split versus process-split coupling of parameterizations and dynamical core. Mon Wea Rev 130:2779–2799
Williamson DL (2007) The evolution of dynamical cores for global atmospheric models. J Meteorol Soc Japan 85B:241–269
Yanenko NN (1971) The method of fractional steps. Springer-Verlag, 160 pp
Zalesak ST (1981) Very high order and pseudospectral flux-corrected transport (FCT) algorithms for conservation laws. In: Vichnevetsky R, Steplman RS (eds) Advances in Computer Methods for Partial Differential Equations IV, International Association for Mathematics and Computers in Simulation, Rutgers University, New Brunswick, N.J.

Chapter 16
Refactoring Scientific Applications for Massive Parallelism

John M. Dennis and Richard D. Loft

Abstract We describe several common problems that we discovered during our efforts to refactor several large geofluid applications that are components of the Community Climate System Model (CCSM) developed at the National Center for Atmospheric Research (NCAR). We stress tested the weak scalability of these applications by studying the impact of increasing both the resolution and core counts by factors of 10–100. Several common code design and implementations issues emerged that prevented the efficient execution of these applications on very large microprocessor counts. We found that these problems arise as a direct result of disparity between the initial design assumptions made for low resolution models running on a few dozen processors, and today's requirements that applications run in massively parallel computing environments. The issues discussed include non-scalable memory usage and execution time in both the applications themselves and the supporting scientific data tool chains.

16.1 Introduction

For the past 30 years, the amount of computing power that could be applied to scientific problems has grown exponentially. This amazing growth rate was a direct result of decreases in transistor sizes, which for decades, directly translated into increases in microprocessor clock frequency and consequently improved single thread performance. The doubling of clock frequency every 18 months became strongly associated with Moore's Law, that actually only describes the underlying rate of improvement in photolithographic techniques. Regardless, for an application developer in the latter part of the twentieth century, Moore's law meant that exponential performance improvements came on a steady schedule with little or no

J.M. Dennis (✉) and R.D. Loft
Computational & Information Systems Laboratory, National Center for Atmospheric Research, Boulder, CO 80307-3000, USA
e-mail: dennis@ucar.edu, loft@ucar.edu

P.H. Lauritzen et al. (eds.), *Numerical Techniques for Global Atmospheric Models*, Lecture Notes in Computational Science and Engineering 80,
DOI 10.1007/978-3-642-11640-7_16,

effort. In this regime there was little incentive to improve application performance by increasing parallelism.

In middle of the 2000s these circumstances began to change, as several fundamental factors began to limit microprocessor frequency. Local (on-chip) interconnect delays began to dominate feature size as the determining factor of clock speed. The heat densities being generated by $\sim$3 GHz microprocessors began hitting thermal design limits. The growing gap between memory and processor speeds increased memory access times, thereby creating the so-called "memory wall", in which faster processor clock speeds no longer guaranteed better performance. In response to these challenges, microprocessor architects began moving toward chip multiprocessor (CMP) designs: under this paradigm, chip performance improvements would come from doubling the number of processors or cores on a silicon die, while clock speed would only increase at a modest rate of 15% per 18 months. It seems clear now that this development is a long-term technology trend, derived from fundamental limitations of the underlying semiconductor technology. For application programmers this means that improved performance must come from speed-ups derived from increased parallelism.

A similar situation has developed in disk subsystem architecture as well: disk spindle parallelism is necessary to match I/O performance with parallel computational performance. Through parallel filesystem technology, modern petabyte filesystems aggregate thousands of rotating disk spindles and access channels to achieve high I/O bandwidth. Thus, as with CPU's, the path to performance for I/O intensive problems is through parallelism.

The impact of the return of massive parallelism is reflected in the increasing number of parallel computing initiatives sponsored by a variety of agencies of the U.S. government. In the High Performance Computing (HPC) arena, much of this effort is now focused around what is known as petascale computing: harnessing $\mathcal{O}(100{,}000)$ or more cores to achieve a petaflop – (10^{15} floating point operations per second). These efforts include funds for the acquisition and deployment of petascale systems, as well as research and development money to develop algorithms and associated applications able to effectively use these systems. For example, the National Science Foundation (NSF) has initiated an ambitious "Track-2" and even larger "Track 1" programs to procure, deploy and operate several petascale systems over the next few years. The first so-called "Track-2" system, Ranger, was installed in Texas Advanced Computing Center in 2006. Ranger is a 62,976 core system, based on the quad-core AMD Barcelona microprocessor with an Infiniband® (IB) interconnect switch designed by Sun Microsystems. A second "Track-2" system, Kraken, a Cray XT5 system currently has 99,072 cores, and was awarded to the National Institute for Computational Sciences (NICS) in 2007. The NSF award for a super-sized "Track-1" system capable of a peak speed of least ten petaflops, was awarded to the Illinois-based Blue Waters Consortium. Additional petascale computers in the United States are being deployed by the Departments of Energy and Defense, and are reportedly planned in Japan, Europe, and China.

Of course, without scalable applications, these large systems can't provide the application acceleration that leads to scientific progress for many important problems, such as climate modeling. To address this issue, U.S. government agencies have also funneled research dollars toward developing new algorithms, frameworks, and applications. Examples of these include the DoE SciDAC and NSF PetaApps programs. Such programs have allowed developers of scientific applications (including ourselves) to make important progress in preparing such codes for operations on increased numbers of cores.

The availability of massively parallel computing systems will place a premium on the scalability of applications. Even so, not every scientific problem needs petascale computing, nor is it the case that every application has a large code base that makes it prohibitively expensive for it to be rewritten from scratch. However, we are interested in the significant class of scientific applications for which improving the parallel performance by refactoring the existing code is the only reasonable option available.

Code refactoring is usually defined as the process of modifying the internal structure of an application without changing the external functional behavior. Such refactoring can be done for a variety of purposes: for example, for readability, performance, or maintainability. The cost, both in terms of human and computational resources, of validating the refactored application is an important determiner of the overall cost of such projects. Refactoring scientific applications for parallel scalability is especially challenging, often requiring new, more suitable algorithms. Using existing parallel programming paradigms, such as Message Passing Interface (MPI) described in Snir et al. (2000) a distributed memory parallel programming library or a shared memory parallel programming standard OpenMP (2005), means that changing the level of parallelism often requires new data structures, and introduces new design issues, unique to parallel execution, that usually have not been considered, let alone addressed, by earlier application developers. Examples of these issues include race conditions, resource contention, and load balancing. The challenge of these restructuring issues is exacerbated by the extraordinary cost of testing and validating many complex applications at scale.

Here we discuss our teams experiences in refactoring six large climate model components for massively parallel execution at significantly increased resolutions. Climate applications represent an especially rigorous test of existing parallel refactoring techniques. First, climate applications, particularly at high resolutions, can be extraordinarily expensive to run: for example, a single simulated year of a high resolution version of the Community Climate System Model requires roughly 80,000 CPU-hours per year on the Kraken XT-5 system at NICS. Second, the climate system is an interacting system of nonlinear PDEs with a multitude of computationally intensive forcings and feedbacks: this makes the system sensitive to initial conditions. Even tiny changes in results at machine-level precision will generate completely different realizations of the climate system after a sufficient amount of time has passed. Third, since the scientific predictions of climate applications are statistical in nature, their results are expressed in monthly and seasonal means and variances of measurable quantities. Thus they require long (perhaps multi-year) runtimes to completely validate.

Previous work by Dennis (2007) and Dennis and Tufo (2008) has revealed numerous structural issues in these codes that prevent successful or efficient execution at higher resolutions and on very large core counts. The reasons for these design issues are complex. Many climate system applications have long histories and have been refactored several times in the past. For example, NCAR's Community Atmosphere Model (CAM) was refactored for multitasking parallelism on vector systems in the 1980s, refactored again for distributed memory (message passing) execution in the early 1990s, and finally modified to support hybrid message passing and OpenMP execution in the late 1990s. For scientific reasons and until recently, climate application resolutions have remained around one to three degrees (100–300 km), as the focus of model development centered on improving the representation of processes and capturing new climate system feedbacks. For these resolutions, a modest level of parallelism 32–128 processors was sufficient to achieve acceptable scientific throughput rates. Now, with a growing sense that the resolution of the climate system must be improved to address remaining model biases, there is a new focus on conducting exploratory research at resolutions increased by one to two orders of magnitude. Not surprisingly perhaps, the need to run efficiently on 10,000–100,000 or greater cores counts, which was never considered by developers, is now needed to provide these high-resolution simulation capabilities.

The essence of the problem we faced in refactoring these applications is that many of their design assumptions, which worked fine on roughly a hundred cores, became problematic on tens or hundreds of thousands of cores. At a high level, the impediments we have uncovered may be classified as either non-scalable memory usage or execution time. Common non-scalable memory usage problems include use of replicated metadata, excessive number of global arrays and serialized I/O. Common problems that impact the scalability of execution time include: non-scalable initialization and communication time. In addition, we describe prevalent scalability issues within the toolchain that also impedes scientific progress and discovery. We present this description of common design issues in the hope that they could be used as a guideline to design future applications for efficient execution at these and even higher levels of parallelism.

16.2 Background

We have refactored six applications. The Parallel Ocean Program (POP) described in Jones (2003) and Jones et al. (2005), developed at Los Alamos National Laboratory is an important multi-agency ocean model used for global ocean modeling. The Community Ice CodE (CICE) described in Hunke and Lipscomb (2008) also developed at Los Alamos National Laboratory is also an important multi-agency code used to model sea ice. The Community Atmosphere Model (CAM) described in Collins et al. (2006), is a widely used atmospheric model, whose development is based at the National Center for Atmospheric Research (NCAR), but has a large international community of contributors and collaborators. The Community Land Model (CLM) described in Hoffman et al. (2005), which was developed at NCAR in collaboration with many national collaborators, models the land surface. The High Order Methods Modeling Environment (HOMME), an atmospheric dynamical core developed at NCAR and described in Dennis et al. (2005) and Chap. 12, is used to evaluate new numerical and computational algorithms. The flux coupler (CPL) combined with POP, CICE, CAM, and CLM form the state of the art climate model, the Community Climate System Model (CCSM). CCSM is one of the most extensively used climate models in the world and has participated in the Intergovernmental Panel on Climate Change (IPCC) Assessment reports by Solomon et al. (2007). POP, CICE, CAM, and HOMME are hydrostatic models that solve the equations of motion on multiple coupled horizontal computational meshes. Without the use of the hydrostatic approximation, the equations and algorithms are considerably more complex. The size of the horizontal computational mesh that is decomposed across cores is significantly larger than the number of levels in the vertical dimension. While CLM does not solve any equations of motion, it in addition to CPL, decomposes the horizontal computational mesh across cores.

Inspired by the work of Shingu et al. (2002), an attempt was made to execute a predecessor to HOMME at 10 km resolution, which represented a significant increase relative to what had previously been attempted. Due to its use of replicated metadata, which will be described in Sect. 16.3.2, HOMME was not even able to complete the initialization of the message passing communication library before it exhausted memory. While a refactoring of the data structures within the communication library enabled it to initialize, a large number of additional issues prevented its successful execution. After concerted effort to improve the scalability of HOMME by a number of developers, CAM based on the HOMME dynamical core (CAM/HOMME) has demonstrated excellent scalability on a range of core counts and resolution combinations. The integration rate of CAM/HOMME on a Aqua-planet simulation courtesy of Taylor et al. (2008) at 0.5° (56 km), 0.25° (28 km), and 0.125° (14 km) is illustrated in Fig. 16.1. Note that CAM/HOMME achieves excellent scalability on the Cray XT4, and the IBM Blue Gene/L (BGL) and Blue Gene/P (BGP) systems.

Our preliminary work with HOMME illustrated that the ultimate scalability of an application was determine not only by the underlying scalability of its numerical algorithms but by the efficiency and quality of its implementation. Subsequent work with other applications revealed that issues that had limited the parallelism or even prevented successful execution of HOMME were also commonly found in other applications. We next describe a set of issues that were discovered in multiple applications that prevented efficient execution on very large core counts.

16.3 Scalability

We base our observations on the work performed preparing these previously mentioned applications for execution at high resolution on very large core counts. Within the group of applications, the one with most limited parallelism has been tested

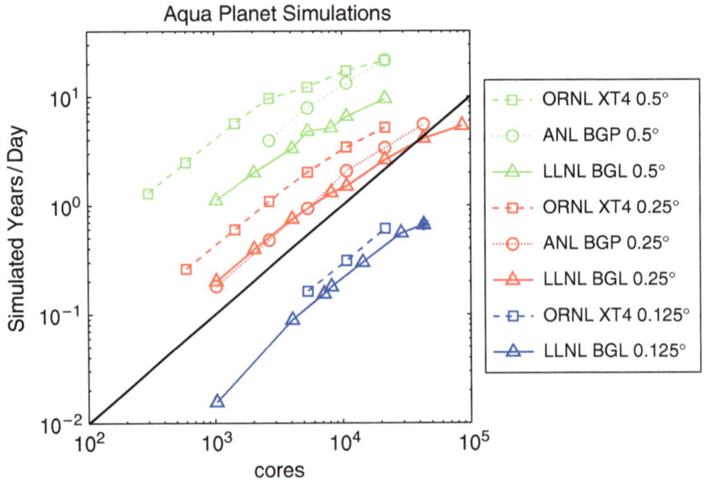

Fig. 16.1 Integration rate of CAM/HOMME on Aqua-planet simulation. Note that scalability is achieved 0.5°, 0.25°, and 0.125° on Cray XT4, IBM Blue Gene/L and Blue Gene/P systems

on a maximum of approximately 3,300 cores, while the most parallelism demonstrated by an application was 96,000 cores. Scalability testing has been performed on several very large systems, including: a 40,960 core IBM Blue Gene/L system described in Adiga and et al. (2002) at Thomas J. Watson Research, a 128,000 core IBM Blue Gene/L system at Lawrence Livermore National Laboratory (LLNL), a 38,912 core IBM Blue Gene/L system at Brookhaven National Laboratory, a 2,048 core IBM Blue Gene/L system at NCAR, a 10,000 core Cray Redstorm system described in RedStorm (2006) at Sandia National Laboratory, a 13,000 core Cray XT3/4 at Oak Ridge National Laboratory, a 9,000 core Appro linux cluster at Lawrence Livermore National Laboratory, and a 99,072 core Cray XT5 at National Institute for Computational Science (NICS). It is useful to evaluate scalability on more than just a single compute platform. The use of multiple compute platforms allows for differentiation between scalability problems in an application and the compute platforms message-passing network. Each application had between five to seven issues. Table 16.1 provides a listing of which applications had which issues.

We found it particularly striking that many of the same design issues were found in applications created by different developers. While the applications were developed by groups of researchers from interrelated scientific disciplines, there is no common origin for all the applications. Therefore we believe that many issues that we have discovered represent inherent stumbling blocks for developers creating parallel applications.

There are two different types of scalability, strong and weak. To test an applications strong scalability, a fixed size problem is executed on variety of processor or core counts. For a code with ideal strong scaling, use of twice the number of cores will reduce the execution time in half. In contrast to strong scaling, weak

Table 16.1 Presence of scalability issues within applications

Problems	Applications					
	POP	CAM	CICE	HOMME	CPL	CLM
Replicated metadata	Yes	Yes	Yes	Yes	Yes	Yes
Excessive global arrays	Yes	Yes	No	Yes	Yes	Yes
Serial I/O	Yes	Yes	Yes	No	Yes	Yes
Non-scalable initialization	Yes	No	Yes	Yes	Yes	Yes
Non-scalable communication	Yes	No	Yes	No	Yes	No
Debugging at scale	Yes	Yes	Yes	Yes	Yes	Yes
Pre/post processing	Yes	Yes	Yes	Yes	Yes	Yes

scaling fixes the size of the part of problem allocated to each core. For a code with ideal weak scaling, use of twice the number of cores will enable the execution of an application with twice the number of grid points in the same amount of time. Because climate modeling if frequently concerned with reducing the time to solution, and resolution changes very infrequently, only the strong scaling characteristics of an application is typically reported. However successful use of very large-scale parallelism will likely involve improvements in both the weak and strong scaling characteristics of the application.

A critical prerequisite to efficiently utilize very large-scale parallel systems is that the underlying numerical method is fundamentally scalable. In other words, scalability is only limited by the quality of implementation of the application or the scalability of the computing platform. The difference between numerical method and application scalability is illustrated by considering the parallelization approach used by POP and an older version of CAM. In the older version of CAM, which decomposed across latitude, a computational grid with 128 grid points in longitude and 64 grid points in latitude could only be parallelized across 64 cores. Additional parallelism within CAM was only possible through either a major change in the computational infrastructure, or an alternative numerical method. The scalability of CAM was limited from a structural perspective. A significant change to the computational structure and numerical method of CAM has since occurred, which has improved its structural scalability. However, CAM still has structural scalability limitations. Alternatively, POP has a much more flexible approach to scalability. The only limit to parallelism within POP is that imposed on it by the compute platform. For example it is possible to place as few as several grid points per core, though with current compute platform characteristics this would not be an efficient configuration. However the flexibility of POP is an important feature and enables it to adapt to computational platforms with different balances between computational and communication costs. The flexibility of POP has allowed it to successfully adapt to both the Cray XT and IBM Blue Gene family of supercomputers. While structural scalability is necessary, it is not sufficient for efficient execution for very large-scale parallelism.

16.3.1 Scalability of Memory Usage

We begin our examination of scalability by discussing the scalability of an applications memory usage. The single most critical, and common issue in the application suite is the presence of non-scalable memory constructs. While problems with the scalability of an applications execution time may just reduce the simulation rate, problems with memory scalability will prevent an application from running at all. Recall that the initial attempt to execute a predecessor of HOMME at 10 km failed because it exhausted memory during initialization. Note that we are referring to memory that is due to the data-structures of an application, not that is devoted to the scientific calculations. Frequently, non-scalable data-structures can limit the type of science questions that can be asked. Even when it is possible to successfully execute an application, non-scalable memory may limit its execution to systems with larger memory. Because of the cost of memory is a considerable component of the overall system cost, systems with a large amount of memory will tend to be rarer and in greater demand than systems with comparatively smaller amount of memory per core. Therefore the ability of an application to run successfully in a small amount of memory gives it a competitive advantage relative to other applications that require larger amounts of memory.

16.3.2 Replicated Metadata

One of the most common design issues discovered in the application suite is the unnecessary use of replicated metadata. We refer to metadata as something that describes the location of something else, like data structures that describe the location of grid cells in the domain decomposition, or the message-passing schedule. An example of message passing metadata is a data structure that indicates that core i (p_i) sends n bytes to core j (p_j). In reality, only the cores p_i and p_j truly require this piece of metadata. Cores p_k, where $k \neq i, k \neq j$ does not need metadata for cores p_i and p_j. The replication of message passing metadata adds an $\mathcal{O}(numNeigh * P)$ sized data structure where P is the total number of cores and *numNeigh* is the number of communication neighbors. If only relevant metadata is stored, the size of the data structure is reduced to $\mathcal{O}(numNeigh)$. The importance of eliminating replicated metadata has been demonstrated by Baker et al. (2006). It should be noted that retaining a single serial algorithm within a parallel application forces replication of some metadata. For example, if an entire application is parallel except for serial I/O, then a data structure of size $\mathcal{O}(P)$ is necessary to describe the domain decomposition.

The most extreme example of the growth of metadata storage was located in CLM. Because of the particular design of CLM's dynamic domain decomposition strategy, $4 * P$ global arrays were present in the initialization subroutines. Global arrays in this case are arrays that are the size of the computational grid replicated on each core. For CLM with $1/6°$ separation between grid points (2,160 × 1,080 total horizontal grid points), these four integer arrays would require 29 TBytes of memory on 10,000 cores. While it was easy to discover large data structures, it is more challenging to discover issues in smaller ones. Thus, long after addressing the blatantly obvious metadata issues, a number of additional replicated data structures continued to emerge.

A good technique for discovering the presence of replicated metadata is to measure the maximum memory usage and plot it as a function of core count. Maximum memory usage is the maximum memory used by a single core at any time during the simulation. While we found the Blue Gene memory usage tool *memmon* developed by Walkup (2007) to be very useful to identify memory usage problems within CLM, it is also possible to develop similar tools that use the Unix system call *getrusage*, or the Linux /proc file system to measure memory usage. The memmon tool provides function calls that print out current memory usage. Calls to the memmon utility were added to CLM at key points in the code. The resulting analysis of memory usage yielded surprising results. Figure 16.2 is a plot of the memory usage as a function of core count for CLM at high-resolution. The maximum memory usage for the original version of CLM (CLM-orig), along with two modified versions of CLM (CLM-mod1) and (CLM-mod2) are shown. Note that CLM-mod2 uses an updated version of the Model Coupling Toolkit (MCT) described in Jacob et al. (2005) and Larson et al. (2005), while CLM-mod1 does not. Note that the memory usage for the CLM-orig was estimated from lower resolution configuration because its memory requirements were too large to successfully execute on Blue Gene. It is interesting to note that memory usage for both the CLM-orig and

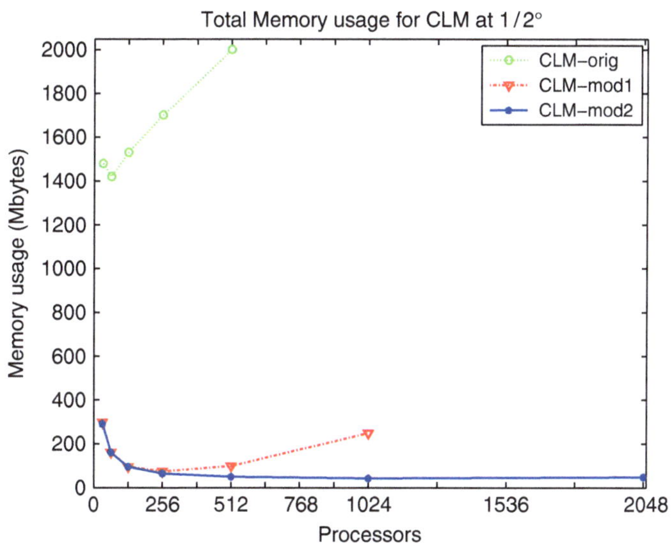

Fig. 16.2 A plot of the memory usage of several different versions of CLM as function of processor count. Elimination of both replicated data structures and global memory reduces memory usage for CLM at high resolution on 512 processors by a factor of 50

CLM-mod1 increase as core count increases. This behavior is a clear indicator of the presence of replicated metadata. The huge difference in memory usage for small core counts between CAM-orig and CLM-mod1 is an indication of another form of non-scalable memory, excessive global arrays, which will be discussed in Sect. 16.3.3.

We discovered $\mathcal{O}(numNeigh * P)$ sized metadata on all six applications that were examined. The simple replication strategy is understandable because most of the applications were initially designed on 32–128 cores. The size of the metadata only become problematic when attempting to scale to much larger core counts.

16.3.3 Excessive Global-Sized Arrays

Most of the applications in this study had an excessive number of arrays that were the size of the entire grid. We discovered that CLM used a large number of persistent global-sized arrays. We differentiate between persistent global arrays that are used for the duration of a run, versus temporary global arrays that are allocated within a subroutine and subsequently deallocated. While the elimination of persistent global arrays will always reduce the necessary memory usage for the application, the reduction of temporary global arrays may or may not reduce the maximum memory usage of the application. Elimination of temporary global arrays will only reduce maximum memory if the subroutine in which they are allocated is the actual location causing the maximum memory usage of the application to be reached.

It is possible to identify the presence of global arrays within an application by looking at memory usage as a function of core count for a fixed resolution. The memory usage of an application with global arrays will not decrease as core count is increased. Another approach is to examine the weak scalability of memory usage by fixing the per core domain size. An increase in memory usage for the larger resolution problem is a sign of the presence of global arrays. Figure 16.2 is a plot of memory usage for a fixed resolution for several different versions of CLM. The reduction in total memory usage for the CLM-mod1 versus the CLM-orig is a result of the reduction in the number of persistent global arrays from approximately 500 to 1. The reduction of excessive global arrays and replicated metadata reduces the memory usage for CLM at high resolution on 512 cores from 2,000 to 42 Mbytes, a reduction of a factor of 50.

16.3.4 Serial I/O

Five of the six applications performed serial I/O. This design decision was likely made when the applications were first refactored for modest levels of distributed memory parallelism. This "triage" was made at the time because (1) parallel I/O to a single file was not supported in the initial implementations of MPI, (2) it is easier to

Table 16.2 Memory usage of the CAM-CICE-CPL part of a high-resolution CCSM configuration running on 480 cores of the Cray XT4

	CCSM component			MPI buffers	Total
	CAM	CICE	CPL		
Memory usage (MB)	250	160	29	500	939
Percentage (%)	27	17	3	53	100

implement and (3) it provided acceptable performance when running on small core counts. Not surprisingly, it becomes problematic when the parallelism of the application is significantly increased. In addition to the creation of a serial performance bottleneck, the serialization of I/O creates several other problems. In particular, it requires the scattering and gathering of data to and from the distributed data representation. The straightforward implementation involves allocating a global-sized array on a single core, which may by itself exhaust memory on a system with limited memory like Blue Gene. Further, the gather/scatter of data may cause MPI to allocate large amounts of buffer memory on the core performing serial I/O. Work with a high resolution (0.1° or 10 km resolution) configuration of POP on Blue Gene revealed that while it was possible to allocate a single global array, the MPI buffer allocation overhead caused the application to fail.

The impact that MPI buffer memory allocation has on overall memory usage is illustrated our by measurements of the memory usage of a high-resolution CCSM configuration on the Cray XT4. This CCSM configuration coupled 0.1° POP and CICE components to 0.5° CAM and CLM. We concentrate on determining memory usage for a single configuration of CCSM where the CICE, CAM, and CPL components executing sequentially on 480 cores, and POP is executing concurrently on a separate set of cores. We estimate each components memory usage by using the resident working set size reported by the Linux/proc file system, and by placing each component on disjoint sets of cores. Note that the MPI buffer space which is a non-separable component of memory usage is easily determined because it is set by environment variable. The results of our analysis for this configuration are shown in Table 16.2, with the memory usage broken down by component model and that used by MPI buffers. The rather large MPI buffers (53% of the total memory usage shown) are necessary to support gathering/scattering related to serial I/O.

The impact of a single gather of a global array, inherent in a serialized I/O design while acceptable on small core counts, has a profound impact on application memory usage at large core counts, and can even prevent the use of certain types of compute systems.

16.4 Scalability of Execution Time

We next describe some examples of poor weak scalability discovered within the six applications. While the previous section concentrated on addressing issues associated with allowing codes to run at all, this section concentrates on reducing the cost to run applications.

16.4.1 Non-scalable Initialization

Several of the applications in this study had non-scalable initialization execution times. All applications in this study are typically run for as long as the queueing system will allow, in which case the cost of initialization is amortized across a 6–48 hours long job. However, the impact of large initialization costs becomes particularly problematic when performing refactoring or development work, and can serious limit the ability to test an application on very large cores counts. For example, an $\mathcal{O}(P^2)$ initialization algorithm, where P is the number of processors, was discovered in POP when it attempted to run it on more than 10,000 cores. What was a modest cost of a few minutes on smaller cores counts grew to 45 min at this scale.

An effective technique for identifying non-scalable initialization is to plot the application's initialization time as a function of core count. Unexpected increases in initialization time at large core counts may indicate an algorithmic problem in the initialization. In POP, the initialization issue turned out to be an $\mathcal{O}(P^2)$ algorithm to calculate the message passing schedule in a single routine: a problem that was easily addressed. The original initialization algorithm contained a nested loop over P tasks that searched for neighbors among a list of $P-1$ tasks. An alternative $\mathcal{O}(P)$ version that computed and stored each tasks neighbors, which eliminated the $\mathcal{O}(P^2)$ search, was developed and reduced the total initialization time at 10,000 processors from 45 to 10 min. Note that at low resolution and core counts the improved $\mathcal{O}(P)$ algorithm only reduced the initialization time from several minutes to 40 s versus the $\mathcal{O}(P^2)$ version, a minor improvement that would not have made sense to pursue at low resolutions and core counts.

16.4.2 Non-scalable Inter-processor Communication

Scalable and efficient boundary exchange strategies are essential to the successful parallel implementation of many commonly-used numerical methods for solving partial differential equations. However, several of the applications we have studied had scalability issues with their boundary exchange routines. These problems can be classified as either unnecessary message serialization or excessive latency sensitivity. Message serialization is typically the result of the serial treatment of special points, edges, or surfaces found in the underlying grids. Latency sensitivity usually involves sending too many small messages, and becomes critical only at large processor counts.

Both types of problem were discovered in the POP ocean model. POP uses curvilinear displaced-pole grids described by Murray (1996) and Smith et al. (1995) to address coordinate singularies at the North Pole. A popular grid variation available in POP is the tripole grid, which provides a quasi-uniform mesh over the Arctic Ocean through the addition of a third pole and a coordinate seam across the Arctic Ocean. An image of the POP tripole grid from Jones (2003) is provided in Fig. 16.3. Because of the complexity of the coordinate seam in the tripole grid at this

Fig. 16.3 The tripole grid used by POP. The coordinate seam in the tripole grid connects a pole in the Yukon Territory of Canada to one near Arkhangelsk, Russia

interface, the boundary exchanges in the parallel implementation of the tripole grid were partially serialized. In particular the POP implementation of the tripole grid duplicate grid points along one logical dimension of the computational grid. POPs update algorithm collects all the duplicated grid points along the tripole boundary on a single core, perform the update coordinate transformation calculation, and then distribute the solution back to all the cores along that edge. Consider the cost to perform a boundary update using a serialized versus a distributed algorithm. For the serialized algorithm, approximately $\sqrt{P}$ cores need to communicate resulting in an $\mathcal{O}(\sqrt{P})$ algorithm. For a distributed algorithm, the cost to update the boundary should be an $\mathcal{O}(1)$, i.e., communicate with one or a small number of neighbors. On 32 cores the serialize algorithm is approximately 6 times as expensive, while on 32,000 it would be 179 times more expensive. While serialization problems in the communication algorithm will show up in both strong and weak scaling experiments, it will be particularly apparent for weak scaling. The elimination of the serialized POP tripole algorithm and replacement with a distributed version reduced the total execution time of POP at 0.1° on a 2,000-core Linux cluster with an Infiniband® interconnect by approximately 15%.

POP also contained a latency-sensitive boundary exchange that underscored the importance of message aggregation. A communication routine may be considered latency sensitive if greater than 50% of the cost to send a message is due to latency cost versus bandwidth cost (message size divided by bandwidth). Because the climate community is primarily concerned with strong scalability, the impact of message latency on application performance is important. Message aggregation is a standard technique to reduce the impact latency has on application scalability, albeit at the cost of increased MPI message buffer sizes. For example, POP's finite difference boundary exchange library was designed to work on two dimensional grid objects and did not provide separate subroutines for boundary exchange of 3-dimensional (3D) or 4-dimensional (4D) variables. Instead, POP implemented the 3D boundary exchange update as a series of 2-dimensional ones, thus sequentially performing a number of 2D boundary exchange updates equal to the number of levels. In POP, which typically uses 60 vertical levels, this strategy has the potential to incur a substantial latency overhead penalty. By writing boundary exchange subroutines specific for both 3D and 4D variables, this latency overhead is paid once per variable, rather than once per vertical level, thus reducing latency sensitivity.

Ideally it is possible to diagnose problems in the scalability of boundary exchanges through both strong and weak scaling experiments on systems with different latency and bandwidth attributes. For example, POP's lack of message aggregation was discovered by comparing the scalability of its boundary exchanges on the Cray XT and Blue Gene systems. It was observed that the scalability of boundary exchanges within POP on the Cray XT was worse than on Blue Gene, a direct result of the Cray XT's higher message latency versus Blue Gene. Further, an application may only exhibit scalability problems on compute platforms with less capable message passing networks that either the Blue Gene or Cray XT systems. The relative cost of the boundary exchange may vary widely depending on the precise system and model configuration. In the HOMME dynamical core, the relative cost of boundary exchanges at low resolution ranges from 1% to 5% to as high as 30%–50% of the total cost at very large core counts.

16.5 Other Impediments

A number of additional impediments were discovered when attempting to significantly increase both the resolution and core counts for these applications. These obstacles include debugging and software development, pre-processing of input, and post-processing of output files. The overarching goal of increased parallelism is to accelerate scientific discovery and thus must therefore involve not only application execution but also a whole host of related tasks.

16.5.1 Debugging at Scale

Apart from memory and performance issues, when applications are run at high resolutions, bugs emerge. This is to be expected, since we are exercising code in a way that has never been tested when we explore higher process counts. However, we were surprised by the large number of bugs in these six applications that emerged when we increased both resolution and core count. Decomposition issues on large core counts were particularly common. Other bugs were discovered in options that engaged model physics not generally used in lower resolution configurations. Experience debugging these applications has led us to make several best practice recommendations. First, the domain decomposition strategies of applications should be subjected to rigorous testing at a variety of scales. Unfortunately, the majority of the domain decomposition bugs would not have been caught using a simplified unit test methodology, but rather involve testing of the entire system by comparing the results obtained using one decomposition to another. Second, while the number of possible combinations of physics options, resolutions, and core count available in scientific applications is daunting, routine testing is critical. Third, routine access to multiple, very large systems is a critical requirement for the testing of applications in order to spot problems introduced during development. In our experience, most large systems, because of their cost, are competitively allocated on scientific merit. As a result, resource providers and scientific users of large systems tend to neglect performance and validation testing. Fourth, successful debugging of large, parallel scientific applications not only requires routine access to large systems, but also routine interactive access, or at the very least, batch access with rapid turn around time. For example, addressing several bugs in POP required daily access to 2,048 cores for a total of a week. A concerted debugging effort of a high-resolution CCSM configuration involved regular access to 1,800 cores, and consumed 600,000 CPU hours in several months time. The development work associated with highly scalable scientific applications frequently require substantial resources. The best way to minimize these costs is to reduce debugging costs through routine testing at a variety of processor counts scales and resolutions.

16.5.2 Pre/Post Processing

The ultimate goal of our refactoring applications for performance is to speed up scientific progress. It does little good to increase the performance of the model, if the tools around it slow down or cease to function. Thus, one must examine the entire scientific workflow for bottlenecks. As in the case of applications themselves, CCSMs pre/post analysis tools were designed to deal with low-resolution simulations executed on small processor counts, indeed many are still serial.

An illustrative example is the serial generation of a river runoff input file for an ultra-high resolution CCSM run. At low resolution, the calculation took 3 h and 2 GBytes of memory. Since the application only needed to be run once, these requirements were acceptable. However, when applied to a high-resolution CCSM configuration, we discovered that the algorithm for generating a river runoff mapping from the 0.5° land model to the ocean would have taken 60 days to execute on a single core, and would have required 128 Gbytes of memory. The existing river runoff algorithm was rewritten by replacing the local search algorithm and by limiting its calculation to be near the coastline. The redesign (still serial) reduced the required resources to 30 min on a system with 5 Gbytes of memory for the high resolution case.

The attempts to analyze a 100-year high-resolution coupled simulation has recently highlighted the lack of scalability within CCSMs post processing workflow and the impact it has on scientific discovery. While we are now able to simulate high-resolution climate at approximately 2 simulated years per day, we are certainly not able to analyze the approximately 3.5 TB of history files per week it generates. A number of the standard post-analysis scripts had to be rewritten to eliminate excessive memory usage in order to even execute successfully. Our inability to analyze this data at a rate commensurate to its generation stems from the fact that most of the analysis processing is serial.

16.6 Conclusions

We have described several common issues gleaned from experience refactoring six scientific computing applications for efficient execution on very large-scale computing systems. The six applications in total represent over a million lines of code that have been developed by multiple scientific communities over the last 25 years. Fortunately, improving their scalability involved addressing a small number of similar issues in $\sim$1% of the source code. Unfortunately, there is no simple solution, or magic bullet. Refactoring, testing and debugging complex scientific applications on large numbers of processors is inherently difficult. However, there is an approach that has been remarkable effective on all six applications that we have worked with. All the problems were discovered by systematically stress testing each application, that is, testing the applications in ways that they have never before been tested, and then systematically studying the way in which they behave or fail. This approach requires routine, and in the case of debugging, rapid turn-around access to very large scalable, parallel systems, and a commitment of resources for this purpose. Routine application testing regimens can then ensure that these issues do not recur.

Our approach of stress testing while refactoring revealed that, despite the unique origin for most of the applications studied, the issues discovered were strikingly similar. At a high level the impediments involved non-scalable memory usage and execution time. Non-scalable memory usage issues were discovered that actually limited the type of science questions that could be asked. However, we also found that a large number of these problems could be addressed by modifying a small amount of source code. All of these issues were the results of design choices made years ago for a single range of resolution and processor counts, and for which they had little impact. The scalability implications of these choices were either ignored or deferred. It is a cautionary tale as we contemplate the exascale systems with millions of processors that loom on the horizon.

Acknowledgments We would like to thank our colleagues Mariana Vertenstein, Tony Craig for all their work addressing the many code design issues discovered during this study. We would like to thank Dr. Mark Taylor for running several of the applications on compute platforms at Sandia National Laboratory, and Lawrence Livermore National Laboratory. We also thank Brookhaven National Laboratory, and Oak Ridge National Laboratory for access to their large compute platforms. We thank Fred Mintzer for access to the Thomas J. Watson Research facility through the 2nd and 3rd Blue Gene Watson Consortium Days event. Significant computational resources were provided through grants by the LLNL 2nd and 3rd Institutional Grant Challenge program. Code development would not have been possible without the access to the Blue Gene system at NCAR, which is funded through NSF MRI Grants CNS-0421498, CNS-0420873, and CNS-0420985 and through the IBM Shared University Research (SUR) Program with the University of Colorado.

The work of these authors was supported through National Science Foundation Cooperative Grant NSF01 which funds the National Center for Atmospheric Research (NCAR), and through the grants: #OCI-0749206 and #OCE-0825754. Additional funding is provided through the Department of Energy, CCPP Program Grant #DE-PS02-07ER07-06.

References

Adiga NR, et al (2002) An overview of the Blue Gene/L supercomputer. In: Proceedings of SC2002, Baltimore, MD

Baker AH, Falgout FD, Yang UM (2006) An assumed partition algorithm for determing processor inter-communication. Parallel Computing 32:394–414

Collins WD, Rasch P, Boville BA, Hack J, McCaa J, Williamson DL, Briegleb BP, Bitz CM, Lin SJ, Zhang M (2006) The formulation and atmospheric simulation of the Community Atmosphere Model version 3 (CAM3). Journal of Climate 19(11):2144–2161

Dennis JM (2007) Inverse space-filling curve partitioning of a global ocean model. In: IEEE International Parallel & Distributed Processing Symposium, Long Beach, CA

Dennis JM, Tufo HM (2008) Scaling climate simulation applications on IBM Blue Gene. IBM Journal of Research and Development: Applications for Massively Parallel Systems 52(1/2)

Dennis JM, Fournier A, Spotz WF, St-Cyr A, Taylor MA, Thomas SJ, Tufo H (2005) High resolution mesh convergence properities and parallel efficiency of a spectral element atmospheric dynamical core. Int J High Perf Comput Appl 19:225–235

Hoffman FM, Vertenstein M, Kitabata H, III JBW (2005) Vectorizing the community land model. International Journal of High Performance Computing Applications 19:247–260

Hunke EC, Lipscomb WH (2008) CICE: the Los Alamos sea ice model documenation and software user's manual version 4.0. Tech. Rep. LA-CC-06-012, Los Alamos National Laboratory, T-3 Fluid Dynamics Group

Jacob R, Larson J, Ong E (2005) MxN communication and parallel interpolation in CCSM3 using the Model Coupling Toolkit. Int J High Perf Comp Appl 19(3):293–307

Jones P (2003) Parallel Ocean Program (POP) user guide. Tech. Rep. LACC 99-18, Los Alamos National Laboratory

Jones PW, Worley PH, Yoshida Y, White JBI, Levesque J (2005) Practical performance portability in the Parallel Ocean Program (POP). Concurrency Comput Prac Exper 17:1317–1327

Larson J, Jacob R, Ong E (2005) The Model Coupling Toolkit: A new Fortran90 toolkit for building multiphysics parallel coupled models. Int J High Perf Comp App 19(3):277–292

Murray RJ (1996) Explicit generation of orthogonal grids for ocean models. J Comp Phys 126:251–273

OpenMP (2005) OpenMP application programing interface. http://www.openmp.org/

RedStorm (2006) The Cray XT3 Supercomputer. Http://www.cray.com/products/xt3/index.html

Shingu S, Y T, Ohfuchi W, Otsuka K, Takahara H, Hagiwara T, Habata S, Fuchigami H, Yamada M, Sasaki Y, Kobayashi K, Yokokawa M, Itoh H (2002) A 26.58 Tflops global atmospheric simulation with the spectral transform method on the earth simulator. In: Proceedings of the 2002 ACM/IEEE Conference on Supercomputing, pp 1–19

Smith R, Kortas S, Meltz B (1995) Curvilinear coordinates for global ocean models. LANL Technical Report LA-UR-95-1146

Snir M, Otto S, Huss-Lederman S, Walker D, Dongarra J (2000) MPI: The Complete Reference: Volume 1, The MPI Core. The MIT Press

Solomon S, Qin D, Manning M, Chen Z, Marquis M, Tignor KAM, Miller H (eds) (2007) Contribution of Working Group 1 to the Fourth Assessment Report of the Intergovernmental Panel on Climate Change. Cambridge University Press, Cambridge United Kingdom and New York, NY, USA

Taylor MA, Edwards J, St-Cyr A (2008) Petascale atmospheric models for the community climate system model: New developments and evaluation of scalable dynamical cores. J Phys Conf Ser 125

Walkup B (2007) Personal Communication

Editorial Policy

1. Volumes in the following three categories will be published in LNCSE:

i) Research monographs
ii) Tutorials
iii) Conference proceedings

Those considering a book which might be suitable for the series are strongly advised to contact the publisher or the series editors at an early stage.

2. Categories i) and ii). Tutorials are lecture notes typically arising via summer schools or similar events, which are used to teach graduate students. These categories will be emphasized by Lecture Notes in Computational Science and Engineering. **Submissions by interdisciplinary teams of authors are encouraged.** The goal is to report new developments – quickly, informally, and in a way that will make them accessible to non-specialists. In the evaluation of submissions timeliness of the work is an important criterion. Texts should be well-rounded, well-written and reasonably self-contained. In most cases the work will contain results of others as well as those of the author(s). In each case the author(s) should provide sufficient motivation, examples, and applications. In this respect, Ph.D. theses will usually be deemed unsuitable for the Lecture Notes series. Proposals for volumes in these categories should be submitted either to one of the series editors or to Springer-Verlag, Heidelberg, and will be refereed. A provisional judgement on the acceptability of a project can be based on partial information about the work: a detailed outline describing the contents of each chapter, the estimated length, a bibliography, and one or two sample chapters – or a first draft. A final decision whether to accept will rest on an evaluation of the completed work which should include

- at least 100 pages of text;
- a table of contents;
- an informative introduction perhaps with some historical remarks which should be accessible to readers unfamiliar with the topic treated;
- a subject index.

3. Category iii). Conference proceedings will be considered for publication provided that they are both of exceptional interest and devoted to a single topic. One (or more) expert participants will act as the scientific editor(s) of the volume. They select the papers which are suitable for inclusion and have them individually refereed as for a journal. Papers not closely related to the central topic are to be excluded. Organizers should contact the Editor for CSE at Springer at the planning stage, see *Addresses* below.

In exceptional cases some other multi-author-volumes may be considered in this category.

4. Only works in English will be considered. For evaluation purposes, manuscripts may be submitted in print or electronic form, in the latter case, preferably as pdf- or zipped ps-files. Authors are requested to use the LaTeX style files available from Springer at http:// www. springer.com/authors/book+authors?SGWID=0-154102-12-417900-0.

For categories ii) and iii) we strongly recommend that all contributions in a volume be written in the same LaTeX version, preferably LaTeX2e. Electronic material can be included if appropriate. Please contact the publisher.

Careful preparation of the manuscripts will help keep production time short besides ensuring satisfactory appearance of the finished book in print and online.

5. The following terms and conditions hold. Categories i), ii) and iii):

Authors receive 50 free copies of their book. No royalty is paid.
Volume editors receive a total of 50 free copies of their volume to be shared with authors, but no royalties.

Authors and volume editors are entitled to a discount of 33.3 % on the price of Springer books purchased for their personal use, if ordering directly from Springer.

6. Commitment to publish is made by letter of intent rather than by signing a formal contract. Springer-Verlag secures the copyright for each volume.

Addresses:

Timothy J. Barth
NASA Ames Research Center
NAS Division
Moffett Field, CA 94035, USA
barth@nas.nasa.gov

Michael Griebel
Institut für Numerische Simulation
der Universität Bonn
Wegelerstr. 6
53115 Bonn, Germany
griebel@ins.uni-bonn.de

David E. Keyes
Mathematical and Computer Sciences
and Engineering
King Abdullah University of Science
and Technology
P.O. Box 55455
Jeddah 21534, Saudi Arabia
david.keyes@kaust.edu.sa

and

Department of Applied Physics
and Applied Mathematics
Columbia University
500 W. 120 th Street
New York, NY 10027, USA
kd2112@columbia.edu

Risto M. Nieminen
Department of Applied Physics
Aalto University School of Science
and Technology
00076 Aalto, Finland
risto.nieminen@tkk.fi

Dirk Roose
Department of Computer Science
Katholieke Universiteit Leuven
Celestijnenlaan 200A
3001 Leuven-Heverlee, Belgium
dirk.roose@cs.kuleuven.be

Tamar Schlick
Department of Chemistry
and Courant Institute
of Mathematical Sciences
New York University
251 Mercer Street
New York, NY 10012, USA
schlick@nyu.edu

Editor for Computational Science
and Engineering at Springer:
Martin Peters
Springer-Verlag
Mathematics Editorial IV
Tiergartenstrasse 17
69121 Heidelberg, Germany
martin.peters@springer.com

Lecture Notes in Computational Science and Engineering

1. D. Funaro, *Spectral Elements for Transport-Dominated Equations.*
2. H.P. Langtangen, *Computational Partial Differential Equations.* Numerical Methods and Diffpack Programming.
3. W. Hackbusch, G. Wittum (eds.), *Multigrid Methods V.*
4. P. Deuflhard, J. Hermans, B. Leimkuhler, A.E. Mark, S. Reich, R.D. Skeel (eds.), *Computational Molecular Dynamics: Challenges, Methods, Ideas.*
5. D. Kröner, M. Ohlberger, C. Rohde (eds.), *An Introduction to Recent Developments in Theory and Numerics for Conservation Laws.*
6. S. Turek, *Efficient Solvers for Incompressible Flow Problems.* An Algorithmic and Computational Approach.
7. R. von Schwerin, ***M**ulti **B**ody System **SIM**ulation.* Numerical Methods, Algorithms, and Software.
8. H.-J. Bungartz, F. Durst, C. Zenger (eds.), *High Performance Scientific and Engineering Computing.*
9. T.J. Barth, H. Deconinck (eds.), *High-Order Methods for Computational Physics.*
10. H.P. Langtangen, A.M. Bruaset, E. Quak (eds.), *Advances in Software Tools for Scientific Computing.*
11. B. Cockburn, G.E. Karniadakis, C.-W. Shu (eds.), *Discontinuous Galerkin Methods.* Theory, Computation and Applications.
12. U. van Rienen, *Numerical Methods in Computational Electrodynamics.* Linear Systems in Practical Applications.
13. B. Engquist, L. Johnsson, M. Hammill, F. Short (eds.), *Simulation and Visualization on the Grid.*
14. E. Dick, K. Riemslagh, J. Vierendeels (eds.), *Multigrid Methods VI.*
15. A. Frommer, T. Lippert, B. Medeke, K. Schilling (eds.), *Numerical Challenges in Lattice Quantum Chromodynamics.*
16. J. Lang, *Adaptive Multilevel Solution of Nonlinear Parabolic PDE Systems.* Theory, Algorithm, and Applications.
17. B.I. Wohlmuth, *Discretization Methods and Iterative Solvers Based on Domain Decomposition.*
18. U. van Rienen, M. Günther, D. Hecht (eds.), *Scientific Computing in Electrical Engineering.*
19. I. Babuška, P.G. Ciarlet, T. Miyoshi (eds.), *Mathematical Modeling and Numerical Simulation in Continuum Mechanics.*
20. T.J. Barth, T. Chan, R. Haimes (eds.), *Multiscale and Multiresolution Methods.* Theory and Applications.
21. M. Breuer, F. Durst, C. Zenger (eds.), *High Performance Scientific and Engineering Computing.*
22. K. Urban, *Wavelets in Numerical Simulation.* Problem Adapted Construction and Applications.

23. L.F. Pavarino, A. Toselli (eds.), *Recent Developments in Domain Decomposition Methods.*

24. T. Schlick, H.H. Gan (eds.), *Computational Methods for Macromolecules: Challenges and Applications.*

25. T.J. Barth, H. Deconinck (eds.), *Error Estimation and Adaptive Discretization Methods in Computational Fluid Dynamics.*

26. M. Griebel, M.A. Schweitzer (eds.), *Meshfree Methods for Partial Differential Equations.*

27. S. Müller, *Adaptive Multiscale Schemes for Conservation Laws.*

28. C. Carstensen, S. Funken, W. Hackbusch, R.H.W. Hoppe, P. Monk (eds.), *Computational Electromagnetics.*

29. M.A. Schweitzer, *A Parallel Multilevel Partition of Unity Method for Elliptic Partial Differential Equations.*

30. T. Biegler, O. Ghattas, M. Heinkenschloss, B. van Bloemen Waanders (eds.), *Large-Scale PDE-Constrained Optimization.*

31. M. Ainsworth, P. Davies, D. Duncan, P. Martin, B. Rynne (eds.), *Topics in Computational Wave Propagation.* Direct and Inverse Problems.

32. H. Emmerich, B. Nestler, M. Schreckenberg (eds.), *Interface and Transport Dynamics.* Computational Modelling.

33. H.P. Langtangen, A. Tveito (eds.), *Advanced Topics in Computational Partial Differential Equations.* Numerical Methods and Diffpack Programming.

34. V. John, *Large Eddy Simulation of Turbulent Incompressible Flows.* Analytical and Numerical Results for a Class of LES Models.

35. E. Bänsch (ed.), *Challenges in Scientific Computing - CISC 2002.*

36. B.N. Khoromskij, G. Wittum, *Numerical Solution of Elliptic Differential Equations by Reduction to the Interface.*

37. A. Iske, *Multiresolution Methods in Scattered Data Modelling.*

38. S.-I. Niculescu, K. Gu (eds.), *Advances in Time-Delay Systems.*

39. S. Attinger, P. Koumoutsakos (eds.), *Multiscale Modelling and Simulation.*

40. R. Kornhuber, R. Hoppe, J. Périaux, O. Pironneau, O. Wildlund, J. Xu (eds.), *Domain Decomposition Methods in Science and Engineering.*

41. T. Plewa, T. Linde, V.G. Weirs (eds.), *Adaptive Mesh Refinement – Theory and Applications.*

42. A. Schmidt, K.G. Siebert, *Design of Adaptive Finite Element Software.* The Finite Element Toolbox ALBERTA.

43. M. Griebel, M.A. Schweitzer (eds.), *Meshfree Methods for Partial Differential Equations II.*

44. B. Engquist, P. Lötstedt, O. Runborg (eds.), *Multiscale Methods in Science and Engineering.*

45. P. Benner, V. Mehrmann, D.C. Sorensen (eds.), *Dimension Reduction of Large-Scale Systems.*

46. D. Kressner, *Numerical Methods for General and Structured Eigenvalue Problems.*

47. A. Boriçi, A. Frommer, B. Joó, A. Kennedy, B. Pendleton (eds.), *QCD and Numerical Analysis III.*

48. F. Graziani (ed.), *Computational Methods in Transport.*
49. B. Leimkuhler, C. Chipot, R. Elber, A. Laaksonen, A. Mark, T. Schlick, C. Schütte, R. Skeel (eds.), *New Algorithms for Macromolecular Simulation.*
50. M. Bücker, G. Corliss, P. Hovland, U. Naumann, B. Norris (eds.), *Automatic Differentiation: Applications, Theory, and Implementations.*
51. A.M. Bruaset, A. Tveito (eds.), *Numerical Solution of Partial Differential Equations on Parallel Computers.*
52. K.H. Hoffmann, A. Meyer (eds.), *Parallel Algorithms and Cluster Computing.*
53. H.-J. Bungartz, M. Schäfer (eds.), *Fluid-Structure Interaction.*
54. J. Behrens, *Adaptive Atmospheric Modeling.*
55. O. Widlund, D. Keyes (eds.), *Domain Decomposition Methods in Science and Engineering XVI.*
56. S. Kassinos, C. Langer, G. Iaccarino, P. Moin (eds.), *Complex Effects in Large Eddy Simulations.*
57. M. Griebel, M.A Schweitzer (eds.), *Meshfree Methods for Partial Differential Equations III.*
58. A.N. Gorban, B. Kégl, D.C. Wunsch, A. Zinovyev (eds.), *Principal Manifolds for Data Visualization and Dimension Reduction.*
59. H. Ammari (ed.), *Modeling and Computations in Electromagnetics: A Volume Dedicated to Jean-Claude Nédélec.*
60. U. Langer, M. Discacciati, D. Keyes, O. Widlund, W. Zulehner (eds.), *Domain Decomposition Methods in Science and Engineering XVII.*
61. T. Mathew, *Domain Decomposition Methods for the Numerical Solution of Partial Differential Equations.*
62. F. Graziani (ed.), *Computational Methods in Transport: Verification and Validation.*
63. M. Bebendorf, *Hierarchical Matrices.* A Means to Efficiently Solve Elliptic Boundary Value Problems.
64. C.H. Bischof, H.M. Bücker, P. Hovland, U. Naumann, J. Utke (eds.), *Advances in Automatic Differentiation.*
65. M. Griebel, M.A. Schweitzer (eds.), *Meshfree Methods for Partial Differential Equations IV.*
66. B. Engquist, P. Lötstedt, O. Runborg (eds.), *Multiscale Modeling and Simulation in Science.*
67. I.H. Tuncer, Ü. Gülcat, D.R. Emerson, K. Matsuno (eds.), *Parallel Computational Fluid Dynamics 2007.*
68. S. Yip, T. Diaz de la Rubia (eds.), *Scientific Modeling and Simulations.*
69. A. Hegarty, N. Kopteva, E. O'Riordan, M. Stynes (eds.), *BAIL* 2008 – *Boundary and Interior Layers.*
70. M. Bercovier, M.J. Gander, R. Kornhuber, O. Widlund (eds.), *Domain Decomposition Methods in Science and Engineering XVIII.*
71. B. Koren, C. Vuik (eds.), *Advanced Computational Methods in Science and Engineering.*
72. M. Peters (ed.), *Computational Fluid Dynamics for Sport Simulation.*

73. H.-J. Bungartz, M. Mehl, M. Schäfer (eds.), *Fluid Structure Interaction II - Modelling, Simulation, Optimization.*

74. D. Tromeur-Dervout, G. Brenner, D.R. Emerson, J. Erhel (eds.), *Parallel Computational Fluid Dynamics 2008.*

75. A.N. Gorban, D. Roose (eds.), *Coping with Complexity: Model Reduction and Data Analysis.*

76. J.S. Hesthaven, E.M. Rønquist (eds.), *Spectral and High Order Methods for Partial Differential Equations.*

77. M. Holtz, *Sparse Grid Quadrature in High Dimensions with Applications in Finance and Insurance.*

78. Y. Huang, R. Kornhuber, O.Widlund, J. Xu (eds.), *Domain Decomposition Methods in Science and Engineering XIX.*

79. M. Griebel, M.A. Schweitzer (eds.), *Meshfree Methods for Partial Differential Equations V.*

80. P.H. Lauritzen, C. Jablonowski, M.A. Taylor, R.D. Nair (eds.), *Numerical Techniques for Global Atmospheric Models.*

For further information on these books please have a look at our mathematics catalogue at the following URL: www.springer.com/series/3527

Monographs in Computational Science and Engineering

1. J. Sundnes, G.T. Lines, X. Cai, B.F. Nielsen, K.-A. Mardal, A. Tveito, *Computing the Electrical Activity in the Heart.*

For further information on this book, please have a look at our mathematics catalogue at the following URL: `www.springer.com/series/7417`

Texts in Computational Science and Engineering

1. H. P. Langtangen, *Computational Partial Differential Equations.* Numerical Methods and Diffpack Programming. 2nd Edition
2. A. Quarteroni, F. Saleri, P. Gervasio, *Scientific Computing with MATLAB and Octave.* 3rd Edition
3. H. P. Langtangen, *Python Scripting for Computational Science.* 3rd Edition
4. H. Gardner, G. Manduchi, *Design Patterns for e-Science.*
5. M. Griebel, S. Knapek, G. Zumbusch, *Numerical Simulation in Molecular Dynamics.*
6. H. P. Langtangen, *A Primer on Scientific Programming with Python.*
7. A. Tveito, H. P. Langtangen, B. F. Nielsen, X. Cai, *Elements of Scientific Computing.*

For further information on these books please have a look at our mathematics catalogue at the following URL: `www.springer.com/series/5151`

MIX
Papier aus verantwortungsvollen Quellen
Paper from responsible sources
FSC® C105338

If you have any concerns about our products,
you can contact us on
ProductSafety@springernature.com

In case Publisher is established outside the EU,
the EU authorized representative is:
Springer Nature Customer Service Center GmbH
Europaplatz 3, 69115 Heidelberg, Germany

Printed by Libri Plureos GmbH
in Hamburg, Germany